第十六届中俄双边
新材料新工艺研讨会论文集

Proceedings of the 16th Sino-Russia Symposium on
Advanced Materials and Technologies

中国有色金属学会 编

Edited by The Nonferrous Metals Society of China

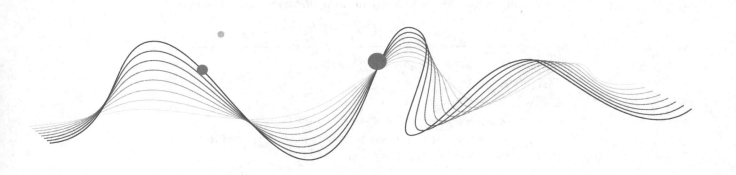

中南大学出版社·长沙
www.csupress.com.cn

图书在版编目(CIP)数据

第十六届中俄双边新材料新工艺研讨会论文集／中国有色金属学会编. —长沙：中南大学出版社，2023.10
ISBN 978-7-5487-5576-0

Ⅰ. ①第… Ⅱ. ①中… Ⅲ. ①新材料应用－国际学术会议－文集 Ⅳ. ①TB3-53

中国国家版本馆CIP数据核字(2023)第186996号

第十六届中俄双边新材料新工艺研讨会论文集
DISHILIUJIE ZHONG-E SHUANGBIAN XINCAILIAO XINGONGYI YANTAOHUI LUNWENJI

中国有色金属学会　编

□责任编辑	刘石年　胡炜　刘小沛　伍华进	
□责任印制	李月腾	
□出版发行	中南大学出版社	
	社址：长沙市麓山南路	邮编：410083
	发行科电话：0731-88876770	传真：0731-88710482
□印　　装	湖南省众鑫印务有限公司	
□开　　本	880 mm×1230 mm　1/16　□印张 32　□字数 1575千字	
□版　　次	2023年10月第1版　□印次 2023年10月第1次印刷	
□书　　号	ISBN 978-7-5487-5576-0	
□定　　价	180.00元	

图书出现印装问题，请与经销商调换

编委会

◇ **总　编**
　　贾明星

◇ **副总编**
　　高焕芝　张洪国　李　芳

◇ **编　辑**
　　尹　普　庞月明　冯　翠　李　勤　张琳琳
　　贺奉先　邹博尧

Editorial Board

◇ **Editor in Chief**
　　Mingxing Jia

◇ **Deputy Editor in Chief**
　　Huanzhi Gao　Hongguo Zhang　Fang Li

◇ **Editorial Board Members**
　　Pu Yin　Yueming Pang　Cui Feng　Qin Li　Linlin Zhang
　　Fengxian He　Boyao Zou

前 言

中俄双边新材料新工艺研讨会是由中国有色金属学会与苏联科学院巴伊科夫研究院、强度物理及材料科学研究所于1991年共同倡议组织的每两年轮流在中国和俄罗斯召开的国际学术会议,由于会议举办不断成功,其规模和范围不断扩大。本届研讨会得到了两国政府和中俄两国共同组建的"二十一世纪和平、友好、发展委员会"的支持,将于2023年11月6—10日在中国海南省海口市举办。

会议旨在为研究人员提供一个自由交流的平台,共同分享新的思路、新的解决方案。主题是"金属、陶瓷与复合材料";内容涵盖了航空航天材料,新能源材料,电子技术材料,生物医用材料,稀有金属、贵金属及高纯材料,节能环保新材料、绿色工程材料、催化剂,表面工程技术与材料,激光信息技术制造复杂形状制品、增材制造,冶金过程新工艺新技术,功能材料(含难熔金属、特硬材料及磁性材料),冶金过程模拟和数字化等领域。此外,会议组织者将邀请知名专家、学者做大会专题报告,所有参会者将有机会与报告人面对面讨论,这对参会者进行更深入的科学讨论和增进相互了解,是非常有益的。

中俄双边新材料新工艺研讨会,历经十五届,积极推动了中俄两国科技界的合作与交流,成为了冶金材料领域的代表性学术盛会。本届研讨会将延续这一优良传统,汇聚各界专家学者,共同探讨金属、陶瓷与复合材料等领域的前沿问题与创新成果。我们由衷感谢两国政府及"二十一世纪和平、友好、发展委员会"的大力支持,相信与会者将共同开启一场思想的盛宴,深化合作,促进学术发展。预祝此次会议圆满成功!

大会主席
中国有色金属工业协会会长

Preface

Ladies and gentlemen, esteemed guests, and esteemed colleagues,

It is an immense honor and with great enthusiasm that I address you today at this prestigious international conference. This gathering of brilliant minds from around the world exemplifies the power of collaboration and the pursuit of knowledge that transcends geographical boundaries.

The China-Russia Symposium on Advanced Materials and Technologies is a remarkable international academic conference, jointly initiated by the Nonferrous Metals Society of China, the Baikov Institute of the Academy of Sciences of the Soviet Union, and the Institute of Strength Physics and Materials Science. Since its inception in 1991, this conference has been held biennially, alternating between China and Russia. Its continued success has led to an expansion in both its scale and scope. The current edition of this symposium has garnered support from both governments and the "21st Century Committee for Peace, Friendship, and Development", a collaborative initiative between China and Russia. It is scheduled to take place in Haikou City, Hainan Province, China, from November 6th to 10th, 2023.

The conference aims to provide a platform for researchers to engage in open exchanges, facilitating the sharing of novel ideas, solutions collaboratively. With the theme "Metals, Ceramics, and Composite Materials", the symposium encompasses a wide array of domains, including aerospace materials; new energy materials; electronic technology materials; biomedical materials; rare metals, precious metals, high-purity materials; energy-saving and environmentally friendly new materials; green engineering materials; catalysts; surface engineering technology and materials; laser information technology for manufacturing complex shapes, additive manufacturing; new processes and technologies in metallurgical processes; functional materials (including refractory metals, superhard materials, and magnetic materials); metallurgical process simulation, and digital technologies. Additionally, the conference organizers will invite renowned experts and scholars to deliver keynote speeches, providing all participants with the opportunity for face-to-face discussions with the speakers. This interaction will facilitate deeper scientific discussions and enhance mutual understanding among attendees, which is highly beneficial.

Through its fifteen editions, the China-Russia Symposium on Advanced Materials and

Technologies has been a driving force in promoting cooperation and exchanges within the scientific communities of both China and Russia. It has rightfully earned its reputation as a preeminent academic event in the domain of metallurgical materials. This edition of the seminar will continue this esteemed tradition, bringing together experts and scholars from various sectors to collectively explore cutting-edge issues and innovative accomplishments in areas such as metals, ceramics, and composite materials. We extend our heartfelt gratitude to the governments of both nations and the generous support rendered by the "21st Century Committee for Peace, Friendship, and Development". We believe that participants will embark on a feast of ideas, deepening collaboration and fostering academic advancement. With fervent anticipation, we wish this conference resounding success.

Honglin Ge

General Chair
The President of China Nonferrous Metals Industry Association

Contents

Diagnostics of the Multi-scale Structure of Amorphous Alloy $Co_{58}Ni_{10}Fe_5Si_{11}B_{16}$
    Vladimir TKACHEV, Anatoliy FROLOV, Galina KRAYNOVA, Vladimir PLOTNIKOV, Sergey DOLZHIKOV, Aleksander FEDORETS, Evgeniy PUSTOVALOV / 1

Acoustic Properties and the Deformability of Materials
    Lev. B. ZUEV / 6

Formation of Carbon Fiber Reinforced ZrB_2-SiC Composites Using Preceramic Slurry
    Aleksei UTKIN, Roman ORBANT, Mikhail GOLOSOV, Denis BANNYKH, Natalia BAKLANOVA / 11

The Role of Magnetic Interactions in the Crystallization of NiO on Graphene Sheets
    Elena TRUSOVA / 13

Low-temperature Synthesis and Luminescence of $La_{0.95}Eu_{0.05}BO_3$:M and $La_{0.95}Eu_{0.05}(BO_2)_3$:M Borates (M-Y, Sm, Tb, Bi)
    Nadezhda STEBLEVSKAYA, Margarita BELOBELETSKAYA, Michail MEDKOV / 18

Research of the Effect of the Vacuum-chamber Design on the Technological Parameters of Steel Processing in a RH-degasser
    O. Y. SHESHUKOV, A. A. METELKIN / 24

Roadmaps "Fractals" "Artificial Intelligence" and "Photonics" Need to Be Combined (Our International Priorities in End-to-end Technologies)
    Alexander A. POTAPOV / 31

Synthesis of a Graphene-containing Composite by Deposition of Oxygen-free Graphene on Nanocrystallite CeO_2
    Ivan PONOMAREV, eLENA TRUSOVA / 37

Solid-state Reaction Kinetics Between Iridium and Zirconium Carbide
    Yaroslav A. NIKIFOROV, Natalya I. BAKLANOVA / 42

Development of the Ion Theory of Slags and Implementation of the Results Obtained at Metallurgical Enterprises in Russia
    A. A. METELKIN, O. Yu. SHESHUKOV, D. K. EGIAZARYAN, V. N. NEVIDIMOV, M. V. SAVELYEV / 44

On the Influence of High Pressures and Boron Nitride on the Processes of Structure Formation and Microhardness of a High-entropy Equiatomic Composition AlNiCoFeCr Alloy
    Svetlana MENSHIKOVA / 48

The Behavior of the CaB_6-Ir System at High Temperature
    Victor V. LOZANOV, Yaroslav A. NIKIFOROV, Mikhail A. GOLOSOV, Tatyana A. GAVRILOVA, Natalya I. BAKLANOVA / 53

Features of Deformation and Destruction of 3D Laser Printed Austenitic Steel
    N. V. KAZANTSEVA, Yu. N. KOEMETS, D. I. DAVIDOV, I. V. EZHOV, M. S. KARABANALOV / 58

Structure and Properties of Electroexplosive Coatings of the TiB_2-Ag System
    Denis ROMANOV, Vasili POCHETUKHA / 68

Catalytically Active Materials for PEM Fuel Cells Applications
    Alexandra KURIGANOVA, Nikita FADDEEV, Marina KUBANOVA / 73

Filler Composite Materials Based on an Al-Si System Alloy for the Synthesis of Wear-resistant Working Layers of Functionally Organized Compositions by Arc Cladding
    Roman MIKHEEV, Igor KALASHNIKOV, Pavel BYKOV, Lubov KOBELEVA / 77

Spinel-Garnet Fibers Based on Organomagnesiumoxane Yttriumoxane Alumoxanes Materials of Aviation and Space Technology
    Anastasiya S POKHORENKO, Galina I SHCHERBAKOVA, Maxim S VARFOLOMEEV, Tatyana L APUKHTINA, Artem A VOROBYEV, Natalia S KRIVTSOVA, Dmitriy V ZHIGALOV, Pavel A STOROZHENKO / 83

Co-processing of Rapeseed Oil and SRGO Blend over Sulfide Catalysts
    Evgeniya VLASOVA, Pavel ALEKSANDROV, Alexey NUZHDIN, Galina BUKHTIYAROVA / 86

Structure and Microhardness of Plasma Composite Layered Coating Ni-(80%WC12%Co+20%Ni)-Ni after Friction Treatment
    Aleksandra MIKHAILOVA, Vasilii KALITA, Dmitrii KOMLEV, Alexey RADYUK, Konstantin DEMIN, Boris RUMYANTSEV / 90

Effect of Zeolite Type on Activity of MoS_2/Al_2O_3-Z Catalysts in Hydroconversion of Methylpalmitate
    Yiheng ZHAO, Evgeniya VLASOVA, Pavel ALEKSANDROV, Ivan SHAMANAEV, Alexey NUZHDIN, Evgeniy SUPRUN, Vera PAKHARUKOVA, Galina BUKHTIYAROVA / 95

Hydroconversion of Methyl Esters over Ni-phosphide Catalyst on Composite Alumina-SAPO-11 Support
    Ivan SHAMANAEV, Evgenia VLASOVA, Galina BUKHTIYAROVA / 100

SiO_2 Precursors for Si-based Negative Electrode Materials for Li-ion Batteries
    Elena ABRAMOVA, Aleksey NEDOLUZHKO, Artem ABAKUMOV / 104

Influence of Graphene Sheet Content on Thermal Expansion of ZrO_2 Nanopowders
    Asya AFZAL, Elena TRUSOVA / 109

Zeeman Splitting of the Absorption Band Edge and Fabry-Perot Oscillations in Materials with Strong Spin-orbit Coupling
    Danil BELYAEV, Mikhail YAKUSHEV, Vladimir GREBENNIKOV, Milan ORLITA, Konstantin KOKH, Oleg TERESHCHENKO, Robert MARTIN, Tatyana KUZNETSOVA / 114

Interaction of Iridium with Silicon Carbide in Diffusion Couples in Wide Temperature Range
    Mikhail GOLOSOV, Aleksei UTKIN, Victor LOZANOV, Natalia BAKLANOVA / 119

Lightweight Ceramic Nanocomosites for Radiation Shielding of Electronics
    Oleg KHASANOV, Edgar DVILIS, Vladimir PAYGIN, Oleg TOLKACHEV / 121

The Synthesis of Triple-lithiated Transition Metal Oxide NCM622 Based on Extracted Cobalt from Spent $LiCoO_2$ Cathode Material
    Alexandra KOSENKO, Konstantin PUSHNITSA, Pavel NOVIKOV, Anatoliy A. POPOVICH / 125

Creep Behavior and Structural Changes in the RE-containing 10% Cr Martensitic Steels
    Alexandra FEDOSEEVA / 130

Achieving High Strength in Cu/graphene Composite Produced by High Pressure Torsion
    Galia KORZNIKOVA, Aynur ALETDINOV, Gulnara KHALIKOVA, Elena KORZNIKOVA / 135

Tissue-specific Decellularized Scaffolds for Pancreas Tissue Engineering
    Anna PONOMAREVA, Natalia BARANOVA, Lyudmila KIRSANOVA, Alexandra KIRILLOVA, Evgeniy NEMETS, Dmitriy KRUGLOV, Yulia BASOK, Victor SEVASTIANOV / 140

Formation of Cartilage-like Tissue from Collagen-based Microparticles in Perfusion Bioreactor
    Yulia BASOK, Alexey GRIGORIEV, Ludmila KIRSANOVA, Alexandra KIRILLOVA, Varvara RYZHIKOVA, Anastasia

SUBBOT, Evgeniy NEMETS, Victor SEVASTIANOV / 145

Development of Nonwoven Hemocompatible Vascular Graft of Small Diameter with Reduced Surgical Porosity
Viktor SEVASTIANOV, Evgenij NEMETS, Vyacheslav BELOV, Lyudmila KIRSANOVA, Alla NIKOLSKAYA, Vyacheslav ZAKHAREVICH, Kirill KIRIAKOV, Varvara RYZHIKOVA, Irina TYUNYAEVA, Artem VYPRYSHKO, Yuliya BASOK / 150

Influence of L-PBF Process Parameters on Mechanical Properties and Nitrogen Content in Nitrogen Steels
Ekaterina VOLOKITINA, Nikolay RAZUMOV, Nikolai OZERSKOI, Anatoly POPOVICH / 154

Observation of Interatomic Auger Transitions in Cu-based Chalcogenides Using Synchrotron Radiation X-ray Photoemission
Vladimir GREBENNIKOV, Tatyana KUZNETSOVA / 159

Possibility of Damping Through the Use of Different Steels and Alloys
Iurii DUBINOV, Oksana ELAGINA, Artem BEREZNYAKOV, Olga DUBINOVA, Victoria FEDOROVA, Alexander KUZNETSOV / 164

Solid-state Reaction Kinetics Between Iridium and Zirconium Carbide
Yaroslav A. NIKIFOROV, Natalya I. BAKLANOVA / 168

Study of the Properties of Damping Materials for Manufacturing of Equipment in the Oil and Gas Industry
A. K. PRYGAEV, Y. S. DUBINOV, O. B. DUBINOVA, M. S. TANASENKO, A. N. DUDKINA, E. A. PRYGAEVA / 172

Gadolinium-based Nanosized Phosphors Activated by Terbium for Medical Applications
Vadim BAKHMETYEV, Polina ZYKOVA, Anna VLASENKO, Olga OSMAK, Nikolay KHRISTYUK, Sergey MJAKIN / 176

X-ray Luminescent Phosphor-photosensitizer Systems for Oncotheranostics
Anna VLASENKO, Vadim BAKHMETYEV, Sergey MJAKIN / 180

Reinforced Plastics for Alternative Wind Power
Natalia KORNEEVA, Vladimir KUDINOV / 185

Thermostable Functional Materials Based on Phosphate Binders
Konstantin N. LAPKO, Alexander N. KUDLASH, Natalia S. APANASEVICH, Aliaksei A. SOKAL / 190

Effect of Short-range Order on Viscosity and Crystallization of Al-Mg Melts
Larisa KAMAEVA, Elizaveta BATALOVA, Nikolay CHTCHELKATCHEV / 194

Pulse Alternating Current Electrosynthesis as an Effective Way to Multifunctional Materials for Hydrogen Energy
Tatyana MOLODTSOVA, Anna ULYANKINA, Daria CHERNYSHEVA, Nina SMIRNOVA / 199

Complex Experimental and Model Study of Non-isothermal Deformation in Al-, Co- and Cu-based Alloys with Different Space Configuration and Structural State
Arseniy BEREZNER, Victor FEDOROV, Michael ZADOROZHNYY, Gregory GRIGORIEV / 204

Development of the Basis for the Industrial Processing of Ilmenite-titanomagnetite Ores of the Basite-ultrabasites Intrusions Deposited in Sikhote-Alin (The Far East of Russia)
Vladimir MOLCHANOV, Fengyue SUN / 208

Gadolinium Doped Tricalcium Phosphates for Hard Tissue Bioimaging: EPR Spectroscopy Control of Chemical Synthesis
Margarita SADOVNIKOVA, Fadis MURZAKHANOV, George MAMIN, Anna FORYSENKOVA, Inna FADEEVA, Marat GAFUROV / 213

Charpy Impact Testing of a Middle-entropy Alloy with an Austenitic Structure Subjected to Cold Rotary Swaging
Dmitrii PANOV, Ruslan CHERNICHENKO, Stanislav NAUMOV / 219

Resonant Photoemission Spectroscopy for Studying Long-lived Excited Atomic States in Multicomponent Rare Earth-transition Metal Materials

Tatyana KUZNETSOVA, Vladimir GREBENNIKOV, Ekaterina PONOMAREVA / 224

Numerical Realization of Helicoidal DNA Model
Maria OSTRIK, Victor LAKHNO / 228

Metal-containing Catalysts for the Hydrogenation Process of Substituted 5-acyl-1, 3-dioxanes
Gul'nara RASKIL'DINA, Yulianna BORISOVA, Simon ZLOTSKII / 233

Solar Technology for Extraction of Metals from Waste
M. S. PAIZULLAKHANOV, O. R. PARPIEV, R. Yu. AKBAROV, A. A. HOLMATOV, N. H. KARSHIEVA, N. N. CHERENDA / 235

Failure of Friction Stir Welded Joint of a Tempformed High-strength Low-alloy Steel
Anastasiia DOLZHENKO, Anna LUGOVSKAYA, Andrey BELYAKOV / 240

Use of Accelerator Technology for Qualitative and Quantitative Analysis of Materials
V. TOVTIN, E. STAROSTIN, S. SIMAKOV, M. PRUSAKOVA, V. VINOGRADOVA / 245

Influence of Pulsed Beams of Helium Ions and Helium High-temperature Plasma on the Ferritic 16Cr-4Al-2W-0.3Ti-0.3Y_2O_3 Steel
E. DEMINA, M. PRUSAKOVA, V. PIMENOV, S. MASLYAEV, N. VINOGRADOVA, A. DEMIN, E. MOROZOV, N. EPIFANOV, S. SIMAKOV / 246

Numerical Simulation's Possibility of Materials' Damping Properties
Robert BAITEMIROV, Iurii DUBINOV, Alexander PRYGAEV, Alexander KUZNETSOV, Dmitriy VISHNIVETSKIY / 250

Amorphous Ferromagnetic Wires for Structural Health Monitoring
Andrey ALPATOV, Vyacheslav MOLOKANOV, Andrey KRUTILIN, Natalia PALII / 253

Changing the Technological Properties of Materials by Conducting Thermomagnetic Treatment
Yu. S. DUBINOV, A. K. PRYGAEV, G. T. BOKOYEV, A. D. KOTOV, M. A. DUBROVIN, O. B. DUBINOVA / 257

Application of Data from Photometric Analysis of Structural Images in the Choice of Heat Treatment Modes for Heat-resistant Alloys
V A ERMISHKIN, N A MININA, D L MIKHAILOV, N A PALII / 259

Fiber Optical Current Sensor — A Method for Online Monitoring of Electrode Current in Electrolytic Metallurgy Process
Yi MENG, Jun TIE, Chun LI, Rentao ZHAO, Hongwei JIANG, Xingzu PENG, Hao XIAO, Dongwei LIU / 264

Mathematical Models and Software for Dynamic Simulation of Ladle Treatment Technology
O. A. KOMOLOVA, K. V. GRIGOROVICH / 271

Spectroscopic Methods for Determination of Gold Content in High-salt Solutions of Complex Composition
Valentina VOLCHENKOVA, Evgeny KAZENAS, Nadezhda ANDREEVA, Boris TAGIROV, Irina ZLIVKO, Alexander ZOTOV, Vladimir REUKOV, Irina NIKOLAEVA / 276

Effect of Growth Rate on Dendritic Morphology and Secondary Dendrite Arm Spacing of Pd-20W Alloy
Jiming ZHANG, Youcai YANG, Ming XIE, Li CHEN, Yongtai CHEN, Jiheng FANG, Saibei WANG, Aikun LI / 283

Investigation into the Effect of Precursor Solution pH on the Phase Structure of the YBCO Superconducting Target
Wenyu ZHANG, Benshuang SUN, Xiaokai LIU, Huiyu ZHANG, Hetao ZHAO, Yongchun SHU, Yang LIU, Jilin HE / 289

A Multi-scale Study of the Thermal Transport Properties of Graphene-reinforced Copper/diamond Composites
Jiarui ZHU, Hui YANG, Shuhui HUANG, Hong GUO, Zhongnan XIE / 301

Application of Rare Earth Metal Ferroalloys in Special Steels
Lifeng ZHANG, Ying REN, Qiang REN / 309

Effect of Ca Treatment on Sulfide Inclusion of High-strength Low-alloy Steel
Lifeng ZHANG, Xiaoyong GAO / 315

Numerical Simulation on Entrapment of Inclusions During Electroslag Remelting Process
 Wei CHEN, Tianjie WEN, Lifeng ZHANG   / 321

Influence of Microstructure on the Flow Boiling Heat Transfer Characteristics of Diamond/Cu Heat Sink
 Mingmei SUN, Nan WU, Hong GUO, Zhongnan XIE, Shuhui HUANG   / 327

Basic Research on Preparation of Al-Si Alloy from Oxides of Al and Si by Molten Salt Electrolysis
 Jiaxin YANG, Zhaowen WANG, Wenju TAO, Youjian YANG   / 335

Study on the Plasma Sphero Process of Rare and Precious Metal Powders
 Kang YUAN, Xiaoxiao PANG, Yubai HOU   / 342

Effects of Co Doping on Martensite Transformation and Thermal Hysteresis in NiTi Alloys
 Baolin PANG, Zhenqiang WANG, Baoxiang ZHANG, Zan LIAO, Jiangbo WANG   / 348

Stabilization of the β'-Cu_4Ti Phase in Cu-Ti Alloys by Micro-alloying with Gd
 Yumin LIAO, Chengjun GUO, Chenyang ZHOU, Weibin XIE, Bin YANG, Hang WANG   / 358

Effect of Process Parameters and Alloy Composition on Residual Stress of Martensitic Stainless Steel Cladding Layer
 Kaiping DU, Ziqiang PI, Xing CHEN, Zhaoran ZHENG, Jie SHEN   / 369

Nano-$Li_{1.3}Al_{0.3}Ti_{1.7}(PO_4)_3$ Modified Ni-Rich Cathode Materials for High Energy Density Lithium-ion Batteries
 Zongpu SHAO, Yafei LIU, Xuequan ZHANG, Yanbin CHEN, Yueguang YU   / 376

The Influence of Feed Rate on the Abradability of AlSi/PHB Seal Coatings
 Xinwo ZHAO, Jiangang SUN, Deming ZHANG   / 381

Single Crystalline $LiNi_{0.65}Co_{0.15}Mn_{0.20}O_2$ for High-energy Li-ion Batteries with Outstanding Cycling Stability
 Jun Wang, Xuequan ZHANG, Shunlin SONG, Yafei LIU, Yanbin CHEN   / 389

A Bulk Oxygen Vacancy Dominating WO_3 Photocatalyst for Carbamazepine Degradation
 Weiqing GUO, Qianhui WEI, Gangrong LI, Feng WEI, Zhuofeng HU   / 394

Application of Acidophilic Bacteria in the Metal Enrichment of Electroplating Sludge Use the External Circulation Bioreactor
 Bingyang TIAN, He SHANG, Wencheng GAO, Jiankang WEN   / 404

Comparative Analysis of Domestic and Foreign Indentation Test Standards
 Qinli Lü, Sitong YE, Shuai XU, Hong JI   / 410

Evolution of S(Al_2CuMg) Phase During Fabrication Process and Its Influence on Mechanical Property in a Commercial Al-Zn-Mg-Cu Alloy
 Kai WEN, Hongwei YAN, Lizhen YAN, Hongwei LIU, Wei XIAO, Ying LI, Guanjun GAO, Rui LIU, Weicai REN   / 419

Tuning the Mechanical and Corrosion Properties of Selective Laser Melted Al-Mg-Sc-Zr Alloy Through Heat Treatment
 Jinglin SHI, Qiang HU, Xinming ZHAO, Yonghui WANG, Jinhui ZHANG, Yingjie LIU   / 429

Effect of Grain Boundary Diffusion of Nano and Micron $Tb_{70}Cu_{15}Al_{15}$ on the Coercivity of Hydrogenation Disproportionation Desorption Recombination $Nd_2Fe_{14}B$ Powders
 Xuhua WANG, Dunbo YU, Yang LUO, Zilong WANG, Yuanfei YANG, Ningtao QUAN, Zhongkai WANG, Weikang SHAN, Wenjian YAN, Wenlong YAN   / 437

Recent Developments of Purification of Rare Earth Metals
 Xiaowei ZHANG, Hongbo YANG, Zhiqiang WANG, Zongan LI, Chuang YU, Xinyu GUO, Jiamin ZHONG, Penghong HU, Fan YANG, Wenli LU, Chenchen XU   / 442

Effect of Pre-deformation Heat Treatment Process on Microstructure and Mechanical Properties of TB3 Titanium Alloy

Baohui ZHU, Feng DU, Xiaofei LI, Lin CHEN, Dongxin WANG, Jingming ZHONG / 452

Prediction Techniques for Shrinkage and Porosity During Solidification of Be-Al Alloy
Yao XIE, Shenmin LI, Junyi LI, Dongxin WANG, Yajun YIN, Jingmin ZHONG / 459

Study on Synthesis of Submicron Polycrystalline Diamond Composite Using Cobalt Acetate as Cobalt Source
Jiarong CHEN, Jinghe ZHU, Qiaofan HU, Haiqing QIN, Peicheng MO, Xiaoyi PAN, Jun ZHANG, Kai LI, Chen CHAO / 466

Wear-resistant CoCrNi Multi-principal Element Alloy at Cryogenic Temperature
Qing ZHOU, Yue REN, Dongpeng HUA / 473

Ce Substitution with La to Improve Magnetic Properties of RE-Fe-B Sintered Magnets
Ming YUE, Hao CHEN, Weiqiang LIU / 474

A Novel Low-density ZrTiNbAl System Multi-principal Element Alloys with Unique Mechanical Properties
Xuehui YAN, Baohong ZHU / 475

A Novel Nb-W-C Alloy with Special Microstructure and Excellent Mechanical Property
Xiaohong YUAN, Yan WEI, Li CHEN, Xian WANG, Changyi HU, Jialin CHEN / 476

Effect of Compaction Pressure on the Sintering Activation Energy and Microstructure of IGZO Targets
Xiaokai LIU, Wenyu ZHANG, Benshuang SUN, Xueyun ZENG, Huiyuyu ZHANG, Zhijun WANG, Chaofei LIU, Yongchun SHU, Yang LIU, Jilin HE / 477

Tribological Properties of MoS_2/a-C: Si Composite Films under High-temperature Atmosphere and Vacuum Environments
Yanjun CHEN, Songsheng LIN, Fenghua SU / 478

Effect of Y Doping on Hot Corrosion Behavior of NiAlHf Coating
Xiaoya LI, Qian SHI, Songsheng LIN, Mingjiang DAI, Kesong ZHOU / 479

The Correlation Mechanism Between the Grain Boundary Characteristics of Sintered Nd-Fe-B Base Magnets and the Heavy Rare Earth Element Diffusion Behaviors
Qingzheng JIANG, Qingfang HUANG, Zhenchen ZHONG, Hang WANG / 480

Thermodynamic Calculation and Experimental Investigation of NiCr-Cr_3C_2 Coating Prepared by Detonation Spray
Song QIN, Donghua LIU, Sen HAN, Jianqiao ZHOU / 481

Influence of Feed Rate on the Abradability of AlSi/PHB Sealing Coatings
Xinwo ZHAO, Jiangang SUN, Deming ZHANG / 482

A New Method of Preparing Hydrophobic Photocatalytic Composite Coatings Based on Nano-TiO_2 and Soluble Polytetrafluoroethylene Prepared by Suspension Plasma Spraying
Chunyan HE, Xiujuan FAN, Shuanjian LI, Sainan CUI, Jie MAO, Chunming DENG, Min LIU, Kesong ZHOU / 483

Challenges and Development Progress of Mn-based Cathode Materials
Jun WANG, Xuequan ZHANG, Yafei LIU, Yanbin CHEN / 484

Activation of Peroxymonosulfate by Single Atom Co-N-C Catalysts for High-efficient Removal of Phenol: Performance, Mechanism and Stability
Jianqun WU, Shangqian ZHAO / 485

Defect-regulated Synthesis of ZrB_2 Powder by Carbon Thermal Reduction and Its Hot-pressing Densification Mechanism
Yuyang LIU, Xiaoning LI, Xingqi WANG / 486

Determination of Elemental Impurities in Nickel-based Superalloys by Glow Discharge Mass Spectrometry
Fangfei HU, Jidong LI, Yingxin ZHANG, Pengyu LIU / 487

Effect of A-EMS Melt Treatment on Microstructure and Mechanical Properties of Al-Zn-Mg-Cu Alloy Castings

Y. T. XU, Z. F. ZHANG, Z. H. GAO, Y. B. WANG, C. S. CHEN, J. Z. FAN / 488

Interface Modulation for Inorganic Perovskite Solar Cell with Efficiency over 16%
Zhenyun ZHANG, Haoyue MA, Hongling ZHANG, Lin CAO, Bo WANG, Chenyang ZHU, Peng SHANG, Hongchun SHI, Yuanfei MA / 489

NiFe Layered Double Hydroxide Supported on Ni Fiber Felt for Oxygen Evolution at High Current Density
Qinglin LIU, Yiwen TANG, Man LUO, Lijun JIANG / 490

Study on Heat Transfer Performance of Mg Based Composite Hydrogen Storage Materials under Working Condition
Liyu ZHANG, Jianhua YE, Zhinian LI, Lijun JIANG / 491

The Effect of Garnet Type Oxide Solid State Electrolyte Coating on the Properties of Separators and the Electrical and Safety Performances of Pouch Cells
Bo WANG, Tianhang ZHANG, Zenghua CHANG, Rennian WANG, Xiaopeng QI, Jiantao WANG, Rong YANG / 492

The Microstructural Evolution and Mechanical Properties after the Solution Treatment of GH4099 Produced by Laser Powder Bed Fusion
Jiahao LIU, Wenqian GUO, Yonghui WANG, Shaoming ZHANG, Qiang HU / 493

Core-shell FeCo@ SiO_2 Nanocomposites with Controllable Shell Thickness and Tunable Electromagnetic Properties
Longxia YANG, Haicheng WANG / 494

In-situ EBSD Investigation on Deformation Mechanism of Room Temperature Superplasticity of Bulk Recrystallized Molybdenum
Wenshuai CHEN, Yan LI, Xueliang HE, Zenglin ZHOU / 495

Diagnostics of the Multi-scale Structure of Amorphous Alloy $Co_{58}Ni_{10}Fe_5Si_{11}B_{16}$

Vladimir TKACHEV, Anatoliy FROLOV, Galina KRAYNOVA, Vladimir PLOTNIKOV, Sergey DOLZHIKOV, Aleksander FEDORETS, Evgeniy PUSTOVALOV

Far Eastern Federal University (FEFU), Vladivostok, 690091, Russia

Abstract: In this work, the multiscale structural inhomogeneities of the amorphous alloy $Co_{58}Ni_{10}Fe_5Si_{11}B_{16}$, obtained by rapid quenching from the melt, were investigated. The presence of a hierarchy of structural inhomogeneities of spatial dimensions from 0.1 nm to 21.8 nm was confirmed by using the methods of scanning and transmission electron microscopy with the subsequent application of the method of spectral analysis of microimages of this amorphous alloy. When using the method of antiplane deformation of the amorphous $Co_{58}Ni_{10}Fe_5Si_{11}B_{16}$ alloy, the presence of multiscale structural inhomogeneities in the range from 10 nm to 10 μm was found in scanning electron microscopy images. The developed technique of spectral analysis of electron microscopic images confirmed the presence of a hierarchical cellular structure in the studied amorphous alloy.

Keywords: Amorphous metal alloys; Electron microscopy; Fractograms; Spectrum of structural inhomogeneities; Hierarchical structures

1 Introduction

Soft magnetic amorphous metal alloys obtained by rapid quenching from the melt have high magnetic permeability, which allows them to be successfully used for the manufacture of magnetic heads, magnetic shields, and secondary power supplies[1-2]. These alloys have high firmness and corrosion resistance, which is important for the exploitation of products made from them in aggressive environments[3]. Amorphous alloys $Co_{58}Ni_{10}Fe_5Si_{11}B_{16}$ also belong to this class of materials. The problem of the structure and morphology of alloys, obtained by rapid quenching, in particular by spinning, has not yet found a reliable solution[4]. Disorder in amorphous materials is not absolute. Experiments indicate the presence of universal structural formations at scales of nearly 1 nm, which can be considered as the correlation radius of the structure[5,6]. Spinning tapes have a contact to the quenching disk and free from its influence surfaces, characterized by significantly different morphology, anisotropy of the distribution of inhomogeneities[4,7]. In this regard, it is not surprising that there is a stratification of rapidly quenched tapes in their thickness[8]. Amorphous bodies, with multiscale structural formations along the thickness of the tape from the contact surface (CS) to the free surface (FS), are referred to as modulation or hierarchical structures. The existence of such a structure of nonequilibrium media with a high level of energy absorption is largely due to its final flux from sources to sinks[9]. Obtaining of amorphous tapes by spinning the melt onto a rotating disk is a classical method that satisfies the dissipation conditions[4].

2 Materials and research methods

X-ray (Bruker D8 Advance diffractometer, K_α-radiation of Fe) amorphous metal alloy (AMS) based on cobalt composition $Co_{58}Ni_{10}Fe_5Si_{11}B_{16}$ (at%) with a thickness of (21.1±4.5) μm was obtained by single-roll sinning in an argon atmosphere on a copper wheel in an installation MeltSpinner SC. Work is aimed at studying structure of tapes of the $Co_{58}Ni_{10}Fe_5Si_{11}B_{16}$ alloy, between the contact and free surfaces using scanning electron microscopy (SEM CarlZeiss Crossbeam 1540XB) and high-resolution transmission electron microscopy (HRTEM). The standard HRTEM (FEI TITAN 300) samples were prepared by ion milling under flowing argon gas.

The method, decorating the internal structure of rapidly quenched alloys, was destruction according to the antiplane deformation scheme. The direction of the kinks matched with the tape rolling axis. The resulting fractograms were used for further analysis of the structure over the thickness of rapidly quenched tapes. In this case, the obtained spectrum of inhomogeneities sizes can have a scale comparable to surface periodicities, but carry fundamentally different information, and the morphology of fractures reflects not only traces of the causes of the fracture process, but also characteristic features of the structural state[10].

Structure of the tapes in terms of thickness was controlled by using an energy dispersive X-ray attachment (Oxford Instruments, Inca software) at 20 kV with preliminary calibration on a cobalt standard.

3 Results and discussions

The first scale level of the alloy under study is represented by the structure, which is decorated by the antiplane deformation method. It covers the size range from 10 nm to 10 μm. The image of fractograms typical of the $Co_{58}Ni_{10}Fe_5Si_{11}B_{16}$ alloy for this production mode is shown in Figure 1(a). A coral-like structure is observed for the FS and gradually transforms into a wave-like structure with a period of nearly 100 nm for the CS. On electron microscopic images, a grid with a size of 20 nm×20 nm is visualized in Figure 1(b).

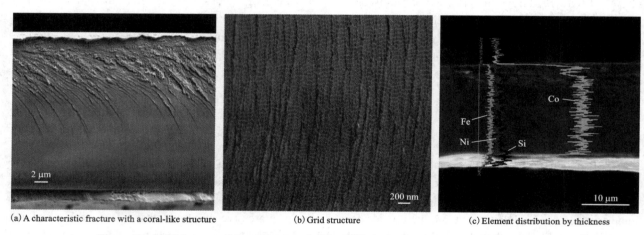

(a) A characteristic fracture with a coral-like structure　　(b) Grid structure　　(c) Element distribution by thickness

Figure 1　SEM images of fractograms of the rapidly quenched alloy $Co_{58}Ni_{10}Fe_5Si_{11}B_{16}$

The observed coral-like structures are specific for tapes, obtained by spinning the melt in air[10]. One of the most probable reasons for the appearance of such a structure is the slip of an antiplane crack in the zone near the free surface of rapidly quenched tapes on structure inhomogeneities. The authors[11] at the ductile-brittle transition observed this effect. Already at this level of structural-morphological inhomogeneities, decorated by the antiplane deformation method, a multiscale structure of tapes from 20 nm to (1-2) μm is observed in Figure 1. A visual surface analysis of the obtained electron microscopic images revealed a structure with a color difference from light to dark. Since the elemental composition is constant across the thickness of the tapes [Figure 1(c)], this is not associated with a change in the concentration of chemical elements. The observed structure was called typical salt-and-pepper-like[11]. It demonstrates inhomogeneities with a period (wavelength) λ about 0.26 nm [Figure 2(a) and (b)]. Analysis of the image makes it possible to visualize the combination of short-wavelength periodicities into larger irregularities with a period of 0.7 nm and 1.3 nm or more [Figure 2(b)].

Visual analysis of images of this kind is very difficult; it also does not allow assessing the anisotropy of the structure. Therefore, in this work, we analyze the structure of the nanoscale in the spectral representation. A typical view of the Fourier spectrum from a microscopic image in Figure 2(a) and the integral frequency characteristic (IFC) and integral spatial characteristic (ISC) calculated from them are shown in Figure 3(a).

Spectral representation in the form of a diffraction pattern (DFP) from an electron microscopic image (Figure 2), has the form of a diffuse halo bounded in reciprocal space with a size of 0.52 nm^{-1} (about 0.26 nm) and a central spot. The halo is isotropic, but the bright reflections on the distinguished ring of the DFP of the sample indicate that there is local anisotropy in the structure. Talking about the spectrum of sizes of

(a) General view and electron diffraction pattern (inset)　　(b) Fine structure and electron diffraction pattern (inset)

Figure 2 Typical HRTEM image of an amorphous alloy of composition $Co_{58}Ni_{10}Fe_5Si_{11}B_{16}$

both long-wavelength and smaller ranges of inhomogeneities, up to periodicities of 0.2 nm, IFC [Figure 3(a)], mid-wave irregularities are also widely represented. The calculated IFC shows the presence of modes: $\lambda \sim 0.26$ nm (intensity 87%), $\lambda \sim 1.08$ nm (intensity 100%) and $\lambda \sim 1.8$ nm (intensity about 100%).

The analysis ISC, a pie chart [Figure 3(a)], allows speaking in general about the isotropy of the structure in the range (0.09-5.4) nm. However, a visual analysis of the electron microscopic image in Figure 2 demonstrates the inhomogeneity of the structure of the rapidly quenched amorphous alloy in the nanoscale. Therefore, further analysis was carried out for local areas of the image in order to establish their differences.

The original image [Figure 2(a)] for the sample had a size of 87 nm×87 nm. For further analysis, a square area 8.7 nm×8.7 nm in size was isolated on the image. The selected area was shifted by half a period in the horizontal direction along the image from the left edge to the right. For all regions, Fourier spectra were obtained and the IFC and ISC were calculated from them. As an example, Figures 3(b) and (c) show the spectral characteristics for two such regions (the first and fourth) out of ten. The spectral characteristics (Figure 3), show that the structure of the regions differs from each other and from the structure of the entire image. The differences are significant both in the intensity of modes of different sizes of IFC and in the anisotropy of the distribution of inhomogeneities [pie charts in Figure 3(a), Figures 3(b) and (c)].

To introduce a quantitative estimate of the difference in the structure of local regions, we will use the methodology for calculating the integral functions of Lebesgue measures (IFLM)[7]. Comparison of IFLM using the concept of linearized Kullback divergence (LivI)[7] was carried out with two basic structures—with linear IFLM, corresponding to IFLM from the spectral characteristic of white noise, and IFLM from the spectral characteristic of the complete image of the sample. The results of comparison of the selected regions for the studied amorphous alloy $Co_{58}Ni_{10}Fe_5Si_{11}B_{16}$ are shown in Figures 4(a) and (b).

The greatest interest for comparative analysis [Figures 4(a) and (b)] lies in area 4 and area 9 of the structure image: structure of area 4 and area 9 is the farthest from the structure of the entire image [Figure 4(b)], but simultaneously approach to the structure of white noise-complete disordering[8]. Thus, in these local areas, the structure is the most disordered relatively to the integral. If we consider the HRTEM image as a projection of the material density (darker structures—increased density, lighter ones—decreased), then area 4 and area 9 are characterized by a reduced density of the structure of the areas under consideration. Area 7 is characterized, on the contrary, by an increase in the density of the structure. Based on Figure 4, it can be concluded that the entire interval from area 5 to area 8 has an increased structure density, although it is not uneven.

In the projection onto the plane, the structure of HRTEM images at the nanoscale can be interpreted as a grid structure. Areas with high density are cells, and areas with low density are the boundaries of cells. In this case, for the $Co_{58}Ni_{10}Fe_5Si_{11}B_{16}$ alloy, a cellular structure about 20 nm×20 nm in size was observed in fractograms (Figure 1). The size of the grid coincides with the distance between the regions of low density of the structure. Here, the distance between the centers of area 4 and area 9 is 21.8 nm, which corresponds to the order of the grid cell size in Figure 1.

The structure of the tape during the spinning of the melt is formed as a result of athermohydrodynamic process. The main role in this process is played by the removal of heat from the melt to the refrigerator. Since

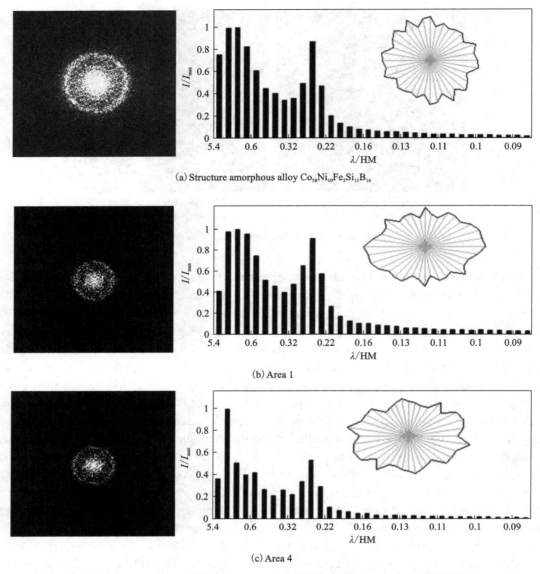

Figure 3 Fourier spectrum, IFC (histogram) and ISC (pie chart) from HRTEM images

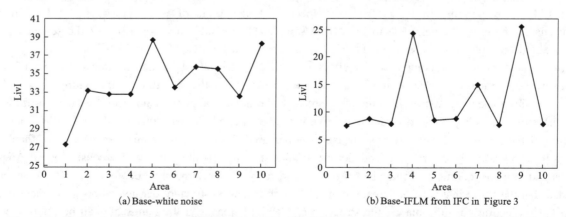

Figure 4 Kullback divergence for IFLM, obtained with TEM image of a spinning tape $Co_{58}Ni_{10}Fe_5Si_{11}B_{16}$

the process is extremely non-equilibrium and rather fast, it should be accompanied by a restructuring of the melt structure during solidification. Heat transfer is most likely of a wave nature[4, 7], and the corresponding structure is cellular. But its visualization on fractograms (Figure 1), can be considered an "ideal case". Cellular structure in fractograms (Figure 1), can be considered an elementary microarea, participating in the phenomenon of wave-like heat transfer along the simplest mesostructure.

The performed procedures for processing HRTEM and TEM microimages of amorphous alloys showed the presence of a hierarchy of structural inhomogeneities in the range of spatial dimensions from 0.09 nm to 21.8 nm, which is consistent with previously published results in [12]. The same hierarchical structure is observed for soft magnetic nanocrystalline alloys[11].

4 Conclusions

Using the antiplane deformation method, the hierarchical structure of fractograms of the rapidly quenched $Co_{58}Ni_{10}Fe_5Si_{11}B_{16}$ alloy in the range from 20 nm to (1-2) μm in the form of coral-like and cellular structures is shown. Analysis of HRTEM images of the structure of the amorphous alloy was carried out in spectral representation. The calculated integral frequency (IFC) and integral spatial characteristics (ISC) showed the presence of modes: short-wave, $\lambda \sim 0.26$ nm, medium-wave, $\lambda \sim 1.08$ nm and long-wave, $\lambda \sim 1.8$ nm ranges, as well as the nature of their ordering.

Insertion of a quantitative assessment of the similarity/difference in the structure of the amorphous alloy $Co_{58}Ni_{10}Fe_5Si_{11}B_{16}$ using the method for calculating the integral function of Lebesgue measures and Kullback divergence confirmed the presence of a grid (cellular) structure with a size of nearly 20 nm×20 nm, which is considered as the distance between areas of reduced material density. In view of the formation of the spinning tape structure as a whole, this cellular structure can be represented as an elementary microarea, participating in the wave-like transfer of heat along the mesostructure-coral branch. This result can be used in the future to control the parameters of the cellular structure of amorphous alloys, controlled by the conditions of technology and subsequent heat treatment of amorphous alloys to obtain the specified service properties.

Acknowledgments: This work was financially supported by FEFU EF No. 22-02-03-005.

References

[1] GUTFLEISCH O, WILLARD M A, BRÜCK E, et al. Magnetic materials and devices for the 21st century: stronger, lighter, and more energy efficient[J]. Advanced Materials, 2011, 23(7): 821-842.
[2] WILLARD M A, DANIIL M. Nanocrystalline soft magnetic alloys two decades of progress[M]//Handbook of Magnetic Materials. Amsterdam: Elsevier, 2013(21): 173-342.
[3] PUSTOVALOV E V, MODIN E B, FROLOV A M, et al. Effect of the process conditions for the preparation of CoNiFeSiB amorphous alloys on their structure and properties[J]. Journal of Surface Investigation: X-Ray, Synchrotron and Neutron Techniques, 2019, 13(4): 600-608.
[4] FILONOV M R, ANIKIN U A, LEVIN U B. Theoretical foundations of the production of amorphous and nanocrystalline alloys by ultrafast quenching[M]. MISIS, 2006: 228.
[5] OSSMURA K, SHIBUE K, SUZUKI R, et al. SAXS study on the structure and crystallization of amorphous metallic alloys[J]. Colloid and Polymer Science, 1981, 259(6): 677-682.
[6] SHENG H W, LUO W K, ALAMGIR F M, et al. Atomic packing and short-to-medium-range order in metallic glasses[J]. Nature, 2006, 439(7075): 419-425.
[7] FROLOV A, KRAINOVA G, DOLZHIKOV S. Anisotropy of the structural inhomogeneities of rapidly quenched alloys[J]. Journal of Surface Investigation X-Ray Synchrotron and Neutron Techniques, 2018, 12(2): 370-376.
[8] FROLOV A M, KRAINOVA G S, DOLZHIKOV S V, et al. Peculiarities of the stratification effect in tapes made of Fe-B and Fe-Cr-B alloys obtained by spinning[J]. Deformation and Destruction of Materials, 2019, 5: 2-5.
[9] KUKLIN V M. Formation of self-similar spatial structures in modulation-unstable media[J]. Electromagnetic phenomena, 2004, 4, (13): 85-100.
[10] BROBERG K B. Cracks and fracture[M]. San Diego: Academic Press, 1999.
[11] TONG X, ZHANG Y, WANG YC, et al. Structural origin of magnetic softening in a Fe-based amorphous alloy upon annealing[J]. Journal of Materials Science & Technology, 2022, 96: 233-240.
[12] YANG Z Z, JIANG S S, YE L X, et al. Nanoscale structural heterogeneity perspective on the ameliorated magnetic properties of a Fe-based amorphous alloy with decreasing cooling rate[J]. Journal of Non-Crystalline Solids, 2022, 581: 121433.

Acoustic Properties and the Deformability of Materials

Lev. B. ZUEV

Institute of Strength Physics and Materials Science, SB RAS, Tomsk, 634055, Russia

Abstract: A new method for non-destructive evaluation of the mechanical properties of materials has been developed. This is based on measurements of the ultrasound propagation velocity in deforming materials. Preliminarily investigations were carried out in order to relate the ultrasound propagation velocity to the mechanical characteristics of the deforming material.

Keywords: Strength; Plasticity; Deformation; Alloys; Ultrasound

1 Introduction: Experimental justification of the method

It was established previously[1,2] that the ultrasound propagation rate measured directly for metal specimens tested in tension would depend on material structure, total deformation and flow stress. Similar data were obtained for small total strains in [3]. Of particular interest is the form of ultrasound propagation rate dependence on flow stress and deformation obtained for the tested pure aluminum specimen (Figure 1). This consists of three linear parts that can be described[1] by the Equation (1)

$$V_S = V_0 + \xi\sigma \quad (1)$$

Here the empirical constants V_0 and σ have different values for the different stages of the flow.

It follows from Figure 1 that σ can be either positive or negative. However, the proportionality $V_S - \sigma$ is always fulfilled within a single stage with the correlation coefficient $|r| \geq 0.9$.

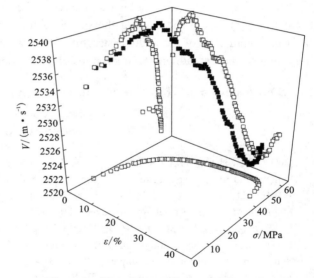

Figure 1 The ultrasound propagation rate dependence on flow stress and deformation obtained for the tested pure aluminum specimen

The aim of the present study is to verify that Equation (1) can be used for the evaluation of mechanical characteristics of materials by the non-destructive method developed. To elucidate the issue, the dependence $V_S(\sigma)$ was obtained for various kinds of alloys (see Table 1). The propagation rate of Rayleigh waves was measured directly for flat specimens tested in tension by the method of sound impulses self-circulation, which is described in detail below.

The dependencies $V_S(\sigma)$ obtained for all the materials above have a similar shape. Using the dimensionless variables V_S/V_0 and σ/σ_B (V_0 is the rate of ultrasound propagation in the material before the deformation and σ_B is the strength limit of the material), one can easily establish the general form of this dependence (Figure 2).

The above normalization permits pooling of the data obtained for all the materials tested; stages 1 and 2 of the dependence $V_S(\sigma)$ are

$$\frac{V_S}{V_0} = k_i + \alpha_i \cdot \frac{\sigma}{\sigma_B} \quad (2)$$

Here i is stage number 1 or 2; the constants k_i and α_i are independent of the kind of material and are evaluated experimentally. It is found that the respective values for stages 1 and 2 are as follows: $k_1 = (1 \pm 2.7) \times 10^{-4}$ and $k_2 = (1.03 \pm 3) \times 10^{-3}$; $\alpha_1 = -6.5 \times 10^{-3} \pm 4.7 \times 10^{-4}$ and $\alpha_2 = -3.65 \times 10^{-2} \pm 3.2 \times 10^{-3}$.

Table 1 Chemical composition of the alloys investigated (mass fraction) %

No.	Material	Symbol	C	Mg	Mn	Li	Cr	Cu	Ni	Zn	Zr
1	Steel		0.12	-	2.0	-	17.0	0.3	10.5	-	-
2	Steel		<0.1	-	1.3-1.7	-	<0.3	<0.3	<0.3	-	-
3	Steel		<0.1	-	0.5-0.8	-	0.6-0.9	0.4-0.6	0.5-0.8	-	-
4	Steel		0.18	-	0.35	-	<0.3	<0.3	<0.3	-	-
5	Duralumin		-	1.5	-	-	-	4.35	<0.1	<0.3	-
6	Al-Mg	+	-	5.8-6.2	0.15	1.8-2.2	-	-	-	-	0.1
7	Al-Li	×	-	-	-	1.8-2.0	-	2.8-3.2	-	-	0.12
8	Brass		-	-	-	-	-	-	-	38.0	-
9	Zr-Nb		-	-	-	-	-	-	-	-	99.0
10	Zr-Nb		-	-	-	-	-	-	-	-	97.5

Equation (2) can be transformed as follows:

$$\sigma_B = \frac{\alpha_i \sigma}{V_S/V_0 - k_i} \tag{3}$$

which can be used for the estimation of strength limit at small total plastic strains precluding specimen failure. To do this, the ultrasound propagation rate, V_S, was measured at the stress $\sigma_{0.2} < \sigma < 0.6\sigma_B$ ($\sigma_{0.2}$ is proof stress), which initiates small plastic deformation only.

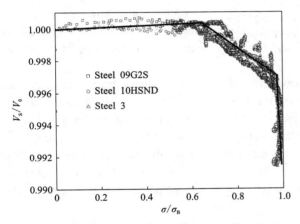

Figure 2 The generalize dependence $V_S/V_0(\sigma/\sigma_B)$ obtained for steels

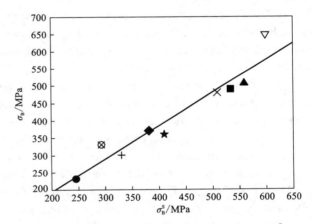

Figure 3 The correlation between countered σ_B^S and experimental obtained σ_B

The strength limit values obtained from Equation (3) (σ_B^S) are matched against those derived conventionally from the curves $\sigma-\varepsilon(\sigma_B)$ in Figure 3. The rate, V_S, was measured at the deformation $\varepsilon \approx 1\%$ (flow stress $\sigma \approx 0.1\sigma_B$). The values of σ_B and σ_B^S are proportional

$$\sigma_B = 0.96\sigma_B^S \tag{4}$$

The correlation coefficient is about 0.96. The above testifies the efficiency of the proposed method for strength limit evaluation in structural materials, which deform at small total plastic strains precluding failure. Thus, it is a promising method for structural integrity monitoring of metalwork and machine parts, produced from various materials.

The nature of the above relation might be addressed on the assumption that material hardening is determined by the magnitude of internal stress fields, which moving dislocations have to overcome[4]. On the other hand, with increasing internal stresses, the ultrasound propagation rate decreases[1,2]. Thus, the above two values are defined by the same factor; therefore, they are closely correlated.

2 Equipment designed for ultrasound method application

The devices designed for ultrasound method implementation and made in small lots are Acoustic Strain Tester Rapid (ASTR) and Acoustic Non-Destructive Analyzer (ANDA). These devices have been elaborated in the institute. These are meant for structural integrity inspection of metals and alloys in metalwork and machine parts during long-term service in both regular and severe conditions. The general principle of operation of the units is self-circulation frequency measurement of Rayleigh wave impulses. They are simple in operation and allow the ultrasound rate to be measured to an accuracy of nearly 3×10^{-5}.

3 Use of the ultrasound method to evaluate residual internal stress level

The applications of the proposed method include the estimation of stressed state in zirconium billets used for the manufacture of nuclear reactor fuel cladding. During the cold rolling of Zr-Nb alloy tubes, an intricate distribution of residual internal macro-stresses would form in the worked billet, which enhances the probability of its failure at one of the process stages. When tackling the problems of process optimization one therefore has to take into account the level and distribution of residual internal macro-stresses in worked billets. Because of their large size, however, this is hardly feasible with the aid of conventional methods, e.g. X-ray techniques, very much so under process conditions.

This investigation was carried on using the ASTR unit to determine internal stress level for finished product samples and worked billets. The measurements were made in a wide range of internal stresses for the deforming Zr-Nb alloy 9 specimens in order to relate the internal stresses to the propagation rate of acoustic wave. These were performed using a head having base length L_b = 30 mm for the same points and in accordance with the same scheme as that of an X-ray technique. The measured self-circulation frequency was converted to ultrasound rate. The most significant results were obtained for the worked billets in which internal stresses varied over a wide range.

The present work is aimed at development of non-destructive methods for the determination of residual stresses, which form in thin-wall Zr tubes manufactured by the process of cold rolling[5]. This would help improve the technologies currently employed for tube production. The investigation was made for a wide range of specimens, i.e. tubes and round billets made from Zr based alloys 9 and 10. The lifetime of materials and constructions is in many ways affected by worked material uniformity and by the stressed state of finished products manufactured from the same material. Therefore, the investigation of residual macro-stresses was performed by both traditional (X-ray) and non-conventional (acoustic) methods; the two sets of data obtained by the above two techniques were matched.

The X-ray investigation was conducted using an X-ray diffractometer. The measurements were performed in Cu K_α radiation using a diffracted monochromatic beam. The X-ray examination allows one to distinguish in the worked material lattice certain regions characterized by variations in the interplanar spacing, which can be measured with a sufficient degree of accuracy. In the case of plane stressed state, one of the principal normal stresses is equal to zero and the sum of the remaining two is given[4, 6] by

$$\sigma_1 + \sigma_2 = \frac{E}{\nu}(\theta_1 - \theta_0)\cot\theta_0 \tag{5}$$

where θ_1 and θ_0 are the Bragg angles determined by the stressed and unstressed (reference) samples on the base of X-ray data respectively; E is the Young modulus and ν is the Poisson ratio. A 30 mm length of tube made from alloy 9 (after re-crystallization annealing) was used as reference sample. The sample investigated was moved along the X-ray beam (scanned) and the local stresses were thus defined for the various points of the generatrix. Then the sample was rotated through $\pi/4$ and scanned over the other generatrix. Thus the local residual stresses were measured and mapped and then matched with the ultrasound rate data.

It has been found that the magnitude of macro-stresses $\sigma_1 + \sigma_2$ is linearly related to the frequency of self-circulation f in alloy 9 (see Figure 4), i.e.

$$\sigma_1 + \sigma_2 = \sigma_0 - \beta f \tag{6}$$

where σ_0 = 420 MPa and β = 0.42 MPa·s are the constants, and the correlation coefficient is about 0.7. The high value of the correlation coefficient allows one to conclude that the relationship is a functional one, so that the self-circulation frequency can be safely converted to stresses using Equation (6). On the base of the above results, a technique has been developed which is intended for internal stress measurement in Zr alloy tubes. In what follows, two sets of data on residual stresses are discussed. These were obtained for round billets and finished items with the aid of the above two techniques.

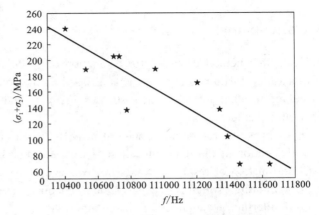

Figure 4 The correlation between the ultrasound rate and the level of stresses in alloy 9

The macro-stresses, i.e. residual stresses resulting from rolling, were measured with the aid of X-ray technique for round zirconium billets ϕ14.8 mm×9.5 mm. The macro-stresses in alloy 10 specimens are found to vary over a wide range from 400 to 900 MPa (Figure 5), especially so near the transition area from the smaller to the larger diameter. It should be noted that regions removed far enough from the transition area reveal sufficiently smooth and uniform distributions of macro-stresses. The level of stresses in alloy 9 is found to be considerably lower relative to alloy 10. The small height of the stress jumps suggests that alloy 9 worked by rolling is in a more homogeneous state, which might be due its greater ductility relative to alloy 10. The use of appropriate die profile enabled one to reduce considerably the jump of stresses in the worked material. To measure the stresses accurately, the test objects shall conform to the following requirements:

①Absence of surface defects;
②Equidistant points marked over the tube envelope;
③Availability of reference sample.

The stress distributions in tubes were determined using a specially designed attachment. This features a stage with two guides, which allow the sample to be aligned in both the beam plane and relative to the goniometer axis. The scanning was performed manually every 20 mm, using the marks over the tube envelope. Figure 6 illustrates the variation in macro-stresses σ_1 over the length of tubes made from alloy 9. It can be seen that homogeneous distributions are observed in the range of 200 MPa. To obtain more detailed distribution patterns, recording was performed for four equidistant points marked over the tube envelope. The measurements were made for the tube ends and the middle lengths of the tubes. As is seen from Figure 6, more uniform distributions of stresses are observed for the middle lengths of the tubes relative to the tube ends where stresses may be due partly to the material non-uniformity and partly to the tube deformation by cutting.

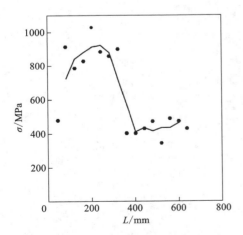

Figure 5 The distribution of internal stresses in the pipe billets tested

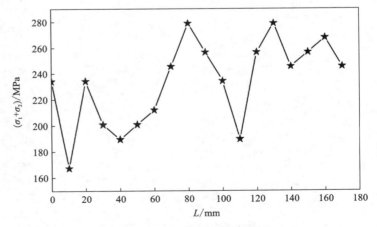

Figure 6 The macro-stresses distributed over the length of pipe from alloy 9

4 Conclusion

Thus, the method designed for estimation of mechanical characteristics facilitates considerably residual stress measurement in real objects. This is based on the correlation between the ultrasound propagation rate and the level of residual internal macro-stresses in tubes and round billets. The applications of the ultrasound method also include:

①Analysis of stress-strained state of heavily loaded large-sized metalwork, e. g. bridges;
②Evaluation of the remaining lifetime of water-tube boiler parts and pipelines;
③Eestimation of residual stresses in steels and alloys by welding;
④Monitoring of high pressure vessels in chemical industry;
⑤Monitoring of cumulative fatigue damages;
⑥Analysis of chemical heat-treatment of surface (carburizing, nitriding, hydrogen saturation);
⑦Monitoring and evaluation of the remaining lifetime of railway transport parts.

It should also be noted that over fifty sheds in Russia are equipped with ASTR units for the inspection of metal state of locomotive trolleys in the course of their reconditioning and overhaul.

Acknowledgments: This study was performed with the help of the grant No. 21-19-00075 of Russian Scientific Foundation.

References

[1] ZUEV L, BARANNIKOVA S, DANILOV V, et al. Plasticity: from crystal lattice to macroscopic phenomena[J]. Uspehi Fiziki Metallov, 2021, 22(1): 3-57.
[2] KOBAYASHI M, TANG S H, MIURA S, et al. Ultrasonic nondestructive material evaluation method and study on texture and cross slip effects under simple and pure shear states[J]. International Journal of Plasticity, 2003, 19(6): 771-804.
[3] LÜTHI B. Physical acoustics in the solid state[M]. 2nd printing of the 1st ed. Berlin: Springer, 2007.
[4] PELLEG J. Mechanical properties of materials[M]. Dordrecht: Springer Netherlands, 2013.
[5] ZUEV L B, ZAVODCHIKOV S Y, POLETIKA T M, et al. Phase composition, structure, and plastic deformation localization in Zr1% Nb alloys[M]//Zirconium in the Nuclear Industry: Fourteenth International Symposium. 100 Barr Harbor Drive, PO Box C700, West Conshohocken, PA 19428-2959: ASTM International, 2008: 264.
[6] MEYERS M A. dynamic behavior of materials[M]. New York: Wiley and sons, 1994.

Formation of Carbon Fiber Reinforced ZrB_2-SiC Composites Using Preceramic Slurry

Aleksei UTKIN, Roman ORBANT, Mikhail GOLOSOV, Denis BANNYKH, Natalia BAKLANOVA

Institute of Solid State Chemistry and Mechanochemistry, Novosibirsk, 630090, Russia

In recent years there has been growing interest in the development and application of ceramics based on zirconium diboride, which is successfully used for the manufacture of cutting tools, plasma-arc electrodes, electrodes for electrical discharge machining, in microelectronics, and solar radiation concentrators[1]. Since ZrB_2 is resistant to molten metals and slags, it can also be used in the steel industry as a refractory protective crucible coating. It has been shown that the mechanical strength, crack resistance, and oxidation resistance of monolithic ZrB_2 ceramics can be improved by introducing a second component, silicon carbide, which slows down high-temperature oxidation due to the formation of a glassy passivation layer during oxidation[2]. To impart increased crack resistance to monolithic ceramics, studies are currently underway on the introduction of various reinforcing components such as carbon or SiC whiskers, nanotubes and fibers[3].

This work is devoted to the study of microstructural features and some mechanical properties of composite materials based on ZrB_2-SiC matrix reinforced with continuous carbon fiber. The morphology, texture, and bending strength of C/ZrB_2-SiC ceramic composites obtained by a new method based on the impregnation of a carbon bundle with a ceramic suspension, the formation of unidirectional ceramic tapes, followed by pyrolysis, and siliconization, have been studied.

It has been shown that when pre-ground ZrB_2 powder is used as a filler, composites are formed with a porosity of less than 4% and a low proportion of closed pores [Figure 1(a)]. The replacement of part of the ground ZrB_2 powder with submicron SiC powder leads to an additional reduction in porosity. The rheological properties of ceramic suspensions have been studied and the compositions of suspensions that are optimal in terms of impregnating ability have been determined [Figure 1(b)]. The developed approach makes it possible to sufficiently evenly distribute the reinforcing component in the ceramic matrix and obtain composites with controlled properties.

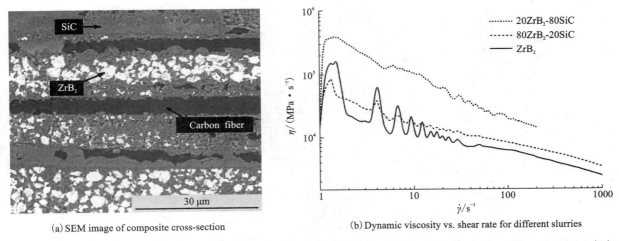

(a) SEM image of composite cross-section (b) Dynamic viscosity vs. shear rate for different slurries

Figure 1 SEM image of composite cross-section (a) and dynamic viscosity vs. shear rate for different slurries (b)

Acknowledgments: This work was supported by the Russian Science Foundation (Project No. 23-19-00212).

References

[1] ASL M S, NAYEBI B, AHMADI Z, et al. Effects of carbon additives on the properties of ZrB_2-based composites: a review[J]. Ceramics International, 2018, 44(7): 7334-7348.

[2] CARNEY C M, MOGILVESKY P, PARTHASARATHY T A. Oxidation behavior of zirconium diboride silicon carbide produced by the spark plasma sintering method[J]. Journal of the American Ceramic Society, 2009, 92(9): 2046-2052.

[3] HU P, GUI K X, HONG W H, et al. High-performance ZrB_2-SiC-C_f composite prepared by low-temperature hot pressing using nanosized ZrB_2 powder[J]. Journal of the European Ceramic Society, 2017, 37(6): 2317-2324.

The Role of Magnetic Interactions in the Crystallization of NiO on Graphene Sheets

Elena TRUSOVA

Baikov Institute of Metallurgy and Materials Science, Russian Academy of Sciences, IMET RAS, Moscow, 119334, Russia

Abstract: We synthesized the composite nanostructured magnetic particles based on graphene and NiO. A combination of sol-gel and sonochemical techniques was used for this purpose. The mixture of graphene suspension and Ni-containing sol was used for the hybrid structure synthesize under anaerobic conditions. This approach excludes oxidized graphene formation and leads to incorporation of the oxygen-free sheets into composite structure without the destruction of sp^2-electron system of graphene. It was shown that composite consists of the fibers with a thickness up to 3 nm and a length of 8-10 nm. We assumed that the graphene sheets direct the anisotropic growth of NiO crystallites along the graphene surface due to the edge magnetic field and promote the formation of composite fibers with a thickness no more than 3 nm. It was found that the synthesized composite was a soft ferromagnet, its specific magnetization was nearly 4.6 emu/g, and the coercive force was 110 Oe.

Keywords: Nanostructured graphene-NiO composite; Sol-gel; Sonochemical exfoliation of graphene; Soft ferromagnetic material

1 Introduction

As determined, the most remarkable electronic features of graphene are due to the chiral character of the charge carriers[1]. A number of recent theoretical publications show the impact of deformations and strains in electronic properties of graphene[2,3]. It was shown by calculations that upon mechanical deformation of a graphene sheet, pseudo magnetic fields appear in it due to nonuniform strain distributions[3,4]. The authors of these theoretical works assume great prospects for the creation of new materials with finely tuned properties for various industries, in particular, for valleytronics. However, there are much fewer practical works on the chemical synthesis of materials where these properties are observed, which is associated with the difficulties of obtaining and using oxygen-free graphene, as well as creating conditions for achieving a uniform distribution of graphene in the bulk of the composite at the nanolevel. The hybrid structures consisting of NiO nanocrystals, wrapped by graphene sheets, are recognized as the most effective for many relevant materials for supercapacitors, anodes of Li-batteries, fuel cells and sensors[5]. This way of packing allows to save high mobility of sp^2-electrons of graphene, while preventing agglomeration of its sheets[6]. It has been proven that graphene addition to material improves its thermal and electrical properties, as well as specific strength[7]. As is known, NiO and graphene react differently to magnetic fields: nanosized NiO is superantiferromagnet/superparamagnet[8], and graphene as nanoclusters has ferromagnetic properties due to high density of uncompensated electrons located at the edges of its sheets[9,10]. Composite nanostructured particles based on NiO and graphene are of particular interest for medicine, environmental protection, electronic, biological and biomedical applications such as biosensing, drug delivery, hyperthermia, magnetic resonance imaging and cellular capture, for vector drug delivery and magnetic resonance imaging[11]. The creation of a reliable method for the synthesis of such hybrids with a controlled structure and properties proved to be very difficult in practice. The production of graphene-NiO composite powders is complicated by the fact that oxygen-free graphene is required to obtain a quality product. Unfortunately, most of the published works are devoted to the creation of NiO composites based on graphene oxide or reduced graphene oxide, which does not solve the problem of obtaining high-quality nanomaterials. An alternative to these methods is the least developed methods for creating composite nanosystems based on oxygen-free graphene, since they are the most interesting to study and apply. Earlier, we reported on a developed low cost and technologically promising method for the synthesis of van der Waals systems based on oxygen-free graphene and nanosized crystalline metal oxides. In the presented paper, we report about the synthesis of magnetic nanostructured powders based on oxygen-free graphene and NiO structure characterization and study of magnetic properties of the composites synthesized in various ways.

2 Experiment

Synthetic graphite powder (NPO UNIHIMTEC, Russia) with a particle size of 600-800 microns was used to obtain graphene suspension. Deionized water was poured into a flask with a portion of graphite powder and then DDA (MERCK Schuchardt) was added; DDA/C molar ratio was equal to 1. The KOH solution was added to the resulting emulsion to achieve a pH of 10. The graphene suspension was obtained by ultrasonic treatment of liquid substrate (Sonoswiss SW1H, power 200 W) for 1 h. The graphene suspension was separated from unreacted graphite after sedimentation for 12 h. Nickel nitrate $Ni(NO_3)_2 \cdot 6H_2O$ (GOST 4055-78, Russia) was used to prepare 0.4 mol/L aqueous solution. An ethanol solution of DDA was used for the formation and stabilization of Ni-containing sol at 85-90 ℃; a DDA/Ni molar ratio was equal to 1. As-prepared Ni-containing sol was added to the selected graphene suspension on a magnetic stirrer at 85-90 ℃ during 30-40 min. The mixture was stirred for 30 min at 90 ℃ in a flask with a reflux condenser. Then the reflux condenser was removed and temperature was raised to 92-95 ℃. A sol was evaporated until the gel state. As-prepared hybrid gel was calcined in a muffle furnace at 500 ℃ for 1 h.

X-ray diffraction (XRD) data were recorded using SHIMADZU XRD-6000 diffractometer with monochromatic copper radiation (λ_{K_α} = 1.54178 Å). An average crystallite size was calculated by the Rietveld method, viz. using an iterative procedure to minimize the experimental diffraction pattern deviations from those calculated. Transmission electron microscopy (TEM) studies of graphene suspension and synthesized composite were carried out with the use of the LEO-912 AB OMEGA electron microscope operating at 100 kV. High resolution transmission electron microscopy (HRTEM) studies were carried out with the use of JEM 2010 instrument (JEOL ltd.). Elemental analysis was carried out on a Leco instrument, model CS-600. Raman spectra were measured in approximately back-scattering geometry using a TRIAX 552, Jobin Yvon spectrometer equipped with CCD Spec-10, 2KBUV, a Princeton Instruments 2048×512 detector and razor edge filters. A 514.5 nm exciting laser STABILITE 2017 was used. Magnetic measurements were performed using VM-23-K vibrating sample magnetometer under normal conditions.

3 Results and discussions

The morphology of graphene suspension obtained for the synthesis of the NiO-based composite with graphene content 0.2 wt.% was characterized using HRTEM. Figure 1(a) clearly shows that it mainly consists of sheets with a thickness of several layers. The length of sheets does not exceed 10 nm, and the thickness was 2-3 nm (some fragments are marked with ovals). In composite, NiO nanocrystals are lined in chains, the direction of which is given by graphene sheets enveloping them [Figure 1(b)]. In agglomerates, the chains are packed parallel to each other. HRTEM microphotos show uniform distribution of NiO crystallites with sizes not exceeding 7 nm along the graphene sheets with its longer side. Comparison of HRTEM images in Figure 1(b) for the composite and TEM ones in Figure 1(c) for a pure NiO nanopowder obtained from the same sol as the composite indicates differences in the morphology, as well as in NiO crystallites stacking order in these two powders. In a pure NiO nanopowder, the crystallites chains are not observed; this powder consists of the agglomerates with sizes of 80-300 nm, moreover, in them the crystallites with sizes of 10-35 nm are randomly arranged. Thus, it can be argued that only graphene sheets can lead to the appearance of a particularly stylized packing of crystallites during the formation of the composite.

In the XRD pattern of composite, there are two species of reflexes corresponded to NiO and Ni^0, while there are no reflexes corresponding to carbon species (Figure 2). The peaks at 37.06°, 43.10°, 62.62° and 75.09° have been identified as peaks of cubic NiO crystallites with the diffracting planes [111], [200], [220] and [311], respectively (JCPDS Card 47-1049). The calculation by the Rietveld method showed that the average crystallite size of NiO did not exceed 8 nm, which is completely concordant with the HRTEM data

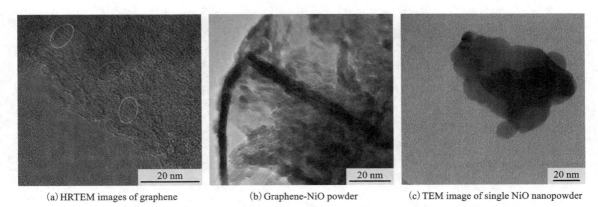

(a) HRTEM images of graphene (b) Graphene-NiO powder (c) TEM image of single NiO nanopowder

Figure 1 HRTEM images of graphene (a), graphene-NiO powder (b) and TEM image of single NiO nanopowder (c)

[Figure 1(a)]. However, along with the main phase of nickel oxide, a completely reduced nickel phase is observed, and its content does not exceed 3 wt.%. The peaks at 44.50°, 51.86° and 77.39° have been identified as peaks of cubic NiO crystallites with the diffracting planes [111], [200] and [220], respectively (JCPDS Card 87-0712). Comparison of XRD diffraction patterns of the composite and pure nano-NiO obtained from the same sol [Figure 2(a), inset] shows that NiO dispersion in the composite is higher than in the pure NiO powder. Previously, it was shown using the Williamson-Hall analysis when estimating the lattice deformation of NiO that the broadening of the reflections is associated with the deformation of the crystal lattice as a result of the formation of a large number of dislocations on the crystal faces[11].

In the Raman spectrum of graphene-NiO composite [Figure 2(b)], there are bands in the area of 490-540 cm^{-1}, 900 cm^{-1}, 1050 cm^{-1} and a weak band in the area of 350-450 cm^{-1} (shoulder band 520 cm^{-1}). An increase in the line half-width (over 90 cm^{-1}) of the band in the region of 490-540 cm^{-1} and its shift from 506 to 495 cm^{-1} in the Raman spectrum of graphene-nickel oxide composite is connected with decrease of NiO crystalline size. Also an increase in the intensity of the band in the region of 350-450 cm^{-1} (left shoulder) compared to single NiO in the graphene-NiO composite spectrum is connected with the higher content of NiO$_x$ phase ($x<1$), which indicates a larger nickel oxide non-stoichiometry in the composite.

To determine the magnetization curves for the graphene-NiO nanocomposite under normal conditions, a vibrating sample magnetometer was used. The hysteresis loops of microwires were measured by a vibrating sample magnetometer producing a maximum magnetic field of 107 A/m. Figure 3 shows that the magnetization curve approaches saturation at a magnetic field strength above ±7.5 kOe. The synthesized composite was a soft ferromagnet, its specific magnetization was about 4.6 emu/g, and the coercive force was 110 Oe.

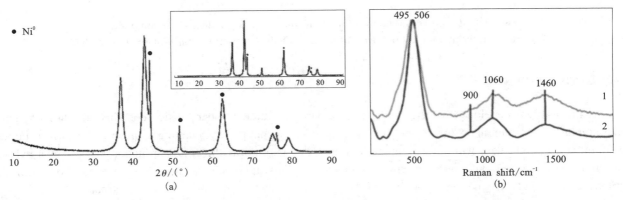

Figure 2 XRD data for graphene-NiO composite (a) and single nano-NiO (inset) and Raman spectra of the composite graphene-NiO (b1) and single nano-NiO (b2)

Figure 4(a) shows the proposed mechanism for the formation of a graphene-NiO composite. The hydrolysis of the starting salt, nickel nitrate, proceeds in two stages with the formation of A and B intermediates with multiple predominance of hydroxyl and nitrate ligands (A). When DDA is added to a salt

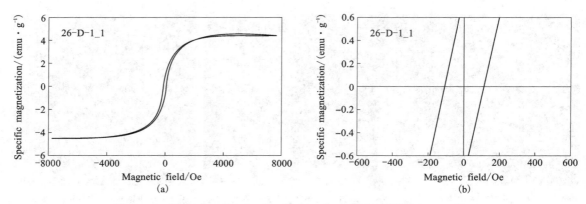

Figure 3 Magnetization hysteresis loop for graphene-NiO composite:
in the maximum magnetic fields ±7.5 kOe (a) and in the small field region (b)

solution at a DDA/Ni molar ratio of 1, a sol is formed. Then the resulting sol was combined with a graphene suspension, as a result of which the sol particles were coordinated on graphene sheets due to Coulomb forces, which led to the formation of a mixed suspension. During subsequent evaporation, the sol → gel transition occurred on the graphene surface [structures C and D in Figure 4(b)] at discretely located centers, which subsequently become crystallization centers during calcination in a furnace (air, 500 ℃) [structure D in Figure 4(b)]. In this case, the forming crystallites are oriented in the form of chains parallel to each other [structure E in Figure 4(b)].

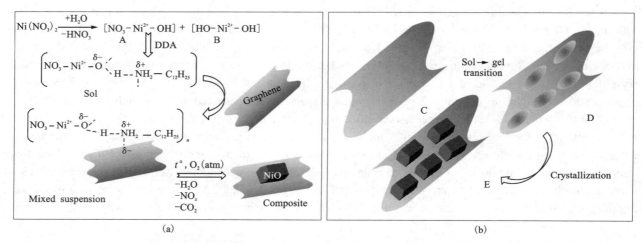

Figure 4 Proposed mechanism for the formation of a graphene-NiO composite:
Processes occurring on the graphene sheets in a colloid (a) and during heat treatment (b)

4 Conclusions

A method was developed for the synthesis of a nanostructured powder, which contains NiO and oxygen-free graphene, by combining the sol-gel and sonochemical techniques. Obtaining graphene sheets in a liquid medium and mixing them with an as-prepared Ni-containing sol allow the entire formation process of a hybrid structure to be carried out under anaerobic conditions. This approach excludes the formation of oxidized graphene and leads to the incorporation of oxygen-free sheets into the composite structure without the destruction of the sp^2-electron system of graphene. In the synthesized composite, NiO nanocrystals are lined in chains, the direction of which is given by graphene sheets enveloping them. It has been suggested that the edge magnetic effect of the graphene sheets controls the self-organization of Ni-containing sol particles on their surface and, during subsequent heat treatment, directs the anisotropic growth of NiO crystals along the graphene sheet surface. Crystallization of NiO took place under the orienting influence of the graphene sheet, and, as a result, all the crystallites were almost the same size, and were strictly oriented in chains. At the same time, the formation of structural discontinuities, similar to the grain boundaries formed during sintering, did not occur. Encapsulation of NiO crystallites in a graphene shell ensures the homogeneity of the composite material at the nano level, while oxygen-free graphene is protected from the formation of functional groups on its surface, and, therefore, retains all its unique electronic properties. Since the behavior of electrons of graphene is very sensitive to the shape and degree of bending of its sheets, the synthesized heterostructure acquired magnetic properties due to the strong interaction of the sp^2-orbitals. The proposed approach to synthesis of the composites based on oxygen-free graphene and nanoscale NiO is promising for a technology of the actual raw products in the development of small-sized high-speed electronic devices.

Acknowledgments: The study was carried out in the Baikov Institute of Metallurgy and Materials Science, RAS in accordance with the State Assignment 075-01176-23-00.

References

[1] GEIM A K, NOVOSELOV K S. The rise of graphene[J]. Nature Materials, 2007, 6(3): 183-191.
[2] AMORIM B, CORTIJO A, DE JUAN F, et al. Novel effects of strains in graphene and other two dimensional materials[J]. Physics Reports, 2016, 617: 1-54.
[3] MAO J H, MILOVANOVIĆ S P, ANĐELKOVIĆ M, et al. Evidence of flat bands and correlated states in buckled graphene superlattices[J]. Nature, 2020, 584(7820): 215-220.
[4] GEORGI A, NEMES-INCZE P, CARRILLO-BASTOS R, et al. Tuning the pseudospinpolarization of graphene by a pseudomagnetic field[J]. Nano Letters, 2017, 17(4): 2240-2245.
[5] ZHOU W, TAN ML, ZHOU X S. Graphene-NiO composite electrode for supercapacitors[J]. Advanced Materials Research, 2011, 345: 75-78.
[6] CHEN YM, HUANG Z D, ZHANG H Y, et al. Synthesis of the graphene/nickel oxide composite and its electrochemical performance for supercapacitors[J]. International Journal of Hydrogen Energy, 2014, 39(28): 16171-16178.
[7] REINA A, THIELE S, JIA X T, et al. Growth of large-area single- and Bi-layer graphene by controlled carbon precipitation on polycrystalline Ni surfaces[J]. Nano Research, 2009, 2(6): 509-516.
[8] RAHDAR A, ALIAHMAD M, AZIZI Y. NiO nanoparticles: synthesis and characterization[J]. J. Nanostruct. (JNS), 2015, 5: 145-151.
[9] ESQUINAZI P, HÖHNE R, HAN K H, et al. Magnetic carbon: explicit evidence of ferromagnetism induced by proton irradiation[J]. Carbon, 2004, 42(7): 1213-1218.
[10] YAZYEV O V, HELM L. Defect-induced magnetism in graphene[J]. Physical Review B, 2007, 75(12): 125408.
[11] BHOWMIK K, MUKHERJEE A, MISHRA M K, et al. Stable Ni nanoparticle-reduced graphene oxide composites for the reduction of highly toxic aqueous Cr(VI) at room temperature[J]. Langmuir: the ACS Journal of Surfaces and Colloids, 2014, 30(11): 3209-3216.

Low-temperature Synthesis and Luminescence of $La_{0.95}Eu_{0.05}BO_3$:M and $La_{0.95}Eu_{0.05}(BO_2)_3$:M Borates (M-Y, Sm, Tb, Bi)

Nadezhda STEBLEVSKAYA, Margarita BELOBELETSKAYA, Michail MEDKOV

Institute of Chemistry, Far-East Department of the Russian Academy of Sciences, Vladivostok, 690022, Russia

Abstract: Extraction-pyrolytic method is proposed for synthesis of orthoborate $LaBO_3$ and metaborate α-$La(BO_2)_3$ co-doped with ions Y^{3+}, Sm^{3+}, Tb^{3+}, Bi^{3+}. The compounds were characterized by X-ray phase analysis, IR and luminescent spectroscopy. The parameters of the crystal lattice of borates of various compositions were calculated. Structures of orthorhombic modification of aragonite for orthoborates and monoclinic modification of α-type for metaborates are preserved during doping. The Eu^{3+} ion-doped compounds show intense luminescence of red in the 400-750 nm region. The introduction of Tb^{3+} and further increase in its concentration leads to a decrease in the luminescence of all phosphors. Introduction of Sm^{3+}, Y^{3+}, Bi^{3+} to 5% (atom fraction) into orthoborates $La_{0.95-x}Eu_{0.05}BO_3$ increases luminescence intensity due to energy transfer from these ions to Eu^{3+}.

Keywords: Lanthanum borates; Doping; Rare earth elements; Luminescence; Pyrolysis

1 Introduction

Eu^{3+}, Tb^{3+}, Sm^{3+}, Dy^{3+}, Gd^{3+} and Ce^{3+}, having high luminescence efficiency, large stock shear and narrow-band radiation in the visible and near-infrared regions when exciting UV light, are used as activators in $LaBO_3$ and $La(BO_2)_3$. In order to increase the luminescence intensity, an ion-sensitizer is added to the luminescence phosphor, which transmits part of the absorbed energy to the activator ions for further glow[1-15]. The energy transfer phenomenon is used both to improve the luminescent characteristics of the phosphors and to extend the excitation spectrum of the ion-activator by transferring energy from the ion-sensitizer, which usually has a more intense absorption at a certain wavelength compared to the activator ion.

The main methods for producing effective ortho-or metaborates of lanthanum are solid phase[4,14], sol-gel[1,15], hydrothermal[3]. Solid-phase synthesis is characterized by high temperatures and the duration of calcination of the initial precursors, granulometric heterogeneity of the products. These inconveniences are partially eliminated when using alternative synthesis methods: hydrothermal or sol-gel. The functional properties of oxide materials, including the phosphorus, are influenced by a number of factors: morphology, structure and microstructure, concentration ratio of alloying ions, etc., which in turn are largely determined by the method of synthesis. It is often possible to obtain functional material with improved properties, including the most economical, only by a certain method. This document presents the synthesis data not previously used for this pyrolytic extraction method (EP) and the research on lanthanum borates, doped together with Eu^{3+}, Sm^{3+}, Y^{3+}, Tb^{3+}, Bi^{3+}.

2 Results and discussion

Synthesis of orthoborates of $La_{1-x}Eu_xBO_3$, $La_{0.95-x}Eu_{0.05}Sm_xBO_3$, $La_{0.95-x}Eu_{0.05}Y_xBO_3$, $La_{0.95-x}Eu_{0.05}Tb_xBO_3$, $La_{0.95-x}Eu_{0.05}Bi_xBO_3$, $La_{0.95-x}Eu_{0.05}Tb_{0.02}Bi_yBO_3$ and metaborates $La_{1-x}Eu_x(BO_2)_3$, $La_{0.95-x}Eu_{0.05}Sm_x(BO_2)_3$, $La_{0.95-x}Eu_{0.05}Y_x(BO_2)_3$, $La_{0.95-x}Eu_{0.05}Tb_x(BO_2)_3$, $La_{0.95}Eu_{0.05}Bi_x(BO_2)_3$, $La_{0.95-x}Eu_{0.05}Tb_{0.02}Bi_y(BO_2)_3$ ($x = 0.005; 0.01; 0.02; 0.025; 0.05; 0.075; y = 0.005; 0.01; 0.02; 0.025; 0.05; 0.075$) were performed by EP method, successfully used earlier to obtain some functional materials[5]. X-ray phase analysis was carried out on a D8 ADVANCE "BrukerAXS" diffractometer

(Germany) in Cu K_α radiation using the EVA search software with the PDF-2 powder data bank. Luminescence excitation and luminescence spectra of luminophores were recorded under the same conditions at 300 K on a Shimadzu RF-5301 PC spectrofluorimeter. IR spectra were recorded at room temperature on a Vertex 70.

For co-doped Y^{3+}, Sm^{3+}, Tb^{3+}, Bi^{3+} orthoborates, the structure of $LaBO_3$ aragonite is conserved, and for metaborates, the structure of α-type monoclinic modification $La(BO_2)_3$ changes, which formation begins in all cases at 550 ℃ and ends at 750 ℃ and 800 ℃ respectively.

When La^{3+} ion in $LaBO_3$ orthoborate and $La(BO_2)_3$ metaborate is replaced with additional ions with smaller ionic radii and when their concentration is increased, the parameters of the unit cell decrease somewhat (Table 1).

Table 1 Elementary cell parameters of orthoborates and metaborates

Phase composition	a/Å	b/Å	c/Å	α/(°)	β/(°)	γ/(°)	wR_p/%
$LaBO_3$	5.872	8.257	5.107	90	90	90	
$LaBO_3$:5%Eu	5.858(2)	8.229(2)	5.100(1)	90	90	90	2.98
$LaBO_3$:10%Eu	5.848(2)	8.202(2)	5.094(1)	90	90	90	3.12
$LaBO_3$:5%Eu+5%Tb	5.834(3)	8.181(3)	5.082(2)	90	90	90	7.81
$LaBO_3$:5%Eu+2.5%Bi	5.850(4)	8.208(6)	5.092(3)	90	90	90	6.51
$LaBO_3$:5%Eu+2.5%Bi+2%Tb	5.841(1)	8.210(2)	5.089(1)	90	90	90	8.21
$LaBO_3$:5%Eu+5%Sm	5.846(2)	8.206(2)	5.089(1)	90	90	90	4.12
$LaBO_3$:5%Eu+10%Sm	5.841(4)	8.197(6)	5.089(3)	90	90	90	6.52
$LaBO_3$:5%Eu+5%Y	5.853(2)	8.209(2)	5.093(1)	90	90	90	4.12
$La(BO_2)_3$	7.956	8.161	6.499	90	93.630	90	
$La(BO_2)_3$:5%Eu	7.942(3)	8.153(3)	6.480(2)	90	93.560(3)	90	3.36
$La(BO_2)_3$:5%Eu+2%Tb	7.928(2)	8.139(2)	6.458(2)	90	93.531(2)	90	2.23
$La(BO_2)_3$:5%Eu+2.5%Bi	7.939(2)	8.149(2)	6.478(2)	90	93.555(2)	90	3.21
$La(BO_2)_3$:5%Eu+2%Tb+2.5%Bi	7.938(2)	8.148(2)	6.466(1)	90	93.521(2)	90	2.84
$La(BO_2)_3$:5%Eu+5%Sm	7.924(2)	8.131(2)	6.454(2)	90	93.553(2)	90	2.23
$La(BO_2)_3$:5%Eu+10%Sm	7.918(2)	8.128(2)	6.443(1)	90	93.527(2)	90	2.84
$La(BO_2)_3$:5%Eu+5%Y	7.941(2)	8.152(2)	6.467(1)	90	93.519(2)	90	2.23

In IR-spectra of orthoborates, one can observe intensive absorption bands 550-1400 cm^{-1}, typical for vibrations of planar trigonal $[BO_3]^{3-}$-groups and (B-O) in $[BO_3]^{3-}$-groups, and for metaborates-absorption bands of tetrahedral $[BO_4]^{5-}$ and trigonal $[BO_3]^{3-}$ groups, from which correspondent crystal structures are constructed.

Excitation spectra of Eu^{3+} in $La_{0.95-x}Eu_{0.05}M_xBO_3$ orthoborates and $La_{0.95-x}Eu_{0.05}M_x(BO_2)_3$ metaborates at the excitation wavelength at the luminescence maximum of Eu^{3+} ion $\lambda_{em}=615$ nm are identical and contain the $O^{2-} \rightarrow Eu^{3+}$ charge transfer band at about 260 nm[2,5,6] (Figure 1). Narrow bands in the region of 310-420 nm correspond to resonance excitation of Eu^{3+} ion and transitions of f-electrons from ground state to excited levels $^5D_1$, $^5D_4$, $^5L_6$, $^5G_{4,5}$. In the excitation spectra of orthoborates and metaborates (Figure 1), registered at $\lambda_{em}=545$ nm and containing Bi^{3+} ion, the band λ_{max} nearly 262 nm of $^1S_0 \rightarrow {}^3P_1$ transition in the Bi^{3+} ion is present in the wavelength range 230-280 nm[5,11,12]. The intensity of this band increases with increasing Bi^{3+} concentration. At the same time, in the region below 260 nm in the excitation spectra $La_{0.95-x}Eu_{0.05}Tb_{0.02}Bi_yBO_3$ and $La_{0.95-x}Eu_{0.05}Tb_x(BO_2)_3$, a less intense band is superimposed on this band

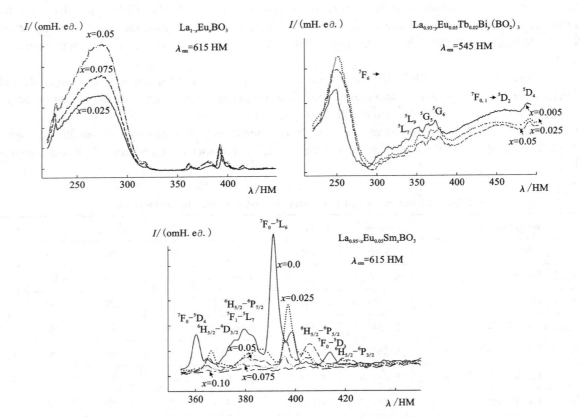

Figure 1 Excitation spectra of borates (300 K)

about 235 nm of the transition $4f^8 \rightarrow 4f^75d^1$ in the Tb^{3+} ion (Figure 1), which clearly manifests itself at $\lambda_{em} = 545$ nm (one of the luminescence bands of the Tb^{3+} ion) in doped Tb^{3+} orthoborates and metaborates[2, 4, 8]. In the excitation spectra of orthoborates and metaborates co-doped with Eu^{3+}, Tb^{3+} and Bi^{3+} ions at wavelength $\lambda_{em} = 545$ nm in the 300-350 nm range we also observe bands of different intensity and width related to the transitions from the ground state of Tb^{3+} ion $^7F_6$ to excited levels $^5D_0$, $^5D_4$, $^5L_7$, $^5L_9$, $^5G_5$, $^5G_6$. In the 360-420 nm region of the excitation spectrum of compounds doped with Sm^{3+} ion, the excitation bands appear at $\lambda_{em} = 615$ nm, related to the transitions of both the Sm^{3+} ion from the ground state $^6H_{5/2}$ to the $^4D_{3/2}$, $^6P_{7/2}$, $^6P_{3/2}$, $^6P_{5/2}$, $^4G_{5/2}$, $K_{11/2}$, $^6I_{13/2}$, $^6I_{11/2}$[7, 13] and the $^7F_0$ level of the Eu^{3+} ion to the excited levels $^5D_1$, $^5D_4$, $^5L_6$, $^5G_{4;5}$ (Figure 1). It should be noted that when Y^{3+}, Sm^{3+}, Bi^{3+} ions are included in the $La_{1-x}Eu_{0.05}BO_3$ or $La_{1-x}Eu_{0.05}(BO_2)_3$ composition, not only does the intensity of the bands in the excitation spectra change, but also a certain shift of the maxima of the excitation bands into the long wave region occurs, which correlates with a decrease in the parameters of the unit cell.

Emission spectra of doped phosphors recorded at excitation wavelengths $\lambda_{ex} = 260$ nm or $\lambda_{ex} = 235$ nm consist of a series of bands in the 450-750 nm region corresponding to transitions between multiplet $^5D_0$-^7F_j ($j = 0, 1, 2, 3, 4$) and characteristic for Eu^{3+} ion (Figure 2). The fact of conservation of Eu^{3+} ion luminescence spectra-position of transition bands and distribution of intensities in the bands at the same excitation wavelengths (λ_{ex}) in the series of doped orthoborates or metaborates at changing the concentration of doping ions proves the identity and conservation of the symmetry of Eu^{3+} ion close environment in the crystal structure of compounds. In the emission spectra of $\lambda_{ex} = 235$ nm $La_{0.95}Eu_{0.05}BO_3$ and $La_{0.95}Eu_{0.05}(BO_2)_3$ doped with Tb^{3+} and Bi^{3+} ions, besides the $^5D_0$-^7F_j transition bands of Eu^{3+} ion the about 545 nm transition band in the Tb^{3+} ion appears (Figure 2), and at $\lambda_{ex} = 260$ nm in the region of 420-450 nm a wide low-intensity band of 3P1-1S0 transitions in the Bi^{3+} ion is observed (Figure 2).

In contrast to orthoborates, a weak band of $^5D_0$-$^7F_0$ transition (λ about 580 nm) of Eu^{3+} (Figure 2) appears in the spectra of metaborates. The main share of the radiation energy of the ion Eu^{3+} doped $La_{1-x}Eu_xBO_3$ accounts for the dominant electrodipole $^5D_0$-$^7F_2$ transition (λ about 625 nm). The band of the

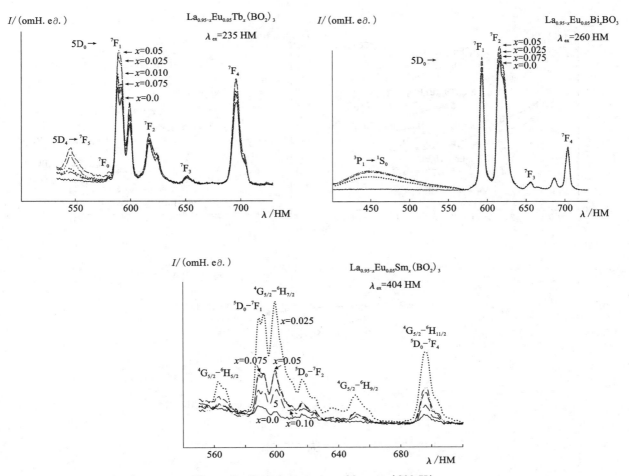

Figure 2 Emission spectra of borates (300 K)

magnetodipole $^5D_0$-$^7F_1$ transition (λ about 595 nm) has a slightly lower intensity. In the emission spectra of $La_{1-x}Eu_x(BO_2)_3$ with the same doping ions, the major part of the emission energy of the Eu^{3+} ion occurs at $^5D_0$-$^7F_1$ (λ about 595 nm) and $^5D_0$-$^7F_4$ (λ about 700 nm) transitions. The ratio of the intensities of the $^5D_0$-$^7F_1$ and $^5D_0$-$^7F_2$ transition bands indicates the lower symmetry of the Eu^{3+} ion nearest environment in doped orthoborates[6].

In the emission spectrum of $La_{0.95-x}Eu_{0.05}Sm_xBO_3$ and $La_{0.95-x}Eu_{0.05}Sm_x(BO_2)_3$ at λ_{ex} = 404 nm, additional broad bands appear with maxima at 602 nm, 648 nm, 703 nm corresponding to $^4G_{5/2} \rightarrow {}^6H_{7/2}$, $^4G_{5/2} \rightarrow {}^6H_{9/2}$, $^4G_{5/2} \rightarrow {}^6H_{11/2}$ transitions of Sm^{3+} ion (Figure 2)[7,11].

The dependence of the luminescence intensity determined by integrating the area under the bands in the emission spectra of phosphors is determined by both the type of doping ion and its concentration and has a complex character (Figure 3). Increasing the concentration of Y^{3+} ion to x = 0.05 leads to an increase in luminescence intensity (λ_{ex} = 615 nm). The introduction of Tb^{3+} and Sm^{3+} into orthoborates results in decreased luminescence of all phosphors. At addition of Bi^{3+} ion up to 5% (atomic fraction) in $La_{0.95-x}Eu_{0.05}BO_3$ and $La_{0.95-x}Eu_{0.05}Tb_xBO_3$ an increase in luminescence intensity is observed, which may be related to the possibility of energy transfer from Bi^{3+} to Eu^{3+} (λ_{ex} = 615 nm).

The decrease of integral luminescence intensity in metaborates at introduction of doping ions Tb^{3+} and Bi^{3+} can be explained by distinctive features of their crystal structure. At excitation of luminescence in band of maximum absorption of ion Sm^{3+} λ_{ex} = 404 nm, luminescence intensity of $La_{0.925}Eu_{0.05}Sm_{0.025}(BO_2)_3$ and $La_{0.925}Eu_{0.05}Sm_{0.025}BO_3$ increases (Figure 3). Considering that at λ_{ex} = 615 nm there is no increase in luminescence intensity of these compounds, the increase in luminescence intensity when excited in the band of maximum absorption of Sm^{3+} ion can be explained by the possibility of efficient transfer of absorbed energy by Sm^{3+} ion to Eu^{3+} ion. The increase of the doping ion concentration above a certain value (Figure 3) reduces the luminescence intensity, which is due to the absence of energy transfer between rare earth ions.

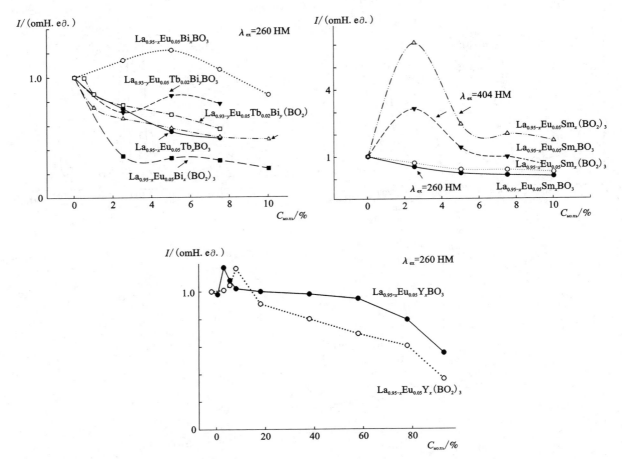

Figure 3 Dependence of integral luminescence intensity on doping ion concentration (300 K)

3 Conclusion

The extraction-pyrolytic method allows accurate introduction of dopants in a wide range of element ratios and producing efficient phosphors based on $LaBO_3$ orthoborate and $La(BO_2)_3$ metaborate doped with Eu^{3+}, Y^{3+}, Sm^{3+}, Tb^{3+}, Bi^{3+} ions at lower temperature and less process time than in solid-phase synthesis.

References

[1] FUCHS B, HUPPERTZ H. β-Eu(BO_2)$_3$-a new member of the β-RE(BO_2)$_3$ (RE=Y, Nd, Sm, Gd-Lu) structure family[J]. Zeitschrift Für Naturforschung B, 2019, 74(9): 685-692.
[2] SZCZESZAK A, KUBASIEWICZ K, LIS S. [J]. Opt Mater, 2013, 35(6): 1297-1303.
[3] ABACI O G H, ESENTURK O, YıLMAZ A, et al[J]. Opt Mater, 2019, 98: 109487-109489.
[4] ZHU Q, FAN Z S, LI S Y, et al. Implanting bismuth in color-tunable emitting microspheres of (Y, Tb, Eu)BO_3 to generate excitation-dependent and greatly enhanced luminescence for anti-counterfeiting applications[J]. Journal of Asian Ceramic Societies, 2020, 8(2): 542-552.
[5] STEBLEVSKAYA N I, MEDKOV M A. Coordination compounds REE. Extraction and obtaining of nanocomposites [J]. Deutschland, Saarbrucken: Palmarium academic publishing, 2012: 371.
[6] HALEFOGLU Y Z. Luminescent properties and characterisation of LaB_3O_6: Eu^{3+} phosphor synthesized using the combustion method[J]. Applied Radiation and Isotopes, 2019, 148: 40-44.
[7] BEIHOUCIF R, VELAZQUEZ M, PLATEVIN O. [J]. Opt Mater, 2017, 73: 658-665.
[8] SHMURAK S Z, KEDROV V V, KISELEV A P, et al. Energy transfer from Ce^{3+} to Tb^{3+} in yttrium and gadolinium orthoborates obtained by hydrothermal synthesis[J]. Physics of the Solid State, 2018, 60(12): 2579-2592.
[9] XU Y W, CHEN J, ZHANG H A, et al. White-light-emitting flexible display devices based on double network hydrogels crosslinked by YAG: Ce phosphors[J]. Journal of Materials Chemistry C, 2020, 8(1): 247-252.

[10] SHMURAK S Z, KEDROV V V, KISELEV A P, et al. [J]. Phys Solid State, (Rus), 2022 64(8): 955-966.
[11] SUN X R, YANG R R, SONG R X, et al. β-$RE_{1-x}Bi_xB_3O_6$(RE = Sm, Eu, Gd, Tb, Dy, Ho, Er, Y): Bi^{3+} substitution induced formation of metastable rare earth borates at ambient pressure[J]. Inorganic Chemistry, 2016, 55(18): 9276-9283.
[12] GAO Y, JIANG P F, GAO W L, et al. Facile synthesis of high-pressure polymorph β-$YB3O_6$ by co-doping Bi^{3+} and RE^{3+}(RE=Tb, Eu) with color-tunable emissions via energy transfer[J]. Journal of Solid State Chemistry, 2019, 278: 120915.
[13] BLASSE G, GRABMAIER B C. Luminescent Materials[M]. Berlin: Springer, 1994.
[14] WEI H W, SHAO L M, JIAO H A, et al. Ultraviolet and near-infrared luminescence of $LaBO_3$: Ce^{3+}, Yb^{3+}[J]. Optical Materials, 2018, 75: 442-447.
[15] YANG R R, SUN X R, JIANG P F, et al. Sol-gel syntheses of pentaborate β-$LaB5O_9$ and the photoluminescence by doping with Eu^{3+}, Tb^{3+}, Ce^{3+}, Sm^{3+}, and Dy^{3+}[J]. Journal of Solid State Chemistry, 2018, 258: 212-219.

Research of the Effect of the Vacuum-chamber Design on the Technological Parameters of Steel Processing in a RH-degasser

O. Y. SHESHUKOV[1,2], A. A. METELKIN[1]

1 Ural Federal University named after first President of Russia B. N. Yeltsin, Yekaterinburg, Russia
2 Federal State Budgetary Institution of Science Institute of Metallurgy of the Ural Branch of the Russian Academy of Sciences, Yekaterinbur, Russia

Abstract: In order to obtain a low residual content of hydrogen, nitrogen and carbon in steel, the metal is processed at RH-degasser. This metallurgical unit is at the last stage of steel processing before casting on continuous casting machines, so it is important to study and improve the technological processes in this unit. Based on the physical model, the main dependencies between the structural and technological parameters of a RH-degasser designed for metal processing in steel ladles with a capacity of 140-180 t were determined. Theoretical calculations of the process were carried out, which were confirmed by practical smelts in a steelmaking unit. It is shown that the wear of the lining of the inlet pipe of the vacuum chamber affects the technological process of vacuuming.

Keywords: Hydrogen removal; Decarbonization; RH-degasser; Out-of-furnace steel treatment; Degassing; Rational parameters

Improving the quality of products is one of the main goals of modern metallurgical enterprises.

The high quality of the manufactured products is ensured by the gradual processing of the melt by various technological operations in various metallurgical units. The RH-degasser is at the last stage of steel processing before casting on continuous casting machines, so it is important to study and improve the technological processes in this unit.

At most enterprises, the tasks of RH-degasser are to obtain a given content of dissolved gases, such as hydrogen and nitrogen. Additionally, it is possible to carry out a technological operation-decarbonization, i. e. removal of carbon, along the way of interaction with oxygen and the formation of gaseous products. To optimize the technological parameters of the RH-degasser installation, it is necessary to study the processes of gas removal and decarbonization in this metallurgical unit.

The lower part of the vacuum chamber has two snorkels, i. e. inlet and outlet. Before vacuuming, both snorkels are immersed in the metal located in the steel ladle. In the vacuum chamber, a vacuum is created and the metal begins to rise into the snorkels, in addition, neutral gas is supplied to one of the snorkels (inlet), the bubbles of which, loosening the liquid steel, reduce its density, which leads to the suction of an additional amount of melt into the RH-degasser.

In the vacuum chamber, the metal becomes denser and flows out through the outlet snorkel[1]. The process of vacuuming steel is shown in Figure 1.

Consider the degassing process.

According to the literature data[2], the final hydrogen content may be determined by the Equation (1)

$$[H]_t = ([H]_0 - [H]_p) \cdot 10^{-0.227 \cdot n} + [H]_p \tag{1}$$

where: $[H]_t$ is the final hydrogen content after time, t; $[H]_p$ is equilibrium hydrogen content, by $p = 0.07$ kPa (residual pressure in the vacuum chamber) $[H]_p = 0.64 \times 10^{-6}$; $[H]_0$ is initial hydrogen content, before processing; η is the multiplicity of circulation, this parameter is determined by Equation (2):

$$\eta = \frac{G \cdot t}{M} \tag{2}$$

where: G is the amount of metal entering the vacuum chamber (circulation speed), t/min; t is processing time, min; M is the mass of metal in the steel ladle, t.

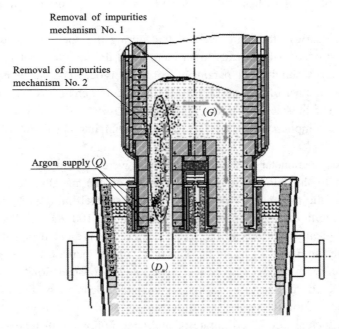

Figure 1 The process of vacuuming steel on a RH-degasser and the mechanisms of removing hydrogen in a vacuum chamber

According to Equation (1), the main indicator for the degassing of steel is the n-multiplicity of circulation, i.e. the greater this parameter, the lower the residual hydrogen content in the metal.

However, it is known that there are several ways to remove gases dissolved in metal[3-6]. In our opinion, two ways are realized in the RH-degasser: the first from the metal surface in the vacuum chamber, the second through the contact surface of the neutral gas bubbles and the melt (Figure 1).

The authors[7-8] showed that the main mechanism of degassing is the contact surface of neutral gas bubbles and metal.

Thus, when studying the degassing process, it is necessary not only to take into account the dependencies (1) and (2), but also to determine the rational technological parameters of the RH-degasser taking into account the literature data[7-8].

It is known that there is a relationship between the amount of neutral gas supplied to the inlet snorkel of the vacuum chamber (Q), the metal circulation speed (G), the inner diameter of the inlet pipe (D_u) and the contact surface of the neutral gas and metal bubbles (S_{puz}). To analyze the process of steel vacuuming, it is necessary to identify the relationships between all the presented parameters and verify the results obtained in practice.

To determine the relationship between the technological parameters of a RH-degasser based on UrFU, a model of a circulating type vacuum cleaner installation was created.

Three experiments were carried out with different diameters of the inlet snorkel, in each of the experiments, the flow rate of gas supplied to the inlet snorkel was varied, in addition, in each experiment, the liquids speed was measured. Water with the addition of aluminum powder was used as a liquid modeling the movement of the melt, aluminum powder was suspended in water and the distance of movement of Al particles for a certain moment of time was determined by its metallic luster.

Based on the experiments conducted to simulate the movement of liquid in a vacuum chamber, the interdependencies between the technological parameters of a RH-degasser designed for metal processing in steel ladles with a capacity of 140-180 tons were determined. It is revealed that the speed of dependence is described by Equation (3)

$$G = \frac{0.0209 \cdot Q + 28.944}{0.6686 \cdot D_u^{-0.8678}} \text{ t/min} \qquad (3)$$

Equation (3) makes it possible to determine the metal circulation speed (G) in the vacuum chamber depending on the gas flow rate (Q) supplied to the inlet snorkel and the diameter of the inlet snorkel (D_u).

Equation (3) in the range of Q values from 800 to 3000 L/min and D_u from 0.4-0.75 m gives an error of no more than 20%.

Additionally, it is necessary to take into account that during the operation of the vacuum chamber, the lining of the snorkels is destroyed, which leads to an increase in the inner diameter of the inlet snorkels and, accordingly, a change in the technological parameters of steel processing. The inner diameter of the inlet snorkels will depend on the number of processed melts in the vacuum chamber (wear resistance of the lining).

Technological parameters of steel processing in a RH-degasser with a capacity of 140-180 t have the following values: argon consumption 800-1000 L/min, D_u initial internal diameter of the inlet pipe of the vacuum chamber 0.42 m.

The main technological parameters of steel processing were determined, with different wear of the snorkel, which correspond to the number of processed melts on the metallurgical unit. At the end of the vacuum chamber operation, the minimum thickness of the inlet snorkel lining is 30 mm, while the number of processed melts or the durability of the lining reaches more than 100 melts. The data, according to the calculations of the technological parameters of the RH-degasser, are presented in Table 1.

From the data presented in Table 1, it can be seen that with an increase in the inner diameter of the inlet snorkel (D_u), for example, with severe wear of the lining, and a constant flow of neutral gas supplied to the inlet pipe (Q), the number of gas bubbles in the gas-metal suspension of the inlet snorkel will change, respectively, the area of interaction of the melt with the surface of the neutral gas (S_{puz}) will increase.

Additionally, with an increase in S_{puz}, the rate of hydrogen removal will increase, which will lead, with the same metal processing time, to a decrease in the content of this gas in the metal.

Table 1 Calculated values of the parameters G, S_{puz}, at a constant value Q
(constant volume of gas flow into the inlet snorkel)

No.	Gas flow rate supplied to the inlet pipe (Q) /(L·min^{-1})	Inlet snorkel diameter (D_u) /m	Number of processed melting /pcs	Metal circulation speed (G) /(t·min^{-1})	The speed of metal movement in the inlet pipe /(m·s^{-1})	The area of interaction of the melt with the neutral gas surface (S_{puz})/m^2
1	800	0.42	0	32.17	0.55	22.77
2	800	0.45	9	34.16	0.51	23.81
3	800	0.48	18	36.12	0.48	24.79
4	800	0.51	27	38.07	0.44	25.71
5	800	0.54	36	40.01	0.42	26.59
6	800	0.57	45	41.93	0.39	27.42
7	800	0.60	54	43.84	0.37	28.21
8	800	0.63	63	45.74	0.35	28.95
9	800	0.66	72	47.62	0.33	29.66
10	800	0.69	81	49.49	0.32	30.34
11	800	0.74	96	52.59	0.29	31.39
12	800	0.75	99	53.21	0.29	31.59

To verify this assumption, process degasser on the vacuum chambers was analyzed. The selected melts were sorted by the state lining of the inlet snorkel, the vacuum chamber and the processing time of the steel at the RH-degasser. Also, these melts measurements were carried out on the hydrogen content after processing the steel on the RH-degasser and in the tundish of the continuous casting machine. The selected data set consisted of 219 melts, the time of processing the melt by at the RH-degasser was 17-18 min. The results obtained are shown in Figure 2.

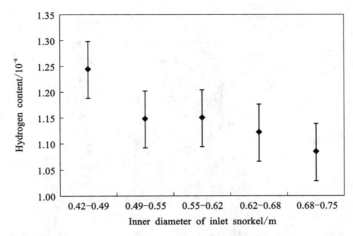

Figure 2 Dependence of the hydrogen content in the metal on the inner diameter of the inlet snorkel, the melt treatment time in a vacuum chamber is 17-18 min

From the data presented in Figure 2, it can be seen that with an increase in the inner diameter of the inlet snorkel the residual hydrogen content in the metal decreases, which corresponds to the assumption that the contact surface of gas bubbles with the melt increases.

Thus, during the operation of the vacuum chamber, it is necessary to take into account the wear of the refractory products of the inlet snorkel and, with an increase in the number of processed melts in the RH-degasser, it must introduce changes in the technological parameters of metal processing.

Consider the decarbonization process.

When studying the mechanism of removing carbon from the melt, in the system of aggregates "RH-degasser-steel ladle", an integrated approach is needed, which consists not only in considering the mechanisms or ways of removing this element from the melt, but also in the need to consider ways of dissolving or getting carbon into the metal during the destruction of the lining of the steel ladle.

The main ways to remove carbon from the melt are the following:

(1) The formation of {CO} gas in the volume of the melt located in the vacuum chamber, and the further ascent of the bubble to the metal surface.

(2) The formation of {CO} gas and its removal by bubbles of inert gas supplied to the inlet pipe of the RH-degasser.

Simultaneously with the removal of carbon, the reverse process occurs—saturation of the melt with carbon. An increase in the carbon concentration in steel occurs due to the destruction of the lining of the steel ladle, which is made of periclase-carbon products (Figure 3).

The first mechanism of carbon removal is the main and predominant, but only during the initial treatment period. With a high content of this element in steel (up to 0.006%-0.007%), the nucleation of {CO} bubbles in the melt can occur at a considerable depth. As the carbon concentration decreases to 0.005%-0.006%, the depth of bubble nucleation decreases, respectively, and the rate of carbon removal by this mechanism decreases.

The second mechanism is the removal from the melt by diffusion, both with bubbles of Ar supplied to the inlet pipe and from the surface of the melt in a vacuum chamber. At low concentrations of carbon in the melt, this mechanism becomes predominant, however, the removal rate, which depends on the contact area of the melt with gaseous phases, is more constant and less dependent on the carbon content in the metal.

The increase in the carbon content in the metal is due to the destruction of the lining of the steel ladle. The rate of carbon saturation of the melt is uneven and depends on the condition of the lining and the contact area of the metal with the lining. In the initial period of vacuum treatment, the saturation rate is not high, because the intensity of metal mixing is less than that during deep vacuum treatment. Over time, the mixing intensity increases, respectively, the wear of the lining of the steel ladle and the rate of saturation of the melt with carbon increase.

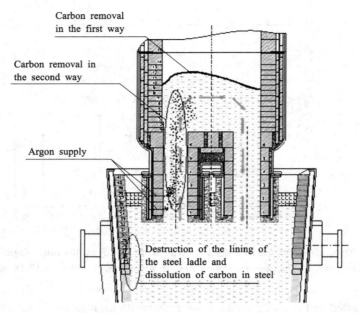

Figure 3 Mechanisms of removal and carbon saturation of the melt in the system "RH-degasser-steel ladle"

From the practical data provided by the metallurgical enterprise, we will calculate the rate of carbon removal from the metal in the "RH-degasser-steel ladle" system. The results are shown in Figure 4.

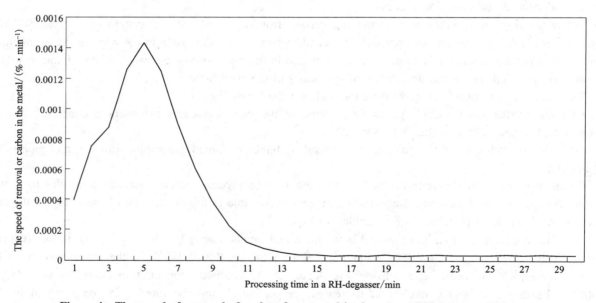

Figure 4 The speed of removal of carbon from metal in the system RH-degasser-steel ladle

We compare the rates of carbon removal by two mechanisms. Additionally, it is necessary to take into account the rate of dissolution of carbon in the metal. The data are shown in Figure 5.

From the data presented in Figure 5, it can be seen that in order to reduce the carbon content in the metal, it is necessary to reduce the speed of dissolution (saturation) of this element of liquid steel. This is achieved by reducing the wear rate of refractory products of the steel ladle.

To reduce the wear of refractory products, it is necessary to apply a protective garnish to the lining surface, which is formed from slag when the metal level in the steel ladle decreases, during the casting of steel at the CCM. In the literature, the conditions for the formation of a protective garnish were determined, the main of which are the content of Al_2O_3 more than 15% and the content of MgO in the range of 6.5%-7.5%. An additional condition is that the slag must be homogeneous.

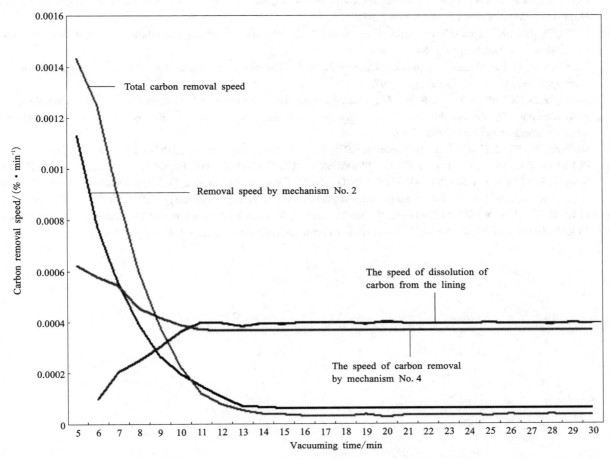

Figure 5 Speeds of removal and dissolution of carbon in metal in the system "RH-degasser-steel ladle"

Thus, to reduce the wear of the lining, it is necessary to bring homogeneous slag of the following chemical composition CaO 52%-54%, Al_2O_3 15%-25%, SiO_2 15%-23%, MgO 6.5%-7.5% at the previous melting. This slag composition will create a protective garnish, reduce the specific wear of the lining of the steel ladle and reduce the rate of dissolution of carbon in steel.

Additionally, it should be noted that the speeds of removal and dissolution of carbon will depend on the technological parameters of the RH-degasser, i.e., on the inner diameter of the inlet pipe (D_u), the flow rate of neutral gas supplied to the inlet pipe (Q), the area of interaction of the melt with the surface of the neutral gas (S_{puz}) and the metal circulation speed (G). To obtain an ultra-low carbon content (less than 0.002%), it is necessary to take into account the interdependencies between all parameters. However, these dependencies are more complex than when removing dissolved gases.

Conclusions

(1) During the operation of the vacuum chamber, the lining of the inlet snorkel is destroyed, which leads to a change in the technological parameters of steel processing in it.

(2) During the operation of the vacuum chamber, it is necessary to take into account the wear of the refractory products of the inlet snorkel and with an increase in the number of processed melts in the RH-degasser, and it is necessary to introduce changes in the technological parameters of metal processing.

References

[1] BIGEEV A M. Metallurgical table. Theory I technology plavki table: uchebnik dlya vuzov/a. M. Bigeev, V. A. Bigeev. -3-e izd. pererab. I DOP. -Magnitogorsk: MGTU, 2000: 544.

[2] GIZATULIN R A. Vnepechnie I Kovshevie prosessi obrabotki Stali: ucheb. posob. dlya vuzov/R. A. Gizatulin, V. I. Dmitrienko[M]. Novokuznesk: Sibgiu, 2006.

[3] KNYUPPEL G. Raskislenie I vakuumnaya obrabotka Stali. Osnovi i technology kovshovoy metallurgii/G. Knyuppel [M]. Moscow: Metallurgy, 1984.

[4] KNYUPPEL G. Raskislenie I vakuumnaya obrabotka Stali. Thermodynamicheskie I kineticheskie zakonomernosti/G. Knyuppel[M]. Moscow: Metallurgy, 1973.

[5] METELKIN COMPILED A A. K voprosu udaleniya trace metal in hydrogen V vacuumatore tsirkulyasionnogo TIPA/ a. A. Metelkn, O. Yu. Sheshukov, I. V. Nekrasov, O. I. Shevchenko, A. Yu. Korogodsky//theory i technology metallurgicheskogo proizvodstva, 2016, 1: 29-33.

[6] DIENER A, STOLTE G. VII. Internationale DH/BV-Vakuumtagung Malente (BRD) 1975, S. 209/36

[7] PLESHIVSEV K. N. IZUCHENIE PROSESSA UDALENIYA HYDROGEN V CIRCULATIONNOM VAKUUMATORE V USLOVIYAX KTS-2 PAO "NLMK"/K. N. Pleshivsev, O. Yu. Sheshukov, A. A. Metelkin, O. I. Shevchenko[C]//Izvestia visshix uchebnix zavedenius. Chernaya metallurgy, 2021, 64(8): 543-549.

[8] PLESHIVTSEV K N, SHESHUKOV O Y, METELKIN A A, et al. Hydrogen removal in circulating vacuum degasser under conditions of PJSC "NLMK"[J]. Izvestiya Ferrous Metallurgy, 2021, 64(8): 543-549.

Roadmaps "Fractals" "Artificial Intelligence" and "Photonics" Need to Be Combined (Our International Priorities in End-to-end Technologies)

Alexander A. POTAPOV[1,2]

1 Kotel'nikov Institute of Radio Engineering and Electronics of Russian Academy of Sciences, Moscow, 125009, Russia

2 Jinan University, College of Information Science and Technology, Department of Electronic Engineering, JNU-IREE RAS Joint Laboratory of Information Technology and Fractal Processing of Signals, Guangzhou, 510632, China)

Abstract: In this article, a brief retrospective analysis of principles of roadmaps "Fractals" "Artificial Intelligence" and "Photonics" is presented. At the present time, promising applications of artificial intelligence in many branches including optics, engineering, medicine, economics, etc. appear. Recently, in many fields of optics and photonics, new perspectives appear because of optical metasurfaces changed rules of manipulation of light. Application of fractal systems, transducers and nods are principally new decisions that essentially changes principles of design of complex intellectual radioengineering systems and devices. We also present our view point concerning selected ideas and perspective directions in the field of fundamental interdisciplinary researches. In the symposium report, the range of concepts and examples under consideration will be essentially expanded.

Keywords: Fractal; Scaling; Photonics; Artificial intelligence; New technologies

1 Introduction

This work is an attempt of briefly presentation of our results in the field of fractals, photonics and artificial intelligence taking into account their synergy. Before the consideration of a basic subject, a brief presentation of essence of these three scientific fields is necessary.

Relevance of these researches is connected with necessity of more precise description of all real processes that occur in modern complex systems, including radiophysic and radioengineering ones, taking into account memory (hereditarity), non-Gaussian and scaling of physical signals and fields[1]. The concept of fractal engineering is being developed.

We point out that consideration of obtained data in the context of new technologies of 21st century is necessary. End-to-end technologies are of key importance to develop a few perspective markets simultaneously. This will undoubtedly make a worthy practical contribution to the developing philosophy of engineering.

As compared with the reports presented by the author at 14th Sino-Russia Symposium "Advanced Materials and Technologies" 2017 (Sanya, China), the new theoretical and experimental results are included.

2 Fractals and fractality

For the first time in the world the author began to investigate the problem of fractals and fractality more than forty years ago at the Institute of Radio Engineering and Electronics of the USSR Academy of Sciences (since 1979)[1-10]. The main distinction of the fractal methods proposed by the author from classical ones is connected with the fundamentally new approach for signals and fields and their constituents. This permitted to transform into the new level of information structure of real non-Markovian signals and fields. Fractals are related to sets with an extremely irregular branched or jagged structure. The theory of fractals considers fractional measures instead of integer ones and it is based on the new quantitative indicators in the form of

---

E-mail: potapov@cplire.ru.

fractional dimensions D and corresponding fractal signatures $D(t, f, r)$. Fractal dimensions D describe both topology of the object and reflect the processes of evolution of dynamic systems and they are connected with their properties.

The main stages and directions of fundamental research on texture and fractal technologies in modern radiophysics and radioelectronics are schematically presented in Figure 1-Figure 5.[1-10]

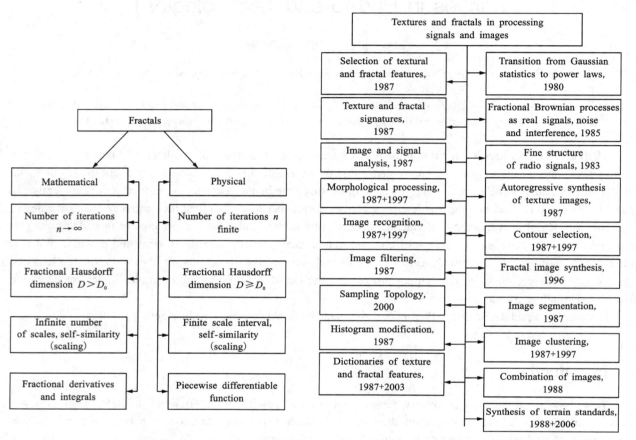

Figure 1 The author's classification of fractal sets and signatures approved by B. B. Mandelbrot

Figure 2 Classification of texture and fractal methods for processing low-contrast images and ultra-weak signals

Fractal geometry is the great merit of B. B. Mandelbrot (1924-2010). But its radio physics/radio engineering and practical implementation is the exclusive merit of known in the world Russian scientific school of fractal methods under the direction of Professor A. A. Potapov (V. A. Kotel'nikov IREE of RAS). The classification of fractals (Figure 1) developed by the author was approved by Professor B. B. Mandelbrot in December 2005 in USA (Figure 4). This fact is of great importance among the world community of scientists dealing with dynamic chaos and fractals.

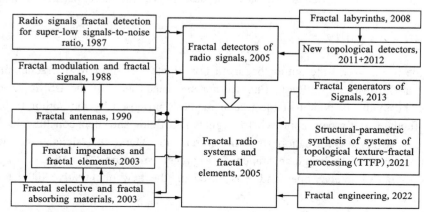

Figure 3 The concept of fractal radiosystems, sensors, devices and radioelements

Figure 4　Prof. B. B. Mandelbrot and Prof. A. A. Potapov: New York, USA, 2005
B. Mandelbrot created the theory of fractals in 1975

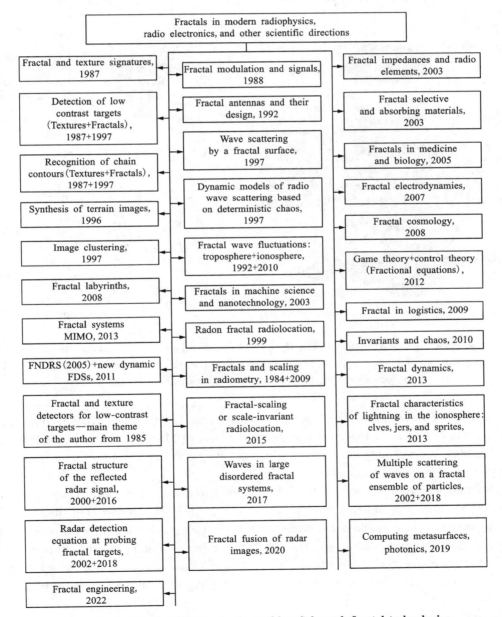

**Figure 5　A sketch of the development of breakthrough fractal technologies:
FNDRS-fractal non-parametric detector of radar signals, FDS-fractal detector of signals**

3 Photonics: Dielectric metamaterials and computational metasurfaces

Continuous improvement of topological texture-fractal processing (TTFP) of signals and fields in modern radiophysics and radioelectronics implies a constant improvement of the speed of information processing and the search for new physical principles for its implementation. Here, undoubtedly, the future belongs to photonic and radiophotonic technologies. The results in the field of photonics, radiophotonics, computational meta-optics and 2D dielectric metamaterials (MM) or computational metasurfaces (MS) are presented below. These results were obtained by the author with Chinese scientists at the Joint Laboratory of Information Technology and Fractal Signal Processing of Jinan University in Guangzhou, China for period 2019-2023. The results have been published in leading international scientific journals.[11] It should be noted that China has a special state program, and in 2015 China became the world leader in the field of production of photonics devices. The proposed Laplace MS is based on the excitation of a bound state in a continuum which has demonstrated exotic optical properties (Figure 6). The highly symmetric mode profile provides an almost isotropic Laplace the optical transfer function (OTF).

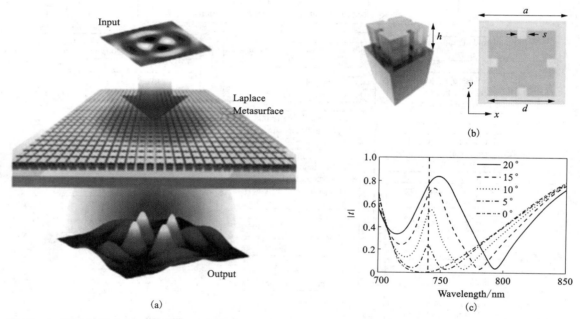

Figure 6 Dielectric Laplace MS transforming the input two-dimensional space function into another function as the Laplace operator (a), Unit cell of the Laplace dielectric MS (b), Transmission spectra of the Laplace MS at various angles of incidence along the x direction for the p-wave[11] (c)

MS is the base of many interesting topological phenomena in physics and exotic manipulations with waves. Development of optical analogue calculations on the basis of MS gives unique possibilities for effective collection of information concerning the contour of images with parallel processing, ultra-fast calculation speed and low and even zero energy consumption.

In view of restricted work volume, other directions of photonics that are investigated and presented in detail in works are briefly mentioned by the author.[11-17] They are: control of light scattering with nanoparticles by means of magnetoelectric connection and zero backscatter (the theory of light scattering with nanoparticles and electromagnetic multipoles, numerical simulation, validation experiments at the range of 4 up to 7.5 GHz); strong optomechanical connection in coupling in chain-like waveguides of silicon nanoparticles with quasi-bound states in the continuum (photon-phonon coupling with microstructures), etc.

4 Artificial intelligence

Our first works concerning a fractal application in problems if artificial intelligence (AI) were appeared at the beginning of XXI century.[6] Now we deal with the application TTFP multidimensional signals and big data (for example, the project Digital Earth). Currently the volume of data processing is incredibly increased and photons currying multidimensional information can effectively extend throughput of information operations. In synergy artificial intelligence and photonics will help to carry out the investigations and develop ment of modern optical chips that will provide implementation of optical devices and systems of next generation. In total it opens great potential possibilities and new perspectives for processing of multidimensional signals in adjacent scientific and technical fields.[18, 19] In other words, a full description of processes of modern signal and field processing is impossible by means of approaches and formulae of classical mathematics. Fractal-scaling methods of signal, wave field and image processing are based on the information that in classical processing methods just could not be taken into account.

5 Conclusion

We presented some of the most interesting current tendencies in considered fields and discussed possible future research directions. Today the obtained results allow us to speak about the most influential (and the only one in Russia) Russian scientific school of fractal engineering in all areas of radioelectronics and radiophysics. Now this school is widely known in the world.[1-19] The global fractal-scaling method and TTFP owing to our pioneer works is developed. It exists and has a known completeness, well-known world priorities and fully deserves serious analysis. As a result in the scientific world the new sense space with unusual for classical sciences properties and problems is formed. All this determines fractal engineering.[18]

We point out that in accordance with the author's monographs lecture courses on fractals in different Russian universities and neighboring countries and in China were created. At the beginning of 2023 the results of fundamental investigation of the author are presented in more than 1200 works and 45 books and their chapters in Russian, English, and Chinese, the reports in 23 countries are made.[3, 4]

In conclusion the author is sure that in the future integration of roadmaps "Fractals" "Photonics" and "Artificial Intelligence" will bring out human civilization on a higher level.

Acknowledgments: The author is grateful to the China grant "Leading Talent Program in Guangdong Province" (No. 00201502, 2016-2022) Jinan University (China, Guangzhou). The author is also sincerely grateful to the big team of Chinese scientists that are co-workers of articles on photonics[11-17] in leading international scientific journals during 2019-2023 and to the Chinese scientists that promptly sent to me via Research Gate their recently released original works.

References

[1] POTAPOV A A. Author's approaches to fractal engineering and the philosophy of fractal engineering: fractal radio systems and international priorities in the study of fractal applications in radio electronics[C]//2022 IEEE Conference on Antenna Measurements and Applications (CAMA). Guangzhou, China. IEEE, 2023: 1-6.
[2] POTAPOV A A, WU H, XIONG S. Fractality of wave fields and processes in radar and control[M]. Guangzhou, South China University of Technology Press, 2020.
[3] POTAPOV A A. Fractals in action: Biography and publication index, Acad[M]. Yu V Gulyaev, Ed., Moscow: Raduga Publ., 2019.
[4] POTAPOV A A. Short scientific biography in International Forum of Industrial Development of New Materials (Jining, China, 11-13 December, 2019), Jining, Jining National High-tech Industrial Development Zone, 2019, 8 (Chinese, Japan, Russian).
[5] POTAPOV A A. Fractals in radio physics and radar[M]. Moscow: Logos Publ., 2002.
[6] POTAPOV A A. Fractals in radio physics and radar: Topology of a sample, 2rd ed. [M]. Moscow: University book, 2005.
[7] POTAPOV A A. Fractals and chaos as a base of new breakthrough technologies in modern radio systems[M].//R. M. Crownover, Introduction to Fractals and Chaos, Moscow: Tekhnosfera, 2006: 374-479.
[8] BUNKIN B V, REUTOV A P, POTAPOV A A. et al. Aspects of perspective radiolocation[M]. Moscow: Radiotekhnika, 2003.
[9] POTAPOV A A. Chaos theory, fractals and scaling in the radar: a look from 2015[M]// The Foundations of Chaos

Revisited: From Poincaré to Recent Advancements. Cham: Springer, 2016: 195-218.
[10] POTAPOV A A. Postulate "The Topology Maximum at the Energy Minimum" for textural and fractal-and-scaling processing of multidimensional super weak signals against a background of noises, in Nonlinearity: Problems, Solutions and Applications, Ludmila Uvarova et al., Ed., New York: Nova Science Publ., 2017, 2: 35-94.
[11] PAN D P, WAN L, OUYANG M, et al. Laplace metasurfaces for optical analog computing based on quasi-bound states in the continuum[J]. Photonics Research, 2021, 9(9): 1758-1766.
[12] WAN L, PAN D P, YANG S F, et al. Optical analog computing of spatial differentiation and edge detection with dielectric metasurfaces[J]. Optics Letters, 2020, 45(7): 2070-2073.
[13] FENG T H, POTAPOV A A, LIANG Z X, et al. Huygens metasurfaces based on congener dipole excitations[J]. Physical Review Applied, 2020, 13(2): 021002.
[14] FENG T H, YANG S F, LAI N, et al. Manipulating light scattering by nanoparticles with magnetoelectric coupling[J]. Physical Review B, 2020, 102(20): 205428.
[15] WAN L, PAN D P, FENG T H, et al. A review of dielectric optical metasurfaces for spatial differentiation and edge detection[J]. Frontiers of Optoelectronics, 2021, 14(2): 187-200.
[16] YANG S F, WAN L, WANG F G, et al. Strong optomechanical coupling in chain-like waveguides of silicon nanoparticles with quasi-bound states in the continuum[J]. Optics Letters, 2021, 46(18): 4466-4469.
[17] WANG F G, YUAN J, YANG S F, et al. Compact ring resonators of silicon nanorods for strong optomechanical interaction[J]. Nanoscale, 2023, 15(10): 4982-4990.
[18] POTAPOV A A. Topological texture-fractal processing of signals and fields in radiophysics, radio engineering and radiolocation: developed methods and technologies (1979-2022)-fractal engineering[J]. Annual Geospatial Almanac 《GeoContext》, 2022, 10(1), 6-56.
[19] POTAPOV A A. Fractal technologies: Problems and prospects. Collection of abstracts of the XXVII Baikal All-Russian Conf. with International Participation "Information and Mathematical Technologies in Science and Management", Baikal session, (Lake Baikal, Olkhon Island, June 29-July 08, 2022), Irkutsk, Melentiev Energy Systems Institute of the Siberian Branch of the RAS, 2022, 19.

Synthesis of a Graphene-containing Composite by Deposition of Oxygen-free Graphene on Nanocrystallite CeO_2

Ivan PONOMAREV, eLENA TRUSOVA

Baikov Institute of Metallurgy and Material Science of the Russian Academy of Sciences, IMET RAS, Moscow, 119334, Russia

Abstract: Method is proposed for the synthesis of hybrid structures based on nano-CeO_2 and oxygen-free graphene, which consist in applying the latter onto oxide powder. As a result, it is possible to achieve an uniform distribution of components in the material bulk. The resulting composite consisted of 20-60 nm agglomerates is formed by CeO_2-crystyllites with sizes 3-11 nm, wrapped by graphene sheets with a thickness of 3-5 nm. A comparative analysis of morphology of pure CeO_2 and graphene-CeO_2 composite using nitrogen adsorption-desorption shows that when graphene is applied to CeO_2 crystallites, the porosity character practically does not change compared to the initial powder. In both cases, these are mesoporous systems with similar pore sizes (4.5 and 4.7 nm) and micropores fraction not exceeding 3% (by volume). However, in the composite, the specific surface area is one-third smaller, which can be explained by the partial blocking of the pores by graphene sheets impenetrable to N_2 molecules. The proposed method for the synthesis of composite nanostructured powders based on graphene and CeO_2 can be used as the basis for an economical and environmentally friendly technology for the production of nanopowders, which are in demand in the development of materials for small-sized electronic devices.

Keywords: Oxygen-free graphene; Nano-CeO_2; Nanostructured composite

1 Introduction

Currently, researchers and developers are interested in materials for supercapacitors, which are in demand for the development of small-sized electronic devices. Graphene due to its high specific surface area and electrical conductivity is an ideal supercapacitor electrode material. However, the most common method of obtaining graphene is reduction of oxidized graphene (GO). The resulting reduced oxidized graphene (RGO) doesn't have the all complex of unique electronic properties of oxygen-free graphene, in addition, graphene sheets prone to agglomeration after reduction, which leads to a significant capacity loss. Last decade the production of hybrid materials based on graphene and various metal oxides, for example, manganese, zinc or stannum, has attached great interest[1, 2]. Graphene-containing composites based on RuO_2 contrast favorably from them[3], but such materials are relatively more expensive than those already used. Graphene-containing materials based on CeO_2 are promising, both from an economic and practical points of view[4, 5].

Nano-CeO_2 is in demand in the various fields of industry, in particular, in production of structural ceramics, adsorbents, catalysts[6]. Addition of graphene to nanocrystalline CeO_2 makes it possible to improve its electronic properties, which is important for increasing the capacity and the number of supercapacitor working cycles[7, 8].

For the effective composite work, the graphene content in it should not be more than 2 wt.%, and it should be uniformly distributed in the bulk of the material and contain a minimum amount of oxygen. The last requirement presents the greatest difficulty. Therefore, currently RGO is most often used in graphene-containing metal-oxide composites, which doesn't have the all complex of properties of oxygen-free graphene. In addition, the methods for obtaining RGO are based on the Hammers method, i.e. requiring the use of strong oxidizing and toxic reducing agents.

In the paper, the method developed at IMET RAS for obtaining of nanocomposites based on oxygen-free graphene and nanocrystalline CeO_2 is presented.

2 Experiment

The synthesis of the composite was carried out in three stages (Figure 1). At the first stage, an oxygen-free graphene suspension was obtained in an isopropanol-water mixture (1:1, vol.) according to a previously developed technique. Graphite powder (600-800 microns) in a mixture of isopropanol-water was subjected to ultrasonic treatment in the Sonoswiss SW1H unit (200 W, 38 kHz) for 1 h. The result graphene-graphite suspension was separated by sedimentation and step decantation within 24 h for obtaining pure graphene suspension.

At the second stage, nano-CeO_2 was synthesized by sol-gel method with the use of $Ce(NO_3)_3$ as metal source and dodecylamine (DDA) for formation and stabilization of sol and acetylacetone (AcAcH) as complexing agent. Resulting sol was evaporated with stirring and heating to 90-95 ℃ to form a gel, which was subjected to heat treatment at 500 ℃ in oven for 1 h. At the third stage an aqua-alcohol suspension of synthesized nano-CeO_2 was mixed with the decanted graphene suspension with intensive stirring and heating to 60-65 ℃. Then an obtained mixed colloid was evaporated with stirring and heating to 90-95 ℃ to a past state, after which it was subjected to heat treatment at 400 ℃ in oven for 1 h.

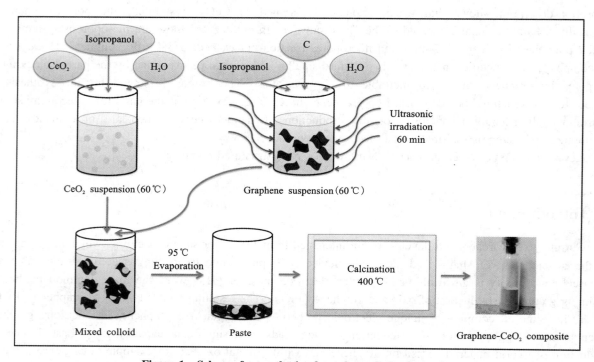

Figure 1 Scheme for synthesis of graphene-CeO_2 composite

Synthesized composite samples were studied using a complex of instrumental methods, including transmission electron microscopy (TEM) with electron diffraction (LEO 912 ab Omega Carl Zeiss), X-ray diffraction(XRD, SHIMADZU XRD-6000 with JCPDS database) and elemental analysis of carbon content ("Leco" device, model CS-600). The study of the surface and porosity of powders was carried out according to the nitrogen adsorption-desorption curves obtained at the TriStar 3000 installation of Micromeritics. The NOVA 2200 specific surface area analyzer was also used. The specific surface area was calculated by the Brunauer-Emmett-Teller (BET) method.

3 Results and discussion

According to TEM data, nano-CeO_2 synthesized by sol-gel method consisted of nano-and submicron sizes agglomerates [Figure 2(a)-(c)]. These agglomerates are formed by differently oriented rod-shaped nanocrystals with sizes $2 \div 10 \times 10 \div 50$ nm. According to electron diffraction pattern for the agglomerate shown at Figure 2, it can be concluded that CeO_2 has high crystallinity degree [Figure 2(d)].

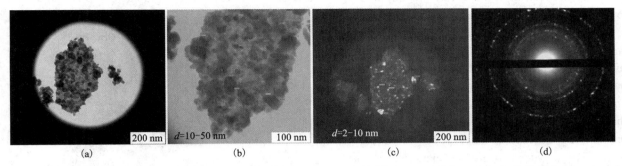

Figure 2 TEM data for the crystalline CeO_2 nanopowder: Light-field [(a) and (b)] and dark-field (c) images and electron diffraction pattern for the sample area shown in Figures (a) (c) and (d)

The XRD pattern of the CeO_2 nanopowder shown at Figure 3 corresponds to fluorite type face-centered cubic lattice (JCPDS card No 34-0394). The average crystallite size of cerium dioxide is 13.1 nm.

According to N_2 adsorption-desorption data, the specific surface area of pure nano-CeO_2, calculated by the BET-method, was 68 m^2/g. The average pore diameter was 4.5 nm which corresponds to a mesoporous system, while micropores provided less than 4% of volume and about 13% of the powder surface.

The morphology of graphene-CeO_2 composite with graphene content 0.61 wt.%, synthesized by mixing suspensions of crystalline nano-CeO_2 and graphene, is shown in Figure 4. The composite powder consisted of agglomerates with sizes not exceeding 100 (20-60) nm, which had graphene shells 3-5 nm thick [Figure 4(a)-4(c)].

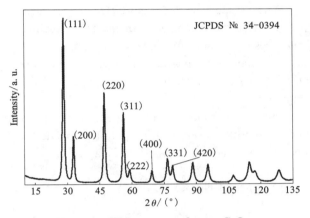

Figure 3 XRD pattern of nano-CeO_2

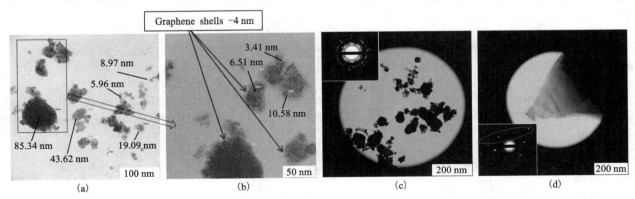

Figure 4 TEM data for the graphene-CeO_2 composite and electron diffraction pattern in the shown areas (insets)

CeO_2 crystallites inside these formations had sizes from 3-11 nm and were randomly packed. Figure 4(b) clearly shows that inside the agglomerates the CeO_2 crystallites are wrapped with the graphene sheets and do not tightly packed. The morphology of pure nano-CeO_2 powder and the composite is different, for example, the agglomerate sizes of the latter is much smaller than the initial oxide nanopowder, since agglomerates of submicron sizes predominate in it (Figures 3 and 4). Electron diffraction pattern in Figure 4(c) for the entire sample indicates the presence in the composite, in addition to graphene, of highly dispersed differently oriented CeO_2 crystallites. Figure 4(d) shows an image of the peripheral part of agglomerate, where graphene sheets mainly appear. The electron diffraction pattern corresponds to multilayer graphene, and the ratio of the characteristic reflection intensities (highlighted by a red oval) indicates that the graphene sheets had variable layering: multilayer sections were adjacent to single-layer ones.

The specific surface of the composite powder was 44.9 m^2/g. The decrease in the BET surface of the composite by 1.5 times compared to pure nano-CeO_2 after the deposition of graphene is apparently due to the fact that the sheets covering the CeO_2 nanocrystals block some of the pores and prevent the adsorption of nitrogen molecules in them. The powder can also be classified as meso-microporous, with an average pore diameter of 4.7 nm. At the same time, micropores provide 3% of the pore space volume of the synthesized powder.

When suspensions of nano-CeO_2 and oxygen-free graphene are combined with intensive stirring, CeO_2 crystallites and graphene sheets supported by stabilizer molecules interact: graphene sheets are attached to CeO_2 crystallites or their agglomerates (Figure 5).

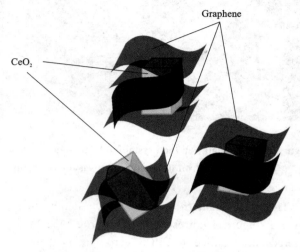

Figure 5 Scheme of the composite formation from nano-CeO_2 crystallites and graphene sheets in a colloid

During subsequent heat treatment, the liquid phase is removed, and composite agglomerates are formed with an almost uniform distribution of graphene sheets and CeO_2 crystallites in the bulk of the material, which was demonstrated above by TEM data.

4 Conclusion

Thus, a method has been developed for the synthesis of hybrid structures based on nanocrystalline CeO_2 and oxygen-free graphene with a content of the latter of no more than 1 wt.%. The developed method makes it possible to obtain the thinnest sheets of oxygen-free graphene and uniformly distribute them in the bulk of nanocrystalline CeO_2. The proposed method for the synthesis of composite nanostructured powders based on graphene and CeO_2 can be used as the basis for an economical and environmentally friendly technology for the nanopowders production, which are in demand in the development of materials for small-sized electronic devices.

Funding: The study was carried out in the Baikov Institute of Metallurgy and Materials Science RAS in accordance with the State Assignment 075-01176-23-00.

Acknowledgments: The authors thank PhD Shelekhov E. V. (National University of Science and Technology MISIS) for the XRD study and PhD Abramchuk S. S. (Lomonosov Moscow State University) for TEM study.

References

[1] TIWARI S K, THAKUR A K, ADHIKARI A D, et al. Current research of graphene-based nanocomposites and their application for supercapacitors[J]. Nanomaterials, 2020, 10, 2046: 48.

[2] SARANYA M, RAMACHANDRAN R, WANG F. Graphene-zinc oxide (G-ZnO) nanocomposite for electrochemical supercapacitor applications[J]. Journal of Science: Advanced Materials and Devices, 2016, 1(4): 454-460.

[3] LIU XR, HUBER T A, KOPAC M C, et al. Ru oxide/carbon nanotube composites for supercapacitors prepared by spontaneous reduction of Ru(Ⅵ) and Ru(Ⅶ)[J]. Electrochimica Acta, 2009, 54(27): 7141-7147.

[4] NEMATI F, REZAIE M, TABESH H, et al. Cerium functionalized graphene nano-structures and their applications; A review[J]. Environmental Research, 2022, 208: 112685.

[5] KUMAR R, AGRAWAL A, NAGARALE R K, et al. High performance supercapacitors from novel metal-doped ceria-decorated aminated graphene[J]. The Journal of Physical Chemistry C, 2016, 120(6): 3107-3116.

[6] BARANIK A, SITKO R, GAGOR A, et al. Graphene oxide decorated with cerium(IV) oxide in determination ofultratrace metal ions and speciation of selenium[J]. Analytical Chemistry, 2018, 90(6): 4150-4159.

[7] BRITTO S, RAMASAMY V, MURUGESAN P, et al. Graphene based ceriananocomposite synthesized by hydrothermal method for enhanced supercapacitor performance[J]. Diamond and Related Materials, 2020, 105: 107808.

[8] CHAITOGLOU S, AMADE R, BERTRAN E. Evaluation of graphene/WO_3 and graphene/CeO_x structures as electrodes for supercapacitor applications[J]. Nanoscale Research Letters, 2017, 12(1): 635.

Solid-state Reaction Kinetics Between Iridium and Zirconium Carbide

Yaroslav A. NIKIFOROV, Natalya I. BAKLANOVA

Institute of Solid State Chemistry and Mechanochemistry SB RAS, Novosibirsk, 630090, Russia

Iridium-based alloys are of growing interest as potential materials for high-temperature applications.[1,2] Here, we focus on the $ZrIr_3$ intermetallics, which characterized by large Young's modulus and yield stress, high thermal conductivity nearly independent of temperature, low coefficient of thermal expansion, and strong bonding at $Ir/ZrIr_3$ interface.[3,4] The reaction of iridium with zirconium carbide can serve as a convenient way to produce $ZrIr_3$, as it is the only intermetallic compound forming in this ternary system.

This paper presents a kinetics study of the solid-state reaction between iridium and zirconium carbide. The reaction proceeds according to the equation (1):

$$ZrC + 3Ir = ZrIr_3 + C \qquad (1)$$

The reaction kinetics is controlled by the interdiffusion of iridium and zirconium atoms through $ZrIr_3$ in the product layer. As $ZrIr_3$ is an ordered alloy, the interdiffusion process in it is strongly correlated. The correlation effect manifests in the higher mobility of iridium, which is experimentally observed in this work. The most peculiar feature of this reaction is a non-parabolic growth of the product layer, a behavior connected to processes near the reaction interface.

The as-received iridium plate and a sintered ZrC pellet m were used as diffusion couples. They were heated at 1600 ℃ for different time under a pressure of 20 MPa to ensure better contact. Morphology was characterized by scanning electron microscopy (SEM), concentration profiles were measured by wavelength-dispersive spectroscopy (WDS), the carbon microstructure was characterized by Raman spectroscopy.

A product layer consisting of intermetallic phase $ZrIr_3$ and graphite-like inclusions forms during the reaction. The Raman spectra from the randomly selected inclusions have peaks at $ca.$ 1330, 1580, and 2660 cm^{-1} corresponding to D, G, and 2D bands of graphite. Carbon particles are distributed through almost the entire product layer thickness, indicating that the reaction proceeds via interaction at $ZrIr_3/ZrC$ interface as a result of the higher mobility of iridium atoms in $ZrIr_3$. The composition of the intermetallic phase is superstoichiometric by Ir, varying from 80 at% Ir (higher boundary of the homogeneity range) near the $ZrIr_3/Ir$ interface to 75%-76% (at.%) Ir near the $ZrIr_3/ZrC$ interface. This also confirms that the diffusion of iridium is faster than that of zirconium. The growth kinetics is non-parabolic, obeying the power law with exponent $n = 0.32$. This

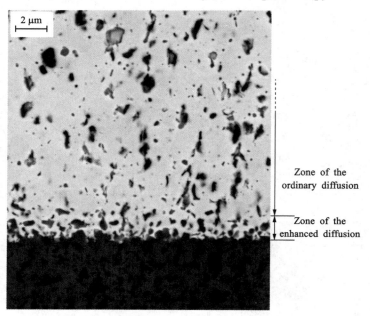

Figure 1 SEM image of the reaction zone

behavior is attributed to the diffusion enhancement in the reaction zone because of the finer $ZrIr_3$ grains and intricate microstructure of the reaction zone (see Figure 1). These findings provide valuable insights into the high-temperature behavior of ZrC, highlighting its potential for advanced applications.

Acknowledgments: This work was supported by the Russian Science Foundation (Project No. 23-19-00212).

References

[1] YAMABE-MITARAI Y, RO Y, HARADA H, et al. Ir-base refractory superalloys for ultra-high temperatures[J]. Metallurgical and Materials Transactions A, 1998, 29(2): 537-549.

[2] CONG X N, CHEN Z F, WU W P, et al. Co-deposition of Ir-containing Zr coating by double glow plasma[J]. Acta Astronautica, 2012, 79: 88-95.

[3] WU J Y, ZHANG B, ZHAN Y Z. Ab initio investigation into the structure and properties of Ir-Zr intermetallics for high-temperature structural applications[J]. Computational Materials Science, 2017, 131: 146-159.

[4] GONG H R, LIU Y, TANG H P, et al. Bond strength and electronic structures of coherent Ir/Ir$_3$Zr interfaces[J]. Applied Physics Letters, 2008, 92(21): 2006-2009.

Development of the Ion Theory of Slags and Implementation of the Results Obtained at Metallurgical Enterprises in Russia

A. A. METELKIN[1], O. Yu. SHESHUKOV[1,2], D. K. EGIAZARYAN[1,2],
V. N. NEVIDIMOV[1], M. V. SAVELYEV[3]

1 Ural Federal University named after first President of Russia B. N. Yeltsin, Yekaterinburg, Russia
2 Federal State Budgetary Institution of Science Institute of Metallurgy of the Ural Branch of the Russian Academy of Sciences, Yekaterinburg, Russia
3 JSC EVRAZ NTMK, Nizhny Tagil, Russia

Abstract: The paper deals with the issues of sulfur removal in the ladle-furnace unit. A formula for calculating the optical basicity is proposed. This formula takes into account the influence of basic, acidic oxides and amphoteric oxide Al_2O_3. It is shown that slags consisting entirely of a homogeneous phase have an increased optical basicity of aluminum oxide. Heterogeneous slags have a reduced optical basicity of Al_2O_3 compared to homogeneous slags. Perhaps this fact can be explained by the fact that in homogeneous slags there is a deficiency of the main CaO oxide and under these conditions Al_2O_3 begins to show more basic properties than acidic ones, so in homogeneous slags the optical basicity of aluminum oxide is increased and approaches the optical basicity of CaO. Calculations performed on real melts show that with an increase in the Al_2O_3 content in the slag, its optical basicity decreases. The known value of optical basicity allows us to determine the sulfide capacity of the slag, the coefficient of sulfur distribution between the metal and the slag, and, accordingly, the final content of sulfur in the metal. The calculations show that it is advisable to use the ionic theory of slags to determine the sulfide capacity.

Keywords: Desulfurization of metal; A ladle-furnace unit; Optical basicity

Slags are the most important component of the technological process of steel smelting. In modern metallurgical production, it is impossible to obtain high-quality metal without processing with refining slags. Therefore, the management of the physicochemical properties of slags is one of the important tasks of modern steelmaking technology.

The study of metallurgical processes allowed us to imagine that slag consists of positively and negatively charged ions. Numerous X-ray structural studies of solidified slags, the electrical conductivity of molten slags, the presence of electric charges in the boundary layers of metal and slag, high values of the surface tension of slags, etc. serve as proof of the ionic structure of slags[1,2].

In molten metallurgical slags, there are ions of the following groups[1-3]:

(1) Cations Ca^{2+}, Mg^{2+}, Mn^{2+}, Fe^{2+}.

(2) Anions O^{2-}, S^{2-}, SiO_4^{4-}, PO_4^{3-}, AlO_2^{1-}, FeO_2^{1-}.

(3) More complex silicon-oxygen anions can be formed in acidic slags $(SiO_3^{2-})_n$, $Si_3O_9^{6-}$, $Si_4O_{12}^{8-}$, $Si_6O_{18}^{12-}$ и т. д.

X-ray diffraction analysis of solid slags shows that their structure can be considered as a dense packing of oxygen ions, in which there are two types of planes[2,3].

Tetrahedral planes are between four oxygen ions whose centers form a regular tetrahedron [Figure 1(a)] and octahedral planes are between six oxygen ions whose centers form a regular octahedron [Figure 1(b)][2,3].

The basic oxides (CaO, MgO, FeO, MnO, etc.) form NaCl-type lattices in which each metal cation is surrounded by six oxygen anions, and each oxygen anion is surrounded by six metal cations, i. e. they form an octahedral coordination [Figure 1(b)][2,3].

This position in the ionic theory of slags was developed in the works on the polymer model (PM) developed at the Department of "Theory of Metallurgical Processes" of UrFU, V. K. Novikov and V. N. Nevidimov. The developed polymer model makes it possible to calculate the structural and thermodynamic

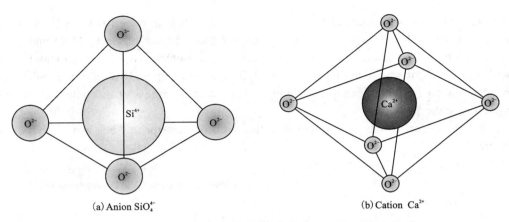

(a) Anion SiO_4^{4-} (b) Cation Ca^{2+}

Figure 1 Tetrahedral and octahedral environment of ions in slag

characteristics of binary silicate melts and to apply the equations to multicomponent oxide melts containing various elements of complexing agents. The polymer model makes it possible to determine the refining properties of slags by calculating the "free" oxygen anions in the oxide melt.

The development of the polymer model by V. K. Novikov and V. N. Nevidimov and, accordingly, the improvement of the ion theory was undertaken by scientists at IMeT UrO RAS and UrFU and went in several directions.

One of the stages was the use of (PM) in determining the boundaries of homogenization of slags formed in the bucket furnace unit (LF). For example, for a temperature of 1585 ℃, the boundary of homogenization of the oxide melt is shown in Table 1 and Figure 2.

Table 1 Boundary composition of slag transition to a heterogeneous state at a temperature of 1585 ℃

Oxide	Mass fraction/%														
CaO	57.6	57.0	56.5	55.9	55.3	54.7	54.0	53.3	52.6	51.9	51.1	50.4	49.6	48.7	47.8
Al_2O_3	4.2	6.0	7.8	9.7	11.6	13.6	15.7	17.8	20.1	22.3	24.7	27.2	29.8	32.4	35.2
SiO_2	28.5	27.3	26.0	24.6	23.2	21.8	20.3	18.8	17.2	15.5	13.8	12.1	10.3	8.4	6.4
MgO	8.9	9.0	9.0	9.1	9.1	9.1	9.2	9.2	9.2	9.3	9.3	9.3	9.4	9.4	9.5
MnO+FeO	0.7	0.7	0.8	0.8	0.8	0.8	0.8	0.9	0.9	1.0	1.0	1.0	1.0	1.0	1.1

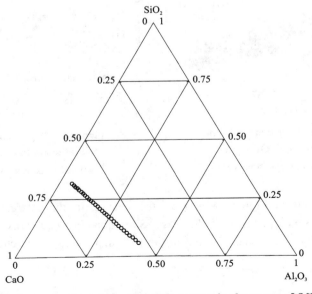

Figure 2 Slag heterogenization zone with MgO content in the range of 8%, 7%-9%, 4%

The next stage in the development of the ion theory was the work of scientists from UrFU and IMeT UrO RAS in the field of the study of slags formed in the LF. If the behavior of CaO, MgO and SiO$_2$ oxides is clear, i. e. they are basic and acidic, then Al$_2$O$_3$ oxide is an amphoteric oxide, it can exhibit both basic and acidic properties. In [4], it was found that with an Al$_2$O$_3$ content of up to 16% in oxide melts corresponding to slags formed in the LF, alumina exhibits basic properties, with an excess of its content of more than 16%, it begins to exhibit acidic properties.

To verify the theoretical results, the smelting at the industrial enterprise "N" was analyzed. The data is shown in Figure 3.

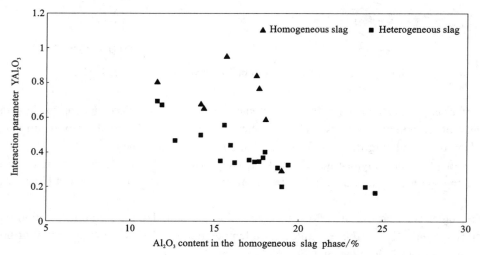

Figure 3 Dependence of the interaction parameter YAl$_2$O$_3$ on the type of slag (homogeneous/heterogeneous) and the content of Al$_2$O$_3$ in it

From the data presented in Figure 3, it can be seen that with an increase in the content of Al$_2$O$_3$ in the slag, its interaction parameter decreases, i. e. this oxide begins to exhibit acidic properties. Additionally, it was determined that slags consisting entirely of a homogeneous phase have an increased interaction parameter of aluminum oxide (Figure 3).

These conclusions were also confirmed at the metallurgical plant "V".

The joint results are presented in Figures 4 and 5.

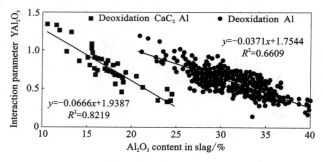 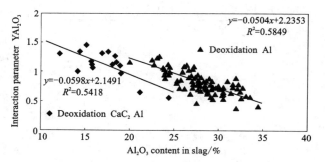

Figure 4 Dependence of the change in the interaction parameter YAl$_2$O$_3$ on the content of alumina in heterogeneous slag

Figure 5 Dependence of the change in the interaction parameter YAl$_2$O$_3$ on the alumina content in homogeneous slag

From the data presented in Figures 4 and 5, the change in the interaction parameter ΛAl_2O_3 was determined depending on the composition of the slag. Subsequently, equations (1)-(4) were proposed to determine the sulfide capacity of the slags formed in the LF, depending on its chemical composition.

$$\Lambda^*_{\text{ГетероAl}} = \sum_{i=1}^{n} (X_i \cdot \Lambda_i)_{\text{ОСН}} - \sum_{i=1}^{n} (X_i \cdot \Lambda_i)_{\text{КИС}} + X_{Al_2O_3} \cdot [-0.0371 \cdot (Al_2O_3) + 1.7544] \tag{1}$$

$$\Lambda^*_{\text{ГетероCaC}_2+\text{Al}} = \sum_{i=1}^{n} (X_i \cdot \Lambda_i)_{\text{ОСН}} - \sum_{i=1}^{n} (X_i \cdot \Lambda_i)_{\text{КИС}} + X_{Al_2O_3} \cdot [-0.0666 \cdot (Al_2O_3) + 1.9387] \tag{2}$$

$$\Lambda^*_{\text{ГомоAl}} = \sum_{i=1}^{n}(X_i \cdot \Lambda_i)_{\text{ОСН}} - \sum_{i=1}^{n}(X_i \cdot \Lambda_i)_{\text{КИС}} + X_{\text{Al}_2\text{O}_3} \cdot [-0.0504 \cdot (\text{Al}_2\text{O}_3) + 2.2353] \quad (3)$$

$$\Lambda^*_{\text{ГомоCaC}_2+\text{Al}} = \sum_{i=1}^{n}(X_i \cdot \Lambda_i)_{\text{ОСН}} - \sum_{i=1}^{n}(X_i \cdot \Lambda_i)_{\text{КИС}} + X_{\text{Al}_2\text{O}_3} \cdot [-0.0598 \cdot (\text{Al}_2\text{O}_3) + 2.1491] \quad (4)$$

According to formulas (1)-(4), it is possible to determine the sulfide capacity of slags and the maximum desulfurizing properties. The data is presented in Table 2.

Table 2 Calculated parameters of sulfide capacity for various types of slags (homogeneous/heterogeneous) and deoxidation technologies

Slag type and deoxidation technology	The content of oxides in the slag/%					Ls
	CaO	Al_2O_3	SiO_2	MgO	MnO+FeO	
Heterogeneous slag, deoxidation of Al	49.56	29.77	10.26	9.38	<1.0	1029.4
Heterogeneous slag, deoxidation of CaC_2+Al	53.34	17.84	18.77	9.19	<1.0	90.53
Homogeneous slag, deoxidation of Al	50.36	27.20	12.09	9.35	<1.0	1795.08
Homogeneous slag, deoxidation of CaC_2+Al	51.88	22.34	15.54	9.27	<1.0	364.4

From the examples presented, it can be seen that the ionic theory allows us to determine the rational composition of the slag having the maximum sulfide capacity.

The next stages in the research of the ionic theory of slags will be the directions for determining the composition of slag having rational properties not only with the maximum sulfide capacity, but also additionally with the best properties for the absorption of non-metallic inclusions, as well as determining the rational composition of slag in steelmaking units (converter, arc steelmaking furnace), for FeO behaviors.

Acknowledgments: The research funding from the Ministry of Science and Higher Education of the Russian Federation (Ural Federal University Program of Development within the Priority-2030 Program) is gratefully acknowledged.

References

[1] BIGEEV A M. Metallurgy of steel. Theory and technology of steel melting [M]. Textbook for Universities/BIGEEV A. M., BIGEEV V. A..-3RD ED., Reprint. and Additional-Magnitogorsk: Mstu, 2000: 544.
[2] POPEL S I. Theory of metallurgical processes [M]. Handbook for Universities/POPEL S. I., SOTNIKOV A. I., BORONENKOV V. N. [M]. Metallurgy, 1986: 463.
[3] NOVIKOV V K. Polymer nature of molten slag [M]. Manual/NOVIKOV V. K., INVISIBLE V. N..-Yekaterinburg: GO VPO UGTU-UPI. 2006: 62.
[4] SHESHUKOV O YU. Issues of utilization of refining slags of steelmaking production/SHESHUKOV O. YU., MIKHEENKOV M. A., NEKRASOV I. V., et al. Ministry of Education and Science of the Russian Federation; Fsaou Vpo "Urfu Named after the First President of Russia B. N. Yeltsin", Nizhny Tagil. Technol. IN-T (Phil.). Nizhny Tagil: NTI (Financial) URFU, 2017: 208.

On the Influence of High Pressures and Boron Nitride on the Processes of Structure Formation and Microhardness of a High-entropy Equiatomic Composition AlNiCoFeCr Alloy

Svetlana MENSHIKOVA

The Udmurt Federal Research Center of the Ural Branch of RAS, Izhevsk, 426034, Russia

Abstract: The structure of equiatomic high-entropy AlNiCoFeCr alloy obtained by arc melting was investigated. The influence of high pressures (5, 8 and 11 GPa), quenching temperature (1650 ℃) and small additions of reinforcing agent-boron nitride (10% of the alloy volume) on the microstructure and microhardness of the alloy after quenching was studied. Depending on the conditions of thermobaric action, structures based on solid solution of the B2 type or mixed phases with structures of the A1, A2 or B2 types are formed in the AlNiCoFeCr alloy, which influences the alloy microhardness that varies in the range of 5-12.5 GPa.

Keywords: Melt; Pressure; Microstructure; Microhardness; Phase

1 Introduction

The alloys of the Al-Ni-Co-Fe-Cr system are dispersion-strengthened composites and refer to high-entropic alloys (HEAs). HEAs contain five or more elements; the amount of each element should not exceed 35 at% and should not be less than 5 at%[1,2]. HEAs belong to a special group of alloys because the processes of the structure and phase formation in them, the diffusion mobility of atoms, the mechanism of the mechanical properties formation and thermal stability essentially differ from similar processes in conventional alloys. In contrast to conventional multi-component alloys, such alloys are characterized by higher values of entropy of mixing. Additions of various chemical compounds, such as carbides, nitrides, borides or oxides having high values of strength, hardness and high chemical stability, are used for strengthening materials. This takes place due to the creation of barriers preventing dislocation displacement similar to that in precipitation-hardened metal alloys. At present, the most commonly used technology for obtaining a dispersion-strengthened composite is powder metallurgy. The main technological processes are the preparation of powder mixtures, powder pressing with subsequent sintering and plastic deformation of the prepared mass. During plastic deformation, the density of a composite increases and its porosity decreases. In composites reinforced by particles of more than 1 μm in size, the most optimal content of the particles is in the range of 20%-25% (throughout the volume) while dispersion-strengthened composites contain from 1% to 15% (throughout the volume) of particles of sizes in the range of 0.01-0.1 μm. The sizes of the particles in the composition of nanocomposites (new class of nanocomposites) are even smaller, in the range of 10-100 nm. The material properties and structure can be changed by not only intensive plastic deformation but also by other extreme actions or their combination such as self-propagating high-temperature synthesis, high pressure and temperatures, etc. Extreme actions permit to obtain unique materials: polymers with high strength, new semiconducting materials, new stable crystalline structures with new unusual physical properties from well-known substances, new superconductors operating at room temperature, and many others.

In the present work, an equiatomic alloy AlNiCoFeCr was chosen as an object for investigation as one of the typical model HEAs of this system and a promising alloy for application. Studies of the influence of high pressures on the formation of structures at quenching liquid alloys Al-Ni-Co-Fe-Cr and the influence of reinforcing agents, in particular boron nitride, on the alloy structure and properties are absent in the literature. Therefore, the present investigation is of importance.

2 Materials and investigation methods

An alloy of the equiatomic composition AlNiCoFeCr was produced in the atmosphere of high-purity argon by arc melting with a nonconsumable electrode from components with purity of 99.999%. To improve the chemical homogeneity, an ingot was remelted five times. The structure of the ingot was studied in the as-obtained state, after its melting and subsequent solidification under high pressure of 5 and 11 GPa. The samples were obtained in a high-pressure chamber of the "toroid" type[3] (Figure 1). As a pressure transmitter, catlinite was used. The sample, placed into a crucible from hexagonal boron nitride, was heated and melted by passing alternating current through it. The value of the temperature was calculated based on the indications of a thyristor with calibration according to the alternating current power. The melts were cooled at the rate of 1000 deg/s; the melt temperature before quenching was 1650 ℃. The scheme of the experiment was as follows: pressure setting → pulse heating → holding at the set pressure and temperature-cooling to room temperature without pressure release → high pressure decrease to atmospheric pressure. The phase composition of the samples was determined by X-ray diffraction analysis on a device DRON-6 at CoK_α-monochromatic radiation. The investigation of the structure and the determination of the chemical and elemental composition, morphology and dimensions of the structural components of each sample were performed using a system Quatro S-scanning electron microscope (SEM) equipped with a standard detector DBS (direct backscattering) ABS/CBS. The error in the determination of the percentage content of elements in the samples was no more than 5%. Durametric measurements (Vickers hardness) were performed on a microhardness tester PMT-3M. A load of 100 g was applied on the indenter for a period of 10 s. The H_v values were averaged over 20 measurements.

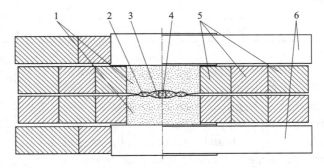

1—solid substance; 2—torus; 3—central part in the form of a lentil;
4—heater and sample; 5—steel rings; 6-support plate

Figure 1　Chamber "toroid"

3 Results and discussion

An initial ingot has a submicrocrystalline structure with the average grain size of 120 nm [Figure 2(a)]. All the alloy elements are present in the grains and intergranular space, though, in different amounts (Table 1). The ingot has packing with components corresponding to the B2 structure based on the distorted bcc-lattice of nickel monoaluminide NiAl with the lattice period of 0.2870-0.2883 nm and the space group Pm3m (Figure 3).

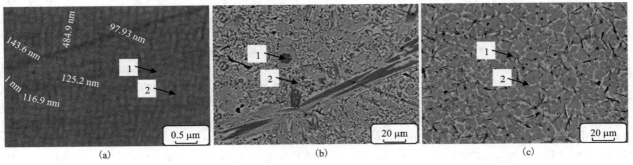

Figure 2　Microstructure of the initial ingot (a), the sample 1650 ℃, 100 deg/min, 5 GPa (b) and the sample 1650 ℃, 100 deg/min, 11 GPa (c) with the addition of BN

Table 1 Elemental composition of the selected areas of the initial alloy

No. of area	Al/at%	N/at%	Co/at%	Fe/at%	Cr/at%
1	20	20	20	20	20
2	22	17	19	20	22

In the sample obtained under the pressure of 5 GPa, the decomposition of the initial solid solution takes place; separate phases with different morphology and eutectic can be seen [Figure 2(b)]. Initially, a phase of a regular geometric shape is formed, which are hexagons in the section in Figure 2(b) that are denoted by 1. The phase is rich in chromium (Table 2). After that the rest of phases and eutectic are formed. All the phases are multicomponent. As can be seen from Table 2, all elements of the alloy are present in the eutectic composition [denoted by 2 in Figure 2(b)]. According to the X-ray diffraction analysis data, a mixed structure of two types: A1 type and A2 type is formed in the alloy (Figure 3). In the chosen conditions of the investigation, the exact stoichiometric composition of all the phases cannot be determined. The analysis of the concentration maps of the element distribution for this sample shows that aluminum is distributed homogeneously, iron and cobalt-quasi-homogeneously and nickel and chromium-heterogeneously. Solidification under pressure of 8 GPa is preformed similarly. The X-ray diffraction patterns of the samples obtained under 5 and 8 GPa do not practically differ (Figure 3).

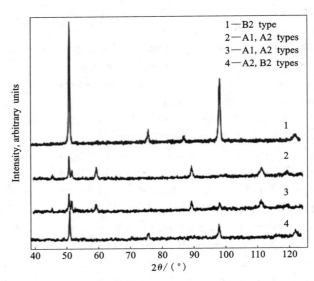

1—initial alloy; 2—alloy obtained under 5 GPa; 3—8 GPa; 4—11 GPa

Figure 3 X-ray diffraction patterns of the alloys

Table 2 Elemental composition of the selected areas of the alloy (1650 ℃, 100 deg/min, 5 GPa)

No. of area	Al/at%	Ni/at%	Co/at%	Fe/at%	Cr/at%
1	-	6	14	20	60
2	21	18	18	22	21

Then, the initial sample was ground to chips; a small amount (10% of the sample volume) of the powder of hexagonal boron nitride (BN)[①][4] was added to the metal chips (Figure 4) and the mixture was blended. The mixture obtained was melted and cooled under pressure of 11 GPa according to the above-mentioned scheme. The microstructure of the obtained sample considerably differs from those considered above. In the structure, there are dendrites denoted by 1 in Figure 2(c). In the interdendritic space [denoted by 2 in Figure 2(c)], mainly along the dendrite boundaries, a needle-like phase with branches and ball-shaped inclusions is located [shown in black color in Figure 2(c)]. The elemental composition of the selected areas in Figure 2(c) is given in Table 3. As seen from Table 3, in the points chosen for analysis, practically all the alloy elements are present in different amounts. Dendrites are aluminum-rich, and in the interdendritic space, the amount of chromium is larger than that of other elements. As the investigations have shown, the needle-like phase of black color with inclusions represents aluminum nitride and boron. According to the X-ray diffraction data, a mixed structure consisting of A2 and B2 types (Figure 3) is formed in the sample. The

---

① Owing to chemical inertness and thermal stability, boron nitride is widely used in many branches of industry. Hexagonal (α)-h-BN form of boron nitride has the largest number of applications owing to low friction coefficient, electric conduction and thermal stability (the compound withstand temperatures up to 3000 ℃). The material represents fine powder of pure white color resembling talc [Figure 4(a)]; 90% of its particles are no more than 20 μm in diameter.

analysis of the concentration maps of element distribution in the sample shows that in dendrites the elements are heterogeneously distributed.

Figure 4 Boron nitride powder (a) and the powder particles (b)

Table 3 Elemental composition of the selected areas of the alloy
(1650 ℃, 100°/min, 11 GPa with BN addition)

No. of area	Al/at%	Ni/at%	Co/at%	Fe/at%	Cr/at%
1	53	20	16	7	4
2	27	6	10	18	39

Figure 5 shows the microhardness of all the samples under study. From Figure 5 it follows that the sample obtained under pressure of 11 GPa has the largest microhardness value 12.5 GPa. The average microhardness of this sample is 2.5 times higher than that of the initial sample, and almost 2.3 times higher than that of the samples obtained under pressure of 5 GPa and 8 GPa. This is explained by the morphological features of the structure and the presence of reinforcing agents. In the studied alloy, under the action of high pressure and temperatures hexagonal boron nitride turns into hard aluminum nitride and boron[5].

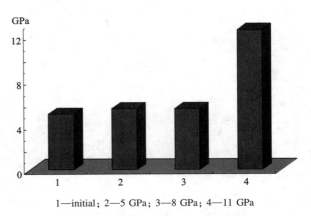

1—initial; 2—5 GPa; 3—8 GPa; 4—11 GPa

Figure 5 Microhardness of the samples

4 Conclusion

Thus, the alloy of the AlNiCoFeCr can have a structure based on a solid solution, complex structure and mixed phases.

Depending on the composition, microstructure and corresponding properties, high-entropy alloys, for example, the alloy considered in the present work, have a great potential for the application as heat-resistant materials; coatings requiring high hardness and high wear-resistance; and corrosion-resistant materials with high strength. At present, many different HEAs have been studied. Despite the fact that some investigations are of purely scientific character and directed to the establishment of the regularities of the influence of different factors such as atomic size, enthalpy of solution, electron concentration, etc. on the properties of obtained HEAs, among the studied alloy there are materials that can compete with the best conventional alloys of special purpose in hardness, high-temperature strength, heat-resistance, corrosion resistance, wear-resistance and thermal stability.

Acknowledgments: The work is performed within the framework of the Project of RNF (No. 22-22-

00674). The electron-microscopic investigations were performed with the use of the equipment of the Center for Collective Usage "The Center of physical and physical-chemical methods of analysis, investigation of properties and surface characteristics, nanostructures, materials and items". UdmFITs UrO RAS, Izhevsk. The samples under high pressure were obtained in IFVD RAS, the city of Troitsk, Moscow region. The author expresses her sincere appreciation to Doctor of Physics and Mathematics, Academician of RAS V. V. Brazhkin for his assistance in obtaining samples in the chamber of the type "toroid".

References

[1] SU Y, LUO SC, WANG Z M. Microstructure evolution and cracking behaviors of additively manufactured Al_xCrCuFeNi$_2$ high entropy alloys via selective laser melting[J]. Journal of Alloys and Compounds, 2020, 842: 155823.

[2] SISTLA H R, NEWKIRK J W, FRANK LIOU F. Effect of Al/Ni ratio, heat treatment on phase transformations and microstructure of Al_xFeCoCrNi$_{2-x}$ (x = 0.3, 1) high entropy alloys[J]. Materials & Design, 2015, 81: 113-121.

[3] BRAZHKIN V V. Influence of high pressure on the solidification of metal melts (Pb, In, Cu, copper-based binary alloys): dissertation for the degree of candidate of physical and mathematical sciences[M]. MIPT, Moscow, 1987: 150.

[4] PEREVISLOV S N. Structure, properties andapplications of graphite-like hexagonal boron nitride[J]. Novye Ogneupory (New Refractories), 2019(6): 35-40.

[5] KURDYUMOV A V, MALOGOLOVETS V G, NOVIKOV N V, et al. Polymorphic modifications of carbon and boron nitride[M]. Moscow: Metallurgiya, 1994: 320.

The Behavior of the CaB$_6$-Ir System at High Temperature

Victor V. LOZANOV[1], Yaroslav A. NIKIFOROV[1], Mikhail A. GOLOSOV[1],
Tatyana A. GAVRILOVA[2], Natalya I. BAKLANOVA[1]

1   Institute of Solid State Chemistry and Mechanochemistry SB RAS, Novosibirsk, 630090, Russia
2   Rzhanov Institute of Semiconductor Physics SB RAS, Novosibirsk, 630090, Russia

Abstract: The reaction of iridium with calcium hexaboride at 1900 ℃ was studied. It was shown that the formation of IrB$_{1.1}$, IrB$_{1.35}$, CaIr$_4$B$_4$ phases, as well as a new phase (NP) was observed. For the NP phase, a variant of the crystal structure was proposed based on the previously discovered Ca$_3$Ir$_8$B$_6$ phase. EDX analysis revealed the presence of CaIr$_4$B$_4$ phase with the near-stoichiometric (CaIr$_{3.4\pm0.2}$B$_{3.7\pm0.1}$) and superstoichiometric (CaIr$_{5.7\pm0.2}$B$_{5.9\pm0.4}$) composition. Additionally, the presence of a boron carbide phase was found by Raman spectroscopy. The presence of boron carbide confirms the participation of the crucible material (graphite) in the reaction.

Keywords: Iridium; Calcium hexaboride; Ternary borides; High temperature synthesis

1   Introduction

Calcium hexaboride is one of the most demanded borides in industry due to its high melting point (2235 ℃), high hardness (HV = 27 GPa), low density (2.45 g/cm^3), etc.[1]. There is also evidence that CaB$_6$ can be used in the preparation of superhard (HV > 40 GPa) ceramic materials[2, 3] and as a perspective component of high temperature oxidation-resistant materials[4, 5]. However, the data on the reaction of CaB$_6$ with other high temperature materials are rare. It was shown that no new phases are formed in the mixtures of CaB$_6$ with refractory carbides and borides, such as SiC[1], TaB$_2$[3], ZrB$_2$, and HfB$_2$[6], heat-treated at 2000-2200 ℃.

Iridium has a considerable interest for high-temperature structural applications where it can act as barrier layer. It was stated that CaB$_6$ reacts with iridium at temperatures 700-1600 ℃, forming binary and ternary compounds, such as IrB$_{1.1}$, CaIr$_4$B$_4$, etc.[4]. Unfortunately, the high-temperature behavior of the CaB$_6$-Ir system was not studied. Thus, the aim of this work was to study the reaction of iridium with CaB$_6$ at temperature as high as 1900 ℃.

2   Experimental part

The CaB$_6$ and Ir powders were mixed in the CaB$_6$:Ir = 1:1, 1:3 and 1:6 molar ratios and placed into graphite crucible. Heat-treatment of the mixtures was carried in home-made equipment UPG (Novosibirsk, Russia) at 1900 ℃ for 1 h in argon atmosphere ($P = 120$ kPa). After heat treatment, the samples were ground in an agate mortar. The diffraction patterns of the products were recorded using a D8 Advance diffractometer (Bruker, USA) and processed by the full-profile Rietveld method using the TOPAS 4.2 program (Bruker, USA). The ISCD database for Ir (#64922), IrB$_{1.1}$(#24364), IrB$_{1.35}$(#43319) was used. Data on the crystal structure of CaB$_6$ was taken from the Springer Materials database[7].

Microstructure was studied by scanning electron microscopy using microscopes TM-1000 (Hitachi Ltd., Japan) at an accelerating voltage of 15 kV, and SU8220 (Hitachi Ltd., Japan) at 6 kV. Elemental microanalysis of the reaction products was performed by energy dispersive X-ray spectroscopy (EDX) at 6 kV on a SU8220 microscope equipped with QUAD and Quantax 60 EDX detectors (Bruker, USA).

Micro-Raman spectroscopy (LabRam HR Evolution, Horiba, Japan) was used to measure Raman spectra of selected areas of the products. The spectra were recorded at excitation wavelength of 633 nm. The acquisition time was 10 s. For each spectrum, 100 scans were accumulated. This was sufficient to obtain the spectra with a low noise/signal ratio required for the fitting procedure. Deconvolution of the spectra was performed using the Fityk 1.3.1 software[8]. Voigt function profiles were used for the peak-fitting procedure.

3 Results and discussion

3.1 X-ray diffraction analysis of reaction products

The X-ray diffraction patterns of products obtained in the CaB_6-Ir mixtures with the different molar ratios after annealing at 1900 ℃ are presented in Figure 1. Main products of the CaB_6-Ir interaction are iridium borides, $IrB_{1.1}$ and $IrB_{1.35}$, which present in all mixtures under investigation. In the 1:1 mixture the emergence of a series of XRD peaks that cannot be ascribed to any known binary or ternary phases in the Ca-Ir-B system was observed [Figure 1(a)]. To identify unknown phase (named as NP), the crystal structure of known ternary $Ca_3Ir_8B_6$ phase was modified (Table 1). Modification included the changes in the atom coordinates and occupations (marked with an asterisk in Table 1). Using the data of Table 1, the XRD peaks of NP phase were calculated. They proved to be in a good accordance with experimental data. Thus, NP phase belongs to $F\,mmm$ space group with atomic parameters presented in Table 1 (lattice parameters a = 5.585 Å, b = 9.939 Å, c = 17.20 Å). With increasing of the iridium concentration in initial powder mixture, the quantity of $CaIr_4B_4$ phase in product increases too, whereas the content of NP phase decreases [Figure 1(b), (c)]. In a mixture with an initial ratio of 1:6, the NP phase was not detected by XRD analysis. Additionally, the presence of reflections of unreacted CaB_6 was observed in products formed from the 1:6 mixture.

Table 1 Atomic coordinates for NP phase

Site	x	y	z	Occupation
Ca_1*	0	0	0.06424	1
Ca_2	0	0	1/2	1
Ca_3*	0	0	0.31664	1
Ca_4*	0.24678	0	0.17067	0.25
Ir_1*	1/4	1/4	0.08360	1
Ir_2	1/4	1/4	1/4	1
Ir_3*	0.24678	0	0.17067	0.33
B_1	0	0.166	0	1
B_2	0	0.161	0.17130	1

3.2 SEM/EDX analysis of reaction products

A sample with an initial ratio of CaB_6:Ir = 1:1 is a well-sintered conglomerate consisting of dark gray faceted CaB_6 crystals and a light gray iridium-containing phase [Figure 2(a)]. As the iridium content increases, the number of dark gray crystals decreases. In a 1:6 mixture, light gray particles are observed [Figure 2(b)]. With an increase in image contrast, patches of inhomogeneous hue are observed. Elemental analysis was carried out to identify Ca-Ir-B containing phases. Elemental mapping [Figure 3(a)] demonstrates the presence of CaB_6 (green particles), iridium borides (blue particles), a ternary compound (light blue area) and boron-enriched particles (red particles) in the sample. The presence of calcium in the light blue areas is confirmed by the EDX spectrum [Figure 3(b)]. To analyze the local composition, six Ca-Ir-B containing points were measured. According to the obtained results, the sample with the initial 1:1 ratio contains two groups of compositions, $CaIr_{3.4\pm0.2}B_{3.7\pm0.1}$ (3 points) and $CaIr_{5.7\pm0.2}B_{5.9\pm0.4}$ (3 points). These compositions are close to those detected earlier by us[4]. They can be considered as the $CaIr_4B_4$ solid solutions. The absence of intense reflections of the $CaIr_4B_4$ phase in the XRD pattern of the product of the 1:1

mixture can be related to the low concentration of the $CaIr_4B_4$ phase.

3.3 Raman spectroscopy analysis of reaction products

Elemental mapping of the samples showed the presence of particles enriched in boron, but not containing calcium and iridium. Note that the analysis of the diffraction patterns did not reveal the presence of elemental boron, which can be due to the low scattering power of boron against the background of heavy elements (atomic form factors: $f_B \sim 5$, $f_{Ca} \sim 20$, $f_{Ir} \sim 73$[9]). The Raman spectroscopy of the selected areas showed that the reaction products of the mixture $CaB_6:Ir = 1:1$ contain the CaB_6[10, 11], boron carbide, and carbon phases (Figure 4). The most intense vibrations appeared at ~264, 321, 718, and 1087 cm^{-1} can be ascribed to boron carbide[12, 13]. Its formation could be resulted from the interaction of iridium-rich borides with material of crucible (graphite).

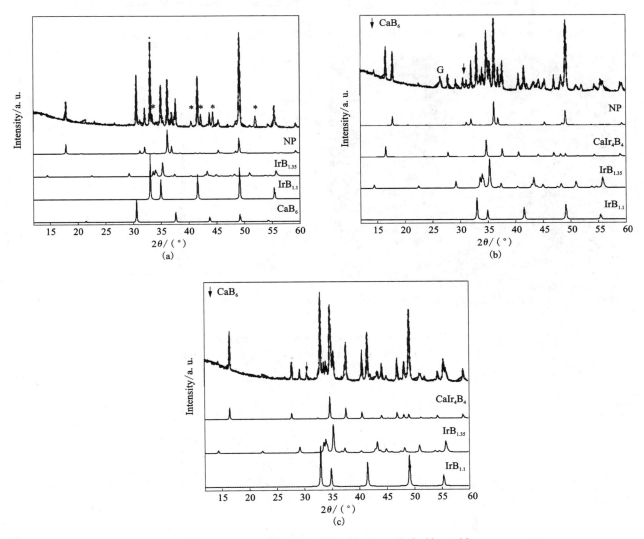

G is the graphite phase; non-identified peaks are marked with asterisks.

Figure 1 The X-ray diffraction patterns of products prepared in the mixtures with the molar ratio $CaB_6:Ir = 1:1$ (a), 1:3 (b) and 1:6 (c) after heat treatment at 1900 ℃

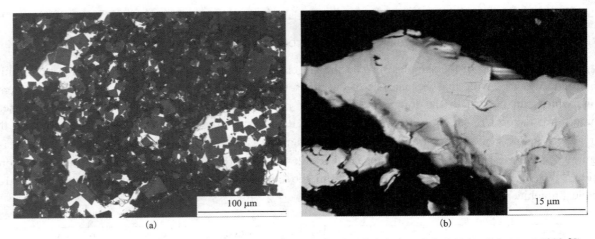

Figure 2 The SEM images of product obtained in the $CaB_6:Ir=1:1$ (a) and $1:6$ (b) mixtures at 1900 ℃

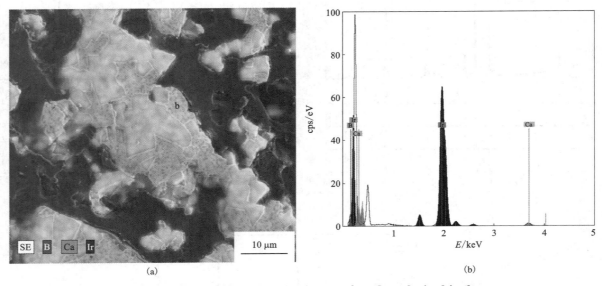

Figure 3 The elemental mapping images of product obtained in the $CaB_6:Ir=1:1$ mixture (a) and EDX analysis of selected areas (b)

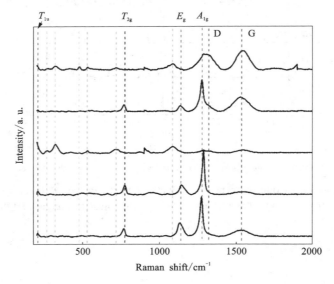

Green dashed lines are pointed with B_4C, blue dashed lines connected with CaB_6 phase, and red dashed lines are pointed with D and G lines of carbon phase

Figure 4 Raman spectra of selected points of the sample powder with ratio $CaB_6:Ir=1:1$ after annealing at 1900 ℃

4 Conclusions

The chemical interaction of iridium with CaB_6 at 1900 ℃ was studied. It was found that iridium borides, $IrB_{1.1}$ and $IrB_{1.35}$, ternary Ca-Ir borides are formed. Not only a recently discovered $CaIr_4B_4$ ternary phase, but also a new ternary Ca-Ir-B phase was detected by different analytical techniques. The crystal structure (F mmm space group) and atomic parameters (coordinates and occupations) of new phase were suggested. Boron carbide was also detected as by-product. Its formation can be explained by the participation of the crucible material (graphite) in the reaction.

Acknowledgments: This work was supported by Russian Science Foundation (Grant No. 22-79-00019). The authors are grateful to Dr. Yulia S. Myasnikova for recording Raman spectra, and Dr. Arina V. Ukhina and Tatyana A. Borisenko for recording of diffraction patterns. We are also thankful to the "Nanostructures" Center, Institute of Semiconductor Physics, SB RAS.

References

[1] WANG Y, ZHANG G, WU Y, et al. [J]. Int. J. Miner. Metall. Mater., 2020 (27), 37-45.
[2] NESMELOV D D, TURTSOVA A I, FEDEROV E V, et al. [J]. Refractories and Industrial Ceramics. 2014 (11-12): 3-8.
[3] SEREBRYAKOVA T I, OCHKAS L F, SHAPOSHNIKOVA T I, et al. [J]. Powder Metall. Met. Ceram. 1998 (37): 507-511.
[4] LOZANOV V V, GAVRILOVA T A, BAKLANOVA N I. Phase formation in the calcium hexaboride-iridium system [J]. Russian Journal of Inorganic Chemistry, 2023(68): 752-759.
[5] LOZANOV V V, BAKLANOVA N I. The effect of calcium hexaboride on the phase formation in the hafnium diboride—iridium and tantalum diboride—iridium systems [J]. Transactions of the Kola Science Centre of RAS Series: Engineering Sciences, 2023, 14(4/2023): 15-20.
[6] CRISCIONE J M, VOLK H F, NUSS J W, et al. High temperature protective coatings for graphite[R]. Technical documentary report ML-TDR-64-173, Part Ⅲ. Parma, Ohio. 1965: 218.
[7] CaB_6 crystal structure: Datasheet from "PAULING FILE Multinaries Edition-2012" in Springer Materials (https://materials.springer.com/isp/crystallographic/docs/sd_1721776)
[8] WOJDYR M[J]. J. Appl. Cryst. 2010(43): 1126-1128.
[9] CHANTLER C T, OLSEN K, DRAGOSET R A, et al. X-Ray form factor, attenuation and scattering tables (version 2.1). [DB/OL] Available: http://physics.nist.gov/ffast [25.08.2016]. National Institute of Standards and Technology, Gaithersburg, MD, 2005.
[10] OGITA N, NAGAI S, UDAGAWA M, et al. Raman scattering study of rare-earth hexaboride[J]. Physica B: Condensed Matter, 2005, 359/360/361: 941-943.
[11] OGITA N, NAGAI S, OKAMOTO N, [J]. J. Solid State Chem., 2004(177): 461-465.
[12] XIE K Y, DOMNICH V, FARBANIEC L, et al. Microstructural characterization of boron-rich boron carbide[J]. Acta Materialia, 2017, 136: 202-214.
[13] ROMA G, GILLET K, JAY A, et al. Understanding first-order Raman spectra of boron carbides across the homogeneity range[J]. Physical Review Materials, 2021, 5(6): 063601.

Features of Deformation and Destruction of 3D Laser Printed Austenitic Steel

N. V. KAZANTSEVA[1,2], Yu. N. KOEMETS[1], D. I. DAVIDOV[1,3], I. V. EZHOV[1], M. S. KARABANALOV[3]

1. Institute of Metal Physics, Ural Branch of the Russian Academy of Sciences, 620108, Ekaterinburg, 18, S. Kovalevskaya Str., Russia
2. Ural State University of Railway Transport, Ekaterinburg, 66, Kolmogorova Str., 620034, Russia
3. Ural Federal University named after the first President of Russia B. N. Yeltsin, Institute of New Materials and Technologies, Ekaterinburg, 19, Mira Str., 620002, Russia

Abstract: Producing the porous materials are necessary for medicine use. The additive laser printing allows one to produce the product with controlling density. In this paper, the features of the deformation and fracture processes in the 316L steel samples with controlling density manufactured with the laser 3D printer are considered. For intentionally manufacture of the porous specimen, laser power of 150 W, point distance of 60 μm, hatch distance of 110 μm, exposure time of 80 μs were used. The density of 92% in the obtained laser powder fusion (L-PBF) samples was measured by the Archimedes' method. The analysis showed that the structure and mechanical properties of steel samples of the austenitic class obtained using a 3D laser printer differ significantly from their cast references. It was found that with an increase in the strain rate under tensile tests, both the strength and ductility of the L-PBF samples simultaneously increase. In contrast to this fact, with the increasing in the strain rate under tensile tests of the conventional samples, both the strength and ductility decrease. Analysis of the deformation process of the L-PBF samples showed existence of different micromechanisms, which associated with the deformation of the porous materials. The strain localization bands, which confirmed the presence of the unstable plastic flow of the material under deformation, were found in the L-PBF sample after compression up to 30% at room temperature. Martensitic α'-phase was found by transmission electron microscopy in the samples after compression at room temperature, with strain rate started from 3×10^{-3} s^{-1}. The obtained results are compared with the literature date for L-PBF and conventional 316L samples.

Keywords: Austenitic steel; Selective laser melting; Deformation; Destruction; Microstructure

Austenitic grade stainless steels are non-magnetic, strong, and corrosion-resistant in a wide range of temperatures and environmental conditions[1]. The temperature of the beginning of the martensitic transformation of austenitic steels (Ms) depends on the chemical composition, because of that corrosion-resistant austenitic steels may be divided into two classes: stable, containing 17%-25% Cr and 14%-25% Ni, and unstable, containing 17%-20% Cr and 8%-12% Ni.[2]. Stable Fe-Cr-Ni austenitic steels have the phase composition of austenite FCC gamma phase, which remains unchanged even after severe deformation or deep cooling. Thermomechanical treatments of unstable chromium-nickel austenitic steels lead to the formation of martensite[2]. A chromium-nickel steel grade 316 (EN 1.4404) is a typical stable austenitic steel. This steel has a very wide application due to its mechanical properties and high corrosion resistance over a wide temperature range. 316L austenitic steel products are widely used in medicine, oil and gas industries, and reactor engineering. Austenitic 316L stainless steel is ideal for implants in dentistry and orthopedics. As well as biomedical titanium alloy Ti-6Al-4V, 316L steel have good biocompatibility and strength properties. However, 316L has better deformability and fracture resistance, and is cheaper than titanium alloy[3]. Usually, austenitic corrosion resistant steels have low strength properties after standard heat treatment. In 316L steel, the main strengthening mechanisms are solid solution strengthening and Hall-Petch strengthening (an increase in hardness associated with the reduction of average grain size)[4]. Figure 1 presents the quasi-binary diagram where one can see that steel 316L at room temperature has a single-phase (fcc, gamma austenite) state. A small amount (up to 2 wt.%-3 wt.%) of the high-temperature bcc (delta-ferrite) phase can be retained in the sample during rapid quenching. The formation of the α'-martensite phase was observed in conventional steel 316L at a strain rate of 1×10^{-3} s^{-1} under mechanical test by compression up to 50%, as well as during low-temperature rolling[5, 6]. The increase in strength in alloys and steel at determined strain rates and temperatures are usually related to unstable plastic flow[7-10]. In the austenitic 316L steel, the unstable plastic flow[4, 7, 9] and negative strain-rate sensitivity (the flow stress decreases with an increase in the rate of deformation)[10] were found under deformation with strain rates of 10^{-4}-10^{-2} s^{-1} in the temperature range of 200-500 ℃.

The active development of additive technologies allows obtaining a finished product with a complex geometric design. Additive laser production used 3D laser printer has a number of advantages and disadvantages. Firstly, the density of the resulting products is comparable to its cast references. However, due to the specifics of the process of

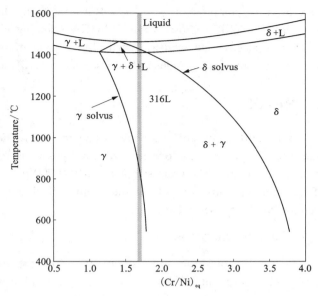

Figure 1　Fe-Ni-Cr quasi-binary diagram[11]

additive laser manufacturing, the structure and phase composition of the resulting 3D product may significantly differ from the structure of the cast reference, which can significantly affect its exploitation properties, deformation and destruction[11-12]. The strength characteristics of austenitic stainless steels manufactured using additive technologies are in the range from 300 to 600 MPa for the yield strength $\sigma_{0.2}$ and from 350 to 760 MPa for the ultimate tensile strength σ_B[13-14]. The high strength and plasticity in L-PBF samples of 316L were suggested to associate with the enhancement of the TWIP effect (twinning induced plasticity)[11]. The appearance of a martensite phase may be considered as one another variant for increasing of the yield strength[3]. There are many problems in producing the porous materials for medicine use. The additive laser printing allows one to produce the product with controlling density. However, literature results on the deformation and destruction of the L-PBF materials with the different controlling density are not too much.

The main aim of this paper is to reveal the features of the deformation and fracture processes in the 316L steel samples with controlling density manufactured with the laser 3D printer.

1　Experiment

Laser printed samples (L-PBF) of austenitic 316L steel were used to study. L-PBF samples were manufactured with a Renishaw AM 400 metal laser printer. Spherical gas-atomized standard ASTM F3184 (316L steel) powder was used for the study. For intentionally manufacture of the porous specimen, laser power of 150 W, point distance of 60 μm, hatch distance of 110 μm, exposure time of 80 μs were used. The density of 92% in the obtained L-PBF samples was measured by the Archimedes' method. Tensile and compression (up to 30%) properties were measured with the standard specimens using a mechanical testing machine Instron at room temperature (Figure 2). Nine samples were used for tensile tests and nine samples

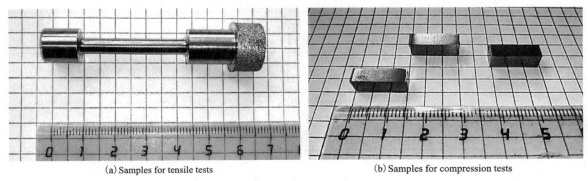

(a) Samples for tensile tests　　　　(b) Samples for compression tests

Figure 2　Shape and size of the L-PBF samples for the test measurements were used for compression tests. A DRON-3 (X-ray diffractometer with CuK$_\alpha$ radiation was used for phase analysis. The ZEISS Cross Beam AURIGA scanning electron microscope (SEM) equipped with an EBSD HKL Inca spectrometer and a Tecnai G2-30 Twin transmission electron microscope were used for structural studies. A conventional industrial cast sample of 316L austenitic steel was used as a reference.

2 Results

X-ray diffraction analysis showed that the conventional and L-PBF samples have austenitic single-phase state (fcc, γ-phase). No one lines of other phases were found. The as-build L-PBF sample has an equiaxed austenite structure. The grain size is about 100-200 μm. A dendritic-cellular structure with an average cell size of 1-3 μm was observed inside the grains. Such a structure is typical of 316L austenitic stainless steel manufactured by laser 3D printing[15].

2.1 Mechanical properties

Figure 3 shows the stress-strain curves of the investigated samples of steel 316L after tensile and compression deformation. Table 1 presents the results of tensile tests of L-PBF samples. It can be seen that with an increase in the strain rate under tensile tests, both the strength and ductility of the L-PBF samples simultaneously increase, that may be associate with the unstable plastic flow in the material under deformation conditions (Table 1).

During the compression tests, cracks appeared in the sample, but the complete destruction of the samples did not occur. The character of the curves at all strain rates is identical (Figure 3). A slight increase in the value of compressive strength is observed at strain rates of 3×10^{-3} and 2×10^{-2} s^{-1}, which may mean an increase in the strength of the studied material (Table 2). It should be mentioned that yield stress in the compress tests is lower than that in the tensile tests. Such observed difference was earlier explained in [5] by the difference in the Schmid factors in dislocation slip systems due to the texture of the initial material.

According to the literature data, the as-build L-PBF sample of 316L usually demonstrates strength characteristics higher than for samples obtained by traditional methods[5, 13, 15]. However, in our case, it can be seen that both the strength and ductility of the L-PBF sample are significantly lower than those of the conventional one. The strength characteristics under tensile test of conventional steel 316L samples are presented in Table 3. Comparison of the results of the tensile deformation L-PBF and conventional samples (Tables 1, 3) shows that the strain rate dependence of the mechanical characteristics is also different. The strain rate dependence of the conventional sample shows the normal (usual) behavior, which means the decreasing of the strength and plasticity characteristics with the increasing of the strain rates (Table 3).

Table 1 Results of tensile tests of the L-PBF samples at different rates

Strain rate/s^{-1}	σ_b/MPa	$\sigma_{0.2}$/MPa	δ/%
3×10^{-4}	535	520	1.7
1×10^{-3}	554	526	5.0
8×10^{-3}	593	562	5.4

Table 2 Results of compression tests of the L-PBF samples at different rates

Strain rate/s^{-1}	$\sigma_{0.05}$/MPa	$\sigma_{0.2}$/MPa
8×10^{-4}	345	436
3×10^{-3}	408	476
2×10^{-2}	437	516

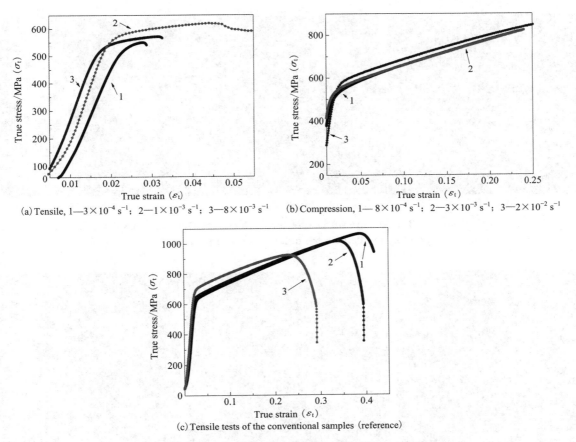

(a) Tensile, 1—3×10^{-4} s^{-1}; 2—1×10^{-3} s^{-1}; 3—8×10^{-3} s^{-1} (b) Compression, 1—8×10^{-4} s^{-1}; 2—3×10^{-3} s^{-1}; 3—2×10^{-2} s^{-1}

(c) Tensile tests of the conventional samples (reference)

Figure 3 Results of the mechanical tests with different strain rates

Table 3 Results of tensile tests of the conventional samples at different rates

Strain rate/s^{-1}	σ_b/MPa	$\sigma_{0.2}$/MPa	δ/%
3×10^{-4}	748	618	49
1×10^{-3}	746	621	44
8×10^{-3}	724	628	34

2.2 Microstructure

Ductile fracture of the conventional samples at all studied strain rates under tension at room temperature is presented in Figure 4.

In opposite to ductile fracture of the conventional samples, the mixed brittle-ductile fracture is observed on the fracture surface of a 316L steel sample after deformation by tensile at a rate of 3×10^{-4} s^{-1} [Figure 5 (a)]. The nucleation of new pores and the merging of existing pores in the microstructure were observed. With an increase in the strain rate to 1×10^{-3} s^{-1}, ductile fracture along the pores is observed, which is unusual for a material with high porosity. At a strain rate of 8×10^{-3}, the nucleation of new pores insignificantly occurs. In this case, a coalescence of existing pores is the main mechanism causing ductile fracture.

Compared with the initial (as-build) state, the internal porosity in the deformed samples is preserved, no complete closure of the technological pores has been detected. An almost non-deformed structure was observed near the technological pores. This fact indicates the heterogeneity of the deformation process in a porous sample. Strain localization bands, which confirmed the presence of the unstable plastic flow in the material under deformation, may be seen on the sample sides after deformation by compression [Figure 5(d)-(f)]. With the increasing of the strain rate, the slip traces become thinner, but the frequency and periodical of the slip traces is retain. The displacement of the material layer in the strain localization bands can be seen in Figure

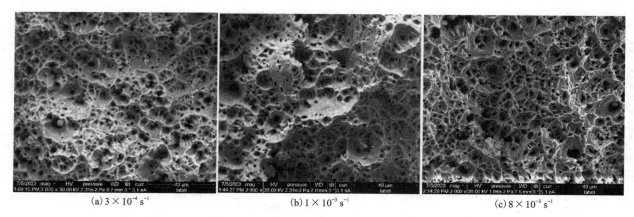

Figure 4 The fracture surfaces of the conventional 316L steel samples deformed by tensile with the different strain rates, SEM images

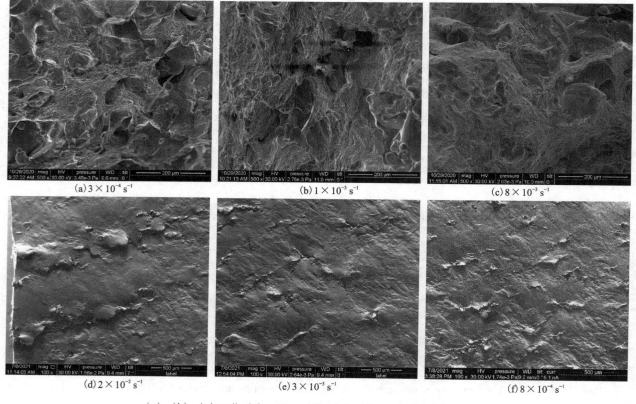

(a), (b), (c) tensile deformation; (d), (e), (f) compression deformation.

Figure 5 The fracture surfaces and sub-fracture microstructures of the L-PBF 316L steel samples deformed by tensile with the different strain rates, SEM images

5(d)(f). The behavior of pores during compression deformation is different from that during tensile tests. The technological pores were closed under compression.

Twins were observed in the L-PBF samples after tensile tests at all studied strain rates. As the strain rate increases, the number of twin packages increases [Figure 6(a)-(c)].

Crystallographic analysis shown that ⟨110⟩{111} dislocation slip prevails at low strain rates; with an increase in the strain rate, dislocation slip occurs both in ⟨110⟩ direction associated with easy dislocation slip and in ⟨112⟩ direction associated with twinning. The twinning process is observed at all strain rates in both uniaxial tension and compression of porous 316L samples. Twinning was more pronounced in specimens deformed at a higher strain rate. As the strain rate increases, the twinning was changed by microtwinning [Figure 6(e)-(f)].

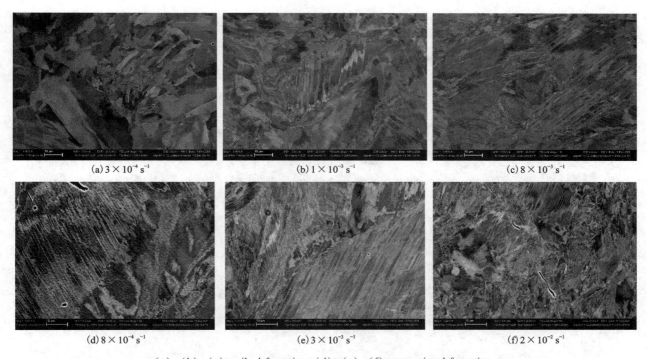

(a), (b), (c) tensile deformation; (d), (e), (f) compression deformation.

Figure 6 Twins in the L-PBF 316L steel samples deformed with the different strain rates, SEM images

TEM study of the initial (as-build) L-PBF sample shows the presence of the sell-dendritic structure of the austenite (Figure 7). High dislocation density was found in the L-PBF sample inside the cells. No one reflexes from carbides or martensite phase were found.

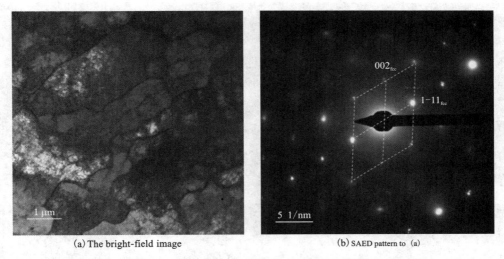

Figure 7 Microstructure of the L-PBF sample (as-built), TEM images

Deformation twins were observed by TEM study in the L-PBF samples after tensile deformation (Figure 8). With the increasing of the strain rate, the number of twins increases. Diffraction patterns taken from twinning regions showed a typical FCC twinning, which had {111} twinning planes.

Figure 9 presents the TEM results of the compression tests. Thin twins were observed in the structure of the sample after deformation with strain rate 8×10^{-4} s^{-1} [Figure 8(a)]. An increase in the deformation rate in the studied samples leads to the martensitic transformation. Deformation martensite α'-phase forms at the boundaries of the intersection of twins [Figure 8(c)-(f)]. The width of the lamellae of the martensite α'-phase is about 20-50 nm.

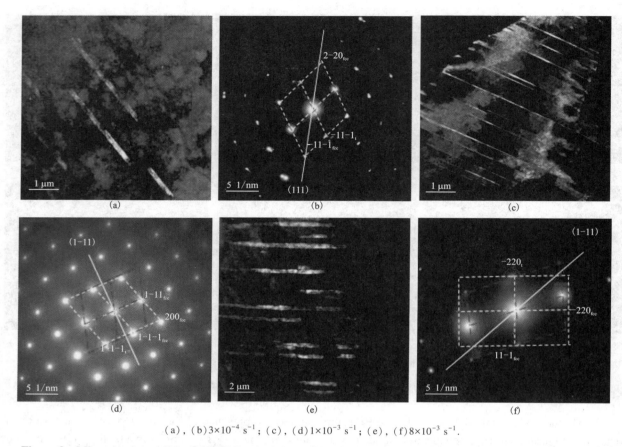

(a), (b) 3×10^{-4} s^{-1}; (c), (d) 1×10^{-3} s^{-1}; (e), (f) 8×10^{-3} s^{-1}.

Figure 8 Microstructure of the L-PBF sample deformed by tension with the different strain rates, TEM images

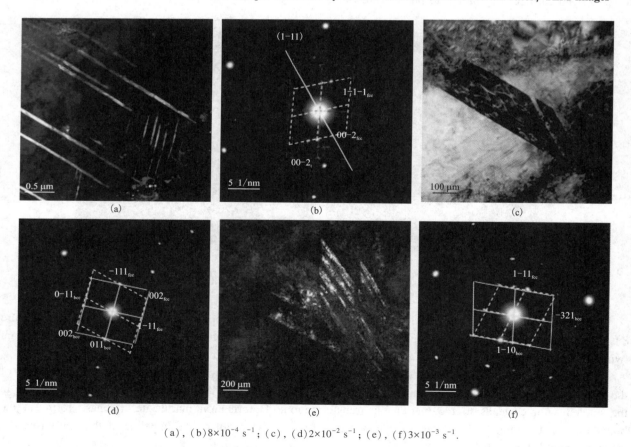

(a), (b) 8×10^{-4} s^{-1}; (c), (d) 2×10^{-2} s^{-1}; (e), (f) 3×10^{-3} s^{-1}.

Figure 9 Microstructure of the L-PBF sample deformed by compression with the different strain rates, TEM images

3 Discussion

According to the literature data, in cold-rolled sheet of 316L common steel, the elongation at break (fracture strain) of the sample was 40%[16-17]. In our case, the elongation at break of the conventional sample at standard strain rate (1×10^{-3} s^{-1}) is 44% (Table 3). Low ductility (2%) and high tensile strength (1400 MPa) were found in the austenitic samples of steel 316L with small grains (30 nm) and with twinning inside them[18]. In the present work, the L-PBF samples in the initial state (as-build) have a cellular structure with an average size of 1-3 μm. Thus, the observed low plasticity at the tensile tests (Table 1) is probably associated with the high porosity of the studied L-PBF samples.

It was shown by some researches that the mechanical properties of L-PBF 316L steel samples depend on both the density of the obtained sample and its orientation in the 3D printer chamber during manufacture. Table 4 summarizes the mechanical properties of L-PBF samples of steel 316L obtained by various authors during tensile tests at room temperature.

Table 4 Summary results of the tensile tests of the L-PBF steel 316L samples

Sample	Density	Orientation	$\sigma_{0.2}$/MPa	σ_b/MPa	E/GPa	δ/%	Ref.
cast	100	-	365	563	190-200	69	[19]
L-PBF	99, 97	Horiz.	503	644	183	51	[20]
L-PBF	99, 97	Vert.	452	548	165	56	[20]
L-PBF	99, 34	Vert.	438	525	164	15	[20]
L-PBF	99, 43	Horiz.	440	573	167	38	[20]
L-PBF	98, 85	Horiz.	-	645	-	25	[21]
L-PBF	92	Horiz.	526	554	-	5	This work

As can be seen in the table, changes in the density and orientation of the samples have a particularly large effect on plasticity, while the strength of the obtained samples does not change so significantly.

The appearance of deformation martensite at room temperature is atypical for samples of steel 316L. Austenitic corrosion — resistant steel 316L (16 wt.%-18 wt.% Cr—10 wt.%-14 wt.% Ni—2 wt.%-3 wt.% Mo) has a high energy of packaging defects and belongs to the category of stable steels with respect to the formation of deformation martensite even at low temperature[6, 23]. No martensitic transformation was observed in conventional 316L steel during deformation at room temperature[6]. It was found that the formation of deformation martensite in conventional 316L steel during cold rolling deformation at 0 ℃ began only after deformation of 30%. The volume fraction of deformation martensite was only 0.4% when a degree of deformation was more than 50%[6]. It was shown in [23] that metastable intermediate phase ε-martensite was observed in the transformation of $\gamma \rightarrow \varepsilon \rightarrow \alpha'$ after deformation at cryogenic temperatures of the massive samples of 316L steel. The detailed study of martensitic transformation in 316L steel samples was carried out at a deformation rate of 5×10^{-5} s^{-1} in the low temperature range of 80-300 K in [22]. In [22] it was shown that the amount of deformation martensite decreased with an increase in the deformation temperature. At a deformation temperature of 300 K, the practical absence of deformation martensite was noted, ε-martensite also not detected. The authors of [24] also confirmed the absence of α'-martensite in the L-PBF 316L steel sample after deformation at room temperature, while after deformation at cryogenic temperature (-77 K), both types of martensite (α' and ε) were found in the structure of the L-PBF sample.

Deformation α' martensite was observed in conventional 316L porous samples, deformed by compression up to 45% at room temperature[25-26]. In [25] authors also did not find the diffraction reflexes of the martensitic phase in X-ray diffraction patterns; presence of the deformation martensite was supported by magnetic studies due to observing the magnetic hysteresis loops in the samples after deformation.

It is known that martensitic α'-phase plates have specific and well-recognized morphology. In low-carbon austenitic steels (class 300), deformation martensite has a cubic body-centered crystal lattice (BCC), which makes it difficult to separate diffraction lines α'-phase and residual ferrite, which also has a BCC lattice. However, the morphology of the regions and the size of these two phases (ferrite and α' martensite) are significantly different, which makes it easy to separate them by electron microscopic studies. Moreover, the amount of δ-ferrite does not change with increasing degree or rate of deformation, unlike deformation martensite, which is formed in the form of thin plates and the amount of which increases with increasing degree of deformation.

The α′-phase deformation martensite plates with a cubic body-centred (BCC) crystal lattice are clearly visible in Figure 8(c)-(f). The location of the reflexes from the planes of austenite and deformation martensite on SAED patterns corresponds to the of Kurdyumov-Sachs orientation relations between FCC and BCC crystal lattices. , the calculated energy of the stacking faults in the studied samples of 316L steel, according to formula suggested in [6], is 180 J/m^2. This fact means that austenite has high stability not only to martensitic transformations, but also to mechanical twinning. However, as our studies show, both martensitic transformations and FCC twinning are observed in the L-PBF 316L steel samples. It allows us to conclude that the stacking faults energy in L-PBF 316L steel samples is significantly reduced. The effect of the deformation rate and grain size on the martensitic transformation in L-PBF samples also differs from that observed in conventional 316L steel samples.

All of the above allows us to conclude that the L-PBF material with controlling density has properties different from its references obtained by the conventional method of casting. Thus, knowledge of the deformation and morphological features of destruction process in steels of the austenitic class manufactured with 3D laser printer is essential value for assessing the level of long-term strength of products obtained from them. The density of the resulting material and the position of the sample in the 3D chamber during growth should also be taken into account.

4 Conclusions

The analysis showed that the structure and mechanical properties of steel samples of the austenitic class manufactured with 3D laser printer with controlling density of the samples differ significantly from their cast references.

(1) Analysis of the deformation process of the L-PBF samples showed existence of different mechanisms, which associated with the deformation of the porous materials.

(2) In L-PBF samples, α′-martensite was detected after high-speed deformation up to 30% compression at room temperature, which is unusual for stable austenitic steel. It is likely that the high level of residual stresses arising in the material during 3D laser printing contributes to the accumulation of crystal lattice defects and thereby facilitates the martensitic transitions.

(3) Unlike HIP, compression deformation at room temperature has little effect on internal gas porosity. The strain localization bands, which confirmed the presence of the unstable plastic flow of the material under deformation, were found in the L-PBF sample after compression at room temperature up to 30%.

Support by Russian Science Found (Project No. 22-29-01514) is acknowledged. Structural studies were done with a Tecnai G2-30 Twin transmission electron microscope and a scanning electron microscope JSM 6490 with the Oxford Inca energy dispersive and wave microanalysis at the Center of Electron Microscopy of the Testing Center for Nanotechnologies and Advanced Materials of the Institute of Metal Physics, Ural Branch, Russian Academy of Sciences.

References

[1] FONTANA M G, STAEHLE R W. Advances in corrosion science and technology[M]. Springer, 2012.
[2] SALUJA R, MOEED K. The emphasis of phase transformations and alloying constituents on hot cracking susceptibility of type 304L and 316L stainless steel welds[J]. International Journal of Engineering Science and Technology, 2012, 4(5): 2206-2212.
[3] SPENCER K, VÉRON M, YU-ZHANG K, et al. The strain induced martensite transformation in austenitic stainless steels: part 1-Influence of temperature and strain history[J]. Materials Science and Technology, 2009, 25(1): 7-17.
[4] DIEPOLD B, NEUMEIER S, MEERMEIER A, et al. Temperature-dependent dynamic strain aging in selective laser melted316L[J]. Advanced Engineering Materials, 2021, 23(10): 2001501.
[5] EL-TAHAWY M, JENEI P, KOLONITS T, et al. Different evolutions of the microstructure, texture, and mechanical

performance during tension and compression of316L stainless steel[J]. Metallurgical and Materials Transactions A, 2020, 51(7): 3447-3460.

[6] SOHRABI M J, NAGHIZADEH M, MIRZADEH H. Deformation-induced martensite in austenitic stainless steels: a review[J]. Archives of Civil and Mechanical Engineering, 2020, 20(4): 1-24.

[7] HONG S G, LEE S B. Mechanism of dynamic strain aging and characterization of its effect on the low-cycle fatigue behavior in type316L stainless steel[J]. Journal of Nuclear Materials, 2005, 340(2/3): 307-314.

[8] KAZANTSEVA N, KOEMETSY, DAVYDOV D, et al. Analysis of unstable plastic flow in the porous 316L samples manufactured with a laser 3D printer[J]. Materials, 2022, 16(1): 14.

[9] SAMUEL K G, MANNAN S L, RODRIGUEZ P. Serrated yielding in AISI-316 stainless steel[J]. Acta Metall, 1988, 36(8): 2323-2327.

[10] LV. X L, CHEN S H, WANG Q Y, et al. Temperature dependence of fracture behavior and 312 mechanical properties of AISI 316 austenitic stainless steel[J]. Metals. 2022, 12: 1421.

[11] SUN Z J, TAN X P, TOR S B, et al. Simultaneously enhanced strength and ductility for 3D-printed stainless steel 316L by selective laser melting[J]. NPG Asia Materials, 2018, 10(4): 127-136.

[12] KAZANTSEVA N, KRAKHMALEV P, ÅSBERG M, et al. Micromechanisms of deformation and fracture in porous l-PBF 316 L stainless steel at different strain rates[J]. Metals, 2021, 11(11): 325.

[13] BAJAJ P, HARIHARAN A, KINI A, et al. Steels in additive manufacturing: a review of their microstructure and properties[J]. Materials Science and Engineering: A, 2020, 772: 138633.

[14] DEBROY T, WEI H L, ZUBACK J S, et al. Additive manufacturing of metallic components - Process, structure and properties[J]. Progress in Materials Science, 2018, 92: 112-224.

[15] TUCHO W M, LYSNE V H, AUSTBØ H, et al. Investigation of effects of process parameters on microstructure and hardness of SLM manufactured SS316L[J]. Journal of Alloys and Compounds, 2018, 740: 910-925.

[16] EN 10088-2: 2005. Stainless steels. Part 2: Technical delivery conditions for sheet/plate and strip of corrosion resisting steels for general purposes. BSI. 30. 06. 2005. 46

[17] LEE W S, CHEN T H, LIN C F, et al. Dynamic mechanical response of biomedical316L stainless steel as function of strain rate and temperature[J]. Bioinorganic Chemistry and Applications, 2011: 1-13.

[18] YAN F K, LIU G Z, TAO N R, et al. Strength and ductility of316L austenitic stainless steel strengthened by nano-scale twin bundles[J]. Acta Materialia, 2012, 60(3): 1059-1071.

[19] SANDER G, THOMAS S, CRUZ V, et al. On the corrosion and metastable pitting characteristics of316L stainless steel produced by selective laser melting[J]. Journal of the Electrochemical Society, 2017, 164(6): 250-257.

[20] RÖTTGER A, BOES J, THEISEN W, et al. Microstructure and mechanical properties of 316L austenitic stainless steel processed by different SLM devices[J]. The International Journal of Advanced Manufacturing Technology, 2020, 108(3): 769-783.

[21] GATÕES D, ALVES R, ALVES B, et al. Selective laser melting and mechanical properties of stainless steels[J]. Materials, 2022, 15(21): 7575.

[22] MAN J, OBRTLÍK K, PETRENEC M, et al. Stability of austenitic 316L steel against martensite formation during cyclic straining[J]. Procedia Engineering, 2011, 10: 1279-1284.

[23] NALEPKA K, SKOCZEŃ B, CIEPIELOWSKA M, et al. Phase transformation in 316L austenitic steel induced by fracture at cryogenic temperatures: experiment and modelling[J]. Materials, 2020, 14(1): 127.

[24] MISHRA P, ÅKERFELDT P, FOROUZAN F, et al. Microstructural characterization and mechanical properties of L-PBF processed 316L at cryogenic temperature[J]. Materials, 2021, 14(19): 5856.

[25] GRĄDZKA-DAHLKE M, WALISZEWSKI J. Analysis of phase transformation of austenitic 316L implant steel during compression[J]. Defect and Diffusion Forum, 2009, 283/284/285/286: 285-290.

[26] MUMTAZ K, TAKAHASHI S, ECHIGOYA J, et al. Magnetic measurements of martensitic transformation in austenitic stainless steel after room temperature rolling[J]. Journal of Materials Science, 2004, 39(1): 85-97.

Structure and Properties of Electroexplosive Coatings of the TiB$_2$-Ag System

Denis ROMANOV, Vasili POCHETUKHA

Siberian State Industrial University, Novokuznetsk, 654007, Russia

Abstract: The structure of TiB$_2$-Ag coatings deposited by electron-ion-plasma sputtering is studied. The hardness, the modulus of elasticity, the wear resistance, and the friction coefficient are determined. The coatings are shown to affect positively the combination of the operating properties of copper electric contacts of switches of high-power networks.

Keywords: Electric contact material; Electron-ion-plasma sputtering; Structure; Properties; Hardness

1 Introduction

In most cases, the physical and mechanical properties of the surfaces of electrical contacts determine the service life of the entire electrical and power facilities[1-3]. Therefore, the surface of the contacts should be designed appropriately and possess the required functional properties[4] rather than making the whole of the article from an expensive material spending much power[5]. In this connection, surface modification with effective electron-ion-plasma beams is a promising direction for creating steady coatings resisting electroerosion[6]. In each specific case, it is hard to obtain an obvious physical picture making it possible to predict the result of such a treatment as a function of the properties of the substrate and of the parameters of the active beams of charged particles[7]. In the first turn, this concerns the complex effect of plasma and accelerated electron beams on the surface treated[8]. The fact that the electron-ion-plasma technology is the most efficient tool of surface hardening is determined by the possibilities and the recent level of development of the equipment for its implementation[9]. Due to the use of this process, the surface layer of the contact acquires a multiphase submicro-and nanocrystalline structure due to the superfast rates of heating and cooling and due to formation of ultimate temperature gradients[10]. Thus, determination of the nature and of the laws of formation of structural and phase states and properties of electroerosion-resistant coatings obtained by an electronion-plasma treatment is an important direction in the physics of condensed state[11, 12]. The aim of the present work was to analyze the elemental and phase composition, the state of the defective substructure, the mechanical and tribological properties of a TiB$_2$-Ag coating formed on copper by the method of electron-ion-plasma sputtering.

2 Methods of study

A TiB$_2$-Ag coating was deposited onto commercially pure copper M100 with the following chemical composition 3: 99.99 Cu, 0.001 Fe, 0.001 Ni, 0.001 S, 0.001 P, 0.001 As, 0.001 Pb, 0.001 Zn, 0.001 Ag, 0.0005 O, 0.0005 Sb, 0.0005 Bi, 0.0005 Sn. The sputtering was conducted by the method of electron-ion-plasma treatment involving successive electroexplosion sputtering in a "EVU 6010M" device (Novokuznetsk) and an electron beam treatment combined with nitriding in a "KOMPLEKS" device for electron-ion-plasma surface engineering (Institute of High-Current Electronics of the Siberian Branch of the Russian Academy of Sciences, Tomsk). The copper samples were shaped as parallelepipeds with sizes 25 mm ×25 mm×5 mm. The electroexplosion sputtering of the samples was performed with the use of a bilayer foil. One of the layers was a titanium diboride foil with a mass of 150 mg; the other layer was a silver foil (99.9% Ag, 0.003% Pb, 0.035% Fe, 0.002% Sb, 0.002% Bi, 0.058% Cu) with a mass of 800 mg. The regime of the thermal and force impact on the irradiated surface was assigned by choosing the charge voltage of the capacitive energy storage of the device, which was then used to compute the absorbed power density. The sputtering parameters were as follows: The time of the plasma action on the surface of the specimen is about 100 s, the absorbed power density on the axis of the jet is about 5.5 GW/m^2, the pressure in the impact-compressed layer near the irradiated surface is about 12.5 MPa, the residual gas pressure in the working chamber is about 100 Pa, the plasma temperature on the silver nozzle section is about 10^4 K. The combination of the parameters of the electron-beam treatment and nitriding was as follows. The parameters of the low-energy strong-current electron beam: the surface energy density $E_s = 40$ J/cm^2, the pulse duration $t = 200$ s, the number of pulses $N = 5$. The nitriding was conducted in the same chamber at 520 ℃ for 5 h. The two processes were combined in a single vacuum cycle, i. e., at first the electron-beam treatment and then the nitriding. We analyzed the elemental and phase compositions and the defective substructure of the coating using Carl Zeiss EVO50 (Germany) and JEOL JEM-210F (Japan) devices for scanning and transmission electron diffraction microscopy of thin foils[13, 14]. The electron diffraction patterns were analyzed by the methods of [13, 15]. The parameters of the defective substructure were determined as described in [13, 15]. The foils for the transmission electron microscopy were prepared by ion thinning of plates using argon ions (Ion Slicer EM-091001S device). The plates were cut from massive samples using an Isomet Low Speed Saw device perpendicularly to the modified surface, which allowed us to trace the changes in the structure and in the elemental and phase compositions of the material over the thickness of the coating.

3 Results and discussion

The tribological tests show that the wear parameter (a quantity inverse to the wear resistance) $P_w = 0.26 \times 10^{-4}$ mm^3/m for the copper sample coated with TiB$_2$-Ag. This is 2.3 times lower than that for the uncoated copper sample ($P_w = 0.59 \times 10^{-4}$ mm^3/m). The friction coefficient of the coated samples is 2.2 times lower than that of the uncoated samples ($\mu = 0.2$ and 0.43 respectively). Consequently, the TiB$_2$-Ag coating deposited on copper by electron-ion-plasma sputtering combined with nitriding possesses higher (by a factor of ≥2) tribological properties than the copper samples without coating.

The friction coefficient in the tribological tests of copper and of coated copper behaves differently. The results presented in Figure 1, show that the sample with the coating exhibits a longer stage before saturation than the uncoated sample. The profiles of the friction groove of the uncoated copper and of the coated copper are smooth and exhibit no flaw or tear, which indicates absence of brittle inclusions in the coatings (Figure 2).

The hardness of the coating was measured in a cross section over a groove going parallel to the surface of the substrate at a distance of 50 μm from the coating substrate interface. The hardness values determined for the coating and for the substrate are close and amount to about 1.3 GPa. The elasticity moduli of the coating (109 GPa) and of the substrate (107 GPa) also differ little.

The thickness of the coating measured in cross sections with the help of scanning electron microscopy (SEM) varies from 120 to 300 μm [Figure 3(a) and (b)]. The coating substrate interface is not homogeneous [Figure 3(b) and (c)]; the elements of the coating penetrate the substrate to a depth of up to 100 μm most probably over the boundaries of copper grains. This indicates good cohesion between the coating and the substrate. Analysis of the structure of the coating in transverse direction [Figure 3(c) and (d)] shows its multiphase nature. In the bulk of the grains of the coating the structure is of a eutectic type and is composed of alternating plates of different phases [Figure 3(d)].

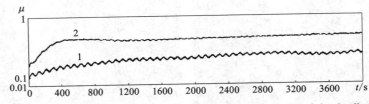

Figure 1 Friction coefficient (μ) as a function of the duration (t) of tribological tests of copper samples without coating (1) and with a TiB$_2$-Ag coating (2)

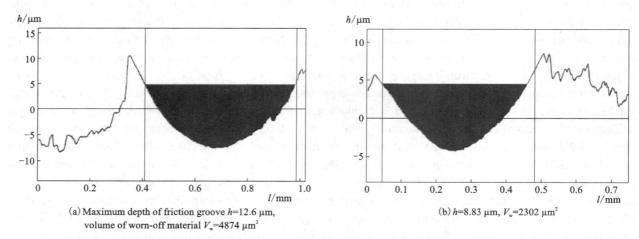

(a) Maximum depth of friction groove $h=12.6$ μm, volume of worn-off material $V_w=4874$ μm²

(b) $h=8.83$ μm, $V_w=2302$ μm²

Figure 2　Profile of the friction groove (h is the depth and l is the width) of copper samples without coating (a) and with a TiB_2-Ag coating (b)

The elemental composition of the coating was studied by microscopic X-ray spectrum analysis. The mapping of a cross section (SEM) shows that the elements are distributed in the coating in quasi-uniform manner. A well manifested interface is observable between the coating and the substrate. The main elements in the coating are titanium diboride and silver; copper atoms are present in the coating in a little amount. The relative content of chemical elements and their distribution in the coating were studied by macroscopic X-ray spectrum analysis using scanning over the chosen line. The results of this study show that the coating consists primarily of silver and titanium diboride and bears comparatively little copper. It addition, the samples contain substrate regions with deep penetration of the elements of the coating.

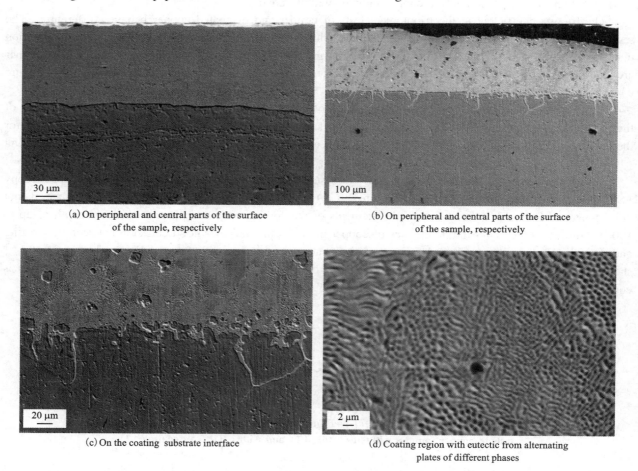

(a) On peripheral and central parts of the surface of the sample, respectively

(b) On peripheral and central parts of the surface of the sample, respectively

(c) On the coating substrate interface

(d) Coating region with eutectic from alternating plates of different phases

Figure 3　Structure of a TiB_2-Ag coating on copper (transverse section, SEM)

A more detailed study of the distribution of elements in the structure of the coating was made by microscopic X-ray spectrum analysis with the help of a transmission electron microscope (TEM). The results show that the base of the coating is a TiB_2-Ag solid solution of titanium diboride in silver. Copper is present in the coating in the form of inclusions of a cylindrical (cigar) shape. Nitrogen atoms do not interact with copper.

The phase composition, the defective substructure of the coating, and the morphology of the constituent phases were studied by transmission electron microscopy, microdiffraction analysis and dark-background imaging.

The results show that the coating is a multiphase formation. The inclusions of the second phase have a lamellar (cigar) shape and are arranged in parallel rows in the volume of the grains. The structure in the vicinity of the coating substrate interface is polycrystalline. Round nanosize particles of the second phase are located over the boundaries and in the bodies of grains.

Structure based on grains of TiB_2, Ti_2B and Ti_3B_4 forms near the coating substrate interface, which means that the main elements of the phases are copper and cadmium. Copper and titanium diboride form nanosize particles of phase TiB_2, Ti_2B, Ti_3B_4, TiB.

4 Conclusion

We have studied TiB_2-Ag coatings with a thickness of up to 300 μm deposited on copper samples by the method of electron-ion-plasma sputtering. It has been shown that the wear parameter (a quantity inverse to the wear resistance) of a copper sample with a coating is less than that of an uncoated sample by a factor of 2.3 and the friction coefficient is less by a factor of 2.2. The hardness of the coating and the modulus of elasticity are virtually the same as in copper. The main chemical elements in the coating are silver and titanium dibride. Copper is present in an inconsiderable amount. The main phases in the structure are solid solutions based on copper and silver. In addition, the coating acquires nanosize intermetallic phases TiB_2, Ti_2B, Ti_3B_4, TiB. The results of the study allow us to conclude that the growth in the strength and in the tribological parameters of the coating with respect to those of the copper substrate is caused by formation of a multiphase structure represented by submicron-size solid solutions based on silver and copper and nanosize inclusions of boride phases.

Acknowledgments: This work was performed within Russian Science Foundation project No. 22-79-10012 (https://rscf.ru/en/project/22-79-10012/).

References

[1] LIU XL, LI Z H, HE J F, et al. Influence of humidity on endurance of electrical contact during fretting wear[J]. Journal of Materials Engineering and Performance, 2022, 31(2): 933-943.

[2] SUN SG, CUI J R, DU T H. Research on the influence of vibrations on the dynamic characteristics of AC contactors based on energy analysis[J]. Energies, 2020, 13(3): 559.

[3] TANG LF, HAN Z P, XU Z H. A sequential adaptive control strategy for the contact colliding speed of contactors based on fuzzy control[J]. IEEE Transactions on Industrial Electronics, 2021, 68(7): 6064-6074.

[4] ISARD M, LAHOUIJ I, LANOT JM, et al. In-depth investigation of a third body formed by selective transfer in a NiCr/AgPd electrical contact[J]. Wear, 2021, 474/475: 203753.

[5] QI L, ZONG M, WANG XJ. Eliminating the contact bounce of AC contactor based on speed feedback[J]. International Journal of Circuit Theory and Applications, 2021, 49(3): 731-745.

[6] KUMAR S P, SENTHIL S M, PARAMESHWARAN R, et al. Fabrication of a novel silver-based electrical contact composite and assessment of its mechanical and electrical properties[J]. Arch. Metall. Mater., 2021(66): 1087-1094.

[7] SANTOSH S, KEVINTHOMAS J, PAVITHRAN M, et al. An experimental analysis on the influence of CO_2 laser machining parameters on a copper-based shape memory alloy[J]. Optics & Laser Technology, 2022, 153: 108210.

[8] HE CY, FENG F, WANG J B, et al. Improving the mechanical properties and thermal shock resistance of W-Y_2O_3 composites by two-step high-energy-rate forging[J]. International Journal of Refractory Metals and Hard Materials, 2022, 107: 105883.

[9] ZHANG M Y, WANG C M, MI G Y, et al. in situ study on fracture behaviors of SiC/2A14Al composite joint: co-construction of microstructure and mechanical properties via laser welding[J]. Composites Part B: Engineering, 2022, 238: 109882.

[10] LU Y, TURNER R, BROOKS J, et al. A study of process-induced grain structures during steady state and non-steady state electron-beam welding of a titanium alloy[J]. Journal of Materials Science & Technology, 2022, 113: 117-127.

[11] ROMANOV D A, POCHETUKHA V V, SOSNIN K V, et al. Structure and properties of composite coatings of the SnO_2-In_2O_3-Ag-N system intended for strengthening the copper contacts of powerful electric network switches[J]. Journal of Materials Research and Technology, 2022, 17: 3013-3032.

[12] ROMANOV D, POCHETUKHA V, GROMOV V, et al. Fundamental research on the structure and properties of electroerosion-resistant coatings on copper[J]. Uspehi Fiziki Metallov, 2021, 22(2): 204-249.

[13] UTEVSKII L M. Diffraction electron microscopy in metals science [J]. Metallurgiya, Moscow, 1973: 584.

[14] TOMAS G, GORINGE M J. Transmission electron microscopy of materials [M]. Moscow: Nauka, 1983.

[15] ANDREWS K, DYSON D, KEOWN S. Electron diffraction patterns and their interpretation [J]. Metallurgiya, Moscow, 1977: 208.

Catalytically Active Materials for PEM Fuel Cells Applications

Alexandra KURIGANOVA, Nikita FADDEEV, Marina KUBANOVA

Platov South Russian State Polytechnic University (NPI), Novocherkassk, 346428, Russia

Abstract: Electrocatalytic systems based on platinum, an alloy of platinum and tin, palladium, tin oxides, and titanium oxides were obtained under the conditions of nonstationary electrolysis in aqueous electrolytes without the use of organic solvents, stabilizers, and structure-forming agents.

Possibilities of non-stationary electrolysis to obtain platinum-tin-containing systems, in which tin is introduced in the form of a doping element, or in the composition of an alloy with platinum, or in the form of tin oxide, as a component of a hybrid carrier, platinum-containing catalysts based on hybrid supports with different contents of the oxyphilic component, palladium-containing catalyst systems with controlled palladium phase particle size and palladium oxide phase amount are shown.

The prospect of using such electrocatalytic systems in the technologies of solid polymer fuel cells, namely, fuel cells with direct oxidation of organic fuel (ethanol, formic acid, dimethyl ether) is shown.

Keywords: Polymer exchange membrane fuel cells; Non-stationary electrolysis; Electrocatalysis

1 Introduction

At present, the study of catalytic systems based on platinum and palladium in electrocatalysis is primarily due to the need to create efficient electrocatalysts for polymer exchange membrane fuel cells (PEMFCs) using hydrogen or organic molecules (alcohols, acids, ethers) as fuel[1]. Despite numerous attempts to replace expensive platinum catalysts in PEMFCs with cheaper materials, platinum group metal nanoparticles deposited on carbon supports (CS) with a developed surface area remain the most effective catalysts for processes occurring in PEMFCs today. Today, the requirements for environmental neutrality are also imposed on technologies for the production of catalytic materials based on platinum group metals. In this sense, the technologies of electrochemical production are very promising. They use electric current as an oxidizing agent, a reducing agent instead of chemical reagents[2].

The basis of non-stationary electrolysis is the use of alternating current for the implementation of various electrochemical processes. In particular, non-stationary electrolysis makes it possible to obtain dispersed materials with catalytic activity.

One of the electrochemical approaches to the production of catalytic materials is non-stationary electrolysis, the advantage of which is the possibility of obtaining in one stage catalytically active nanoparticles of platinum group metals deposited on the surface of carbon or hybrid (metal oxide-carbon) supports. This article will present the results of studies of the physicochemical properties of mono-and polymetallic catalytic systems based on platinum, palladium obtained under conditions of non-stationary electrolysis and their electrocatalytic properties in anodic processes implemented in PEMFCs: oxidation of hydrogen, CO, ethanol, formic acid, dimethyl ether.

2 Experiment

2.1 Catalysts preparation

The process of catalytically active materials based on Pt and Pd particles under conditions of non-stationary electrolysis is carried out according to the following procedure (Figure 1): at the first stage, two metal electrodes (Pt or Pt_3Sn or Pd) with the same geometrical surface area were placed in a support suspension in an aqueous electrolyte solution. Carbon black Vulcan XC-72 or hybrids Vulcan XC-72+MO_x (M = Sn, Ti) was used as a support. As an electrolyte, aqueous solutions of sodium hydroxide with a concentration of 1 mol/L were used as an alternating pulsed current with a frequency of 50 Hz and an average density of 1.0 A/cm^2 was applied to the electrodes. Under the action of an alternating pulsed current, the metal electrodes were dispersed to nanoparticles and deposited on the support, as described in [3, 4]. The synthesis process was carried out with constant stirring and cooling of the electrolyte.

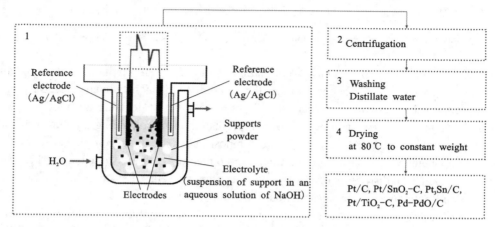

Figure 1 Scheme for the synthesis of catalytic materials under conditions of non-stationary electrolysis

At the next stages, the catalyst was separated from the electrolyte by centrifugation, washed many times with distilled water, and dried in air to constant weight.

2.2 Catalysts characterization

To determine the composition and microstructural characteristics, all synthesized catalysts were studied by X-ray phase analysis and electron microscopy.

For electrochemical measurements, "catalytic inks" were prepared based on the synthesized catalysts, isopropyl alcohol, and an aqueous suspension of Nafion (10 wt.%).

Calculation of the electrochemically active surface area (Pt_{ECSA}) of the catalysts was carried out by CO-stripping method according to the charge spent on CO oxidation, with the exception of the charge that went to charge the double layer:

$$Pt_{ECSA} = \frac{Q_{CO}}{Q_m \times g_{Pt}} \qquad (1)$$

where: $Q_{CO}(\mu C)$ is the charge passed during CO oxidation, g_{Pt} is the mass of Pt on the working electrode, and $Q_m = 420\ (\mu C/cm^2)$ [5] is the charge required for desorption of the CO_{ads} monolayer from 1 cm^2 of platinum.

Studies of the electrocatalytic activity of catalysts in the anodic oxidation of methanol, ethanol, formic acid, and dimethyl ether were carried out both under model conditions (standard three-electrode electrochemical cell) using the methods of cyclic voltammetry (CV), chronoamperometry (CA), a rotating disk electrode (RDE), and as part of a membrane-electrode assembly (MEA).

3 Results and discussion

An important characteristic of electrocatalytic materials is their electrochemically active surface area (ECSA). The method of oxidative desorption of CO[6] is considered as the most accurate method for determining the ECSA of Pt-or Pd-containing materials, that is, the surface on which electrochemical processes can occur in PEMFCs. It should be noted that the ECSA value of Pt-containing catalysts was practically independent of the composition of the catalyst or support and was about 13-14 m^2/g, while the area of Pd-PdO/C was determined by the radius of the hydrated electrolyte cation used in the synthesis of catalytic systems (Table 1).

Table 1 Characterization of Pt-and Pd-containing electrocatalysts

Sample	Electrolyte for synthesis	Microstructural characteristics	Electrochemical measurements		Ref.
			ECSA /($m^2 \cdot g^{-1}$)	Catalytic activity	
Pt/C (25% Pt)	1 mol/L NaOH+ Vulcan XC-72	Pt particles (6.0 ± 0.7) nm deposited onto CS	13.1 ± 1.3	0.5 mol/L H_2SO_4+0.5 mol/L C_2H_5OH $j_{peak}=0.41$ mA/cm^2	[7]
Pt_3Sn/C (25% Pt_3Sn)	1 mol/L NaOH+ Vulcan XC-72	Pt_3Sn particles (5.6 ± 0.6) nm deposited onto CS	13.5 ± 1.5	0.5 mol/L H_2SO_4+0.5 mol/L C_2H_5OH $j_{peak}=0.64$ mA/cm^2	[7]
$PtSn_{doped}$/C (25% $PtSn_x$)	1 mol/L NaOH+ 1 mol/L $SnSO_4$+ Vulcan XC-72	$PtSn_{doped}$ particles (5.7 ± 0.8) nm deposited onto CS	13.5 ± 1.5	0.5 mol/L H_2SO_4+0.5 mol/L C_2H_5OH $j_{peak}=0.84$ mA/cm^2	[7]
Pt/SnO_x-C (30% SnO_x)	1 mol/L NaOH+ Vulcan XC-72 +SnOx	Pt particles (6.0 ± 0.7) nm deposited onto hybrid support	13.1 ± 1.5	0.5 mol/L H_2SO_4+0.5 mol/L C_2H_5OH $j_{peak}=1.19$ mA/cm^2	[7]
Pt/TiO_2-C (10% TiO_2)	1 mol/L NaOH+ Vulcan XC-72+TiO_2	Pt particles (6.0 ± 0.7) nm deposited onto hybrid support	14.0 ± 1.0	Deaerated with DME 0.5 mol/L H_2SO_4 $j_{peak}=0.042$ mA/cm^2	[8]
Pt/TiO_2-C (30% TiO_2)	1 mol/L NaOH+ Vulcan XC-72+TiO_2	Pt particles (6.0 ± 0.7) nm deposited onto hybrid support	14.0 ± 1.0	Deaerated with DME 0.5 mol/L H_2SO_4 $j_{peak}=0.049$ mA/cm^2	[8]
Pt/TiO_2-C (60% TiO_2)	1 mol/L NaOH+ Vulcan XC-72+TiO_2	Pt particles (6.0 ± 0.7) nm deposited onto hybrid support	14.0 ± 1.0	Deaerated with DME 0.5 mol/L H_2SO_4 $j_{peak}=0.055$ mA/cm^2	[8]
Pd-PdO/C (28% PdO)	1 mol/L KCl+ Vulcan XC-72	Pd-containing particles ~11.0 nm deposited onto CS	52.5 ± 4.0	1) 0.5 mol/L C_2H_5OH+1.0 mol/L NaOH $j_{at0.7V}=0.66$ mA/cm^2 2) 0.5 mol/L HCOOH+0.5 mol/L H_2SO_4 $j_{at0.35V}=0.66$ mA/cm^2	[9]
Pd-PdO/C (42% PdO)	1 mol/L NaCl+ Vulcan XC-72	Pd-containing particles ~7.9 nm deposited onto CS	77.5 ± 5.0	1) 0.5 mol/L C_2H_5OH+1.0 mol/L NaOH $j_{at0.7V}=0.55$ mA/cm^2 2) 0.5 mol/L HCOOH+0.5 mol/L H_2SO_4 $j_{at0.35V}=0.70$ mA/cm^2	[9]
Pd-PdO/C (55% PdO)	1 mol/L LiCl+ Vulcan XC-72	Pd-containing particles ~8.5 nm deposited onto CS	31.2 ± 2.0	1) 0.5 mol/L C_2H_5OH+1.0 mol/L NaOH $j_{at0.7V}=0.96$ mA/cm^2 2) 0.5 mol/L HCOOH+0.5 mol/L H_2SO_4 $j_{at0.35V}=0.42$ mA/cm^2	[9]

Studies of a series of catalysts in the ethanol electrooxidation reaction showed that, regardless of the state in which tin is in platinum-tin-containing materials, in which both the platinum and tin components were obtained under pulsed electrolysis conditions, such materials demonstrate increased electrocatalytic activity in the ethanol electrooxidation reaction in comparison with Pt-containing materials obtained by a similar method and in which tin was not included. The best electrocatalytic activity was demonstrated by the material characterized by the predominance of the tin component in its composition in the form of SnO_2[7] (Table 1).

Platinum-containing catalysts based on hybrid TiO_2-C supports can be recommended for increasing the efficiency of PEMFCs with direct oxidation of dimethyl ether (DME)[8] (Table 1). An increase in the catalytic activity of the Pt/TiO_2-C materials compared to the commercial Pt/C catalyst was due to the influence of the size effect during oxidation of dimethyl ether on larger platinum particles in the composition of the catalysts, the structural sensitivity of the reaction of chemisorption and oxidation of DME, which predominantly occurs on the Pt(100) terraces, which mainly represented by cubic particles of platinum obtained under conditions of non-stationary electrolysis, as well as the presence of hydrophilic particles of titanium dioxide in the composition of the hybrid catalyst carrier, which provides a higher rate of the limiting stage of oxidation of strongly chemisorbed intermediate particles due to the presence of active surface oxygen compounds, and also facilitates the activation of C-H-bonds in the methyl groups of the DME.

It has been established that the electrocatalytic activity of a series of Pd-PdO/C materials is determined by both the size effect and the amount of the oxide phase in the material[4, 9] (Table 1). With an increase in the amount of PdO phase in the material, the electrocatalytic activity of Pd-PdO/C increases in the processes of electrooxidation of CO and ethanol, but negatively affects the electrooxidation of formic acid, which is due to differences in the mechanisms of electrooxidation of CO, ethanol, and formic acid on platinum group metals. Thus, CO and ethanol are oxidized according to the Lenmuir-Hinshelwood mechanism, the limiting step of which is the stage of chemical interaction of Pd_{Rads} organic particles adsorbed on the catalyst surface with adsorbed oxygen-containing particles, for example, Pd_{OHads}. For these processes, the presence of a hydrophilic PdO oxide phase on the catalyst surface facilitates the adsorption of reactive oxygen-containing particles, which generally reduces the overvoltage of the process. Formic acid is oxidized through an active intermediate weakly bound to the surface. For this process, the presence of an oxide phase in the composition of the catalyst has a negative effect. This is probably why catalysts with a lower PdO content are more active in the oxidation of formic acid.

4 Conclusion

Thus, non-stationary electrolysis, namely the electrochemical dispersion of metals under the action of an alternating pulsed current, is an effective approach to obtaining dispersed materials with electrocatalytic activity in the oxidation reactions of alcohols, acids, and esters. In particular, non-stationary electrolysis makes it possible to obtain (i) platinum-tin-containing systems, in which tin is introduced in the form of a doping element, or in the composition of an alloy with platinum, or in the form of tin oxide, as a component of a hybrid carrier; (ii) platinum-containing catalysts based on hybrid supports with different contents of the oxophilic component; (iii) palladium-containing catalyst systems with controlled palladium phase particle size and palladium oxide phase amount.

Acknowledgments: This work was carried out within the framework of the strategic project "Hydrogen Energy Systems" of the SRSPU (NPI) Development Program during the implementation of the program of strategic academic leadership "Priority-2030".

References

[1] IOROI T, et al. Electrocatalysts for PEM Fuel Cells[J]. Advanced Energy Materials, 2019, 9(23): 1801284.
[2] PETRII O A. Electrosynthesis of nanostructures and nanomaterials[J]. Russian Chemical Reviews, 2015, 84(2): 159-193.
[3] LEONTYEV I, KURIGANOVA A, KUDRYAVTSEV Y, et al. New life of a forgotten method: Electrochemical route toward highly efficient Pt/C catalysts for low-temperature fuel cells[J]. Applied Catalysis A: General, 2012, 431/432: 120-125.
[4] KURIGANOVA A B, FADDEEV N A, LEONTYEV I N, et al. New electrochemical approach for the synthesis of Pd-PdO/C electrocatalyst and application to formic acid electrooxidation[J]. ChemistrySelect, 2019, 4(29): 8390-8393.
[5] BRIMAUD S, PRONIER S, COUTANCEAU C, et al. New findings on CO electrooxidation at platinum nanoparticle surfaces[J]. Electrochemistry Communications, 2008, 10(11): 1703-1707.
[6] WATT-SMITH M J, FRIEDRICH J M, RIGBY S P, et al. Determination of the electrochemically active surface area of Pt/C PEM fuel cell electrodes using different adsorbates[J]. Journal of Physics D: Applied Physics, 2008, 41(17): 174004.
[7] KURIGANOVA A, KUBANOVA M, LEONTYEV I, et al. Pulse electrolysis technique for preparation of bimetal tin-containing electrocatalytic materials[J]. Catalysts, 2022, 12(11): 1444.
[8] KUBANOVA M S, KURIGANOVA A B, SMIRNOVA N V. Electrooxidation of dimethyl ether on Pt/TiO_2-C catalysts[J]. Russian Journal of Electrochemistry, 2022, 58(10): 916-926.
[9] FADDEEV N A, KURIGANOVA A B, LEONT'EV I N, et al. Palladium-based electroactive materials for environmental catalysis[J]. Doklady Physical Chemistry, 2022, 507(1): 139-146.

Filler Composite Materials Based on an Al-Si System Alloy for the Synthesis of Wear-resistant Working Layers of Functionally Organized Compositions by Arc Cladding

Roman MIKHEEV[1], Igor KALASHNIKOV[2], Pavel BYKOV[3], Lubov KOBELEVA[4]

1　BMSTU, Moscow, 105005, Russian
2　IMET RAS, Moscow, 119334, Russian
3　IMET RAS, Moscow, 119334, Russian
4　IMET RAS, Moscow, 119334, Russian

Abstract: Filler composite materials based on the Al-Si system alloy (grade $AlSi_{12}Cu_2MgNi$) reinforced with ortho-phase (Ti_2NbAl) particles and modifying additives of chromium carbide (Cr_2C_3) have been produced by extrusion process. By means of optical, electron microscopy, and using energy dispersive X-ray spectroscopy method it has been shown that the established extrusion process parameters allow to prevent interaction between the matrix material and reinforcements, and to ensure a uniform composition and a satisfactory reinforcements distribution.

The wear-resistant working layers were obtained on functionally organized compositions by arc cladding with a tungsten electrode in an argon atmosphere using produced the filler composite materials. In addition to the working layer such functionally organized compositions include a base made of low-carbon steel 20, a technological layer of pure tin, and an intermediate layer of babbit B83. The results of wear tests under dry sliding friction showed that the cladded working layers of composite materials provide the friction coefficient and wear rate reduction by a factor of 3-4 compared to the cladded coatings from antifriction matrix alloy $AlSi_{12}Cu_2MgNi$.

Keywords: Filler composite material; Al-Si system; Arc cladding; Functionally organized compositions; Wear rate

1 Introduction

The failure of machines and mechanisms parts and working bodies under normal operating conditions is rarely the result of their insufficient strength. In 85%-90% of cases, the main reason for this is the wear of sliding bearings, journal bearings etc.[1-3]. Therefore, new solutions are required both to increase wear resistance and to reduce the friction coefficients of such bearings.

In recent years, significant progress has been made in the development of particle reinforced composite materials based on Al-Si system alloys which have the best combination of tribological, mechanical, and other characteristics[3, 4]. However, of greatest interest is the task of using such materials as those subjected to intense physical and chemical effects during the operation of the working layers of functionally organized compositions. Moreover, in addition to the working layer of composite material, such composition may include a substrate that perceives the load, technological and intermediate layers, providing the required values of adhesive strength, as well as layers formed directly in the synthesis process[5, 6].

Different processes can be applied for the functionally organized composition synthesis: solid-phase methods (explosion welding, friction surfacing, etc.); liquid-phase methods (foundry technologies, arc and laser cladding, etc.); methods for applying coatings from the gas phase[6-8]. Among them, arc cladding process is the most economical and promising for use in industry. Due to layer-by-layer synthesis arc cladding process makes it possible to form the necessary structure and desired properties of functionally organized compositions, as well as to perform repairing without equipment decommissioning[6, 9].

The filler composite material is required for arc cladding process using in the synthesis and repairing of functionally organized compositions working layers. The filler material has to ensure the reinforcement presence and its uniform distribution in the cladded working layer and can be obtained in the form of a flux-cored wire, as well as solid rods, made by casting technology or by extrusion[6, 10]. The using of extruded rods is preferable. The extrusion process can provide a given content and uniform distribution of the reinforcement in the matrix alloy, as well as to obtain rods of a given size[11]. The work purpose was to obtain filler composite materials based on an Al-Si system alloy and test them for applying wear-resistant working layers of functionally organized compositions by arc cladding.

2 Materials and methods

Widespread antifriction aluminum alloy $AlSi_{12}Cu_2MgNi$ (Al-Si system) was used as a matrix for filler composite materials. Ortho-phase intermetallics (Ti_2NbAl) with an average size of not more than 100 μm, as well as chromium carbide particles (Cr_2C_3) with an average size of 1.5 μm, were used as reinforcements. An important criterion in choosing these reinforcements was the high proportion of the metal component of the chemical bond in them, which determines the thermodynamic and kinetic compatibility with the matrix material[6]. The proportion of reinforcing particles Ti_2NbAl and modifying additives Cr_2C_3 was 10 and 1 wt.%, respectively.

Filler materials in the form of rods were obtained by extrusion of composite mixtures on an OMA 650 B press with an axial force of 115-120 kN. The composite mixtures were preheated to a temperature of 873 K in an electric furnace SNOL 10/11-V. To obtain a composite powder mixture, chips obtained by processing cast ingots of the $AlSi_{12}Cu_2MgNi$ alloy were mixed with reinforcing particles in a Retsch PM100 planetary mill according to the technological parameters presented in Table 1. The main criterion in choosing the values of the technological parameters of the regimes was the possibility of extruding the entire volume of the prepared composite powder mixture and the compliance of the obtained filler rods with the requirements of GOST 21449.

Table 1 Technological parameters for obtaining of composite powder mixtures

Composite powder mixture composition/wt.%	Rotation frequency, ω/min^{-1}	Time, t/min
$AlSi_{12}Cu_2MgNi$ (matrix)	300	20
$AlSi_{12}Cu_2MgNi^* + 10Ti_2NbAl_{(<100)}$	350	60
$AlSi_{12}Cu_2MgNi^* + 10Ti_2NbAl_{(<100)} + 1Cr_2C_{3(1, 5)}$	300	20

* $AlSi_{12}Cu_2MgNi$ particles obtained after treatment according to the parameters: $\omega = 300\ \text{min}^{-1}$, $t = 20\ \text{min}$

The working layers were obtained by arc cladding with a tungsten electrode in an argon atmosphere on functionally organized compositions, which are a substrate of low-carbon steel 20 (GOST 1050), a technological layer of pure tin, and an intermediate layer of B83 babbitt (GOST 1320). Selected technological parameters of the arc cladding process: welding current 140 A; arc voltage 18 V; deposition rate 4.2 m/h; an argon flow rate of 12 L/min ensured satisfactory formation of cladded layers and the absence of unacceptable defects.

The study of the structure and composition of the filler composite rods and cladded working layers is carried out by means of optical and electron microscopy, as well as using the method of energy dispersive X-ray spectroscopy (EDS). Leika DMILM optical microscope equipped with the Qwin software package for image analysis and a JEOL JSM-6000PLUS (NEOSCOPE II) scanning electron microscope equipped with secondary and backscattered electron detectors, as well as an energy-dispersive microanalysis system were used. It is important to note that, in order to ensure the reliability of the obtained results, the structure and phase composition were studied in different areas of the cross section of the manufactured samples.

Wear tests of the working layers of the fabricated functionally organized compositions under dry sliding friction were carried out on the CETR UMT Multi-Specimen Test System according to the test scheme " a rotating sleeve (counterbody, hardened steel 40Cr, GOST 4543, HRC > 45) along a fixed disk (test sample)". The test procedure included continuous stepwise loading of samples at a constant sliding velocity (V) of 0.5 m/s in a wide range of specific loads (p): 0.5; 1.0; 1.5; 2.0 and 2.5 MPa[12]. The tests were carried out for 600 s at each of the proposed specific loads.

The friction coefficient values were determined by using the UMT TestViewer software. The wear of the samples was evaluated by the value of the mass wear rate, I_m, g/m:

$$I_m = \Delta/L \quad (1)$$
$$\Delta = m_b - m_a \quad (2)$$

where: Δ is the mass loss of the sample; m_b is the mass of the sample before testing; m_a is the mass of the sample after testing; L is the total friction path. Additionally, the values of the friction stability parameter (α) were determined, which was dimensionless value of the friction coefficient standard deviation.

The obtained testing results of the working layers of the fabricated functionally organized compositions were compared with those for a coating from the antifriction matrix alloy $AlSi_{12}Cu_2MgNi$.

3 Results and discussion

The produced filler composite materials had a diameter of 3.0 and a length of 350 mm, which met the requirements for rods intended for cladding a wear-resistant layer on parts of machines and equipment operating under conditions of abrasive wear, shock loads, corrosion, erosion at elevated temperatures or in aggressive environments according to GOST 21449.

The results of the analysis of fracture surfaces of filled rods made of composite materials indicate the absence of reinforcements degradation [Figure 1(a)]. The presence of Ti_2NbAl reinforcing particles and Cr_2C_3 modifying additives in the filler composite rods composition was confirmed by the EDS [Figure 1(b)-(e)]. In addition, the difference in the quantitative results of EDS in different areas does not exceed 3%. This allows to consider the filler composite rods composition produced by the extrusion process to be homogeneous, and the distribution of reinforcing particles and modifying additives to be satisfactory.

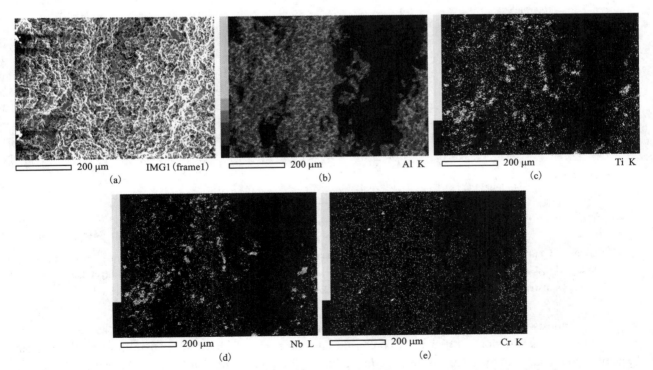

Figure 1 Characteristic fracture surface (a) and EDS results (b), (c), (d), (e) of the filler composite rod, wt.%: $AlSi_{12}Cu_2MgNi + 10Ti_2NbAl_{(<100)} + 1Cr_2C_{3(1.5)}$

The cladded working layers made of composite materials have a thickness of 2.0-3.0 mm and are characterized by the continuity of the fusion line, as well as the pores absence [Figure 2(a)]. The EDS results indicate the reinforcement (reinforcing particles and modifying additives) retention in the cladded layer [Figure 2(b), (c)]. Thus, the produced wear-resistant working layers of functionally organized compositions retain the given reinforcement code, which confirms the possibility of their manufacture by the arc cladding process.

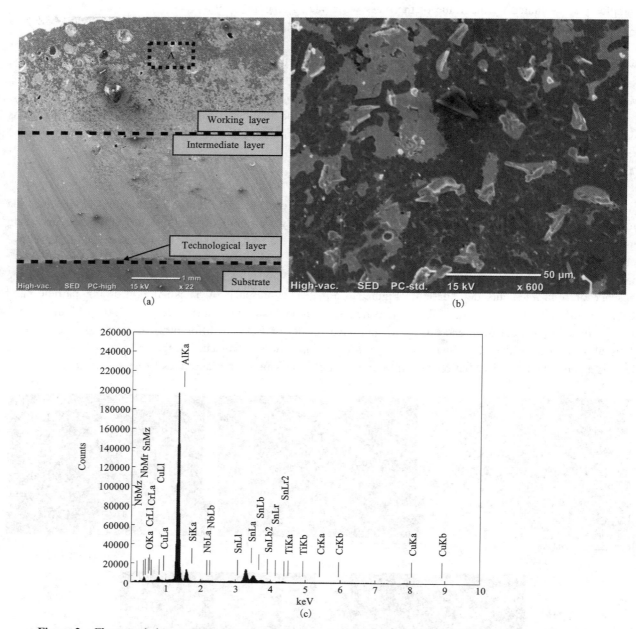

Figure 2　Characteristic macrostructure (a) of the produced functionally organized composition: (working layer-AlSi$_{12}$Cu$_2$MgNi+10Ti$_2$NbAl$_{(<100)}$+1Cr$_2$C$_{3(1.5)}$) - (intermediate layer-B83) - (technological layer-tin) - (substrate-steel 20), microstructure (b) and EDS (c) the cladded working layer in area A

The wear tests results of the cladded working layers are presented in Table 2. It can be seen that the values of the average friction coefficient of composite layers are up to 3 times lower than those of layers from the antifriction matrix alloy. This is due to the presence of reinforcing particles in the composite layers composition, which help to retain the "third body" at the interface of the contacting surfaces[7, 12]. The "third body" formation on friction surfaces prevents direct contact of metal surfaces and, therefore, not only helps to reduce the friction coefficient, but also protects the sample from wear[13].

Table 2 The friction and wear tests results of cladded working layers

The cladded working layer composition/wt. %	Average friction coefficient, f/stability parameter of the friction process, α, depending on the specific load, p/MPa					Wear rate, I_m /(10^{-5} g·m^{-1})
	0.5	1.0	1.5	2.0	2.5	
AlSi$_{12}$Cu$_2$MgNi (matrix)	0.401/0.07	0.171/0.11	0.169/0.07	0.164/0.07	0.192/0.07	9.38
AlSi$_{12}$Cu$_2$MgNi+10Ti$_2$NbAl$_{(<100)}$	0.34/0.07	0.285/0.04	0.179/0.08	0.173/0.08	0.148/0.08	4.16
AlSi$_{12}$Cu$_2$MgNi+10Ti$_2$NbAl$_{(<100)}$+1Cr$_2$C$_{3(1.5)}$	0.388/0.09	0.257/0.05	0.107/0.04	0.054/0.02	0.052/0.03	2.12

A comparison of the wear rate values shows that the working layers from composite materials make it possible to increase the wear resistance by up to 4 for the Al-Si system alloys. In addition, the dry sliding friction process of such layers is characterized by better stability (the stability parameters presented in Table 2 have values of no more than 0.09) in the entire range of specific loads compared to that for matrix alloy layers. This is due not only to the "third body" formation, but also to better strength characteristics due to the presence of high-modulus reinforcements in the working layers of composite materials.

4 Conclusion

Filler composite materials based on the Al-Si system alloy (grade AlSi$_{12}$Cu$_2$MgNi) reinforced with orthophase (Ti$_2$NbAl) particles and modifying additives of chromium carbide (Cr$_2$C$_3$) have been produced by extrusion process. By means of optical, electron microscopy, and using energy dispersive X-ray spectroscopy method it has been shown that the established extrusion process parameters allow to prevent interaction between the matrix material and reinforcements, and to ensure a uniform composition and a satisfactory reinforcements distribution. The wear-resistant working layers were obtained on functionally organized compositions by arc cladding with a tungsten electrode in an argon atmosphere using produced the filler composite materials. In addition to the working layer such functionally organized compositions include a base made of low-carbon steel 20, a technological layer of pure tin, and an intermediate layer of babbit B83. The results of wear tests under dry sliding friction showed that the cladded working layers of composite materials provide the friction coefficient and wear rate reduction by 3-4 times compared to the cladded coatings from antifriction matrix alloy AlSi$_{12}$Cu$_2$MgNi.

Acknowledgments: This research was supported by a grant from the Russian Science Foundation (Project No. 22-29-00366), https://rscf.ru/project/22-29-00366/.

References

[1] BABU M V S, RAMA KRISHNA A, SUMAN K N S. Review of journal bearing materials and current trends[J]. American Journal of Materials Science and Technology, 2015 (2), 72-83.
[2] SANTOS N D S A, ROSO V R, FARIA M T C. Review of engine journal bearing tribology in start-stop applications [J]. Engineering Failure Analysis, 2020, 108: 104344.
[3] CHERNYSHOVA T A, MIKHEEV R S, KALASHNIKOV I E, et al. Development and testing of Al-SiC and Al-TiC composite materials for application in friction units of oil-production equipment[J]. Inorganic Materials: Applied Research, 2011, 2(3): 282-289.
[4] GARG P, JAMWAL A, KUMAR D, et al. Advance research progresses in aluminium matrix composites: manufacturing & applications[J]. Journal of Materials Research and Technology, 2019, 8(5): 4924-4939.
[5] MIKHEEV R S, KALASHNILOV I E, BOLOTOVA L K, et al. Research of the intermetallics formation mechanism during the synthesis of functionally graded layered steel-aluminum compositions[J]. IOP Conference Series: Materials Science and Engineering, 2020, 848(1): 012056.
[6] MIKHEEV R S. Innovative processes of production functional gradient layered compositions with enhanced tribological properties[J]. IOP Conference Series: Materials Science and Engineering, 2020, 934(1): 012036.
[7] ZHANG Y P, WANG Q, CHEN G, et al. Mechanical, tribological and corrosion physiognomies of CNT-Al metal

matrix composite (MMC) coatings deposited by cold gas dynamic spray (CGDS) process[J]. Surface and Coatings Technology, 2020, 403: 126380.
[8] BHADAURIA N, VASHISHTHA P, MISHRA S, et al. Fabrication of aluminium MMCs & associated difficulties-A review[J]. Materials Today: Proceedings, 2022, 64: 1276-1282.
[9] KUMAR DAS A, VERMA O. Development of Ni+TiB$_2$ metal matrix composite coating on AA6061 aluminium alloy substrate by gas tungsten arc cladding process[J]. Materials Today: Proceedings, 2022, 51: 248-253.
[10] KALASHNIKOV I E, KOLMAKOV A G, BOLOTOVA L K, et al. Technological parameters of production and properties of babbit-based composite surfacing rods and deposited antifriction coatings[J]. Inorganic Materials: Applied Research, 2019, 10(3): 635-641.
[11] JAYASEELAN V, KALAICHELVAN K, RAMASAMY N, et. al. Influence of SiC$_p$ volume percentage on AA6063/SiC$_p$ MMC extrusion process: An experimental, theoretical and simulation analysis, International Journal of Lightweight Materials and Manufacture, 2023(3): 357-366.
[12] BYKOV P A, KALASHNIKOV I E, KOBELEVA L I, et al. Mapping wear modes of composite materials with intermetallic reinforcing based on antifrictional alloy of system Al-Sn-Cu[J]. Letters on Materials, 2021(2): 181-186.
[13] DU A D, LATTANZI L, JARFORS A E W, et al. Role of matrix alloy, reinforcement size and fraction in the sliding wear behaviour of Al-SiCp MMCs against brake pad material[J]. Wear, 2023, 530/531: 204969.

Spinel-Garnet Fibers Based on Organomagnesiumoxane Yttriumoxane Alumoxanes Materials of Aviation and Space Technology

Anastasiya S POKHORENKO[1], Galina I SHCHERBAKOVA[1], Maxim S VARFOLOMEEV[1,2],
Tatyana L APUKHTINA[1], Artem A VOROBYEV[1], Natalia S KRIVTSOVA[1],
Dmitriy V ZHIGALOV[1], Pavel A STOROZHENKO[1]

State Research Institute for Chemistry and Technology of Organoelement Compounds, Moscow, 105118, Russia

Abstract: Researches of GNIIChTEOS have developed a method for producing ceramic fiber of mixed spinel-garnet ($MgAl_2O_4/Y_3Al_5O_{12}$) composition based on fiber-forming organomagnesiumoxane yttriumoxane alumoxanes oligomers by melt spinning.

Keywords: Ceramic-forming and fiber-forming organomagnesiumoxane yttriumoxane alumoxanes; Magnesium aluminate spinel; Yttrium aluminum garnet

1 Introduction

Ceramic fibers of mixed oxide composition, for example, α-Al_2O_3 and t-ZrO_2, YAG-ZrO_2, α-Al_2O_3 and MgO[1,2] are in demand for the production of high-temperature ceramic composites with improved mechanical properties which are required for the manufacture of parts of aircraft and ground-based gas turbine engines, hypersonic missiles and aircraft as well as systems for thermal protection of spacecraft and hypersonic vehicles[3].

The oxides of the spinel and garnet structure have not only a high melting point (2135 and 1940 ℃, respectively) but also a complex crystal structure that prevents the movement and propagation of cracks[1,2].

Chinese scientists described a method for producing oxide fibers of mixed composition: spinel-garnet (MAS/YAG) using sol-gel technology[4].

We have developed a method for producing fibers of mixed oxide composition: alumomagnesium spinel ($MgAl_2O_4$) and yttrium aluminum garnet ($Y_3Al_5O_{12}$) from a melt of a pre-ceramic polymer-organomagnesiumoxane yttriumoxane alumoxanes[5].

2 Material and experimental procedures

Magnesium acetylacetonates and yttrium acetylacetonates hydrates were purchased at JSC *Spectr TT*. Used solvents were purchased at JSC *Component-Reaktiv*.

Using a procedure described previously, we synthesized organomagnesiumoxane yttriumoxane alumoxanes with given molar ratio of Al/Mg and Al/Y[6].

Polymer fibers were produced from organomagnesiumoxane yttriumoxane alumoxanes by melt spinning. The fibers were formed by SmartRheo 20 (SR20) capillary viscometer ("CEAST").

Curing of the formed polymer fibers was performed in Nabertherm 50/500/11 tube furnace in the air environment at the rate of 1°/min up to 500 ℃.

Pyrolysis of the cured fiber was performed in an upgraded electric resistance furnace SNOL 12/16 at 1300-1500 ℃ in atmospheric air.

The surface morphology of the polymer was studied on a Quanta 250 and elemental composition of the polymer fibers and samples of ceramic fibers was studied on a Philips SEM 505 equipped with a SapphireSi (Li) SEM10 energy dispersive detector and a Micro Capture SEM3.0 M image capture system.

Diffractometric studies were carried out in a divergent Zeeman-Bolin beam on Shimadzu XRD-6000 vertical X-ray diffractometer at ambient temperature with CuK$_\alpha$ radiation ($\lambda K_\alpha cp = (2\lambda K_{\alpha 1} + \lambda K_{\alpha 2})/3 = 1.54178$. The crystalline phases were identified using ICDD PDF Release 2003 data.

The characteristic temperatures — softening point (T_1), fiberization temperature (T_2), melting or solidification temperatures (T_3) were determined by a procedure developed at GNIIChTEOS (State Research Institute for Chemistry and Technology of Organoelement Compounds)[7].

3 Results and discussion

The production of fibers of mixed oxide composition $MgAl_2O_4/Y_3Al_5O_{12}$ is carried out as follows: 200 g of fiber-forming organomagnesiumoxane yttriumoxane alumoxanes with a molar ratio of Al:Y≈6 and Al:Mg ≈2 are loaded in small portions into the extruder of the molding machine preheated to 110 ℃. The speed of rotation of the receiving spool is set at 250 r/min for pulling and winding the polymer fiber. Then the wound polymer fiber [Figure 1(a)] is removed from the receiving spool, placed on a corundum mat and placed in an oven for further heat treatment [Figure 1.2(b)]. Heating is carried out in an air atmosphere according to the following mode: from room temperature to 500 ℃ at a speed of 1 ℃/min — fiber curing [Figure 1(c)], from 500 ℃ to 1300-1500 ℃ at a speed of 10 ℃/min with exposure for 10 min. Heat treatment is carried out in an atmosphere of air. As a result, ceramic fibers of mixed oxide composition $MgAl_2O_4/Y_3Al_5O_{12}$ are obtained [Figure 1(d)].

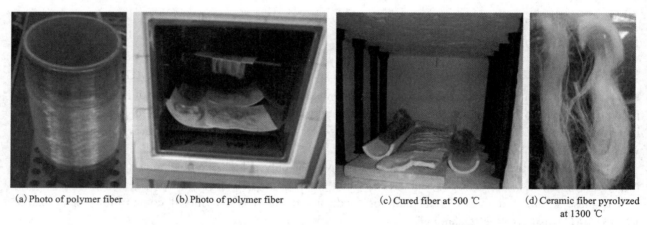

Figure 1　The process of obtaining ceramic fibers of mixed oxide composition $MgAl_2O_4/Y_3Al_5O_{12}$

The elemental composition of the polymer fiber is shown in Figure 2(a), and the ceramic fiber in Figure 2(b).

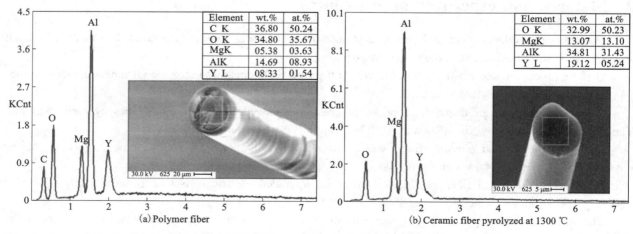

Figure 2　SEM image and X-ray elemental microanalysis

The phase composition of $MgAl_2O_4/Y_3Al_5O_{12}$ ceramic fibers based on organomagnesiumoxane yttriumoxane alumoxanes with a molar ratio of $Al:Y \approx 6$ and $Al:Mg \approx 2$ obtained at different temperatures was proved by the XRD method:

① At 1300 ℃: $MgAl_2O_4$ 79 wt%; $Al_2Y_4O_9$ 17 wt%; $Al_5Y_3O_{12}$ 4 wt%; trace amount Al_3Y_5 [Figure 3(a)];
② At 1500 ℃: $MgAl_2O_4$ 77 wt% and $Y_3Al_5O_{12}$ 23 wt% [Figure 3(b)].

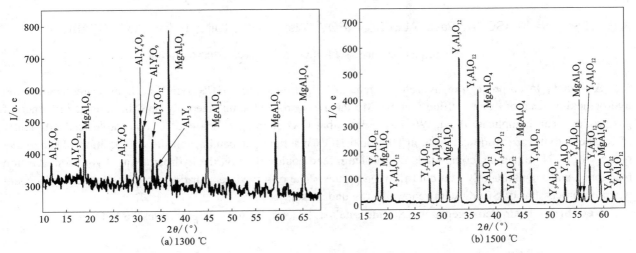

Figure 3　Diffractogram of ceramic fiber $MgAl_2O_4/Y_3Al_5O_{12}$ pyrolyzed at 1300 ℃ and 1500 ℃

Tensile tests show that the fibers that cure at 500 ℃ have strength of 150-300 MPa, however, further heat treatment of the cured fibers up to 1300 ℃ leads to an increase in strength up to 800 MPa.

4　Conclusion

The method for producing ceramic fibers of mixed spinel-garnet composition $MgAl_2O_4/Y_3Al_5O_{12}$ by melt spinning of fiber — forming oligomers to produce polymer fibers at 80-180 ℃ with a molar ratio of $Al:Y \approx 6$ and $Al:Mg \approx 2$ and further stepwise heat treatment in air at 500 ℃, and 1300-1500 ℃ is developed. It is found that polymer fibers heat-treated up to 1300 ℃ have a tensile strength of about 800 MPa.

Acknowledgments: This work was carried out with financial support of State Research Institute for Chemistry and Technology of Organoelement Compounds.

References

[1] AKRAM M Y, FERRARIS M, CASALEGNO V, et al. Joining and testing of alumina fibre reinforced YAG-ZrO_2 matrix composites[J]. Journal of the European Ceramic Society, 2018, 38(4): 1802-1811.
[2] CHANDRADASS J, BALASUBRAMANIAN M. Effect of magnesium oxide on sol-gel spun alumina and alumina-zirconia fibres[J]. Journal of the European Ceramic Society, 2006, 26(13): 2611-2617.
[3] ARMANI C J, RUGGLES-WRENN M B, FAIR G E, et al. Creep of Nextel™ 610 fiber at 1100 ℃ in air and in steam [J]. International Journal of Applied Ceramic Technology, 2013, 10(2): 276-284.
[4] MA XL. Preparation and grain-growth of magnesia-alumina spinel/yttrium aluminum garnet composite fibers[J]. Ceramics-Silikaty, 2018: 279-284.
[5] SHCHERBAKOVA G I, et al. 2022 RF Patent 2776286 C1.
[6] SHCHERBAKOVA G I, et al. 2018 RF Patent 2644950 C1.
[7] SHCHERBAKOVA G I, APUKHTINA T L, KRIVTSOVA N S, et al. Fiber-formingorganoyttroxanealumoxanes[J]. Inorganic Materials, 2015, 51(3): 206-214.

Co-processing of Rapeseed Oil and SRGO Blend over Sulfide Catalysts

Evgeniya VLASOVA, Pavel ALEKSANDROV, Alexey NUZHDIN, Galina BUKHTIYAROVA

Boreskov Institute of Catalysis SB RAS, Novosibirsk, 630090, Russia

Abstract: In co-processing hydrodeoxygenation, hydrotreating (HDS, HDN), hydroisomerization and hydrocracking reactions over sulfide Co(Ni)Mo catalysts proceed simultaneously and the mutual influence of gas products can complicate them. We have seen, that O-containing compounds in renewable feed differently affect the activity of sulfide catalysts in HDS and HDN reactions, depending on the composition of the active phase. S-and N-containing compounds in petroleum feed could effect on the hydrocracking/hydroisomerization reactions, and, consequently, on cold flow properties. The observed pecularities should be taken into account when co-processing of SRGO with renewables is undertaken.

Keywords: Hydrodeoxygenation; Sulfide catalysts; Isomerization; Triglycerides, SAPO-11

1 Introduction

The involvement of renewables in the production of motor fuels is motivated by several reasons such as the challenge of CO_2 emission reduction, legislative regulation increasing demand for environmentally friendly motor fuels, depletion of oil recourses[1-2]. Triglyceride-based feedstocks, especially waste oil, are the attractive source for production of normal and iso-alkanes-valuable components of motor fuels. Oxygen can be removed via different pathways as water, CO or CO_2. At present hydroprocessing of triglycerides is realized in the stand-alone units using the sulfide catalysts, but the deactivation of sulfide phase in the sulfur-free feed is the problem.

Co-processing of renewables with petroleum feeds helps to maintain the sulfide state of catalysts due to sulfur-containing compounds in petroleum feed, provides better control of reactor temperature, improves the economic efficiency due to use of existing refinery and distribution infrastructure. One more problem is the pour cold flow properties of hydrodeoxygenation products, long chain normalalkanes[3-4], that requires the employment of catalysts with isomerization/cracking functionality, for example sulfide phase on zeolite-containing support.

The aim of this paper is to study the effect of active phase (Mo, CoMo and NiMo) and support (Al_2O_3 or Al_2O_3-SAPO-11) composition on the behavior of sulfide catalysts in co-processing of rapeseed oil (RSO) with straight run gas oil.

2 Experimental part

2.1 Support preparation

For preparation of catalysts supported on alumina we used commercial Al_2O_3 extrudates with the diameter of 1.2 mm (specific surface area-206 m^2/g, total pore volume-0.66 cm^3/g, average pore diameter 12.8 nm).

Support Al_2O_3-SAPO-11 was synthesized by mixing of AlOOH (Sasol GmbH, Germany) and SAPO-11 powders (Zeolyst International Co.) in a Z-shaped blade mixer followed by peptization by nitric acid solution and extrusion. Content of SAPO-11 in final composite support was 30 wt.%.

2.2 Catalyst preparation

The Mo/Al_2O_3 catalyst was synthesized by incipient wetness impregnation of alumina granules by aqua solution containing molybdenum trioxide MoO_3 (grade ≥99.0%, Vekton, Russia), citric (grade ≥99.8%, Vekton, Russia) and phosphoric (grade ≥85%, Vekton, Russia) acids.

The $CoMo/Al_2O_3$-SAPO-11 and $NiMo/Al_2O_3$-SAPO-11 catalysts were prepared by incipient wetness impregnation of Al_2O_3-SAPO-11 extrudates using an aqueous solution of active metals precursors (MoO_3, grade ≥99.0%, from Vekton, Russia; $Co(OH)_2$, 95%, Sigma-Aldrich, USA, or $Ni(OH)_2$, 99%, Acros Organics), H_3PO_4 (grade ≥85%, from Vekton, Russia) and diethylene glycol (Acros Organics). After impregnation the catalysts were dried in nitrogen flow at room temperature overnight followed by drying at 110 ℃ for 4 h. The list of catalysts and their characteristics are presented in Table 1 (the contents of Mo, Co, Ni were determined after calcination at 550 ℃ for 4 h).

Table 1 The list of catalysts and their characteristics

Catalyst	Mo/wt.%	Co(Ni)/wt.%
Mo/Al_2O_3	13.3	-
$CoMo/Al_2O_3$	12.5	3.5
$NiMo/Al_2O_3$	12.1	3.4
$CoMo/Al_2O_3$-SAPO-11	13.4	3.4
$NiMo/Al_2O_3$-SAPO-11	14.0	3.3

2.4 Catalyst characterization

The textural properties of the catalyst were determined using nitrogen physisorption with an ASAP 2400 instrument (USA). The elemental analysis was performed using Optima 4300 DV (Perkin Elmer, France). The morphology of sulfide phase was studied using JEM-2010 transmission electron microscope (JEOL, Japan) with accelerating voltage of 200 kV and resolution of 0.14 nm. The sample was applied to copper gauze in alcoholic suspension prepared with an ultrasonic disperser. To obtain statistical information, the structural parameters of ca. 500 particles were measured.

2.5 Catalytic experiments

The experiments were performed in a trickle-bed down-flow reactor in wide range of temperature, hydrogen pressure, H_2/feed ratio 600 Nm^3/m^3, LHSV and RSO content (0, 15 or 30 wt.%) in SRGO after in-situ sulfidation of granulated catalysts (20-60 mL) with DMDS-SRGO mixture.

The liquid and gas products were quantified using gas chromatographs; Lab-X 3500SCl, ANTEK 9000NS and Vario EL Cube were used to follow total S, N and O contents. The quality of products (aromatics, density, cold flow properties, distillation) was detected using the corresponding ASTM methods.

3 Discussion

3.1 Co-processing of RSO-SRGO mixtures over sulfide $CoMo/Al_2O_3$ and $NiMo/Al_2O_3$ catalysts

The comparison of catalyst's performance in RSO-SRGO blends hydrotreatment let us to conclude, that an addition of rapeseed oil (5-15 wt.%) inhibits the HDS and HDN activities of $CoMo/Al_2O_3$ catalyst, whereas activities of $NiMo/Al_2O_3$ catalyst are not affected (Figure 1). In accordance, over $NiMoS/Al_2O_3$ catalyst ULSD can be produced from SRGO and RSO-SRGO blend at nearly the same conditions, whereas the temperature increase is needed if the $CoMoS/Al_2O_3$ catalyst is used for ULSD production from RSO-SRGO blends.

It was observed that addition of CO (by-products of RSO conversion) to the hydrogen flow had the same effect on the HDS and HDN activities of the sulfide $CoMo/Al_2O_3$ and $NiMo/Al_2O_3$ catalysts as the addition of RSO to the SRGO feed. It should be noted that activity of sulfide CoMo returned to the initial level after carbon monoxide or RSO was stopped to feed. Carbon monoxide is proposed to be the main inhibitor of sulfide

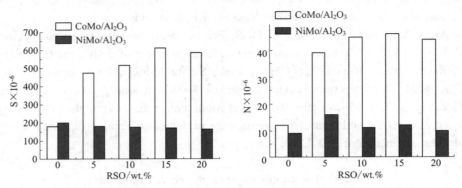

Figure 1 Effect of RSO additives on sulfur and nitrogen content in the product in co-processing of RSO-SRGO mixtures over CoMo/Al$_2$O$_3$ and NiMo/Al$_2$O$_3$ catalysts

CoMo/Al$_2$O$_3$ catalyst's activity, stronger adsorption of CO molecules on CoMoS centers is the cause of the inhibiting effect of carbon monoxide on HDS activity.

Taking into account the discovered peculiarities the dual-bed catalytic system was proposed for the hydrotreating of RSO-SRGO mixture. The use of sulfide Mo/Al$_2$O$_3$ catalyst in the front layer for the RSO conversion diminished the CO$_x$ formation and allowed to use the sulfide CoMo/Al$_2$O$_3$ catalyst in the second layer without significant decrease of HDS activity. Absence of CO$_x$ in the gas phase gives one some technological advantages: CH$_4$ formation through CO$_x$ hydrogenation is escaped and necessity of costly and energy-intensive purification of recycle hydrogen in co-processing of RSO-SRGO mixtures is avoided.

3.2 Co-processing of RSO-SRGO mixtures over sulfide CoMo/Al$_2$O$_3$-SAPO-11 and NiMo/Al$_2$O$_3$-SAPO-11 catalysts

To improve the cold flow properties of the products, we used dual bed catalytic systems (Mo/Al$_2$O$_3$+Ni(Co)Mo/Al$_2$O$_3$-SAPO-11) in the hydroprocessing of 30 wt.% of RSO-SRGO mixture. The reaction was performed using a pilot plant with a trickle-bed reactor at 4.0-7.0 MPa, 350-380 ℃, LHSV 1-2 h^{-1}, H$_2$/feed ratio 600-1000 Nm^3/m^3. It should be noted that in whole range of process conditions complete conversion of oxygen-containing compounds was observed and sulfur content in final products did not exceed 10 ppm.

It was found that the temperature increase from 350 to 380 ℃ led to increasing of n-C$_{18}$ alkanes conversion and, consequently, decreasing of cloud point of the final products (Figure 2 and Figure 3).

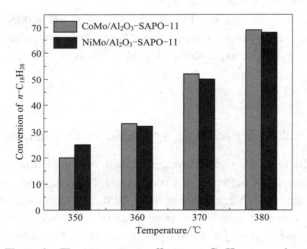

Figure 2 The temperature effect on n-C$_{18}$H$_{38}$ conversion in hydroprocessing of RSO-SRGO mixture

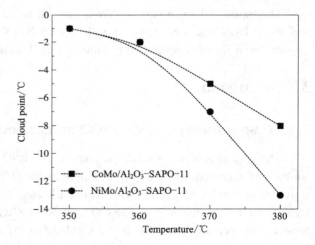

Figure 3 The temperature effect on cloud point of final products

In addition, lower LHSV and higher H/C ratio resulted in increasing of alkanes conversion, but in lower extent than temperature increase. Pressure increase resulted in decreasing of C$_{18}$+C$_{17}$ alkanes conversion and we supposed that by-products of SRGO hydrotreating (H_2S, NH_3) can inhibit hydrocracking/hydroisomerization reactions due to their higher solubility in liquid phase at higher pressure. To verify this observation, additional experiments were performed using mixture of 30 wt. % of RSO in hydrotreated SRGO (HT SRGO, 6 ppm of N and 8 ppm of S) and different additives: dimethyldisulfide as source of H_2S, and dihexylamine as sourse of NH_3. The sulfur and nitrogen contents in the feed RSO-HTSRGO were adjusted to their contents in RSO-SRGO mixture. It was observed that $C_{18}+C_{17}$ alkanes conversion decreased with the increase of nitrogen content in the feed, while effect of hydrogen disulfide on hydrocracking/hydroisomerization reactions was less noticeable.

4 Conclusion

Selectivity of rapeseed oil HDO and conversion of S-, N-containing components of SRGO in mixture with RSO depend on active phase composition. Cooperation of MoS_2/Al_2O_3 catalyst (providing selective HDO of RSO) and traditional hydrotreating catalysts (providing deep hydrotreating of SRGO) makes it possible to produce ULSD from SRGO-RSO mixture without CO formation, which increases the yield of product and softens a technological problems associated with the purification of recycle hydrogen.

When using Ni(Co)Mo/Al_2O_3-SAPO-11 catalysts, an improvement in the low-temperature properties of the products is observed. An increase in temperature leads to an increase in the conversion of n-octadecane and a decrease in the cloud point of the product; at the same time, an increase in pressure negatively affects these values. N-containing compounds was shown to be an inhibitor of hydroisomerization/hydrocracking of alkanes in the course of hydroprocessing of RSO/SRGO mixture.

References

[1] FIVGA A, GALILEU SPERANZA L, MUSSE BRANCO C, et al. A review on the current state of the art for the production of advanced liquid biofuels[J]. AIMS Energy, 2019, 7(1): 46-76.

[2] VAN DYKJ, SU J P, MCMILLAN J, et al. Potential synergies of drop-in biofuel production with further co-processing at oil refineries[J]. Biofuels Bioproducts and Biorefining, 2019, 13(10): 750-760.

[3] MITTELBACH M. Fuels from oils and fats: Recent developments and perspectives (Review)[J]. Eur. J. Lip. Sci. Tech., 2015, 117(11): 1832-1846.

[4] YAKOVLEV V A, KHROMOVA S A, BUKHTIYAROV V I. Heterogeneous catalysts for the transformation of fatty acid triglycerides and their derivatives to fuel hydrocarbons (Review)[J]. Russ Chem Rev 2011, 80(10): 911-925.

Structure and Microhardness of Plasma Composite Layered Coating Ni-(80%WC12%Co+20%Ni)-Ni after Friction Treatment

Aleksandra MIKHAILOVA, Vasilii KALITA, Dmitrii KOMLEV,
Alexey RADYUK, Konstantin DEMIN, Boris RUMYANTSEV

A. Baikov Institute of Metallurgy and Materials Science, Moscow, 119334, Russia

Abstract: The results of studies of friction treatment (FT) effects on the structure and microhardness of plasma composite coatings based on WC-12%Co composition and deposited on cylindrical steel substrates are presented. The coatings were obtained from a mechanical mixture of 80 wt. % (88 wt. % WC-12 wt. % Co) + 20 wt. % Ni powders and consisted of a base layer and Ni buffer layers. After friction treatment deformation bands formed in the coatings structure and at the boundaries with the steel substrate. The microhardness of the 80 wt. % (88 wt. % WC-12 wt. % Co) +20 wt. % Ni layer was on the level: 16.89±4.05 GPa. Changes in the phase composition, substructure, and microstructure of coatings after spraying and friction treatment have been studied by phase and structural analysis methods: RFA, optical microscopy and SEM.

Keywords: Plasma coatings; Friction treatment; Structure; Cylindrical surface; Microhardness

1 Introduction

The mechanism of plasma spraying technology makes it possible to form wear-resistant, corrosion-resistant and heat-shielding coatings with uneven shape of the surface. After collision with the substrate, sprayed melted particles 10-100 μm in size are deformed into disks 1-10 μm thick, and their surface increases ~ in 4 times, interacting with the surrounding gas atmosphere[1]. The coatings could be considered as a porous powder body formed from particles deposited separately, solidified on the substrate and partially oxidized on the surface. Applying thermoplastic treatment, welding the sprayed particles between themselves and with the substrate, including friction treatment[3,4], it is possible to eliminate porosity, increase cohesion and adhesion and result in the mechanical properties of coatings. The tool-analogue of the cutter rotates over the coating, presses on it with its end at the same time linearly moving along the coating, which heats up to ~ 1300 ℃ and are plastically deformed under the cutter. In[5,6] after (FT) the microstructure of the tungsten plasma coatings is modified, becoming identical to the bulk wolfram samples, and equiaxed grains with a size of 3-5 μm are formed[5]. The microhardness of the (WC-12 wt. % Co) cermet coatings increase from 13 GPa to 20 GPa[6] as a result of an increase in the coating density and simultaneous refinement of WC, reaching the hardness of a bulk hard alloys. Friction treatment of coatings on cylindrical surfaces and complex surfaces like to industrial products with complex shape has the greatest practical interest. The purpose of this study is to establish the relationship between the parameters of (FT) and the structure, microstructure, and mechanical properties of (WC-12 wt. % Co) coatings on a cylindrical substrate.

2 Experimental methods

The coatings were sprayed using a plasma torch with a nozzle for protection from the surrounding atmosphere, the plasma gas—a mixture ($Ar + N_2$). Coating substrates were made of steel 45 and has a cylindrical shape with diameter of 30 mm and length of 300 mm. The main coating layer was sputtered with a mechanical mixture of standard powder 80% (88 wt. % WC-12 wt. % Co), fractions 20-53 μm+20 wt. % Ni powder. Coating layers from Ni were used as a sublayer, an intermediate layer with a medium thickness and a top layer. The lower relaxed Ni layer on the substrate let it possible to reduce the stress concentration between the coating and the substrate, which could arise due to the temperature difference during (FT) and the difference in thermal expansion coefficients. The Ni top layer is a "sacrificial layer" for a softer interaction of the coating with the ends of the tools.

The friction treatment of the plasma coated samples was carried out on an installation assembled on the basis of a 1E61PM lathe, placing them in the machine chuck and then, due to its kinematics, they were rotated and the working tools moved along the generatrix of the substrate. The tools were pressed to the coating by force (P), using pneumatic cylinders, mounted on the machine carriage instead of the tool holder. The microstructure of the coatings was studied by JENAVERT optical microscope and a JEOL JAMP-9500F OJE scanning microscope. Structural changes and phase composition of the coatings were studied sequentially before deposition, after deposition, and after (FT), using RFA software packages (Sieve and MAUD programs). The diffraction profiles were recorded on an Ultima IV Rigaku X-ray diffractometer in CuK_α. The microhardness of the coatings was determined on a PMT-3 device at a load of 200 G.

3 Results and discussion

The main objective of the research was aimed at obtaining high mechanical properties, because the microhardness of modern WC-Co coatings is lower than the microhardness of sintered hard alloys. The obtained coatings in the state after deposition consisted of disk-shaped particles solidified separately on the substrate. In the main layer of the WC coating, both carbides of the initial powder and phases formed from the liquid phase with the initial dissolved WC carbides were fixed during deposition [Figure 1(a), (b)].

Figure 1 Structure of (a) (WC-12 wt.% Co) sputtering powder, (b) coating after sputtering, (c) cross section of the coating after friction treatment, (e) cross section after argon etching at 3 kV in OGE microscope

To obtain a strong connection between the deposited particles and the substrate and to achieve physical contact between the areas to be joined along the juvenile surfaces the mechanisms of deformation action of (FT) were used. One of the signs of such deformations is the deformation bands in the coating, starting from the boundary with the substrate. Detailed studies of the substructure of the coatings after (FT) were performed by XRF and SEM [Figure 1(d)].

The investigated (88 wt.% WC-12 wt.% Co) cermet belongs to the hypereutectic composition with a solidus temperature of 1340 ℃ and a liquidus temperature of about 2000 ℃. Taking into account the overheating of the sprayed particles relative to the melting point by 1000 ℃, most of the WC carbides can dissolve in the liquid phase before the impact with the substrate. When particles solidify on a substrate, firstly nanosized carbides solidify from the liquid phase and at a solidus temperature of 1340 ℃ the amorphous phase also solidifies. During (FT), the amorphous phase crystallizes at temperatures above 645 ℃. In the general case, nanosized carbides are formed both during sputtering and during (FT).

The nominal composition of the sprayed powder: 70.4 wt.% WC, 9.6 wt.% Co, 20 wt.% Ni. According to XRD data, the powder for sputtering has an equilibrium state, and the phase peaks are not broadened [Figure 2(a)].

For a coating in the state after deposition, broadening of the peaks at the corners with a close arrangement of the matrix and carbide phases, an amorphous halo is fixed, which is formed as a result of quenching from the liquid state upon solidification of the eutectic phase[7]. The thermal effect of the transition from the amorphous phase to the crystalline phase in DSC studies at 645 ℃ was observed earlier for (88 wt.% WC-12 wt.% Co) coating[7]. After sputtering, the coatings contain: 18 wt.% WC, 5 wt.% W_2C, 68 wt.% NiCo, 4 wt.% NiO, 5 wt.% $NiO_{0.96}$. The formation of W_2C is determined by the dissolution of WC in the liquid phase of the sprayed particle, followed by the interaction of carbon with oxygen dissolved in this liquid phase with the formation of gaseous oxide, which leads to the loss of carbon from the coating[8]. Oxygen also interacts with the matrix phases to form nickel oxides.

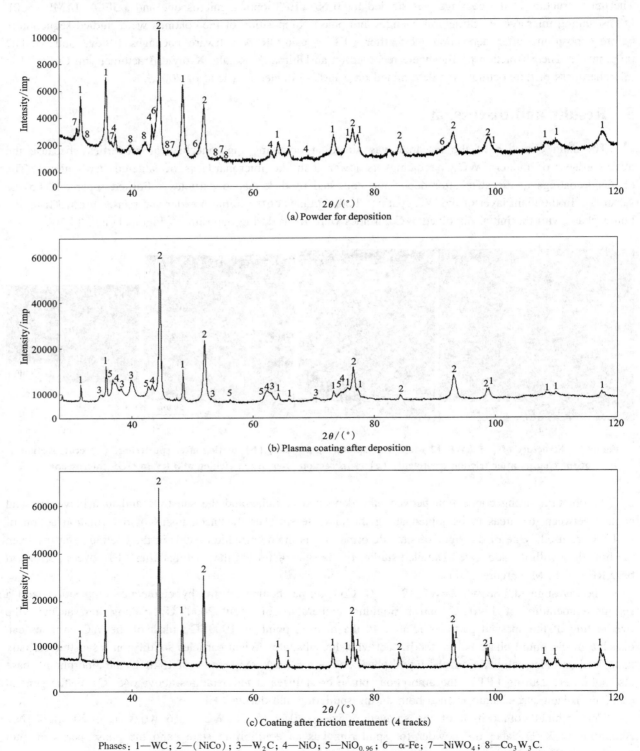

Phases: 1—WC; 2—(NiCo); 3—W_2C; 4—NiO; 5—$NiO_{0.96}$; 6—α-Fe; 7—$NiWO_4$; 8—Co_3W_3C.

Figure 2 X-ray diffraction patterns of samples of nominal composition Ni+(80%WC-12Co-20%Ni)+Ni

At (FT) (track 4), the coating was at a temperature of 1176 ℃ for 30 s, which increased the equilibrium of the coating: 20 wt.% WC, 46 wt.% NiCo, 6 wt.% NiO, 6 wt.% α-Fe [Figure 2(c)]. It can be assumed that in this case the amorphous phase decomposes with the formation of Co_3W_3C and WC. The formation of the Co_3W_3C phase in small amounts increases the strength of the bond between the WC carbide and the matrix Co phase. In the process of (FT), phases of the $NiWO_4$ type are additionally formed in the coating, which implies the occurrence of diffusion processes under conditions of a stressed state and significant strains in the air atmosphere. The Fe phase got into the sample when the coating was removed from the steel substrate by turning.

The microhardness under an indenter of 200 G load was determined in the main coating layer. The value of the microhardness of the coating of 1 track after (FT) is (16.89±4.05) GPa. In[9], a similar WC-36% Co coating with a hardness of up to 9.6 GPa was formed by cold spraying with a helium jet at a temperature of 760-980 ℃. Higher hardness was obtained by sputtering a WC-10Co-4Cr coating with two sizes of carbides-micro and nano[10] and using a solid matrix WC-Co/NiSiCrFeB-14 GPa[11]. The wear resistance of the high-entropy NiCoCrAlYTa coating was improved by increasing its hardness from 0.6 GPa to 0.9 GPa by introducing WC-17 wt.% Co particles into the sprayed powder[12]. Coatings deposited from agglomerated powder with different additional nickel powder content were obtained: (WC-Co)-5 wt.%, 15 wt.%, 25 wt.% and 75 wt.% Ni. The microhardness of the coatings according to the measurements is 0.4-0.7 GPa, the composition of the surface layer is WC, W_2C, Co, Ni and Co_6W_6C. The wear of the coating occurred due to knocking out of WC carbides and delamination of particles during plastic deformation. The high content of Ni particles in the powder made it possible to reduce porosity and increase adhesion, cohesion, and fracture toughness[13]. Coatings with an additional content of the Ni phase are similar to the coatings studied in this article, but have a lower hardness, because most of them were not subjected to subsequent thermoplastic processing or (FT).

4 Conclusions

(1) Studies of plasma coatings based on WC-12 wt.% Co, deposited on cylindrical steel substrates after frictional treatment, showed that (FT) can significantly improve the microstructure of coatings, compacting it and forming a coating with higher mechanical properties compared to coatings after deposition.

(2) The processing modes of calculated power and the value of work performed on the coating by (FT), the ratio of this value of work to the area of the coating, and the dependence of the coating temperature on this ratio were measured and selected.

(3) Deformation bands in the structure of the coatings and at its boundary with the steel substrate were studied after (FT). For the main coating layer, a microhardness of (16.89±4.05) GPa was obtained.

(4) The phase composition of the coatings after friction treatment changed insignificantly due to the formation of Co_3W_3C and $NiWO_4$ phases, suggesting partial decomposition of the coating structure hardened from the liquid state after deposition. The relative loss of carbon during plasma spraying and subsequent friction treatment amounted to 2.6%.

References

[1] GAN J A, BERNDT C C. Review on the oxidation of metallic thermal sprayed coatings: a case study with reference to rare-earth permanent magnetic coatings[J]. Journal of Thermal Spray Technology, 2013, 22(7): 1069-1091.

[2] SEMENOV A P. Investigation of the seizure of metals during joint plastic deformation[M]. Publishing House of the Academy of Sciences of the USSR, 1953: 123.

[3] VILL' V I. Svarka metallov treniem [M]. Mashgiz. Friction welding of metals//Leningrad. 1970: 175.

[4] GEL'MAN A S. Osnovy' svarki davleniem. Fundamentals of pressure welding. Engineering. [M], 1970: 312.

[5] MISHRA R S, MA Z Y. Friction stir welding and processing[J]. Materials Science and Engineering: R: Reports, 2005, 50(1/2): 1-78.

[6] SHARMA V, PRAKASH U, MANOJ KUMAR B V. Surface composites by friction stir processing: a review[J]. Journal of Materials Processing Technology, 2015, 224: 117-134.

[7] TANIGAWA H, OZAWA K, MORISADA Y, et al. Modification of vacuum plasma sprayed tungsten coating on

reduced activation ferritic/martensitic steels by friction stir processing[J]. Fusion Engineering and Design, 2015, 98/99: 2080-2084.
[8] MORISADA Y, FUJII H, MIZUNO T, et al. Modification of thermally sprayed cemented carbide layer by friction stir processing[J]. Surface and Coatings Technology, 2010, 204(15): 2459-2464.
[9] BARTULI C, VALENTE T, CIPRI F, et al. A parametric study of an HVOF process for the deposition of nanostructured WC-co coatings[C]//Thermal Spray 2003: Proceedings from the International Thermal Spray Conference", "International Thermal Spray Conference. Orlando, Florida, USA. ASM International, 2003: 283-290.
[10] KEAR B H, SKANDAN G, SADANGI R K. Factors controlling decarburization in HVOF sprayed nano-WC/co hardcoatings[J]. Scripta Materialia, 2001, 44(8/9): 1703-1707.
[11] MEGHWAL A, BERNDT C C, LUZIN V, et al. Mechanical performance and residual stress of WC-Co coatings manufactured by Kinetic Metallization™[J]. Surface and Coatings Technology, 2021, 421: 127359.
[12] GAO P H, CHEN B Y, WANG W, et al. Simultaneous increase of friction coefficient and wear resistance through HVOF sprayed WC-(nano WC-Co)[J]. Surface and Coatings Technology, 2019, 363: 379-389.
[13] KHUENGPUKHEIW R, WISITSORAAT A, SAIKAEW C. Wear behaviors of HVOF-sprayed NiSiCrFeB, WC-Co/NiSiCrFeB and WC-Co coatings evaluated using a pin-on-disc tester with C45 steel pins[J]. Wear, 2021, 484/485: 203699.
[14] HAO EK, ZHAO X Q, AN Y L, et al. WC-Co reinforced NiCoCrAlYTa composite coating: effect of the proportion on microstructure and tribological properties[J]. International Journal of Refractory Metals and Hard Materials, 2019, 84: 104978.
[15] FU W, CHEN Q Y, YANG C, et al. Microstructure and properties of high velocity oxygen fuel sprayed (WC-Co)-Ni coatings[J]. Ceramics International, 2020, 46(10): 14940-14948.

Effect of Zeolite Type on Activity of MoS_2/Al_2O_3-Z Catalysts in Hydroconversion of Methylpalmitate

Yiheng ZHAO[1], Evgeniya VLASOVA[2], Pavel ALEKSANDROV[2], Ivan SHAMANAEV[2], Alexey NUZHDIN[2], Evgeniy SUPRUN[2], Vera PAKHARUKOVA[2], Galina BUKHTIYAROVA[2]

1　Novosibirsk National Research University, Novosibirsk, 630090, Russia
2　Boreskov Institute of Catalysis SB RAS, Novosibirsk, 630090, Russia

Abstract: The MoS_2 bifunctional catalysts supported on different zeolite composite supports were prepared by vacuum impregnation method. The prepared catalysts and their performance were analyzed by a series of characterization methods. Under the conditions of temperature 270-350 ℃, pressure 3.0-5.0 MPa, LHSV 36 h^{-1}, the activity of prepared catalysts in hydrodeoxygenation and isomerization reactions was investigated in a trickle-bed reactor. Experimental results showed that the selectivity of methyl palmitate conversion through direct hydrodeoxygenation route is about 90% in the presence of MoS_2 catalyst. In hydrodeoxygenation reaction, the complete transformation of oxygenated compounds can only occur at a temperature of 310 ℃ or above; in the hydroisomerization reaction, the yield of isomerized hexadecane on different supports is ranked as Al_2O_3<Al-Z12<Al-Z5<Al-SAPO-11<Al-Z22. In addition to this, the activity of sulfided catalysts decreases with increasing pressure and temperature.

Keywords: Bio-jet fuel; MoS_2 catalyst; Hydrodeoxygenation; Hydroisomerization

1　Introduction

With the continuous development of industry, the demand for energy is increasing. The large-scale use of traditional fossil fuels has led to a series of environmental problems, such as global warming. Therefore, people began to search for new alternative energy sources, and electricity and biofuels are one of them. At this stage, the use of electricity or hydrogen to replace hydrocarbons (gasoline and diesel) is one of the main development trends.

Currently, two-stage processes are used industrially to obtain bio-jet fuel components from fatty acid triglycerides (TFA), the most common of which are NEXBTL (Neste, Finland) and Ecofining (UOP/Honeywell, USA and Eni, Italy)[1]. In the first stage the feedstock is hydrodeoxygenated over a sulfide catalyst (Ni(Co)Mo/Al_2O_3) and in the second stage (after thorough purification from sulfur compounds, CO and CO_2) the alkanes obtained are hydroisomerizied/hydrocracked in presence of noble metal based catalysts (platinum on composite supports, including zeolites ZSM-22, ZSM-23, SAPO-11). Using a one-step process to transform triglycerides into bio-jet fuel components would reduce capital and operating costs, avoid intermediate purification stages, simplify process control, and reduce energy and hydrogen consumption. Therefore, the development of multifunctional catalysts that ensure the simultaneous occurrence of multiple reactions (HDO, hydroisomerization, and hydrocracking) to obtain products of a given composition is a promising research direction that has been actively developed in recent years.

For the preparation of bio-jet fuel from triglycerides, sulfide nanoparticles or noble metals (Pt, Pd, Ru) on supports containing zeolites, silicoaluminophosphates, or other materials with pronounced acidity are most commonly used[1-4]. The most commonly used materials include zeolites ZSM-5, ZSM-22, zeolite, SAPO-11 silicoaluminophosphate[1-4]. At the same time, ZSM-5 molecular sieve has a two-dimensional channel system, which promotes aromatization reaction, while materials with one-dimensional channel system (ZSM-22, SAPO-11) promote isomerization and mild hydrocracking reaction[1,3]. The main obstacle to large-scale use of noble metal-containing bifunctional catalysts on zeolite-containing media is high cost.

It is well known that fatty acid triglycerides are converted mainly through two pathways: the direct hydrodeoxygenation pathway (alkanes and water in the product) and the decarboxylation/decarbonylation pathway (short-chain alkanes and carbon oxides). Related literature shows that in the presence of MoS_2/Al_2O_3 catalyst, the hydrodeoxygenation of methyl palmitate proceeds through the route of direct hydrodeoxygenation, and the main product of the reaction, hexadecane, has high selectivity[5].

Therefore based on the literature analysis, it can be concluded that the current trend in the production of biojet fuel from aliphatic lipids and triglycerides of fatty acids is to develop multifunctional catalysts for the one-step process and to replace them with more readily available and cheaper hydrogenation components metals.

2 Material and methods

2.1 Support preparation

Alumina support was prepared by HNO_3 peptization of pseudoboehmite (Disperal 20, Sasol GmbH). Zeolite-containing granular supports were prepared by mixing of pseudoboehmite (Disperal 20, Sasol GmbH) and zeolite powders (ZSM-5, ZSM-12, ZSM-22, SAPO-11 from Zeolyst) followed by peptization with nitric acid and then piston extrusion through a trefoil-shaped die. After extruding, support granules were dried at 110 ℃ during 12 h and then were calcined at 550 ℃ in air flow during 6 h. Zeolite content was 30 wt.% in all calcined composite supports. Synthesized supports were denoted as Al-Z5, Al-Z12, Al-Z22, Al-SAPO-11.

2.2 Catalyst preparation

Mo catalysts were prepared by incipient wetness impregnation of synthesized alumina and zeolite-containing extrudates by aqua solution containing ammoniumheptamolybdate ($(NH_4)_6Mo_7O_{24} \cdot 4H_2O$ from Vekton) and citric acid monohydrate ($C_6H_8O_7 \cdot H_2O$ from Vekton). Mo content was about 7.0 wt.% after calcination of the catalysts at 550 ℃ for 4 h.

2.3 Support and catalyst characterization

When characterizing catalysts and supports, we use a variety of characterization techniques. The textural properties of the supports were determined by low-temperature (-196 ℃) adsorption of N_2 using an Autosorb-6B-Kr instrument ("Quantachrome Instruments", USA). The chemical analysis of Mo in the catalyst was performed by atomic adsorption spectrometry using spectrometer "Optima 4300 DV" (Perkin Elmer, France). X-ray diffraction (XRD) analysis of supports and catalysts was carried out using MoK_α radiation ($\lambda = 0.7093$ Å) on a STOE STADI MP instrument (STOE, Germany) equipped with a MYTHEN2 1K detector. The structure and microstructure of the sulfide samples were studied by a high-resolution transmission electron microscope (HRTEM)-electron microscope Themis Z (Thermo Fisher Scientific, USA), with an accelerating voltage of 200 kV and a limit resolution of 0.07 nm. Experiments for morphological studies were performed on a scanning electron microscope (SEM) Hitachi Regulus SU8230 FESEM (Hitachi, Japan).

2.4 Catalytic experiments

The catalytic experiments were performed using an experimental setup with a trickle-bed reactor with an inner diameter 12 mm and length 370 mm. In each experiment 0.5 mL of catalyst (0.25-0.50 mm size fraction) was diluted with inert material, carborundum (0.1-0.25 mm size fraction) in a 1:8 volume ratio. Prior to the catalytic experiments the catalysts were activated by in-situ sulfidation with dimethyldisulfide in dodecane (0.6 wt.% sulfur) at H_2 pressure-3.5 MPa, H_2/feed ratio-300 Nm^3/m^3 and LHSV-20 h^{-1}. Sulfidation was performed at temperature 340 ℃ during 4 h with a heating rate of 25 ℃/h. Hydroprocessing of methylpalmitate (MP) was carried out at temperature range 230-350 ℃, H_2 pressure 3.0 and 5.0 MPa, H_2/feed ratio-630 Nm^3/m^3 and LHSV-36 h^{-1}. The feed was 10 wt.% of MP in dodecane (1.17 wt.% O). The duration of the each step was 6 h.

2.5 Product analysis

The products of methylpalmitate (MP) conversion were analyzed using an Agilent 6890N gas chromatograph ("Agilent Technologies", USA) equipped with a flame ionization detector and an HP-1MS quartz capillary column (30 m×0.32 mm×1 μm). Methylpalmitate conversion was calculated as Equation (1):

$$X_{MP} = \frac{C^0_{MP} - C_{MP}}{C^0_{MP}} \times 100\% \tag{1}$$

where: C^0_{MP} is the chromatogram peak area of MP in the feed; C_{MP} is the chromatogram peak area of MP in the final product.

The total oxygen content in liquid samples was determined using a Vario EL Cube elemental CHNSO analyzer ("Elementar Analysensysteme GmbH", Germany).

Oxygen conversion was calculated as Equation (2):

$$X_O = \frac{C^0_O - C_O}{C^0_O} \times 100\% \tag{2}$$

where: C^0_O is total oxygen content in the feed; C_O is total oxygen content in the final product.

Gas phase during the MP hydroprocessing was analyzed online using a gas chromatograph Chromos 1000 ("Chromos", Russia) equipped with a methanator and a flame ionization detector.

3 Results and discussion

3.1 Catalyst characterization

Firstly, the textural characteristics of the synthesized supports and the molybdenum content of the catalysts were characterized (Table 1). Through the analysis of the data in the table, it can be concluded that the introduction of 30% by weight of zeolite (samples Al-Z5, Al-Z12, Al-Z22, Al-SAPO-11) into the alumina support leads to a small increase in specific surface area, while the pore volume is reduced. It should be noted that the zeolite additive does not affect the average pore size.

Table 1 Textural characteristics of supports and molybdenum content in the synthesized catalysts

Catalyst	Mo/wt.%	Support	Textural characteristics		
			$S/(m^2 \cdot g^{-1})$	$V_{pore}/(cm^3 \cdot g^{-1})$	D_{pore}/nm
Mo/Al$_2$O$_3$	6.95	Al$_2$O$_3$	142	0.66	25.1
Mo/Al-Z5	6.90	Al$_2$O$_3$-ZSM-5	202	0.48	25.6
Mo/Al-Z12	6.96	Al$_2$O$_3$-ZSM-12	165	0.49	22.8
Mo/Al-Z22	6.90	Al$_2$O$_3$-ZSM-22	175	0.53	25.5
Mo/Al-SAPO-11	6.97	Al$_2$O$_3$-SAPO-11	177	0.42	22.6

It was found by X-ray diffraction analysis that the zeolite phase in the prepared composite supports Al-Z5, Al-Z12, Al-SAPO-11 and Al-Z22 remained unchanged (Figure 1). By SEM-EDX it was shown that in the process of preparation zeolite-containing supports a uniform distribution of zeolite in the structure of the support was obtained.

According to HRTEM data dispersed sulfide phase is presented on the surfaces of the sulfided catalysts. The average size of nanoparticles was varied from 4 to 6 nm; stacking number was 1.5-1.7 for all catalysts.

3.2 The effect of zeolite type on hydrodeoxygenation and hydroisomerization of methypalmitate

At temperatures above 310 ℃, complete conversion of oxygenates was achieved. Methane and negligible carbon monoxide were detected in gas phase analysis. The selectivity for the conversion of MP via the direct hydrodeoxygenation route in the presence of MoS$_2$ catalyst was about 90%. It was found that the zeolite-containing catalysts had a higher conversion of the intermediate oxygenates than the alumina-containing sample, while the alumina-containing catalyst had a higher conversion of methyl palmitate. It can be explained by a rate increase of the acid-catalyzed ester hydrolysis reaction in presence of zeolite-containing catalysts. Moreover, temperature increase leads to a decrease in the selectivity of the formation of C$_{16}$ alkanes due to

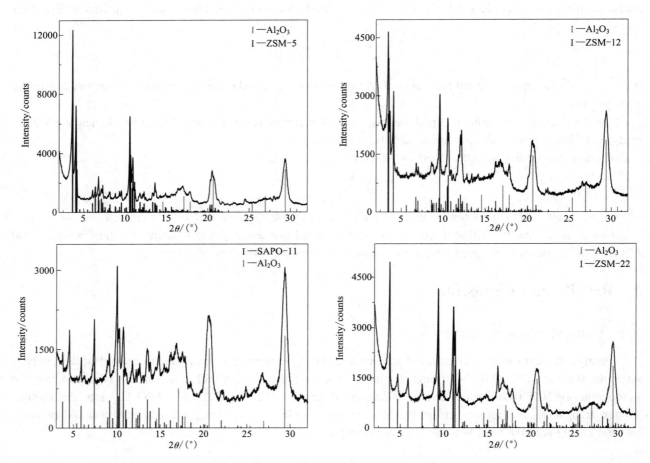

Figure 1 X-ray Diffractograms of Prepared Composite Supports

decarbonylation reaction.

It was found that the yield of isohexadecanes in hydroisomerization of MP increased in the following order: Al_2O_3 < Al-Z12 < Al-Z5 < Al-SAPO-11 < Al-Z22. The most active sample in the hydroisomerization reaction was the catalyst supported on a composite containing zeolite ZSM-22 with yield of isomerized $C_{16}H_{34}$ of 40%. Furthermore, at temperature increase from 310 ℃ to 350 ℃ the yield of iso-$C_{16}H_{34}$ decreased from 7% to 4% in over Mo/Al-Z12; from 40% to 24% over Mo/Al-Z22; from 23% to 14% over Mo/Al-SAPO-11.

In addition, the effect of pressure (3.0 and 5.0 MPa) on the hydroisomerization of methyl palmitate in the presence of Mo/Al-Z22 and Mo/Al-SAPO-11 catalysts was also investigated. The reaction was carried out at a temperature of 350 ℃, LHSV 36 h^{-1}, and a H_2/feed ratio of 600 Nm^3/m^3. According to the data, at pressure increase from 3.0 to 5.0 MPa the yield of isohexadecanes decreases. The yield of isohexadecanes decreased from 26% to 14.5% over MoS_2/Al-Z22 and from 15% to 10% over MoS_2/Al-SAPO-11.

4 Conclusion

Composite supports, containing 30 wt.% zeolite (ZSM-5, ZSM-12, ZSM-22, SAPO-11) and 70% Al_2O_3, and catalysts were characterized by XRD, HRTEM and SEM. According to XRD data structure of zeolites was preserved in synthesized supports and catalysts. Uniform distribution of zeolite crystallites in composite materials was confirmed by SEM-EDX. According to HRTEM data, the average size of sulfide nanoparticles was 4.5-5.5 nm.

A comparative study of sulfide Mo catalysts showed that higher conversion of the intermediate oxygenates was obtained over zeolite-containing catalysts than the alumina-containing sample, while the alumina-containing catalyst had a higher conversion of methyl palmitate. 100% conversion of oxygen-containing compounds is achieved at a temperature of 310 ℃.

The yield of isohexadecanes in hydroisomerization of MP increased in the following order: Al_2O_3<Al-Z12

<Al-Z5<Al-SAPO-11<Al-Z22. The highest yield of iso-alkanes was observed for ZSM-22 containing catalyst.

Acknowledgments: This work was supported by the Russian Science Foundation (Grant No. 22-13-00371).

References

[1] LONG F, LIU W G, JIANG X, et al. State-of-the-art technologies for biofuel production from triglycerides: a review [J]. Renewable and Sustainable Energy Reviews, 2021, 148: 111269.

[2] GUTIÉRREZ-ANTONIO C, GÓMEZ-CASTRO F I, DE LIRA-FLORES J A, et al. A review on the production processes of renewable jet fuel[J]. Renewable and Sustainable Energy Reviews, 2017, 79: 709-729.

[3] YELETSKY P M, KUKUSHKIN R G, YAKOVLEV V A, et al. Recent advances in one-stage conversion of lipid-based biomass-derived oils into fuel components-aromatics and isomerized alkanes[J]. Fuel, 2020, 278: 118255.

[4] MÄKI-ARVELA P, MARTÍNEZ-KLIMOV M, MURZIN D Y. Hydroconversion of fatty acids and vegetable oils for production of jet fuels[J]. Fuel, 2021, 306: 121673.

[5] MAGHREBI R, BUFFI M, BONDIOLI P, et al. Isomerization of long-chain fatty acids and long-chain hydrocarbons: a review[J]. Renewable and Sustainable Energy Reviews, 2021, 149: 111264.

Hydroconversion of Methyl Esters over Ni-phosphide Catalyst on Composite Alumina-SAPO-11 Support

Ivan SHAMANAEV, Evgenia VLASOVA, Galina BUKHTIYAROVA

Boreskov Institute of Catalysis, Novosibirsk, 630090, Russia

Abstract: Ni-phosphide catalyst on composite Al_2O_3-SAPO-11 support was studied in the hydrodeoxygenation-isomerization (hydroconversion) of methyl esters with different amount of carbon atoms in the chain and different number of double bounds: C16:0-methyl palmitate, C18:0-methyl stearate, C18:1-methyl oleate, C18:2-methyl linoleate, and C18:3-methyl linolenate. The catalyst was synthesized by impregnation of the support with aqueous solution of Ni hypophosphite with subsequent reduction in H_2 flow. The catalyst was characterized by ICP-AES analysis, N_2 physisorption, H_2-TPR, NH_3-TPD, XRD, and ^{27}Al MAS NMR. The hydroconversion experiments were carried out in a flow reactor at 310-340 ℃, 2.0 MPa, 5.3 h^{-1}. 100% conversion of all esters was achieved. The number of carbon atoms was shown to influence the selectivity to iso-alkanes (at 340 ℃ for C16 ester it was ~22%, and for C18 esters it was ~30%), but the number of double bonds did not show any impact on the selectivity of iso-alkanes.

Keywords: Hydrodeoxygenation; Isomerization; Nickel phosphide catalysts; SAPO-11

1 Introduction

Due to ecological and economic problems renewable sources attract a lot of attention in production of motor fuels and chemicals. Hydrodeoxygenation (HDO) of triglycerides and fatty acids in vegetable oils, animal fats, and tall oils leads to formation of n-alkanes. HDO is carried out over conventional sulfided Ni(Co)Mo/Al_2O_3 catalysts. The second step is isomerization over noble metal catalysts supported on silicoaluminophosphates (SAPO-11) or zeolites (ZSM-23) to meet the requirements of commercial fuels. It is interesting to intensify hydroconversion (HC) process by combining HDO and isomerization into one step over bifunctional catalysts[1].

SAPO-11 is a well-known hydroisomerization component of industrial catalysts. It has AEL topology and belongs to a member of one-dimensional molecular sieves. It has a medium-sized (0.4-0.65 nm) pore structure and mild acidity[2]. But to the date there are only a few works devoted to composite alumina-SAPO-11 supported catalysts[3].

Ni-phosphides are attractive active components for HDO of esters[3], but there is no information in literature about influence of the carbon chain length and the number of double bonds on the HC products over these catalysts. Thus, the aim of this work is to prepare and test Al_2O_3-SAPO-11-supported Ni-phosphide catalyst in HC of different methyl esters.

2 Experimental part

2.1 Materials

SAPO-11 (ZD09004, $SiO_2/Al_2O_3/P_2O_5$ = 0.25/1.0/0.80) was purchased from "Zeolyst International" (USA). AlOOH "Disperal 20" was purchased from "Sasol" (South Africa).

Following reagents were used for the preparation of Ni-phosphide catalysts: nickel(Ⅱ) acetate tetrahydrate (Ni(CH_3COO)$_2$·$4H_2O$, "Reachim", 99%), hypophosphorous acid (H_3PO_2, "Sigma-Aldrich", 50 wt.% in H_2O). Following reagents were used directly for HC experiments or in methyl esters synthesis for HC experiments: methyl palmitate ($C_{15}H_{31}COOCH_3$, "Sigma-Aldrich", 97%), stearic acid ($C_{17}H_{35}COOH$, "BLDpharm", 95%), oleic acid ($C_{17}H_{33}COOH$, "Fisher Chemical", 70%), linolic acid

($C_{17}H_{31}COOH$, "Leap Chem", 80%), linoleic acid ($C_{17}H_{29}COOH$, "Leap Chem", 80%), methanol (CH_3OH, "J. T. Baker", 99.9%), sulfuric acid (H_2SO_4, "Sigma Tech", 99%), n-dodecane ($C_{12}H_{26}$, "Acros Organics", 99%) as a solvent, n-decane ($C_{10}H_{22}$, "Komponent-reaktiv", 99%) as an internal standard.

Methyl esters (methyl stearate $C_{17}H_{35}COOCH_3$, methyl oleate $C_{17}H_{33}COOCH_3$, methyl linoleate $C_{17}H_{31}COOCH_3$, and methyl linolenate $C_{17}H_{29}COOCH_3$) synthesis was carried out by Fischer-Speier method in a Rotavapor© R-215 using reflux mode.

2.2 Support and catalyst preparation

The support was prepared by mixing of SAPO-11 powder (30 wt.%) with AlOOH in a Z-shaped blade mixer. HNO_3 solution was used as a peptizator. The obtained paste was formed into trefoil-shaped extrudates using laboratory spritz. The extrudates were dried at 110 ℃, and calcined at 550 ℃ in air flow for 6 h. 0.25-0.5 mm particles of the supports were prepared by crashing and sieving the extrudates.

For the catalyst preparation $Ni(CH_3COO)_2 \cdot 4H_2O$ was dissolved in H_3PO_2 solution. The obtained green solution was used to impregnate the support. The precursor was dried overnight, then at 80 ℃ for 24 h. The precursor was reduced in H_2 flow according to the following program: heating to 600 ℃ at a ramp of 1 ℃/min, and holding at 600 ℃ for 1 h. The reduced sample was stored in inert atmosphere. For catalytic experiments the catalysts were reduced in situ in catalytic reactor to avoid oxidation in air[4].

2.3 Catalyst characterization

The content of Ni, P, Al, and Si was determined by inductively coupled plasma atomic emission spectrometer (ICP-AES, Optima 4300 DB, Perkin-Elmer, USA). N_2 physisorption was carried out on ASAP-2400 (Micromeritics, USA). Temperature-programmed reduction in hydrogen (H_2-TPR) and temperature programmed desorption of ammonia (NH_3-TPD) were carried out on Chemosorb (Neosib Ltd, Russia). X-ray diffraction (XRD) analysis was carried out on Bruker D8 Advance (Bruker, Germany). Transmission electron microscopy (TEM) images were obtained on Themis Z (Thermo Fisher Scientific, USA). Energy dispersive X-ray (EDX) analysis was conducted on SuperX spectrometer (Thermo Fisher Scientific, USA). ^{27}Al magic angle spinning (MAS) NMR experiments were carried out using Bruker Avance 400 ("Bruker", Germany) pulsed Fourier spectrometer at the Larmor frequency of 104.31 MHz (constant magnetic field 9.4 T).

2.4 Catalytic experments

MP HC was carried out in a continuous-flow fixed-bed reactor (i.d. 12 mm). 1.5 mL of the catalyst precursor was mixed with SiC ($V_{cat}:V_{SiC} = 1:2$) and loaded to the reactor. The precursor was reduced according to the temperature program described above in H_2 flow (100 mL/min, 0.1 MPa). After cooling the reactor was pressurized (2.0 MPa) and heated to the reaction temperature (310-340 ℃), the liquid feed (0.281 mol/L of methyl ester in n-dodecane with 1% n-decane as an internal standard) was supplied using Gilson 305 HPLC pump (Gilson Inc., USA). The reaction was conducted at LHSV = 5.3 h^{-1} and H_2/feed = 600 Ncm^3/cm^3 until stationary level of product concentrations (6-10 h).

Samples of liquid products were analyzed every hour using Agilent N6890 gas chromatograph (Agilent, USA) equipped with HP-1MS capillary column (30 m × 0.32 mm × 1 μm) and FID detector. Gas-phase products were analyzed using Chromos 1000 chromatograph (Chromos engineering, Russia) equipped with a packed column (HayeSep, "Sigma-Aldrich", USA), methanator with Ni catalyst, and FID detector.

Ester conversions were calculated as follows:

$$X = \frac{C_0 - C}{C_0} \cdot 100\% \tag{1}$$

where: C_0 is concentration of the ester in the feed; C is concentration of the ester in the products.

Selectivity to i^{th} compound was calculated as follows:

$$S_i = \frac{C_i}{C_0 - C} \cdot 100\% \tag{2}$$

where: C_i is concentration of the ith compound in the products.

3 Results and discussion

3.1 Physicochemical properties of Ni_2P/Al_2O_3-SAPO-11

Results of chemical analysis of the support and the reduced catalyst are summarized in Table 1. The catalyst contains 4.10 wt.% Ni and 10.2 wt.% P. Ni/P molar ratio in the sample (0.47) is close to the initial Ni/P ratio in the impregnation solution (0.5). H_2-TPR showed a characteristic peak of H_2 and/or phosphines formation which starts at 190 ℃ and has a maximum at 280 ℃ (1.3 mmol/g H_2 formed, 1.8 mmol/g is approximate theoretical value).

Table 1 Results of physicochemical characterization of Al_2O_3-SAPO-11 and Ni_2P/Al_2O_3-SAPO-11

Sample	Ni /wt.%	P /wt.%	Ni/P	Al /wt.%	Si /wt.%	S_{BET} /($m^2 \cdot g^{-1}$)	V_p /($cm^3 \cdot g^{-1}$)	D_p /nm	NH_3-TPD /μmol-NH_3/g	D_{XRD} /nm	D_{TEM} /nm	Al in $AlPO_4$ form /at%
Al_2O_3-SAPO-11	-	5.58	-	36.6	0.97	175	0.417	22.8	138	-	-	14
Ni_2P/Al_2O_3-SAPO-11	4.10	10.2	0.47	33.7	0.69	96	0.305	22.6	139	45	11.9	21

S_{BET} and V_p decrease after Ni_2P supporting due to blocking pores of the Al_2O_3-SAPO-11 support by the active component and unreduced phosphates (PO_x)[4]. D_p and acidity remain almost unchanged after Ni_2P supporting. XRD analysis showed presence of Ni_2P phase, and TEM images confirmed presence of particles with interplanar distances characteristic for Ni_2P. The mean particle size according to TEM is 11.9 nm, but the particle size distribution includes a few particles with sizes 30-70 nm, which resulted in the D_{XRD} = 45 nm. ^{27}Al MAS NMR analysis showed formation of some amount of $AlPO_4$ (~5 at%) due to strong interaction of P precursor with Al_2O_3 after reduction.

3.2 Catalytic properties of Ni_2P/Al_2O_3-SAPO-11

Esters HC was first conducted at 340 ℃ and 2 MPa. Full conversion of the esters was achieved at these conditions. No oxygen-containing compounds were observed in the products. Product selectivities are shown in Figure 1. The main products for methyl palmitate (C16:0) are n-C_{15}, iso-C_{15}, n-C_{16}, and iso-C_{16} alkanes, and for C18 esters-n-C_{17}, iso-C_{17}, n-C_{18}, and iso-C_{18} alkanes. Selectivity to iso-alkanes does not depend on the number of double bonds in C18 esters and is ~30%. But for C16 ester the selectivity to iso-alkanes is slightly lower (~22%). At lower HC temperatures the difference in selectivity towards iso-alkanes is also observable. At 330 ℃-20 and 8%, and at 310 ℃-5 and 2% for C18 and C16 esters correspondingly.

Most probably, the number of double bonds did not influence the selectivity due to rapid hydrogenation over metal sites of Ni_2P/Al_2O_3-SAPO-11 catalyst. The esters can be hydrogenated even before HDO giving the same initial compounds for HC-C18:0. The difference in iso-alkanes selectivity between C18 and C16 esters may be related to different stability of C18 and C16 carbocations[5]. The formation of carbocations in esters HC can be closely associated with HDO reaction because the main reaction of carbocations formation is protonation of alkenes over acid sites of SAPO-11. Alkenes can be formed in alkanes dehydrogenation, but there is a more probable way of alkenes formation in consecutive transformation of methyl ester over catalyst surface: ester → acid → aldehyde → alkene (decarbonylation), or ester → acid → aldehyde → alcohol → alkene (direct HDO). Thus, the amount of intermediate alkenes may be similar for C18 and C16 esters, but the amount of iso-alkanes can be different due to higher stability of C18 carbocations.

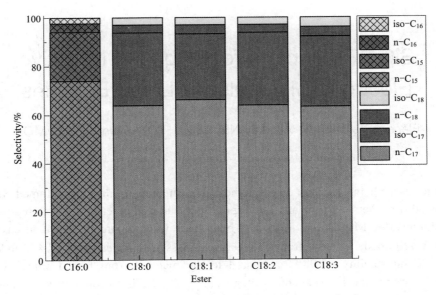

Figure 1 Product selectivities over Ni$_2$P/Al$_2$O$_3$-SAPO-11 in methyl esters HC

4 Conclusion

Ni$_2$P/Al$_2$O$_3$-SAPO-11 catalyst was prepared by reduction of hypophosphite in H$_2$ and studied by a range of physicochemical methods. According to H$_2$-TPR the reduction and decomposition of precursor starts at ~ 190 ℃. Chemical analysis showed similar Ni/P ratio in the precursor and in the catalyst after reduction. N$_2$ physisorption revealed surface area reduction after Ni$_2$P supporting. NH$_3$-TPD did not show significant differences in acidity between the support and the catalyst. XRD and TEM confirmed formation of Ni$_2$P phase. ^{27}Al MAS NMR showed formation of small amounts of AlPO$_4$.

Hydroconversion experiments of C16:0, C18:0, C18:1, C18:2, and C18:3 methyl esters were carried out in continuous-flow reactor at 310-340 ℃, 2 MPa, LHSV = 5.3 h^{-1}, and H$_2$/feed = 600 Nm^3/m^3. The esters and oxygen conversion were complete. The selectivity to iso-alkanes was shown to depend on the carbon chain length. At 340 ℃ C18 esters gave ~30% iso-alkanes, and C16 ester gave 22% iso-alkanes, which can be related to different stabilities of the carbocations. The number of double bonds in C18 esters did not influence the selectivity to iso-alkanes, probably, due to high rate of esters hydrogenation.

Acknowledgments: This work was supported by the Russian Science Foundation (Grant No. 22-13-00371).

References

[1] MÄKI-ARVELA P, MARTÍNEZ-KLIMOV M, MURZIN D Y. Hydroconversion of fatty acids and vegetable oils for production of jet fuels[J]. Fuel, 2021, 306: 121673.
[2] SHAMANAEVA I A, PARKHOMCHUK E V. Variability of molecular sieve SAPO-11 crystals: acidity, texture, and morphology[J]. Journal of Porous Materials, 2022, 29(2): 481-492.
[3] SHAMANAEV I V, VLASOVA E N, SCHERBAKOVA A M, et al. Hydroconversion of methyl palmitate over Ni-phosphide catalysts on SAPO-11 and ZSM-5 composite supports[J]. Microporous and Mesoporous Materials, 2023, 359: 112667.
[4] SHAMANAEV I V, DELIY I V, ALEKSANDROV P V, et al. Effect of precursor on the catalytic properties of Ni$_2$P/SiO$_2$ in methyl palmitate hydrodeoxygenation[J]. RSC Advances, 2016, 6(36): 30372-30383.
[5] CNUDDE P, WISPELAERE KD, VANDUYFHUYS L, et al. How chain length and branching influence the alkene cracking reactivity on H-ZSM-5[J]. ACS Catalysis, 2018, 8(10): 9579-9595.

SiO₂ Precursors for Si-based Negative Electrode Materials for Li-ion Batteries

Elena ABRAMOVA, Aleksey NEDOLUZHKO, Artem ABAKUMOV

Skolkovo Institute of Science and Technology, Moscow, 121205, Russia

Abstract: The research investigates magnesiothermic reduction of silica from fumed SiO_2 and the SiO_2 precursor synthesized from Tetraethoxysilane. Both poorly crystallized precursors develop crystal structure after the reduction. Additionally, Mg_2SiO_4 is formed in both cases with its larger content in case of fumed silica precursor. The Si-based anode material from TEOS reveals less specific capacity of 477 mA · h/g at the 1st cycle and better cycling stability of 98% compared to the material from fumed silica. It reveals specific capacity of 580 mA · h/g at the 1st cycle and cycling stability of 72% in Li-ion half-cell. Thus, it is shown the influence of different SiO_2 precursors on the magnesiothermic reduction reaction and produced products properties.

Keywords: Lithium-ion batteries; Anode material; Silicon oxide; Material characteristics; Electrochemical properties

1 Introduction

The development of next-generation lithium-ion batteries (LIBs) is driven by the improvements of electrode materials[1]. Among promising negative electrode (anode) materials, silicon-based ones attract much attention due to high theoretical specific capacity of Si (up to 4200 mA · h/g), which is >10 times higher compared to that of graphite. However, there are still challenges, including dramatic volume expansion (up to 280%[2]) and small Initial Coulombic efficiency (ICE)[3] that impede the practical application of the Si-based anodes. Many preparation techniques have been researched to obtain Si-based anode materials with attractive electrochemical properties at low cost[4]. Among them the magnesiothermic reduction of silica is a rather simple technique to produce Si-based materials. Kim et al.[5] reported that Si-based materials with general formula SiO_x, where $x \leqslant 1$, demonstrated improved cycling stability, but lowers ICE with the increase of x value.

The magnesiothermic reduction of SiO_2 has been known for decades. The critical process parameters include temperature, heating rate, atmosphere, duration, reagents ratio, but the nature of the precursor influences the final product of magnesothermic reduction as well. The aim of this study is to investigate and compare impact of silica precursors on the obtained silicon-based materials and their electrochemical properties using silica synthesized from tetraethoxysilane and commercial fumed silica as examples.

2 Experiment

2.1 Materials

The silica precursor was obtained with hydrolysis of tetraethoxysilane (TEOS) in presence of ammonia. 25 mL of TEOS was mixed with 132 mL of ethanol, then 4 mL of HCl (0.1 mol/L) and 1.5 mL of ammonia (4 mol/L) were added and stirred for a night. The obtained product was dried at 80 ℃ overnight under vacuum to remove residual solvent and further at 600 ℃ in air for 1 h. The obtained silica was ball-milled in a zirconia bowl. The commercial fumed silica was purchased from Sigma Aldrich. The magnesiothermic reduction of silica (SiO_2:Mg:MgO = 1:0.8:0.4) was conducted in the muffle furnace at 550 ℃ in air for 1 h. The reaction products were washed in the mixture of 40 mL of HCl (38 wt.%) with 200 mL of deionized (DI) water and in DI-water. The Si-based products was separated with a centrifuge. The product was dried under vacuum at 70 ℃ overnight. The final material prepared from the synthesized precursor is further marked as "S" and that from the commercial precursor as "C".

2.2 Material characterization

The morphology of the silica precursors and Si-based materials were investigated by scanning electron microscopy (ThermoFisher Quattro S FEG-SEM). Energy-disersive X-ray (EDX) spectroscopy was used to analyze their chemical compositions on the samples mounted on a SEM holder with carbonaceous sticky tape. Powder X-ray diffraction (PXRD) patterns were collected with a Bruker D8 ADVANCE diffractometer (reflection geometry, CuK$_\alpha$ radiation, LYNXEYE XE position-sensitive detector). Fourier-transformed infrared (FTIR) spectroscopy measurements were performed at a Bruker Alpha II spectrometer in the 4000 to 400 cm^{-1} region with 2 cm^{-1} resolution.

2.3 Electrode preparation and electrochemical characterization

The electrode slurry consisting of Si-based anode material, carbon black Super P (Timcal), single-walled carbon nanotubes (CNT; TUBALL), carboxymethyl cellulose (CMC; Sigma-Aldrich) and styrene-butadiene rubber (SBR) in 65:18:3:7:7 mass ratio was prepared by mixing with deionized water in a SPEX 8000M high-energy ball miller. The resulting electrode slurry was spread on the carbon-coated Cu-foil (MTI, 18 μm) with a Doctor Blade method and subsequently dried at 120 ℃ overnight. After calendaring, the disk electrodes of 16 mm in diameter were cut and dried at 120 ℃ under vacuum for 20 h. An average electrode active mass was 1.2 mg.

Galvanostatic charge/discharge measurements were performed in coin-type electrochemical cells. Half-cells were assembled with Si-based working electrode (WE), polymer separator, Li metal as a counter electrode (CE) and carbonate-based electrolyte (ethylene carbonate (EC), ethyl methyl carbonate (EMC), dimethyl carbonate (DMC) with vinylene carbonate (VC) and fluorine ethylene carbonate (FEC)) of 1 M LiPF$_6$ in a glovebox filled with Ar (MBraun, H$_2$O <0.1 ppm, O$_2$<0.1 ppm). Electrochemical measurements were performed with a BTS3000 Neware battery tester at room temperature in a galvanostatic regime. The voltage range was 0.01÷2.0 V vs Li/Li$^+$ at the current of 0.1 mA (current density of ~83 mA/g).

3 Results and discussion

3.1 Precursors and Si-based materials characterization

Silica precursors for the magnesiothermic reduction were studied with PXRD. They both demonstrate poorly crystalline nature. The synthesized silica has one clear broad peak at 2θ~23° [Figure 1(a)]. This peak can be attributed to (101) plane of cristobalite with d_{101} = 3.95 Å (a smaller value compared to the cristobalite modification of silica with d_{101} = 4.04 Å). Its crystallite size calculated with the Sherer equation from the FWHM of (101) peak (constant K=0.94) is 1.9 nm. The commercial fumed silica reveals a broad peak at 2θ~21° [Figure 1(b)]. It can be attributed to the cristobalite plane of (101) as well with d_{100}=4.26 Å, which is some larger compared to the value of cristobalite. The crystallite size from the FWHM of (101) peak is about 1.2 nm. The chemical composition of SiO$_2$ precursors was confirmed with EDX. The synthesized sample has 66.7 at% of Si and 33.3 at% of O, the fumed silica has 70 at% of Si and 30 at% of O. Particles of the synthesized and commercial silica precursors are of irregular shape, but the commercial sample has smaller agglomerates [Figure 1(b), (d)].

After the magnesiothermic reduction and washing both samples demonstrate high degree of crystallinity [Figure 2(a), (b)], with similar PXRD patterns. Peaks of crystal Si and Mg$_2$SiO$_4$ are identified for both samples. Based on the Rietveld refinement the quantitative content of the Si and Mg$_2$SiO$_4$ were calculated. Sample C has 33.5 wt.% of Si and 66.5 wt.% of Mg$_2$SiO$_4$. Sample S has 69.0 wt.% of Si and 31.0 wt.% of Mg$_2$SiO$_4$. Mg$_2$SiO$_4$ forms due to the reaction between MgO and SiO$_2$. MgO is a by-product of the silica magnesiothermic reduction and it is added to the reaction mixture to decrease the intensity of the reaction and prevent a possible explosive eruption. The magnesiothermic reduction of fumed silica comes amid powder sintering and larger interaction of MgO and SiO$_2$ compared to the TEOS-derived synthesized precursor.

Additionally, samples were studied with FTIR spectroscopy [Figure 2(c)]. Si-Si mode from the crystal

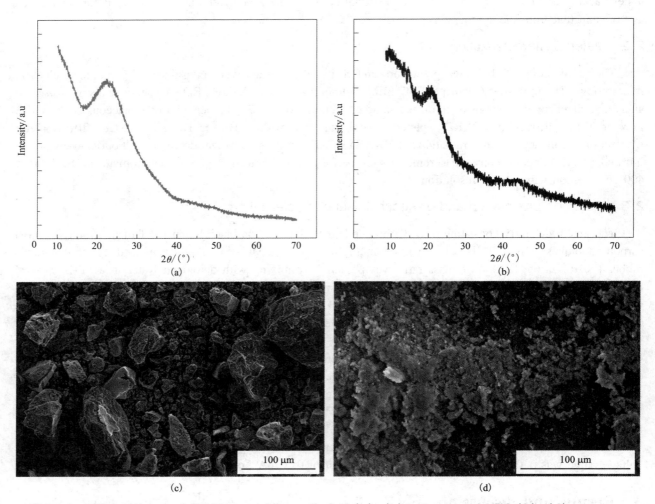

Figure 1 PXRD patterns and SEM images of the synthesized (a), (c) and commercial SiO_2(b), (d) precursors

silicon is identified at 609 cm^{-1} (green arrow). Vibrations of Si-O-Si ν_{as} modes of SiO_2 (red arrow) are observed at 1070 and 1075 cm^{-1} for both samples. Absorbance at 804 cm^{-1} and 793 cm^{-1} for both samples (blue arrow) is attributed to the ν_s(Si-O-Si) modes. Sample C also has obvious band at 455 cm^{-1} (orange arrow), corresponding to δ_r(Si-O-Si)[6]. The absence of a clear absorbance peak at this value for S sample can originate from defects of crystal lattice. Absorbance at ~880 cm^{-1} for both samples can be attributed to symmetric stretching of SiO_4 from Mg_2SiO_4 (violet arrow)[7, 8].

3.2 Electrochemical properties

The galvanostatic charge/discharge curves of the S and C anodes in the potential range of 0.1÷2 V are presented in Figure 3.

Anodes from materials S and C show similar ICE of 74.9% and 72.7%. Sample C has the specific discharge capacity of 580 mA·h/g on the 1st cycle and shows significant degradation with the capacity retention of about 72% after 50 cycles. Its capacity drops to 416 mA·h/g at 50th cycle. The drastic capacity drop of sample C can originate from large volume expansion of the material during cycling resulting in disconnection with the current collector and continuous SEI (solid electrolyte interphase) formation.

The specific discharge capacity of sample S is noticeably lower being about 477 mA·h/g on the 1st cycle, but it demonstrates much better stability compared to the sample C with capacity retention of about 98% after 50 cycles.

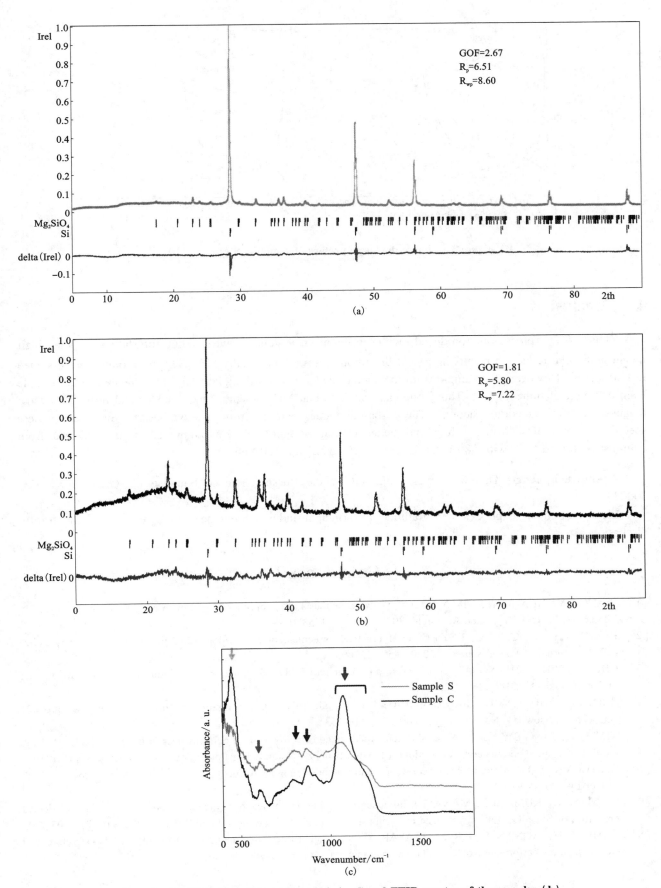

Figure 2　PXRD patterns of samples: S (a); C and FTIR spectra of the samples (b)

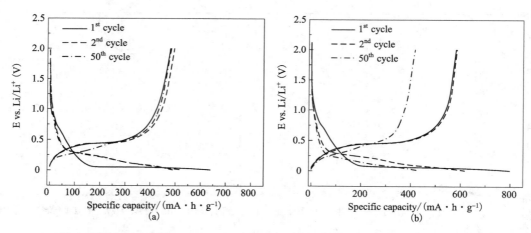

Figure 3　Charge-discharge profiles of samples: S (a), C (b)

4　Conclusion

Thus, SiO_2 precursors produced with different techniques reveal poor crystallinity, but their magnesiothermic reduction results in crystal Si in both cases. Additionally, Mg_2SiO_4 by-products are obtained in both cases. However, the sample from the synthesized silica shows higher Si yield compared to the sample from commercial fumed silica. Thus, one can conclude that SiO_2 precursor has fundamental influence on the magnesiothermic reduction reaction. The Si-based anode material from the synthesized SiO_2 reveals less specific capacity of 477 mA·h/g and better cycling stability of 98% compared to the material from commercial fumed SiO_2 with 580 mA·h/g and 72% in Li-ion half-cell.

Acknowledgments: This work was supported by the Russian Science Foundation (Grant No. 23-73-30003).

Elena Abramova expresses her gratitude to principal industrial engineer Egor Pazhetnov for fruitful discussion.

References

[1] WANG J Y, HUANG W, KIM Y S, et al. Scalable synthesis of nanoporous silicon microparticles for highly cyclable lithium-ion batteries[J]. Nano Research, 2020, 13(6): 1558-1563.
[2] REYNIER Y, VINCENS C, LEYS C, et al. Practical implementation of Li doped SiO in high energy density 21700 cell[J]. Journal of Power Sources, 2020, 450: 227699.
[3] CHEN T, WU J, ZHANG Q L, et al. Recent advancement of SiO_x based anodes for lithium-ion batteries[J]. Journal of Power Sources, 2017, 363: 126-144.
[4] ZUO X X, ZHU J, MÜLLER-BUSCHBAUM P, et al. Silicon based lithium-ion battery anodes: a chronicle perspective review[J]. Nano Energy, 2017, 31: 113-143.
[5] KIM M K, JANG B Y, LEE J S, et al. Microstructures and electrochemical performances of nano-sized SiO_x ($1.18 \leq x \leq 1.83$) as an anode material for a lithium (Li)-ion battery[J]. Journal of Power Sources, 2013, 244: 115-121.
[6] EPHIMOVA A I. Infrared spectroscopy of nanostructures semiconductors and dielectrics[M]. Moscow State University, 2014: 41.
[7] MONDAL K, KUMARI P, MANAM J. Influence of doping and annealing temperature on the structural and optical properties of $Mg_2SiO_4:Eu^{3+}$ synthesized by combustion method[J]. Current Applied Physics, 2016, 16(7): 707-719.
[8] TAMIN S H, ADNAN S B R S, JAAFAR M H, et al. Effects of sintering temperature on the structure and electrochemical performance of Mg_2SiO_4 cathode materials[J]. Ionics, 2018, 24(9): 2665-2671.

Influence of Graphene Sheet Content on Thermal Expansion of ZrO$_2$ Nanopowders

Asya AFZAL, Elena TRUSOVA

Baikov Institute of Metallurgy and Materials Science Russian Academy of Sciences, IMET RAS, Moscow, 119334, Russia

Abstract: The choice of optimal technological modes is especially important for the production of nanostructured powders and fine-grained ceramics. The study of sintering conditions of graphene-ceramic nanostructured powders is relevant and in demand by developers of new functional materials. In this study, a method for the synthesis of composite nanostructures based on oxygen-free graphene and ZrO$_2$, including sol-gel and sonochemical techniques, is proposed. The influence of graphene content on the sintering dynamics of synthesized composite powders was studied using the dilatometry method.

Keywords: Graphene based hybrid nanostructures; Graphene-ZrO$_2$ composite; Dilatometry of nanopowder; Oxygen-free graphene

1 Introduction

Knowledge of the regularities of thermal expansion, optimal compaction temperatures of powders is very important in the development of compaction and sintering modes for obtaining materials for a specific purpose, in order to obtain durable materials capable of operating in a wide temperature range without destroying the structure and losing performance. One of the priority directions in the development of materials science is the preparation of initial powder mixtures for obtaining new functional materials based on graphene and ZrO$_2$. However, ZrO$_2$ ceramics are quite fragile and characterized by low electrical conductivity. Oxygen-free graphene, which has a range of unique properties is used as an additive in a ceramic matrix, contributes to a significant improvement in material performance[1,2]. From the analysis of open scientific publications it follows that nanocomposites based on graphene and ZrO$_2$ play an important role in the development and production of materials used in a wide range of technologies for the production of NO$_2$ sensors[3], lithium-ion batteries[4], (photo)catalysts[5,6], restoration materials[7], supercapacitors, solid oxide fuel cells[8,9] and other electronic application[10]. However, the synthesis of hybrids with controlled structure and properties is difficult to implement in practice. The choice of optimal technological modes for obtaining nanostructured powders and fine-grained ceramics is most often carried out empirically and is aimed at solving a specific practical problem. Compaction and sintering of nanostructured composite powders require completely new technological approaches based on the results of dilatometric studies and careful selection of temperature increase modes during sintering.

2 Experiment

The preparation of hybrid graphene-ZrO$_2$ nanopowders was carried out by the interaction of Zr-containing sol and graphene suspension. The last one was obtained by ultrasonic irradiation of synthetic graphite (purity 99.99%, particle size 600-800 microns) in an isopropanol-water mixture (1:1, vol.). The molar ratio of isopropanol/graphite was 20. Graphite powder was irradiated for 3 h in an ultrasonic bath Sonoswiss SW1H (200 W). The graphene suspension formed by ultrasonic irradiation was separated from unreacted graphite during sedimentation for 20-22 h by step-by-step decantation.

Sol was synthesized using zirconyl nitrate ZrO(NO$_3$)$_2$ as an aqueous 0.3 mol/L solution. Hexamethylenetetramine (GMTA) in the form of a solution in isopropanol (GMTA/Zr = 1, mol.) was used for sol stabilization, and acetylacetone (AcAc/Zr = 2, mol.) was used as a complexing agent. Sol was combined with graphene suspension and mixture was evaporated at 95-98 ℃ and stirred (500 r/min) to form gel, which was subjected to heat treatment in air for 1 h at 500 ℃.

The synthesized objects were characterized using transmission electron microscopy (TEM) and electron diffraction (LEO 912 Ab Omega Carl Zeiss), X-ray diffraction (XRD) (DRONE-3M with CuK$_\alpha$ radiation), high-resolution TEM (JEM 2010 JEOL Ltd. with prefix GIF Quantum, Gatan Inc. for EELS analysis) and elemental analysis (LECO SC-400). The investigation of the surface and porosity of the powders was carried out at the TriStar 3000 installation of Micromeritics using nitrogen adsorption-desorption curves, as well as using the NOVA 2200 specific surface analyzer. The specific surface area was determined by the Brunauer-Emmett-Teller (BET) method, the pore size distribution was determined by the Barrett-Joyner-Halend (BDX) method at a temperature of -196 ℃. Dilatometric study of synthesized nanostructured powders was carried out on a dilatometer DIL 402 C Netzsch (Netzsch, Germany). Cylindrical green bodies had a diameter of 5 mm and a height of 2.5 mm. The flow rate of argon fed into the furnace was 70 mL/min, and the heating rate was 10 ℃/min. Heat treatment was carried out up to 1750 ℃, and then the sample was cooled at a rate of 20 ℃/min.

3 Results and discussion

The TEM data for graphene suspension in an isopropanol-aqua mixture, shown in Figures 1(a) and (b), indicate that the resulting graphene is represented by relatively large multilayer particles (packets) with sizes up to several microns. Electron diffraction indicates that the particles consist of many differently oriented sheets, the thickness of which does not exceed several nanometers (Figure 1(b), insert). The moiré, clearly visible in Figure 1(a), confirms that the packages consist of 1-2-layer sheets of graphene. According to EELS analysis [Figure 1(c)], the resulting flakes consisted only of oxygen-free graphene without admixture of oxides, as evidenced by the absence of a peak of 532 eV, characteristic of graphene oxide. At the same time, a wide peak with a center corresponding to 284 eV indicates the transition of 1s to π^* and indicates the presence of carbon atoms in the system in the state of sp^2 hybridization.

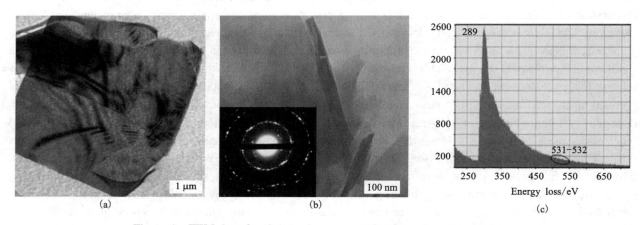

Figure 1　TEM data for the graphene suspension in an isopropanol-aqua mixture (a), electron diffraction (insert (b)), EELS analysis datum (c)

Figure 2 shows TEM data for the graphene-ZrO$_2$ composite with a carbon content of 0.706 wt.% ±0.002 wt.%, according to elemental analysis. Figures 2(a) and (b) show that the powder particles are formed by the graphene sheets with linear dimensions of several hundred nm, into which crystallites with sizes of 7-10 nm are discretely incorporated [Figure 2(b)]. The electron diffraction in Figure 2(c) indicates that the graphene sheets are formed by the chaotically oriented layers.

Figure 3 shows XRD pattern of the powdery graphene-ZrO$_2$ composite obtained using CoK$_\alpha$ and CuK$_\alpha$ sources. The turquoise spectrum was obtained used a copper source, the lilac spectrum was obtained used a cobalt source and recalculated by an angle of 2θ to CuK$_\alpha$. Neither the carbon phase nor the zirconium carbide was detected, which indicates the absence of 3D carbon in the composite. Analysis of XRD data on the composite powder shows that it consists of two modifications of ZrO$_2$: tetragonal phase tP6 and monoclinic mP12 (84∶16 wt.%), identified using JCPDS cards No. 24-1164 and No. 05-0543, respectively, with an

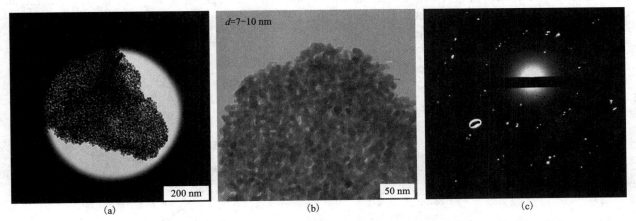

Figure 2　TEM data for the graphene-ZrO$_2$ composite [(a)(b)] and electron diffraction (c)

average crystallite size of 10 nm for both modifications. The elemental composition corresponds to the gross formula ZrO$_2$, which indicates the absence of partially reduced zirconium and a nonstoichiometric phase.

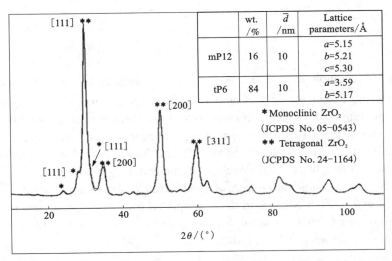

Figure 3　XRD pattern, phase composition and crystallographic data for the graphene-ZrO$_2$ composite powder (insert)

According to the results of nitrogen adsorption-desorption, the nanostructured graphene-ZrO$_2$ powder is mesoporous with a pore size of 4-7 nm. The BET calculation showed that the specific surface area of the composite is 47 m^2/g.

Figure 4 shows the shrinkage and shrinkage rates curves for a series of ZrO$_2$-based composites with different graphene content: green-0.33 wt.%, red-0.82 wt.%, blue-1.60 wt.%. This series was synthesized using Zr-containing sol and various volumes of graphene suspension. Analysis of the curves shows noticeable differences in the behavior of composites at temperatures above 600 ℃: if the graphene content in the composite increases, its ability to shrink decreases. The differences are especially pronounced in the range of 600-800 ℃: with an increase in the graphene content, the shrinkage rate decreases by more than 2 times and the extrema on the shrinkage rate curves shift to the region of lower temperatures.

With a further increase of temperature (1000-1700 ℃), differences are also observed, that affect the change in shrinkage rate, and in this case, the sample with the highest graphene content shrinks at the highest rate, and the extremum falls at a temperature higher than for samples with a lower graphene content. The shrinkage curves also differ for analogs with different graphene content, and this difference begins at temperatures above 600 ℃. In the range of 700-1600 ℃, the curves diverge, and the one for the sample with a lowest graphene content looks steeper; in this area, its shape is similar to linear. An increase in carbon content leads to difficulty in shrinkage, for example, a relative shrinkage value of 0.04 is observed for the sample with a content of 1.6 wt.% at a temperature of ~250 ℃ higher than for the sample with 0.822 wt.%. However, in all cases, the total shrinkage is 17.0%-17.5% in the region of 1700 ℃.

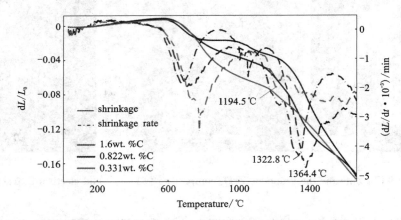

Figure 4　The shrinkage and shrinkage rate curves for the graphene-ZrO$_2$ composites with different graphene content

4　Conclusion

It is shown that the addition of 0.3 wt.%-1.6 wt.% sheets of oxygen-free graphene into the ZrO$_2$ nanopowder leads to a change in its rheological properties. In the range from room temperature to 1700 ℃, the shrinkage and shrinkage rate curves have different shapes at temperatures above 600 ℃: the higher the graphene content, the slower the shrinkage. Apparently, graphene sheets prevent material compaction due to their rigidity. The results of the work will be able to influence the development of methods for the synthesis of nanocomposites and the production of functional graphene-ceramic hybrid materials for a wide range of purposes. The study of the consolidation process of synthesized nanostructured powders based on ZrO$_2$ using the dilatometric method will make it possible to develop optimal conditions for sintering fine-grained ceramics.

Funding: The study was carried out in the Baikov Institute of Metallurgy and Materials Science RAS in accordance with the State Assignment 075-01176-23-00.

Acknowledgments: The authors thank PhD Shelekhov E. V. (National University of Science and Technology MISIS) for the XRD study.

References

[1] PORWAL H, GRASSO S, REECE M J. Review of graphene-ceramic matrix composites[J]. Advances in Applied Ceramics, 2013, 112(8): 443-454.

[2] DHAND V, RHEE K Y, KIM H J, et al. A comprehensive review of graphene nanocomposites: Research status and trends[J]. J. Nanomater., 2013: 14.

[3] KODU M, BERHOLTS A, KAHRO T, et al. Highly sensitive NO$_2$ sensors by pulsed laser deposition on graphene [J]. Appl. Phys. Lett., 2016, 109(11): 113108.

[4] XU H P, YUAN S, WANG Z Y, et al. Graphene anchored with ZrO$_2$ nanoparticles as anodes of lithium ion batteries with enhanced electrochemical performance[J]. RSC Advances, 2014, 4(17): 8472-8480.

[5] LAI J H, ZHOU S L, LIU X X, et al. Catalytic transfer hydrogenation of biomass-derived ethyl levulinate into gamma-valerolactone over graphene oxide-supported zirconia catalysts[J]. Catalysis Letters, 2019, 149(10): 2749-2757.

[6] USHARANI B, MURUGADOSS G, RAJESH KUMAR M, et al. Reduced graphene oxide-metal oxide nanocomposites (ZrO$_2$ and Y$_2$O$_3$): fabrication and characterization for the photocatalytic degradation of picric acid [J]. Catalysts, 2022, 12(10): 1249.

[7] MUTLU Ö, MAZIERO V C A, LUCAS H, et al. Graphene for zirconia and titanium composites in dental implants:

significance and predictions[J]. Current Oral Health Reports, 2022, 9(3): 66-74.
[8] MUDILA H, RANA S, ZAIDI M G H. Electrochemical performance of zirconia/graphene oxide nanocomposites cathode designed for high power density supercapacitor[J]. Journal of Analytical Science and Technology, 2016, 7(1): 1-11.
[9] SINHA S, SINGH W I, NONGTHOMBAM S, et al. Optical properties, electrochemical analysis and corrosion resistance studies of polyaniline/reduced graphene Oxide/ZrO_2 for supercapacitor application[J]. Journal of Physics and Chemistry of Solids, 2022, 161: 110478.
[10] MARKANDAN K, CHIN J K, TAN M T T. Enhancing electroconductivity of yytria-stabilised zirconia ceramic using grapheneplatlets[J]. Key Engineering Materials, 2016, 690: 1-5.

Zeeman Splitting of the Absorption Band Edge and Fabry-Perot Oscillations in Materials with Strong Spin-orbit Coupling

Danil BELYAEV[1], Mikhail YAKUSHEV[1,2,3], Vladimir GREBENNIKOV[1], Milan ORLITA[4], Konstantin KOKH[5,6,7], Oleg TERESHCHENKO[7,8], Robert MARTIN[9], Tatyana KUZNETSOVA[1,3]

1. M. N. Miheev Institute of Metal Physics of UB RAS, Ekaterinburg, 620108, Russia
2. Institute of Solid State Chemistry of the UB RAS, Ekaterinburg, 620990, Russia
3. Ural Federal University, Ekaterinburg, 620002, Russia
4. LNCMI, Grenoble, 38042, France
5. V. S. Sobolev Institute of Geology and Mineralogy of SB RAS, Novosibirsk, 630090, Russia
6. Kemerovo State University, Kemerovo, 650000, Russia
7. Novosibirsk State University, Novosibirsk, 630090, Russia
8. Rzhanov Institute of Semiconductor Physics of SB RAS, Novosibirsk, 630090, Russia
9. Strathclyde University, Glasgow, G4 0NG, UK

Abstract: Zeeman splitting of the absorption band edge was studied in Mid-infrared (MIR) magneto-transmission (MT) at magnetic fields, applied in Faraday configuration, up to 11 T in $Bi_{1.1}Sb_{0.9}Te_2S$. The absorption band edge in the MT spectra is strongly modified by magnetic fields acquiring an S-shape so the bottom half of the edge is shifting towards higher (+) whereas the top one shifts towards lower (−) energies. It was found some saturation of energy shift and dependence of g-factor can be described with linear function $g(B) = 22.5 - 0.745B$. Also, suppression of transmitted wave interference (Fabry-Perot oscillations) due to Faraday rotation of the polarization plane in a magnetic field was found.

Keywords: Spin-orbit coupling; Zeeman splitting; Fabry-Perot oscillations

1 Introduction

A novel quantum state of matter has been revealed in a number of materials called topological insulators (TIs). TIs have insulator like bulk whereas on the surface strong spin-orbit coupling produces topologically protected surface states (SSs) with Dirac-type dispersion cone[1,2]. TIs have exotic electronic properties and high potential to be used spintronics[3] and quantum electronics[4]. Understanding of the electronic properties in TIs is at a very early stage of exploration and experimental studies can reveal novel physical effects giving an opportunity for applications of TIs in new generation devices.

Magneto-optical spectroscopy is amongst the most efficient techniques to examine electronic properties of materials. One of the reasons for this is high concentration of charge carries in a typical TI due to excessive level of doping by intrinsic defects. But such doping shifts the bulk Fermi level to the conduction/valence band reducing the quality and the quantity of information which can be gained from optical spectra. $Bi_{1.1}Sb_{0.9}Te_2S$, which is a derivative of Bi_2Se_3, has a wider band gap so Fermi level is believed to keep easier within the bulk band gap which provides opportunities for magneto-optical studies.

In this paper Zeeman splitting and Fabry-Perot oscillations of $Bi_{1.1}Sb_{0.9}Te_2S$ single crystals are examined by Fourie MIR transmission at magnetic fields up to 11 T.

2 Experiment

An ingot of $Bi_{1.1}Sb_{0.9}Te_2S$ single crystal (a cylinder of 5 mm diameter and 7 mm long) was grown as described in detail in [5]. Flat samples with sizes about 3 mm×3 mm and a thickness of 0.5 mm were cleaved perpendicular to the c-axis from the initial ingot. The samples were previously studied by X-ray microanalysis, transport measurements, Raman spectroscopy and ARPES as described in [5].

The transmission measurements were carried out over the frequency range from 0 to 5000 cm^{-1} at 4.2 K using a Bruker IFS 66v/S FTIR spectrometer with Globar light source and silicon bolometer as a detector at The Laboratoire National des Champs Magnétiques Intenses in Grenoble, France. Films with thicknesses of 25 μm were exfoliated for the transmission measurement from the samples and then placed on copper foils with 1 mm×1 mm holes cut at the centre. Magnetic fields up to 11 T, applied in Faraday configuration (B parallel to the c axis of the samples), were generated by a superconducting magnet at The Laboratoire National des Champs Magnétiques Intenses in Grenoble, France. Light from Globar source with the wave vector parallel to the magnetic field was delivered through a light tube. Each transmission spectrum was normalized by that of Globar source light measured through an aperture attenuating the intensity to take in account variations in the bolometer response induced by the magnetic fields.

3   Results and duscussion

Figure 1(a) demonstrates absorption band edge obtained in magnetic fields from 0 to 11 T. The spectra show significant Fabry-Perot oscillations indicating a high parallelism of the top and bottom surface. It can be clearly seen the division of one edge into two. It is natural to suppose that absorption edge splits due to Zeeman shift energy of spin states in magnetic field. So, we can obtain an equation for calculating the splitting value from the experimental spectra.

Transmission spectrum $I(E)$ is described by a complex function:

$$I(E) = F[n(E), E] \tag{1}$$

where: E is energy of photons. We have clearly distinguished the value $n(E)$ (imaginary part of refractive index), which strongly depends on magnetic field B:

$$n_B(E) = 0.5[n(E+s) + n(E-s)] \tag{2}$$

where: $s = g\mu_B H = (0.0579 \text{ meV/T})gB$ (for free electrons $g = 1$). Thus, we can represent the transmission spectrum in a magnetic field as a series expansion:

$$(E) = F[n(E), E] + 0.5F'_n[n(E), E][nB(E) - n(E)] \tag{3}$$

The change of the imaginary part of refractive index has been found from zero field spectrum shifted by ± s in energy:

$$I(E \pm s) = F[n(E), E \pm s] + F'_n[n(E), E \pm s][n(E \pm s) - n(E)] \tag{4}$$

So, we can find connection between spectra measured with and without magnetic field I_0 substituting the change of imaginary part in I_B:

$$I_B(E) = C_+ I_0(E+s) + c_- I_0(E-s) \tag{5}$$

Coefficients $c\pm = 0.5F'_n[n(E), E]/F'_n[n(E), E\pm s]$ are almost independent from energy and are considered along. The result of the fitting procedure of absorption edge in the magnetic field of 11 T is shown in Figure 1(b). It can be clearly seen that Equation (5) describes the relation of experimental curves with and without magnetic field satisfactorily. We have calculated value of g-factor for different magnetic fields B Figure 1(c) using Equation (5). With increasing field, there is some saturation of energy shift and dependence $g(B)$ can be described with linear function $g(B) = 22.5 - 0.745B$. Thus, g-factor in $Bi_{1.1}Sb_{0.9}Te_2S$ is 15-20, which is unusually high for semiconductor without any magnetic elements. For example, in [6] estimated g-factor being 22.5 or 15.8 depending on the splittings in $ZrTe_5$. We suggest that this can be hidden magnetism due to the strong correlation between the direction of the spin and wave vector of electrons in topological insulators. Measurement of the g-factor of Zeeman splitting in other topological insulators and appropriately theoretical description would be very helpful for understanding nature of electron states in this type of compounds.

Second effect of the applied magnetic field to the sample under examination is decreasing Fabry-Perot oscillations with increasing field. This is shown in Figure 2(a). $P(\nu) = I(\nu)/I_{sm}(\nu) - 1$, normalized to the spectrum without oscillations $I_{sm}(\nu)$, obtained by smoothing the experimental spectra $I(\nu)$ by frequency ν. Note that oscillation amplitude decreases in magnetic field more pronounced in high frequency region. We have connected this with rotation of polarization plane of light when passing through the sample. Although experiment has been conducted without polarized light, Faraday's rotation influences on the signal intensity because of self-interference effect: transmitted wave is added with the same wave that has experienced

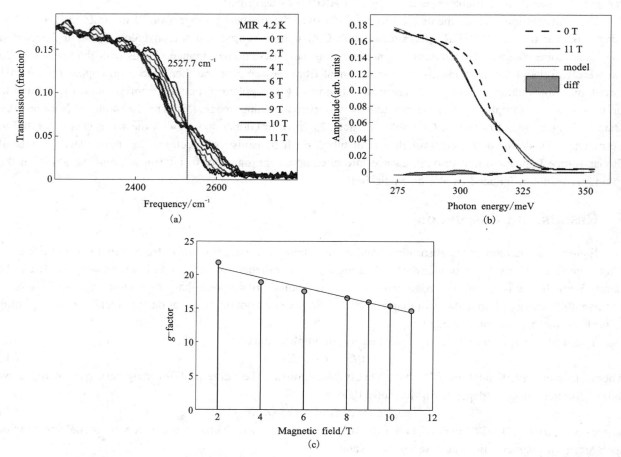

Figure 1 Effects of the magnetic fields on MIR transmission spectra at the absorption edge (a), smoothed absorption edges without magnetic field and in magnetic field (Model (5) is a red solid line and area difference show at the bottom) (b), g-factor of spin splitting energy for magnetic field from 0 to 11 T ($g = 22.5 - 0.745B$) (c)

additional reflection at the two boundaries of the plate. In case of Faraday's rotation second wave differs from the first not only in phase, but also by the rotation of the polarization vector. Consider the well-known formula for light transmittance coefficient (right+, left−) of circular polarization through plate of thickness d:

$$T_\pm = \frac{(1-r^2) e^{i(k \pm b)d}}{1 - r^2 e^{i2(k \pm b)d}} \quad (6)$$

where: $r = |r|e^{i\phi}$ is reflection coefficient from boundary surface; $k = k' + ik'' = n\omega/c$ is complex wave vector (n is refraction coefficient; ω is angular frequency of light; c is speed of the light in vacuum); $bd = \rho Bd$ is angle of rotation of the polarization plane during the passage of the plate, proportional to the magnitude of the field B.

The denominator is obtained by adding the direct wave (amplitude of this wave in taken as 1) with wave experiencing double reflection r^2 from the boundaries and received an additional phase shift from the optical path difference. Transmittance intensity of linearly polarized light Equation (7) is given by half-sum of the moduli square of the transmittance coefficient Equation (6):

$$I = \frac{1}{2} \sum_\sigma = \frac{|(1-r)^2|^2 e^{-2k'd}}{1 + |r^2 e^{-2k'd}|^2 - 2|r^2 e^{-2k'd}| \cos2[(k' + \sigma b)d + \varphi]} \quad (7)$$

The intensity of the spectrum will be determined by the factor in the main order. Thus, rotation of the polarization plane in magnetic field B should weaken interference and decrease the amplitude A of the Fabry-Perot oscillations according to Equation (8):

$$A_B = A_0 \cos 2\rho Bd \quad (8)$$

To check whether Equation (8) holds, we have isolated the main harmonic in oscillations, determined its amplitude for 8 values of the magnetic fields and fitted with Equation (8) contained only Verde constant ρ. The result is shown in Figure 2(b). Main Fourie harmonic contains 26 periods in the frequency range from

404 to 2528 cm^{-1}. It corresponds to the optical thickness of the film equal to $nd = 61$ μm. Fourie-spectrum contains also other harmonics due to the inhomogeneity thickness d of the film or changing in the refractive index n at the edges.

Amplitudes of the main harmonic of Fabry-Perot oscillations obtained from the experiment data in magnetic fields from 0 to 11 T are shown in Figure 2(c). Experimental data are described by Equation (8) with $\rho/n = 7.13$ rad(cm · T). This is the average value of Verde constant in frequency range from 404 to 2528 cm^{-1}. The effect of the magnetic field is greater for higher frequencies as seen at Figure 2(a). Indeed, similar calculations on the frequency interval 1200-2500 cm^{-1} give the value $\rho/n = 9.42$ rad(cm · T), on the frequency interval 2015-2500 cm^{-1} Verde constant is $\rho/n = 13.75$ rad(cm · T). It is worth mentioning that the Equation (8) is well satisfied in a large range of rotation angles of the polarization plane.

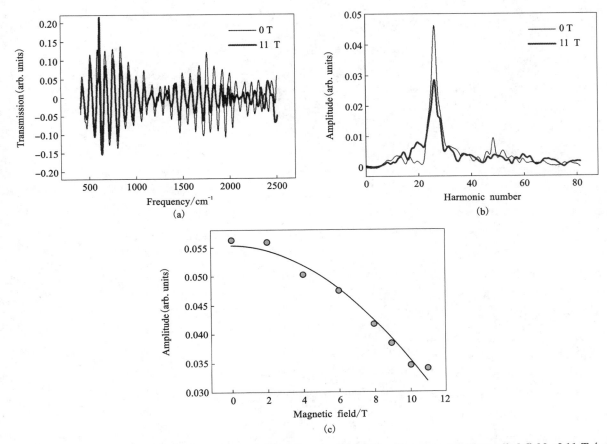

Figure 2 Normalized Fabry-Perot oscillations in MIR transmission spectra in zero and with applied field of 11 T (a), the complex amplitude modulus of the Fourier harmonic depending on its number (b), amplitude of the main harmonic of the Fabry-Perot oscillations versus magnitude of the applied magnetic field (dots) and curve described by Eq. (8) (c)

4 Conclusion

Strong effects of the magnetic field on the electronic structure and optical spectra of the topological insulator $Bi_{1.1}Sb_{0.9}Te_2S$ are found: Splitting of the transmission (absorption) edge due to the Zeeman shift of the spin states energy in a magnetic field with an unusually large g-factor (15-20); Suppression of transmitted wave interference (Fabry-Perot oscillations) due to Faraday rotation of the polarization plane in a magnetic field.

Acknowledgments: The research was funded by the Russian Science Foundation (Project No. 23-72-00067).

References

[1] HASAN M Z, KANE C L. [J]. Rev. Mod. Phys., American Physical Society, 2010: 3045-3067.
[2] QI X L, ZHANG S C. [J]. Rev. Mod. Phys., American Physical Society, 2011: 1057-1110.
[3] FU L., KANE C L. [J]. Phys. Rev. Lett., American Physical Society, 2008: 096407.
[4] MOORE J E. Nature, Nature Research, 2010: 194-198.
[5] KHATCHENKO Y E, et al. [J]. Journal of Alloys and Compounds, Elsevier, 2022: 161824.
[6] CHEN R Y, et. al. [J]. Phys. Rev. Lett, American Physical Society, 2015: 176404.

Interaction of Iridium with Silicon Carbide in Diffusion Couples in Wide Temperature Range

Mikhail GOLOSOV, Aleksei UTKIN, Victor LOZANOV, Natalia BAKLANOVA

Institute of Solid State Chemistry and Mechanochemistry SB RAS, Novosibirsk, 630090, Russia

Silicon carbide is a crucial component of high-temperature structural materials due to unique properties. With a high melting point of 2730 ℃, high specific strength, high hardness and oxidative stability, it can operate effectively for extended periods at temperatures up to 1500 ℃ in an oxidizing environment. However, when exposed to even higher temperatures, the oxide passivation layer undergoes significant evaporation, which leads to significant ablation of the material[1].

To address this limitation and enable SiC applications under more extreme conditions, the development of protective high-temperature oxidation-resistance coatings become necessary. Extensive study has led researches to explore iridium and iridium-containing compounds as potential coating materials. Indeed, iridium is a noble metal which has a high melting point of 2446 ℃ and a low recession in oxygen up to 2300 ℃[2].

Hence, the Ir-SiC system proves to be of great interest in various high-temperature applications. However, literature data on this system are very limited. Therefore, the primary objective of this work is to study a behavior of the Ir-SiC diffusion couple in the 1300-1800 ℃ temperature range.

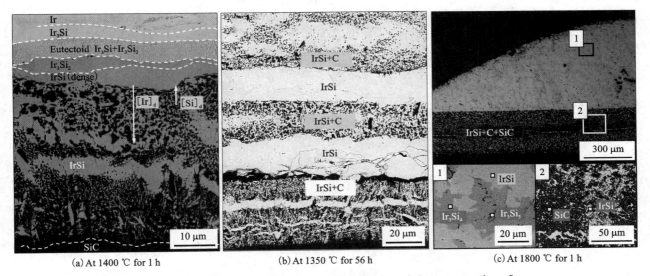

Figure 1 Morphology and elemental composition of the cross-section of the Ir/SiC diffusion couples heat-treated under different conditions

Diffusion couples consisting of iridium foil in thickness of 100 μm and silicon carbide plate in thickness of 5 mm were heat-treated at 1300-1800 ℃ for different exposure time in an argon atmosphere.

It was found that at 1300-1400 ℃, the sequence of silicide phases on the cross-section of the Ir-SiC couple is as follows: Ir_3Si/Eutectoid $Ir_3Si+Ir_3Si_2$/Ir_3Si_2/IrSi/IrSi+C [Figure 1(a)]. The evolving carbon does not dissolve in iridium silicides, therefore, it can be a marker of the Kirkendall plane for IrSi layer. A comparison of the sublayer sizes on both sides of the Kirkendall plane led to the conclusion that the diffusion rate of iridium atoms is much higher than the counter diffusion rate of silicon atoms [Figure 1(a)]. The dependences of the silicide layer thickness on the holding time is linear, indicating that the interaction is controlled by kinetics in experimental conditions under investigation. A morphological feature of the samples treated with long exposure times at 28-56 h is the appearance of periodically located sublayers of dense iridium silicide (IrSi) and IrSi sublayers with a high content of carbon grains [Figure 1(b)]. It seems that the formation of these bands is associated with the tendency of the system to decrease the surface energy between the silicide and carbon phases.

At $T \geqslant 1417$ ℃ (the eutectic temperature between the Ir_3Si_2 and Ir_2Si phases), the interaction of iridium with silicon carbide changes its character from the solid state to the liquid phase interaction, the drop-like layer on the surface of the Ir/SiC diffusion couple being formed[3,4]. The phase composition is presented by Ir_2Si, Ir_3Si_2, IrSi, and carbon. The stabilization of the high-temperature Ir_2Si phase seems to be associated with a high cooling rate (~75 ℃/min). Heat treatment at 1600 ℃ for 4 h leads to the formation of final products IrSi and C. With time, graphitization of the evolved carbon phase occurs that confirmed by Raman spectroscopy analysis. Heat treatment at temperature in the vicinity of IrSi melting point (1707 ℃) leads to supersaturation of the Ir-Si melt with silicon (above 50 at%), therefore, higher silicides, Ir_3Si_4 and Ir_3Si_5, are observed [Figure 1(c)].

These findings provide valuable insights into the high-temperature behavior of the Ir-SiC system, highlighting its potential for advanced applications.

Acknowledgments: This work was supported by the Russian Science Foundation (Project No. 23-19-00212).

References

[1] KIM Y W, MALIK R. SiC ceramics, structure, processing and properties[M].//Encyclopedia of Materials: Technical Ceramics and Glasses/ed. Pomeroy M. Oxford: Elsevier, 2021: 150-164.

[2] WU W P, CHEN Z F. Iridium coating: processes, properties and application. part I[J]. Johnson Matthey Technology Review, 2017, 61(1): 16-28.

[3] GOLOSOV M A, et al. Toward understanding the reaction between silicon carbide and iridium in a broad temperature range[J]. Journal of the American Ceramic Society, 2021, 104(12): 6653-6669.

[4] GOLOSOV M A, UTKIN A V, LOZANOV V V, et al. Microstructural patterning of the reaction zone formed by solid-state interaction between iridium and SiC ceramics[J]. Materialia, 2023, 27: 101647.

Lightweight Ceramic Nanocomosites for Radiation Shielding of Electronics

Oleg KHASANOV, Edgar DVILIS, Vladimir PAYGIN, Oleg TOLKACHEV

National Research Tomsk Polytechnic University, Tomsk, 634050, Russia

Abstract: Full dense lightweight nanostructured metal-ceramic composite have been developed for the purpose of radiation shielding of electronic components from the combined effects of electrons, ions, gamma radiation, neutrons. The composite consisted of low-melting AlMg alloy powder (77.9 vol%), refractory B_4C submicron powder (18.8 vol%) and refractory W nanopowder (3.3 vol%). The composite was sintered by SPS technique up to 100% density at 490 ℃/40 MPa. The achieved mechanical properties of the composite were as follows: microhardness 419.9 HV; elastic modulus 98.62 GPa; creep under indentation 1.45%. The attenuation coefficient of gamma radiation by the composite was 1.34 times grater than for a pure AlMg alloy; the attenuation coefficient of thermal and superthermal neutrons by the composite-2.2 times greater.

Keywords: Nanocomposites; Spark plasma sintering; Consolidation; Radiation shielding

1 Introduction

An urgent task is to ensure the protection of electronic components intended for operation in radiation fields from harmful radiation effects. Concerning the radiation protection of the electronics on the spacecraft it is important to provide the light weight of the protective material.

To provide radiation protection from the combined effects of radiation fluxes of electrons, ions, gamma rays, neutrons, it is necessary to consolidate a composite consisting of several components having different thermophysical properties, but the necessary cross-sections of absorption of these types of radiation.

There are known technologies for manufacturing radiation-protective composites using the mechanochemical alloying and homogeneous mixing of various powders, followed by their sintering by hot extrusion or hot pressing[1]. For hot extrusion, powerful and expensive presses with a force of more than 500 tons and having tools for preheating the green powder mixtures up to 300-500 ℃ are used. The density of metal-matrix composites produced in this way reaches 95%-96% of the theoretical one. The finish machining of the extruded workpiece requires additional milling and turning.

The goal of this work was to study the consolidation processing of mixed low-melting and refractory powders to produce the full-dense and lightweight bulk composite that provides effective radiation shielding of electronic components from the combined influence of several types of radiation fluxes: electrons, ions, gamma rays, neutrons.

We have developed the metal-ceramic composite consisting of powder of low-melting and lightweight Al/Mg matrix (ensuring protection against electrons, ions) with inclusions of refractory B_4C submicron powder (protection against neutrons), and refractory W nanopowder (protection against gamma rays).

2 Experimental procedure

The equipment of the Center for Sharing Use "Nanomaterials and Nanotechnologies" of TPU was applied for this work: JSM-7500F JEOL (SEM analysis); SALD-7101 Shimadzu (particle size distribution determination by the method of laser diffraction-LD); Sorbi META (specific surface measurement by BET technique); XRD-7000S Shimadzu (XRD analysis). Also other devices for characterization of studied powders and consolidated composite samples have been used.

The AlMg alloy powder (77.9 vol%, AMg6 grade, Redmetsplav LLC, Russia; mean particle size of 6.8 micron) with additions of B_4C submicron powder (18.8 vol%, OKB-BOR LLC, Russia; mean particle size of 275 nanometers) and W nanopowder (3.3 vol%; it was produced by TPU using method of electric explosion of wires[2]; mean particle size of 115 nanometers) were mixed in the specified ratio by ball milling using grinding balls made of YSZ for at least 2 h.

The raw powders were mixed in the specified ratio by ball milling using grinding balls made of YSZ for at least 2 h.

Modeling of particle packages of the studied powder mixtures was performed by discrete element analysis using the S3D PorouStructure code by the Ichikawa algorithm (central packaging). The simulation of compaction of a powder mixture was performed using S3D Evolution code.

2.1 Consolidation and testing of AlMg/B_4C/W composite

Experimentally, the compaction curves of mixed dry AlMg/B_4C/W composite powders were plotted by the method of cyclic unloading of the powder during pressing with the use of precision hydraulic press IP-500M-auto ZIPO and computer controller[3].

The method of dry powder compaction using powerful ultrasound assistance (PUA) has been applied[4] using ultrasonic generator IL10-5.0 (power or 4 kW, smooth adjustment of the frequency in the range 16-22 kHz), magnetostrictive converters PMS-15; hydraulic press WK 18 (pressing force up to 100 tons).

Consolidation of the mixed composite dry powders was carried out by the spark plasma sintering (SPS) technique using SPS 515S SYNTEX installation. Disks with a diameter of 15 mm and a thickness of 3 to 6 mm were sintered.

The density of sintered samples was determined by direct measurements of their mass and volume using VLTE-150 digital balance with an accuracy of ±0.001 g, and by Mitutoyo ID-F150 digital thickness gauge with an accuracy of ±1 μm. Mechanical properties of sintered composites (microhardness, elastic modulus, creep under indentation load) were measured by the indentation of the Vickers pyramids using nanohardness tester DUH-2115 Shimadzu, microhardness tester PMT-3M LOMO.

Radiation shielding tests for the developed composites have been performed using facilities of the IRT-T research nuclear reactor at TPU. Experiments to determine the attenuation coefficients of thermal and superthermal neutrons were carried out using the experimental channels of collimated neutron beams HEC-1. The attenuation coefficients of gamma rays were investigated using isotopes ^{57}Co (gamma ray energy E from 14.4 keV up to 136.47 keV), ^{60}Co (1173.2 keV and 1332.5 keV), ^{137}Cs (661.7 keV), and using gamma rays from IRT-T nuclear reactor.

3 Results and discussions

3.1 Optimization of the AlMg/B_4C/W powder mixture composition

The simulation of particle packaging of the mixed AlMg/B_4C/W composite powder was performed using the determined mean particle sizes and particle size distributions obtained by LD experiments for each component. Optimization of the ratio of the mixture components was carried out according to the criteria for achieving the highest average number of interparticle contacts (coordination number N_c) and the highest package density. For this purpose, the particle size distributions were approximated by the corresponding functions. The obtained parameters of these functions were used to define the corresponding particle classes for discrete element modeling by the S3D PorouStructure code. The total number of particles of the representative set in such models ranged from 20000 to 72000 (Figure 1).

The simulation results showed that the optimal ratio of the mixture components according to the criterion for achieving the maximum packing density of all particles has optimal composition AMg6 89.2 vol%+B_4C 7.5 vol%+W 3.3 vol% [Figure 1(c)]. In this case, the partial coordination number N_c of matrix alloy particles ranges from 3 [in the freely filled state-Figure 1(c)] to 4.29 [after deformation of plastic material particles-Figure 1(d)], and the partial packing density of its particles is from 43% [Figure 1(c)] to 67% [Figure 1(d)], respectively.

Deformed composite powder with optimal composition will have the higher packing density 89%, and the coordination number N_c=7.23.

Analysis of the packaging models (Figure 1) shows that studied composites can not be consolidated to a non-porous state by free sintering. After packing the pressed particles of the mixed components, further consolidation with decreasing the free surface at temperatures less than the melting point of the AMg6 alloy (600 ℃) is limited by the presence of the dense packed refractory B_4C and W particles in the interparticle space of the low-melting matrix. This requires significantly higher temperatures for sintering.

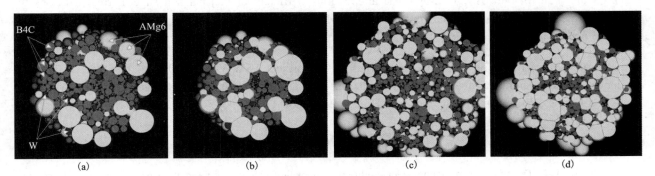

Figure 1 Simulation of the packing of particles of composite components with the minimum permissible (a), (b) and optimal content (c), (d) of matrix alloy AMg6 and dispersed fillers B_4C, W in the free-laying state (a), (c) and after deformation of 25% (b), (d)

Densification of the composite at temperatures about 600 ℃ is possible only due to plastic deformation with the forced flow of the matrix alloy under pressure (intrusion) into the pore space of the dispersed refractory filler, up to filling all the pores.

Therefore, for the complete consolidation of the studied composite, the most promising method is SPS-vacuum sintering atpulsed electric current through the powder particles and at the continuous pressing.

3.2 Properties of consolidated AlMg/B_4C/W composites

Mixtures of composite powders which were compacted using powerful ultrasound assistance at a compaction pressure of 800 MPa without following sintering had the green compact density of 95.1%.

Obtained density of 95.1% of the composite green compact directly after compaction of the dry mixed powder using PUA, but without following sintering, is comparable with the density of similar composite produced by the hot extrusion method (95%-96%)[1]. This indicates the prospects and cost-effectiveness of the ultrasonic compacting method for powder mixtures of studied composites.

Subsequent vacuum free sintering of these green compacts at 590 ℃ for 2 h resulted in a slight increase in density: up to 93.4% for green compacts pressed without ultrasound, but up to 96.8% for green compacts pressed using PUA.

The AMg6/B_4C/W composite was sintered by the SPS technique up to 100% density (3.13 g/cm^3) at temperature of 490 ℃ and pressure of 40 MPa. The microhardness of this composite was 419.9 HV; elastic modulus was 98620 N/mm^2 (98.62 GPa); creep under constant indentation load was 1.45%[5].

Obtained full-dense AMg6/B_4C/W nanocomposites had the following attenuation coefficients of thermal and superthermal neutrons ($K_{t.n}$ and $K_{st.n}$ respectively) and K_{gamma} of gamma radiation:

- $K_{t.n} = 3.34$ and $K_{st.n} = 3.26$, which is 2.2 times greater than for pure AMg6 alloy.
- $K_{gamma} = 1.34$ for gamma rays having $E \leqslant 137$ keV, which is 1.34 times greater than for pure AMg6 alloy.

4 Conclusion

Lightweight metal-ceramic nanocomposites AlMg/B_4C/W were consolidated by SPS technique up to 100% density (3.13 g/cm^3) at temperature of 490 ℃ and pressure of 40 MPa.

Free vacuum sintering does not provide consolidation to full density of the composite having low-melting AlMg alloy matrix and refractory B_4C, W nanopowder fillers.

The optimal component ratio for the studied composite according to criterion of the higher density is

AMg6 89.2 vol%+B_4C 7.5 vol%+W 3.3 vol%.

Static one-side pressing of the dry mixed composite powders at 800 MPa allows to achieve the green density of 92.6%, but compaction using powerful ultrasound assistance provides the green density increasing up to 95.1%.

Developed full-dense AlMg/B_4C/W nanocomposites had attenuation coefficient of gamma radiation 1.34 times greater than for pure AMg6 alloy. The attenuation coefficient of thermal and superthermal neutrons for this composite was 2.2 times greater than for pure AMg6 alloy.

The processing of the developed AlMg/B_4C+W nanocomposite has been patented in Russia[6].

Acknowledgments: The work was supported by the project FSWW-2023-0011 of the State assignment "Science". The equipment of the Center for Sharing Use "Nanomaterials and Nanotechnologies" of TPU, supported by the RF Ministry of Science and Higher Education (grant #075-15-2021-710).

References

[1] GULBIN V N. Development of nanopowder-modified composition materials for radiation protection in nuclear power industry[J]. Nuclear Physics and Engineering (*in Russian*), 2011, 2, 3: 272-286.

[2] KHASANOV O L, POKHOLKOV Y P, BONDARENKO A L, et al. Radiation-induced desorption from the nanopowder, MRS symposia proceedings: Nanostructured powders and their industrial applications[J]. Warrendale, PA, 1998, 520: 89-93.

[3] KHASANOV O L, DVILIS E S, SOKOLOV V M. Compressibility of the structural and functional ceramic nanopowders, J. Eur. Ceram. Soc., 2007, 27: 749-752.

[4] KHASANOV O L, DVILIS E S. Net shaping nanopowders with powerful ultrasonic action and methods of density distribution control[J]. Advances in Applied Ceramics, 2008, 107(3): 135-141.

[5] DVILIS E S, KHASANOV O L, GULBIN V N, et al. Spark plasma sintering of aluminum-magnesium-matrix composites with boron carbide and tungsten nano-powder inclusions: modeling and experimentation[J]. JOM, 2016, 68(3): 908-919.

[6] DVILIS E S, TOLKACHEV O S, PETYUKEVICH M S, et al. Method for manufacturing an aluminum-matrix composite material, Patent of Russian Federation, No. 2616315, 14.02.2017.

The Synthesis of Triple-lithiated Transition Metal Oxide NCM622 Based on Extracted Cobalt from Spent LiCoO$_2$ Cathode Material

Alexandra KOSENKO, Konstantin PUSHNITSA, Pavel NOVIKOV, Anatoliy A. POPOVICH

Peter the Great Saint-Petersburg Polytechnic University, Saint Petersburg, 195221, Russia

Abstract: The environmentally friendly closed cycle of regeneration process of spent LiCoO$_2$ was successfully developed and the following synthesis of triple lithiated transition metal oxides was carried out. The hydrometallurgy recycling route with 1.5 M malic acid and 3 vol.% of H$_2$O$_2$ as a leaching solution for cobalt extraction was chosen. The efficiency of cobalt extraction reached 95%. The obtained material was investigated by X-ray diffraction analysis, EDX and SEM methods. Electrochemical behavior of synthesized NCM622 was analyzed and compared to the commercially available material. Also, the cyclic life and cyclic voltammograms were studied.

Keywords: Regeneration; Cobalt extraction; Spent LiCoO$_2$; Triple-lithiated transition metal oxides

1 Introduction

The use of lithium-ion batteries has increased significantly in recent years, as due to their high energy density, they have found applications in many areas, such as electronic devices, energy storage systems and electric vehicles. However, after several thousand charge-discharge cycles, the service life comes to an end, they become unsuitable for the proper operation of portable devices, and then they are subject to disposal/burial[1]. If battery waste is not treated properly, it poses a serious environmental hazard due to toxic chemicals such as heavy metals (Co, Ni, Cu, etc.) contained in cathode/anode, organic solvents and fluorinated compounds contained in the electrolyte[2-4]. In addition, when using LIBs in large quantities with high growth rates of the use of this technology, significant depletion of raw materials and limited natural resources of metals (for example, Co) threaten the sustainable production of batteries[5]. Recycling of used batteries can reduce the load from both environmental pollution and lack of resources in the long term. Due to the limited resources of materials used in the cell and environmental problems after the termination of the use of LIB, the post-processing of valuable metals from spent batteries is essential. As the most valuable metals are concentrated in the cathode materials, the focus of recycling spent LIBs is the recovery of metals in waste cathode materials. Traditional technologies for recycling metals from waste cathode materials include pyrometallurgy, hydrometallurgy and biometallurgy. All these methods have their own dignities and drawbacks[6-11]. For this research the hydrometallurgy strategy was chosen, which allows to get the closed loop recycling process and the final products of high purity. The hydrometallurgical route comprises leaching (acid or alkali), purification and separation steps, and product recovery using the sol-gel method. This work aims at the methodology development of Co extraction from spent Li-ion batteries by hydrometallurgical processing with organic acid leaching solution, which allows to use the cobalt-containing leachate with additional salts of Li, Ni and Mn as a precursor for following synthesis of triple-lithiated transition metals oxide NCM622.

2 Experiment

2.1 Cobalt extraction and following NCM synthesis

As the initial material the cathode with aluminum current collector from spent lithium-ion battery was used as cobalt source. The spent battery was fully discharged in 5% NaCl solution for 24 h, then disassembled, and the cathode was separated from another parts of battery. The cathode plates were immersed in acetone solution for 24 h and washed distilled water in sequence to remove residual electrolyte and dried in vacuum at 95 ℃.

The cathode material was leached in 1,5M malic acid solution with 3 vol.% H_2O_2 at 90 ℃ with stirring for 1 h. The reaction involved in the leaching is as follow:

$$4LiCoO_2 + 3C_4H_6O_5(aq) + H_2O_2 \longrightarrow 2CoC_4H_4O_5(aq) + Li_2C_4H_4O_5(aq) + H_2O + O_2 \qquad (1)$$

The undissolved residue was removed by filtering to obtain the leachate solution containing Co ions. After that, Ni, Mn and Li acetates were added to the leachate solution in stoichiometric ratio in order to obtain subsequently the NCM material. The calculation was carried out in the following way: after the end of the leaching process the filtrated residue was dried for 24 h and weighted. The obtained mass was subtracted from the initial cathode mass—that was the theoretical dissolved cathode mass in the solution. This value was multiplied to 0,92 to obtain the active mass in the solution. Afterwards, the active mass was multiplied to 0.59—this is theoretical cobalt mass in the solution. Then this value was divided on 59 (Co molar mass) to obtain the amount of Co. Furthermore, to calculate the NCM111 amount, the cobalt amount was multiplied to 3. Then the masses of NiAc, MnAc and LiAc were calculated based on the substances amounts (NiAc and MnAc amounts are equal to Co amount, LiAc value equals to NCM111 amount) and respective molar masses. The next step was water evaporation from the resulting solution with forced removal of gases. The obtained precipitate was used as the precursor for following synthesis of NCM cathode material through heat treatment.

2.2 Elemental analysis and material characterization

The morphology and microstructure of the synthesized sample were studied using a Mira 3Tescan scanning electron microscope in the SE (secondary electrons) mode. The chemical composition of the powder was studied on an electron microscope by X-ray spectroscopy using an EDX Oxford Instruments X-Max 80 energy dispersion detector. X-ray phase analysis of the obtained samples was performed on a Bruker D8 Advance diffractometer ($CuK_\alpha = 1.5406$ Å) in the angle range of 15-85 degree with a step of 0.030 and an exposure of 0.8 s at each step. Phase identification was carried out using the DIFFRAC. EVA V5.0 program. Structural parameters were refined by the Rietveld method using TOPAS5 software. The granulometric composition of the sample was analyzed using Fritsch Analysette 22 NanoTec equipment. To determine the electrochemical characteristics of the synthesized material, it was smeared and assembled into CR2032 layouts. Electrochemical measurements of the charge and discharge capacity of the samples were carried out using the Neware coin cell-5 V 10 mA Battery Test System. The cyclovoltammrograms were obtained using a scanning rate of 0.1 mV/s at potentials 0-3.0 V.I.

2.3 Results and discussion

Figure 1 shows an X-ray diffractogram of a synthesized cathode material of the NCM622 obtained based on extracted cobalt, as well for comparison also diffractogram of commercial NCM material is added. The patterns of synthesized NCM material are well indexed to layered structure with R-3m space group without any impurity phase. The distinct peak split corresponding to (006)/(102) planes suggests a well-ordered layered structure.

Also, it is highly recommended for cathode materials to analyze following parameters[12-14]:

(1) Split of the peaks $I006/I102$ ($2\theta \sim 36°$): This parameter indicates a good ordering of the hexagonallattice, peak separation is present both in the synthesized and commercial NCM.

(2) R-factor, $R = (I006 + I102)/I101$: The

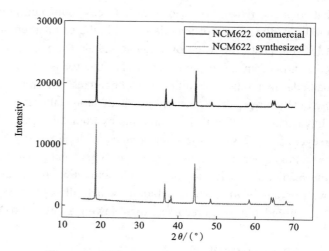

Figure 1 Diffractograms of the synthesized and commercial NCM622

lower the value of this parameter, the higher the ordering of the structure (for good materials <0.5), for both materials this parameter is higher than should be: for commercial is 0.597, for synthesized is 0.739.

(3) The ratio $I003/I104$ ($2\theta\ 003 \sim 20°$, $2\theta\ 104 \sim 45°$). The smaller this ratio, the lattice is closer to cubic. In the context of these materials, this parameter characterizes cationic mixing and for good materials should be > 1.2. For commercial material it is a little bit higher than for obtained (1543 and 1896 respectively).

Obviously, the R-factor of obtained NCM materials after cobalt extraction is extremely high for such type of materials. Probably, it could be a result influence of sol-gel synthesis on powder structure. The ratio of 003 and 104 peaks is higher than ratio of commercial material, hence the crystal lattice is hexagonal and well-ordered.

In the Table 1 particle size distribution is presented.

Table 1 Particle size distribution if synthesized material

	D10	D50	D90
Average particle size/μm	6.6	18.3	35.3

Figure 3 demonstrates the results of the morphology study of the synthesized sample obtained by scanning electron microscopy in the secondary electron mode. Apparently, the powder has rounded particles with an average particle size of 300 nm^{-1} microns, and has a uniform particle size distribution. In addition, the agglomeration of particles in the sample is quite high. It is well known that the small particle size and their uniform size distribution contributes to the rapid migration of Li+ ions, which can contribute to obtaining excellent electrochemical characteristics.

Figure 2 demonstrates the results of the morphology study of the synthesized sample obtained by scanning electron microscopy in the secondary electron mode and the chemical composition of the powder obtained by the EDX. Apparently, the powder has rounded particles with size of 0.5-1 micron, also the particles are agglomerated. Material has a uniform particle size distribution. It is well known that the small particle size and their uniform size distribution contributes to the rapid migration of Li^+ ions, which can contribute to obtaining excellent electrochemical characteristics.

Figure 2 The SEM images of the obtained NCM622 with the chemical composition of the powder

Thus, the atomic ratio of transition metals in obtained materials for calculated NCM622 is actual $LiNi_{0.6}Co_{0.17}Mn_{0.22}O_2$, for calculated NCM811 is $LiNi_{0.80}Co_{0.10}Mn_{0.10}O_2$. The obtained results are close to the theoretically calculated ratios, there is a slightly reduced, relative to the theoretically calculated, cobalt content, which is likely to affect the initial capacitance characteristics. There is also the presence of more than 1% aluminum, which probably got into the precursor during the extraction of cobalt from the aluminum current collector. The introduction of the Al ion increases the Ni^{3+} ratio and limits the mixing of cations. Compared with unmodified samples, the cathode materials NCM811 doped with Al^{3+} demonstrated excellent cycling performance with a capacity retention of 70% at 10 ℃ after 1000 charge-discharge cycles[15]. In addition, doping in the precursor is useful for reducing the energy potential, which leads to a more uniform distribution of elements and a reduction in binding deficiencies during the subsequent calcination process. It is noteworthy that the alloying of Al can suppress the effect of cationic mixing and reduce the destruction of the structure of the material during charging[16].

Figure 3 illustrates the electrochemical behavior of synthesized NCM622.

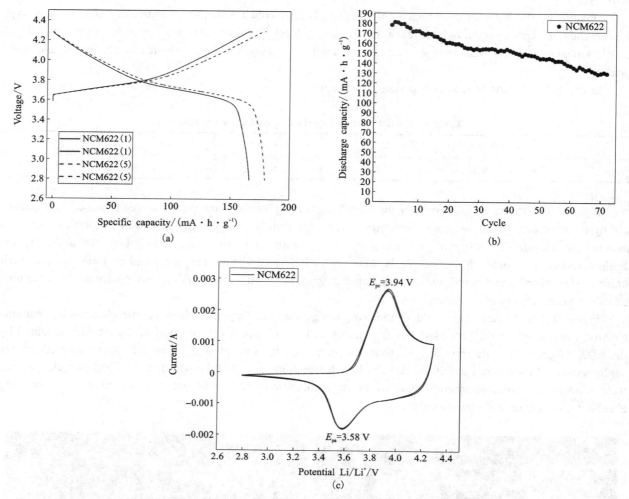

Figure 3 The electrochemical behavior of synthesized NCM622

On Figure 3(a) charge-discharge curves on 1st and 5th cycles are presented. The initial charging and discharge capacities of the NCM622 were 179 and 178 mA·h/g, Coulomb efficiency for 622 is 99%, Coulomb efficiency of the material close to 100% indicates reversible intercalation/deintercalation of lithium indicating sufficient stability of the structure. On cycle 5, the charging and discharge capacities for NCM622 were 181 and 180 mA·h/g. On Figure 3(b) a cyclic resourse study is presented. The capacity drop in NCM622 for 72 cycles was 27%. Such a large drop in the capacity is due to high R-factor value, which has an extremely negative effect on the capacitance characteristics of the material. The synthesized material was also investigated by cyclic voltammetry, the polarization curves are shown on Figure 3(c). This method of analyzing the electrochemical characteristics of a material allows you to analyze the reversibility of electrode reactions, whether side reactions are present on the electrode, and also allows you to compare the kinetics of electrode reactions in different materials. In the cathode and anode regions, peaks of the potentials of the reversible process of deintercalation/intercalation of lithium ions are clearly observed. The peak of the cathode potential is 3.94 V, and the anode potential is 3.58 V. Peaks of side electrode reactions are not observed.

3 Conclusion

This paper describes the synthesis of triple-lithiated transition metal oxide NCM622 from spent $LiCoO_2$ with use of the extracted cobalt malate and Ni, Mn and Li acetates by the sol-gel method followed by thermal treatment. The obtained material was investigated by different methods, such as XRD, SEM, EDX. The initial capacity of the synthesized NCM622 was equal to 179 mA·h/g and 180 mA·h/g on the 5th cycle, under current rate of 0.1 C. The resulting values correspond to requirements to commercial materials of the same type. The cycle life study of material showed the drop by 27% after 72 cycles, which could be explained by the high value of R-factor and disorder of the layered structure. On the polarization curves of NCM622 symmetric peaks of reversible redox reaction were observed at 3.6-3.9 V. There are no signs of adverse reactions.

References

[1] HAN X, LU L, ZHENG Y, et al. A review on the key issues of the lithium ion battery degradation among the whole life cycle E[J]. Transportation. (2019)1: 100005. https: //doi.org/10.1016/j.etran.2019.100005.

[2] YU JD, HE Y Q, LI H, et al. Effect of the secondary product of semi-solid phase Fenton on the flotability of electrode material from spent lithium-ion battery[J]. Powder Technology, 2017, 315: 139-146.

[3] ASSEFI M, MAROUFI S, YAMAUCHI Y, et al. Pyrometallurgical recycling of Li-ion, Ni-Cd and Ni-MH batteries: a minireview[J]. Current Opinion in Green and Sustainable Chemistry, 2020, 24: 26-31.

[4] QU X, CAI M, ZHANG B, et al. A vapor thermal approach to selective recycling of spent lithium-ion batteries[J]. Green Chem. 2021(23): 8673-8684.

[5] ZENG X, LI J, On the sustainability of cobalt utilization in China[J]. Resour. Conserv. Recycl. 2015(104): 12-18.

[6] ZHOU SY, ZHANG Y J, MENG Q, et al. Recycling of $LiCoO_2$ cathode material from spent lithium ion batteries by ultrasonic enhanced leaching and one-step regeneration[J]. Journal of Environmental Management, 2021, 277: 111426.

[7] SONG D, WANG T, LIU Z, et al. Characteristic comparison of leaching valuable metals from spent power Li-ion batteries for vehicles using the inorganic and organic acid system[J]. J. Environ. Chem. Eng. 2022(10): 107102.

[8] YANG H, DENG B, JING X, et al. Direct recovery of degraded $LiCoO_2$ cathode material from spent lithium-ion batteries: Efficient impurity removal toward practical applications[J]. Waste Manag 2021(129): 85-94.

[9] TAO R, XING P, LI H, et al. Recovery of spent $LiCoO_2$ lithium-ion battery via environmentally friendly pyrolysis and hydrometallurgical leaching[J]. Resour. Conserv. Recycl. 2022(176): 105921.

[10] YU JC, MA B Z, SHAO S, et al. Cobalt recovery and microspherical cobalt tetroxide preparation from ammonia leaching solution of spent lithium-ion batteries[J]. Transactions of Nonferrous Metals Society of China, 2022, 32(9): 3136-3148.

[11] BOTELHO A B Jr, STOPIC S, FRIEDRICH B, et al. Cobalt recovery from Li-ion battery recycling: a critical review [J]. Metals, 2021, 11(12): 1999.

[12] REIMERS J N, LI W, DAHN J R. Short-range cation ordering in $Li_xNi_{2-x}O_2$[J]. Physical Review B, 1993, 47(14): 8486-8493.

[13] ZYBERT M, RONDUDA H, DĄBROWSKA K, et al. Suppressing Ni/Li disordering in $LiNi_{0.6}Mn_{0.2}Co_{0.2}O_2$ cathode material for Li-ion batteries by rare earth element doping[J]. Energy Reports, 2022, 8: 3995-4005.

[14] SKVORTSOVA I, SAVINA A A, ORLOVA E D, et al. Microwave-assisted hydrothermal synthesis of space fillers to enhance volumetric energy density of NMC811 cathode material for Li-ion batteries[J]. Batteries, 2022, 8(7): 67.

[15] LI Y C, XIANG W, WU Z G, et al. Construction of homogeneously Al^{3+} doped Ni rich Ni-Co-Mn cathode with high stable cycling performance and storage stability via scalable continuous precipitation[J]. Electrochimica Acta, 2018, 291: 84-94.

[16] CHENG L, ZHANG B, SU S L, et al. Al-doping enables high stability of single-crystalline $LiNi_{0.7}Co_{0.1}Mn_{0.2}O_2$ lithium-ion cathodes at high voltage[J]. RSC Advances, 2021, 11(1): 124-128.

Creep Behavior and Structural Changes in the RE-containing 10% Cr Martensitic Steels

Alexandra FEDOSEEVA

Belgorod National Research University, Belgorod, 308015, Russian

Abstract: The present research was devoted to revealing the reasons of different creep behavior of the Re-containing 10% Cr steels. Two Re-containing 10% Cr steels with different Cu content were normalized at 1323 K with following tempering at 1043 K for 3 h. Creep tests at a temperature of 923 K under the applied stress of 140 MPa were carried out. An increase in the Cu content provides the 8-fold increment in rupture time due to the increase in duration of the primary creep stage. Moreover, the minimum creep rate was reduced by 2 orders of magnitude. Strong degradation of the tempered structure after creep test was found in both steels. Widening of the martensitic laths, reduction in the dislocation density and coarsening of grain boundary particles were typical structural changes after creep test. The Laves phase formation along the lath boundaries was observed in both steels, but constant growth rate of Laves phase was higher in the St0.8Cu steel. On the other hand, precipitation of Laves phase determines the creep behavior during primary creep stage, and the additional nucleation of Laves phase on "Cu" clusters in the St0.8Cu steel provides the improvement of creep behavior.

Keywords: High-chromium steels; Creep; Structural degradation; Rupture.

1 Introduction

9%-12% Cr martensitic steels demonstrate high creep resistance and strength properties as well as a low thermal expansion at elevated temperatures that provides the application of these materials for pipes and tubes of fossil power plants, which are able to work at ultra-supercritical steam parameters (873-893 K, 20-25 MPa)[1,2]. Improved creep resistance of these steels is accompanied with a stability of tempered lath structure of steel, although this structure contains high dislocation density[3,4]. High strength properties are provided by linear sum combination of strengthening mechanisms due to the martensitic laths, dislocations, solid solution and secondary particles[4]. To stabilize tempered lath structure during creep/ageing, $M_{23}C_6$ carbides and MX carbonitrides prevent the migration of dislocations and low-angle boundaries, as well as W, Mo, Co, and Re slow down diffusion-controlled processes[3]. In the modern 9%-12% Cr steels, the N content decreased to 0.001 wt% to prevent the formation of Z-phase particles instead of MX phase, and B content increased up to 0.015 wt.% to provide the low coarsening rate of $M_{23}C_6$ carbides during creep[4]. Moreover, W and Mo deplete from the solid solution with the formation of the Laves phase particles[3]. The Laves phase plays contradictory role[5,6]. On the one hand, the fine Laves phase particles precipitated during primary creep stage decreased creep rate[5]. On the other hand, Laves phase is prone to quick growth that leads to the formation of cracks and voids[6]. An addition of Re provides the retaining fine size of Laves phase along low-angle boundaries up to 10000 h of creep that positively affect creep resistance[7]. Moreover, Re slows down of Fe and W diffusion along grain boundaries. Addition of Cu is considered to be an effective method to control the nucleation of Laves phase[8]. In [8] it was noticed that Cu atoms forms "Cu"-rich clusters with BCC lattice at 923 K, which act as nucleation sites for the Laves phase particles. The aim of the present research is to report on creep behavior and structural changes in the Re containing steels with the different Cu content after creep test at a temperature of 923 K and an applied stress of 140 MPa.

2 Experiment

2%-10% Cr martensitic steels with different Cu content devoted here as St0.8Cu and St0.3Cu were investigated. The chemical compositions represented in Table 1 were controlled via optical emission spectrometer FOUNDRY-MASTER-UVR. These steels were solution treated at 1423 K for 16 h and forged at 1423 K to a true strain of ~1, cooled in air, then normalized at 1323 K for 1 h, cooled in air, and finally tempered at 1043 K for 3 h, air cooling. Flat specimens with a gage length of 25 mm and a cross section of 7 mm×3 mm were crept until rupture at 923 K in air under an applied stress of 140 MPa using computer-controlled ATS Lever Arm Testers. The transmission electron microscope, JEM JEOL-2100, (TEM) with an INCA energy dispersive X-ray spectrometer (EDXS) was used for structural investigations. The specimens for TEM and SEM were prepared by electropolishing using the solution of 10% perchloric acid in glacial acetic acid at a room temperature using Struers Tenupol-5 machine. The transverse size of the martensitic laths was estimated by the linear intercept method. The dislocation density within the lath interiors was evaluated as a number of intersections of individual dislocations with foil surfaces per unit area on at least five selected typical TEM images for each data point. The mean size of the secondary phase particles was evaluated by counting from 100 to 150 particles per specimen. Identification of the secondary phase particles was based on the combination of EDXS composition measurements and indexing of electron diffraction patterns by TEM. The error bars were given according to the standard deviation.

Table 1 The chemical compositions of the steels studied (wt. %)

Ingot	Fe	C	N	B	Cr	Co	W	Mo	Cu	V	Nb	Re	Si	Ni	Mn
St0.8Cu	Bal	0.09	0.002	0.015	9.6	2.9	2.1	0.6	0.8	0.2	0.05	0.17	0.1	0.2	0.03
St0.3Cu	Bal	0.13	0.001	0.015	9.4	3.1	2.1	0.6	0.3	0.2	0.05	0.17	0.1	0.2	0.03

3 Results and discussion

3.1 Tempered structures

After tempering, the tempered martensite lath structure was found in both steels (Figure 1). The mean PAG size and width of the martensitic laths were similar for both steels (Table 2). Dislocation density reduced by 32% in the St0.3Cu steel (Table 2). The grain/lath boundaries were found to be dominant nucleation sites for Cr-enriched $M_{23}C_6$ carbides. The mean size of $M_{23}C_6$ carbides was about 70 nm for both steels, whereas the volume fraction estimated using Thermo-Calc software was by 50% higher in the St0.3Cu steel. Note that higher volume fraction of $M_{23}C_6$ carbides provides the smaller lath width in a case of the same mean size of $M_{23}C_6$ carbide (Table 3). Moreover, NbX carbonitrides with a mean size of 40 nm were observed in both steels. These particles were randomly distributed in ferritic matrix. The volume fraction of these particles was the same in both steels. The formation of "Cu"-clusters with mean size of (3±1) nm was observed in the St0.8Cu steel.

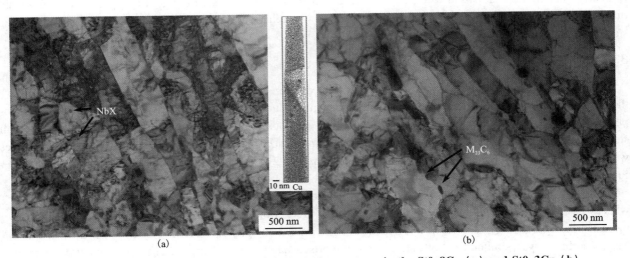

Figure 1 The formation of tempered martensite lath structures in the St0.8Cu (a) and St0.3Cu (b)

Table 2 The structural parameters of the steels studied in the initial states

Ingot	Structure			Particles			
	PAG size /μm	Lath width /nm	Dislocation density /(10^{14} m^{-2})	$M_{23}C_6$		NbX	
				Size/nm	Fraction/%	Size/nm	Fraction/%
St0.8Cu	62±5	370±30	2.1±0.1	72±7	1.56	36±4	0.054
St0.3Cu	65±5	322±30	1.4±0.1	69±7	2.35	29±3	0.056

3.2 Creep behavior

The comparison of creep behavior of the steels studied after creep test at a temperature of 923 K under an applied stress of 140 MPa is represented in Figure 2. An increase in Cu content to 0.8 wt.% positively affected creep resistance. The increment in rupture time was 8 times for the St0.8Cu due to the increase in duration of primary creep stage. Moreover, the minimum creep rate reduced by two orders of magnitude [Figure 2(a)].

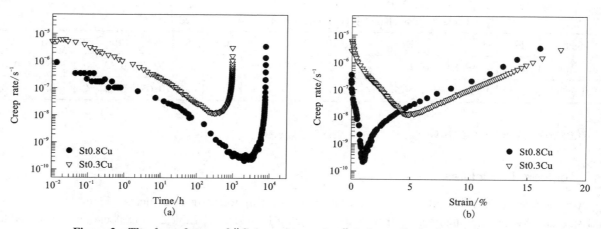

Figure 2 The dependences of "Creep rate vs. time" (a) and "Creep rate vs. strain" (b) obtained after creep tests at 923 K under an applied stress of 140 MPa

During the first 200 h of creep, the behavior of both steels was similar, but at 200 h of creep, the minimum creep rate was attained for the St0.3Cu. On the other hand, for the St0.8Cu after 200 h of creep, the rate of exhaustion of primary creep reduced; the time, at which minimum creep rate was attained, was about 2000 h [Figure 2(a)]. The rate of acceleration of tertiary creep was the same for both steels [Figure 2(b)]. The strain to minimal creep strain comprised 1% and 5% for the St0.8Cu and St0.3Cu steels, respectively, whereas the strain to rupture exceeded 15% for both steels [Figure 2(b)]. So, the significant difference in creep behavior between the steels studied was related to structural evolution during primary creep stage.

3.3 Structure after creep test

Tempered martensite lath structure strongly evolved after creep test in both steels (Figure 3, Table 3): the lath width remarkably increased, dislocation density reduced, secondary phase particles coarsen. Laves phase particles (Fe_2W) located along the boundaries of blocks and martensitic laths were observed in both steels (Figure 3, Table 3). Cu-rich particles with a mean size of 181 nm located along the lath boundaries were found in the St0.8Cu, only. To estimate the constant growth rate of grain boundary particles, Lifshitz-Slyozov-Wagner theory was used[9, 10]:

$$d^n - d_0^n = K_{growth} t$$

where: d_0 is initial particle size; d is particle size for creep time t; K is constant growth rate, coefficient n indicates the coarsening mechanism and equal to 4 for grain boundary diffusion. The constant growth rate for $M_{23}C_6$ carbides was similar for both steels and comprised about 2×10^{-36} m^4/s. The constant growth rate of Laves phase and Cu-rich particles in the St0.8Cu steel was 3.5×10^{-35} m^4/s that was three times higher than growth of Laves phase in the St0.3Cu steel. It is possible that significant coarsening of Laves phase and Cu-rich particles in the St0.8Cu steel occurred during tertiary creep stage[6].

The creep behavior during the primary creep stage is considered to be the nucleation and precipitation of Laves phase[5,6]. The "Cu"-clusters act as the additional nucleation sites for Laves phase that provides to the change in the dispersion of this phase decreasing their mean size and increasing particle number density along the martensitic laths[8]. $M_{23}C_6$ carbides and Laves phase located along the martensitic laths are found to be main sources of back-stress in the steels with low N content[11]. The change in the precipitation of Laves phase could lead to an increase in duration of primary creep stage and reduction of minimum creep rate.

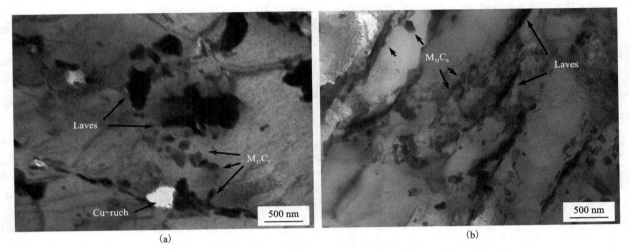

Figure 3 Structures after creep tests in the St0.8Cu (a) and St0.3Cu (b)

Table 3 The structural parameters of the steels studied after creep tests

Ingot	Structure		Particles					
	Lath width /nm	Dislocation density /(10^{14} m^{-2})	$M_{23}C_6$		Laves		NbX	
			Size/nm	Fraction/%	Size/nm	Fraction/%	Size/nm	Fraction/%
St0.8Cu	900±30	1.8±0.1	93±7	1.54	177±7	1.65	45±4	0.054
St0.3Cu	600±30	1.1±0.1	79±7	2.30	78±7	1.65	40±3	0.056

4 Conclusion

The creep behavior and structural changes in two Re-containing 10% Cr steels with the different Cu content after creep test at a temperature of 923 K under an applied stress of 140 MPa was investigated. After tempering, tempered martensite lath structure was observed in both steels. The significant difference between the steels studied in the tempered state comprised the presence of "Cu"-clusters with mean size of 3 nm in the steel with high Cu content. High Cu content in the 10% Cr steels positively affects creep behavior increasing rupture time by 8 times and decreasing minimum creep rate by 2 orders of magnitude. Both steels demonstrate the significant structural degradation after creep test: the martensitic laths widen, dislocation density reduces, grain-boundary particles coarsen. After creep test, the constant coarsening rate of Laves phase in the St0.8Cu steel was three times higher than that in the St0.3Cu steel that could occur during tertiary creep stage. The change in the precipitation of Laves phase during the primary creep via an additional nucleation of this phase on "Cu" clusters could lead to an increase in duration of primary creep stage and reduction of minimum creep rate.

Acknowledgments: This work was supported by Russian Science Foundation, under grant No. 19-73-10089-П. https://rscf.ru/en/project/19-73-10089/.

References

[1] ABE F, KERN T U, VISWANATHAN R. Creep-resistant steels[M]. Cambridge: Woodhead Publishing, 2008, 678.
[2] BHADESHIA H K D H. Design of ferritic creep-resistant steels[J]. ISIJInternational, 2001, 41(6): 626-640.
[3] ABE F. Precipitate design for creep strengthening of 9% Cr tempered martensitic steelfor ultra-supercritical power plants[J]. Science and Technology of Advanced Materials, 2008, 9(1): 013002.
[4] NIKITIN I, FEDOSEEVA A, KAIBYSHEV R. Strengthening mechanisms of creep-resistant 12%Cr-3%Co steel with low N and high B contents[J]. Journal of Materials Science, 2020, 55(17): 7530-7545.
[5] ABE F. Effect of fine precipitation and subsequent coarsening ofFe_2W laves phase on the creep deformation behavior of tempered martensitic 9Cr-W steels[J]. Metallurgical and Materials Transactions A, 2005, 36(2): 321-332.
[6] PRAT O, GARCIA J, ROJAS D, et al. The role of Laves phase on microstructure evolution and creep strength of novel 9%Cr heat resistant steels[J]. Intermetallics, 2013, 32: 362-372.
[7] FEDOSEEVA A, NIKITIN I, DUDOVA N, et al. Superior creep resistance of a high-Cr steel with Re additives[J]. Materials Letters, 2020, 262: 127183.
[8] KLUEH R L. Elevated temperature ferritic and martensitic steels and their application to future nuclear reactors[J]. International Materials Reviews, 2005, 50(5): 287-310.
[9] CARL W. Theorie der alterung von niederschlägen durch umlösen (ostwald-reifung)[J]. Zeitschrift Für Elektrochemie, Berichte Der Bunsengesellschaft Für Physikalische Chemie, 1961, 65(7/8): 581-591.
[10] LIFSHITZ I M, SLYOZOV V V. The kinetics of precipitation from supersaturated solid solutions[J]. Journal of Physics and Chemistry of Solids, 1961, 19(1/2): 35-50.
[11] FEDOSEEVA A, TKACHEV E, KAIBYSHEV R. Advanced heat-resistant martensitic steels: long-term creep deformation and fracture mechanisms[J]. Materials Science and Engineering: A, 2023, 862: 144438.

Achieving High Strength in Cu/graphene Composite Produced by High Pressure Torsion

Galia KORZNIKOVA, Aynur ALETDINOV, Gulnara KHALIKOVA, Elena KORZNIKOVA

Institute for Metals Superplasticity Problems, RAS, Ufa, 450001, Russia

Abstract: Cu/graphene metal matrix composite is promising for a wide range of applications. However, the important problems in the synthesis of the Cu/graphene composite exist, associated with the poor dispersion of the graphene in the matrix and weak interfacial bonding. High pressure torsion processing supplies a possibility to obtain bulk samples with a nanocrystalline structure, without pores and contamination from dissimilar materials. This processing was successfully used for fabrication of the Cu/graphene composite out of thin copper foils coated with a monolayer of graphene. Microstructural characterization of the processed disks demonstrated microstructure with an equiaxed grain size of about 300 nm. The process significantly increased microhardness of Cu/graphene composite. The tensile tests showed the value of the tensile strength reached 670 MPa at room temperature,

Keywords: Cu/graphene composite; High pressure torsion; Microhardness; Tensile strength

1 Introduction

The problem of expanding the range of materials used with high energy efficiency (for example, metal matrix composites with hardening inclusions) and demand for multifunctional composites is growing in modern technologies[1]. This is due to the fact that the progress in modern electronics leads to the miniaturization of high-performance electronic devices to improve accessibility and convenience user. These smart devices need efficient heat dissipation to avoid untimely failures, which is difficult due to the limitations of the compact design. To solve these problems, metal-matrix composites with carbon fillers have been actively studied. Among them, graphene is of the greatest interest due to its exceptional mechanical, electrical and thermal properties[2, 3]. Many promising results have been obtained on thin metal films and powders coated with a layer of graphene. Attempts to scale up and obtain bulk metal-matrix composites with graphene filler suitable for practical use have shown that the mechanical and thermal properties of the consolidated samples, in general, are lower than expected[4]. The reasons for this discrepancy are primarily methods of obtaining the original components, methods of consolidation associated as a rule, not only with high pressure, but also with heating to high temperatures, as well as graphene agglomeration, the interfacial interaction of graphene-metal during processing, anisotropy of the properties of graphene itself in the longitudinal and transverse directions. It was found that the average size and the alignment of graphene flakes are more important parameters defining the heat conduction than the mass density of the graphene laminate[5]. The solution to these problems, at least for individual metal matrix composites of specific compositions, will allow significant progress in the development of materials with high electrical and thermal conductivity.

One of the methods for manufacturing bulk metal-matrix composites is severe plastic deformation by high pressure torsion (HPT)[6]. This method allows without destruction and without change workpiece shapes to deform at room temperature to a significant total degree of deformation. HPT has been successfully used not only for nanostructuring of a wide range of metals and alloys, but also for obtaining heterogeneous and gradient structures, creating "in situ" metal matrix composites from initial plates of pure metals[7], consolidation of amorphous ribbons and powders with preservation amorphous structure and density close to 100%. The advantage of HPT is the absence of introduced contamination and the possibility of varying the deformation conditions (pressure, deformation rate, number of revolutions). This method makes it possible to purposefully change the microstructure of materials, thereby achieving of them a significant increase in the complex of properties.

The aim of the present paper assumes the development of HPT processing for fabrication of Cu-matrix composite reinforced with graphene by compaction of thin copper foils coated with a monolayer of graphene. This study includes not only microstructure examination but also microhardness studies and tensile tests. For the comparison, consolidated samples of pure copper foils with no graphene addition were investigated as well.

2 Experimental methods

The constituent layers for the future HPT processing were cut from copper foils coated on both sides with monolayer graphene produced by Rusgraphene-R & D company[8]. The thickness of the foils was 25 μm and the grain size about 20 μm. Round-shaped plates with a diameter of 12 mm were cut by spark erosion from the foils. Finally, 18 round plates were stacked and placed between the flat anvils, pressed to 5 GPa and processed at room temperature with 5 turns, with a rotation speed of 2 rotation per minute. High quality disc shaped samples after HPT processing have a thickness about 0.1 mm, do not contain cracks, pores and other defects. Stacks of copper foils without graphene coating were processed at the same conditions.

Microstructural analysis was conducted using a Tescan Mira 3LMH scanning electron microscope with backscattered electron detectors at 20 kV accelerating voltage. EBSD analysis was carried out on the same scanning electron microscope using the OXFORD HKL Channel 5 microanalysis system. Due to the limited angular accuracy of EBSD, boundaries with misorientations below 2° were excluded from consideration. A 15° threshold was applied to differentiate low-angle boundaries (LAB) and high-angle grain boundaries (HAB). LAB and HAB on EBSD maps are depicted as red and black lines, respectively. The microstructure was examined at a site 3 mm away from the center of specimens.

The microhardness was measured by the Vickers method, using the instrument AFFRI DM8A "Micromet 5101" with a diamond indenter (a pyramid with an apex angle of 136°), under a load of 1 N (100 g) and with the load holding time of 10 s. Measurements were taken along the radius, the distance between rows was 300 μm, with 4 measurements in each row.

Tensile tests were carried out using specially elaborated locks mounted in an CMT-5 (Liangong Testing Technology Co) tensile testing machine[7]. The gauge length and width were 3 and 1 mm respectively and corresponded to a portion of the disk sample located approximately 3 mm from its edge. The tests were carried out at room temperature at a tensile speed of 1 mm/min until fracture of the sample.

3 Results and discussion

3.1 Microstructure examination

EBSD measurements from HPT processed Cu/graphene and pure Cu samples provided insight into microstructure. Figure 1 presents typical examples of EBSD maps, grain-size distribution, misorientation-angle distribution of Cu/graphene and pure Cu specimens. The microstructures appeared to be nearly equiaxed with a mean grain size about 0.3 μm in both Cu/graphene and pure Cu samples. The distribution of misorientation angles differs from random which is indicated by solid black line on Figure 1(e), (f). The fraction of HAB is rather high, which is associated with dynamic recrystallization during HPT and attributed to the intensive strain imposed by HPT[9]. It should be kept in mind that pure copper can recrystallize even at a temperature of 200 ℃.

3.2 Microhardness distribution

Vickers microhardness measurements for the samples after constrained HPT were carried out on the surface of disc shaped samples along their radius (Figure 2). The distribution of the microhardness was found to increase gradually from the center to the edge in both samples. The results of microhardness measurements are consistent with the well-known fact that shear strain in the deformation by the HPT increases along the radius and achieves minimum value in the center and maximum at the edge of the sample.

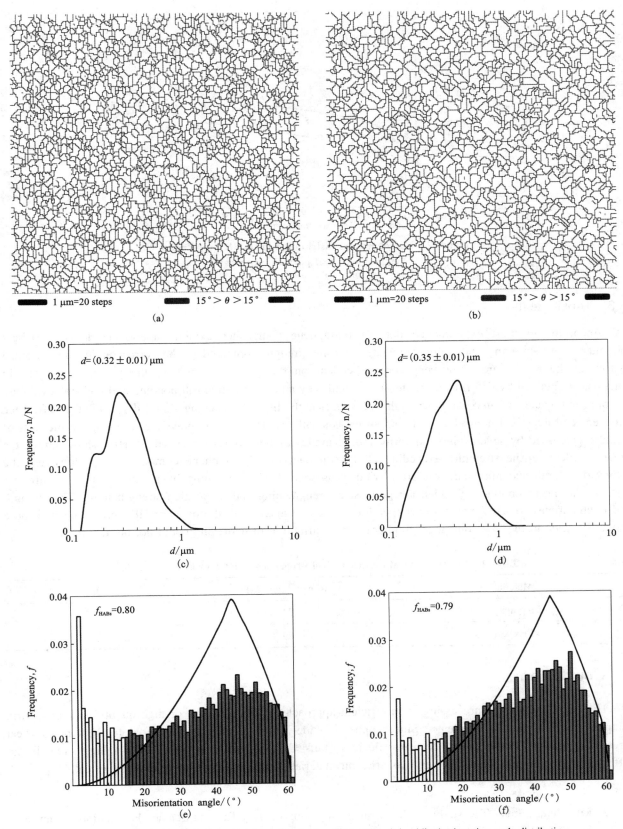

(a), (b) EBSD grain-boundary maps; (c), (d) grain-size distribution; (e), (f) misorientation-angle distribution.

Figure 1 Typical microstructure of HPT processed (a), (c), (e) Cu/graphene and pure Cu (b), (d), (f) samples

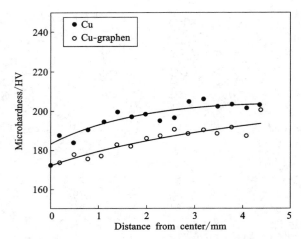

Figure 2 Microhardness distribution in the HPT processed Cu/graphene and pure Cu samples

3.3 Tensile tests

According to most experimental data, Cu/graphene composites exhibit superior hardness, Young's modulus, yield strength, and room temperature tensile strength compared to the corresponding unreinforced matrices. The results are, however, very dependent on the graphene content, the processing route. In composites obtained by deformation route, in particular by an accumulative roll bonding and high-ratio rolling, the obtained values of maximum strength are significantly higher, reaching 485 MPa, and the difference between reinforced and non-reinforced composites is small[10]. The table 1 shows the results of tensile tests of samples processed by severe plastic deformation. Samples of both types processed by HPT show almost the same excellent tensile strength, exceeding all known data for Cu/graphene composites[10]. However, the achieved strength and microhardness are obviously associated with hardening by a superposition of different factors, the most important of which are grain size strengthening and dislocation strengthening. The strength and microhardness of HPT processed copper foils generally agree with the data for HPT processed bulk pure copper[9]. Due to the low content of graphene, its contribution to hardening could not be identified.

Table 1 Room temperature mechanical properties of HPT processed samples

Material	Tensile strength/MPa	Ductility/%
Cu/graphene	674	13.7
Pure Cu	689	12.1

4 Conclusion

High pressure torsion processing is an efficient approach for the synthesis of high-quality bulk Cu-matrix composite reinforced with graphene. Since copper is widely used as structural material for heat pipes, heat spreader and heat sink for smart electronic devices, bulk Cu/graphene composites with a high strength and sufficient plasticity can potentially serve as efficient materials for high power electronic equipment.

Acknowledgments: The study has been financially supported by Russian Science Foundation (Grant No. 23-29-00863). HPT was performed using the shared services center of IMSP RAS and partially supported by State Task of IMSP RAS.

References

[1] LU K. The future of metals[J]. Science, 2010(328): 319-320. https://doi: 10.1126/science.118586.

[2] ALI S, AHMAD F, YUSOFF P S M M, et al. A review of graphene reinforced Cu matrix composites for thermal management of smart electronics[J]. Composites Part A: Applied Science and Manufacturing, 2021, 144: 106357.

[3] LI J, ZHANG P, HE H, et al. Enhanced the thermal conductivity of flexible copper foil by introducing graphene[J]. Materials& Design, 2020, 187: 108373.

[4] WEJRZANOWSKI T, GRYBCZUK M, CHMIELEWSKI M, et al. Thermal conductivity of metal-graphene composites[J]. Materials& Design, 2016, 99: 163-173.

[5] MALEKPOUR H, CHANG K H, CHEN J C, et al. Thermal conductivity of graphene laminate[J]. NanoLetters, 2014, 14(9): 5155-5161.

[6] ZHILYAEV AP, LANGDON T G. Using high-pressure torsion for metal processing: Fundamentals and applications [J]. Progress in Materials Science, 2008, 53(6): 893-979.

[7] KORZNIKOVA G, KABIROV R, NAZAROV K, et al. Influence of constrained high-pressure torsion on microstructure and mechanical properties of an aluminum-based metal matrix composite[J]. JOM, 2020, 72(8): 2898-2911.

[8] KONDRASHOV I, KOMLENOK M, PIVOVAROV P, et al. Preparation of copper surface for the synthesis of single-layer graphene[J]. Nanomaterials, 2021, 11(5): 1071.

[9] LUGO N, LLORCA N, CABRERA J M, et al. Microstructures and mechanical properties of pure copper deformed severely by equal-channel angular pressing and high pressuretorsion[J]. Materials Science and Engineering: A, 2008, 477(1/2): 366-371.

[10] HIDALGO-MANRIQUE P, LEI XZ, XU R Y, et al. Copper/graphene composites: a review[J]. Journal of Materials Science, 2019, 54(19): 12236-12289.

Tissue-specific Decellularized Scaffolds for Pancreas Tissue Engineering

Anna PONOMAREVA[1], Natalia BARANOVA[1], Lyudmila KIRSANOVA[1], Alexandra KIRILLOVA[1], Evgeniy NEMETS[1], Dmitriy KRUGLOV[1], Yulia BASOK[1], Victor SEVASTIANOV[1,2]

1 Shumakov National Medical Research Center of Transplantology and Artificial Organs, 123182, Moscow, Russia
2 The Institute of Biomedical Research and Technology, 123557, Moscow, Russia

Abstract: One of the actual tasks of tissue engineering is to obtain a decellularized scaffold with preserved architectonics and composition features of the native extracellular matrix (ECM) components, which allows simulating the conditions for the prolonged viability of functionally active insulin-producing cells when creating a pancreas tissue equivalent. The aim of the study is the development of a decellularization protocol for fragments of a porcine pancreas for obtaining tissue-specific scaffold. A tissue-specific scaffold was obtained as a result of the physicochemical decellularization method of porcine pancreas fragments (DP scaffold). Morphological analysis and scanning electron microscopy of the DP scaffold were performed, nuclei were stained with fluorescent dye DAPI, the amount of DNA in native and decellularized porcine pancreas was determined, and a cytotoxic study of the resulting scaffolds was performed. Morphological and biological studies have shown that the physicochemical decellularization method of a porcine pancreas using hypotonic and hypertonic solutions (osmotic shock method) is effective to obtain a tissue-specific scaffold. The proposed method for decellularization of porcine pancreatic tissue can be considered for creation of pancreatic tissue engineering as it makes it possible to obtain a tissue-specific fine-fibrous scaffold which is well-purified from cellular components, non-cytotoxic, and non-immunogenic relative to the residual amount of DNA.

Keywords: Pancreatic tissue engineering; Porcine pancreas; Decellularization; Tissue-specific scaffold

1 Introduction

Type I diabetes mellitus is an autoimmune disease with a critical loss of insulin-producing β-cells, for the treatment of which the use of tissue engineering cell technologies seems promising. The development and creation of pancreatic tissue engineering is hindered by problems related to the maintenance of viability of functionally active isolated pancreatic islets. During isolation, the islets are affected by a number of damaging factors such as ischemia, oxidative stress, or possible enzyme cytotoxic action[1,2]. Current research is aimed at creating of pancreas tissue equivalent, consisting of insulin-producing cells (β-cells, differentiated stem cells or isolated Langerhans islets) and a scaffold that provides longer survival and efficient function of transplanted cells. When creating a pancreas equivalent, materials of various nature are used to obtain scaffolds, which have certain physical, mechanical, biological, and functional properties, such as biocompatibility, lack of immunogenicity, mechanical strength and elasticity, biodegradability, etc.[3] Pancreas extracellular matrix (ECM) contains type I, III, IV, V, and VI collagens that provide structural rigidity and tissue adhesion, support for integrity and shape, organ structure; laminin, fibronectin, fibrillin are involved in cytoskeletal remodeling, contractility, and differentiated cell adhesion; elastin provides strength, elasticity and extensibility to the tissue[2,4]. The pancreatic tissue ECM regulates major aspects of islet biology, including development, morphology, and cell differentiation, intracellular signaling, gene expression, adhesion, and migration, proliferation, secretory function, and survival[5]. The development of scaffolds, in optimal cases, acting as a carrier of islets or other insulin-producing cells and mimicking the properties of native ECM, is a key problem in creating tissue engineering of pancreas. Scaffolds with ECM components that prevent cellular stress and preserve the viability and function of β-cells are universal platforms that provide structural and mechanical support to islets, serve as a reservoir of growth factors, cytokines, antioxidants, and transmit signals to islet cells through integrins[5]. Such scaffolds mimic the native microenvironment for islet cells *in vitro* and *in vivo*[2].

Lately, as a prospective approach, the development of tissue equivalent based on tissue-specific scaffolds made from decellularized tissue with their subsequent recellularization has begun[6]. The development of effective protocols of pancreatic decellularization is directed at the utmost possible retention of structural, biochemical and mechanical properties of a native ECM with the maximally complete removal of cell material, including DNA and native tissue cell surface antigens[7]. The presence of main ECM components in decellularized pancreatic scaffolds, such as structure proteins (Type Ⅰ, Ⅲ, Ⅳ, Ⅴ and Ⅵ collagen, elastin, fibronectin and laminin), glycoproteins, and cell adhesion factors, allows to create conditions for a prolonged functional activity of islet (insulin-producing) cells and maximally imitate ECM properties[3,7]. The most complete removal of cellular material from tissue during decellularization leads to a minimized immune response during the further implantation of pancreas tissue equivalent[3].

Due to the acute shortage of donor organs, the possibility of obtaining a tissue-specific scaffold from the porcine pancreas is being considered as it is an accessible material that has been successfully used in clinical practice[6].

The aim of the study is the development of a decellularization protocol for fragments of a porcine pancreas with the subsequent study of morphological features and biological properties of the obtained tissue-specific scaffold.

2 Materials and methods

2.1 Pancreas decellularization

A porcine pancreas was obtained at a slaughterhouse (OOO APK PROMAGRO, Russia) and stored at -80 ℃ until the decellularization. The defrosted porcine pancreas was ground to a size of fragments no more than 1 mm×1 mm×2 mm. The ground tissue was treated at room temperature under continuous stirring on a roller system at a speed of 0.5 r/min sequentially in hypotonic solution (0.1% sodium dodecyl sulfate (SDS) solution in distilled water for 3 h), hypertonic solution (0.1% SDS solution in 1N NaCl for 3 h) and in 0.1% SDS solution in phosphate-buffered saline (PBS, pH = 7.35) for 18 h. An important step in the decellularization process is the washing of decellularized pancreatic tissue fragments from the remnants of surface-active agents in a PBS containing an antibiotic/antimycotic for 72 h. The porcine decellularized pancreas scaffold (DP scaffold) was brought into cryovials, then frozen and sterilized (γ-sterilization, 1.5 Mrad). Sterile DP scaffold was stored at $-20°$.

2.2 Scanning electron microscopy (SEM)

The microstructure of the DP scaffold samples was investigated with a JSM-6360LA scanning electron microscope (Jeol, Japan) at an accelerating voltage of 25 kV. The samples were fixed for 60 min in a 2.5% glutaraldehyde solution and dehydrated in ethyl alcohol solutions of increasing concentrations (from 50% to 100%), followed by air drying. The conductive coating was obtained by ion gold sputtering for 8-10 min at a constant current of 5-7 mA on a JFC-1100 setup (Jeol, Japan).

2.3 Histology staining

Samples of native porcine pancreas and the obtained DP scaffold samples were subjected to morphological examination using routine histological (hematoxylin and eosin staining, Masson's staining, and DAPI fluorescent staining). Additionally, immunohistochemical staining of DP scaffold sections for insulin and glucagon was performed. Analysis and photography of the obtained histological preparations were carried out using a Nikon Eclipse microscope equipped with a digital camera.

2.4 DNA quantification

To determine the immunogenicity of the decellularized material by the residual amount of nuclear material in the DP scaffold, DNA isolation and fluorescent staining were performed. DNA isolation from the DP scaffold samples was performed using the DNeasy Blood & Tissue Kit (QIAGEN, Germany) according to the manufacturer's instructions. The quantitative DNA determination with a fluorescent dye Picogreen Quant-iT™ (Invitrogen, USA) was used according to the protocol.

2.5 Cytotoxicity

The cytotoxicity of the DP scaffold samples *in vitro* was evaluated by direct contact according to the ISO 10993-5: 1999 international standard on the mouse embryonic fibroblast line NIH/3T3. The 10% fetal calf serum culture medium (HyClone, SV30160.03, USA) served as the negative control. The positive control sample was a single-element aqueous standard of 10000 μg/mL zinc (Sigma-Aldrich, USA). Culture monitoring was performed with the Eclipse TS100 (Nikon, Japan) inverted microscope. The metabolic activity of fibroblasts after contact with scaffold samples was determined after 24 h with the presto Blue™ Cell Viability Reagent (Thermo Fisher Scientific, USA) according to the protocol recommended by the manufacturer. The changes in media absorption were recorded using the Spark 10M microplate reader (TecanTrading AG, Switzerland) with Spark Control™ Magellan V1.2.20 software at a wavelength of 570 nm and 600 nm. The data quantitative and statistical processing was performed with SPSS26.0. All results are presented as a mean±standard deviation. The differences were considered significant at $p<0.05$.

3 Results

3.1 Morphological analysis and SEM analysis

The samples of the native porcine pancreas and decellularized pancreatic tissue were evaluated by histological analysis. In the pancreas, a pronounced lobulation of the pancreatic tissue was observed [Figure (a), (b)].

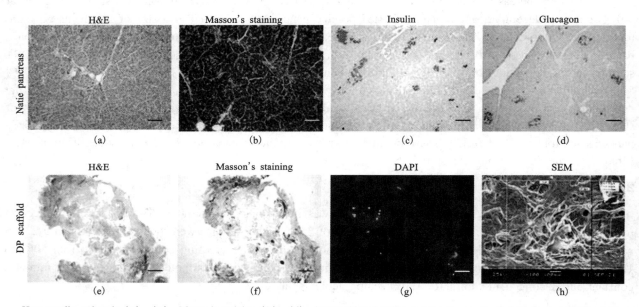

Hematoxylin and eosin (a), (e); Masson's staining (b), (f); immunohistochemical staining for insulin (c); immunohistochemical staining for glucagon (d); fluorescent staining with DAPI (g); electron micrograph of a tissue-specific scaffold from a decellularized porcine pancreas (h). Bar 100 μm.

Figure 1 Porcine pancreas morphology before and after decellularization

A few irregularly shaped Langerhans islets were well identified by immunohistochemical staining [Figure 1(c), (d)]. After decellularization of the porcine pancreatic tissue, the obtained tissue-specific scaffold samples were characterized by a completely fibrous cellular structure with pronounced porosity [Figure 1(e), (f)]. Preserved cells and cell nuclei were not found in the scaffold samples [Figure 1(g)]. According to the results of scanning electron microscopy, a tissue-specific scaffold sample was characterized by a porous structure of various sizes (pore sizes from 37.9 to 102.0 μm). In some areas, granularity was determined, which is associated with the residual presence of grains of cellular detritus. Preserved cells and large cell fragments were not detected in the DP scaffold [Figure 1(h)].

3.2 DNA Content in the DP Scaffold

DNA content in the samples of native porcine pancreatic tissue and the DP scaffold is presented in Figure 2. In the pancreatic tissue samples, DNA amounted to (1159±111) ng/mg of tissue, in the samples of the tissue-specific scaffold-(42±4) ng/mg of tissue, no more than 0.1% of DNA was retained and less than 50.0 ng/mg of tissue was retained, which indicates the high effectiveness of the developed decellularization protocol of pancreatic tissue and the low immunogenicity of the obtained scaffold.

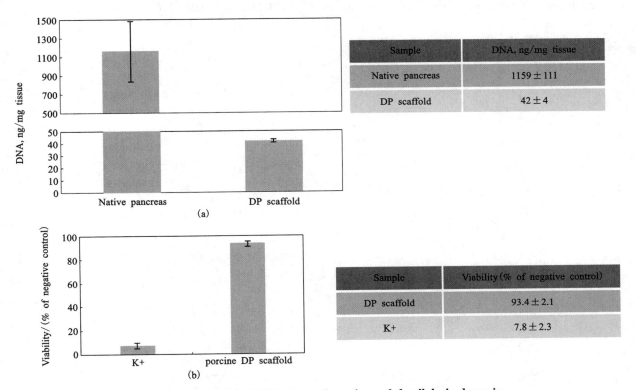

Figure 2 Quantitative DNA content in native and decellularized porcine pancreas (a); cytotoxicity of the tissue-specific DP scaffold (b)

3.3 Cytotoxicity of the DP scaffold

The metabolic activity of fibroblasts after contact with the DP scaffold amounted to (93.4±2.1)% relative to the negative control sample (growth medium), which indicates that the DP scaffolds had no cytotoxic effect [Figure 2(b)].

4 Conclusion

The proposed method for decellularization of porcine pancreatic tissue makes it possible to obtain a fine-fibrous, well-purified from cellular components, non-cytotoxic, and non-immunogenic relative to the residual amount of DNA tissue-specific scaffold, which can be considered for creation of pancreatic tissue engineering.

References

[1] EGUCHI N, KIMIADAMYAR K, ALEXANDER M, et al. Anti-oxidative therapy in islet cell transplantation[J]. Antioxidants, 2022, 11(6): 1038.
[2] AMER L D, MAHONEY M J, BRYANT S J. Tissue engineering approaches to cell-based type 1 diabetes therapy [J]. Tissue Engineering, 2014, 20(5): 455-467.

[3] SALG G A, GIESE N A, SCHENK M, et al. The emerging field of pancreatic tissue engineering: A systematic review and evidence map of scaffold materials and scaffolding techniques for insulin-secreting cells[J]. Journal of Tissue Engineering, 2019, 10: 1-25.

[4] HALPER J, KJAER M. Basic components of connective tissues and extracellular matrix: Elastin, fibrillin, fibulins, fibrinogen, fibronectin, laminin, tenascins and thrombospondins[J]. Adv Exp Med Biol, 2014, 802: 31-47.

[5] RIOPEL M, WANG K. Collagen matrix support of pancreatic islet survival and function[J]. Frontiers in Bioscience, 2014, 19: 77-90.

[6] RANA D, ZREIGAT H, BENKIRANE-JESSEL N, et al. Development of decellularized scaffolds for stem cell-driven tissue engineering[J]. J Tissue Eng Regen Med, 2017, 11(4): 942-965.

[7] SEVASTIANOV V I, PONOMAREVA A S, BARANOVA N V, et al. Decellularization of human pancreatic fragments with pronounced signs of structural changes[J]. Int. J. Mol. Sci, 2023, 24(1): 119.

Formation of Cartilage-like Tissue from Collagen-based Microparticles in Perfusion Bioreactor

Yulia BASOK[1], Alexey GRIGORIEV[1], Ludmila KIRSANOVA[1], Alexandra KIRILLOVA[1], Varvara RYZHIKOVA[1], Anastasia SUBBOT[2], Evgeniy NEMETS[1], Victor SEVASTIANOV[1, 3]

1 Shumakov National Medical Research Center of Transplantology and Artificial Organs, 123182, Moscow, Russia
2 The Research Institute of Eye Diseases, 119021, Moscow, Russia
3 The Institute of Biomedical Research and Technology, 123557, Moscow, Russia

Abstract: The technology of obtaining an injectable form of cartilage-like structures (CLSs) based on collagen-rich microparticles of decellularized porcine cartilage is considered to be promising. The aim of this work was to obtain and comparatively study cartilage-like structures based on microparticles of decellularized cartilage and mesenchymal stromal cells (MSCs) under static conditions and in a perfusion bioreactor. The decellularization process included freeze-thaw cycles ($-196\ °C/37\ °C$), detergents (Triton X-100 and sodium dodecyl sulphate), and DNase treatment. The morphology of the surface and the nearest subsurface layer of samples was examined with SEM. Each CLS consisted of 5×10^5 MSCs on 5 mg DACp. The selected flow mode for the creation of CLS cartilage in the developed perfusion bioreactor 0.5 mL/min allowed to maintain the proliferation and chondrogenic differentiation of MSCs during cultivation on DACp within 2 weeks. It has been found that the cultivation of MSCs on DACp microparticles in bioreactor allows to increase the proliferative activity of cells in comparison with static conditions of 3D cultivation while maintaining the ability to synthesize ECM characteristic of cartilage tissue, histochemical analysis of which revealed the presence of collagen, and GAG.

Keywords: Collagen; Microparticles; Articular cartilage; Decellularization

1 Introduction

One of the formidable problems in an industrial society are joint damage and degeneration, which are associated with the limited ability of cartilage tissue to regenerate.[1]. Mesenchymal stromal cells (MSCs) seem to be a promising cellular material for the restoration of joint defects, since by their nature these cells have the ability to differentiate into various mesenchymal tissues, including cartilage. On the other hand, the therapeutic effect of MSCs is based on the paracrine secretion of a wide range of growth factors and cytokines that stimulate the proliferation of chondrocytes and the synthesis of extracellular matrix (ECM) when injected into the joint[2].

Bioresorbable scaffolds, among which collagen is most often used, are used to deliver and retain MSCs at the site of a cartilage defect, as well as to ensure the vital activity of cells for a time sufficient to start the processes of cartilage tissue repair[3]. The procedure of tissue decellularization ensures the destruction of cells with maximum preservation of ECM (the main component of which is collagen), whose proteins, unlike cells, do not carry antigens that cause a rejection reaction, due to molecular evolutionary conservatism[4]. The technology of obtaining an injectable form of cartilage-like structures (CLSs) based on collagen-rich microparticles of decellularized porcine cartilage is promising. Currently, the clinic is already actively using decellularized skin, bladder, small intestine and heart valves of a porcine, but not cartilage tissue.

To improve the quality of CLSs, perfusion is used, which simulates natural conditions, not only providing nutrition to cells, transport of gases to them and excretion of metabolic products due to constant replacement of the culture medium, but also triggering physiological signals of mechanotransduction[5, 6].

The aim of this work was to obtain and comparatively study cartilage-like structures based on microparticles of decellularized cartilage and MSCs under static conditions and in a perfusion bioreactor.

2 Materials and methods

2.1 Decellularization

Porcine femurs and knee joints were obtained at a slaughterhouse (OOO APK PROMAGRO, Russia). The tissues were decellularized according to the technique published earlier[7]. The high density and the absence of pores in the articular cartilage determined the preference for micronization of native tissue and the need to include physical methods in addition to a detergent (three washes of sodium dodecyl sulfate (0.1% w/v) and Triton X-100 (1% v/v, 2% v/v, 3% v/v for the first, the second, and the third wash, respectively) for 72 h) and DNase treatment in decellularization protocols: 3 freezing/thawing cycles.

2.2 Scanning electron microscopy (SEM)

The morphology of the surface and the nearest subsurface layer of samples was examined with SEM using lanthanoid contrasting solution BioREE (JSC Glaucon, Russia), which allows to observe unfixed biological samples in a low vacuum mode after holding them in a saturated solution of rare earth metal. Thus, the maximum native state of the object under study was preserved, and the image obtained in the mode of detection of backscattered electrons carried expanded information about cellular structures. The observations were made using EVO LS10 (Zeiss, Germany) in a low vacuum mode (EP, 70 Pa), at accelerating voltage of 20 kV.

2.3 Cartilage-like structures formation

The possibility of cartilage-like structures (CLSs) formation in a bioreactor, including MSCs and microparticles of decellularized porcine articular cartilage (DACp), was investigated. To carry out experiments on the cultivation of 5×10^5 MSCs on 5 mg DACp under flow conditions, a modified version of the perfusion bioreactor was used[8]. The flow rate was 0.02 mL/min. The chondrogenic differentiation medium included high-glucose DMEM (Gibco, Billings, MT, USA), 10% ITS+ (Corning Inc, Corning, NY, USA), 1% sodium pyruvate (Sigma-Aldrich Corp, St. Louis, MO, USA), 0.25% ascorbate-2-phosphate (Sigma-Aldrich, St. Louis, MO, USA), 100 nM dexamethasone (Sigma-Aldrich, St. Louis, MO, USA), 0.002% TGF-β1 (PeproTech, Cranbury, NJ, USA), and 1% penicillin-streptomycin-glutamine (Gibco, Billings, MT, USA). On the 14th day, the culture chambers with CLSs were removed from the bioreactor. For comparison, similar CLSs were cultivated under static conditions.

Adipose tissue samples weighing 3-5 g ($n=3$) were obtained with the informed consent of living healthy donors during liver transplantation under general anesthesia. The study was conducted in accordance with the guidelines of the Helsinki Declaration and approved by the Local Ethics Committee at the Shumakov National Medical Research Center of Transplantology and Artificial Organs, Moscow, Russia (15 November 2019, Protocol No. 151119-1/1e). The tissue was incubated in 0.1% collagenase solution type I at 37 ℃ for 20 min. The MSCs were cultured in growth medium (DMEM/F12 (1:1)) with the addition of 10% fetal cattle serum, 100 U/mL penicillin, 100 μg/mL streptomycin sulfate, and 2 mM L-glutamine (all listed reagents-Gibco Inc, Billings, MT, USA). In the experiment 3rd passage cells were used.

2.4 Histological staining

The samples were fixed in formalin and poured into paraffin. The sections were stained with hematoxylin and eosin, alcyan blue, and with Masson's staining. The analysis and photography of the obtained preparations were carried out using a Nikon Eclipse microscope.

3 Results and discussion

3.1 Characteristics of microparticles of decellularized articular cartilage

The average microparticle diameter distribution in the suspension was determined by laser diffraction using SALD-7101 (Shimadzu, Japan). It was shown that the diameter of the obtained microparticles did not exceed

220 μm [Figure 1(a), 1(b)].

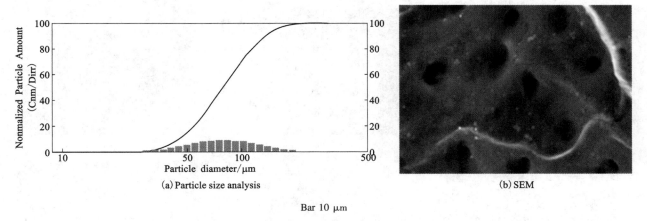

Figure 1 Decellularized porcine articular cartilage microparticles

A significant part of lacunae on the surface of the particles was free of cells [Figure 1(b)]. Quantitative analysis showed that the matrix of decellularized porcine articular cartilage was significantly ($p<0.05$) cleared of DNA: the DNA content decreased from (366.9±53.0) ng/mg of tissue to (9.1±1.1) ng/mg of tissue.

3.2 Cultivation of cartilage-like structures based on decellularized tissue particles in a perfusion bioreactor

On the 14th day of cultivation, the formation of large conglomerates was observed in all samples, which were matrix microparticles connected by cells and their accumulated ECM (Figure 2). The growth of fibroblast-like cells on the DACp surface was visualized in the sample, and the ratio of cells and cell carrier in CLS when cultured in a bioreactor [Figure 2(a), 2(c), 2(e)] was higher than under static conditions [Figure 2(b), 2(d), 2(f)]. Note that the number of cells per unit area of the scaffold calculated using ImageJ (National Institutes of Health, USA) for a sample obtained in a bioreactor was 7 times higher compared to a sample cultured under static conditions [(0.003±0.001) cells/μm^2 and (0.020±0.004) cells/μm^2 for samples obtained under static conditions and perfusion bioreactor, respectively].

ECM formation was observed in all the studied samples against the background of cell proliferation. In the sample obtained in the bioreactor, there were visualized areas in which the ECM fraction significantly exceeded the cell content [Figures 2(a), 2(c), 2(e)]. Blue collagen fibers were determined in ECM [Figures 2(c), 2(d)]. In this case, the ECM was uniformly positively colored with alcyan blue on the glycosaminoglycans [GAG, Figures 2(e), 2(f)]. The results obtained indicate the differentiation of MSCs in the composition of CLS in the chondrogenic direction.

It should be noted that the main experimentally determined difference between CLS obtained during cultivation under dynamic conditions was active cell proliferation while maintaining the ability to synthesize ECM rich in collagen and GAG.

The contradictory studies described above indicate a high value of optimizing the flow regime, which should be selected individually for each bioreactor, since their designs, especially the culture chamber, differ significantly. It should be noted that not only increased cell proliferation while maintaining the ability to synthesize a specific ECM, but also a decrease in the proportion of dead cells compared to cultivation under static conditions, testify in favor of the regime we have chosen.

We emphasize that in all the work described, porous or hydrogel carriers were used as scaffolds when creating CLS. This study showed for the first time the possibility of forming a CLS in a perfusion bioreactor based on DACp microparticles, which indicated the versatility of the culture chamber for the use of various scaffolds. Moreover, the obtained CLS with DACp were characterized by a uniform distribution of cells and their production of a specific ECM containing collagen and GAG.

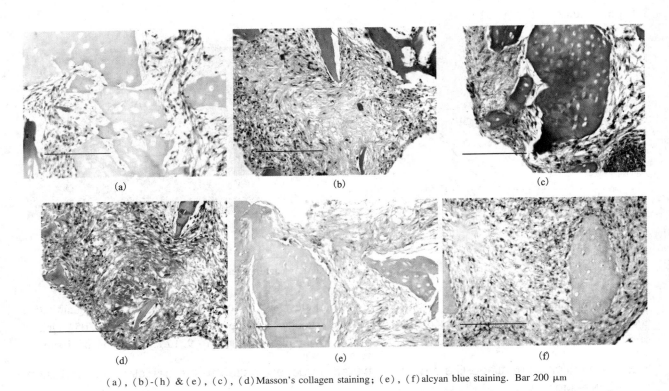

(a), (b)-(h) & (e), (c), (d) Masson's collagen staining; (e), (f) alcyan blue staining. Bar 200 μm

Figure 2 Growth of MSCs when cultured on a tissue-specific matrix from DACp when cultured under static conditions [(a), (c), (e)] in a flowing bioreactor at a flow rate of 0.5 mL/min [(b), (d), (f)] in a differentiating chondrogenic medium for 14 d

4 Conclusion

The selected flow mode for the creation of CLS cartilage in the developed perfusion bioreactor 0.5 mL/min allows to maintain the proliferation and chondrogenic differentiation of MSCs during cultivation on DACp. It has been found that the cultivation of MSCs on DACp microparticles in the selected mode allows to increase the proliferative activity of cells in comparison with static conditions of 3D cultivation while maintaining the ability to synthesize ECM characteristic of cartilage tissue, histochemical analysis of which revealed the presence of collagen and GAG.

Acknowledgments: The research was carried out at the expense of the Russian Science Foundation (Grant No. 21-15-00251), https://rscf.ru/project/21-15-00251/(accessed on 07 july 2023).

References

[1] CARNEIROD C, ARAÚJO L T, SANTOS G C, et al. Clinical trials with mesenchymal stem cell therapies for osteoarthritis: challenges in the regeneration of articular cartilage[J]. International Journal of Molecular Sciences, 2023, 24(12): 9939.
[2] SONG YJ, JORGENSEN C. Mesenchymal stromal cells in osteoarthritis: evidence for structural benefit and cartilage repair[J]. Biomedicines, 2022, 10(6): 1278.
[3] LAM A T L, REUVENY S, OH S KW. Human mesenchymal stem cell therapy for cartilage repair: review on isolation, expansion, and constructs[J]. Stem Cell Research, 2020, 44: 101738.
[4] CRAPO P M, GILBERTT W, BADYLAK S F. An overview of tissue and whole organ decellularization processes [J]. Biomaterials, 2011, 32(12): 3233-3243.
[5] RAVICHANDRAN A, LIU YC, TEOH S H. Review: bioreactor design towards generation of relevant engineered tissues: focus on clinical translation[J]. Journal of Tissue Engineering and Regenerative Medicine, 2018, 12(1): e7-e22.

[6] APRILE P, KELLY D J. Hydrostatic pressure regulates the volume, aggregation and chondrogenic differentiation of bone marrow derived stromal cells[J]. Front Bioeng Biotechnol, 2021, 8: 619914.

[7] SEVASTIANOV V I, BASOK Y B, GRIGORIEV A M, et al. Decellularization of cartilage microparticles: effects of temperature, supercritical carbon dioxide and ultrasound on biochemical, mechanical, and biological properties[J]. Journal of Biomedical Materials Research. Part A, 2023, 111(4): 543-555.

[8] SEVASTIANOV V I, BASOK Y B, GRIGORIEV A M, et al. Application of tissueengineering technology for formation of human articular cartilage in perfusion bioreactor[J]. Russian Journal of Transplantology and Artificial Organs, 2017, 19(3): 81-92.

Development of Nonwoven Hemocompatible Vascular Graft of Small Diameter with Reduced Surgical Porosity

Viktor SEVASTIANOV[1,2], Evgenij NEMETS[1], Vyacheslav BELOV[2], Lyudmila KIRSANOVA[1], Alla NIKOLSKAYA[1], Vyacheslav ZAKHAREVICH[1], Kirill KIRIAKOV[1], Varvara RYZHIKOVA[1], Irina TYUNYAEVA[1], Artem VYPRYSHKO[1], Yuliya BASOK[1]

1 Shumakov Federal Research Center of Transplantology and Artificial Organs, Ministry of Health of the Russian Federation, Moscow, 123182, Russia
2 ANO "Institute of Biomedical Research and Technology", Moscow, 123557, Russia

Abstract: Three-layer frameworks of vascular grafts (VGs) with an inner diameter of 2 mm were obtained by electrospinning. The inner and outer layers were formed from polycaprolactone, and the middle (blocking) layer was made from a mixture of polycaprolactone with gelatin. The obtained VGs demonstrate physical and mechanical characteristics close to those of natural blood vessels. To increase the biocompatibility of VGs, their surface was modified by a biologically active coating of covalently immobilized heparin and human platelet lysate. During intramuscular implantation of VG fragments in rats, it was found that against the background of a mild inflammatory reaction, their rapid vascularization was observed. It has been established that implantation of an VG fragment with low surgical porosity (length 10 mm, inner diameter 3 mm and wall thickness 350 μm) in situ into the infrarenal aorta of a rat ensures the patency of the vascular prosthesis and the absence of parietal thrombi for 16 weeks.

Keywords: Polycaprolactone; Electrospinning; Bioactive coating; Vascular grafts; Implantation

1 Introduction

Since the 1960s, researchers have been actively looking for ways to combine properties such as zero surgical porosity (SP) and high biological porosity that the cells can migrate into the thickness of vascular grafts (VGs). The high permeability of the implant walls contributes to the formation of hematomas, which, when organized, causes fibrosis and a decrease in the lumen of the VGs. Unfortunately, an increase in biological porosity due to an increase in pore size is accompanied by an increase in SP, which increases the risk of bleeding, the duration and severity of which depends on the characteristics of VGs[1].

Studies have shown that there is a correlation ($R^2 > 0.9$) between water permeability and blood loss, while due to the higher viscosity and the presence of uniform elements, blood loss is about 10 times less than water permeability[2], which is explained by a higher viscosity blood. At the same time, water permeability of more than 50 mL/(cm^2·min) is a criterion that determines the need for additional efforts to reduce SP[3].

The very first whole unstabilized blood was used as a natural sealant, and the method was called "preclotting"[4]. In addition, fibrin glue[5] and hydrogels from proteins are used to reduce CP: albumin, collagen, gelatin, fibroin, etc. [1, 6, 7].

A distinctive feature of the electrospinning method is the possibility of obtaining frameworks with a small pore size by varying the parameters of the procedure[8], which ensures low SP. However, biological porosity also decreases at the same time, which is undesirable due to cell migration into the thickness of the matrix. As a solution to the emerging contradiction, it has been proposed to form multilayer scaffolds[9], one of the layers of which provides low SP.

The aim of this work was to develop three-layer small diameter vascular grafts with reduced surgical porosity, the necessary physical and mechanical properties, and high biocompatibility.

2 Materials and methods

2.1 Electrospinning method to obtain small diameter VG

VG samples with an inner diameter of 2 mm were prepared by electrospinning from a 10% solution of polycaprolactone (MM 80000, Sigma-Aldrich, USA) or from solution of polycaprolactone (PCL) with the addition of a 10%-30% (by weight) of gelatin (Sigma-Aldrich, USA) in hexafluoroisopropanol (JSC NPO "PIM-INVEST", Russia), code of PCL and PCL-G samples, respectively, on a NANON-01A electrospinning unit ("MECC CO", Japan). Conditions for the formation of PCL and PCL-G samples: voltage between the electrodes 25 kV, solution supply rate 4 mL/h, distance to the collector 100 mm, rotation speed of the substrate rod 1000 rad/min, needle 18 G. After the end of the solution application process, the obtained samples were dried in a thermostat at a temperature of 37 ℃ for 2 h, followed by evacuation to remove traces of the solvent at a residual pressure of 10-20 millimeters of mercuri and a temperature of 37 ℃ for 24 h.

2.2 Applying a bioactive coating

To form a bioactive coating, VG samples were sequentially incubated in aqueous solutions of bovine serum albumin (1.0 mg/mL), heparin (1.0 mg/mL), glutaraldehyde (1.0%), and platelet lysate. The obtained VGs were dried and subjected to sterilization by gamma radiation at a dose of 1.5 MPa.

2.3 Study of the mechanical properties of VG

Mechanical testing of the samples was carried out on a Shimadzu EZ Test EZ-SX testing (tensile) machine (Shimadzu Corporation, Japan) with TrapeziumX software, version 1.2.6, at a tensile rate of 5 mm/min.

2.4 Determination of the surgical porosity of the VG

SP was determined by measuring the amount of water flowing in one minute through one square centimeter of the phosphate-salt buffer surface at a pressure of 120 mm of mercury.

2.5 Study of the inflammatory response and resorption of the VG

Sterile VGs were implanted into the thigh muscle tissue of adult Wistar rats (male). Implantation was performed under sterile conditions under the action of Zoletil 100 anesthetic (Virbak, France, dose 15 mg/kg). An incision was made in the skin and fascia of the right thigh from the outside by 1 cm, the muscle fibers were moved apart, and experimental samples were placed deep into the wound, after which the wound was sutured with single sutures and treated with an antiseptic. For histological examination, the implanted materials were biopted, together with the surrounding tissues, were fixed in a 10% formalin solution. Sections 4-5 μm thick, obtained using a Leica RM3255 microtome, were stained with hematoxylin and eosin to identify connective tissue elements and Masson's method for protein visualization.

2.6 VG implantation in small laboratory animals

The prosthesis of the infrarenal part of the abdominal aorta of the rat was performed with a 10 mm long VG with an inner diameter of 2 mm. After anesthesia (Xylazine hydrochloride 20 mg/mL + Ketamine 50 mg/mL), access to the abdominal aorta was performed through a median laparotomy, resection of the infrarenal aorta was performed, and proximal and then distal anastomoses with a vascular prosthesis were formed using an atraumatic 8-0 polypropylene suture 6-8 interrupted sutures. At the final stage, layer-by-layer suturing of the surgical wound was performed, followed by treatment of the sutures with an antiseptic agent and an anesthetic solution.

ACL patency was monitored using ultrasound on a Vivid E9 machine (General Electric, USA) with Window Blinds NMM OCX software (Stardock, USA). Zoletil (Virbac Sante Animale, France) at a dose of 0.4 mL/kg was used for sedation of the test animals before Doppler sonography.

2.7 Statistical processing

The obtained data were quantitatively and statistically processed using the Microsoft Excel 2019 application. All results were presented as the mean value ± standard deviation. Differences were considered significant at $p<0.05$ with the number of samples (n) from 3 to 5.

3 Results and discussion

Previously, VGs with an internal diameter of 3 mm, with a minimum CP of $(30.4±1.5)$ mL/(cm^2·min) were obtained[10]. An attempt to apply the parameters that are optimal for the formation of an VG with a diameter of 3 mm to prostheses with a diameter of 2 mm did not lead to success, since the SP in this case was $(47.8±6.6)$ mL/(cm^2·min), which made such an VG inapplicable for implantation without additional sealing. As an internal water-blocking layer, we proposed to use a mixture of PCL with gelatin (PCL-G). It was found that the addition of gelatin at a concentration of 10% (by weight of the polymer) is accompanied by a five-fold decrease in SP to $(9.6±2.5)$ mL/(cm^2·min), and an increase in gelatin concentration to 30% is accompanied by a further decrease in SP to $(2.6±0.3)$ mL/(cm^2·min). Despite the lower values of SP in the case of the introduction of 30% gelatin, the variant with a 10% additive was chosen as optimal, since samples of three-layer prostheses formed on the basis of a sealing layer with a gelatin concentration of 30% were prone to delamination upon contact with an aqueous medium. The SP of three-layer VG with 10% gelatin addition was $(8.4±0.8)$ mL/(cm^2·min), which makes it possible to use VG without additional sealing.

Table 1 summarizes the physical and mechanical characteristics of single-layer and three-layer VG. The addition of a sealing layer of PCL-G is accompanied by only minor changes in the physical and mechanical characteristics, however, both VGs are significantly stronger than natural blood vessels of the same diameter.

Table 1　Mechanical characteristics

	Young's modulus/MPa	Tensile strength/N	Maximum elongation/%
VG from PCL	4.8±0.8	13.4±2.3	448±28
Three-layer VG	5.4±0.9	18.6±3.6	344±30
Rat aorta	8.5±2.2	2.0±0.3	93±16
Sheep artery	1.8±0.3	11.1±1.4	107±23

Figure 1 illustrates the VG surface structure. The fiber thickness is sufficient to maintain cell adhesion, and the pore size allows predicting cell migration into the scaffold volume. In addition, as we have shown earlier, the application of a bioactive coating based on heparin and platelet lysate enhances cell adhesion and proliferation in vitro[11].

Biodegradation and tissue response were studied in a model of VG implantation in the rat thigh muscle. The histological picture (Figure 2) was characterized by the presence of a thin capsule around the perimeter of the implanted sample.

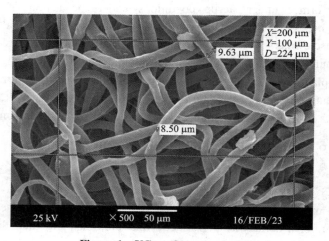

Figure 1　VG surface structure

Pilot experiments on the implantation of a VG fragment into the infrarenal aorta of a rat showed that the prosthesis was passable for up to 16 weeks. The prosthesis lumen diameter was 3 mm, peak systolic velocity 0.61 m/s, maximum end diastolic velocity 0.34 m/s, and the resistivity index calculated from the data obtained $RI=0.44$.

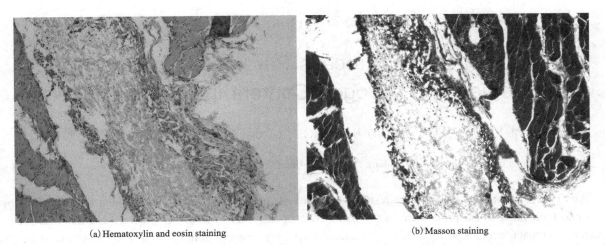

(a) Hematoxylin and eosin staining (b) Masson staining

Figure 2 Histological picture of VG. The term is 4 weeks. Magnification 100×

4 Conclusion

A three-layer design of small-diameter vascular grafts with reduced surgical porosity and mechanical characteristics similar to natural blood vessels has been developed. Experiments in vivo confirmed the high biocompatibility of VGs with bioactive coating and the ability to rapidly vascularize. Prostheses implanted in the infrarenal abdominal aorta of rats remained patent after 16 weeks of the experiment.

References

[1] WESOLOWSKI S A, FRIES C C, KARLSON K E, et al. Porosity: Primary determinant of ultimate fate of synthetic vascular grafts[J]. Surgery, 1961, 50: 91-96.

[2] YATES S G, BARROS D'SA A A, BERGER K, et al. The preclotting of porous arterial prostheses[J]. Ann Surg, 1978, 188: 611-622.

[3] GUAN G, YU C, FANG X, et al. Exploration into practical significance of integral water permeability of textile vascular grafts[J]. J Appl Biomat Funct Mater, 2021, 22808000211014007

[4] XIAO L, SHI D. Role of precoating in artificial vessel endothelialization[J]. Chin J Traumatol, 2004, 7: 312-316.

[4] Guan G, Yu C, Fang X, et al. Exploration into practical significance of integral water permeability of textile vascular grafts[J]. Journal of Applied Biomaterials & Functional Materials, 2021, 22808000211014007.

[5] WEADOCK K S, GOGGINS J A. Vascular graft sealants[J]. J Long Term Eff Med Implants, 1993, 3: 207-22.

[6] DRURY J K, ASHTON T R, CUNNINGHAM J D, et al. Experimental and clinical experience with a gelatin impregnated Dacron prosthesis[J]. Ann Vasc Surg, 1987, 1: 542-547.

[7] HUANG F, SUN L, ZHENG J. In vitro and in vivo characterization of a silk fibroin-coated polyester vascular prosthesis[J]. Artif Organs, 2008, 12: 932-941. DOI: 10.1111/j.1525-1594.2008.00655.x.

[8] Azimi B, Nourpanah P, Rabiee M, et al. Poly (ε-caprolactone) fiber: An overview[J]. Engineered Fibers Fabrics, 2014: 74-90.

[9] VALENCE S D, TILLE J C, GILIBERTO J P, et al. Advantages of bilayered vascular grafts for surgical applicability and tissue regeneration[J]. Acta Biomaterialia, 2012: 3914-3920.

[10] NEMETS E A, SURGUCHENKO V A, BELOV V Y, et al. Porous tubular scaffolds for tissue engineering structures of small diameter blood vessels[J]. Inorganic Materials: Applied Research, 2023, 14: 400-407. DOI: 10.1134/S20751133323020338

[11] SURGUCHENKO V A, NEMETS E A, BELOV V Y, et al. Bioactive coating for tissue-engineered small diameter vascular grafts[J]. Rus J Transpl Artif Organs, 2021, 23: 119-131. DOI: 10.15825/25/1995-1191-2021-4-119-131.

Influence of L-PBF Process Parameters on Mechanical Properties and Nitrogen Content in Nitrogen Steels

Ekaterina VOLOKITINA, Nikolay RAZUMOV, Nikolai OZERSKOI, Anatoly POPOVICH

Peter the Great St. Petersburg Polytechnic University, St. Petersburg, 195251, Russia

Abstract: This paper presents the results of studies of the dependence of the nitrogen content in nitrogen steels and their mechanical properties on the parameters of the laser powder bed fusion (L-PBF). During the research, compact samples were obtained, characterized by a relative density of 0.8%, a tensile strength of 780-1100 MPa, a relative deformation of 4%-10%. In the obtained samples, the nitrogen content was (0.13-0.44) wt.%.

Keywords: Selective laser melting; Additive manufacturing; Nitrogen steel

1 Introduction

Currently, nitrogen is a widely used alloying element along with Cr, Ni, Mn, Mo, etc.[1] It allows to produce steels with a unique combination of strength, ductility and corrosion resistance. The most important advantage of nitrogen over other alloying elements is the availability of nitrogen in almost unlimited quantities.

An analysis of the literature has shown interest on the part of scientists in exploring the possibility of using nitrogen steels in additive manufacturing. Currently, there are publications on the printing of nitrogen-containing steels in the technologies of L-PBF[2,3], LP-DED[4-7], wire-arc additive manufacturing (WAAM)[8,9] and electron beam additive manufacturing (EBAM)[10].

In [2, 3], the authors noted that nitrogen emission occurs in the process of L-PBF of nitrogen stainless steels, which depends on the energy density, it is proportional to the laser power and inversely proportional to the scanning speed. The possibility of using nitrogen steel powders in LP-DED is shown in [4-6]. It is noted that regardless of which process gas is used to feed the powder into the melt pool, the nitrogen content decreases. High temperatures of the melt pool cause a decrease in the solubility of nitrogen in the melt and lead to its degassing with the formation of gas pores. It was shown in [7] that in the LP-DED process it is possible to influence nitrogen emission due to local changes in the geometry of the melt pool, which is a function of energy density. An increase in the energy density leads to an increase in the lifetime of the melt pool along with higher maximum temperatures of the liquid phase. If we assume that nitrogen is mainly lost at the melting stage, then both an increase in the maximum temperature and an increase in the lifetime of the melt pool contribute to the loss of nitrogen. In addition to increasing the temperature and lifetime of the melt pool, the size of the melt pool, i.e. depth and area, increased with increasing energy density.

This article is aimed at identifying the dependence of the nitrogen content in the nitrogen steel and mechanical properties on the parameters of the L-PBF process.

2 Materials and methods

In this work, compositions of nitrogen steel powders were used: Fe-(0.5Cr-0.5Cr2N)-Ni-Mn-Mo, Fe-Cr2N-Ni-Mn-Mo, Fe-(0.5Cr-0.5FeCrN)-Ni-Mn-Mo and Fe-FeCrN-Ni-Mn-Mo obtained by mechanical alloying followed by plasma spheroidization.

Samples of cubic shape with sides of 10 mm were obtained by the L-PBF method. In the process of 3D printing, the laser power, scanning speed, and energy density varied. The thickness of the applied powder layer was 0.05 mm, the distance between the laser passes was 0.12 mm. The parameters of the L-PBF are given in Table 1.

Table 1 L-PBF parameters

No	Laser power/W	Scanning speed/(mm·s^{-1})	E/(J·mm^{-3})
1	240	650	61.54
2	300	800	62.50
3	300	650	76.92
4	360	650	92.31
5	300	500	100.00
6	300	650	115.38

3 Results and discussion

As a result of 3D printing by L-PBF, compact samples with a minimum relative porosity of 0.8% were obtained (Figure 1). It is shown that with an increase in the nitrogen content in the initial powder, the minimum relative porosity of the alloy increases to 11.5%. It was found that the nitrogen content in the alloy obtained by L-PBF is (0.13-0.44)wt.%, which exceeds the critical concentration of nitrogen during crystallization by 2 times. The solubility of nitrogen in the liquid metal, α-phase and γ-phase are significantly different. Consequently, one of the technological problems is that during the crystallization of steel, nitrogen is released into the gas phase, forming nitrogen bubbles and porosity. During crystallization, the composition of the liquid phase and the released solid phases continuously changes depending on changes in temperature and the amount of liquid phase. In this case, the local solubility of nitrogen in the residual liquid phase also changes. The nature of this change depends on the type of crystallization (austenitic, ferritic or mixed) and the ratio of the amounts of phases.

The critical concentration of nitrogen during crystallization of the alloy under study does not exceed 0.2 wt.%. The actual nitrogen content in the liquid phase during printing exceeds its equilibrium solubility at a pressure in the chamber of 1 atm (1 atm = 10^5 Pa), nitrogen is released as a gas. It is worth noting that the porosity of alloys obtained from powders in which FeCrN was used as a nitrogen source is higher than that of alloys with CrN, which is due to the higher decomposition temperature of the latter. The study of the distribution of elements showed that the alloying elements are evenly distributed over the cross-section of the alloys [Figure 2].

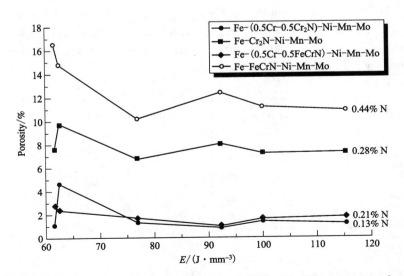

Figure 1 Relative porosity of alloys obtained by L-PBF with different energy densities

The obtained alloys were subjected to heat treatment according to the mode: quenching with (1040 ± 10) ℃ with oil, tempering at 640-680 ℃ for 2 h. Mechanical tests performed at room temperature and 500 ℃ showed (Figure 3, Table 2) that the alloys obtained by L-PBF do not meet the technical requirements for

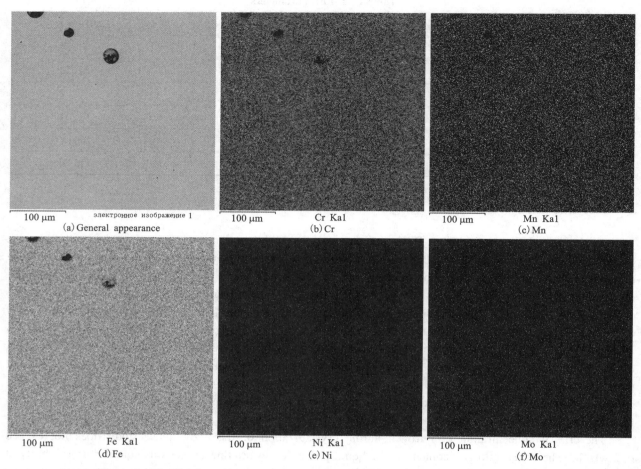

Figure 2　Distribution of components in the composition Fe-FeCrN-Ni-Mn-Mo after N

similar nitrogen steel in terms of elongation, which is due to high porosity. To eliminate porosity, the obtained alloys were subjected to hot isostatic pressing (HIP) at a temperature of 1160 ℃ and a pressure of 150 MPa for 3 h. After the HIP, the relative porosity of the alloys does not exceed 0.2%, which made it possible to increase the plasticity of the material (Figure 4, Table 3). Heat treatment was carried out according to the standard regime for a similar alloy, however, it is worth noting that the compositions under study differ in chemical composition, as a result of which, it can be concluded that standard heat treatment does not allow achieving the maximum possible properties.

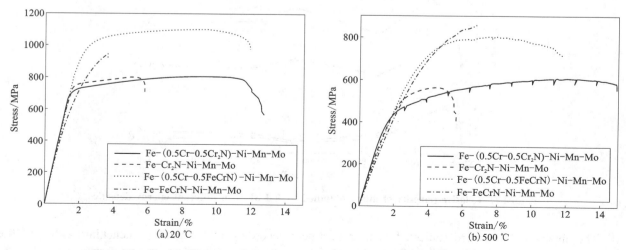

Figure 3　Stretch diagram of alloys obtained by L-PBF at different test temperatures

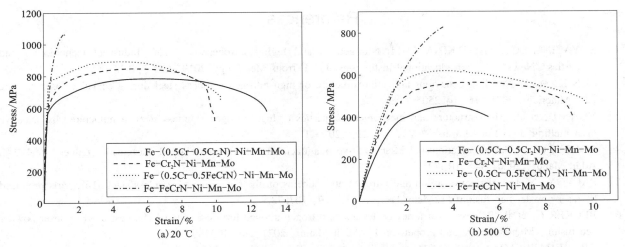

Figure 4 Stretch diagram of alloys obtained by L-PBF followed by HIP

Table 2 Mechanical properties of alloys after L-PBF

Composition	20 ℃			500 ℃		
	$\sigma_{0.2}$/MPa	σ_B/MPa	δ/%	$\sigma_{0.2}$/MPa	σ_B/MPa	δ/%
Fe-Cr_2N-Ni-Mn-Mo	730	780	5.5	480	560	5
Fe-FeCrN-Ni-Mn-Mo	-	980	-	600	820	4
Fe-(0.5Cr-0.5Cr_2N)-Ni-Mn-Mo	700	790	12	430	610	10
Fe-(0.5Cr-0.5FeCrN)-Ni-Mn-Mo	960	1100	10.5	560	800	9
Fe-FeCrN-Ni-MN-Mo	740	935	14	540	640	-

Table 3 Mechanical properties of alloys after L-PBF and HIP

Composition	20 ℃			500 ℃		
	$\sigma_{0.2}$/MPa	σ_B/MPa	δ/%	$\sigma_{0.2}$/MPa	σ_B/MPa	δ/%
Fe-Cr_2N-Ni-Mn-Mo	620	840	16.0	400	560	8.2
Fe-FeCrN-Ni-Mn-Mo	800	1070	1.2	600	830	1.4
Fe-(0.5Cr-0.5Cr_2N)-Ni-Mn-Mo	610	780	21.0	370	460	4
Fe-(0.5Cr-0.5FeCrN)-Ni-Mn-Mo	770	890	16.0	530	620	8.5
Fe-FeCrN-Ni-MN-Mo	740	935	14	540	640	-

4 Conclusion

This paper shows the influence of the parameters of the L-PBF process on the nitrogen content in alloys, porosity and mechanical properties. With an increase in the nitrogen content in the alloy, the minimum relative porosity increases to 11.5 vol.%. The nitrogen content in the alloy obtained by L-PBF is (0.13-0.44)wt.%, which exceeds the critical concentration of nitrogen during crystallization by 2 times. Mechanical tests have shown that the alloys obtained by L-PBF are not inferior in their characteristics to alloys obtained by classical metallurgical technologies. It is assumed that changing the heat treatment mode will increase the mechanical properties of the obtained alloys.

Acknowledgments: This research was funded by the Ministry of Science and Higher Education of the Russian Federation (State Assignment for basic research No. 075-03-2023-004).

References

[1] SVYAZIN A G, KAPUTKINA L M. Nitrogenous and highly nitrogenous steels. Industrial technologies and properties//News of higher educational institutions[J]. Ferrous Metallurgy. 2019, 62, 3: 173-187.

[2] SUN X, et al. Nitriding behaviour and microstructure of high-nitrogen stainless steel during selective laser melting [J]. Materials, 2023, 16, 6: 2505.

[3] YUAN L, et al. Microstructure and mechanical properties of high-nitrogen stainless steel manufactured by selective laser melting[J]. Chin J Lasers, 2022, 49, 22: 2202021.

[4] BOES J, et al. Gas atomization and laser additive manufacturing of nitrogen-alloyed martensitic stainless steel[J]. Addit Manuf, 2020, 34: 101379.

[5] PAUZON C, et al. Effect of argon and nitrogen atmospheres on the properties of stainless steel 316L parts produced by laser-powder bed fusion[J]. Mater Des, 2019, 179: 107873.

[6] BECKER L, et al. Processing of a newly developed nitrogen-alloyed ferritic-austenitic stainless steel by laser powder bed fusion: Microstructure and properties[J]. Addit Manuf, 2021, 46: 102185.

[7] ARABI-HASHEMI A, et al. 3D magnetic patterning in additive manufacturing via site-specific in-situ alloy modification[J]. Appl Mater Today, 2020, 18: 100512.

[8] ZHANG X, et al. Study on microstructure and tensile properties of high nitrogen Cr-Mn steel processed by CMT wire and arc additive manufacturing[J]. Mater Des, 2019, 166: 107611.

[9] ZHANG X, et al. Precipitation characteristics and tensile properties of high-nitrogen chromium-manganese steel fabricated by wire and arc additive manufacturing with isothermal post-heat treatment[J]. Mater Des, 2023, 225: 111536.

[10] ASTAFUROV S, et al. Electron-beam additive manufacturing of high-nitrogen steel: Microstructure and tensile properties[J]. Mater Sci Eng A, 2021, 826: 141951.

Observation of Interatomic Auger Transitions in Cu-based Chalcogenides Using Synchrotron Radiation X-ray Photoemission

Vladimir GREBENNIKOV[1,2], Tatyana KUZNETSOVA[1,3]

1 M. N. Miheev Institute of Metal Physics of UB RAS, Ekaterinburg, 620108, Russia
2 Ural State University of Railway Transport, Ekaterinburg, 620034, Russia
3 Ural Federal University, Ekaterinburg, 620002, Russia

Abstract: The interatomic Auger transitions in compounds containing atomic components with core levels close in energy are studied. The Coulomb transitions of a hole between such levels lead to a resonant enhancement of the Auger spectra (with respect to the energy difference between the levels). Interatomic Auger transitions involving high lying levels are formed by shaking up electrons due to the dynamic field of holes produced during the transition. These effects were observed experimentally in XPS and Auger spectra of $CuInSe_2$ type solar materials.

Keywords: Auger transition; Cu-based chalcogenides; Synchrotron radiation

1 Introduction

The absorption of an X-ray quantum in matter leads to a photoelectron ejection from the core level of the atom, and the resulting photo-hole is then filled with an electron from the overlying shell with the emission of a second electron from the atom due to the Coulomb interaction and conservation of energy (the so-called Auger transition). The kinetic energy distribution of the emitted electrons is recorded by the detector as an X-ray photoemission/Auger spectrum (XPS/Auger). Usually the whole process takes place within one central atom. The influence of neighboring elements is reduced only to a change in the energies of the levels of an absorbing atom during the transfer of valence electrons between atoms due to a chemical bond. We intend to tackle the unusual—to get an answer to the fundamental question: Is the photo effect possible with the ejection of electrons from the core levels of neighboring elements surrounding the atom that has absorbed the quantum? There are certain reasons to believe that such an interatomic photo effect exists and has a significant probability, especially in compounds with elements containing highly localized 4 d shells with binding energy 15-50 eV. These states experience an anomalously strong effect of the dynamic field of the photo hole, forming long-lived excited atomic states.

Our previous XPS experiments[1] showed that interatomic Auger transitions in the X-ray energy range are observed in semiconductor compounds based on $CuInSe_2$. In this case, the final state contains holes at the core levels of two neighboring copper and indium atoms.

Interatomic transitions in condensed matter were discussed many years ago[2]. Theoretical estimates of the direct matrix elements of such transitions between deep levels gave a very low probability. As a result, experiments on their detection were reduced mainly to the study of the spectra of valence states[3,4]. The researchers tried to determine in which of the two valence states a hole is formed on the central atom or on its neighbor. Since the energies of the valence states of the elements forming a chemical bond are approximately the same, it is very difficult to separate these states. A change in the population of these states leads only to a change in the shape of the single Auger line, and model calculations are required to separate two contributions from the experimental line.

The study of interatomic transitions is of fundamental importance for understanding the evolution of excited states in matter. It is appropriate to mention here the effect of multiatomic resonant photoemission (MARPE)[5]. Photoemission associated with a particular electronic level of a given atom "A" can vary significantly in intensity when the photon energy passes through the absorption edge of the core level of the neighboring atom "B". The effect is explained by the change in the complex dielectric constant of the substance when the frequency passes through the corresponding resonance[6,7], as a result, the propagation and focusing of X-rays changes. Another manifestation of interatomic interactions is the so-called interatomic Coulomb Decay (ICD), in which the excitation of the valence electrons of one atom leads to the ionization of the valence shell of the second atom. The ICD is considered as a new source of low-energy electrons in slow ion-dimer collisions[8]. Inelastic scattering of Auger electrons and interatomic Coulomb decay are assumed as mechanisms of population and depopulation, respectively, of excited states in nanoplasma, which are formed when an X-ray free electron laser pulse interacts with a nanometer-sized substance, for example, with a cluster of xenon atoms. ICD can serve as the main mechanism for the rapid emission of hollow atoms resulting from the neutralization of a highly charged ion in solids.[9]

As already mentioned, most of the research in the field of condensed matter has been aimed at studying atomic transitions with the participation of valence electrons. We predict that interatomic Auger transitions should have significant intensity in compounds with atoms having d-type core levels with a binding energy of 10-30 eV. Electrons of these levels are capable of quite easily passing into the valence band when a photo hole is appeared in a neighboring atom, creating interatomic excitation. In this article, we report on the observation of isolated interatomic Auger transitions in the Cu_2SnS_3 compound with the participation of Sn 4d electrons with energies two tens of eV below the energy of the corresponding intra-atomic Cu Auger transitions. Observations of this kind are not described in the literature (except for our first report[1]).

In this work the interatomic Auger transitions $CuL_3M_{2,3}InN_{4,5}$; $CuL_3InN_{4,5}V$; $CuL_3CuM_{2,3}SnN_{4,5}$ and $CuL_3SnN_{4,5}V$ are experimentally observed and studied in semiconductors Cu-based chalcogenides, like a $CuInSe_2$ and Cu_2SnS_3, which used as absorbers in photovoltaics.

2 Results and discussion

Our experiments were carried out on the BESSY-II synchrotron at the Russian-German Laboratory in Berlin. A high-quality working surface was prepared by sample splitting directly in the ultra-high vacuum chamber of the spectrometer. Figure 1 schematically shows possible mechanisms of interatomic transitions in the Cu_2SnS_3 compound, which are further investigated experimentally. The process begins with the formation of a hole at the internal 2p level of copper (binding energy 933 eV) during photon absorption. Figure 1(a) shows the Cu LMV autoionization transition with the formation of two holes at the 75 eV Cu 3p level and in the valence band VB. If we consider the described state as the final state of the autoionization process, then we get the usual intra-atomic Auger spectrum. However, a

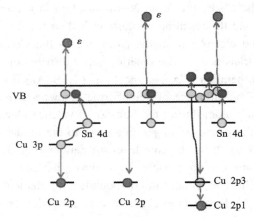

(a) $CuL_3CuM_{2,3}SnN_{4,5}$ (b) $CuL_3SnN_{4,5}V$ (c) $CuL_3SnN_{4,5}V$
Auger transition following formation of the inner CuL_2 core hole

Figure 1 Diagrams of the interatomic Auger process

scenario is possible in which an electron from the Sn 4d level of a neighboring tin atom (binding energy 24 eV) fills the valence state, resulting in a final state corresponding to the $CuL_3CuM_{2,3}SnN_{4,5}$ interatomic Auger transition. In this case, the kinetic energy of the ejected Auger electron will be less than the energy of the corresponding intra-atomic line by approximately the binding energy of the Sn 4d level. The reason for the excitation of the Sn 4d level is the action of the Coulomb field of holes, which are suddenly produced during the photoionization of the CuL_3 level and the Auger transition. Figure 1(b) shows a similar interatomic transition in $CuL_3SnN_{4,5}V$ with finishing holes on the tin atom and in the valence band. If the initial hole is born at a deeper CuL_2 level, then an additional channel of excitation of the electronic system arises associated with the Koster-Kronig transition $L_3 \rightarrow L_2$, which enhances the shaking of the Sn 4d electron [Figure 1(c)].

Let us consider the scheme [Figure 2(a)] of the interatomic Auger transition using the $CuInSe_2$ compound as an example. The photoionized Cu 2p hole (binding energy 933 eV) is filled by an electron from the overlying Cu 3p state (75 eV) due to the Coulomb interaction, and the energy released in this process goes to the ejection of an electron located on the neighboring indium atom In 4d (17 eV) into a free state f. The detector measures the intensity of the outgoing electrons and their kinetic energy or Auger spectrum of $CuL_3M_{2,3}InN_{4,5}$. Usually the intensity of the interatomic transition is vanishingly small in comparison with the interatomic transition, say $CuL_3M_{2,3}V$ (V denotes the valence state), since the distance determining the energy of the Coulomb interaction $e^2/|r_1-r_2|$ between electrons inside the atom is much smaller than the interatomic distance.

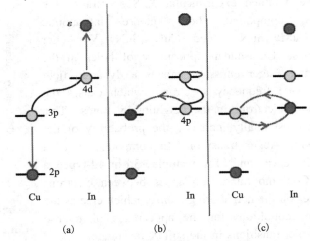

Figure 2 Diagrams of the interatomic Auger process Cu 2p, Cu 3p, In 4d in the CuInSe$_2$ compound (a) considering the real (b) and virtual (c) transitions between Cu 3p and In 4p levels

We will show that the intensity of the interatomic Auger transition substantially increases if the neighboring element has a core level close in energy to the level of the central atom. In our example, this is an In 4p level with a binding energy of 73.5 eV, which is shifted by only 1.5 eV from the energy of the Cu 3p level. In this case, the possibility of a real (in the final state) transition of a hole from one atom to another [Figure 2(b)] appears, and the probability of virtual hole transitions between levels also increases [Figure 2(c)]. Figure 3 shows the CuL_3 Auger spectra obtained at photon energy of 950 eV, which is above the L$_3$ edge but below the L$_2$ edge of the spin-orbit doublet. The background of inelastic electron scattering is drawn by a thin smooth line. The figure also shows the corresponding spectrum of pure metallic copper for comparison. Both curves are formed mainly by intra-atomic Auger lines Cu L$_3$VV (kinetic energy of the maximum is 917.7 eV) and by Cu L$_3$M$_{2,3}$V line split into two peaks 837.9 and 846 eV. However, the main difference between the curves is that an additional extended line is observed at the compound. Its maximum has an energy 22 eV below the peak of the main CuL$_3$VV line. In addition, two more maxima appear in the compound at 23 eV below the CuL$_3$M$_{2,3}$V Auger line. These are interatomic CuL$_3$SnN$_{4,5}$V and CuL$_3$CuM$_{2,3}$SnN$_{4,5}$ Auger excitations.

The interatomic Auger transition occurs as a result of the shaking of Sn 4d electrons. Therefore, let us look for similar processes in other spectra, in particular, in the direct photoemission spectrum. Figure 4 shows the XPS spectra of the Sn 3d doublet in Cu$_2$SnS$_3$. The main lines are accompanied by 28.5 eV satellites arising due to the transfer of additional 4d electrons into unoccupied valence states during the sudden creation of a 3d photohole. The energy shift is determined by the binding energy of the Sn 4d level and the energy of intraatomic repulsion of two holes $U = 1.8$ eV. The Sn 4d electron is shaken by one photo hole in direct photoemission.

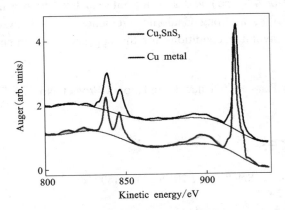

Figure 3 Auger spectrum of the Cu$_2$SnS$_3$ compound, obtained at a photon energy of 950 eV

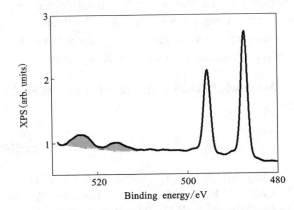

Figure 4 XPS spectra of the Sn 3d spin-orbital doublet and intrinsic loss on excitation of a Sn 4d electron to the valence band during photoemission from Cu$_2$SnS$_3$

The obtained experimental XPS spectra of the Cu_2SnS_3 compound show intense interatomic $CuL_{2,3}CuM_{2,3}SnN_{4,5}$ and $CuL_{2,3}SnN_{4,5}V$ Auger transitions. The sudden appearance of holes in the photo and Auger emission creates a dynamic field with a wide frequency spectrum, which causes the shaking of electrons in the neighboring atoms. This process substantially increases the probability of the interatomic Auger transition. In compounds with a narrow valence band (for example, 3d-or 4d-type) a strong Coulomb interaction arises between electrons and holes on the neighboring atoms, which creates the favorable conditions for the appearance of intense interatomic transitions in the soft X-ray range.

Figure 5 shows the Auger spectra of copper in $CuInSe_2$ and $CuIn_{0.9}Ga_{0.1}Se_2$ compounds, obtained at

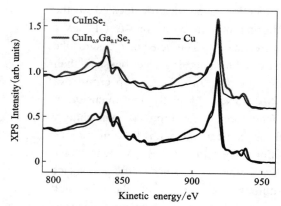

Figure 5   Intra-atomic $CuL_{2,3}VV$ and interatomic $CuL_{2,3}M_{2,3}InN_{4,5}$ Auger bands in compounds in $CuInSe_2$ and $CuIn_{0.9}Ga_{0.1}Se_2$ (thick lines) and pure copper spectrum[8] (thin line)

a photon energy of 1200 eV. The Auger spectrum of metallic copper obtained on MgK_α radiation of 1253.6 eV is also shown[10] for comparison. The intra-atomic CuL_3VV Auger transitions (the maximum of the kinetic energy of 918 eV) and the triple Auger line formed by the $CuL_3M_{2,3}V$ transition (the main maximum of 838 eV, the multiplet splitting because of the addition of the moments of the two holes 3p and 3d) are clearly visible on all the curves. At ~20 eV above the main lines, one can observe their replicas originating from the CuL_2 hole. At the same time, Auger spectra of compounds differ markedly from the spectrum of metallic copper. To better understand the nature of the differences, we first consider the electronic structure of the top-filled states. Figure 5 shows an additional signal in the compounds with indium at 15 eV below the peak CuL_3VV, which is absent in the pure copper spectrum. Its energy corresponds to the interatomic $CuL_3InN_{4,5}V$ Auger transition. Note that the $CuM_{2,3}$ hole (binding energy 75 eV) can be supplemented by an $InN_{2,3}$ hole whose binding energy is only 1.5 eV lower. The intensity of the transitions is enhanced due to the resonant interaction of the levels of Cu 3p and In 4p, which are close in energy.

4 Conclusion

The interatomic Auger transitions $CuL_3M_{2,3}InN_{4,5}$, $CuL_3InN_{4,5}V$, $CuL_3CuM_{2,3}SnN_{4,5}$ and $CuL_3SnN_{4,5}V$ are experimentally observed and studied in Cu-based chalcogenides, like a $CuInSe_2$ and Cu_2SnS_3. The sudden appearance of holes in the photo and Auger emission creates a dynamic field with a wide frequency spectrum, which causes shaking of electrons in neighboring atoms. This process substantially increases the probability of the interatomic $CuL_3InN_{4,5}V$ and $CuL_3SnN_{4,5}V$ Auger transitions, respectively. In compounds with a narrow valence band (for example, 3d-or 4d-type) and a strong internal level with a not very high binding energy (In 4d, 17 eV or Sn 4d, 24 eV), a strong Coulomb interaction arises between electrons and holes on neighboring atoms, which creates favorable conditions for the appearance of intense interatomic transitions in the soft X-ray range.

Acknowledgments: The research was funded by the Russian Science Foundation (Project No. 23-72-00067).

References

[1] GREBENNIKOV V I, KUZNETSOVA T V[J]. Physica Status Solidi (A), 2019(216): 1800723.
[2] RAO C N R, SARMA D D[J]. Phys. Rev. B, 1982, 25: 2927.
[3] JÈRÔME D, et al[J]. Phys. Rev. B, 2001, 64: 045110.
[4] WERTHEIM G K, ROWE J E, BUCHANAN D N E, et al. [J].Phys. Rev. B, 1995, 51: 13669.
[5] MANNELLA N, et al[J]. Phys. Rev. B, 2006, 74: 165106.
[6] KAY A W, et al[J]. Phys. Rev. B, 2001, 63: 115119.

[7] ARAI H, FUJIKAWA T[J]. Phys. Rev. B, 2005, 72: 075102.
[8] ISKANDAR W, et al[J]. Phys. Rev. Lett., 2015, 114: 033201.
[9] RICHARD A, et al[J]. Phys. Rev. Lett., 2017, 119: 103401.
[10] Chastain J. Handbook on electron spectroscopy[M]. Minnesota: Perkin-Elmer Corporation, 1995.

Possibility of Damping Through the Use of Different Steels and Alloys

Iurii DUBINOV[1], Oksana ELAGINA[1], Artem BEREZNYAKOV[1],
Olga DUBINOVA[1], Victoria FEDOROVA[1], Alexander KUZNETSOV[2]

1 Gubkin Russian State University of Oil and Gas (National Research University), Moscow, 119991, Russia
2 Gazprom Gaznadzor LLC, Moscow, 117418, Russia

Abstract: This paper considers the possibility of damping by means of materials with different damping mechanisms: alloy with structural deformation High quality steel 40CrNi2Mo, alloy with thermoelastic martensite NiTi, steel with magnetic properties 01Al5Ti. Tests were carried out to model the vibration conditions under friction and on a fixed support. Damping was analyzed on the basis of vibration voltage, vibration amplitude, vibration velocity and vibration acceleration. The most effective materials for different frequency ranges have been identified.

Keywords: Damping; Vibrations; Alloys

Most of the processes observed in nature and technology are oscillatory. Vibration is a complex oscillatory process where the amplitude is relatively small and the frequency is large.

Vibration can occur as a side effect of equipment operation, most often in the case of reciprocating or impact motion of parts, flow of fluids, synchronized operation of various equipment. An example of vibrations in the oil and gas complex is the vibration that occurs when fluid is transported through a pipe[1, 2].

Vibration has various effects on technological equipment. In the negative case there is acceleration of wear, growth of mechanical fatigue, increase in backlash of moving joints, reduction of the overhaul period[3]. It is impossible to exclude mechanical vibrations, the problem is solved by selecting special designs for vibration damping and improving the design of machines, but the latter is not always possible due to the design features, so recently it is becoming popular to use optimal materials with the effect of vibration damping.

One of the main causes of equipment failure is fatigue failure caused by cyclic loading[4, 5, 6]. This type of failure can be described using the Snyder-Mackormack sinusoidal law. Sinusoidal oscillations occurring in equipment have a similar nature of effect as the load is applied alternately, leading to failure in pumping equipment, drilling equipment. Damping of random oscillations will increase the time until failure of process equipment.

Metal alloys of high damping can provide reduction of such parameters as frequency and amplitude of vibrations in a wide range, which, in turn, can improve the efficiency of equipment[7, 8], but it should be noted that many damping materials have low strength values, damping and mechanical properties are usually inversely proportional.

Examples of alloys of high damping alloy with structural deformation (high quality steel 40CrNi2Mo)[8], alloy with thermoelastic martensite (NiTi)[9] and steel with magnetic properties (01Al5Ti)[10, 11] are considered in this paper.

The mechanism of realization of damping properties using materials with magnetic properties is as follows: under the action of stresses in the material, the domain boundary (the boundary of ordered macroformations) moves against the reactive force trying to return it to its original position, and at the moment when the reactive force has the maximum value, the domain boundary loses its equilibrium and becomes unstable and there is a reverse spontaneous jump to a new position, called the Barkhausen jump. At this moment, the domain structure changes and the energy dissipation process takes place.

The mechanism of damping in alloys with thermoelastic martensite occurs due to reversible diffusionless phase processes and the effect of elastic twinning under the application of external loads.

In Ni-Ti alloys, absorption of large vibrations occurs due to mechanical hysteresis in the pseudoelastic deformation region.

In alloys with structural deformation, damping occurs due to the mobility of dislocations when external loads are applied. The disadvantage of these materials is relatively low strength characteristics.

The purpose of this work is to investigate the effectiveness of damping properties of materials when changing various parameters characterizing the vibration process: frequency, amplitude, vibration velocity, vibration acceleration.

In the first stage, to simulate passive damping, a methodology was developed and tests were conducted on the finger-disk system, at a fixed load and rotational speed of the MT-3 unit (Figure 1).

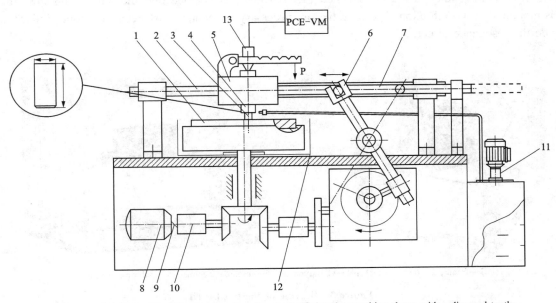

1—abrasive disk; 2—guides; 3—sample; 4—reference for fixing the transition sleeve with a diamond tooth; 5—carriage; 6—rocker mechanism; 7—guides; 8—electric motor; 9—coupling; 10—reducer; 11—electric pump (PA-22); 12—housing; 13—vibration sensor with vibrometer

Figure 1 Scheme of installation MT-3

During the experiment the value of axial load applied to the samples was varied in the range from 5 to 1000 N to determine the best damping ranges. The tests were carried out on the bit mockup, with installed cutters with diameter 13.44 mm and height 13 mm, which were fixed in the mandrel, friction was carried out on the surface of fine granite.

To compare the friction damping properties, we calculate the stresses using Formula (1) and the vibration force using Formula (2).

$$\sigma = \frac{N}{S} \tag{1}$$

where: S is contact surface area, mm^2; N is static clamping load, N.

$$F = N + m \cdot a \tag{2}$$

where: m is the mass acting on the cutter, kg, a is root-mean-square acceleration, m/s^2, N is static clamping load, N.

As a result of the experiment we obtained the dependence presented in Figure 2, which shows that steel 01Al5Ti and NiTi have the highest value of vibration stress in the whole range of tests, which indicates better damping. Comparing Ni-Ti and 01Al5Ti, we can note that 01Al5Ti steel performs better in the range from 400-600 N and 800-1000 N, the frequency for the steel was 47-

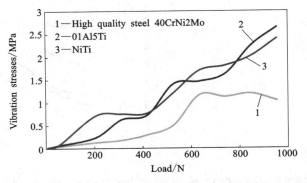

Figure 2 Dependence of vibration stress on the applied load

53 Hz and 59-61 Hz, the amplitude was 27.2-28.3 μm and 26.7-26.3 μm, vibration acceleration 2.4-3.2 mm/s² and 3.7-3.8 mm/s². For the alloy, frequency has the value of 48-58 Hz and 50-54 Hz, amplitude has the value of 27.9-28.5 μm and 30.9-33.1 μm, vibration acceleration has the value of 2.5-3.7 mm/s² and 3.3-3.5 mm/s².

In the second stage, a bench simulating random forced oscillations on a fixed Figure 3 support is designed and a test methodology is developed. The vibration dynamo generates sinusoidal oscillations in the range from 1 to 1000 Hz, which are set using an oscilloscope connected through an amplifier. The stand includes a frame to which a table is attached on vibration-damping ties. The vibration speaker is mounted on the table. The material sample under study is placed between the vibrodynamic device and the table, and the vibrometer sensor is mounted on it. Vibration absorbing ties are necessary to separate surfaces from induced vibrations and to exclude the resonance effect.

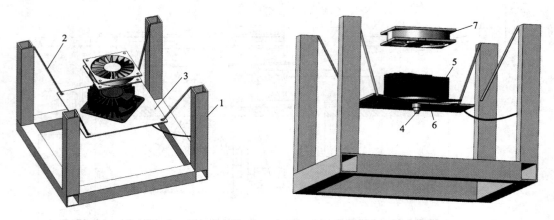

1—frame; 2—ties; 3—table; 4—vibrodynamic; 5—research specimen; 6—vibrometer sensor

Figure 3 Scheme of the test bench

According to the obtained data, the dependencies presented in Figure 4 are plotted.

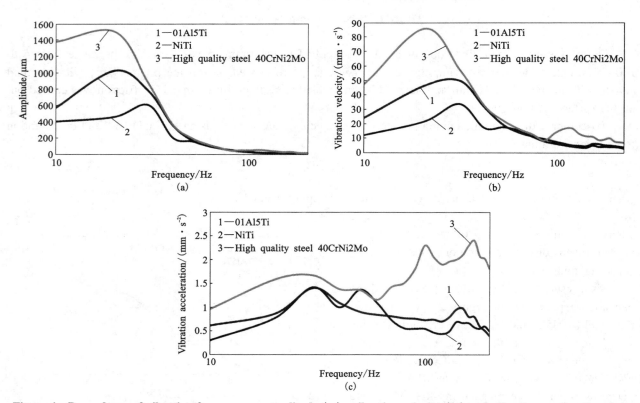

Figure 4 Dependence of vibration frequency on amplitude (a), vibration velocity (b) and vibration acceleration (c)

NiTi damping is better than other materials in the range from 10 to 40 Hz. At frequency from 40 to 200 Hz 01Al5Ti steel shows insignificant difference of vibration indices with NiTi, which indicates that 01Al5Ti steel can be used in this range.

The following conclusions are obtained by the authors:

(1) The most effective way to reduce vibrations when this is not structurally possible is to use materials with magnetic properties, the next most effective being alloys with thermoelastic martensite.

(2) For each of the materials the ranges of frequencies and loads were obtained, on the basis of which recommendations for the use of different materials for the manufacture of technological equipment are given: NiTi dampens better in the frequency range from 10-200 Hz, in the range from 40-200 Hz, steel 01Al5Ti can serve as an analog of NiTi, because it has similar damping parameters, high quality steel 40CrNi2Mo can serve as an analog of the materials presented above in the frequency range from 50-80 Hz.

(3) A bench was constructed and material testing methods were developed to determine the damping capacity.

(4) The application of materials presented in this work in the specified frequency range will increase the operating time before failure of technological equipment.

References

[1] FROLOV K V. Vibrations in engineering: Reference book: in 6. M.: Mashinostroenie. T. 6. Vibration and shock protection, 1995: 456.

[2] SHVINDIN A I. Vibration reliability[J]. Business Journal Neftegaz, 2016 RU, 4: 90-93.

[3] DUBINOV Y S, DUBINOVA O B, ELAGINA O Y. Methods of combating vibration and its negative effects on pipelines and equipment. New technologies in the gas industry: gas, oil, energy: XIV All-Russian Conference of Young Scientists, Specialists and Students: abstracts, Moscow, November 14-18, 2022.

[4] KUSHNARENKO B M, CHIRKOV Y A, REPYAKH V S, et al. Fatigue fractures of oil and gas equipment parts [J]. Herald of Orenburg State University, 2012, 4: 271-279.

[5] RUDENKO A L, MISHAKIN A L, FOMIN A E, et al. Fatigue failure of hydraulic power equipment in the process of long-term operation[J]. Repair Restoration Modernization, 2021, 9: 18-24.

[6] SIMSIVE D C, SIMSIVE J V, KUTYSHKIN A V. Prediction of fatigue fracture of carbide cutting tools during machining[J]. Metal processing: Technology, equipment, tools, 2012, 3: 52-55.

[7] GOLOVIN I S. Internal friction and mechanical spectroscopy of metallic materials[J]. Moscow Izd. Dom MISIS, 2012: 247 с.

[8] GOLOVIN I S. Damping mechanisms in high damping materials[J]. Key Engineering Materials, 2006, 319: 228-230.

[9] PRYGAEV A K, DUBINOV Y S, ELAGINA O Y, et al. Damping properties of titanium nickelide as one of the key features in the creation of oil and gas equipment[J]. Equipment and technologies for oil and gas complex, 2020, 3: 8-13.

[10] CHUDAKOV I B, ALEKSANDROVA N M, MAKUSHEV S Y. Peculiarities of consumer properties of new high-demanding steels[J]. Steel, 2014, 8: 92-95.

[11] CHUDAKOV I B. New industrial high-demanding steels[C]. Proceedings of the IV All-Russian Youth School-Conference "Modern Problems of Metallurgy", Sevastopol, 2016.

[12] DUBINOV Y S, DUBINOVA O B, ELAGINA O Y, et al. Numerical and experimental modeling of damping properties of materials under friction[J]. Promising Materials, 2022, 4: 80-86.

Solid-state Reaction Kinetics Between Iridium and Zirconium Carbide

Yaroslav A. NIKIFOROV, Natalya I. BAKLANOVA

Institute of Solid State Chemistry and Mechanochemistry SB RAS, Novosibirsk, 630090, Russia

Abstract: The kinetics of the solid-state reaction between iridium and zirconium carbide was studied using Ir/ZrC diffusion couples. Samples after annealing at 1600 ℃ during the time from 2 to 16 h were characterized with scanning electron microscopy, wavelength-dispersive spectroscopy, and Raman spectroscopy. The product layer consists of the $ZrIr_3$ matrix phase [(75-80)at% Ir] with small inclusions of graphite-like carbon particles. The experimental results confirm that iridium has a higher self-diffusion coefficient than zirconium in the $ZrIr_3$ phase, which is typical of the $L1_2$ ordered alloys. The growth behavior of the product layer is non-parabolic, obeying the power law with exponent $n = 0.32$, indicating an enhanced diffusion near the reaction zone.

Keywords: Solid-state reaction; Intermetallics; Diffusion; Kinetics

1 Introduction

Iridium-based alloys are of growing interest as potential materials for high-temperature applications.[1-3] Here we focus on the $ZrIr_3$ intermetallics, characterized by large Young's modulus and yield stress, high thermal conductivity nearly independent of temperature, low coefficient of thermal expansion, and strong bonding at $Ir/ZrIr_3$ interface.[4-7] The main focus of the existing works is on the physical properties of this and other intermetallics, meanwhile, the production of $ZrIr_3$ is yet to be investigated. Direct interaction between metallic Ir and Zr is an ineffective way of $ZrIr_3$ formation because other Zr-Ir intermetallic compounds ($ZrIr_2$, $ZrIr$ and Zr_2Ir) form alongside.[8] The reaction of iridium with zirconium carbide can serve as a convenient way to produce $ZrIr_3$, as it is the only intermetallic compound forming in this ternary system.[9]

This paper presents a kinetics study of the solid-state reaction between iridium and zirconium carbide. The reaction proceeds according to the equation:

$$ZrC + 3Ir = ZrIr_3 + C \tag{1}$$

The reaction kinetics is controlled by the interdiffusion of iridium and zirconium atoms through $ZrIr_3$ in the product layer. As $ZrIr_3$ is an ordered alloy, the interdiffusion process in it is strongly correlated. The correlation effect manifests in the higher mobility of iridium, which is experimentally observed in this work. The most peculiar feature of this reaction is a non-parabolic growth of the product layer, a behavior connected to processes near the reaction interface.

2 Experiment

The as-received iridium plate (99.97% purity, GOST 55084-2012, Russia) sized 5 mm×5 mm×0.1 mm and a sintered ZrC pellet sized ϕ12 mm×3 mm were used as diffusion couples. ZrC pellets were produced from the powder (99.8% purity, $D_{50} = 2$ μm, technical specifications TU 6-09-408-75) by hot pressing at 1900 ℃ under 20 MPa for 30 min with subsequent polishing. The density was 6.4 g/cm³ representing 97% of the theoretical; the lattice parameter was 4.693 Å, corresponding to a stoichiometric carbide. Figure 1 shows the microstructure of the sintered ZrC. One can see that the particles are sintered, but still, there are numerous micron-sized pores.

Diffusion couples were hot pressed using a UGP laboratory hot press (IA & E SB RAS, Russia) at 1600 ℃ from 2 to 16 h under a pressure of 20 MPa to ensure better contact. After annealing diffusion couples were cross-sectioned, packed with epoxy, and polished using 6, 3 and 1 μm diamond suspension Aquapol-P

(Kemet International Limited, UK). Morphology was characterized by scanning electron microscopy (SEM) using TM-1000 (Hitachi Ltd, Japan) and MIRA 3 LMU (TESCAN, Czech Republic). Concentration profiles were measured by wavelength-dispersive spectroscopy (WDS) using JXA-8230 electron probe microanalyzer (Jeol Ltd, Japan) at an accelerating voltage of 20 kV; PET (Zr) and LiF (Ir) were used as analyzer crystals, pure iridium (Ir) and zirconium carbide (ZrC) were used as standards to calculate concentrations. The structure of the formed carbon was characterized by Raman spectroscopy using a microscope LabRam HR Evolution (Horiba, Japan).

3 Results and discussion

3.1 Morphology and elemental composition

Let us first consider the morphology of the product layer formed during the reaction. Figure 1 shows an SEM image of a cross-section of the diffusion couple annealed for 8 h. The product layer consists of a continuous matrix of $ZrIr_3$ with small carbon inclusions. The Raman spectra from the randomly selected inclusions have peaks at ca. 1330 cm^{-1}, 1580 cm^{-1} and 2660 cm^{-1} corresponding to D, G and 2D bands of graphite. This evidence with the facts of axial loading during the reaction and of larger molar volume of products (4% compared to stoichiometric $ZrC+3Ir$) rules out a formation of pores in the product layer.

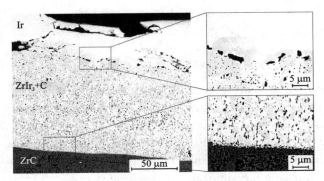

Figure 1 SEM image of the diffusion couple cross-section

It can be seen from the SEM images that carbon inclusions do not significantly change their shape and size, hence one can regard them as inert markers. As carbon forms in a reaction at the $ZrIr_3/ZrC$ interface and $ZrIr_3/Ir$ interface only $ZrIr_3$ forms, the carbon inclusions indicate the reaction by which the layer is formed. These inclusions are observed through almost the entire width of the product layer, with only a small region of carbon-free $ZrIr_3$ near the $ZrIr_3/Ir$ interface. Thus, we conclude that the product layer grows predominantly through reaction with ZrC and iridium is the dominant diffusing species.

Figure 2 shows iridium concentration profiles inside the product layer from $ZrIr_3/Ir$ to $ZrIr_3/ZrC$ boundary. Concentrations are above stoichiometric 75 at%, varying from 80 at% (higher boundary of the homogeneity range) near the $ZrIr_3/Ir$ interface to (75-76) at% near the $ZrIr_3/ZrC$ interface. Thus, the mobility of iridium is high enough to saturate the entire product layer so that zirconium-rich $ZrIr_3$ solid solutions are not observed.

3.2 Growth kinetics

To describe the kinetics of the reaction between iridium and zirconium carbide, the mean product layer thickness was measured for

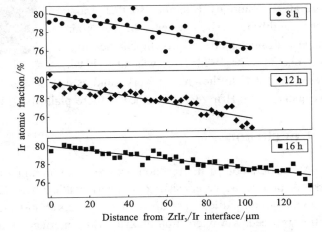

Figure 2 Iridium concentration profile in the $ZrIr_3$ phase

each sample from SEM images. A plot of the layer thickness against time is shown in Figure 3 both in linear and logarithmic scale. One can observe fast growth during the first two hours with subsequent growth deceleration. Growth kinetics obeys power law:

$$l=k(t/t_0)^n \qquad (2)$$

Where: l is layer thickness; t_0 is unit time (here $t_0 = 1$ h); k is a proportionality constant with units of length.

An exponent $n = 0.5$ indicates ordinary diffusional, or parabolic, growth. Our experimental data is described by a power law with exponent $n = 0.32$ indicating a more peculiar growth mechanism. Similar growth kinetics was reported in earlier works[10-14] on solid-solid and solid-liquid interactions with the formation of an intermetallic phase. Existing models[15, 16] relate non-parabolic growth with the diffusion enhancement near a reaction zone achieved through grain boundary diffusion, poor crystallization, or residual strain. These effects are taken into account using a variable diffusion coefficient $\overline{D}$:

$$\overline{D} = D\left[1 + \frac{D'}{D}\exp(-\lambda l)\right] \quad (3)$$

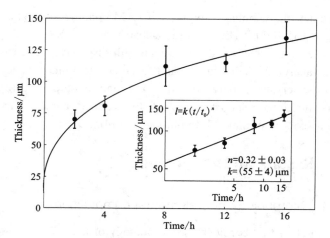

Figure 3 Total thickness of product layer vs annealing time at 1600 ℃

where: D' is a diffusion coefficient in the reaction zone, D is a bulk diffusion coefficient; l is the layer thickness and λ is a decay constant. Numerical analysis showed[15] that such treatment more adequately describes experimental data. In the case of a grain boundary diffusion with uniform grain coarsening with time, it was shown[16] that layer growth will obey the power lower with the exponent $n = 1/3$.

The growth of the product layer is approximately proportional to $t^{1/3}$ indicating that diffusion enhancement takes place in the reaction zone. One can assume that grain structure and tortuous microstructure of the reaction zone (see Figure 4) are the main sources of the observed behavior. In a similar system, HfC-Ir diffusion couples, the parabolic growth of a layer HIr_3-C was observed.[17] Experiments in that work were carried out at temperatures in the range of 1900-2200 ℃, where atomic mobility is higher so that grains reach equilibrium size faster, hence grain boundary diffusion affects growth kinetics to a lesser extent. The used HfC was also better sintered than ZrC in the present work, and as a consequence, the $HfIr_3$/HfC interface is more plain-like, so its morphology affects kinetics less.

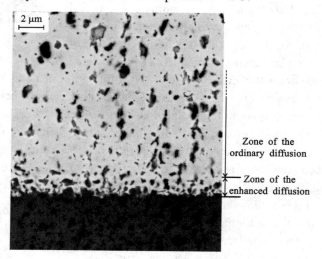

Figure 4 SEM image of the reaction zone

4 Conclusion

A kinetic study of the reaction between iridium and zirconium carbide at 1600 ℃ was carried out in this work. A product layer consisting of intermetallic phase $ZrIr_3$ and graphite-like inclusions forms during the reaction. Carbon particles are distributed through almost the entire product layer thickness, indicating that the reaction proceeds via interaction at $ZrIr_3$/ZrC interface as a result of the higher mobility of iridium atoms in $ZrIr_3$. The composition of the intermetallic phase is superstoichiometric by Ir also confirming that the diffusion of iridium is faster than that of zirconium. The growth kinetics is non-parabolic, obeying the power law with exponent $n = 0.32$. This behavior is attributed to the diffusion enhancement in the reaction zone because of the finer $ZrIr_3$ grains and intricate microstructure of the $ZrIr_3$/ZrC interphase. These findings provide valuable insights into the high-temperature behavior of ZrC, highlighting its potential for advanced applications.

Acknowledgments: This work was supported by the Russian Science Foundation (Project No. 23-19-00212).

References

[1] YAMABE-MITARAI Y, et al. Ir-base refractory superalloys for ultra-high temperatures[J]. Metall Mater Trans A, 1998, 29(2): 537-549.

[2] CONG X, et al. Co-deposition of Ir-containing Zr coating by double glow plasma[J]. Acta Astronautica, 2012, 79: 88-95.

[3] TAGUETT A, et al. Comparison between Ir, $Ir_{0.85}Rh_{0.15}$ and $Ir_{0.7}Rh_{0.3}$ thin films as electrodes for surface acoustic waves applications above 800 ℃ in air atmosphere[J]. Sensors and Actuators A: Physical, 2017, 266: 211-218.

[4] GONG H R. Ideal mechanical strengths of Ir and Ir_3Zr[J]. Scr Mater, 2008, 59(11): 1197-1199.

[5] WU J, ZHANG B, ZHAN Y. Ab initio investigation into the structure and properties of Ir-Zr intermetallics for high-temperature structural applications[J]. Comput Mater Sci, 2017, 131: 146-159.

[6] TERADA Y, et al. Thermal conductivity and thermal expansion of $Ir_3 X$ (X=Ti, Zr, Hf, V, Nb, Ta) compounds for high-temperature applications[J]. Mater Chem Phys, 2003, 80(2): 385-390.

[7] GONG H R, et al. Bond strength and electronic structures of coherent Ir/Ir_3Zr interfaces[J]. Appl Phys Lett, 2008, 92(21): 2006-2009.

[8] LI F Y, et al. Oxidation behaviours of Ir-Hf and Ir-Zr coatings under different air pressures at 1800 ℃. Available at SSRN: https://ssrn.com/abstract=4260049.

[9] HOLLECK H. Binäre und Ternäre Cabide und Nitride der Übergangsmetalle und ihre Phasenbeziehungen. Hochschulschrift. Universität Karlsruhe. 1981.

[10] VIANCO P T, ERICKSON K L, HOPKINS P L. Solid state intermetallic compound growth between copper and high temperature, Tin-rich solders — Part I: Experimental analysis[J]. J Electron Mater, 1994, 23(8): 721-727.

[11] MITA M, et al. Growth behavior of Ni_3Sn_4 layer during reactive diffusion between Ni and Sn at solid-state temperatures[J]. Mater Sci Eng A, 2005, 403: 269-275.

[12] SUZUKI K, et al. Reactive diffusion between Ag and Sn at solid state temperatures[J]. Mater Trans, 2005, 46(5): 969-973.

[13] SAKAMA T, KAJIHARA M. Influence of Ag on kinetics of solid-state reactive diffusion between Pd and Sn[J]. Mater Trans, 2009, 50: 266-274.

[14] LIS A, et al. Early stage growth characteristics of Ag_3Sn intermetallic compounds during solid-solid and solid-liquid reactions in the Ag-Sn interlayer system: Experiments and simulations[J]. J Alloys Compd, 2014, 617: 763-773.

[15] ERICKSON K L, HOPKINS P L, VIANCO P T. Solid state intermetallic compound growth between copper and high temperature, Tin-rich solders — Part II: Modeling[J]. J Electron Mater, 1994, 23(8): 721-727.

[16] SCHAEFER M, FOURNELLE R A, LIANG J. Theory for intermetallic phase growth between Cu and liquid Sn-Pb solder based on grain boundary diffusion control[J]. J Electron Mater, 1998, 27(11): 1167-1176.

[17] KWON J W. Formation and growth of Ir_3Hf layers at Ir/HfC interfaces between 1900 ℃ and 2200 ℃[D]. The Ohio State University, 1989.

Study of the Properties of Damping Materials for Manufacturing of Equipment in the Oil and Gas Industry

A. K. PRYGAEV, Y. S. DUBINOV, O. B. DUBINOVA,
M. S. TANASENKO, A. N. DUDKINA, E. A. PRYGAEVA

Gubkin Russian State University of Oil and Gas (National Research University), Moscow, 119991, Russia

Abstract: All equipment in the oil and gas industry is subject to vibrations that have a negative impact on the condition of wells, formations and the equipment itself, which is why any research aimed at preventing the influence of external vibrations on structures is relevant. When manufacturing equipment for the oil and gas industry, different grades of steel are used depending on the operating conditions. Recently, many companies have been developing equipment using structural materials with additional functional properties, for example, allowing to dampen or change vibration parameters. Examples of such materials are: high quality steel (HQS) 40CrNi2Mo, NiTi alloy with heat treatment (NiTi HT), as well as steel 01Al5Ti. In this paper, the dependences between acoustic and mechanical vibrations occurring in the process of equipment operation are obtained, as well as dependences for estimation of damping capacity from mass and dimensional parameters.

Keywords: Acoustic methods; Vibrations; Highly damping alloys; Damping steels; Acoustic and mechanical vibrations

According to the statistics of oil and gas equipment failures: units of gas distribution stations, surface drilling equipment, drilling rigs, separators, compressor and pumping equipment, more than 35% of failures are caused by random vibrations of high intensity[1]. Therefore, the development of new and better methods to prevent the influence of third-party vibrations on structures is a very urgent issue for most companies.

Dampener (from German "Dämpfer"-"muffler, shock absorber", "dämpfen"-"to muffle") is a special device for damping or damping vibrations or preventing random vibrations occurring in various technological installations while performing its functional purpose.

Damping ability of materials is explained by imperfect elasticity of the body material, which manifests itself in a non-linear dependence of mechanical stresses on the resulting deformations[1].

The authors propose a method for evaluating the damping ability of materials by acoustic vibrations. The method consists in recording the maximum and minimum peak readings of a noise meter (either by arithmetic mean readings or by octave readings, depending on the measurement requirements) installed inside a noise-and vibration-isolated cylindrical device shown in Figure 1.

Inside the device, a sample of material under investigation is placed, which is subjected to a mechanical impact, namely an impact load of a fixed mass from a measured height, causing vibrations that are measured by an acoustic device.

The authors, using the proposed scheme for measuring the damping ability of materials, have carried out studies of the following materials: high quality steel 40CrNi2Mo, steel 01Al5Ti[2], NiTi alloy[3] after heat treatment according to [4], and also investigated the influence of mass-size index on the process of damping vibrations of different frequency.

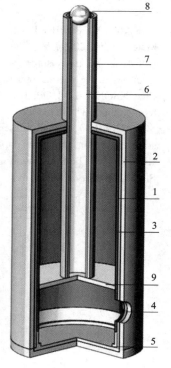

1—housing; 2—sound insulation;
3—vibration isolation;
4—hole for the sound meter;
5—bottom cover; 6—tube;
7—sound insulation of the tube;
8—ball; 9—sample.

Figure 1 3D model of the proposed installation

The evaluation is carried out by installing different number of plates of fixed thickness of 5 mm.

The results of measuring the sound pressure above and below the plate formed when a ball of known mass falls from a fixed height onto a plate, 5 mm thick, made of different materials, are presented in Figure 2.

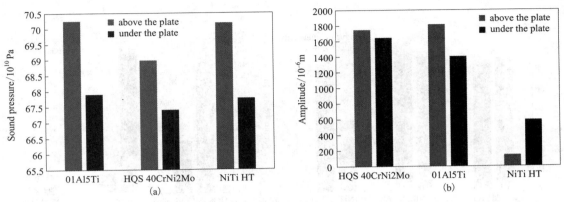

Figure 2 Sound pressure (a) and amplitude of vibrations (b) above and below the plate formed when a ball of known mass falls from a fixed height onto a 5 mm thick plate made of different materials

Analysing the graphs, we can say that the sound pressure level above the plate is approximately the same for steel 01Al5Ti and NiTi alloy with heat treatment, and for high quality steel 40CrNi2Mo is lower, which indicates the amplification of sound pressure, but under the plate there is a significant damping of sound pressure, the best result is again shown by steel 01Al5Ti and NiTi alloy with treatment (2.35 dB and 2.41 dB respectively).

In terms of vibration amplitude, it can be seen that the NiTi alloy with heat treatment shows the best result, although it should be noted that the amplitude above the plate is lower than below the plate.

The results of measuring the sound pressure and vibration amplitude above and below the plate formed when a ball of known mass falls from a fixed height onto a plate, 10 mm thick, made of different materials, are presented in Figure 3.

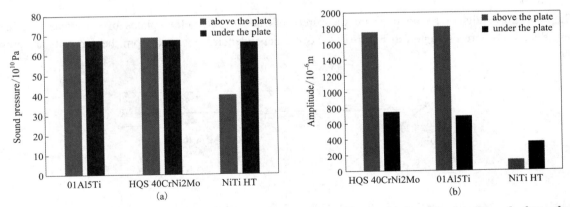

Figure 3 Sound pressure (a) and amplitude of oscillations (b) above and below the plate formed when a ball of known mass falls from a fixed height onto a 10 mm thick plate made of different materials

Analysing the graphs, we can say that the sound pressure level above the plate is very different and in comparison with tests with a plate thickness of 5 mm, differ only for NiTi alloy with heat treatment by 30 dB, which indicates the influence of mass-dimensional indicators for this type of material. The same value of sound pressure above and below the plate at different plate thicknesses shows high quality steel 40CrNi2Mo and NiTi alloy with heat treatment, i.e. mass-dimension indicators negatively influenced vibration damping. NiTi alloy with heat treatment shows not damping, but amplification of sound pressure under the plate, i.e. there is a process of generation of random sound vibrations.

According to the amplitude of vibrations it can be seen that the best result shows NiTi alloy with heat treatment, although it should be noted that the amplitude above the plate is lower than under the plate. An increase in mass dimension results in a significant decrease in vibration amplitude.

The results of sound pressure and vibration amplitude measurements above and below the plate formed when a ball of known mass falls from a fixed height onto a plate, 15 mm thick, made of different materials, are presented in Figure 4.

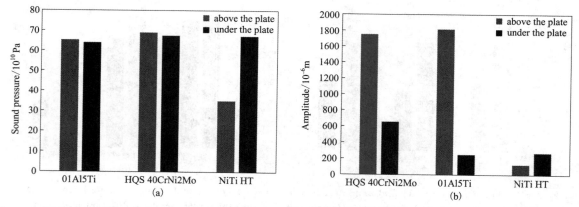

Figure 4　Sound pressure (a) and amplitude of oscillations (b) above and below the plate formed when a ball of known mass falls from a fixed height onto a 15 mm thick plate made of different materials

Analysing the graphs, we can say that the sound pressure level above the plate is very different and in comparison with tests with a plate thickness of 5 mm, differ only for NiTi alloy with heat treatment by 35 dB, which indicates the influence of mass-dimensional indicators for this type of material. The same value of sound pressure above and below the plate at different plate thicknesses shows high quality steel 40CrNi2Mo and NiTi alloy with heat treatment, i.e. mass-dimension indicators negatively influenced vibration damping. NiTi alloy with heat treatment shows not damping, but amplification of sound pressure under the plate, i.e. there is a process of generation of random sound vibrations.

The amplitude of vibrations shows that the best result shows NiTi alloy with heat treatment and 01Al5Ti steel. Increasing the mass dimensions of 01Al5Ti steel leads to the same values as shown by NiTi alloy with heat treatment.

Summarised results of sound pressure measurements above and below plates of different materials and different thicknesses are presented in Figure 5, and the dependencies obtained from these results are presented in table 1.

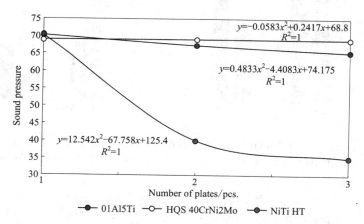

Figure 5　Variation of sound pressure above and below a plate formed when a ball of known mass falls from a fixed height onto plates of different thicknesses

Table 1 Dependencies derived from the results

Material	Dependencies
NiTi HT	$y=12.542x^2-67.758x+125.4$
HQS 40CrNi2Mo	$y=-0.0583x^2+0.2417x+68.8$
01Al5Ti	$y=0.4833x^2-4.4083x+74.175$

The obtained dependences allow to determine the level of sound pressure arising at the application of impact load, using different materials and mass-dimension indicators.

The test results show that for high quality steel 40CrNi2Mo sound pressure values are preserved at change of mass-dimension indicators, for steel 01Al5Ti sound pressure decreases at increase of mass-dimension indicators, for NiTi alloy with heat treatment sound pressure sharply decreases at increase of mass-dimension indicators and remains in one range at further increase of mass-dimension indicators.

The occurrence of a large value of sound pressure indicates the process of conversion of mechanical energy into sound energy and, thus, there is a damping of vibrations in the plate body.

The following conclusions can be drawn based on the results of the work performed:

(1) According to the results of the analysis of literature sources it is revealed that at present the field of application of structural steels and alloys with damping properties to combat random vibrations is expanding in the direction of the oil and gas industry.

(2) A methodology has been developed and a laboratory bench has been set up to evaluate the damping ability of various materials.

(3) It's determined that NiTi alloy with heat treatment converts mechanical vibrations into acoustic vibrations and thus dampens vibrations. It is most effective at low mass and dimensions. High quality steel 40CrNi2Mo retains its low damping ability when mass and dimensions are changed. 01Al5Ti steel has the same damping ability as NiTi with heat treatment when mass dimensions increase. It should be noted that the cost of 01Al5Ti is significantly lower than that of NiTi with heat treatment.

(4) Dependences allowing to calculate the damping capacity of the presented materials on the basis of mass and dimensional indices are obtained.

References

[1] DUBINOVA Y S, ELAGINA O Y, DUBINOVA O B, et al. Influence of the vibration parameters on the destruction of abrasives under sliding friction[J]. Inorganic Materials: Applied Research, 2021, 12(2): 576-580.

[2] CHUDAKOV I B, ALEXANDROVA N M, MAKUSHEV S Y. Features of consumer properties of new high-damping steels[J]. Steel, 2014(8): 92-95.

[3] SHEVCHENKO A D, DEVIN L N, OSADCHY A A. Features of damping ability of composite material based on titanium nickelide[J]. Modern Problems of Natural Sciences, 2014, 1(2): 63-68.

[4] Application 036600 Russian Federation. Method of thermomagnetic processing of steel products; applicant Gubkin Russian State University of Oil and Gas (NIU)-No. 2023117137; application 29.06.2023: 33.

Gadolinium-based Nanosized Phosphors Activated by Terbium for Medical Applications

Vadim BAKHMETYEV, Polina ZYKOVA, Anna VLASENKO,
Olga OSMAK, Nikolay KHRISTYUK, Sergey MJAKIN

Saint-Petersburg State Institute of Technology (Technical University), Saint-Petersburg, 190013, Russia

Abstract: GdF_3:Tb and Gd_2O_2S:Tb nanosized phosphors are synthesized by sol-gel and hydrothermal methods and characterized by the study of their phase composition, particle size and luminescence spectra. In GdF3:Tb phosphors, a directed growth of crystallites the (020) and (210) crystallographic planes is observed. The hydrothermal synthesis is shown to provide phosphors with smaller particles featuring twice higher luminescence intensity compared to the phosphor prepared by the sol-gel method. According to the luminescence performances and particle sizes GdF_3:Tb and Gd_2O_2S:Tb phosphors obtained using the hydrothermal procedure are suitable for the use in pharmaceutical drugs for X-ray photodynamic therapy of cancer in combination with photosensitizers Rose Bengal and Radachlorin.

Keywords: Nanosized phosphors; Sol-gel synthesis; Hydrodynamic synthesis; X-ray photodynamic therapy

1 Introduction

Terbium-activated phosphors based on gadolinium compounds provide an effective X-ray luminescence in the green spectral region, affording their use in medical drugs for X-ray photodynamic therapy (XRPDT) of cancer in combination with Rose Bengal photosensitizer. However, commercial Gd_2O_2S:Tb phosphors have particle size up to 10 μm, whereas the considered medical application requires nanosized phosphors (below 100 nm)[1] which cannot be obtained by conventional synthetic approaches based on a high temperature calcination of the charge mixture but can be prepared using such methods as sol-gel precipitation and hydrothermal synthesis[2]. This study is aimed at the development and optimization of procedures for obtaining rare earth element based nanosized GdF_3:Tb and Gd_2O_2S:Tb phosphors by sol-gel and hydrothermal syntheses, as well as the characterization of the prepared materials.

2 Experiment

As considered above, the objectives of this research are GdF_3:Tb and Gd_2O_2S:Tb nanosized phosphors. GdF_3:Tb cphosphor with the activator concentration 30 mol. % was synthesized using two methods:

(1) Sol-gel precipitation in aqueous medium at ambient temperature;
(2) Hydrothermal synthesis in ethylene glycol medium.

The first approach was implemented using gadolinium and terbium chlorides as precursors. Ammonium fluoride aqueous solution was used as a precipitating agent added dropwise to the precursors aqueous solution at stirring without any dispersing agent. The resulting precipitate was separated from the solution by centrifugation, then washed with ethanol and water followed by dispersing in bidistilled water at ultrasonic treatment to obtain a stable colloid.

The second synthetic procedure was carried out using the same precursors and precipitating agent dissolved in ethylene glycol with the addition of poly(ethylene glycol) with molecular weight 2000 as a dispersing agent. The precipitating agent was added at stirring, then the solution was sonicated, followed by placing into an autoclave and maintaining at 200 ℃ within 24 h. Subsequently, the precipitate was centrifugated, washed with ethanol and dispersed in bidistilled water similar to the first method.

Gd_2O_2S:Tb was prepared by hydrothermal synthesis in ethylenediamine. The activator concentration was

0.2 mol.%, as optimized in our earlier studies[3]. The synthesis was carried out using gadolinium and terbium nitrates dissolved in ethanol as precursors, sulfur dissolved in ethylenediamine as a precipitating agent and poly (vinyl pyrrolidone) Kollidon 30 (MW 44000-54000) as a dispersing agent. An ethanol solution of the precursors and dispersing agent was added dropwise to the sulfur solution in ethylenediamine. Then the prepared solution was placed into an autoclave and maintained at 220 ℃ within 24 h. The resulting precipitate was centrifugated, washed with ethanol and dispersed in bidistilled water similar to the synthesis of GdF_3:Tb phosphors described above.

The luminescence spectra of the prepared phosphors were studied using spectrofluorimeters AvaSpec-3648 (Avantes B. V., Netherlands) and CM 2203 (JSC SOLAR, Belarus). XRD characterization of the obtained samples was performed using a Rigaku SmartLab 3 (Rigaku Corporation, Japan) diffractometer and their dispersion was studied by SEM technique using a Tescan MIRA 3 (Tescan Orsay Holding, a. s., Czech Republic) electron microscope.

3 Results and discussion

XRD spectra of the synthesized GdF_3:Tb nanosized phosphors (Figure 1) show that their phase composition corresponds to the PDF card 12-788 (orthorhombic GdF_3) with different width of the peaks relating to different crystallographic planes. The most narrow and intensive peaks are observed for the planes (020) and (210), while other peaks are much broader. This difference can be accounted for a predominant growth of the phosphor crystals in the planes of (020) and (210). The calculations according to Scherrer equation indicated that the crystallite sizes of the synthesized samples towards the planes (020) and (210) are in the ranges of 36-42 nm and 19-21 nm, respectively, while in other directions is much lower (e. g. 6-7 nm for (111) plane).

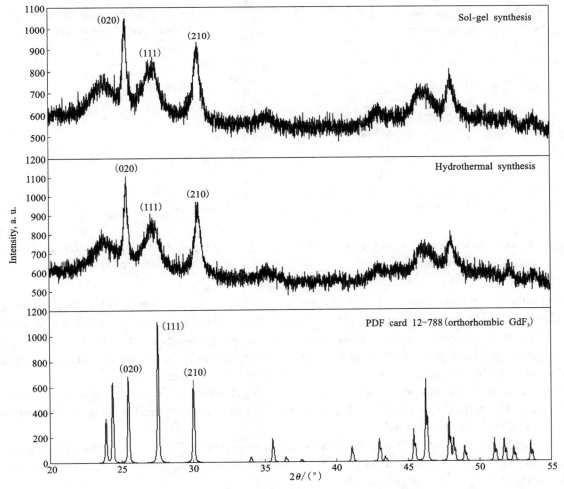

Figure 1 XRD spectra of GdF_3:Tb nanosized phosphors

The luminescence spectra of the synthesized GdF_3:Tb nanosized phosphors are shown in Figure 2. All the samples feature the presence of 4 luminescence bands are observed with the peaks at $\lambda_{max}=489$ nm, $\lambda_{max}=543$ nm, $\lambda_{max}=588$ nm and $\lambda_{max}=620$ nm. The highest intensity is observed for the green band at $\lambda_{max}=543$ nm, well corresponding to the absorption maximum of the Rose Bengal photosensitizer (Figure 3), which should provide an effective energy transfer from the phosphor to photosensitizer within a pharmaceutical drug for XRPDT. The intensity of this band for the phosphor prepared via the hydrothermal synthesis is twice higher compared with the sample obtained by sol-gel method.

The luminescence spectrum of the prepared Gd_2O_2S:Tb phosphor (Figure 4) also features a narrow green band with the peak at $\lambda_{max}=543$ nm intrinsic to Tb^{3+}, but, in addition, involves a broad blue band with the maximum at $\lambda_{max}=434$ nm, close to the absorption maximum of Radachlorin photosensitizer (Figure 5). The observed spectral characteristic suggest that the obtained Gd_2O_2S:Tb phosphor can be used in pharmaceutical drugs for XRPDT in combination with either Rose Bengal or Radachlorin photosensitizers.

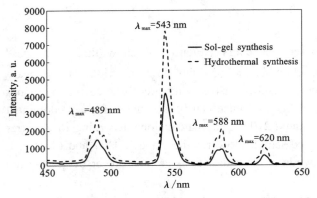

Figure 2 Luminescence spectra of GdF_3:Tb nanosized phosphors

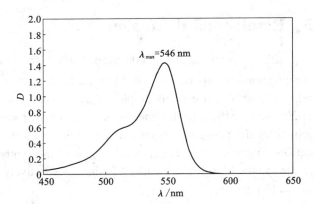

Figure 3 Absorption spectrum of Rose Bengal photosensitizer

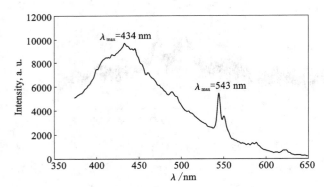

Figure 4 Luminescence spectrum of Gd_2O_2S:Tb nanosized phosphors

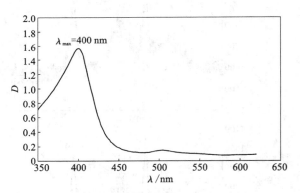

Figure 5 Absorption spectrum of Radachlorin photosensitizer

SEM images of the synthesized phosphors are shown in Figure 6. The sample GdF_3:Tb prepared by sol-gel synthesis comprised oblong aggregates with the length up to 400 nm and width up to 150 nm, which in turn consist of thin needle-shaped crystals with the length 100-200 nm and width no more than 20 nm. These data are in a good agreement with XRD results (Figure 1). Evidently, the needle-like crystals forming the aggregates are mostly elongated in the directions of (020) and (210) planes.

On the contrary, the phosphors prepared by hydrothermal synthesis consist of smaller particles with a regular shape. The particle size of GdF_3:Tb and Gd_2O_2S:Tb phosphors obtained by hydrothermal method are 40-60 nm and 20-40 nm, respectively.

Thus, the nanosized GdF_3:Tb and Gd_2O_2S:Tb phosphors prepared by hydrothermal synthesis are more applicable for XRPDT compared with GdF_3:Tb phosphor obtained using sol-gel precipitation, in respect to both luminescence brightness and particle size.

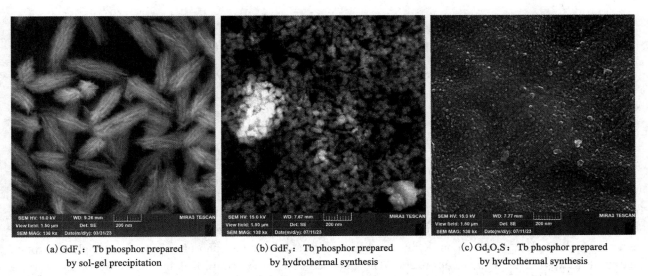

(a) GdF$_3$:Tb phosphor prepared by sol-gel precipitation

(b) GdF$_3$:Tb phosphor prepared by hydrothermal synthesis

(c) Gd$_2$O$_2$S:Tb phosphor prepared by hydrothermal synthesis

Figure 6 SEM images of the synthesized nanosized phosphors

4 Conclusion

The performed studies resulted in the hydrothermal synthesis of GdF$_3$:Tb and Gd$_2$O$_2$S:Tb nanosized phosphors according to their spectral characteristics and particle size suitable for the application in pharmaceutical drugs for X-ray photodynamic therapy of cancer in combination with Rose Bengal and Radachlorin photosensitizers.

References

[1] BAKHMETYEV V V, MINAKOVA T S, MJAKIN S V, et al. Synthesis and surface characterization of nanosized Y$_2$O$_3$:Eu and YAG:Eu luminescent phosphors which are useful in photodynamic therapy of cancer[J]. European Journal of Nanomedicine, 2016, 8(4): 173-184.

[2] BAKHMETYEV V V, DOROKHINA A M, KESKINOVA M V, et al. Synthesis and characterization of nanosized phosphors for enhanced photodynamic therapy of cancer[J]. Chemical Papers, 2020, 74(3): 787-797.

[3] RODIONOVA A V, KUULAR V I, MINAKOVA T S, et al. Acid-base and luminescent properties of Gd$_2$O$_2$S:Tb luminescent phosphors synthesized in a reducing atmosphere[J]. Key Engineering Materials, 2020, 854: 57-63.

X-ray Luminescent Phosphor-photosensitizer Systems for Oncotheranostics

Anna VLASENKO, Vadim BAKHMETYEV, Sergey MJAKIN

Saint-Petersburg Institute of Technology (Technical University), Moskovsky pr., 26, St. Petersburg, 190013, Russia

Abstract: X-ray luminescent phosphor-photosensitizer systems Y_2O_3:Eu-Radachlorin and Gd_2O_2S:Tb-Rose Bengal useful for X-ray photodynamic therapy (XRPDT) of cancer are comparatively studied by measuring their spectral characteristics and acid-base properties of the surface. The system based on terbium-doped gadolinium oxysulfide in combination with Rose Bengal photosensitizer provides the most efficient generation of singlet oxygen upon X-ray irradiation. In addition, this phosphor features a high adsorption activity towards the used photosensitizers. The obtained results are confirmed by the study of the surface functional composition via the adsorption of acid-base indicators indicating that Gd_2O_2S:Tb features a higher surface activity, particularly increased Lewis acidity and absolute value of ζ-potential compared with Y_2O_3:Eu. Furthermore, Gd_2O_2S:Tb phosphors obtained by hydrothermal synthesis are the most finely dispersed among the studied materials which is promising for XRPDT application.

Keywords: Biomedical applications; ζ-potential; X-ray photodynamic therapy; Oncotheranostics; Rare earth metals

1 Introduction

Rare earth metals based luminescent phosphors Y_2O_3:Eu and Gd_2O_2S:Tb are known for a long time and advantageously feature a high efficiency of photo-, cathodo-and X-ray luminescence, reduced toxicity and high stability and resistance to various impacts, particularly including cathode radiation[1].

The issues of obtaining and subsequent study of phosphors doped with various rare earth elements are becoming increasingly important, which is particularly determined by their possible use for medical purposes, such as oncotheranostics, both for the treatment and diagnostics of cancer[2-7]. Such rare earth based phosphors can be obtained by hydrothermal synthesis widely used for preparing materials of various compositions and purposes[8].

In addition to all the requirements for luminescent phosphors used for medical purposes, particularly for the development of pharmaceutical drugs, the study of their surface properties is also highly important. Surface properties largely determine the interaction of phosphor particles with body tissues and other components of drugs, and can affect the luminescence process, since surface active groups can serve as nonradiative recombination centers reducing the luminescence yield. Another important characteristic of the surface of particles responsible for the stability of nanosized phosphor suspensions depends is ζ-potential. The aim of this study was a comparative study of the target characteristics (particularly efficiency of active oxygen generation) of rare earth based phosphor-photosensitizer systems promising for use in photodynamic therapy, in comparison with the characteristics of their surface.

2 Experiment

The following systems mostly promising for XRPDT application in respect to many parameters were selected for comparative studies:

(1) Y_2O_3:Eu phosphor+photosensitizer Radachlorin;
(2) Gd_2O_2S:Tb phosphor+photosensitizer Rose Bengal.

The most important parameter of luminescent phosphor-photosensitizer systems used in photodynamic therapy is the generation of active (singlet) oxygen characterized in this study using the "chemical trap" method. According to this technique a substance (chemical trap) turning from a colored to the colorless state upon reacting with the singlet oxygen is added to a colloid solution containing a phosphor and photo-or Z-ray sensitizer. Since the amount of the discolored chemical trap is equal to the amount of yielded singlet oxygen, its generation can be estimated according to the change of the solution optical density. In this research, 1, 3-Diphenylisobenzofuran[9] was used as a chemical trap.

For all the studied systems the active oxygen generation was studied under a "hard" X-ray irradiation in comparison with control experiments involving a similar irradiation of the systems free of the phosphors and X-ray sensitizers, followed by measuring the absorption spectra of all the samples using a SF-56 spectrophotometer (LOMO, St. Petersburg, Russia).

The luminescence spectra of the studied materials were measured using a spectrofluorimeter AvaSpec-3648 (Avantes B.V., Netherlands) upon excitation with X-ray radiation supplied by the X-ray installation based on RPD-200 apparatus with the voltage at X-ray tube 160 kV, current 4.5 mA and exposure time 16 min.

The surface properties of the phosphors were characterized by the adsorption of acid-base indicators with different pK_a values in the range from -5 to 15 based on optical density measurements for standard indicator solutions before and after the contact with the studied materials using the spectrophotometer SF-56 according to the procedure described in detail in [10]. Based on the obtained results, the content of adsorption centers with different pK_a values on the surface of the phosphors was calculated and distributions of adsorption centers by pK_a values were comparatively analyzed for different samples.

The dispersion of the synthesized nanosized phosphors was studied by SEM technique using a Tescan MIRA 3 (Tescan Orsay Holding, a.s., Czech Republic) electron microscope.

3 Results and discussion

The luminescence spectra of Y_2O_3:Eu-Radachlorin and Gd_2O_2S:Tb-Rose Bengal systems is shown in Figure 1. The maximum of X-ray luminescence for Y_2O_3:Eu-Radachlorin system is close to the peak of this photosensitizer absorption at 660 nm but does not coincide with it. For the system Gd_2O_2S:Tb-Rose Bengal the photosensitizer absorption band completely overlaps the peak of the phosphor luminescence. Consequently, according to spectral characteristics the system Gd_2O_2S:Tb-Rose Bengal provides a more efficient energy transfer from the phosphor to the photosensitizer.

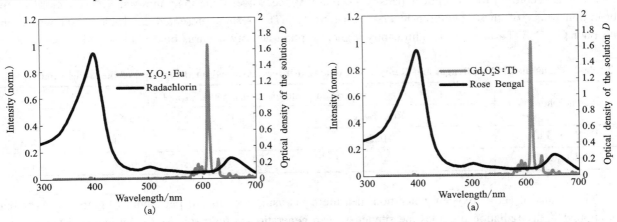

Figure 1 Luminescence spectra of the systems Y_2O_3:Eu-Radachlorin (a) and Gd_2O_2S:Tb-Rose Bengal (b)

The distribution of adsorption centers on the surface of the studied materials is presented in Figure 2. The overall content of active surface centers and ζ-potential values are summarized in Table 1.

Table 1 Total content of active surface centers and ζ-potential values of the studied phosphors

Phosphor	Total content of active surface centers/($\mu mol \cdot g^{-1}$)	ζ-potential/mV
Y_2O_3:Eu	51.38	-12.07
Gd_2O_2S:Tb	71.23	-16.85

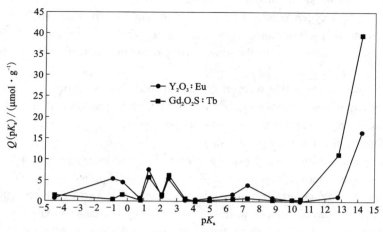

Figure 2 Distribution of adsorption centers on the surface of Y_2O_3:Eu and Gd_2O_2S:Tb phosphors

The data in Figure 2 show that the surface of the studied phosphors is predominantly occupied with Lewis acidic centers with pK_a 14.2 corresponding to positively charged rare earth metal ions (yttrium and europium in Y_2O_3:Eu, gadolinium and terbium in Gd_2O_2S:Tb). According to the Table 1, the phosphor Gd_2O_2S:Tb features a higher content of active centers and more negative ζ-potential compared with Y_2O_3:Eu. The negative ζ-potential of the studied phosphors in the analyzed solutions are probably determined by the attraction of hydroxyl ions from the solution to Lewis acidic centers, resulting in the formation of a negatively charged coating on the surface of the particles. Since the Gd_2O_2S:Tb phosphor surface contains more acidic centers compared with Y_2O_3:Eu, its hydroxyl coating has a stronger negative charge providing a higher stability of the Gd_2O_2S:Tb phosphor colloid. Furthermore, higher total content of active surface centers should provide an increased adsorption capacity towards the photosensitizer.

It was confirmed in another experiment on the comparison of the studied phosphors in respect of the adsorption of Rose Bengal and Radachlorin photosensitizers from aqueous solutions with the concentration 0.18 mol/L according to the optical density decrease at the wavelengths 548 nm and 400 nm close to the absorption peaks of these photosensitizers, respectively. The results presented in Table 2 show that the phosphor Gd_2O_2S:Tb features a significantly higher sorption activity towards both photosensitizers.

Table 2 Amounts of Rose Bengal and Radachlorin photosensitizers adsorbed on the surface of the studied phosphors

Phosphor	Adsorbed photosensitizer amount/(μmol·g^{-1})	
	Rose Bengal	Radachlorin
Y_2O_3:Eu	0.20	0.74
Gd_2O_2S:Tb	0.53	0.77

The results of active oxygen generation measurements using the studied phosphors (Figure 3) show that the applied X-ray radiation provides the singlet oxygen generation in their solutions even in the absence of the phosphors and photosensitizers. The addition of Gd_2O_2S:Tb phosphor to the system containing Rose Bengal photosensitizer results in a 3.7-fold increase in the active oxygen production, while the phosphor addition in the system Y_2O_3:Eu+Radachlorin almost does not affect the singlet oxygen generation.

SEM images of the synthesized phosphors (Figure 4) indicate that the particle size of Y_2O_3:Eu phosphor prepared by hydrothermal synthesis is about 100 μm and more which requires the further optimization to obtain the nanosized material, while in the case of Gd_2O_2S:Tb the particle size is about 20-40 nm that is optimal for biomedical applications.

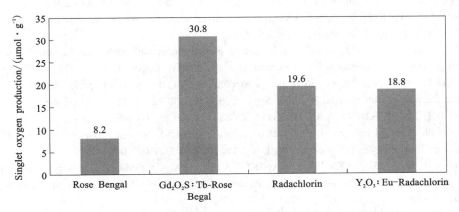

Figure 3　Comparison of active oxygen generation by the studied systems

Figure 4　SEM images of Y_2O_3:Eu (a) and Gd_2O_2S:Tb (b) phosphors prepared by hydrothermal synthesis

4　Conclusion

　　The obtained results show that among the studied rare earth metal based material the phosphor Gd_2O_2S:Tb in combination with Rose Bengal photosensitizer provides the best performances in singlet oxygen generation and adsorption of photosensitizers which makes this system promising as a component of drugs for photodynamic therapy of cancer. A high efficiency of this material is largely determined by the nanoscale size of its particles and high activity of its surface featuring an increased content of Lewis acidic centers.

References

[1] SYCHOV M M. Optimization of low-voltage cathodoluminescence of electron-beam-evaporated Y_2O_3:Eu thin film phosphor[J]. J. of Appl. Phys. Japan, 2008, 47, 9: 7206-7210.
[2] YU X J, LIU X Y, WU W J, et al. MRI-guided synergistic radiotherapy and X-ray inducible photodynamic therapy using Tb-doped Gd-W-nanoscintillators[J]. Angew. Chem. -Int. Edit, 2019, 58(7): 2017-2022.
[3] OUYANG Z J, LI D, XIONG Z J, et al. Antifouling dendrimer-entrapped copper sulfide nanoparticles enable photoacoustic imaging-guided targeted combination therapy of tumors and tumor metastasis[J]. ACSAppl. Mater. Interfaces, 2021, 13(5): 6069-6080.
[4] YU X J, LIU X Y, WU W J, et al. MRI-guided synergistic radiotherapy and X-ray inducible photodynamic therapy

using Tb-doped Gd-W-nanoscintillators[J]. Angew. Chem. -Int. Edit, 2019, 58(7): 2017-2022.

[5] LI B, ZHAO M, ZHANG, F. Rational design of near-infrared-II organic molecular dyes for bioimaging and biosensing[J]. ACS Mater. Lett, 2020, 2: 905-917.

[6] FENG Y S, LIU X M, LI Q Q, et al. A scintillating nanoplatform with upconversion function for the synergy of radiation and photodynamic therapies for deep tumors[J]. J. Mater. Chem. C, V, 2022, 10(2): 688-695.

[7] MAITI D, YU H, KIM B S, et al. Rose bengal decorated NaYF4:Tb nanoparticles for low dose X-ray-induced photodynamic therapy in cancer cells[J]. ACS Appl. Bio Mater, 2022, 5(11): 5477-5486.

[8] VLASENKO A B, DOROKHINA A M, BAKHMETYEV V V, et al. Hydrothermal synthesis and characterization of nano-sized phosphors based on rare-earth activated yttrium compounds for photodynamic therapy[J]. Journal of Sol-Gel Science and Technology, 2023. https://doi.org/10.1007/s10971-022-06013-6.

[9] VLASENKO A B, BAKHMETYEV V V, NIANIKOVA G G. Investigation of the activity of phosphorphotosensitizer systems for oncoteranostics by chemical and microbiological methods[J]. Bulletin of the St. Petersburg State Institute of Technology (Technical University), 2022, 62(88): 12-15.

[10] SYCHOV M M, MINAKOVA T S. Kislotmo-osnovnye scoistba poverhnosti tverdykh tel I upravlenie svoistvami materialov I kompozitov (Acid-base properties of solids and control over properties of materials and composites) [M]. St. Petersburg: Khimizdat Publishers, 2022: 288.

Reinforced Plastics for Alternative Wind Power

Natalia KORNEEVA[1], Vladimir KUDINOV[2]

1 Semenov Federal Research Center of Chemical Physics, Russian Academy of Sciences, Moscow, 119334, Russia
2 Baikov Institute of Metallurgy and Materials Science, Russian Academy of Sciences, Moscow, 119334, Russia

Abstract: A study on the development and production of ultra-light CMs with a density of 1.1 g/cm³ for wind energy is presented. Polyethylene plastics reinforced with woven fabrics and non-woven materials from UHMWPE-fibers activated by non-equilibrium low-temperature plasma were obtained and their properties were studied.
Keywords: Ultra-light composites; Wind energy; Turbine blades; UHMWPE-fiber

1 Introduction

Traditional sources of energy such as coal, peat, oil, shale and natural gas when burned emit carbon dioxide into the atmosphere which cause a greenhouse effect and global warming. Alternative energy provides the opportunity to humankind without environmental degradation use non-conventional energy sources, which make it possible to obtain the energy necessary for its existence. The main reason for the search for alternative energy sources is the need to obtain it from renewable or virtually inexhaustible natural resources.

In this connection, wind energy has a special place among the alternative energy sources, because it is available and easily converts kinetic energy of the atmospheric air into the electrical, mechanical, thermal, or any other form of energy for use in the national economy. Wind energy is a renewable kind of energy. The use of wind energy does not lead to irreversible or critical changes in the atmosphere, hydrosphere and lithosphere. The main environmental benefit of wind energy is the reduction in greenhouse gas emissions.

To obtain electricity and mechanical energy, the transformation of atmospheric air energy is carried out using wind generators and windmills.

To date wind energy is developing rapidly. In 2022 the total installed wind power capacity worldwide was approximately 906 GW[1]. The power of modern wind turbines (WT), such as SWT-7.0-154 and Enercon 126, reached 7 MW and 7.58 MW, respectively[2]. By the end of 2022, the installed capacity of wind turbines was 365420 MW in China, 144226 MW in the USA and 66322 MW in Germany[3,4].

The leader in annual capacity commissioning is China: in 2022, 37631 MW were installed in the country (32579 MW onshore and 5052 MW off shore). China's renewable energy capacity will continue to grow as the country seeks to meet half of its electricity demand from renewable sources[3].

The development of wind energy depends on solving the fundamental problem associated with the severity of wind turbine blades. The power of a wind farm is critically dependent on the size of the turbine blades. A wind farm can provide competitive electricity generation as a result of an increase in the radius of the wind turbine blade. The increase in electricity generation by factor 2 is achieved by lengthening the turbine blade. It is known that the power of a wind generator depends on the power of the wind flow (N), which is determined by the wind speed and the area swept by the generator blades according to the following equation:

$$N = \rho S v^3 / 2$$

where: v is the wind speed; ρ is the air density; $S = \pi R^2$ is the swept area; R is the radius of the blade.

The traditional materials from which the blades are made are metals and composite materials (CMs) based on carbon, glass and basalt fiber. However, their use is associated with increased noise from wind farms. Without the use of new generation CMs, wind farms will remain a noisy and low-power segment of the energy sector. In Germany, there is an active renovation process, called "repowering", when old wind turbines are replaced with more powerful and less noisy ones.

New materials such as advanced carbon fiber reinforced plastic (CFRP) and glass fiber reinforced plastic (GFRP) make it possible to manufacture twice as long blades, which brings wind farms to an efficiency comparable to and superior to that of traditional sources for generating energy. That is why in countries where laws have been adopted to limit harmful emissions into the atmosphere, an increased demand for blades made from the latest high-strength CMs has been recorded. In Russia as well as in China we also see this growth in demand linked with the program for the development of alternative energy.

The production of wind turbine blades twice and three times longer than traditional ones is possible due to the use of ultra-light composites with high physical and mechanical properties based on both carbon, glass and basalt fibers as well as high-strength high-modulus fibers from ultra-high molecular weight polyethylene (UHMWPE) as well as their hybrids[5, 6].

Due to the nanocrystalline structure, multifilament UHMWPE-fiber has a unique combination of main properties[7, 8]. Its density is less than unity (0.97 g/cm^3), and its specific strength is 15-20 times higher than that of steel fibers. UHMWPE-fibers allow one to get rid of the noise of wind farms, since they have the necessary acoustic properties, their dielectric loss tangent is -2×10^{-4}.

The aim of the work was to develop a method for obtaining ultra-light composite materials so-called "polyethylene plastics" reinforced with woven and non-woven materials from UHMWPE-fibers activated by non-equilibrium low-temperature plasma (NLT), and to study their properties.

2 Experiment

The reinforcing fillers were unidirectional non-woven materials made of Dutch high-strength high-modulus UHMWPE-fiber of trademark Dyneema ® SK-75 and fabric made of Chinese UHMWPE-fiber 800D (Table 1). Epoxy and epoxyurethane binders based on ED-20, ED-22, and Epicot-828 epoxy resins cured with aliphatic and aromatic amines were used as matrices[9].

Table 1　Properties of UHMWPE-fibers

Fiber brand	Density, $\rho/(g \cdot cm^{-3})$	Tensile strength, σ_t/GPa	Specific strength, σ_t/ρ/km	Elastic modulus, E/GPa	Elongation at break /%
SK-75	0.97	3.4	350	110	3.8
800D	0.97	2.48	256	85	2.94

To enhance the interfacial interaction during the production of CMs, fibrous fillers were activated by non-equilibrium low-temperature argon radio-frequency (RF) plasma at the reduced pressure from 1.33 Pa up to 660 Pa. The thermal component of such a plasma can be minimized due to the low ion current density of $j_i = 0.5$-1 A/m^2 and the short duration of plasma exposure to the fiber[10].

In order to control the CM properties and increase the realization of the properties of UHMWPE-fibers in the composite, the following technological methods were developed: ①NLT-plasma pretreatment of the fiber and obtaining CM in air; ②impregnation and production of CM in vacuum from raw fiber; ③preliminary plasma treatment of the fiber and obtaining CM in vacuum[11-13].

Strength properties of polymer CMs reinforced with multifilament fibers such as Dyneema, Spectra, etc., are mainly determined by two parameters. First, this is the degree of multifilament fibers impregnation with the matrix. The better is the fiber impregnation, the larger is the contact area between the filaments and matrix. Second, this is the strength of the fiber/matrix joint at the interface. CM fracture is often initiated by poor bond at the fibers/matrix interface. We compared vacuum impregnation and air impregnation. The strength of the fibers/matrix joint was estimated by the wetted-pull-out method (W-P-O). The W-P-O method allows one to estimate the wettability and multifilament fibers impregnation with the liquid matrix and measure fiber capillary rise h and joint strength after curing process finished[12].

To obtain non-woven CMs, non-woven reinforcing materials were produced by winding fibers on a polypropylene mandrel frame. The fibers on the frames were treated with the help of NLP-plasma. The non-woven materials consisted of parallel-laid unidirectional fibers with longitudinal-transverse laying (1:1) fibers. Plasma-activated fibrous fillers were impregnated with a binder to form prepregs. In the manufacture of CMs, a prepreg blank was assembled from alternating layers of non-woven materials or fabrics. The composite laminates were molded by pressing prepreg blanks[14].

The properties of laminated CM polyethylene plastics were evaluated using a three-point bending loading scheme for samples according to breaking stresses during bending and shear. Specimens with width, length and thickness (δ) equal to 15 mm, 300 mm and 3.2-3.3 mm, respectively, were cut from the laminate in order to determine the CM shear strength (interlaminar shear strength ILSS) and the CM bending strength (ultimate flexural strength, UFS).

The studies were carried out on an Instron 3382 universal testing machine at a cross-head speed of 5 mm/min. The properties were compared for the CMs reinforced with the plasma-treated and untreated fibers.

3 Results and discussion

In previous works[11-13] using the W-P-O method it was found out that NLT-plasma treatment of UHMWPE-fiber improved its wetting in air and in vacuum with an epoxyurethane matrix (EPUR)[9] by 35% and 173%, respectively, compared to fiber without plasma treatment impregnated in air. Plasma treatment of UHMWPE-fiber improved its wetting in air and in vacuum with an EPUR matrix by 86% and 141%, respectively, compared to untreated fiber impregnated in air. At the same time, it was shown that the joint strength between the fiber and the matrix increased by a factor of 2 and 3 at atmospheric pressure and in vacuum[10-12]. Coupled action of vacuum and plasma treatment of the fibers improved the strength of the fiber/matrix joint by 200% or by factors of 3.

Thus, for improving the properties of composite materials, reinforced with UHMWPE-fibers, one should use three following methods: First, it is the impregnation of initial fibers in vacuum and the production of CMs in vacuum; second, it is plasma treatment of the fibers and the production of CMs by the impregnation of plasma-activated fibers in air; third, it is plasma treatment of the fibers and the production of CMs by the impregnation of plasma-activated fibers in vacuum. The technology of producing CMs in vacuum is complicated. That is why in our further experiments we used only plasma treatment of the fibers and the impregnation of plasma-activated fibers in air.

It was previously established[14, 15] that after the plasma treatment of non-woven material made of SK-75 fibers the CM bending strength (σ_b) was increased by a factor of 1.65 from 265 MPa to 436 MPa. The CM shear strength (τ_{sh}) was increased by the factor of 1.68 from 19 MPa to 33 MPa (Table 2). New experiments confirm these data.

Table 2 Properties of non-woven polyethylene plastics[14, 15]

Material characteristics	Non-woven reinforcement, cross-ply lay-up, SK-75/EPUR laminates (1:1)	
	Untreated	Plasma treated
Relative (volumetric) content of fibers, V_f/%	72	56
Specimen thickness, δ/mm	3.2	3.3
Ultimate flexural strength UFS (σ_b)/MPa	265	436
Interlaminar shear strength ILSS(τ_{sh})/MPa	19	33

The properties of the obtained polyethylene plastics from Chinese UHMWPE-fiber 800D are given in Table 3. After plasma treatment of woven fabric the CM bending strength σ_b was increased by a factor of 2.7 from 165 MPa to 450 MPa. The CM shear strength τ_{sh} was increased by the factor of 1.9 from 13 MPa to 25 MPa.

Table 3 Properties of woven fabric polyethylene plastics

Material characteristics	Woven fabric reinforcement, cross-ply lay-up, 800D/EPUR laminates (1:1)	
	Untreated	Plasma treated
Relative (volumetric) content of fibers, V_f/%	60	60
Specimen thickness, δ/mm	3.2	3.3
Ultimate flexural strength UFS (σ_b)/MPa	165	450
Interlaminar shear strength ILSS (τ_{sh})/MPa	13	25

Thus, it has been established that the flexural and shear strengths for CMs made of non-woven materials and fabrics activated by NLT-plasma increase by a factor of nearly 2 when the fiber is cross-laid. Table 4 shows the properties of polyethylene plastic in comparison with the properties of used materials.

Table 4 Properties of materials[16]

Materials	ρ/(g·cm^{-3})	σ_t/(kg·mm^{-2})	σ_t/ρ/km	E/GPa	E/ρ/(GPa·t·m^{-3})
UHMWPE-FRP	1.1	154	140	50	45
CFRP	1.55	180	116	150	97
GFRP	2.5	155	62	51	20
Aluminum	2.8	80	29	73	26
Titanium alloys	4.5	160	36	120	27
Steel	7.8	220	28	210	27

Plasma treatment improves interfacial interaction, wetting and impregnation of UHMWPE-fibers with polymer matrices, which makes it possible to obtain lightweight polyethylene plastics with a density approximately of 1.1 g/cm^3, which are superior in specific strength to metals, GFRP and CFRP. In terms of strength, CM polyethylene plastic is at the level of known and widely used materials[16].

4 Conclusion

The high specific strength of new polyethylene plastics makes it possible to use them for the manufacture of lighter wind turbine blades. Such CMs allows one to double and even triple the length of the blades of wind turbines, which will increase the efficiency of existing wind farms and reduce their noise.

Acknowledgments: The authors are grateful to Prof. Abdullin I. Sh. & Prof. Shaekhov M. F. for plasma treatment.

References

[1] www.statista.com
[2] www.enercon.de//E-126
[3] eprussia.ru
[4] global-wind-report-2022
[5] KORNEEVA N V, KUDINOV V V, ABDULLIN I S. Plasma technology for the production of composite materials for alternative energy, the Collection of materials of the All-Russian Physics of low-temperature plasma, Kazan, Publ. House KNRU, 2014(2): 25.
[6] KORNEEVA N V, KUDINOV V V, KRYLOV I K. Regulation of carbon plastic properties upon impact by hybridization of the reinforcing fibers[J]. IOP Conf. Ser., Institute of Physics Publ., 2019(1347): 012057. DOI: 10.1088/1742-6596/1347/1/012057

[7] PEREPELKIN K E. Past, present and future of chemical fibers[J]. Moscow: Kosygin's MSTU, 2004: 208.
[8] www.dsm.com
[9] Russian Patent No. 2277549
[10] Russian Patent No. № 2467101
[11] KUDINOV V V, SHAECHOV M F, KORNEEVA N V. Effect of plasma treatment and impregnation mode on the strength of joining of polyethylene fibers with epoxy resin matrix in the process of CMs production[J]. Phys. & Chem. Mater. Treat., 2004(3): 18-24.
[12] KUDINOV V V, KORNEEVA N V. Using plasma-activated high performance fibers with nanocrystalline structure in producing new reinforced composite materials[J]. Macromol. Symp., 2009(286): 187-194. DOI: 10.1002/masy.200951223
[13] KUDINOV V V, KORNEEVA N V. Using plasma-activated UHMPE-fibers with nanocrystalline structure in producing textile composite materials[J]. Proc. of the Int. Joint Sheffield-Cambridge-Manchester Conf. (DFC12/SI6): Deformation and Fracture of Composites (DFC-12) & Structural Integrity and Multi-scale Modeling (SI-6). Cambridge: Univer. of Sheffield, 2013: S12T3.
[14] KORNEEVA N V, KUDINOV V V, KRYLOV I K, et al. [J]. Encyclopedia of the Chemical Engineer, 2012(6): 7-8.
[15] KUDINOV V V, KORNEEVA N V. Composite materials reinforced with plasma-activated ultra-high molecular weight polyethylene fibers with nanocrystalline structure[M]. Proc. of the 12th China-Russia Symp. on Advanced Mater. and Technol.: Advanced Metals, Ceramics and Composites (CRS AMT 2013). Kunming, Yunnan Publ. Group Corp. & Yunnan Scie. & Tech. Press, 2013: 345.
[16] KUDINOV V V, KORNEEVA N V, KRYLOV I K. Influence of components on the properties of polymer composite materials[M]. Moscow: Science, 2021: 134. DOI: 10.7868/9785020408654.

Thermostable Functional Materials Based on Phosphate Binders

Konstantin N. LAPKO[1,2], Alexander N. KUDLASH[1], Natalia S. APANASEVICH[2], Aliaksei A. SOKAL[1]

1. Belarusian State University, Faculty of Chemistry, Minsk, 220030, Belarus
2. Research Institute for Physical Chemical Problems of the Belarusian State University, Minsk, 220006, Belarus

Abstract: The paper considers the use of liquid and solid phosphate binders to obtain a variety of functional thermostable composite materials. The use of microwave radiation for curing and obtaining of thermostable phosphate composites is shown to be promising.

Keywords: Thermostable materials; Phosphate composites; Liquid and solid binders; Dry building mixtures; Microwave treatment

1 Introduction

The use of phosphate binders allows to obtain a wide range of thermally stable materials with operation temperatures in a quite broad range from −200 ℃ to +2000 ℃[1-4].

Specific application area of the materials, conditions of their use, and attaining definite physical, chemical and mechanical characteristics are determined by the composition and proportions of the main components: the phosphate binders and the fillers. Thus, depending on the filler's type, composites may perform as radio-transparent or radio-absorbing materials, dielectrics or electric conductors, heat-conducting or heat-insulating materials, electro-conductive or neutron-absorbing composites as well as materials for ionizing radiation protection.

Modification of well-known phosphate materials and imparting new properties to them is possible in two ways: changing the formulation of composites by introducing new functional fillers and improving the technology for producing composites.

The manufacture of phosphate materials is waste-free and does not need any complicated equipments. It can be successfully organized at any enterprise of building industry. The phosphate composite materials are non-combustive, nontoxic, and environmentally friendly.

The available application areas of the thermostable materials based on phosphate binding composition are: aviation and space industry, metallurgy, ceramics industry, production of glass, refractories and building materials.

2 Experiment

With the appearance of a new type of materials that can act as different functional fillers, the development of new phosphate composites with specific properties is of great interest. Thus, the production on an industrial scale and the subsequent use of inorganic hollow microspheres (HMS) led to the creation of highly efficient heat-insulating materials[5,6]. The occurrence of carbon nanotubes (CNTs) as a nanosized functional filler made it possible to develop thermally resistant and electrically conductive phosphate materials that also have shielding properties with respect to electromagnetic radiation[7,8]. When using such nanosized fillers as $BaTiO_3$, Fe_3O_4, SiC and introducing them into phosphate matrices, functional materials with special electromagnetic properties were obtained[9-11].

In most cases, liquid binders, for example, aluminum-phosphate, aluminum-chromium-phosphate, etc., are widely used in the preparation of phosphate composites. However, production of materials based on liquid phosphate binders involves a number of difficulties. Such binders are usually unstable during storage with crystallization processes occurring over time, which leads to formation of rather strong deposits that changes the atomic ratio of phosphorus and metal (P/Me) in the remaining liquid phase. The latter one requires adjustment of the filler/binder ratio in the preparation of thermally and hydrolytically stable composites.

The transition to the use of solid binders, such as magnesium dihydrogen phosphate tetrahydrate and calcium dihydrogen phosphate monohydrate, made it possible to improve the functional properties of composites[5, 6, 12]. The use of solid binders proved to be especially effective when working with relatively weak filler components, such as hollow microspheres, expanded vermiculite and perlite, CNTs and whiskers. When using phosphate binders in solid form, the mixing of the initial reaction mixture is carried out in an air-suspended state. This method of composite materials preparation allows better preservation of the integrity of these fillers, which ensures the manifestation of the functional properties of the composites to the maximum extent.

For example, the results of electron microscopic studies of heat-insulation phosphate composites with hollow microspheres (Figure 1) have shown, the use of proposed method of air-weighted mixing of HMS with other components for the preparation of heat-insulating compositions practically does not lead to the destruction of microspheres, as evidenced by the images of the surfaces and side split of the obtained composites based on solid magnesium phosphate binder (S-MPB)[6]. The micrographs have shown that the microspheres in the prepared composite remain predominantly in an undestroyed state, which determines the effectiveness of the applied technique for obtaining heat-insulating phosphate composites.

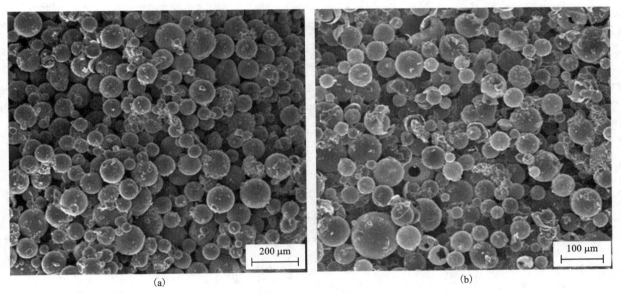

Figure 1 SEM images of the surface (a) and side split (b) of the prepared heat-insulation composites based on S-MPB with fly-ash (a) and glass (b) hollow microspheres[6]

Research results[5] show that using microwave radiation is a promising method of materials treatment. Volume absorption of microwave radiation enables uniform and much faster heating of the products as compared to traditional methods. This provides for economic and technical benefits including energy saving, shorter treatment time, high level of automation and control of sintering processes, and a significant increase of the dispersion ability and homogeneity of the sintered materials.

The results of measurement of the compressive strength of the samples have been shown that thermostable composites based on a solid magnesium phosphate binder $Mg(H_2PO_4)_2$

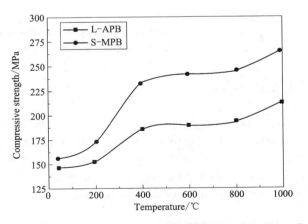

Figure 2 The dependence of phosphate composite strength on the thermal treatment temperature for the samples obtained on the base of liquid or solid binders[5]

· $4H_2O$ (S-MPB) have the improved strength properties as compared to ones based on a traditional liquid aluminum phosphate binder $Al(H_2PO_4)_3$ (L-APB). The resulting data (Figure 2) have been clearly illustrated that the strength of S-MPB-based composites is 40-60 MPa higher than that of L-APB[5].

A similar effect of increasing strength was observed for phosphate composites containing multi-walled CNTs as a functional filler[12]. Composites were obtained using a solid magnesium phosphate binder in comparison with a liquid aluminum phosphate binder. The results showed (Figure 3) that the strength (σ_{comp}) of composites based on S-MPB is 1.3-3.0 times higher than the strength of composites based on L-APB in the entire studied range of CNT content from 0 to 5 wt. %.

The examination of the power and duration of microwave treatment[5] on the strength of the phosphate composites, obtained on the base of S-MPB, revealed (Figure 4) that increase of the microwave treatment power from 140 W to 700 W leads to a double successive increase of the sample strength. Furthermore, high strength properties of the composites can be achieved as soon as after 3-5 min of microwave treatment. Based on this experiment, one can conclude that microwave treatment provides materials with high compressive strength properties at low energy consumption and significantly shorter samples treatment time.

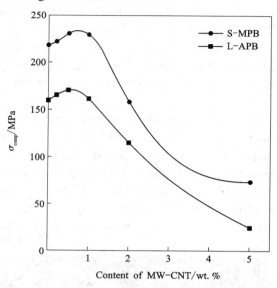

Figure 3 Compressive strength of phosphate composites based on solid and liquid binders versus multi-walled CNT content[12]

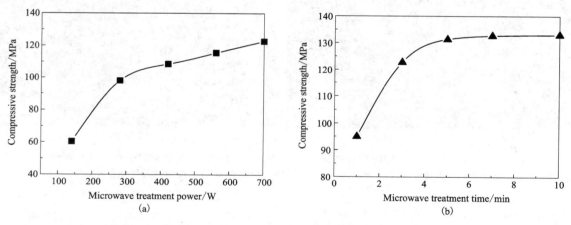

Figure 4 The dependence of magnesium-phosphate sample strength on the power (a) and duration (b) of microwave treatment[5]

4 Conclusion

Thus, based on the studies carried out by the authors, it has been shown that phosphate binder systems used in liquid and solid form are an universal matrix for obtaining various functional thermostable composite materials. The use of solid phosphate binders as well as introducing of microwave treatment of composites is of great interest and is a promising direction in the development of a modern efficient technology for obtaining thermostable composites for various purposes.

Acknowledgments: This research work was supported by the Ministry of Education of the Republic of Belarus (assignment 2.1.07.2 of the state program of scientific research "Chemical processes, reagents and technologies, bioregulators and bioorgchemistry")

References

[1] KOPEIKIN V A, KLEMENT'EVA V S, KRASNYI B L. Refractory solutions based on phosphate binders[M]. Moscow: Metallurgiya, 1986: 240.

[2] KINGERY W D. Fundamental study of phosphate bonding in refractories: Part IV, Mortars bonded with monoaluminum and monomagnesium phosphate[J]. Journal of the American Ceramic Society, 1952, 35(3): 61-63.

[3] SUDAKAS L G. Phosphate binding systems[J]. Saint Petersburg: Kvintet, 2008: 260.

[4] WAGH A S. Chemically bonded phosphate ceramic [M]. 2nd ed. Amsterdam: Elsevier, 2016.

[5] LAPKO K N, APANASEVICH N S, SHULGA T N, et al. Dry building mixtures based on solid phosphate binders for thermostable functional composite materials. ALITinform: Cement. Concrete[J]. Dry Mixtures, 2015, 2, 39: 78-83.

[6] APANASEVICH N S, SOKAL A A, KUDLASH A N, et al. Thermostable heat-insulating composite materials based on hollow microspheres and solid phosphate binders: Development and research[J]. Journal of the Belarusian State University Chemistry, 2022, 2: 70-82.

[7] Patent RU № 2524516 dated 06.06.2014. Russian.

[8] PLYUSHCH A, BYCHANOK D, KUZHIR P, et al. Heart-resistant unfired phosphate ceramics with carbon nanotubes for electromagnetic application[J]. Physical Status Solidi A, 2014, 211, 11: 2580-2585.

[9] PLYUSHCH A, MACUTKEVIC J, SOKAL A, et al. The phosphate-based composite materials filled with nano-sized $BaTiO_3$ and Fe_3O_4: Toward to the unfired multiferroic materials[J]. Materials, 2021, 14, 1, 133: 1-8.

[10] PLYUSHCH A, MACUTKEVIC J, KUZHIR P, et al. Synergy effects in electromagnetic properties of phosphate ceramics with silicon carbide whiskers and carbon nanotubes[J]. Applied Sciences, 2019, 9, 4388: 1-8.

[11] PLYUSHCH A, MACUTKEVIC J, SVIRSKAS S, et al. Silicon carbide/phosphate ceramics composite for electromagnetic shielding applications: Whiskers vs particles[J]. Applied Physics Letters, 2019, 114, 18, 183105: 1-5.

[12] APANASEVICH N, SOKOL A, LAPKO K, et al. Phosphate ceramics-carbon nanotubes composites: Liquid aluminum phosphate vs solid magnesium phosphate binder[J]. Ceramics International, 2015, 41: 12147-12152.

Effect of Short-range Order on Viscosity and Crystallization of Al-Mg Melts

Larisa KAMAEVA[1], Elizaveta BATALOVA[1], Nikolay CHTCHELKATCHEV[2]

1 Udmurt Federal Research Center, Ural Branch of Russian Academy of Sciences, Izhevsk, 620990, Russia
2 Vereshchagin Institute for High Pressure Physics, Russian Academy of Sciences, Troitsk, Moscow, 108840, Russia

Abstract: In this work, using the methods of viscosimetry and thermal analysis, the concentration changes in the values of the supercooling viscosity of Al-Mg melts with Mg content from 2.5 at% to 95 at% are studied. It is shown that the temperature dependence of viscosity is well described by an exponential dependence. The change in the chemical short-range order in the liquid phase can be seen in the concentration dependence of viscosity, which is not monotone. The type of solid phase formed during solidification determines the concentration dependence of the supercooling of Al-Mg melts. The concentration behavior of supercooling is also sensitive to the most significant changes in the chemical short-range order in the liquid phase, which are observed at 20 at% and at 80 at% Mg. Al-Mg alloys with Mg contents of (0-10)at%, (40-50)at% and (90-100)at% are susceptible to non-equilibrium crystallization, the formation of quasi-eutectics, and solidification without intermediate intermetallic phases.

Keywords: Al-Mg alloys; Viscosity; Crystallization; Chemical short-range order

1 Introduction

Al-based metal systems have a complex structure in the liquid state[1-3]. However, these alloys are actively used as structural materials. In particular, the Al-Mg alloys are widely used in aircraft, shipbuilding, and mechanical engineering. This is due to the fact that adding a small amount of Mg to Al leads to a simultaneous increase in the strength and ductility of the obtained alloys[4]. On the other hand, Al-Mg alloys are characterized by susceptibility to intergranular corrosion, which is due to the appearance of β-phase (Al_3Mg_2) crystals in the structure of the alloys; therefore, the conditions for its formation are intensively studied[4]. Understanding the mechanisms of the formation of the complex compounds γ-$Mg_{17}Al_{12}$, ζ-$Al_{52}Mg_{48}$[5] and ε-$Al_{30}Mg_{23}$[6] for the range of Mg-rich alloys is necessary for a wider practical application of Al-Mg alloys. An analysis of the electronic structure and vibrational entropy of intermetallic compounds showed that the description of the formation of such compounds requires an analysis of the short-range order in the arrangement of atoms[7-9]. The short-range order in the liquid and supercooled Al-Mg melts with Mg content from 0 to 100 at% using ab initio molecular dynamics (AIMD) was carried out in [3, 10]. The authors showed through an analysis of the Warren-Cowley parameters, that in the concentration ranges with Mg concentrations of (20-25) at% and (70-80) at%, respectively, the development of extended bonds between Al atoms and the fragmentation of Mg atomic groups. Our studies[11-13] for Al-based ternary systems showed that concentration changes in the chemical short-range order of melts affect not only the concentration behavior of properties, but also the crystallization process, which confirms the structural heredity between the liquid and solid states. Therefore, taking into account the identified features of the chemical short-range order in [3, 10], in this work we have studied the kinematic viscosity and crystallization processes of melts of the Al-Mg system with Mg concentration ranging from 2.5 at% to 95 at%.

2 Materials and methods

The kinematic viscosity (ν) of liquid Al-Mg alloys was measured on an automated setup[14] using the method of damped torsional vibrations of a crucible with a melt. Samples for research weighing 8-12 g were obtained by melting electrolytic aluminum and high-purity magnesium in the furnace of the installation[14] in a protective atmosphere of purified He at a temperature of 800 ℃ for 30 min. Differential thermal analysis (DTA) was implemented on an automated high-temperature thermal analyzer (VTA-983). The method of operation on this setup is described in [15]. The concentration dependences of undercooling (ΔT) show all the data obtained for each investigated alloy under various cooling conditions. This is necessary to obtain objective values. The structural characteristics of the studied melts were determined using first-principles molecular dynamics modeling (AIMD). The calculations were carried out using the Vienna ab initio simulation program (VASP)[16]. A detailed calculation procedure is presented in [11, 13].

3 Results and discussion

At the first stage of the work, the temperature dependences of the viscosity (polytherms) of the Al-Mg melts (from 2.5 at% to 95 at% Mg) were measured and plotted in the coordinates $\nu(T)$ and $\ln\nu(1/T)$. In the studied temperature range, significant anomalies in the temperature dependence of the viscosity of Al-Mg melts were not detected. Viscosity polytherms have monotonic dependencies and are well described by the Arrhenius relation. Table 1 shows the parameters of the approximating equation:

$$\ln\nu = a\frac{1}{T} - b \quad (1)$$

Table 1 Parameters of the approximating equation (1) of the temperature dependence of the viscosity of Al-Mg melts

Composition/at%	a	b	Composition/at%	a	b
$Al_{97.5}Mg_{2.5}$	1309	15.99	$Al_{55}Mg_{45}$	2076	16.21
$Al_{95}Mg_5$	1365	15.90	$Al_{50}Mg_{50}$	2180	16.31
$Al_{92.5}Mg_{7.5}$	1559	16.06	$Al_{40}Mg_{60}$	2134	16.22
$Al_{90}Mg_{10}$	1348	15.79	$Al_{32}Mg_{68}$	1953	16.03
$Al_{87.5}Mg_{12.5}$	1547	15.97	$Al_{30.5}Mg_{69.5}$	1820	15.90
$Al_{85}Mg_{15}$	1577	15.96	$Al_{28.4}Mg_{71.6}$	1919	16.06
$Al_{80}Mg_{20}$	1708	15.94	$Al_{25}Mg_{75}$	1526	15.66
$Al_{75}Mg_{25}$	1561	16.11	$Al_{20}Mg_{80}$	1216	15.31
$Al_{70}Mg_{30}$	1902	16.16	$Al_{10}Mg_{90}$	1804	15.86
$Al_{60}Mg_{40}$	2115	16.31	Al_5Mg_{95}	1751	15.87

According to the previously obtained polytherms, Figure 1(b) shows the concentration dependence of the viscosity of the Al-Mg melts for two temperatures. It can be seen from the figure that with an increase in the magnesium concentration in Al-Mg melts, their viscosity increases. Moreover, the increase occurs nonmonotonically: there are 5 intervals with different types of dependence $\nu(T)$. Viscosity increases sharply at a magnesium content of 2.5 at% to 5 at%; when its concentration reaches 7.5 at%, the same sharp decrease occurs; as a result, a pronounced maximum appears on the concentration dependence at 5 at% Mg. With an increase in the magnesium concentration in the melt to 20 at%, an increase in viscosity occurs; with a magnesium content of 22.5 at%, a pronounced minimum is observed, and then a monotonous increase in viscosity occurs until the concentration reaches 72 at% Mg. Then there is a minimum at 75 at% Mg, which corresponds to the eutectic concentration, after which the viscosity increases to (80-85) at% Mg, and then the viscosity values begin to decrease again.

The concentration features of the viscosity of Al-Mg melts that we found are in good agreement with the data obtained in [3, 10] on the concentration change in the chemical short-range order in these melts. The authors[3] used the Warren-Cowley parameters, which show how much the real chemical environment nearest to the selected atom differs from the random one at a certain concentration of components. In order to study in more details the processes occurring in the regions with magnesium content of 20 at% and 80 at% (the largest

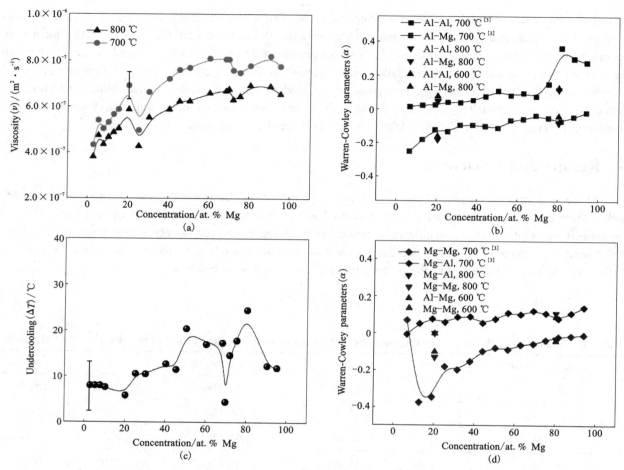

Figure 1　Concentration dependences of viscosity (a), Warren-Cowley parameters (b, d) and undercooling (c)

changes in viscosity, which manifest themselves in its concentration dependence), we also analyzed the chemical local ordering by the method of AIMD for two temperatures.

　　The results were compared with the data obtained in [3] and are shown in Figure 1(b). Considering the advanced calculation methods we use, which allow us to carry out MD calculations for low concentrations of elements in alloys with ab initio accuracy, the parameters calculated by us are more accurate than the results obtained in 2017. The Cowley parameters obtained by us confirm the main regularities established in this article but show that the chemical short-range order does not have such sharp changes as they were presented in [3, 10]. Comparing the obtained concentration dependences of viscosity and Warren-Cowley parameters[3] [Figure 1(a), (b)], one can observe that the concentration dependence of viscosity reflects changes in the chemical short-range order in melts. A sharp change in the chemical interaction in the melt between Mg atoms is observed in the concentration range from 10 at% to 20 at% Mg [Figure 1(a)], while the conglomeration of Mg atoms occurs, as a result of which the interaction between Al-Mg atoms weakens, which was pronounced at a minimum Mg content. A similar feature with respect to Al atoms is also observed at 80 at% Mg. Both of these features are manifested in the concentration dependence of viscosity.

　　The DTA thermograms were used to calculate the values of the undercooling required to initialize the crystallization process and plot the concentration dependence of ΔT [Figure 1(c)]. Despite the scatter in the values of ΔT, the concentration dependence of undercooling shows concentrations at which a fundamental change in the nature of the concentration behavior of ΔT occurs, which is not associated with a change in the type of crystals formed. In the concentration range from 2.5 at% to 40 at% Mg at undercoolings up to 20 ℃, crystallization begins with the formation of an Al-based solid solution, and up to 20 at% Mg, the effect of concentration on the magnitude of undercooling is not observed [Figure 1(c)]. When the Mg content is between 20 at% and 40 at%, the undercooling value increases [Figure 1(c)], which indicates a decrease in the crystallization ability of the Al-based solid solution. The formation of Al_3Mg_2 intermetallide [(40-45) at% Mg] also proceeds at low undercoolings of 10 ℃. The $Al_{12}Mg_{17}$ intermetallic compound is first formed from the melt in the concentration range from 50 at% to 70 at% Mg; its formation requires greater undercooling than for the Al_3Mg_2 intermetallic compound; therefore, at 50 at% Mg, a jump is observed in the concentration dependence of ΔT [Figure 1(c)]. A further increase in the Mg content in the melt increases the crystallization ability of $Al_{12}Mg_{17}$. The minimum undercooling values for the Al-Mg system (5 ℃) are observed for the $Al_{30}Mg_{70}$ alloy; this is an eutectic alloy, but under the selected cooling conditions, its crystallization begins with the growth of primary Mg dendrites. Thus, the Mg-based solid solution is the first from the melt to solidify in the concentration range of 70 at% Mg and more. In the concentration range from 70 at% to 80 at% Mg, undercooling increases, and then, with an increase in the Mg concentration to 95 at%, it decreases. From the analysis of the concentration dependence of undercooling, two concentrations, 20 at% and 80 at% Mg, can be distinguished at which there is a significant change in undercooling but no change in the type of crystallization is observed. The found concentration features of undercooling are in good agreement with the data on the concentration change in viscosity and chemical short-range order in Al-Mg melts[3] (Figure 1).

The nonequilibrium crystallization of the melt that we observed, in which the liquid phase kept for eutectic crystallization, can be explained by the significant effective attraction between Mg atoms and the subsequent local segregation. Features of phase formation in Al-Mg alloys at Mg concentrations of more than 20 at% are mainly determined by the interaction between Al atoms. Al atoms, as they are replaced by Mg, tend to maintain interaction with each other, as a result of which they shift to more distant distances than the first coordination sphere[3], which leads to a decrease in their number both around Al and Mg atoms. This interaction is maximally manifested at an equiatomic ratio of Al and Mg and in the region of 80 at% Mg. During crystallization, this interaction prevents the formation of disordered phases (solid solutions based on Al and Mg) and contributes to an increase in undercooling, under which their crystallization is observed.

4 Conclusion

As a result, the concentration dependence of the viscosity of Al-Mg melts from 2.5 at% to 95 at% Mg is not monotonic and has pronounced features in the region of 20 at% and 80 at% Mg. The concentration dependence of undercooling has features in the regions of 20 at%, 35 at%, 50 at%, 70 at% and 80 at% Mg. Features at 35 at%, 50 at% and 70 at% Mg are associated with a change in the crystal structure formed from the melt. The bends in concentration dependence of undercooling and viscosity at 20 at% and 80 at% Mg correspond to the concentrations at which the maximum chemical interactions in melts are observed (at 20 at% Mg, effective attraction of Mg-Mg, at 80 at% Mg, effective repulsion of Al-Al).

Acknowledgments: The investigation has been financed by the grant of the Russian Science Foundation No. 22-22-00912, https://rscf.ru/project/22-22-00912/.

References

[1] LIN W, LI S S, LIN B, et al. Liquid-liquid phase separation and solidification behavior of Al-Bi-Sn monotectic. alloy[J]. Journal of Molecular Liquids, 2018, 254: 333-339.

[2] ROIK O S, SAMSONNIKOV O V, KAZIMIROV V P, et al. Medium-range order in Al-based liquid binary alloys [J]. Journal of Molecular Liquids, 2010, 151: 42-49.

[3] WANG J, LI X X, PAN S P, et al. Mg fragments and Al bonded networks in liquid Mg-Al alloys [J]. Computational Materials Science, 2017, 129: 115-122.

[4] YANG Y K, TODD A. Direct visualization of β phase causing intergranular forms of corrosion in Al-Mg alloys[J]. Materials Characterization, 2013, 80: 76-85.

[5] KASPER J S, WATERSTRAT R M. Ordering of atoms in the r phase[J]. Acta Crystallographica, 1956, 9(3): 289-295.

[6] SAMSON S, GORDON E K. The crystal structure of ε-$Mg_{23}Al_{30}$[J]. Acta Crystallographica B, 1986, 24(8): 1004-1013.

[7] ZHONG Y, YANG M, LIU Z K. Contribution of first-principles energetics to Al-Mg thermodynamic modeling[J]. Calphad, 2005, 29(4): 303-311.

[8] VRTNIK A, JAZBEC S, et al. Stabilization mechanism of $Mg_{17}Al_{12}$ and Mg_2Al_3 complex metallic alloys[J]. Journal of Physics: Condensed Matter, 2013, 25(42): 425703.

[9] SHIN D, WOLVERTON C. The effect of native point defect thermodynamics on off stoichiometry in $Mg_{17}Al_{12}$[J]. Acta Materially, 2012, 60(13): 5135-5142.
[10] DEBELA T T, ABBAS H G. Role of nanosize icosahedral quasicrystal of Mg-Al and Mg-Ca alloys in avoiding crystallization of liquid Mg: Ab initio molecular dynamics study[J]. Journal of Non-Crystalline Solids, 2018, 499: 173-182.
[11] KAMAEVA L V, RYLTSEV R E, et al. Viscosity, undercoolability and short-range order in quasicrystal-forming Al-Cu-Fe melts[J]. Journal of Molecular Liquids, 2020, 299: 112207.
[12] KAMAEVA L V, STERKHOVA I V, et al. Phase selection and microstructure of slowly solidified Al-Cu-Fe alloys [J]. Journal of Crystal Growth, 2020, 531: 125318.
[13] KAMAEVA L V, RYLTSEV R E, et al. Effect of copper concentration on the structure and properties of Al-Cu-Fe and Al-Cu-Ni melts[J]. Journal of Physics: Condensed Matter, 2020, 32: 224003(9).
[14] BEL'TYUKOV A L, LAD'YANOV V I. An automated setup for determining the kinematic viscosity of metal melts [J]. Instruments and Experimental Techniques, 2008: 51.
[15] STERKHOVA I V, KAMAEVA L V. The influence of Si concentration on undercooling of liquid Fe[J]. Journal of Non-Crystallin Solids, 2014, 401: 250-253.
[16] KRESSE G, FURTHMULLER J. Efficiency of ab-initio total energy calculations for metals and semiconductors using a plane-wave basis set[J]. Computational Materials Science, 1996, 6(1): 15-50.

Pulse Alternating Current Electrosynthesis as an Effective Way to Multifunctional Materials for Hydrogen Energy

Tatyana MOLODTSOVA, Anna ULYANKINA, Daria CHERNYSHEVA, Nina SMIRNOVA

Platov South-Russian State Polytechnic University (NPI), Novocherkassk, 346428, Russia

Abstract: The electrochemical synthesis of nanomaterials for diverse applications ranging from energy conversion to environmental protection is a growing research field due to its simplicity, scalability, and eco-friendliness. Besides, it provides materials with tailored properties, which are, in most cases, unattainable by conventional chemical routes. The use of pulse alternating current (PAC) provided the one-step fabrication of various materials for hydrogen energy-related electrocatalytic reactions. In this work, we summarize recent progress in the pulse alternating current electrosynthesis of photo-and electrocatalytic materials and materials for electrochemical energy storage systems. Emphasis is placed on the straightforwardness of the electrochemical route-in contrast to more conventional synthesis-in fabricating the wide range of highly efficient catalytic materials, which otherwise often require harsh conditions.

Keywords: Electrosynthesis; Pulse alternating current; Photocatalyst; Electrocatalyst

1 Introduction

Transition metal oxides (TMOs) have been widely investigated for recent decades due to their huge potential in magnetic, electronic, optical as well as energy conversion and storage applications including supercapacitors (SCs), lithium-ion batteries (LIBs), electrocatalysts, which can effectively combat the existing energy and environment crisis[1]. However, precise synthesis of photo-and electroactive TMOs-based nanostructures is one of the key challenges that hinder the practical application of many important hydrogen energy-related electrocatalytic reactions. Therefore, the multitude of studies focusing on the catalyst fabrication make the immense effort for the rational synthesis of active and stable catalysts.

Various methods exist to yield TMOs-based materials, such as plasma-assisted, chemical precipitation, solvo/hydrothermal, etc[2]. While these methods can yield TMOs, oftentimes such methods are complex and involve the use of toxic and expensive reagents. Thus, the development of alternative synthetic protocols, which are more environmentally benign, simple, cost-effective, high-yield, and scalable, is of high interest. Compared with conventional chemical synthesis routes, electrochemical synthesis has been considered as a sustainable and economically attractive method to produce highly efficient catalysts. Furthermore, electrochemical methods can be used to synthesize TMOs-based composite structures with carbon support, which are not easily accessible, via embedding carbon additives.

Our research group is dedicated to advancing the field of electrochemistry by developing innovative techniques for synthesizing TMOs-based nanostructures using pulse alternating current(PAC) electrochemical method. Additionally, we are interested in studying the relationship between the structural properties of these materials and their photo-and electrocatalytic performance. By leveraging these materials, we aim to gain a deeper understanding of how to optimize their performance for a variety of electrochemical applications. Therefore, this work will summarize recent progress in the pulse alternating current electrosynthesis of photo-and electrocatalytic materials for hydrogen energy-related electrocatalytic reactions and electrochemical energy storage systems.

2 Experiment

High purity metal plates (Zn, Ti, Cu, Fe, Co, Ni, W or In) used as electrodes were firstly polished using sandpaper and then washed with distilled water. The electrodes were immersed in the aqueous solutions of salts, acids or alkaline used as the electrolytes and connected to a home-designed pulse alternating current source (Figure 1). Other synthesis conditions as well as the resulting products and their applications are presented in detail in Table 1.

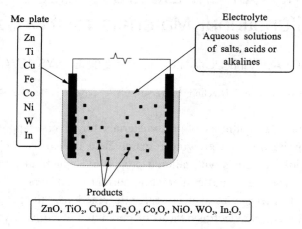

Figure 1 Schematic of the electrochemical synthesis using pulse alternating current

Table 1 PAC electrosynthesis conditions, the resulting synthesis products and their applications

Metal	Synthesis conditions			Product/s	Application/s	Ref.
	$j_a:j_k$ /(A·cm^{-2})	Electrolyte	Others			
Zn	2.4:1.2	2 mol/L NaCl, 2 mol/L KCl, 2 mol/L LiCl and 1 mol/L Na$_2$SO$_4$	stirring and cooling	ZnO	PC degradation of azo dyes and antibiotics	[3, 4]
In	2.5:2.5		stirring and cooling; annealing in air for 1 h at 400 ℃	c-and c/rh-In$_2$O$_3$	PEC use	[5]
Ti	0.2:1	2 mol/L NaCl	stirring and cooling; annealing in air for 3 h at 400-600 ℃	TiO$_2$	PC degradation of azo dyes and oxidation of biomass	[6, 7]
Cu	0.5:0.5 1:1 1.5:1.5		stirring and cooling	CuO or Cu$_2$O-CuO	Methanol electrooxidation and SC	[8]
Fe	3:3		stirring and cooling	γ-Fe$_2$O$_3$/ Fe$_3$O$_4$	ORR and environmental pollutant detection	[9]
	1.2:2.4			γ-Fe$_2$O$_3$/ δ-FeOOH		[10]
Co	0.5:0.5	2 mol/L NaOH	-	Co$_3$O$_4$/ CoOOH	LIB and SC	[11]
Ni	0.25:0.5		carbon powder as support material	NiO/C	SC	[12]
	0.5:1		multilayer graphene as support material	NiO/MLG		[13]
	0.13:0.47		multi-walled carbon nanotube as support material	NiO$_x$/ MWCNTs		[14]
W	3:3	0.5 mol/L C$_2$H$_2$O$_4$, 0.5 mol/L H$_2$SO$_4$ and 0.5 mol/L HNO$_3$	stirring and cooling; annealing in air for 3 h at 500 ℃	WO$_3$	PEC reforming of organic substances	[15]

3 Results and discussion

3.1 Photocatalytic applications of electrochemically synthesized materials

The energy conversion efficiency of photocatalysis is greatly influenced by light absorption, separation and transport of the charge carrier, the number of surface-active sites, and band structures. TMOs have important applications in photocatalysis, primarily because they can act both as active phases (i.e., bulk or self-supported catalyst) and as supports. Considering the unique properties of electrochemically synthesized materials mentioned above, Zn-, Ti-, W- and In-containing oxide materials are viewed as promising photocatalysts with good performances. In this section, the recent progress of oxide materials obtained by the PAC electrosynthesis in some important applications of photocatalysis, including photocatalytic degradation of pollutants and oxidation of biomass as well as photoelectrochemical use is reviewed.

TiO_2 and ZnO are among the most often used photoactive materials in heterogeneous catalysis, owing to their reasonably high surface area, higher electron mobility, accelerated electron transfer and higher quantum efficiency. Electrochemical synthesis using PAC has been shown to provide easy access to the nanoparticulate form of these phases[3, 4, 6]. These oxide nanostructures provide a large surface area with full contact between catalysts and organic molecules and thus inspires the study of its environmental applications. Moreover, electrochemically synthesized TiO_2 nanoparticles were used in the oxidation of HMF as a photocatalyst to produce valuable DFF[7].

Apart from the photodegradation of pollutants in water, metal oxide can also be used in PEC applications. For example, WO_3 and In_2O_3 nanostructures were prepared using PAC electrosynthesis and characterized by high photoelectrochemical performances caused by the optimal morphological, electronic, and charge-transfer properties[5, 15].

3.2 Electrochemical applications of electrochemically synthesized materials

The electrochemical activity is mainly determined by the number of active sites, the configuration states of the atoms, and the conductivity of the active materials. Excellent electrochemical performances are expected for (oxihydr)oxide-based materials due to their unique structural advantages, charge storage capacity with outstanding charge-discharge performance, long cycle life and higher power density. In this section, we mainly focus on the recent progress of oxide materials obtained by the PAC electrosynthesis in the different fields of electrochemistry including supercapacitors, lithium-ion batteries, alcohol electrooxidation reaction (AEOR) applications and non-enzymatic sensors.

Metal oxide nanoparticles are studied for the AEOR due to their fascinating features of high exposed surface area for abundant catalytic active sites and large contact electrolyte area for rapid electron transfer. Recently, we reported the electrochemical synthesis of a Cu_2O-CuO bilayered polyhedra and investigated its electrocatalytic performance in the methanol oxidation reaction[8]. In addition, this approach also was successful in the synthesis of Cu_2O octahedra with suitable specific capacitance for SCs.

Moreover, other TMOs-based nanostructures are obtained using PAC electrosynthesis for energy storage applications. For example, the interaction between NiOx material and various carbon-based supports (carbon black, multi-walled carbon nanotube or multilayer graphene) improved their electrochemical properties indicating potential applications as high-performance supercapacitor electrode materials[12-14]. In addition to Ni-based oxides, electrochemically prepared Co_3O_4/CoOOH nanocomposite material also demonstrated as a promising candidate for high-performance SCs and LIBs applications[11].

On the other hand, TMOs and their composites have been used as efficient materials for non-enzymatic electrochemical sensing applications. Our group's recent works have demonstrated electrochemically synthesized Fe-based electrocatalysts for use in EC sensors. For instance, the finding γ-Fe_2O_3/δ-FeOOH and γ-Fe_2O_3/Fe_3O_4 nanocomposites achieved an excellent analytical performance for amperometric determination of acetaminophen and hydrogen peroxide respectively[9, 10]. Therefore, electrosynthesis using pulse alternating current holds great promise for the development of the single and mixed-phased oxide catalysts for multifunctional electrochemical applications (Figure 2).

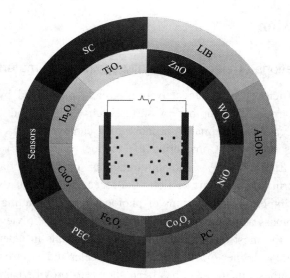

Figure 2　The unique roles of electrochemically synthesized materials in hydrogen energy-related electrochemical applications

4　Conclusion

This work provides valuable insights into the electrochemical synthesis of transition metal oxide-based materials and their potential use as highly effective catalysts for photo-and electrocatalytic reactions as well as electrochemical energy storage systems. By using pulse alternating current, it is possible to obtain nanostructures with unique phase composition and tailored properties, which can be optimized for hydrogen energy-related electrocatalytic reactions. With further research, pulse alternating current electrochemical synthesis may enable the development of new and improved transition metal oxide-based materials with even higher electrochemical activity, making them ideal candidates for use in a wide range of industrial processes.

Acknowledgments: This work was carried out within the framework of the strategic project "Hydrogen Energy Systems" of the SRSPU (NPI) Development Program during the implementation of the program of strategic academic leadership "Priority-2030".

References

[1] LEI Z, LEE J M, SINGH G, et al. Recent advances of layered-transition metal oxides for energy-related applications [J]. Energy Storage Mater, 2021, 36: 514-550.
[2] TAWALBEH M, KHAN H A, AL-OTHMAN A. Insights on the applications of metal oxide nanosheets in energy storage systems[J]. J Energy Storage, 2023, 60: 106656.
[3] ULYANKINA A, LEONTYEV I, AVRAMENKO M, et al. Large-scale synthesis of ZnO nanostructures by pulse electrochemical method and their photocatalytic properties[J]. Mater Sci Semicond Process, 2018, 76: 7-13.
[4] ULYANKINA A, MOLODTSOVA T, GORSHENKOV M, et al. Photocatalytic degradation of ciprofloxacin in water at nano-ZnO prepared by pulse alternating current electrochemical synthesis[J]. Journal of Water Process Engineering, 2021, 40: 101809.
[5] MOLODTSOVA T, GORSHENKOV M, KOLESNIKOV E, et al. Fabrication of nano-In_2O_3 phase junction by pulse alternating current synthesis for enhanced photoelectrochemical performance: Unravelling the role of synthetic conditions[J]. Ceram Int., 2023, 49: 10986-10992.
[6] ULYANKINA A, AVRAMENKO M, KUSNETSOV D, et al. Electrochemical synthesis of TiO_2 under pulse alternating current: Effect of thermal treatment on the photocatalytic activity[J]. Chemistry Select, 2019, 4: 2001-2007.
[7] ULYANKINA A, MITCHENKO S, SMIRNOVA N. Selective photocatalytic oxidation of 5-HMF in water over electrochemically synthesized TiO_2 nanoparticles[J]. Processes, 2020, 8: 647.

[8] ULYANKINA A, LEONTYEV I, MASLOVA O, et al. Copper oxides for energy storage application: Novel pulse alternating current synthesis[J]. Mater Sci Semicond Process, 2018, 73: 111-116.

[9] MOLODTSOVA T, GORSHENKOV M, SALIEV A, et al. One-step synthesis of γ-Fe_2O_3/Fe_3O_4 nanocomposite for sensitive electrochemical detection of hydrogen peroxide[J]. Electrochim Acta, 2021, 370: 137723.

[10] MOLODTSOVA T, GORSHENKOV M, KUBRIN S, et al. One-step access to bifunctional γ-Fe_2O_3/δ-FeOOH electrocatalyst for oxygen reduction reaction and acetaminophen sensing[J]. J Taiwan Inst Chem Eng, 2022, 140: 104569.

[11] CHERNYSHEVA D, VLAIC C, LEONTYEV I, et al. Synthesis of Co_3O_4/CoOOH via electrochemical dispersion using a pulse alternating current method for lithium-ion batteries and supercapacitors[J]. Solid State Sci, 2018, 86: 53-59.

[12] LEONTYEVA D V, LEONTYEV I N, AVRAMENKO M V, et al. Electrochemical dispergation as a simple and effective technique toward preparation of NiO based nanocomposite for supercapacitor application[J]. Electrochim Acta., 2013, 114: 356-362.

[13] CHERNYSHEVA D V, LEONTYEV I N, AVRAMENKO M V, et al. One step simultaneous electrochemical synthesis of NiO/multilayer graphene nanocomposite as an electrode material for high performance supercapacitors [J]. Mendeleev Communications, 2021, 31: 160-162.

[14] SHMATKO V, LEONTYEVA D, NEVZOROVA N, et al. Interaction between NiO_x and MWCNT in NiO_x/MWCNTs composite: XANES and XPS study[J]. J Electron Spectros Relat Phenomena, 2017, 220: 76-80.

[15] TSARENKO A, GORSHENKOV M, YATSENKO A, et al. Electrochemical synthesis-dependent photoelectrochemical properties of tungsten oxide powders[J]. Chem Engineering, 2022, 6.

Complex Experimental and Model Study of Non-isothermal Deformation in Al-, Co- and Cu-based Alloys with Different Space Configuration and Structural State

Arseniy BEREZNER[1], Victor FEDOROV[1], Michael ZADOROZHNYY[2], Gregory GRIGORIEV[1]

1 Derzhavin Tambov State University, Tambov, 392000, Russia
2 The National University of Science and Technology "MISiS", Moscow, 119049, Russia

Abstract: A generalization of non-isothermal creep, dynamo-mechanical or thermo-mechanical experiments has been performed in frames the same physical model. Good agreement between the model approach and experiment is shown that permits estimation of main applied parameters such as the linear thermal expansion coefficient and others. Necking forms and critical thickness of corrugation for ribbon and rod specimens are also calculated. The main size fractal analysis of corrugation folds has been carried for polycrystalline and amorphous ribbon specimens.

Keywords: Creep; Thermomechanical processes; Metallic materials

1 Introduction

Study of elastic and plastic deformation is essential, useful and well-known approach at the different fields of fundamental and applied science[1-5]. Depending on initial and boundary deformation conditions, a specimen has a personal response[6].

Non-isothermal creep[7] is qualitatively identical to thermal mechanical analysis (TMA)[8] because of an similar deformation curve, appearing at static external loading. For plastic and fracture deformation stages of creep, fatigue and other tests, universal description of necking[9] or corrugation[10] (in thinner specimens), observing in different materials, is an important task. In case of alternating temperature[11], mentioned model analysis is complicated by non-linear boundary conditions which define a search of novel and optimal methods.

Taking into account descripted peculiarities at alternating temperatures, the main goal of this work is the study of non-isothermal creep behaviour of Al- and Co-based amorphous or polycrystalline specimens in comparison with TMA. Generalization of creep tests, dynamic mechanical analysis (DMA) and TMA will be performed by using thermodynamic and deformation models. Moreover, the plane and cylinder specimen configuration (a ribbon or a rod, consequently) will be considered, and their deformation or fracture features (such as necking and corrugation) must be determined. An estimation of scale sizes, under which those effects occur, will be also provided.

2 Materials and methods

As the specimens for this work, Al-Y-Ni-Co, Cu-Pd-P and Co-Fe-Si-Mn-B-Cr ribbon amorphous alloys (metallic glasses, MGs), whose percent element composition, manufacturing and amorphous structure were thoroughly described in previous articles[12-14]. Except the MGs, we used industrial polycrystalline Al-Fe (Al-98 wt.%, Fe-2 wt.%) ribbons and Cu rods. The structural similarity between the investigated copper or Al-Fe alloys and common polycrystalline analogs[15-16] was estimated with a Bruker D2 Phaser (CuK$_\alpha$ spectrum) X-Ray diffractometer (see Figure 1).

TMA of Al-Y-Ni-Co and Cu-Pd-P ribbon metallic glasses was carried out with a Q800 (TA Instruments) test complex under 3 MPa fixed stress at 5 K/min heating rate up to 673 K. Work sizes of Al-Y-Ni-Co amorphous specimens (in length, width and thickness, consequently) were 16.5 mm, 1.4 mm, 0.02 mm with 16.5 mm, 2.5 mm and 0.025 mm ones for copper-based MGs. For comparison, we used the DMA data from works [12-13], and non-isothermal creep of polycrystalline Al-Fe or Cu was studied (and compared

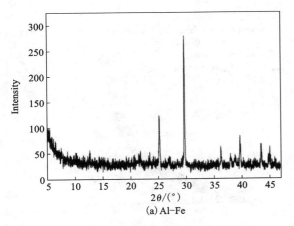
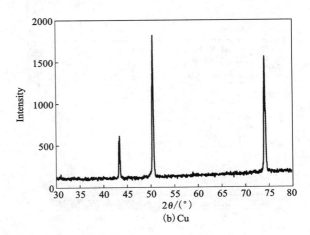

Mainly useful parts (with diffraction peaks) of the spectra are depicted here. Observed spectral lines of specimens correspond to standard crystal directions from literature.

Figure 1 X-ray scattering plots of polycrystalline alloys

with Co-Fe-Si-Mn-B-Cr MG) at the same conditions as in [14], i. e. at 1 N of static load (or 14 MPa stress) together with 1 K/s of heating rate. However, the sizes of Al-Fe crystalline alloys were 50 mm in length, 6 mm in width, and 0.01 mm of thickness while the copper specimens occupied 50 mm of length and 0.625 mm of diameter.

Configuration (ribbons or rods) was chosen differently for estimation of personal mechanical response in specimens at the same testing conditions. By using of differential scanning calorimetry (at 5 K/min heating rate), glass-transition (T_g) and crystallization (T_x) temperatures were estimated for Al-Y-Ni-Co, Cu-Pd-P and Co-Fe-Si-Mn-B-Cr MGs as 533 K and 545 K, 528 K and 568 K, 716 K and 743 K consequently, which testifies about glass or crystal transition in some specimens during the TMA or creep tests. Nevertheless, this testing approach (at non-isothermal creep or TMA) is necessary for a noticed plastic deformation and fracture of the specimens. Also, comparison between the deformation of crystalline and amorphous specimens during the whole experiment is required for better understanding the dependence of atomic reorganizations on non-isothermal impact.

For a surface relief estimation of finally tested specimens (after their two-fragment fracture), a Femto Scan (Advanced Technologies Center) scanning probe microscope (worked at the atomic force-AFM mode) or Vega3 (Tescan) apart with JCM-7000 (Jeol) scanning electron microscopes (SEM) were used.

3 Results and discussion

3.1 Primary experimental analysis and generalization

In Figure 2, experimental TMA and non-isothermal creep curves for ribbon and rod samples with different composition and interatomic structure have been presented.

Analysis of the graphs from Figure 2 permits a conclusion about the identity [by $x(t)$ curve view] between the investigated case, DMA[12] and creep[14]. Here with noticeable difference in the deformation rate (during TMA and creep) can be caused both by structural features (MG or crystal)[13] and by size-effect (copper rods are 25 times thicker than Cu-Pd-P ribbons). Despite the mentioned differences, interpolation by the fractional function is possible for all experimental data in the form[14]:

$$x(t) = x_0 + \frac{Ct}{B^2 - Bt} \tag{1}$$

with high correlation coefficient $r(0.9 < r < 1)$ and x_0 (or l_0), C, B, t parameters, having physical meaning and stable dimension[12-14]. For TMA of Al-Y-Ni-Co amorphous alloys, Equation (1) has to be added by the linear $-Dt$ term [$-D$ (m/s) is a process rate], which is related by reversible shrinkage of the specimen under the static load (relatively lesser than at creep tests). All parameters, determined for the experiments, have been listed in the Table 1.

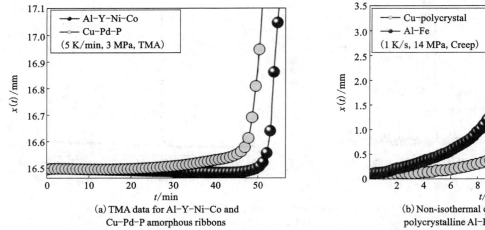 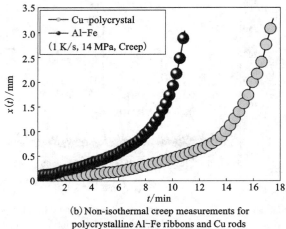

(a) TMA data for Al-Y-Ni-Co and Cu-Pd-P amorphous ribbons

(b) Non-isothermal creep measurements for polycrystalline Al-Fe ribbons and Cu rods

Figure 2 Deformation curves of investigated specimens

Table 1 Experimental parameters and their physical meaning in equation (1)

Composition, alloy type, experiment	Value			
	C multiplier $/(10^{-3} \cdot 60 \text{ m} \cdot \text{s}^{-1})$	Fracture time $B/(60 \text{ s})$	Shrinkage rate $D/[10^{-3} \cdot (60 \text{ m} \cdot \text{s})^{-1}]$	Linear correlation coefficient, r
Al-Y-Ni-Co (MG, ribbon, TMA)	0.43	55.2	0.00099	0.97
Co-Fe-Si-Mn-B-Cr (MG, ribbon, non-isothermal creep)	3.9	15.5	0	0.997
Cu-Pd-P (MG, ribbon, TMA)	1.5	52.84	0	0.99
Al-Fe (polycrystal, ribbon, non-isothermal creep)	6	12.6	0	0.998
Cu (polycrystal, rod, non-isothermal creep)	8	19.5	0	0.996

3.2 Calculation of some material parameters at non-isothermal experiment

To continue the previous calculations[17], the numerical value of the linear thermal expansion coefficient (CLTE) at TMA and non-isothermal creep can be calculated by differentiation of (1) by temperature. For Al-Y-Ni-Co amorphous alloy (at TMA and $T = 507$ K) $\alpha_L = 8.6 \times 10^{-6}$ K^{-1}, and for Al-Fe crystalline one (at creep at $T = 507$ K) $\alpha_L = 24 \times 10^{-6}$ K^{-1} that agrees by magnitude order with work[18] on ternary Al-based MG at the same temperature. The same good agreement with literature data[19] takes place for amorphous and crystalline copper-based alloys, and CLTE is 4×10^{-6} K^{-1} (for MG) or 7×10^{-6} K^{-1} (for polycrystal) in the 300-500 K temperature range. For Co-based MG, $\alpha_L = 3.4 \times 10^{-6}$ K^{-1} (for non-isothermal creep at $T = 300$-400 K), that is also approved by the work[20]. Note that CLTE increases for alloys with the same first component (such as Al-Y-Ni-Co and Al-Fe) because of the different predetermined heating rate (as C and B depend on V_T), and geometry factor can impact [α_L(Cu-Pd-P) < α_L(Cu) as rods elongate more intensive than wider ribbons]. In this work, we calculate the deformation necking form of the specimens, and either linear or more complicated fracture dynamics can be predicted that is experimentally approved. By using the fractal analysis, the critical thickness for presence (or absence) of corrugating in ribbons is estimated together with typical (mean) length (nearly 10^{-2} m) of a corrugation fold. It is found that relationship like the Paris-Erdogan equation takes place for non-isothermal crack growth. All model expressions and numeric calculations are experimentally confirmed with AFM, SEM and optical microscopy.

4 Conclusion

The similarity of deformation dynamics in TMA, DMA, and nonisothermal creep takes place and is described within the framework of a general physical model, regardless of the sample configuration and structural state. A quantitative characteristic is achieved by selecting the C and B values, which give the maximal correlation between the model and the experiment. Calculation of the main thermodynamic parameters and material characteristics is possible in frames of the proposed model, and deformation analysis of the sample necking together with the corrugation conditions can be performed. The obtained numerical data agree with the experiment by the magnitude order.

Acknowledgments: This work was supported by the Russian Science Foundation (Grant No. 22-22-00226). The results were partially obtained using the equipment of the Center for Collective Use of Scientific Equipment at Derzhavin Tambov State University.

References

[1] KHAN A S, ZHANG H. Finite deformation of a polymer: Experiments and modeling[J]. Int. J. Plast., 2001, (17): 1167-1188.
[2] PRASAD G V S S, GOERDELER M, GOTTSTEIN G. Work hardening model based on multiple dislocation densities[J]. Mater. Sci. Eng., A 2005, 400-401: 231-233.
[3] SANDITOV D S, OJOVAN M I, DARMAEV M V. Glass transition criterion and plastic deformation of glass[J]. Physica B: Condensed Matter, 2020, 582: 411914.
[4] LIU X, WANG T, WANG Q, et al. Shear band evolution related with thermal annealing revealing ductile-brittle transition of $Zr_{35}Ti_{30}Be_{27.5}Cu_{7.5}$ metallic glass under complex stress state[J]. Intermetallics, 2022, 140: 107378.
[5] LOUZGUINE-LUZGIN D V. Structural changes in metallic glass-forming liquids on cooling and subsequent vitrification in relationship with their properties[J]. Materials, 2022, 15: 7285.
[6] MEYERS M, CHAWLA K. Mechanical behavior of materials 2nd edition[M]. Cambridge University Press, 2008: 882.
[7] VLASAK G, SVEC P, DUHAJ P. Application of isochronal dilatation measurements for determination of viscosity of amorphous alloys[J]. Mater. Sci. Eng. A, 2001, 304-306: 472-475.
[8] GYUROV S, CZEPPE T, DRENCHEV L, et al. Thermo-mechanical study of rapidly solidified NiNbZrTiAl amorphous metallic alloys[J]. Mater. Sci. Eng. A, 2017, 684: 222-228.
[9] CRIST B, METAXAS C. Neck propagation in polyethylene[J]. J. Polym. Sci., Part B: Polym. Phys., 2004, 42(11): 2081-2091.
[10] WANG M, FOURMEAU M, ZHAO L, et al. Self-emitted surface corrugations in dynamic fracture of silicon single crystal[J]. Proc. Natl. Acad. Sci. U.S.A., 2020, 117(29): 16872-16879.
[11] LOUZGUINE-LUZGIN D V, INOUE A. Structure and transformation behaviour of a rapidly solidified Al-Y-Ni-Co-Pd alloy[J]. J. Alloys Compd., 2005, 399: 78-85.
[12] BEREZNER A D, FEDOROV V A, YU M. Zadorozhnyy, Relaxation behavior of an Al-Y-Ni-Co metallic glass in as-prepared and cold-rolled state[J]. J. Alloys Compd., 2022, 923: 166313.
[13] BEREZNER A D, FEDOROV V A, YU M, et al. Deformation of Cu-Pd-P metallic glass under cyclic mechanical load on continous heating[J]. Theor. Appl. Fract. Mech., 2022, 118: 103262.
[14] FEDOROV V A, BEREZNER A D, BESKROVNYI A I, et al. Determining the form of a hydrodynamic flow upon creep of an amorphous cobalt-based metal alloy in a variable temperature field[J]. Tech. Phys. Lett., 2018, 44: 678-680.
[15] ZHANG T, ZHAO N, LI J, et al. Thermal behavior of nitrocellulose-based superthermites: Effects of nano-Fe_2O_3 with three morphologies[J]. RSC Adv., 2017, 7: 23583.
[16] NARUSHIMA T, TSUKAMOTO H, YONEZAWA T. High temperature oxidation event of gelatin nanoskin-coated copper fine particles observed by in situ TEM[J]. AIP ADVANCES, 2012, 2: 042113.
[17] BEREZNER A, FEDOROV V, GRIGORIEV G. A few fracture features of Al-based and Cu-based ribbon metallic glasses under non-isothermal and oscillating loading[J]. Lecture Notes in Mechanical Engineering, 2022.
[18] LI G H, WANG W M, BIAN X F, et al. Correlation between thermal expansion coefficient and glass formability in amorphous alloys[J]. Mater. Chem. Phys., 2009, 116: 72-75.
[19] WANG Y, BIAN X, JIA R. Effects of cooling rate on thermal expansion of $Cu_{49}Hf_{42}Al_9$ metallic glass[J]. Trans. Nonferrous Met. Soc. China, 2011, 21(9): 2031-2036.
[20] TESLIA S, SOLODKYI I, YURKOVA O, et al. Phase compatibility in (WC-W2C)/AlFeCoNiCrTi composite produced by spark plasma sintering[J]. J. Alloys Compd., 2022, 921: 166042.

Development of the Basis for the Industrial Processing of Ilmenite-titanomagnetite Ores of the Basite-ultrabasites Intrusions Deposited in Sikhote-Alin (The Far East of Russia)

Vladimir MOLCHANOV[1], Fengyue SUN[2]

1 Far East Geological Institute, Far East Branch, Russian Academy of Sciences, Vladivostok, 690022, Russia
2 Jilin University, Changchun, 130061, China

Abstract: The results of studying the opening of the ilmenite-titanomagnetite ores of the ultrabasic rocks occurring in the Koksharovsky massif (south of the Far East of Russia) conducted with using both ammonium hydrofluoride NH_4HF_2 and its mixture with ammonium sulfate $(NH_4)_2SO_4$ are presented. The usage of mixture of these reagents has proved to facilitate better opening of titanium-containing mineral raw materials. The experience of deep processing of ilmenite-titanomagnetite ores will allow outlining the ways for the industrial development of complex deposits of the Far East.

Keywords: Titanomagnetite; Ilmenite; Ammonium hydrodifluoride; Ammonium sulfate; Heat treatment; Water leaching; Koksharovky massif of ultramafic rocks; Primorye

1 Introduction

There has been a sharp increase in interest in deposits of ilmenite-titanomagnetite ores all over the world, and especially in Russia recently because the last ones are an important industrial sources of titanium and iron. At the same time, the technologies used for processing mineral raw materials are most often suitable only for certain types of ores with strict requirements to the quality of their enrichment[1]. Thus, decrease in the concentration of titanium dioxide is an important condition for the usage of ilmenite-titanomagnetite ores as iron ore raw materials, in view of the fact that the amount of TiO_2 should not exceed 4% to maintain the normal course of blast-furnace smelting. The titanium-rich ores are the only ones used to obtain commercial titanium slag. They can be processed as high-quality titanium raw materials with an ilmenite content in the ore being equal to more than 5%. Processing titanium ores with the sulfuric acid is traditionally the most common method which is due to the simplicity of usage of equipment and the availability and low cost of reagents[2]. On the other hand this method is characterized by a high degree of environmental impact with increased amounts of harmful liquid and solid wastes being released. Besides, the structure of domestic titanium-containing raw materials is distinguished by a large proportion of associated components, which makes in inferior to traditional sources of mineral raw materials, such as ilmenite and rutile, if the titanium content is considered. All this requires creating new technical solutions that will allow conducting complex processing of mineral raw materials in compliance with the principles of rational use of natural resources and environmental protection.

The attention of researchers to the ways of processing mineral raw materials by fluoride methods using such reagents as ammonium fluoride and ammonium hydrodifluoride[3, 4] has increased recently. It is especially advantageous to use this approach in the processing of complex mineral raw materials, since it allows expanding the range and the degree of extraction of useful components[5]. According to the data from [6] obtained, due to the comparing of the technical and economic indicators of existing methods for the production of pigment titanium dioxide, the use of fluoride technology appeared to provide significant reduction of the pigment cost. The difference between the fluorination of samples with dry salts and the acid opening, in which the process temperature is limited by the boiling point of the solution, is that the first one can be carried out at temperatures of 230-250 ℃ in open-ended vessels[7], which allows increasing the speed and completeness of fluorination significantly. The opening process is carried out with the usage of solid-phase interaction of ammonium fluoride/hydrofluoride with elements of mineral raw materials. The fluorinated products are dissolved in nitric acid with a subsequent analysis of the solutions by instrumental methods, such as using inductively coupled plasma mass spectrometry[8]. The interaction of minerals, which rocks are composed of results in the formation of a number of the water-soluble complex fluoroammonium compounds of metals (such as iron, aluminum, titanium, zirconium, etc.), silicon and practically insoluble fluorides of calcium, magnesium and rare earth elements (REE).

On the other hand, as shown in [9, 10], sparingly soluble fluorides, such as calcium and rare-earth elements, can be transformed into more soluble sulfates by solid-phase interaction with ammonium sulfate.

This approach is especially beneficial in the processing of complex mineral raw materials, since it allows expanding the range and degree of extraction of valuable components[5]. In addition, a comparison of the technical and economic indicators of existing methods for the production of pigment titanium dioxide showed that the use of fluoride technology allows reducing the prime cost of the pigment by several times[6].

The purpose of our work is to study the possibilities of opening ilmenite-titanomagnetite ore with a mixture of ammonium sulfate and ammonium hydrofluoride and to find out if the subsequent complete transformation of valuable elements into solution is possible.

2 Materials and methods

The ore-bearing intrusions of mafic-ultramafic rocks of the Sikhote-Alin orogenic belt which are new sources of high-tech metals (titanium, platinum group metals, rare earth elements, niobium, tantalum, hafnium, vanadium, cobalt, antimony) have been discovered in the south of the Russian Far East. The Koksharovsky ultramafic massif located in a densely populated center of Primorye with a developed infrastructure facilitating the usage of up-to-date methods of production and extraction of minerals is a prominent representative of this group. The formation of the massif occurred in several phases, with pyroxenites relating to the earliest one and compositing the main body of the intrusion. Titanomagnetite and ilmenite are the essential minerals constituting up to 30% and 5%-10% of the rock mass, respectively. Titanomagnetite $[n\text{FeTiO}_4 \cdot (1-n)\text{Fe}_3\text{O}_4]$ is found in the form of small rounded grains of disturbed octahedral and sharply-angular fragments and represents a solid solutio with isomorphic inclusion of titanium into the magnetite lattice. Ilmenite (FeTiO_3) often develops grains with well-defined crystallographic forms resembling thin plates. According to X-ray fluorescence analysis calculated in terms of oxides, the content of the main components in the studied rock sample was as follows (wt. %): SiO_2-33.0; TiO_2-8.8; Al_2O_3-3.6, Fe_2O_3-17.7; FeO-11.7; MgO-11.3; CaO-14.2. The industrial development of Koksharov titanium ores is constrained by the lack of technology for extracting useful components, which predetermined the line of our research.

The opening of the studied mineral raw materials was carried out by means of using ammonium bifluoride NH_4HF_2, ammonium sulfate (NH_4) and concentrated nitric acid HNO_3, each belonging to "chemically pure" grade. The previously mentioned components were mi) in various mass proportions to study the possibility of interaction of ore samples with a mixt of ammonium sulfate with ammonium hydrofluoride. The resulting mixture poured in glassy carbon or platinum crucibles was placed in a muffle furnace-controller produced by Nabertherm GmbH (Germany) and equipped with an electronic controller with a digital display. The mixture was heated at the rate of 2.5 K/min to a predetermined temperature kept for 4-6 h. The quantities weighted 10-40 g.

The changes that the sample underwent to sample during heating were controlled by the weight loss of the initial mixture, X-ray phase and X-ray fluorescence analyses of the product obtained during processing, and atomic absorption analysis of leaching solutions.

The leaching process of the samples treated with a mixture of ammonium sulfate and ammonium hydrodifluoride was carried out at room temperature by using 4-fold dissolution of the resulting product in water at the ratio of T:L equal to 1:10 with duration of 15-30 min and subsequent filtration through a blue ribbon filter. Leaching of samples obtained by the interaction with NH_4HF_2 was carried out in fluoroplastic cups and the ones obtained by the interaction with the mixture of NH_4HF_2 and $(NH_4)_2SO_4$ underwent the procession in glass cups.

The content of the main components of the fractions at various stages of processing was determined by X-ray fluorescence analysis using a Shimadzu EDX 800 HS spectrometer (tube with a rhodium anode, vacuum) at the room temperature with a tablet containing polytetrafluoroethylene (PTFE) being used. The quantities weighing 1 g were ground in an agate mortar with 0.5 g of PTFE, then placed in a mold of 20 mm in diameter, and pressed for 2 min at the pressure of 20 MPa.

The X-ray diffraction patterns of the samples were taken on a D-8 ADVANCE automatic diffractometer with sample rotation in CuK_α radiation. The X-ray phase analysis was performed using the EVA search program with the PDF-2 powder database.

The induced activity was measured on a spectrometric box monitor on GC2018 coaxial Ge detector basis manufactured by Canberra. The energy resolution of the detector was 1.8 keV with the radiation energy being equal to 1332 keV. The relative detection efficiency in the 1332 keV peak was 20%. In addition to the detector, the spectrometric box monitor consists of SBS-75 information processing unit and a 2002CSL preamplifier with a cooled head stage and 3 m cables. The eSBS Version 1.6.7.0 program and the Gamma Analyzer for semiconductor detectors (SPD)" Version 1.0 were used, measuring the gamma spectra and processing the results of the measurements, respectively.

The thermogravimetric studies were performed on a Q-1000 derivatograph in platinum crucibles in air at a heating rate of 2.5 K/min and weighed quantities of 100-200 mg.

3 Results and discussion

While interacting with ammonium hydrodifluoride, the oxygen-containing compounds of transitional and most of non-transitional elements form ammonium fluoro-or oxofluorometallates, which facilitate the solubility of the products due to their physicochemii properties and allow transforming the main part of the complex salts into solution upon subsequent leaching of the product with water, which makes the use of NH_4HF_2 very perspective for the usage with the purpose of opening of mineral raw materials.

According to [7], the procedure of preparing samples for analysis includes the following sequence of actions: mixing a weighed quantities of the analyzed sample with NH_4HF_2 in a rate of 1 : (1-6), heating the resulting mixture for 1-6 h, dissolving the resulting product in nitric acid, evaporating the solution to dryness and re-dissolving in nitric acid when heated for 6 h. There were cases of incomplete dissolution of fluorinated samples containing aluminum by the passivation of the gibbsite surface by fluorine in this case, the same happened to alkaline earth and rare earth elements[3]. As noted above, the solid phase interaction of the calcium fluoride and REE with the ammonia sulphate at the temperature of 35-400 ℃ leads to the formation of sulfates of these elements. It should be mentioned that these substances are more soluble than fluorides[9, 10]. Accordingly, the comparison between the degree of decomposition of the studied geological samples and the transformation of their constituent components into solution was made using the hydrodifluoride of NH_4HF_2 and the mixture of last one with ammonium sulfate $(NH_4)_2SO_4$.

According to the data obtained as a result of the experiment, the fluorinated titanium-containing ore proved to dissolve much worse in nitric acid without preliminary leaching witl water in case of processing the mineral raw material by NH_4HF_2. Better solution is achieved to the processing of the ore by the mixture of hydrodifluoride and ammonium sulphate firstl at the temperature of 190 ℃ followed by heating up to 350 ℃. This mode was chosen based on the analysis of the thermal behavior of a mixture of hydrodifluoride and ammonium sulfate.

The thermogravimetric study showed that heating the mixture of NH_4HF_2 and $(NH_4)_2SO_4$ is accompanied by both thermal decomposition in the temperature range from 125 ℃ to 430 ℃ and almost complete transformation of the products into the gas phase. The first endothermic effect on the given thermogram refers to the melting of NH_4HF_2 ($t_m = 126.2$ ℃). The weight loss observed during the further heating of the mixture, which was caused by insignificant evaporation of NH_4HF_2 at the temperature kept within the range from 126 ℃ to 200 ℃. The subsequent heating over 200 ℃ leads to the decomposing of ammonium sulfate accompanied by the release of ammonia NH_3 and the formation of ammonium hydrosulfate NH_4HSO_4. According to X-ray phase analysis the product released at 220 ℃, appeared to be the mixture of $(NH_4)_2SO_4$, NH_4HF_2 and NH_4HSO_4. Two endothermic effects overlapping each other occur within the temperature range of 220-280 ℃, which are as follows: boiling of NH_4HF_2 ($t_{boil} = 238°$) accompanied by decomposition into NH_3 and HF, and melting of NH_4HSO_4 ($W_{it} = 251$ ℃). The further increase in temperature is accompanied by two more endothermic effects, proceeding at the maximum rate at the temperature of 330 ℃ and 425 ℃ being caused by the stepwise decomposition of ammonium hydrosulfate NH_4HSO_4 into sulfuric anhydride, ammonia and water.

The process of solid-phase interaction of ilmenite-titanomagnetite ore of the Koksharovsky massif with both NH_4HF_2 and the mixture of NH_4HF_2 and $(NH_4)_2SO_4$ and subsequent leaching o' the resulting product with water was used to illustrate the above said. A comparative analysis of the data obtained under various processing conditions of titanium-containing mineral raw materials indicates that the use of a mixture of hydrodifluoride with ammonium sulfate allows better opening of the mineral raw material compared with the usage of ammonium hydrofluoride alone. Accordingly, better transformation of the resulting product into the solution will be obtained during the water leaching. The good response of using a mixture of NH_4HF_2 with $(NH_4)_2SO_4$ is achieved due to NH_4HF_2 capability to destroy effectively the crystal lattice of minerals, including silicate ones, breaking Si-O bonds, but in this case, poorly soluble fluorides of some elements can be formed and $(NH_4)_2SO_4$ transfers them into sulfates. If silicate rocks are processed, the bulk of silicon is removed into sublimation, reducing the mass of soluble solids in the final solution.

The usage of the mixture of NH_4HF_2 and $(NH_4)_2SO_4$ for water leaching of the product of the interaction of titanium-containing mineral raw materials allows transferring almost the whole of titanium and the bulk of iron into the solution of highly water soluble double salts. When the leaching solution having pH equal to 1-2 is heated (up to 50-60 ℃), the process of hydrolysis o the titanium salt $(NH_4)_2TiO(SO_4)_2$ starts accompanied by the formation of titanium dioxide in the form of anatase. This method allows complete separation of titanium and iron from the leaching solution. The resulting filtrate, which is a mixture of phases of NH_4HSO_4 and $NH_4Fe(SO_4)_2$, can be separated by stepwise neutralization.

4 Conclusion

Thus, according to the results of the conducted experiments for the decomposition of Koksharovsky ilmenite-titanomagnetite ores, the mineral raw material proved to open better in case of using the mixture of ammonium sulfate with ammonium hydrofluoride in comparis on with the usage of ammonium hydrofluoride alone. Combining the processes of fluorination and sulfatization allows effective opening of rocks, including silicate ones, due to the breaking of Si-O bonds with the participation of NH_4HF_2 and the transformation of insoluble fluorides into soluble sulfates $(NH_4)_2SO_4$. The proposed technical solutions for the extraction of useful components from titanium ores worked out in compliance with the principles of environmental management and environmental safety are only the first step in the development of minerals in the south of the Russian Far East. It's obvious that further research needs to be carried out in the direction of the quality improvement of the processing degree, which will allow reducing the cost of obtaining individual products and ensure higher production efficiency.

Acknowledgments: This work was supported by the Russian Science Foundation (Project No. 23-17-00093), under international agreement No. 1136 between the Far East Geological Institute of the Far East Branch of the Russian Academy of Sciences and Jilin University.

References

[1] BYKHOVSKY L Z, PAKHOMOV F P, TURLOVA M A. The mineral resource base and prospects for the integrated use of titanomagnetite and ilmenite magmatogenic deposits in Russia[J]. Minir Information and Analytical Bulletin, 2008(1), 209-215.

[2] KOROVIN M. Rare and trace elements. Chemistry and technology[M]. MISIS, 1996: 461.

[3] ZHANG W, HU Z[J]. Spectrochimica acta Part B, 2019, 160, 105690. https://doi:10.1016/j.sab.2019.105690.

[4] KHANCHUK A I, MOLCHANOV V P, MEDKOV M A[J]. Doklady Chemistry, 2020, 491(2): 65-67. https://doi:10.1134/S0012500820040011.

[5] KARELIN V A, KARELIN A I. The fluoride technology for the processing of the rare metals concentrates[J]. Tomsk: NTL Publishing House, 2004: 221.

[6] ANDREEV A A. Fluoride technology for the production of the pigment titanium dioxide. Fluoride Technologies[M]. Tomsk: TPU Publishing House, 2009: 27.

[7] O'HARA M J, KELLOGG C M, PARKER C M, et al. Decomposition of diverse solid inorganic matrices with molten ammonium bifluoride salt for constituent elemental analysis[J]. Chemical Geology, 2017, (466): 341-351. https://doi:10.1016/j.chemgeo.2017.06.023

[8] POTTS P J, WEBB P C, THOMPSON M. Bias in the determination of Zr, Y and rare earth element concentrations in selected silicate rocks by ICP-MS when using some routine acid dissolution procedures: Evidence from the GeoPT proficiency testing program[J]. Geoanal, 2015(39): 315-327. https://doi:10.1111/j.1751-908X.2014.00305.x.

[9] KRYSENKO G F, EPOV D G, MERKULOV E B. Studying the possibility for defluorination of calcium and rare-earth fluorides by ammonium sulfate[J]. Theoretical Foundations of Chemical Engineering, 2021, 55(5): 996-1001. https://doi:10.31044/1684-5811-2020-21-9-395-402.

[10] ZHUMASHEV K, NAREMBEKOVA A, KATRENOV B B. Determination of the reaction mechanise of the calcium fluoride interaction with ammonium sulphate[J]. Bulletin of the Karaganda University "Chemistry" series, 2019, 3(95): 83-87. https://doi:10.31489/2019ch3/83-87.

Gadolinium Doped Tricalcium Phosphates for Hard Tissue Bioimaging: EPR Spectroscopy Control of Chemical Synthesis

Margarita SADOVNIKOVA[1], Fadis MURZAKHANOV[1], George MAMIN[1], Anna FORYSENKOVA[2], Inna FADEEVA[2], Marat GAFUROV[1]

1　Kazan Federal University, Kazan, 420008, Russia
2　A. A. Baikov Institute of Metallurgy and Material Science, Moscow, 119334, Russia

Abstract: Tricalcium phosphate (TCP) is a calcium phosphate bioceramic preferred by orthopedic clinics for bone repair owing to its high bioactivity and biocompatibility. The introduction of impurity ions instead of Ca^{2+} positions leads to significant favorable changes in the biochemical characteristics of TCP, which makes the current biomaterial more multifunctional and suitable for orthopedics and dentistry. Intrinsic magnetic and optical features of Rare Earth (RE) elements caused by their 4f-electronic configuration, make RE ions appropriate as magnetic resonance (MR) agents and highly sensitive diagnostic bioassays. The successful introduction of impurity centers, precise control of the concentration values and the formation of side phases are a determining factor for the creation of suitable biocompatible ceramics based on calcium phosphates. The most suitable method for analyzing the Gd^{3+} ions in TCP materials is the technique of electron paramagnetic resonance spectroscopy. In this article, it was unequivocally established that Gd^{3+} is incorporated into the TCP crystal lattice and occupies two structurally nonequivalent positions of Ca^{2+} during synthesis by precipitation from aqueous solutions of salts. Whereas mechano-chemical activation leads to a weak incorporation of Gd^{3+} ions into the TCP structure, despite the high declared concentration. Wherein Gd^{3+} ions don't lead to significant distortion of the crystal lattice of the material for both chemical synthesis methods.

Keywords: Rare earths; Tricalcium phosphate; Electron paramagnetic resonance

1　Introduction

The bone regenerative materials are gaining increasing attention in an alternative role to autologous bone, where the latter is still considered as the gold standard material for hard tissue transplantation[1]. The finiteness suitable parts, donor-site morbidity, long-term surgical intervention and high failure rates at difficult healing lead to significant restrictions in application of autologous bones for various age groups[2]. The bone tissue engineering has emerged as a promising biomedical direction for treatment of large bone defects where intrinsic bone repairing is insufficient[3,4]. The similarity of the mineral and elemental composition with inorganic components of bone tissue, as well as the ability to freely introduce various impurity ions into the crystal lattice, calcium phosphates have become widely known and attractive materials for design bone implants[5]. Beta-tricalcium phosphate (β-TCP) is a calcium phosphate bioceramic preferred by orthopedic clinics for bone repair owing to its high bioactivity and biocompatibility. Implants based on calcium phosphates contribute to adhesion and proliferation of osteoblastic cells in contact with the damaged area which leads to a robust biological contact between bone tissue and implant[6]. However, TCP has poor mechanical properties, low antibacterial activity and uncontrollable bio resorption which requires an improvement procedure to create a more perfect composite[7]. The favorable osteogenic environment can be achieved by modified of TCP in a number of ways including adjustment physical properties (e.g., pore sizes or porosity) using special conditions of chemical synthesis, combining with ionic inclusions in the crystal lattice. The introduction of impurity ions in small amounts instead of Ca^{2+} positions leads to significant favorable changes in the biochemical characteristics of TCP, which makes the current biomaterial more multifunctional and suitable for orthopedics and dentistry[8]. Recent experiments establish that lanthanides or Rare Earth (RE) elements are also appropriate for Ca^{2+} substitution[9,10]. Intrinsic magnetic and optical features caused by their 4f-electronic configuration, make RE elements suitable as magnetic resonance (MR) agents and highly sensitive diagnostic bioassays. Hard tissue regeneration must be accompanied optimized by imaging techniques, such as X-ray radiography, computed tomography (CT), single-photon emission computerized tomography (SPECT), and magnetic resonance imaging (MRI). At the same time, rare earth ions should not lead to toxicity and, as a consequence, rejection of the implant by the human body. Thus, the successful introduction of impurity centers, precise control the concentration values and the formation of side phases are a determining factor for the creation of suitable biocompatible ceramics based on calcium phosphates[11]. The chemical preparation of tricalcium phosphate can be carried out in two main ways: mechano-chemical activation synthesis due to mechano-activation and by wet precipitation. Among rare earth atoms, gadolinium ions are the well-known and widely used contrast agent for MRI due to their magnetic and optical features. The gadolinium ion Gd^{3+} in its ground state $^8S_{7/2}$ has an electron spin $S=7/2$, which leads to the presence of paramagnetic properties. The most suitable method for analyzing the presence of side phases (by-products), establishing the Gd^{3+} position and mechanisms of interaction of Gd^3 ions with surrounding atoms is the technique of electron paramagnetic resonance (EPR) spectroscopy. The high sensitivity and spectral resolution make it possible to carry out comprehensive studies of gadolinium-containing samples in a wide temperature range. The use of special pulse sequences allows to study the local nuclear environment and draw conclusions regarding the position of the Gd^{3+} ion and the effect on the degree of crystallinity/amorphism of the biomaterial. Thus, the goal of this work is to determine favorable synthesis conditions with the successful incorporation of Gd^{3+} ions into the sample structure and the avoidance of side toxic phases by monitoring the obtained samples by EPR spectroscopy.

2 Materials and methods

2.1 Synthesis

Tricalcium phosphates doped with gadolinium ions were synthesized by two methods: precipitation from aqueous solutions of salts (pc)(1) and mechano-chemical activation (ma)(2). More details see in the article [12].

The raw materials for precipitation from aqueous solutions of salts were calcium nitrate (chemical grade, Chimmed, Moscow, Russia), diammonium phosphate (analytical grade, Chimmed, Moscow, Russia), gadolinium chloride (chemical grade, Chimmed, Moscow, Russia), and 25% aqueous ammonia solution (analytical grade, Chimmed, Moscow, Russia).

$$(3-x)Ca(NO_3)_2 + 2x/3GdCl_3 + 2(NH_4)_2HPO_4 + 2NH_4OH \longrightarrow$$
$$Ca_{3-x}Gd_{2x/3}(PO_4)_2 + 0.2NH_4Cl + 5.8NH_4NO_3 + 2H_2O \quad (1)$$
$$(3-x)CaO + 2x/3GdCl_3 + 2(NH_4)_2HPO_4 \longrightarrow$$
$$Ca_{3-x}Gd_{2x/3}(PO_4)_2 + 3.8NH_3 + 0.2NH_4Cl + 2.9H_2O \quad (2)$$

The mechano-chemical activation was carried out in the planetary mill container PM-1, (Vibrotechnik, St. Petersburg, Russia) at a 1500 min^{-1} rotation rate. The ratio between materials mixture: zirconium oxide balls was 1:5 and the activation time was 30 min.

2.2 Electron paramagnetic resonance spectroscopy

The EPR spectra in continuous wave (CW) mode were recorded at $T=297$ K in the X-band microwave range ($v_{MW}=9.6$ GHz) on a Bruker spectrometer Elexsys E580. The modulation amplitude 0.01 mT at 100 kHz, microwave power $P=2$ μW were set in a such way to avoid over-modulation, distortion, or saturation of the EPR signal, respectively.

To measure the low-temperature ($T=25$ K and $T=12$ K) EPR spectra in the pulsed mode, the method of detecting the integral intensity of the electron spin echo (ESE) during the sweep of the magnetic field B_0 was used. The spin-spin relaxation time T_2 was measured by tracking the primary amplitude of the ESE with the same pulse durations $\pi/2$-τ-π-ESE with a change of delay time τ. The spin-lattice relaxation time T_1 was extracted from inversion-recovery studies by applying the pulse sequence π-T_{delay}-$\pi/2$-τ-π-ESE, while the delay time T_{delay} varied. The electron-nuclear interactions were analyzed using a three-pulse Electron Spin Echo Envelope Modulation (ESEEM) sequence ($\pi/2$-τ-$\pi/2$-T-$\pi/2$-ESE), with a change T from 180 ns to 1204 ns.

To investigate the radiation-induced paramagnetic species, X-ray irradiation of the materials was performed using a URS-55 source ($U = 50$ kV, $I = 15$ mA, W-anticathode) at $T = 297$ K for 1 h with a calculated dose of 15 kGy.

3 Results and discussion

3.1 Conventional EPR Spectroscopy

Figure 1 shows the EPR spectra recorded for a TCP-Gd sample with different concentrations of gadolinium ions at room temperature. The top two intense signals was obtained for TCP-Gd synthesized using mechanochemical activation, and the bottom one is for a sample synthesized by wet precipitation way. The EPR signals consist of an asymmetric broad line (from 0 to 600 mT) with a weakly resolved structure. In the region of a weak magnetic field, one can single out resonant absorptions corresponding to an effective g factor of 2.8 and additional unresolved components at $g_{eff} = 5.9$. The main contribution of intensity to the EPR spectrum is concentrated at $g = 2.0$, associated with typical 3d/4f metal ions in the powder form of the samples[13]. As can be seen, the blue line 0.01Gd(ma) in addition to gadolinium includes the spectrum of manganese ions (by-product)[14], which is not in the structure of the sample and probably located on the surface of the TCP particles.

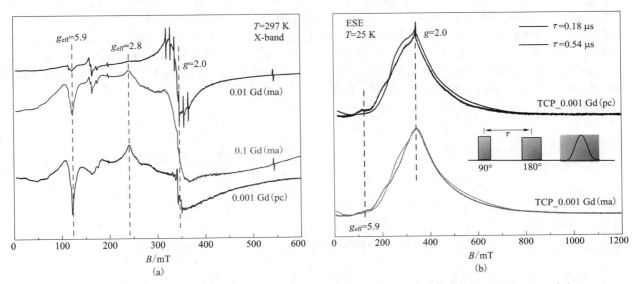

Figure 1 Concentration dependence of CW EPR spectra in the X-band (a) and EPR spectra of the TCP-0.001Gd in pulsed mode (X-band) for two different τ delay time values (b)

Also, analyzing the intensity of the spectrum lines, one can notice that during synthesis by the pc method, gadolinium is better integrated into the TCP lattice. For a more detailed study of the influence of the synthesis method on the incorporation of gadolinium into the structure, EPR experiments were carried out in a pulsed mode.

3.2 Pulse EPR Spectroscopy

The EPR spectrum was recorded using a two-pulse Hahn sequence with different τ times between pulses $\pi/2$ and π to determine the presence of other different contributions [the time between pulses is shown in Figure 1(b)]. Measurements of the transverse magnetization decay revealed two exponential curves. To confirm the presence of two different processes, the EPR spectrum was recorded with an increased time τ between pulses. As can be seen, with increasing time τ, a strong decrease in the spectrum component with a short relaxation time is observed against the background of the spectrum component with longer times, which confirms the existence of two types of Gd^{3+} centers in TCP structure. Thus, the difference between the EPR spectra arises due to the redistribution of intensities. The spin-lattice relaxation rate was also measured for both centers, at a magnetic field value of $B_0 = 344.3$ mT, in which the contribution of both types of centers, and $B_0 = 118.2$ mT, where there is a contribution from only the second Gd^{3+} center. To distinguish the dynamic characteristics of both centers to each other, the additional measurements of the relaxation rates were carried out at $T = 12$ K (Table 1).

Table 1 Relaxation times at $T = 12$ K of Gd^{3+} ions in TCP

Magnetic field/mT	$T_1/\mu s$	T_2/ns
TCP-0.001Gd (pc)		
$B_0 = 344$	157±3	173±5
$B_0 = 118$	173±5	1000±50
TCP-0.001Gd (ma)		
$B_0 = 344$	82±1	711±5
$B_0 = 118$	58±1	706±3

The ESEEM method is important for determining the anisotropic hyperfine structure from the modulation of the spin-spin (transverse) relaxation decay curve[15]. The integral intensity of the electron spin echo is recorded depending on the time interval (τ) between two pulses in a fixed magnetic field (B_0). The Fourier transform of modulations gives a frequency spectrum (Figure 2) in which the frequencies of nuclear-spin transitions are visible, which is necessary for the interpretation of hyperfine interaction mechanisms. Harmonic oscillations in the transverse magnetization decay curve occur when Gd^{3+} is bound to the surrounding nuclei (phosphorus) via an anisotropic dipole-dipole interaction[13].

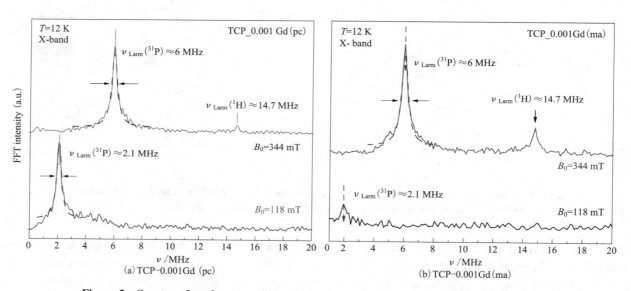

Figure 2 Spectra of nuclear transitions in the frequency range for different values of B_0

The spectra shows the ^{31}P signal centered on the Larmor frequency $\nu_{Larm}(^{31}P) = 6$ MHz for the magnetic field $B_0 = 344$ mT, and $\nu_{Larm}(^{31}P) = 2.1$ MHz for $B_0 = 118$ mT, respectively (Figure 2). These results indicate that the paramagnetic center (Gd^{3+}) is located in the crystal structure of the sample and can be successfully used as a spin probe in further investigations. In the studied TCP-0.001Gd (pc) samples include at least two distinguishable types of Gd^{3+} centers with different nuclear environments. Thus, we revealed the presence of two structurally non-equivalent positions of Gd^{3+} in the TCP lattice. In the case of TCP-0.001Gd (ma) sample at $B_0 = 118$ mT, this effect is weak and almost invisible at the frequency $\nu_{Larm}(^{31}P)$ of the signal, which indicates that Gd^{3+} ions are weakly incorporated into the TCP structure and occupy one structurally non-equivalent position Ca^{2+}.

A well-known way to study paramagnetic centers is to create radiation defects and study their spectroscopic properties. Earlier, several radiation-induced paramagnetic particles (radical anions) located in hydroxyl or phosphate centers were identified in calcium phosphates[16]. The samples under study were irradiated with an X-ray source under ambient conditions. EPR spectra (Figure 3) were recorded in pulsed mode using the Hahn sequence at $T = 297$ K.

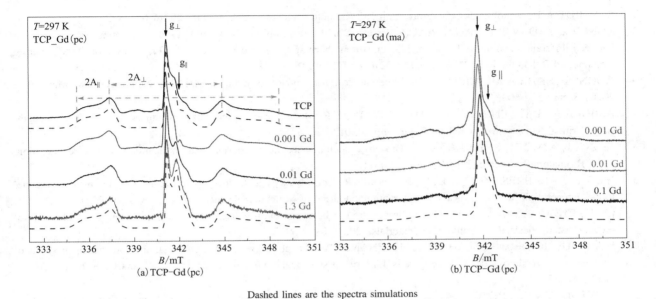

Dashed lines are the spectra simulations

Figure 3 EPR spectra of irradiated TCP samples as a function of Gd^{3+} concentration in pulsed mode

TCP-Gd(pc) contains impurities of the nitrate anion (a by-product of precipitation synthesis), which is used as a probe for analyzing the local environment. At different concentrations of gadolinium, the spectrum is not distorted, so we can conclude that TCP retains the original symmetry group. Simulation was also carried out using the EasySpin program[17], the main spectroscopic parameters are shown in Table 2. It has been shown that during synthesis using mechanochemical activation, nitrate salts are absent in the synthesis scheme, which subsequently leads to their disappearance. Thus, this method shows that it is possible to obtain TCP without NO_3, which in turn are toxic.

Table 2 Electron relaxation times of the nitrogen radical at $T = 297$ K

TCP-Gd (pc)				
	$g_\parallel$	$g_\perp$	$A_\parallel$/mT	$A_\perp$/mT
NO_3^{2-}	2.005	2.009	6.68	3.75
TCP-Gd (ma)				
CO_3^{2-}	2.006	2.00	-	-

4 Conclusion

Using the functionality of the X-band EPR spectroscopy in continuous and pulsed modes, it has been unequivocally established that Gd^{3+} is incorporated into the TCP crystal lattice and occupies two structurally nonequivalent positions of Ca^{2+} during synthesis by precipitation from solutions. Whereas mechano-chemical activation leads to a weak incorporation of Gd^{3+} ions into the TCP structure, despite the high declared concentration. Sufficiently intense signals from side synthesis ions (Mn^{2+}) on the surface of TCP particles were also revealed. The mechano-chemical activation way don't consist of nitrate salts in the synthesis scheme, which subsequently leads to the disappearance of toxic nitrate anions in TCP bioceramics confirmed by EPR measurements. Gd^{3+} ions don't lead to significant distortion of the crystal lattice of the material for both chemical synthesis methods.

Acknowledgments: This research was funded by RSF grant No. 23-63-10056.

References

[1] SCHMIDT A H. Autologous bone graft: Is it still the gold standard? [J]. Injury, 2021: S18-S22.
[2] TONK G, YADAV P K, AGARWAL S, et al. Donor site morbidity in autologous bone grafting—a comparison between different techniques of anterior iliac crest bone harvesting: A prospective study[J]. Journal of Orthopaedics, Trauma and Rehabilitation, 2022, 29(1), 22104917221092163.
[3] KOONS G L, DIBA M, MIKOS A G. Materials design for bone-tissue engineering[J]. Nature Reviews Materials, 2020, 5(8): 584-603.
[4] AMINI A R, LAURENCIN C T, NUKAVARAPU S P. Bone tissue engineering: Recent advances and challenges [J]. Critical Review in Biomedical Engineering, 2012, 40(5).
[5] JEONG J, KIM J H, SHIM J H, et al. Bioactive calcium phosphate materials and applications in bone regeneration [J]. Biomaterials Research, 2019, 23(1): 1-11.
[6] IELO I, CALABRESE G, DE LUCA G, et al. Recent advances in hydroxyapatite-based biocomposites for bone tissue regeneration in orthopedics[J]. International Journal of Molecular Sciences, 2022, 23(17): 9721.
[7] LU H, ZHOU Y, MA Y, et al. Current application of beta-tricalcium phosphate in bone repair and its mechanism to regulate osteogenesis[J]. Frontiers in Materials, 2021, 8: 698915.
[8] WANG G, ROOHANI-ESFAHANI S I, ZHANG W, et al. Effects of Sr-HT-gahnite on osteogenesis and angiogenesis by adipose derived stem cells for critical-sized calvarial defect repair[J]. Scientific Reports, 2017, 7(1): 41135.
[9] CAWTHRAY J F, CREAGH A L, HAYNES C A, et al. Ion exchange in hydroxyapatite with lanthanides[J]. Inorganic Chemistry, 2015, 54(4): 1440-1445.
[10] FADEEVA I V, LAZORYAK B I, DAVIDOVA G A, et al. Antibacterial and cell-friendly copper-substituted tricalcium phosphate ceramics for biomedical implant applications[J]. Materials Science and Engineering C, 2021, 129: 112410.
[11] BOHNER M, SANTONI B L G, DÖBELIN N. β-tricalcium phosphate for bone substitution: Synthesis and properties [J]. Acta Biomaterialia, 2020, 113: 23-41.
[12] FADEEVA I V, DEYNEKO D V, BARBARO K, et al. Influence of synthesis conditions on gadolinium-substituted tricalcium phosphate ceramics and its physicochemical, biological, and antibacterial properties[J]. Nanomaterials, 2022, 852.
[13] SADOVNIKOVA M A, MURZAKHANOV F F, FADEEVA I V, et al. Study of tricalcium phosphate ceramics doped with gadolinium ions with various EPR techniques[J]. Ceramics, 2022, 5(4): 1154-1166.
[14] MURZAKHANOV F F, FORYSENKOVA A A, FADEEVA I V, et al. Incorporation of manganese (II) in beta-tricalcium phosphate from EPR and ENDOR measurements for powders[J]. Ceramics, 2022: 318-329.
[15] VAN DOORSLAER S, GOLDFARB D, STOLL S. Hyperfine spectroscopy-ESEEM [J]. EPR Spectroscopy: Fundamentals and Methods, 2018: 377-400.
[16] MURZAKHANOV F F, GRISHIN P O, GOLDBERG M A, et al. Radiation-induced stable radicals in calcium phosphates: Results of multifrequency epr, ednmr, eseem, and endor studies[J]. Applied Sciences, 2021, 11(16): 7727.
[17] STOLL S, SCHWEIGER A. EasySpin, a comprehensive software package for spectral simulation and analysis in EPR [J]. J Magn Reson, 2006, 178: 42-55.

Charpy Impact Testing of a Middle-entropy Alloy with an Austenitic Structure Subjected to Cold Rotary Swaging

Dmitrii PANOV, Ruslan CHERNICHENKO, Stanislav NAUMOV

Belgorod State University (BSU), Belgorod, 308015, Russia

Abstract: The middle-entropy alloy with austenitic structure was subjected to cold rotary swaging with a reduction of up to 85%. In the cross-section of the cold-swaged rod, the non-uniform structure was attained. The dynamic mechanical behavior of the program material was studied during Charpy V-notch impact testing at ambient temperature. It was found that impact toughness and impact plane-strain fracture toughness decreased dramatically after a reduction of 20%. However, with following swaging reduction, impact toughness and impact plane-strain fracture toughness was stabilized at 0.6 MJ/m^2 and 70 MPa·m$^{1/2}$, respectively. Yet, maximum stress (σ_m) increased with swaging reduction. The fracture mechanisms of the program material during Charpy impact V-notch impact testing were also studied.

Keywords: Austenitic structure; Cold rotary swaging; Middle-entropy alloy; Charpy impact V-notch impact test

1 Introduction

Medium-entropy alloys (MEAs) possess broad prospects for obtaining unique properties and their combinations due to the infinite variety of possible compositions[1-3]. Recently, such alloys have been significantly developed by the addition of interstitial elements (carbon, nitrogen, etc.)[4-7] that resulted in a attractive combination of strength and ductility. However, MEAs perform low yield strength. To increase strength properties, cold deformation hardening is usually applied[8-10]. Meanwhile, cold rotary swaging might be considered as one of the most promising deformation technics due to obtaining non-uniform microstructure throughout the rod cross-section during processing[11]. In spite of the large body of works[12], how cold rotary swaging influence impact toughness of MEAs at room temperature is not studied properly. Therefore, this work aimed to study the effect of the cold rotary swaging mode on the impact toughness of MEA.

2 Experiment

MEA with the chemical composition 49.5Fe-30Mn-10Co-10Cr-0.5C (wt.%) was used as the program material. The program material was subjected to water quenching at 1050 ℃ with a dwelling time of 2 h. Then the rod underwent cold rotary swaging with a reduction of up to 85%. The microstructure characterization was performed by a JEOL JEM-2100 transmission electron microscope (TEM) with an accelerating voltage of 200 kV. Templets with a thickness of 0.3 mm were cut by a Sodik AQ300L electric discharge machine, followed by mechanical grinding and polishing up to 0.1 mm. The obtained foils were perforated by TenuPol-5 equipment in an electrolyte consisting of 10% perchloric acid in acetic acid. V-notch specimens with dimensions 10 mm×5 mm×55 mm were cut along the rod axis. The Charpy impact testing was carried out by an Instron 450 J machine according to ISO 148-1: 2009. The fracture after the Charpy V-notch impact testing was studied by a Nova Nano SEM 450 scanning electron microscope.

3 Result and discussion

3.1 Microstructure characterization

The results of microstructure characterization after different cold rotary swaging modes are shown in

Figure 1. Uniaxial austenitic grains with a diameter of 52 μm were found in the initial condition. A few annealing twins and single dislocations were also observed herein [Figure 1(a)]. Furthermore, the microstructure of the initial rod was uniform over the cross-section. However, with swaging reduction, the microstructure along the rod cross-section became non-uniform. Cold rotary swaging with a reduction of up to 40% resulted in increasing twin and dislocation density. So, in the rod center, mechanical twins of one system were observed inside the grains [Figure 1(b)]. Meanwhile, at the rod edge, the lamellar microstructure was formed due to twinning development [Figure 1(c)]. Apparently, after 80%-85% swaging reduction, twins of various systems were derived in the rod center [Figure 1(d)]. Thereby, a block-shaped microstructure was obtained due to the crossing of twins of different slid systems. At the rod edge, the fragmented microstructure was found because of shear banding and dislocation slip development [Figure 1(e)]. Thus, the morphology of the microstructure in the edge and center was dramatically different.

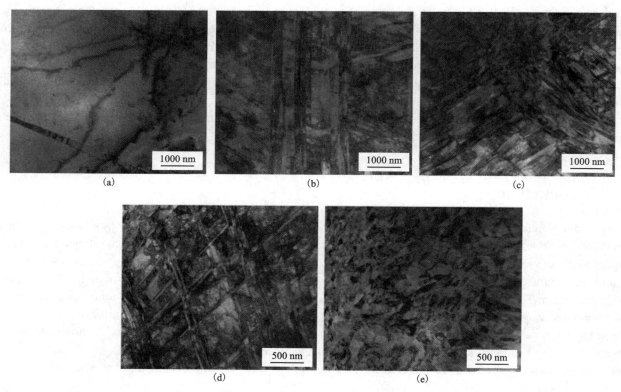

Figure 1 Microstructure of the program material in the initial condition (a) and after cold rotary swaging with a reduction of (b), (c) 40% and (d), (e) 80%: (a), (b), (d) rod center; (c), (e) rod edge

3.2 Charpy V-notch impact testing

The fracture diagrams and impact toughness value depending on cold rotary swaging mode are presented in Figure 2. In the initial condition, the yield plateau was found after the elastic part [Figure 2(a)]. However, after cold-swaged samples did not demonstrate the yield plateau. Meanwhile, with swaging reduction, increasing the applied load was observed during Charpy V-notch impact testing [Figure 2(a)]. On the other hand, well-defined inflections were observed on the diagrams after swaging with a reduction of 60%-85% reduction. Yet, no inflections were found on similar diagrams of the program material with the uniform microstructure. Hence the mechanical behavior can be ascribed to non-uniform microstructure produced by cold rotary swaging[13]. Obviously, crack propagation throughout various structure layers can be associated with additional energy adsorption and caused the inflections.

According to the obtained results [Figure 2(b)], in the initial condition, the program material possessed the KCV value at 1.37 MJ/m^2. After cold rotary swaging with a reduction of 20%, the KCV value decreased drastically to 0.6 MJ/m^2. Further swaging with a reduction of up to 85% did not change the KCV value.

The effect of swaging reduction on impact plane-strain fracture toughness and dynamic strength might be

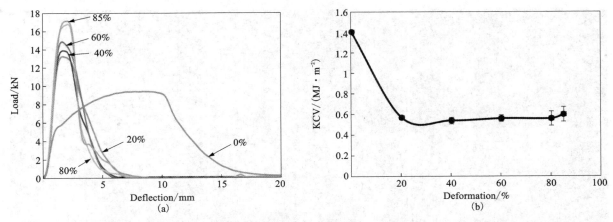

Figure 2 Fracture diagrams (a) and effect of impact toughness on swaging reduction (b)

defined by the analysis of the diagrams [Figure 2(a)]. The values of the maximum stress (σ_m) after different swaging modes were also estimated (Table 1) using the following equation[14]:

$$\sigma = \frac{\beta LP}{2C_f(w-a)^2 B} \quad (1)$$

where: L is the distance between supports ($=40$ mm); P is the load, kN; w is the sample's width ($=10$ mm); B is the thickness of sample ($=5$ mm); a is the notch depth ($=2$ mm); C_f is the constant factor ($=1.363$ for ASTM tup); β is the Tresca criterion ($=2$)[15].

To evaluate the crack resistance of the MEA, impact plane-strain fracture toughness (K_{Id}) was also calculated. The K_{Id} values were estimated by the following equation[16]:

$$\frac{(K_{Id})^2}{E} = 6.46 \times 10^{-4}(\text{CVN}) \quad (2)$$

where: E is Young's modulus ($=200$ GPa); CVN is the shock absorption energy, J.

Table 1 Dynamic parameters of the program material

Parameters	Deformation/%					
	0	20	40	60	80	85
Maximum stress (σ_m)/MPa	850	1210	1090	1270	1530	1620
Impact plane-strain fracture toughness (K_{Id})/(MPa·m$^{1/2}$)	108	69	66	67	69	70

The increase in a reduction of up to 85% doubled the maximum stress (Table 1). So, the maximum stress increased from 850 MPa for the initial condition to 1620 MPa for the 85%-swaged condition. Apparently, the attained strengthening might be ascribed to the deformation hardening. Interestingly, an increase in reduction resulted in a decrease in the K_{Id} values (Table 1). The K_{Id} values decreased from 108 MPa·m$^{1/2}$ to 70 MPa·m$^{1/2}$.

The fracture overviews and microfracture photographs are shown in Figure 3. In the initial condition, the fraction of the shear lips on the fracture overviews reached 95% [Figure 3(a)]. However, after a reduction of 40%, the fraction of shear lips decreased to 62% [Figure 3(b)]. Following swaging with a reduction of 80% resulted in an increase in the fraction of the shear lips to 92% [Figure 3(c)]. Apparently, the increased fraction of the shear lips was associated with an increase in impact toughness[17]. In the initial condition and after swaging with a reduction of 40%, the program material was destroyed by the ductile fracture micromechanism as a result of the formation and coalescence of microvoides [Figures 3(d), (e)]. However, the fraction of large dimples decreased from 70% to 55% that resulted in decreasing the value of impact toughness. Interestingly, in the 80%-swaged condition, the main fracture micromechanism was stepwise [Figure 3(f)] probably because of the development of crack both along and across the fibrous microstructure of the rod.

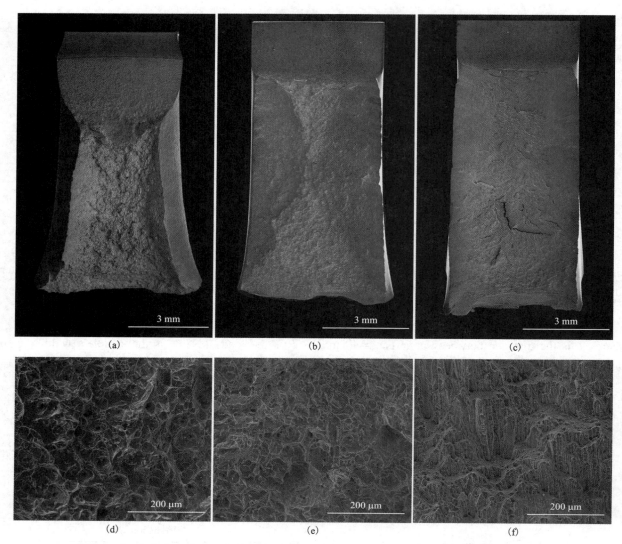

Figure 3 (a)-(c) Fracture overviews and (d)-(f) microfracture photographs in (a), (d) the initial condition, (b), (e) after 40% and (c), (f) 85% swaging reduction

Generally, the value of impact toughness relied on both the fraction of shear lips and big dimples. Comparing the 40%-swaged material to the initial condition, the structure refinement caused decreasing the fraction of big dimples and shear lips on the fracture surfaces. In 80%-swaged condition, the alteration of fracture micromechanism from formation and coalescence of microvoides to stepwise resulted in changing the crack development direction and, therefore, an increase in the fraction of shear lips on the fracture surfaces. Thereby, with swaging reduction, strengthening (Table 1) was accompanied stabilization of impact toughness and impact plane-strain fracture toughness.

4 Conclusion

The effect of cold rotary swaging on the impact toughness of the middle-entropy alloy 49.5Fe-30Mn-10Co-10Cr-0.5C was studied. The following conclusions were obtained:

(1) Cold rotary swaging resulted in the non-uniform microstructure formation due to intersection of mechanical twins of different systems in the center and fragmentation of a lamellar structure at the edge after a reduction of 80%.

(2) With swaging reduction, significant strengthening was accompanied by stabilization of impact toughness and impact plane-strain fracture toughness at the level of 0.6 MJ/m^2 and 70 MPa·m$^{1/2}$,

respectively.

(3) In the initial and 40%-swaged conditions, the fracture micromechanism was formation and coalescence of microvoids. However, in the 80%-swaged condition, the main fracture micromechanism was the crack development both along and across the fibrous microstructure.

Acknowledgments: This work was funded by the Russian Science Foundation Grant No. 20-79-10094. The authors are grateful to the personnel of the Joint Research Center, "Technology and Materials", Belgorod National Research University, for their assistance.

References

[1] CANTOR B, CHANG I T H, KNIGHT P, et al. Microstructural development in equiatomic multicomponent alloys [J]. Mater. Sci. Eng. A, 2004: 213-218.
[2] YEH J W, CHEN S K, LIN S J, et al. Nanostructured high-entropy alloys with multiple principal elements: Novel alloy design concepts and outcomes[J]. Adv. Eng. Mater., 2004: 299-303.
[3] MIRACLE D B, SENKOV O N. A critical review of high entropy alloys and related concepts[J]. Acta Mater., 2017: 448-511.
[4] LI Z, TASAN C C, SPRINGER H, et al. Interstitial atoms enable joint twinning and transformation induced plasticity in strong and ductile high-entropy alloys[J]. Nature Publishing Group, 2017: 1-7.
[5] SONG R, PONGE D, RAABE D, et al. Overview of processing, microstructure and mechanical properties of ultrafine grained bcc steels[J] Mater. Sci. Eng. A, 2006: 1-17.
[6] SU J, WU X, RAADE D, et al. Deformation-driven bidirectional transformation promotes bulk nanostructure formation in a metastable interstitial high entropy alloy[J]. Acta Mater., 2019: 23-39.
[7] SU J, RAABE D, LI Z. Hierarchical microstructure design to tune the mechanical behavior of an interstitial TRIP-TWIP high-entropy alloy[J]. Acta Mater., 2019: 40-54
[8] VALIEV R Z, ISLAMGALIEV R K, ALEXANDROV I V. Bulk nanostructured materials from severe plastic deformation[J]. Progress in Materials Science, 2000: 103-189.
[9] PANOV D O, SMIRNOV A I, PERTSEV A S. Formation of structure in metastable austenitic steel during cold plastic deformation by the radial forging method[J]. Phys. Met. Metallogr, 2019: 184-190.
[10] PANOV D O, CHERNICHENKO R S, KUDRYAVTSEV E A, et al. Cold swaging on the bulk gradient structure formation and mechanical properties of a 316-type austenitic stainless steel[J]. Materials, 2022: 2468
[11] PANOV D O, KUDRYAVTSEV E A, NAUMOV S V, et al. Gradient microstructure and texture formation in a metastable austenitic stainless steel during cold rotary swaging[J]. Materials, 2023: 1-16.
[12] MAO Q, LIU Y, ZHAO Y. A review on mechanical properties and microstructure of ultrafine grained metals and alloys processed by rotary swaging[J]. J. Alloys Compd., 2022: 163122
[13] PANOV D O, CHERNICHENKO R S, NAUMOV S V, et al. Excellent strength-toughness synergy in metastable austenitic stainless steel due to gradient structure formation[J]. Mater. Lett., 2021: 130585.
[14] CHAOUADI R, FABRY A. On the utilization of the instrumented Charpy impact test for characterizing the flow and fracture behavior of reactor pressure vessel steels[J]. European Structural Integrity Society, 103-117.
[15] DUDKO V A, FEDOSEEVA A E, KAIBYCHEV R O. Ductile-brittle transition in a 9% Cr heat-resistant steel[J]. Mater. Sci. Eng. A, 2017: 73-84.
[16] BARSOM J M, ROLFE S T. Fracture and fatigue control in structures: Applications of fracture mechanics[J]. ASTM, West Conshohocken, PA, 1999.
[17] LI J, QIN W, PENG P. Effects of geometric dimension and grain size on impact properties of 316L stainless steel [J]. Mater. Lett., 2021: 128908

Resonant Photoemission Spectroscopy for Studying Long-lived Excited Atomic States in Multicomponent Rare Earth-transition Metal Materials

Tatyana KUZNETSOVA[1,2], Vladimir GREBENNIKOV[1,3], Ekaterina PONOMAREVA[1]

1　M. N. Miheev Institute of Metal Physics of UB RAS, Ekaterinburg, 620108, Russia
2　Ural Federal University, Ekaterinburg, 620002, Russia
3　Ural State University of Railway Transport, Ekaterinburg, 620034, Russia

Abstract: An investigation of the interplay of d- and f-elements and effect on the formation of the electronic structure in different multicomponent R-T compounds is presented. The method of resonant X-ray photoemission spectroscopy allows to select the contributions of the various components in the valence bands (VB). We discuss about possibility of study not only the ground state, but also the lifetime of the excited (a core-level hole-VB electron) state, determine energies of the VB single-particle states and two-hole states at selected atoms, and see reactions to sudden appearance of the core-level photo-hole.

Keywords: Multicomponent rare earth-transition metal materials; resonant photoemission spectroscopy; Excited atomic states

1 Introduction

Intermetallic compounds formed by rare earth (R) and transition (T) metals are of great interest for research due to their unique magnetic properties and features of the electronic structure[1-2]. Traditionally, to study the electronic structure of such compounds, theoretical calculations and photoelectron spectra are performing. The use of synchrotron radiation makes it possible to obtain additional information about long-lived atomic states in multicomponent R-T compounds.

Among the rare-earth intermetallic compounds, there are groups of ternary intermetallic compounds RT_2Si_2 and RT_2T, in which, depending on the interatomic distances of transition metals T and the choice of a rare-earth element, the crystal structure and magnetic properties change[3]. RMn_2Si_2 (R = La, Sm, Tb) compounds have a layered structure and are characterized by a large magnetocaloric effect, and the type of magnetic ordering in them depends on the intralayer distance between manganese atoms[4-6]. In intermetallics with the Laves phase RNi_2Mn (R = Tb, Er, Dy), giant magnetostriction[7] and spontaneous magnetization at a temperature of 5 K[8] are observed. Manganese addition in RCo_2Mn_x compounds (R = Tb, Er, Dy) leads to an increase in the Curie temperature (T_c) and the appearance of a magnetocaloric effect[9-10]. With an increase in the manganese concentration, the magnetic moment of the 3d sublattice increases, which probably caused with changing in the electronic structure of the compounds. The magnetic properties of most ternary rare-earth intermetallic compounds has been studied in detail, in contrast to the electronic structure, information about which can supplement the results of available magnetometric measurements. Another group of R－T compounds, for example, the intermetallic RT5 compounds where R is a rare earth and T is a 3d transition metal has been extensively studied due to their interesting magnetic properties[11-13]. One of the most popular is the $GdNi_5$ compound which indicates a ferromagnetic behaviour with ordering temperature T_c = 32 K and magnetic moment 6.2B/(f.u.) in the $GdCu_5$ compound which are antiferromagnetic arrangement below T_N = 26 K and a unusual behaviour of $\rho(T)$ below 30 K. These features are probably connected with its helimagneticlike structure[14]. The Ni/Cu substitution in the $RNi_{5-x}Cu_x$ is reflected in the different magnetic properties depending on the type of R atom[15]. These properties are connected with crystal field effects, band magnetism and electronic structure.

One of the most informative methods for studying the electronic structure is resonant photoelectron spectroscopy, which can be used to determine the localization of 4f electrons of rare earth elements and 3d-electrons of transition metals in the valence band (VB). R-T and T-T interactions have a significant effect on the magnetic properties of both layered intermetallic compounds and compounds with the Laves phase. The method used is surface-sensitive and requires careful preparation of the surface of the samples under study.

The aim of this work is an experimental study of the interaction of f-and d-elements and its influence on the formation of the electronic structure of intermetallic compounds in a wide energy range. The used method of resonant photoemission spectroscopy makes it possible to separate the contributions of various components to the valence bands. In this case, it is possible to study not only the ground state, but also the characteristics of the excited states, the response to an external action, and the processes of relaxation of the electronic system.

2 Results and discussion

The interaction of X-rays with matter provides unique opportunities for measuring local atomic characteristics that are selective in terms of elements and their depth, which are widely used by researchers. Traditionally, the interaction of X-ray quant with matter considered in the first order of the theory of perturbations. Second-order processes are much less studied: the transition of an atom to an excited state upon absorption of a photon and the decay of this state with the birth of a new photon (RIXS-resonant inelastic X-ray scattering) or photoelectron (RXPS-resonant X-ray photoemission). Our experiments carried out on the BESSY-II synchrotron at the Russian-German Laboratory in Berlin. All samples cleaved directly in the ultra-high vacuum chamber of the spectrometer.

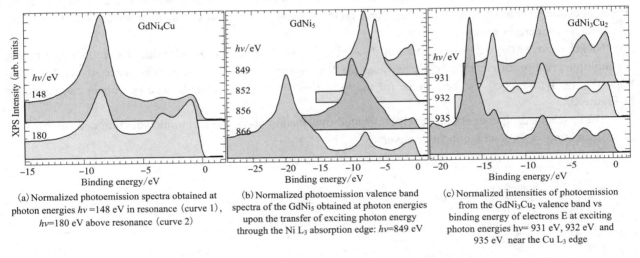

(a) Normalized photoemission spectra obtained at photon energies $h\nu$ =148 eV in resonance (curve 1), $h\nu$=180 eV above resonance (curve 2)

(b) Normalized photoemission valence band spectra of the GdNi$_5$ obtained at photon energies upon the transfer of exciting photon energy through the Ni L$_3$ absorption edge: $h\nu$=849 eV

(c) Normalized intensities of photoemission from the GdNi$_3$Cu$_2$ valence band vs binding energy of electrons E at exciting photon energies $h\nu$= 931 eV, 932 eV and 935 eV near the Cu L$_3$ edge

Figure 1 Normalized photoemission spectra of ternary intermetallic compounds GdNi$_{5-x}$Cu$_x$

Resonance photoemission in narrow-band materials described by the sum of the first and the second order transitions, their quantum mechanical interference leads to an increase in the spectrum from the valence bands and the appearance of an asymmetric dependence on the photon energy. These effects experimentally studied on the example of ternary intermetallic compounds GdNi$_{5-x}$Cu$_x$ (Figure 1). The competition between elastic and inelastic photoemission channels leads to the dependence of the photoemission spectra of nickel and manganese on the photon energy. The elastic channel realized on atoms with large magnetic moments, while the inelastic Auger decay occurs on atoms with small moments. The VB XPS spectra obtained at different photon energies $h\nu$ crossing the core-level excitation thresholds. A comparative analysis is gave for the two enhancement channels (elastic and Auger electron) of the valence band photoemission under resonant excitation of the 2p core-level. For example, on Gd N$_{4,5}$ edges (148 eV) the 8.5 eV peak in valence band increases in GdNi$_4$Cu. The Gd 4d core-level electron transfers to a long-lived Gd 4f-state, and then the reverse transition occurs, accompanied by emission of an electron detected. Note that the peak is enhanced much more (350 times) on the Gd M$_5$ absorption edge (1184 eV). The nickel and copper spectra behave otherwise. The VB signal (0-9 eV) in GdNi$_3$Cu$_2$ does not change at the Cu L$_3$ excitation edge $h\nu$ = 932 eV, but the Auger line appears with a starting binding energy of 14 eV. This means that the Cu 2p-3d excited state very quickly leaves the parent atom and then the Auger decay occurs, forming two holes in the Cu 3d-states. Subtracting energy of two single-hole states (2×3.5 eV) from 14 eV, we find experimentally the interaction energy for two holes on the copper atom 7 eV. Nickel behaves the same as copper, but its Auger line is significantly broader than the appropriate copper line. Broadening is due to shaking or the multiple creation of electron-hole pairs near the Fermi energy upon the sudden occurrence of the photo-hole. The probability of this process on a nickel atom is very high, since the Ni has high density of 3d-states at the Fermi level. Shaking on the copper atom is much smaller, since the Cu 3d-states are deepened by 3.5 eV.

The Figure 1 shows the change in the shape of X-ray photoemission spectra (XPES) obtained at different photon energies. When the threshold of excitation of the internal level is reached, an additional electron emission channel opens, which is associated with the filling of the formed hole with a valence electron and the ejection of a second electron from the sample. Combined with direct photoemission from the valence band, this channel increases the yield of photoelectrons from an excited atom, clearly indicating the energy position of its states. It can be seen in the GdNi$_4$Cu figure that at the Gd 4d excitation threshold of the 148 eV level, a peak with a binding energy of 8 eV intensifies in the spectrum of the valence band, which means that it is formed by gadolinium states. The intensity of the 8 eV peak at the Gd 3d excitation threshold ($h\nu$ = 1184 eV) increases even more. Thus, the unoccupied 4f states on the rare earth atom form a long-lived excited atomic state (trap) for the excited electron, which it leaves through the reverse transition to the initial level with the simultaneous ejection of a valence electron from the rare earth atom, which is fixed by the detector as a photoelectron. The excited states of the transition atoms Ni and Cu, but not Mn, behave quite differently. There is no trap here, the excited electron leaves the parent atom, and the core hole evolves by Auger decay. Auger lines are visible on nickel (GdNi$_5$) and copper (GdNi$_3$Cu$_2$) atoms, but their shape is different, for which there are certain reasons.

Nickel behaves the same as copper but its Auger line is significantly broader than the copper line due to the multiple creation of electron-hole pairs near the Fermienergy upon the sudden occurrence of the 2p photo-hole [Figure 1(b), (c)]. The scheme for the formation of the final two-hole state on copper can be shown in Figure 2.

Thus, resonant photoemission provides detailed information about the electronic structure and mechanisms of its formation in three-component compounds.

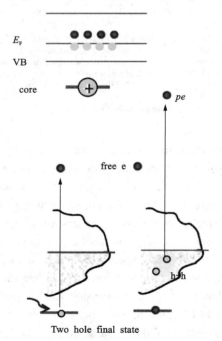

Figure 2 Scheme of formation of the final two-hole state on copper

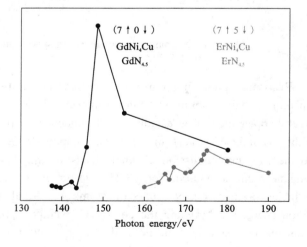

Figure 3 The maximum value of the XPS intensity versus excitation energy near R M$_5$ edges in Gd(Er)Ni$_4$Cu

Figure 3 shows the maximum value of the XPS intensity versus excitation energy near R M_5 edges in Gd(Er)Ni_4Cu. Especially high gain and narrow resonance is observed in atoms with half-filled f-shell. With increasing number of electrons in the elastic channel resonant photoemission adds a competing channel with excitation of the electron-hole pair in VB. This leads to a broadening of the XPS spectrum and the weakening of the resonant amplitude.

4 Conclusion

Photoemission from the VB 4f-states increases manifold on the threshold of excitation the REM 3d and 4d levels. The magnetic moments (spin) of the atoms determine the magnitude and width of the resonance spectra. Especially high gain and narrow resonance is observed in atoms with half-filled f-shell. The competing channel with excitation of the electron-hole pair in VB adds with increasing number of electrons in the elastic resonant photoemission. This leads to a broadening of the XPS spectrum and the weakening of the resonant amplitude. Transition metals RXPS are roughly the same dependence on the atomic magnetic moments, but their behavior is more complex and diverse. This is due to greater spatial and energy delocalization of d-states. The d-band does not have a clear high-energy boundary. The absorption intensity far above the threshold significantly exceeds its value before the threshold. REM have a narrow absorption band, and TM rather broad. Resonance XPS study not only the ground state, but also the lifetime of the excited states, determine energy of the VB single-particle states and two-hole states at selected atoms, and see reactions to sudden appearance of the core-level photo-hole.

Acknowledgments: The research was funded by the Russian Science Foundation (Project No. 23-72-00067).

References

[1] SZYTUłA A. Handbook of magnetic materials[M]. Amsterdam: Elsevier, 1991: 667.
[2] SZYTUłA A. Handbook of crystal structures and magnetic properties of rare earth intermetallics[M]. Florida: Boca Raton, CRC Press, 1994: 281.
[3] GSCHNEIDNER K A J. Handbook on the physics and chemistry of rare earths[M]. Amsterdam: Elsevier, 1989: 434.
[4] GSCHNEIDNER K A J, PECHARSKY V K, TSOKOL A O. [J]. Reports on progress in physics, 2005, 68: 1479-1539.
[5] GERASIMOV E G, KURKIN M I, KOROLYOV A V, et al. [J]. Physica B, 2002, 322: 297-305.
[6] MUSHNIKOV N V, GERASIMOV E G, GAVIKO V S, et al. [J]. Phys. Solid State, 2018, 60: 1082-1089.
[7] ENGDAHL G. Handbook of giant magnetostrictive materials[M]. Amsterdam: Elsevier, 1999: 386.
[8] WANG J L, MARQUINA C, IBARRA M R, et al. [J]. Physical Review B, 2006, 73: 094436.
[9] MAJI B, SURESH K G, NIGAM A K. [J]. Journal of Magnetism and Magnetic Materials, 2010, 322: 2415.
[10] GERASIMOV E G, et al. [J]. Journal of Alloys and Compounds. 2016, 680: 359-365.
[11] BAJOREK A, et al. [J]. Journal of Alloys and Compounds, 2011, 509: 578-584.
[12] PECHEV S, BOBET J L, CHEVALIER B, et al. [J]. Solid State Chem, 2000, 150: 62.
[13] CHEłKOWSKA G, BAJOREK A, KATOLIK A, et al. [J]. Magn, Mater, 2004, 272-276(Suppl.) E455.
[13] BARANDIARAN J M, GIGNOUX D, RODRIGUEZ-FERNANDEZ J, et al. [J]. Physica B, 1989, 154: 239.
[14] KUCHIN A D, ERMOLENKO A S, KULIKOV Y A, et al. [J]. Mater, 2006, 303: 119.
[15] BURZO E, CHIUZBAIAN S G, NEUMANN M, et al. [J]. J. Phys.: Condens. Matter, 2000, 12: 5897.

Numerical Realization of Helicoidal DNA Model

Maria OSTRIK[1], Victor LAKHNO[2]

1 Leonov University of Technology, Korolev, 141074, Russia
2 The Institute of Mathematical Problems of Biology RAS, Pushchino, 142290, Russia

Abstract: The numerical realization of the helicoidal mechanical model of the DNA molecule by the Runge-Kutta method of the fifth order of accuracy with an automatically selectable variable step in time is proposed. The model is a further development of the well-known helicoidal DNA model developed by M. Barbi et al. Molecular dynamic modeling of DNA denaturation was carried out on the basis of the proposed model. Agreement of the calculated temperature of denaturation with its experimental values is obtained.

Keywords: DNA molecule; Biomolecule conductivity; Denaturation; Runge-Kutta method; Molecular dynamics method

1 Introduction

DNA molecules can be used as structural elements of promising electronic devices[1]. Electronic nanobiochips have a number of advantages over modern silicon chips (miniature, high-speed performance and accuracy). It is also possible to use DNA molecules in memory and logic devices[1, 2]. The success of the use of DNA in electronics depends on the possibility of ensuring its conductivity, which is largely determined by the properties of open states. The formation of open states (denaturation bubbles) and their propagation can be numerically modeled based on various mechanical models. Interest in description of the physical and mechanical behavior of DNA molecules began with work[3] and increased in part of mathematical modeling of this behavior after work [4-6]. A fairly general dynamic helicoidal model for describing the mechanical behavior of a DNA molecule was developed in the works of M. Barbi et al.[7-9] This work is devoted to the further development and numerical realization of the M. Barbi model.

2 Mechanical model of DNA molecule

The helicoidal mechanical DNA model of M. Barbi et al.[7-9] is a further development of the well-known Peyrard-Bishop flat model (PB model)[5]. It correctly describes the helical structure of the molecule and the relationship between rotation deformations and the opening of hydrogen bonds. Unlike works [7-9], our model does not use Tailor series expansion of interaction potentials. This makes it possible not to discuss issues related to the possibility of such expansion at large amplitudes of radial and angular oscillations of bases near the DNA denaturation region.

Figure 1 shows the stages of transitioning from the actual helical geometry of a DNA molecule to a helicoidal DNA model. In it, each pair of nitrogenous bases rotates in a relatively rigid backbone of the molecule and its configuration is given by two generalized coordinates ($n=1, \ldots, N$; N is the number of base pairs in DNA): r_n is a radial variable associated with the break of hydrogen bonds and Φ_n is the rotation angle of each of the base pairs (the combination of these angles sets the current helical structure of DNA; the initial twist of the molecule is given by the constant difference between the angles of rotation of adjacent pairs θ). In this case, the bases themselves are considered as indivisible and non-deformable objects (point masses).

As in the PB model of DNA, hydrogen bonds between bases are described by Morse potential with depth D and width a

$$U_{Hn} = D[e^{-a(r_n - R)} - 1]^2 \qquad (1)$$

where: $2R$ is the equilibrium distance between the bases in the pair; D, a are the depth and width characteristics of the Morse potential well, respectively. The helical structure of DNA is formed due to two

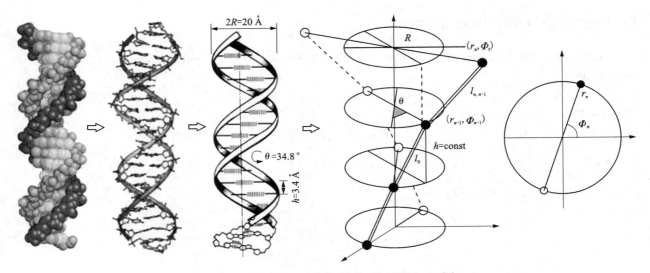

Figure 1 Geometry of the helicoidal DNA model

factors: the stacking interaction that brings together base pairs, and the stiffness of the sugar-phosphate backbone of the molecule that prevents this convergence. The stiffness of the skeleton is so great that the distance between adjacent base pairs along the DNA double helix is practically fixed.

The helical structure of DNA is formed due to two factors: the stacking interaction that brings together base pairs and the stiffness of the sugar-phosphate backbone of the molecule that prevents this convergence. The stiffness of the skeleton is so great that the distance between adjacent base pairs along the DNA double helix is practically fixed. To approach the base pairs, they can only turn, tilting the sugar-phosphate skeleton in a spiral structure[10]. In the model of M. Barbi et al.[7-9], to simulate the described features of DNA twisting, base pairs remain in planes, the distance between which remains constant and equal to h. The connection between the planes of the adjacent base pairs is modeled by elastic rods with an equilibrium length of $l_0 > h$ (the fulfillment of this inequality is possible only with the inclined position of the rods, see Figure 1. Equilibrium l_0 and current $l_{n,n-1}$ (between the planes of pairs $n-1$ and n) of rod lengths are determined from the model geometry

$$l_0 = \sqrt{h^2 + 4R^2 \sin(\theta/2)}, \quad l_{n,n-1} = \sqrt{h^2 + r_n^2 + r_{n-1}^2 - 2 r_n r_{n-1} \cos(\Phi_n - \Phi_{n-1})} \qquad (1)$$

Then the potential energy of each pair of rods between the planes of pairs $n-1$ and n is written as (K is the stiffness coefficient of the elastic bond)

$$U_{\text{rod }n} = K(l_{n,n-1} - l_0)^2 \qquad (2)$$

The potential energy of the stacking interaction between adjacent pairs $n-1$ and n is[9]

$$U_{Sn} = St(r_n - r_{n-1})^2 e^{-b(r_n + r_{n-1} - 2R)} \qquad (3)$$

where: St and b are model parameters characterizing the energy of the stacking interaction connection.

The kinetic energy of the n-th pair is composed of radial motion and rotation (m is the mass of the base)

$$T_n = m(\dot{r}_n^2 + r_n^2 \dot{\Phi}_n^2) \qquad (4)$$

where: the point above the variable means time differentiation, m is the mass of the base.

Summing the energy over all bases and interactions, we get from Equation (1) to (4), the dimensionless Lagrange function for the considered mechanical system ($1/a$ is the length scale; D is energy scale; $\sqrt{m/(Da^2)}$ is time scale)

$$\tilde{L} = \sum_n y_n^{*2} + (y_n + r)^2 \varphi_n^{*2} - [(e^{-y_n} - 1)^2 + k(\tilde{l}_{n,n-1} - \tilde{l}_0)^2 + s(y_n - y_{n-1})^2 e^{-\beta(y_n + y_{n-1})}] \qquad (5)$$

where: a wave over a variable means its dimensionless value, an asterisk over a variable means differentiation over dimensionless time $\tilde{t} = t\sqrt{Da^2/m}$, and the rest of the variables are calculated from the relations

$$y_n = (r_n - R)a, \quad r = Ra, \quad \varphi_n = (\Phi_n - n\theta), \quad k = K/(Da^2), \quad s = St/(Da^2), \quad \beta = b/a$$

3 Lagrange equations of motion

According to the obtained dimensionless Lagrange function (5), we find a system of 2N equations of base motion

$$\frac{d}{d\tilde{t}}\frac{\partial \tilde{L}}{\partial y_n^*}=\frac{\partial \tilde{L}}{\partial y_n}: y_n^{**}=(y_n+r)\varphi_n^{*2}+(e^{-y_n}-1)e^{-y_n}-k\left[(\tilde{l}_{n,n-1}-\tilde{l}_0)\frac{\partial \tilde{l}_{n,n-1}}{\partial y_n}+(\tilde{l}_{n,n+1}-\tilde{l}_0)\frac{\partial \tilde{l}_{n,n+1}}{\partial y_n}\right]- \quad (6)$$

$$-s[(y_n-y_{n-1})e^{-\beta(y_n+y_{n-1})}+(y_n-y_{n+1})e^{-\beta(y_n+y_{n+1})}]+\beta s[(y_n-y_{n-1})^2e^{-\beta(y_n+y_{n-1})}+(y_n-y_{n+1})^2e^{-\beta(y_n+y_{n+1})}],$$

$$\frac{d}{d\tilde{t}}\frac{\partial \tilde{L}}{\partial \varphi_n^*}=\frac{\partial \tilde{L}}{\partial \varphi_n}: (y_n+r)^2\varphi_n^{**}=-2(y_n+r)\varphi_n^* y_n^*-k\left[(\tilde{l}_{n,n-1}-\tilde{l}_0)\frac{\partial \tilde{l}_{n,n-1}}{\partial \varphi_n}+(\tilde{l}_{n,n+1}-\tilde{l}_0)\frac{\partial \tilde{l}_{n,n+1}}{\partial \varphi_n}\right] \quad (7)$$

where: $\tilde{l}_0=al_0=\sqrt{\tilde{h}^2+4r^2\sin(\theta/2)}$, $\tilde{h}=ah$,

$$\tilde{l}_{n,n-1}=al_{n,n\pm1}=\sqrt{\tilde{h}^2+(y_n+r)^2+(y_{n-1}+r)^2-2(y_n+r)(y_{n-1}+r)\cos(\varphi_n-\varphi_{n-1})}$$

$$\frac{\partial \tilde{l}_{n,n\pm1}}{\partial y_n}=\frac{y_n+r}{\tilde{l}_{n,n\pm1}}\left[1-\frac{y_{n\pm1}+r}{y_n+r}\cos(\varphi_n-\varphi_{n\pm1})\right]$$

$$\frac{\partial \tilde{l}_{n,n\pm1}}{\partial \varphi_n}=\frac{(y_n+r)(y_{n\pm1}+r)}{\tilde{l}_{n,n\pm1}}\sin(\varphi_n-\varphi_{n\pm1})$$

Each of the ordinary differential equations of the system (6)-(7) has a second order and to obtain a single solution to this system. It is necessary to set the 4N of initial conditions (2N initial values of generalized coordinates y_{n0}, φ_{n0} and 2N initial values of generalized velocities v_{n0}, ω_{n0})

$$y_n(0)=y_{n0}, \quad y_n^*(0)=v_{n0}, \quad \varphi_n(0)=\varphi_{n0}, \quad \varphi_n^*(0)=\omega_{n0} \quad (8)$$

The system of nonlinear equations (6), (7) with initial conditions (8) is written in the form of a system of 4N equations of the first order and is integrated numerically by the Runge-Kutta method[11] of the fifth order of accuracy with variable and automatically selectable time steps. The time step is selected from the conditions providing stability and the required accuracy.

4 Numerical results

To study the statistical properties of DNA and its denaturation, the method of direct molecular dynamics modeling (DMDM) was used. The study was carried out for a microcanonical ensemble, i.e. a set of microstates of the DNA molecule was statistically averaged at constant external thermodynamic parameters [the number of particles, its total energy and volume (no external work under the molecule)]. The dynamic behavior of the homogeneous DNA molecule from $N = 128$ A-T base pairs with the parameters from [9] presented in Table 1 was numerically modeled.

Table 1 Homogeneous DNA molecule parameters

Parameter	Value	Dimension
Characteristic of Morse potential well width, a	6.3	1/Å
Characteristic of well width stack-interactions, b	0.5	1/Å
Morse potential well depth, D	0.15	eV
Stack-interaction bond parameter, S	0.65	eV/Å^2
Interplanar distance, h	3.4	Å
Elastic bonding stiffness coefficient, K	0.04	eV/A^2
Equilibrium distance of base to axis, R	10	
Initial turning angle, θ	0.607	radians
Mass of base, m	300	a.m.u.

The setting of each DMDM microstate was generated by setting the initial radial velocities of the pairs randomly according to a normal distribution with a mathematical expectation of $\mu=0$ and a standard deviation σ

$$y_n=0,\ y_n^*=\zeta_n,\ \varphi_n=0,\ \varphi_n^*=0,\ \zeta\sim G(\mu=0,\ \sigma)$$

Performing a statistically representative series of calculations, we obtain the average values of the total E and kinetic E_{kin} energy of the system of interacting base pairs. To establish a microstate (redistribution between potential and kinetic energies) according to the equations of motion (6), (7) at a given mean square deviation of the σ and averaging over the number of draws in the series, it was necessary to perform at least 10^4 implementations of random DNA states obtained by numerical inegration with a dimensionless time step $\Delta t \leqslant 0.02$ up to $t=10^5$.

Temperature is determined by the average kinetic energy of the system. Each base pair has two degrees of freedom and, according to the equipartition theorem of kinetic energy over degrees of freedom, it corresponds to the energy in dimensionless variables $2k_B T/(2D)$ (k_B is the Boltzmann constant), from where (division by D is preserved in equations to use dimensionless values obtained in the calculation) $k_B T/D = E_{kin}/(ND)$. The total dimensionless energy per freedom degree is $e = E/(2ND)$.

Having carried out the above series of calculations of DMDM for various σ, we obtain the dependence of temperature on energy $T=T(e)$ in parametric form (value σ plays the role of a parameter): $k_B T/D = f(\sigma)$, $e=\psi(\sigma)$, where f, ψ are functions defined from DMDM.

DMDM results are presented in Figure 2. A solid line shows the work data[9], crosses are the values obtained in our work. There is agreement on the results, but in the denaturation zone (horizontal line on the graph $T=T_D=$ const) the temperature values in the case of our DMDM do not remain strictly constant.

The results show that during denaturation, the temperature is almost constant $(k_B T_D/D)_{num} \approx 0.2$, and like the phase transition of the first order, this T_D temperature can be taken as one of the main characteristics of the formation of open states in a DNA molecule. Then we have $(k_B T_D/D)_{num} \approx 0.2 \Rightarrow T_D \approx 0.2 D/k_B \approx 350$ K ≈ 77 ℃. Experimental data give close values of $(T_D)_{exp} = 75\text{-}85$ ℃.

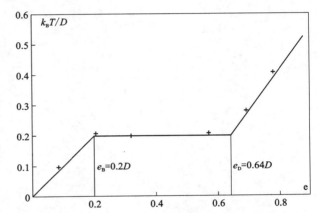

Figure 2　Temperature versus DNA internal energy per freedom degree

5　Conclusions

(1) Model M. Barbi et al.[7-9] well describes the denaturation of homogeneous DNA with single T-A bonds between bases. Satisfactory DMDM results require at least 10^4 implementations of DNA random states with dimensionless time increments $\Delta t \leqslant 0.02$ up to $t=10^5$.

(2) Good agreement was obtained with the experimental data and calculations[9] for the denaturation temperature.

(3) DMDM with a large number of base pairs requires the development and implementation of a parallel numerical algorithm.

(4) It is necessary to generalize the model for a heterogeneous DNA molecule and continue DMDM.

References

[1]　LAKHNO V, VINNIKOV A. Molecular devices based on DNA, Preprint IPM named after M. V. Keldysh RAS, Num, 2018, 137.

[2]　GOLDHABER-GORDON D, MONTEMERLO M, LOVE J, et al. Overview of nanoelectronic devices[J]. IEEE, 1997, 85, 4: 521-540.

[3]　SCHRÖDINGER E. What is life? The physical aspect of the living cell[M]. Cambridge University Press, 1944.

[4]　FRANK-KAMENETSKII M. Unraveling DNA: The most important molecule of life [M]. Cambridge, VCH Publishers (UK) Ltd., 1993.

[5] PEYRARD M, BISHOP A R. Statistical mechanics of a nonlinear model for dna denaturation[J]. Phys. Rev. Letters, 1989, 62: 2755.
[6] DAUXOIS T, PEYRARD M. Entropy-driven transition in a one-dimensional system[J]. Phys. Rev. E, 1995, 51: 4027.
[7] BARBI M, COCCO C, PEYRARD M. Helicoidal model for DNA opening[J]. Phys. Lett. A, 1999, 253, 5-6: 358-369.
[8] BARBI M, COCCO C, PEYRARD M, et al. A twist opening model for DNA[J]. Journal of Biological Physics, 1999, 24: 97-114.
[9] BARBI M, LEPRI S, PEYRARD M, et al. Thermal denaturation of a helicoidal DNA model[J]. Phys. Rev. E, 2003, 68: 061909.
[10] CALLADINE C, DREW H. Understanding DNA[M]. London, Academic Press, 1992.
[11] HAIRER E, NШRSETT S P, WANNER G. Solving ordinary differential equations I: Nonstiff problems[M]. New York: Springer, 1987.

Metal-containing Catalysts for the Hydrogenation Process of Substituted 5-acyl-1, 3-dioxanes

Gul'nara RASKIL'DINA, Yulianna BORISOVA, Simon ZLOTSKII

Ufa State Petroleum Technological University, Ufa, 450064, Russia

Abstract: Hydrogenation of substituted 5-acyl-1, 3-dioxanes was used to synthesize heterocyclic alcohols in the presence of metal-containing catalysts. It has been established that the best catalyst for the reduction of substituted 5-acyl-1, 3-dioxanes is Pd/C, which makes it possible to achieve a high selectivity for the formation of the corresponding heterocyclic alcohols at a conversion of the initial ketones of 60%-90%.

Keywords: Hydrogenation; 5-acyl-1, 3-dioxanes; Metal-containing catalysts

1 Introduction

Oxymethyl-1, 3-dioxacycloalkanes and their derivatives—ethers and esters, thioethers and others are used as corrosion inhibitors, plant protection chemicals, and also exhibit various biological activities[1-3].

The main method for obtaining alcohols containing acycloacetal fragment is the condensation of 1, 1, 1-trioxymethylalkanes with carbonyl compounds[4, 5]. However, in some cases, secondary 1, 3-dioxacycloalkane alcohols are necessary, so it was proposed to obtain them by reduction of the keto group in 5-acyl-1, 3-dioxanes with metal hydrides[6]. At the same time, this hydrogenation method is of little use for preparative synthesis under industrial conditions.

In this regard, we studied the heterogeneous catalytic reduction of substituted 5-acyl-1, 3-dioxanes in the presence of various metal-containing catalysts (Pd/C, Ni on kieselguhr, Pt/Re, Ni/Mo) in this work.

2 Result and discussion

Previously[7] we showed that in a hydrogen flow in the presence of a Pd/C catalyst, 5-acyl-1, 3-dioxanes are reduced to the corresponding heterocyclic alcohols. Continuing this work, we studied the hydrogenation of heterocyclic ketones 1-5 in the presence of a number of industrial metal-containing catalysts: Pd/C, Ni/kieselguhr, Pt/Re, or Ni/Mo.

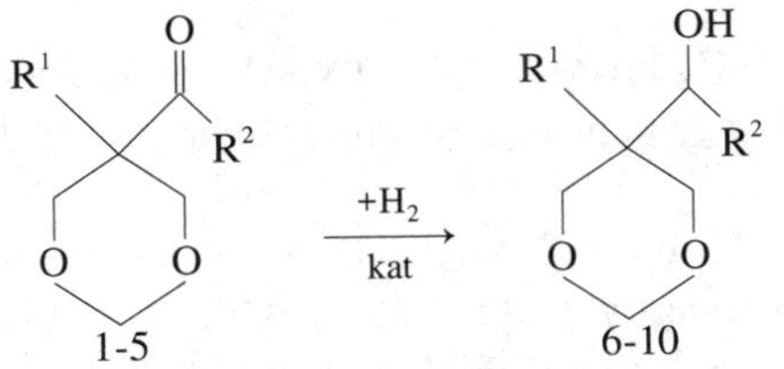

$R^1 = CH_3$, $R^2 = CH_3$(1, 6), $R^1 = C_2H_5$, $R^2 = CH_3$(2, 7)
$R^1 = i\text{-}C_3H_7$, $R^2 = CH_3$(3, 8), $R^1 = CH_3$, $R^2 = C_2H_5$(4, 9)
$R^1 = CH_3$, $R^2 = Ph$(5, 10).

Figure 1 Hydrogenation of 5-acyl-1, 3-dioxanes

The best result was shown by Pd/C among the studied catalysts (Table 1), which is used in the reduction of unsaturated and carbonyl compounds[8, 9]. The conversion on Pt-and Ni-containing catalysts is 1.5-2.5 times lower, while the selectivity in all cases is more than 70%.

Table 1 Hydrogenation of substituted 5-acyl-1, 3-dioxanes 1-5 in the presence of various catalysts synthesis conditions: 200 ℃, reaction time=1 h, molar ratio ketone: H_2 =1:6.

Reagents	Products	Catalyst							
		Pd/C		Pt/Re		Ni on kieselguhr		Ni/Mo	
		C^*/%	S^*/%	C/%	S/%	C/%	S/%	C/%	S/%
1	6	80	98	70	95	50	85	40	95
2	7	90	95	50	95	40	80	40	90
3	8	80	95	40	95	30	80	20	95
4	9	60	95	50	80	30	60	30	80
5	10	65	95	40	70	25	75	20	70

* C—conversion, %; S—selectivity, %.

The conversion of ketones 1-5 is also affected by substituents with different structures at the carbonyl group and in the 5th position of the 1, 3-dioxane ring. The ethyl and phenyl radicals at the C═O group reduce the conversion of compounds 4, 5. The activity of ketones 2, 3, containing ethyl or isopropyl groups in the 5th position, slightly decreases compared to the methyl ethyl ketone (MEK) derivative 1. Note that during the hydrogenation of the ketone 5 no products of complete or partial reduction of the aromatic nucleus were found.

3 Conclusion

The heterogeneous Pd/C catalyst makes it possible to reduce 5-acyl-1, 3-dioxanes to the corresponding alcohols with a selectivity of more than 95%. Catalysts containing Ni are substantially less active in this process.

Acknowledgment: This work was performed in the framework of the state assignment of the Ministry of Education and Science of the Russian Federation in the field of scientific activity, publication number FEUR-2022-0007 "Petrochemical reagents, oils and materials for thermal power engineering."

References

[1] GAFAROV N A, KUSHNARENKO V M, BUGAI D E. Corrosion inhibitors[J]. M.: Chemistry, 2002, 2: 367.
[2] YAKOVENKO E A, RASKIL'DINA G Z, ZLOTSKII S S, et al. Synthesis and herbicidal and antioxidant activity of a series of hetero-and carbocyclic derivatives of monochloroacetic acid[J]. Russian Journal of Applied Chemistry, 2020, 93, 5: 712-720.
[3] KUZ'MINA U S, RASKIL'DINA G Z, ISHMETOVA D V. Cytotoxic activity against SH-SY5Y neuroblastoma cells of heterocyclic compounds containing gem-dichlorocyclopropane and/or 1, 3-dioxacycloalkane fragments[J]. Pharm Chem J, 2022, 55, 1293-1298.
[4] MAXIMOV A L, NEKHAEV A I, RAMAZANOV D N. Ethers and acetals, promising petrochemicals from renewable sources[J]. Petroleum Chemistry, 2015, 55, 1: 1-21.
[5] SAMOILOV V, GONCHAROVA A, ZAREZIN D, et al. Bio-based solvents and gasoline components from renewable 2, 3-butanediol and 1, 2-propanediol: synthesis and characterization[J]. Molecules, 2020, 25(7): 17-23.
[6] ZLOTSKIJ S S, LESNIKOVA E T, RACHMANKULOV D L, et al. Synthesis of 5-hydroxyalkyl-1, 3-dioxane and 5-alkenyl-1, 3-dioxane[J]. Z. Chem, 1990, 30, 8: 281.
[7] BORISOVA Y G, MUSIN A I, YAKUPOV N V, et al. Pd/C-catalyzed hydrogenation of substituted 5-acyl-1, 3-dioxanes[J]. Russian Journal of General Chemistry, 2021, 91(9): 1619-1622.
[8] MAO Z, GU H, LIN X. Recent advances of Pd/C-catalyzed reactions[J]. Catalysts, 2021, 11: 1078.
[9] DU R, ZHU C, ZHANG P, et al. Selective hydrogenation of aromatic aminoketones by Pd/C catalysis[J]. Synthetic Communications, 2008, 38(17): 2889-2897.

Solar Technology for Extraction of Metals from Waste

M. S. PAIZULLAKHANOV[1,*], O. R. PARPIEV[1], R. Yu. AKBAROV[1],
A. A. HOLMATOV[2], N. H. KARSHIEVA[3], N. N. CHERENDA[4]

1　Material Sciences Institute of the Academy of Sciences Republic of Uzbekistan, Tashkent, 100084, Uzbekistan
2　Fergana Polytechnical Institute, Fergana, 150100, Uzbekistan
3　I. A. Karimov Tashkent State Technical University, Tashkent, 100095, Uzbekistan
4　Belarusian State University, Physics Faculty, Minsk, 220030, Belarus

Abstract: The possibilities of solar installations based on mirror-concentrating systems for processing of waste materials in a stream of concentrated high-density solar radiation were analyzed. It was proposed to use mobile compact solar installations located near metallurgical plants for processing mining and metallurgical waste. The geometric and optical-energy parameters of the concentrator for processing were calculated in order to extract metals from mining wastes. It was shown that a system of mirrors consisting of a heliostat (100 m^2) and a paraboloid-shaped concentrator with a diameter of 10 m can focus a solar radiation flux with a density sufficient to melt metallurgical waste from the Almalyk Mining and Metallurgical Plant. It was shown that ultrasonic treatment of waste materials stimulates an increase in the amount of copper containing phase in the melt by 8 times compared to the initial state of the material.

Keywords: Solar concentrators; Mirror concentrating systems; High flux densities; Material processing; Metal recovery

1　Introduction

The usage of techniques (lasers, plasmatrons, cathode-beam or arc sources) that create high-density quantum or particle fluxes for surface modification and material processing lead to the formation of non-equilibrium microstructures that can be used to fabricate materials with higher corrosion resistance, high temperature oxidation and wear resistance. Solar installations based on mirror concentrating systems have unique capabilities for processing of metal (welding and surfacing, surface treatment and surface hardening), powders and non-metal (ceramics, fullerenes, carbon nanotubes) materials (Herranz, 2010; Fernández-González and Ruiz-Bustinza, 2018). A number of researchers (Fernández-González and Prazuch, 2018; Ruiz-Bustinza, 2013; Sibieude, 1982; Steinfeld, 1991; Akbarov, 2017; Faiziev, 2008) considered the possibility of using a stream of concentrated solar energy instead of burning hydrocarbons at high temperatures.

At present, the search for new energy sources is being intensively carried out along with the investigations on the efficient use of existing sources. Especially great attention is paid to renewable energy sources because of their ability to be regenerated by natural processes. In this regard, solar energy is one of the most promising renewable energy sources (Parpiev, 2021; Paizullakhanov, 2021). Various designs of melting units installed in the focal area of solar installations make it possible to implement technologies for energy-intensive processes in the ceramic, glass and metallurgical industries (Figure 1).

As can be seen from Figure 1, the melting unit contains a graphite cavity which absorbs the energy of concentrated high-density solar radiation and becomes a heat source. Due to the high thermal conductivity of graphite, the material in the reaction chamber is heated and melted. However, it seems to us that the efficiency of such melting unit design will not be high in terms of the amount of melted material per unit of time, as well as the full melting of the loaded material. This work is aimed for the development of a solar installation and a melting unit for processing waste from metallurgical production in order to extract metal alloys from them. The influence of ultrasonic treatment of industrial waste from the metallurgical production of the Almalyk Mining and Metallurgical Plant (AMMP) on the process of metal recovery will also be studied.

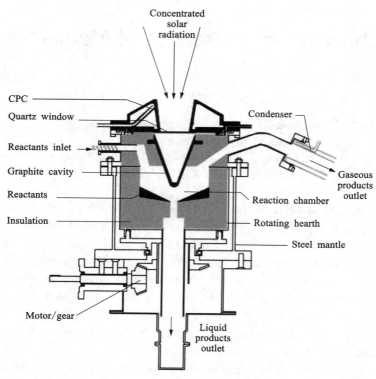

Figure 1 Solar furnace, for extracting metals from ore (PSI Spain)[12]

2 Experiment

At the first stage, the chemical composition of waste from the metallurgical production of the AMMP was analyzed (Table 1).

Table 1 Chemical composition of waste from the metallurgical production of the AMMP

Component	SiO_2	Fe_2O_3	CaO	K_2O	ZnO	MgO	CuO	PbO	MnO	MoO_3
Content/wt. %	52.29	38.58	3.28	2.57	1.07	1.13	0.40	0.24	0.22	0.22

As follows from Table 1, silicon and iron oxides were the predominant components in industrial waste. The analysis showed that for such a composition of AMMP wastes, a low melting point is characteristic, which was $T=1750$ K with a material dispersion of no more than 100 μm.

At the second stage of the study, the flux density of concentrated solar radiation (Q) was calculated using the Stefan-Boltzmann equation, which describes the radiation of heated bodies:

$$Q = \sigma \varepsilon T^4$$

where: ε is the emissivity of the material; $\sigma = 5.67 \times 10^{-8}$ W/(m²·K⁴) is the Stefan-Boltzmann constant, T is the body temperature (K). When the value of the materials emissivity is 0.85, the required flux density for melting AMMP ($T=1750$ K) is $Q=50$ W/cm².

From the analysis of the previously shown estimations it follows that the energy density $E=100$ W/cm² at the focus of the solar furnace will be more than sufficient for the melting of waste with the extraction of metal alloys from it. Obviously, the larger the focal spot size, the greater the amount of processed waste and the greater efficiency of this process. Let us estimate what optical-geometric (dimensional) parameters a solar concentrator should have in order to provide the required technological mode. For calculations, we use the following data: the flux density of concentrated solar radiation $E=100$ W/cm², preferably with a flat distribution over the spot with the diameter $d=30$ cm. To do this, we first determine what power will be in a circle with a diameter of 30 cm at a uniform density of 100 W/cm² in the focal zone:

$$W_f = \frac{E_f \pi d^2}{4} = 70650 \text{ W}$$

On the other hand, such power should be provided by the concentrator midsection area. Let us consider a round concentrator. If we denote the diameter of the midsection of the concentrator as D_c, then:

$$\frac{E_0 R_g E_c \pi D^2}{4} = W_f$$

where: R_g, R_c are the reflection coefficients of the heliostat and concentrator mirrors; E_0 is direct solar radiation density. Thus:

$$\frac{E_f \pi d^2}{4} = \frac{E_0 R_g R_c \pi D_c^2}{4}$$

From here, for the midsection of the concentrator, we get:

$$D_c = d_f \sqrt{\frac{E_f}{E_0 R_c R_g}}$$

This equation makes it possible to calculate the diameter of the concentrator corresponding to the given values of optical and energy parameters, taking into account the conditions of technological processes.

The proposed scheme of the solar installation, which can be used for the processing of materials, is shown in Figure 2.

As can be seen from Figure 2, the solar furnace is designed according to the bi-mirror scheme of mirror concentrating systems of the heliostat-concentrator type. Such a scheme of mutual arrangement of reflecting elements (mirrors) allows concentrating solar radiation into a focal region with a vertical flux vector that is more convenient for technological purposes of processing materials in the focal plane.

At the third stage of the experiments, technogenic wastes were irradiated with ultrasonic pulses in an ultrasonic bath of the DSA50-Ski-1.8 L brand (manufactured in China) in order to identify the effect of preliminary ultrasonic exposure on the material before melting in a solar furnace.

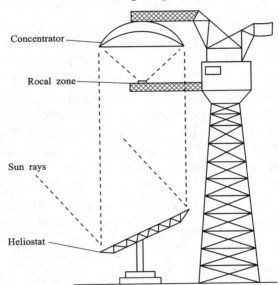

Figure 2 Scheme of the solar installation for materials recycling

3 Results and discussion

Assuming the flux density of solar radiation to be 0.07 W/cm² and the reflection coefficient of the mirrors to be 0.9 (the same for the concentrator and heliostat), we obtain D_c = 17.81 m. This value of the concentrator diameter is quite large. Therefore, it is necessary to clarify experimentally the smallest value of the concentrated radiation density sufficient to melt the material being processed. Figure 3 shows the dependence of the concentrator diameter on the concentrated radiation flux density. The calculations were performed for two values of the focal spot diameter, 30 cm and 20 cm.

Figure 3 shows that if the diameter of the concentrator is 10 m, then it should provide a spot with a diameter of 30 cm with an average radiation flux of 65 W/cm² in the focal zone of such a concentrator. For the processing of materials in the focal region, we have developed a special design of a graphite melting unit in the form of a truncated cone (Figure 4).

The process of metallurgical waste processing from the AMMP in a solar furnace consisted of material melting and it quenching in the water. Analysis of the chemical composition of the fused material showed the presence of substances of metallic (approximately 22 wt.% Fe-Cu based alloy) and ceramic (approximately 71 wt.% $CaMaSi_2O_6$) compositions separately.

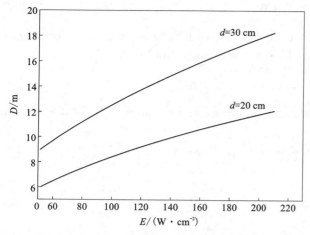

Figure 3 Dependence of the concentrator diameter on the concentrated radiation flux density

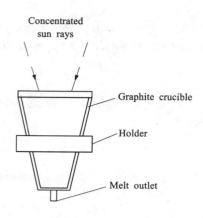

Figure 4 Graphite melting furnace

In a carbon medium, the process of metal reduction from their oxide states proceeded according to the reaction MeO+C === Me+CO. However, before the melt drops enter the water, the metals in the drop have time to oxidize in the air atmosphere, according to their chemical affinity for oxygen. For example, at a temperature of 1600 ℃, the chemical affinity of elements to oxygen decreases in the following line Be, Ca, Zr, Mg, Al, Ti, C, Si, V, B, Mn, Cr, Sb, Zn, Fe, W, Mo, Co, Ni, Cu, As. Due to the fact that the elements located to the left of iron, in comparison with it, have a higher chemical affinity for oxygen, they were quickly oxidized, that was observed in the experiment.

The chemical composition of the metallic part of the melted AMMP waste material subjected to preliminary ultrasonic treatment is given in Table 2.

Table 2 Main chemical composition of the metallic part of the melted AMMP waste material subjected to preliminary ultrasonic treatment

Component	Fe	Cu	Mo	SiO_2	CaO	Sb_2O_3	MgO	MnO	PbO	ZnO
Content/wt. %	88.04	3.28	0.63	2.81	3.32	0.12	1.13	0.22	0.24	0.21

An analysis of the composition of the metallic part of industrial waste melted in a solar furnace showed that their preliminary ultrasonic treatment led to an 8-fold increase in the amount of copper containing phase compared to the initial state of the material. This may be due to the acceleration of physical processes, based on the absorption of the energy of high-intensity ultrasonic frequency mechanical vibrations by the particles of the substance. In particular, as noted in [6], the use of high-intensity ultrasonic vibrations accelerates technological processes and increases the yield of useful products, and also makes it possible to obtain a material with new properties.

The capacity of the developed melting unit is 2 kg of material. Such an amount of substance can be melted out in 3 s of irradiation in the focal area of a solar installation. In case of providing a continuous supply of material to the melting unit a total weight of processed material for an 8 h sunny day will be 2400 kg. Based on the fact that the minimum number of sunny days in the Republic of Uzbekistan is 220 t, then 500 t of man-made waste can be processed in a working year. At the same time, the yield of iron will be 110 t, and copper-about 16 t.

4 Conclusions

The findings showed that a system of mirrors consisting of a heliostat (100 m²) and a paraboloid-shaped concentrator with a diameter of 10 m could focus the solar radiation flux with a density sufficient to melt the metallurgical waste of the Almalyk Mining and Metallurgical Plant. It was revealed that in case of heating of the material in a carbon medium, the process of metal reduction from their oxide states proceeded according to the reaction MeO + C === Me + CO. Some metals oxidized when cooled in air according to their chemical affinity for oxygen. The continuous supply of material to the melting unit would make it possible to melt the material in the amount of 500 t of industrial waste per 1 year with the extraction of 110 t of iron and 16 tons of copper on one solar furnace. Solar extraction of metals can become an alternative in metallurgical processes. It was shown that preliminary ultrasonic treatment of waste materials stimulated an 8-fold increase in the amount of copper containing phase in the melt compared to the initial state of the material.

Acknowledgments:

The authors are grateful to the researchers R. Yu. Akbarov and Sh. R. Nurmatov for participation in the experiments and discussion of the results.

This work was carried out within the framework of the research program of the laboratory "Synthesis and processing of materials" of the Institute of Materials Science.

References

[1] AKBAROV R Y, PAIZULLAKHANOV M S. Characteristic features of the energy modes of a large solar furnace with a capacity of 1000 kW[J]. Applied Solar Energy, 2017, 54(2): 99-109.

[2] FAIZIEV S A, PAIZULLAKHANOV M S, NODIRMATOV E Z. Synthesis of pyroxene pyroceramics in large solar furnace with ZrO_2 crystallization nucleator[J]. Applied Solar Energy, 2008, 44(2): 139-141.

[3] FERNÁNDEZ-GONZÁLEZ D, PRAZUCH J, RUIZ-BUSTINZA I, et al. Solar synthesis of calcium aluminates[J]. Sol. Energy, 2018, 171: 658-666.

[4] FERNÁNDEZ-GONZÁLEZ D, RUIZ-BUSTINZA I, GONZÁLEZ-GASCA, C, et al. Concentrated solar energy applications in materials science and metallurgy[J]. Sol. Energy, 2018, 170: 520-540.

[5] HERRANZ G, RODRíGUEZ G P. Uses of concentrated solar energy in materials science//Radu D Rugescu Solar Energy, Croatia: INTECH, 2010: 145-170.

[6] KHMELEV V N. Ul'trazvukovyye mnogofunktsional'nyye i spetsializirovannyye apparaty dlya intensifikatsii tekhnologicheskikh protsessov v promyshlennosti (Ultrasound multifunctional installations for intensification of technological processes in industry), Barnaul: AltGTU, 2007.

[7] PAIZULLAKHANOV M S, SHERMATOV Z Z, NODIRMATOV E Z. Synthesis of materials by concentrated solar radiation[J]. High Temperature Material Processes, 2021, 25(2): 25-34.

[8] PARPIEV O R, PAIZULLAHANOV M S, NODIRMATOV E Z. Peculiarities of processing metallurgical waste in a large solar furnace[J]. Theory and technology of the Metallurgical Process, 2021, 36(1): 15-20.

[9] RUIZ-BUSTINZA I, CAÑADAS I, RODRíGUEZ J, et al. Magnetite production from steel wastes with concentrated solar energy[J]. Steel Res. Int., 2013, 84: 207-217.

[10] SIBIEUDE F M, TOFIGHI A, AMBRIZ J. High temperature experiments with a solar furnace: The decomposition of Fe_3O_4, Mn_3O_4, CdO[J]. Int. J. Hydrogen Energy, 1982, 7: 79-88.

[11] STEINFELD A, FLETCHER E A. Theoretical and experimental investigation of the carbothermic reduction of Fe_2O_3 using solar energy[J]. Energy, 1991, 16: 1011-1019.

[12] BADER R, LIPIńSKI W. Solar thermal processing//Advances in Concentrating Solar Thermal Research and Technology, 2017: 403-459.

Failure of Friction Stir Welded Joint of a Tempformed High-strength Low-alloy Steel

Anastasiia DOLZHENKO, Anna LUGOVSKAYA, Andrey BELYAKOV

Belgorod National Research University, Belgorod, 308015, Russia

Abstract: The failure of friction stir welded joint of a high-strength low-alloy steel Processed by tempforming was analyzed. The stir zone was characterized by the specific microstructure with an average grain size of 800 nm. The yield strength of the welded joint was 1220 MPa, whereas the yield strength of the base material was 1350 MPa. The fracture of the welded joint occurred in the heat affected zone between the stir zone and the base material.

Keywords: Low-alloy steel; Tempforming; Friction-stir welding; Mechanical properties; Fractography

1 Introduction

High-strength low-alloy steels with a nanocrystalline lamellar microstructure formed as a result of thermo-mechanical treatment (tempforming) have a unique combination of high strength (the yield strength above 1000 MPa) and high impact toughness at low temperatures (KCV more than 100 J/cm^2 at 77K)[1-3]. Such steels are promising materials for replacing price-limited maraging steels in products produced in large series, designed for structures operating under shock loads at low temperatures. The high-strength steels are also used to create lightweight welded structures. Various technologies are used to join parts made of high-strength steels: automatic, argon-arc welding, submerged arc welding, etc. The main difficulty in welding the high-strength structural steels is their increased susceptibility to hardening, which leads to a sharp increase in metal hardness in the near-weld zones, which adversely affects the mechanical properties of welded joints. In addition, the weld and heat-affected zones of welded joints obtained by traditional steel welding methods are characterized by a coarse-grained microstructure, which is an inevitable consequence of the melting of the welded material. This is completely unacceptable for steels after tempforming, whose outstanding mechanical properties are due to the formation of a specific microstructure with a transverse grain size significantly less than one micrometer. Currently, a new welding method is being developed in the world, that is friction stir welding developed at the British Institute of Welding[4,5]. With this method of welding, the welding tool is twisted and introduced into the joint of the sheets in such a way that the pin penetrates deep into the material and mixes it, while pressing and rotating the shoulders provides heating and softening of the material around the immersed pin. It is known that welds obtained by this method have a number of advantages compared to traditional types of welding. Namely, the absence of hot cracks and porosity, welding in air, low requirements for the quality of the surface of welded elements, high welding speed, etc[6,7].

2 Material and methods

A low-alloy steel with a chemical composition of Fe-0.36C-0.4Si-0.56Cr-0.57Mn-0.54Mo-0.0067P-0.0034S (wt.%) was hot rolled at 1123 K followed by water quenching. Then, the steel samples were tempered for 1 h at a temperature of 873 K followed by rolling at the same temperature to a total strain of 1.5 (tempforming). After tempforming, the steel sample was cut along the RD-TD section (RD means the rolling direction, TD means the transverse direction). Thus, two sheets of 4 mm thick were obtained. Next, the sheets were friction-stir welded (FSW) with a FSW seam parallel to TD using the AccuStir 1004 FSW machine (Figure 1). Attempting to provide a full-penetration FSW, a double-side technique of FSW was applied in mutually opposite directions such that the advancing side and retreating side reversed from the upper to the bottom surfaces of the weld. The welding process was performed using a tool with shoulder diameter of

12 mm and cylindrical pin of 3.5 mm in length and of 5 mm in diameter. A tool rotation speed and a travel speed were 400 r/min and 100 mm/min, respectively. During FSW, the tool was tilted by 2.5° from the sheet normal so that the rear of the tool was lower than the front. The microstructural observation was performed on the RD-ND sections (ND is the normal direction), using a Quanta 600 FEG scanning electron microscope (SEM) incorporating an orientation imaging microscopy (OIM) system. The SEM specimens were electro-polished using an electrolyte containing 10% perchloric acid and 90% acetic acid at a voltage of 20 V at room temperature. The OIM images were subjected to a cleanup procedure using the Grain Dilation method with Grain Tolerance Angle of 2, Minimal Grain Size of 3. The mean

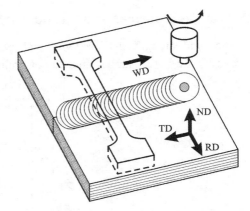

Figure 1 Scheme of the friction stir welding and the tensile specimen orientation

grain size was evaluated on the OIM images as average distances between high-angle grain boundaries (HAB) with misorientations of $\theta \geqslant 15°$. The tensile tests were carried out by using an Instron 5882 testing machine on specimens with a gauge length of 25 mm and cross-section of 7 mm×3 mm at ambient temperature and a crosshead rate of 2 mm/min with the tensile direction parallel to the rolling direction. The tensile specimens were machined transverse to welding direction (WD) and included all FSW microstructural zones (Figure 1). The fractures of the specimans after tensile tests were examined using a Quanta 600 FEG SEM.

3 Results and discussion

3.1 Microstructure

Figure 2 shows the microstructures evolved in the base material, in the stir zone and the heat-affected zone (HAZ) of FSW sample. The base material [Figure 2(a)] is characterized by a typical lamellar-type microstructure consisting of highly elongated grains with the transverse grain size of 330 nm, which corresponds to that developed by previous tempforming at 873 K[1]. In contrast, the microstructure evolved in the stir zone [Figure 2(b)] consists of fine irregular grains with frequently wavy/serrated boundaries. The mean grain size in the stir zone comprises 900 nm. Such difference in the grain sizes suggests some recrystallization/grain coarsening processes taking place in the stir zone during FSW. The heat-affected zone [Figure 2(c)] is characterized by a mixed-type microstructure, since both highly elongated grains and fine grains of irregular shape are observed.

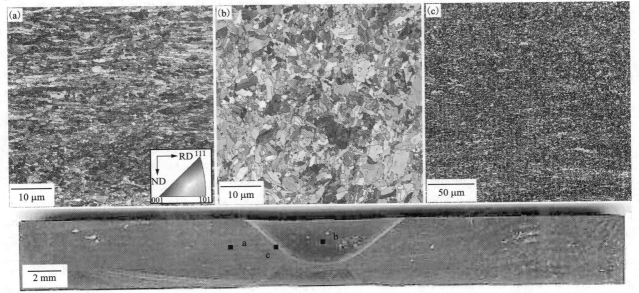

Figure 2 Microstructures in base material (a), stir zone (b) and heat-affected zone (c) of FSW sample

3.2 Tensile properties and fractography

The yield strength of FSW joint comprises 1220 MPa that is just below the yield strength of 1350 MPa as recorded by the base material after tempforming at 873 K (Figure 3)[1].

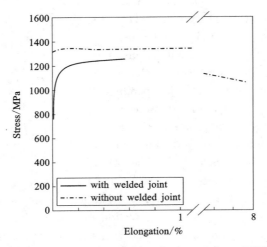

Figure 3 Tensile stress-elongation curve for a FSW joint specimen (with welded joint) and tensile stress-elongation curve for the base material (without welded joint)[1]

Therefore, the FSW joint exhibits rather high strength comparable with that of the base material, which was work hardened by tempforming treatment. However, in contrast to the base material, the FSW sample is characterized by a quite small plasticity. Total elongation for the specimen made of FSW joint sample does not exceed 0.6%. This specimen experienced rapid fracture localized in the heat affected zone, i.e., between the FSW seam and the base material.

Typical SEM images of the tensile fracture surfaces of the FSW joint specimen after tensile test at 293 K are shown in Figure 4.

The fracture surfaces in Figure 4 are shown for 3 zones including side zones [Figures 4(a), (c)] and centre portion of the specimen [Figure 4(b)]. On the fracture surface, dimples/cells of different sizes and shapes are observed irrespective of the fracture zone. The shallow dimples of a more elongated shape are in dominant in the side zones in Figure 4(a) and Figure 4(c). It is interesting to note that there is also a zone of the flat terraces formed parallel to RD-TD plane on the surface [Figure 4(d)] that is typical of the destruction of steels after tempforming[1-3]. This suggests that the failure occurred in the heat-affected zone between the stir zone and the base material, where a mixed structure is present.

4 Conclusion

The microstructures developed during FSW and the FSW joint failure were studied for a high-strength low-alloy steel after tempforming. The main results can be summarized as follows.

The developed microstructures in the stir zone are characterized by the mean grain size of 800 nm. On the other hand the heat-affected zone is characterized by a mixed-type microstructure consisting of elongated grains developed by previous tempforming and irregular-shaped grains evolved by FSW.

The yield strength of the FSW joint was 1220 MPa, whereas the yield strength of the base material was 1350 MPa. The fracture surface has a pitted/cellular character. The fracture of the FSW joint occurred in the heat affected zone between the stir zone and the base material.

Acknowledgments: This work was funded by Russian Science Foundation under Agreement No. 20-19-00497-П. The work was carried out using the equipment of the Joint Research Center, Technology and Materials, Belgorod National Research University.

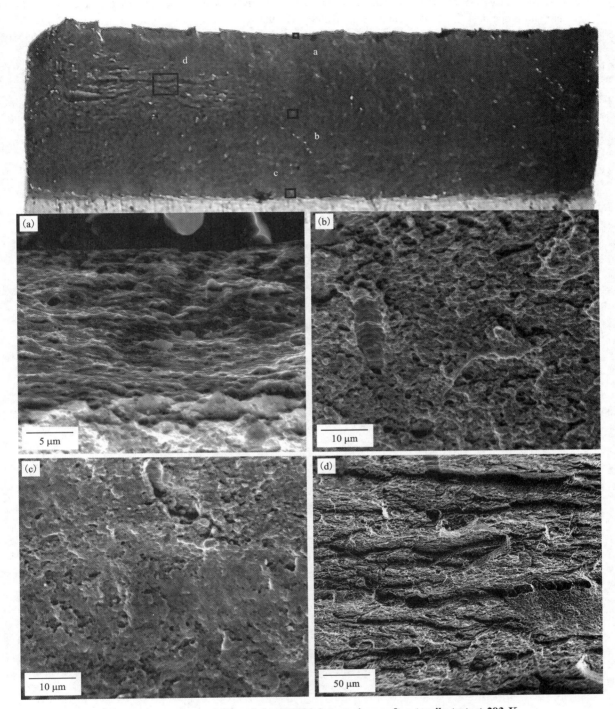

Figure 4 Fracture surface of the FSW joint specimen after tensile test at 293 K. General view of tested sample and high magnifications of indicated portions (a) to (d)

References

[1] DOLZHENKO A, KAIBYSHEV R, BELYAKOV A. Tempforming strengthening of a low-alloy steel[J]. Materials, 2022(15): 5241.
[2] DOLZHENKO A, BELYAKOV A. Mechanical properties of high-strength low-alloy steel after tempforming[C]. AIP Conference Proceedings, AIP Publishing LLC, 2022(250), C. 020056.
[3] DOLZHENKO A S, DOLZHENKO P D, BELYAKOV A N, et al. Microstructure and impact toughness of high-strength low-alloy steel after tempforming[J]. Phys. Met. Metall., 2021(122): 1014-1022.

[4] NANDAN R, DEBROY T, BHADESHIA H K D H. Recent advances in friction-stir welding-process, weldment structure and properties[J]. Prog. Mater. Sci., 2008(53): 980-1023.
[5] MISHRA R S, MA Z Y. Friction stir welding and processing[J]. Materials Science and Engineering R, 2005(50): 1-78.
[6] REYNOLDS A P, LOCKWOOD W D, SEIDEL T U. Processing-property correlation in friction stir welds[J]. Mater. Sci. Forum, 2000 (331-337): 1719-1724.
[7] MURR L E, LI Y, FLORES R D, et al. Intercalation vortices and related microstructural features in the friction-stir welding of dissimilar metals[J]. Mater. Res. Innovat., 1998(2): 150-163.

Use of Accelerator Technology for Qualitative and Quantitative Analysis of Materials

V. TOVTIN, E. STAROSTIN, S. SIMAKOV, M. PRUSAKOVA, V. VINOGRADOVA

Baikov Institute of Metallurgy and Materials Science of Russian Academy of Sciences,
Moscow, 119334, Russia

The cyclic accelerator "Microtron-sT" with electron energy of 21 MeV at the beam current of 5-6 μA is used as a powerful source of electron irradiation for the subsequent production of hard bremsstrahlung gamma radiation and its use in gamma-activation analysis. Gamma-activation analysis (GAA) is based on the interaction of high-energy gamma quanta (5-20 MeV) with the nuclei of the substance[1]. This leads to photonuclear reactions, probability of which depends on the energy of gamma rays. The reactions with the interaction threshold in the range of electron energy 8-20 MeV are of practical importance. This is usually the reaction: (γ, n), (γ, p), $(\gamma, 2n)$, (γ, pn). As a result neutron-deficient radionuclides (isotopes) are obtained with different characteristics i. e. half-lives time, quantum radiation intensities, energy spectrum. The presence of an element in the test sample is identified by the type of emitted particles, their energy, and the lifetime of the resulting radionuclide[2, 3].

After irradiation of the studied samples, the quantum gamma spectrum of radionuclides is analyzed using a detection unit made of high-purity germanium, a gamma spectrometer and spectrum processing software "GENIE 2000". On the basis of nuclear physical data, a library of radionuclides formed by photonuclear reactions has been developed. Gamma-activation analysis is used to study high-purity metals, alloys, steels, semiconductors, single crystals, ceramics and other multicomponent compounds. Multi-element GAA allows to simultaneously determine a number of impurities in one sample with the sensitivity of (10.4-10.7)wt. %. The activation method is used to determine short-lived radionuclides, including gas impurities such as C, N, O. We have carried out a series of experiments on gamma-activation analysis, such as pure vanadium and its alloys V-Ga-Cr, a number of steels etc.

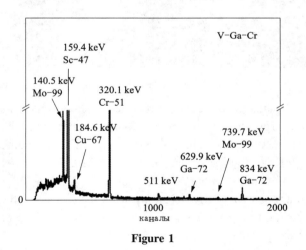

Figure 1

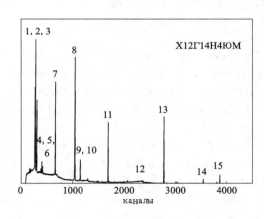

Figure 2

References

[1] Ю. М. Ципенюк. Фундаментальные и прикладные исследования на микротроне. Москва, физматлит, 2009: 424 с.
[2] RANDA Z, KREISINGER F J. Radioanalytical chemistry, Akademia Kiado, Dudapest, Elsevier sequoia S. A., 1983, 77: 502.
[3] http://www.nnds,bnl.gov/nudat2.

Influence of Pulsed Beams of Helium Ions and Helium High-temperature Plasma on the Ferritic 16Cr-4Al-2W-0.3Ti-0.3Y$_2$O$_3$ Steel

E. DEMINA, M. PRUSAKOVA, V. PIMENOV, S. MASLYAEV,
N. VINOGRADOVA, A. DEMIN, E. MOROZOV, N. EPIFANOV, S. SIMAKOV

Baikov Institute of Metallurgy and Material Science RAS, Moscow, 119991, Russia

Abstract: Oxide dispersion strengthened (ODS) ferritic and ferritic/martensitic steels, which produced by mechanical alloying of the elemental metallic powder with yttrium oxide (Y$_2$O$_3$) powder and consolidated by hot extrusion or hot isostatic pressing, are a class of advanced structural materials for fusion applications. The material used for this investigation was 16Cr-4Al-2W-0,3Ti-0,3Y$_2$O$_3$(K3) ODS ferritic steel after final cold rolling operation.

Keywords: 16Cr-4Al-2W-0.3Ti-0.3Y$_2$O$_3$ Steel; Pulsed Beams; High-temperature Plasma

A study of the radiation-thermal resistance of this steel was made. The "Vikhr" Plasma Focus installation was used for the production of powerful pulsed flows of helium ions and helium plasma. The power density of a beam of fast helium ions and high-temperature helium plasma flows was nearly 10^8 W/cm^2 and 10^7 W/cm^2 at exposure times of nearly 50 ns and 100 ns, respectively. The number of pulses N varied in the range from 10 to 30. The irradiated samples were examined by scanning electron microscopy, X-ray phase analysis, atomic force microscopy and precision weighing.

The rate of evaporation and radiative sputtering changed slightly with an increase in the number of pulses of energy flows acting on the

Figure 1

material and amounted to $h \approx (0.01\text{-}0.02)$ μm/imp. The irradiated surface after repeated melting under the action of a pulsed radiation-thermal load with powerful energy flows acquired a wave-like character with inclusions of dispersed micro particles of the second phase, containing mainly yttrium, oxygen, aluminum, iron, and titanium. At the same time, there is a grinding of the grain structure, a decrease in the parameter and the formation of micro-distortions of the crystal lattice. In contrast to the refractory metals (W, Mo, Nb) earlier under similar radiation conditions studied, no micro and macro cracks were formed on the surface of the material facing the plasma. "Vikhr" Plasma Focus setup proved to be an effective tool for simulation testing of candidate materials for fusion plant with magnetic and inertial plasma confinement.

Oxide dispersion strengthened (ODS) ferritic and ferritic/martensitic steels, which produced by mechanical alloying of the elemental metallic powder with yttrium oxide (Y$_2$O$_3$) powder and consolidated by hot extrusion or hot isostatic pressing, are a class of advanced structural materials for fusion applications. The material used for this investigation was 16Cr-4Al-2W-0,3Ti-0,3Y$_2$O$_3$(K3) ODS ferritic steel after final cold rolling operation.

A study of the radiation-thermal resistance of this steel was made. The "Vikhr" Plasma Focus installation was used for the production of powerful pulsed flows of helium ions and helium plasma. The power density of a beam of fast helium ions and high-temperature helium plasma flows was nearly 108 W/cm^2 and 107 W/cm^2

at exposure times of nearly 50 ns and 100 ns, respectively. The number of pulses N varied in the range from 10 to 30. The irradiated samples were examined by scanning electron microscopy, X-ray phase analysis, atomic force microscopy and precision weighing.

The rate of evaporation and radiative sputtering changed slightly with an increase in the number of pulses of energy flows acting on the material and amounted to $h \approx (0.01\text{-}0.02)\ \mu m/imp$. The irradiated surface after repeated melting under the action of a pulsed radiation-thermal load with powerful energy flows acquired a wave-like character with inclusions of dispersed micro particles of the second phase, containing mainly yttrium, oxygen, aluminum, iron, and titanium. At the same time, there is a grinding of the grain structure, a decrease in the parameter and the formation of micro-distortions of the crystal lattice. In contrast to the refractory metals (W, Mo, Nb) earlier under similar radiation conditions studied, no micro and macro cracks were formed on the surface of the material facing the plasma. "Vikhr" Plasma Focus setup proved to be an effective tool for simulation testing of candidate materials for fusion plant with magnetic and inertial plasma confinement.

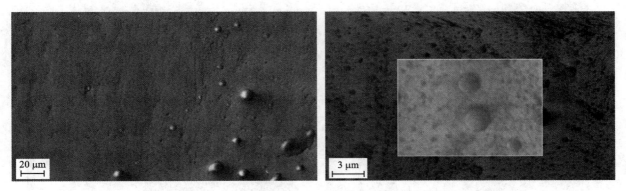

Figure 2　SEM structure of areas of the irradiated surface of ODS steel samples after beam-plasma exposure in the PF installation at $q = 10^7\text{-}10^8\ W/cm^2$, $\tau = 50\text{-}100$ ns of pulses 30

Table 1

Element	O	Al	Ti	Cr	Fe	Cu	Y	W	Ir	Total
Spectrum	23.10	3.94	0.22	1.16	4.42	0.00	61.20	1.87	4.09	100.00

Conclusions

With the use of the PF "Vikhr" installation, tests were carried out of ferritic steel 16Cr-4Al-2W-0.3Ti-0.3Y_2O_3, which is promising for use in DUO nuclear fusion installations, under conditions of exposure to powerful pulsed fluxes of helium ions and helium plasma at a power density of a beam of fast

Tests of steel DUO, carried out with varying the number of pulsed loads in the range from 10 to 30 impacts, showed its sufficiently high thermal and radiation resistance and the absence of micro-and macrocracks on the surface of the material facing the plasma.

The irradiated surface after repeated melting under the action of a pulsed radiation-thermal load by powerful beams of helium ions and helium plasma acquired a wave-like character with inclusions of spherical dispersed microparticles of the second phase, containing mainly yttrium, oxygen, aluminum, iron, titanium. The size of visible particles of the second phase lies within the micron and submicron scales. In this case, a refinement of the grain structure, a decrease in the parameter, and the formation of microdistortions of the crystal lattice are observed.

The Vortex Plasma Focus setup proved to be an effective tool for simulation testing of candidate materials for fusion facilities with magnetic and inertial plasma confinement.

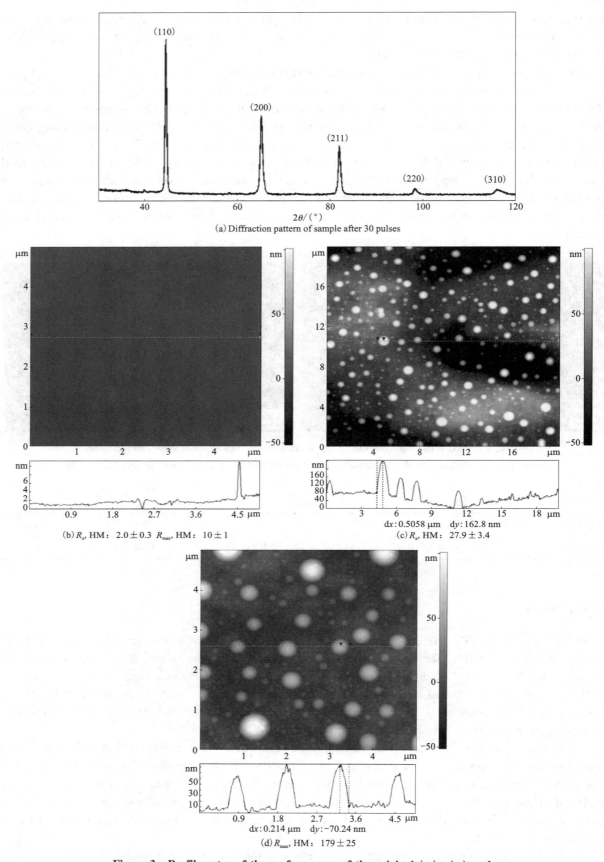

Figure 3 Profilometry of the surface areas of the original (a), (c) and irradiated (b), (d) samples of the investigated ODS steel at $N=30$.
An increased concentration of spherical micro particles is visible, located mainly on the surface of wave crests

References

[1] LINDAU R. Mechanical and microstructural properties of a hipped RAFM ODS-steel/LINDAU R, MÖSLANG A, SCHIRRA M, et al. Journal of Nuclear Materials, 2002, 307-311: 769-772.

[2] KLUEH R L. Tensile and creep properties of an oxide dispersion-strengthened ferritic steel/KLUEH R L, P. J. MP. J. MAZIASZ, KIM I S, et al. [J]. Journal of Nuclear Materials, 2002, 307-311: 773-777.

[3] KASADA R. Pre-and post-deformation microstructures of oxide dispersion strengthened ferritic steels/KASADA R, N. TODA, YUTANI K et al[J]. Journal of Nuclear Materials, 2007, 367-370: 222-228.

[4] GRIBKOV V A, DEMIN A S, DEMINA E V, et al. Specifics of damageability of the silicon single crystal under exposure of powerful plasma streams and fast helium ions[J]. Inorganic Materials: Applied Research, 2020, 11, (2): 349-358.

[5] SCHOLZ M, MIKLASZEWSKI R, GRIBKOV V A, et al. PF-1000 device[J]. Nukleonika, 2000, 45, 3: 155-158.

[6] GRIBKOV V A. Physical processes taking place in dense plasma focus devices at the interaction of hot plasma and fast ion streams with materials under test[J]. Plasma Phys. Control. Fusion, 2015, 57: 065010. DOI: 10.1088/0741-3335/57/6/065010.

[7] GRIBKOV V A, BOROVITSKAYA I V, DEMIN A S, et al. The Vikhr plasma focus device for diagnosing the radiation-thermal resistance of materials intended for thermonuclear energy and aerospace engineering[J]. Instruments and experimental techniques, 2020, 63, 1: 68-76.

[8] PIMENOV V, BOROVITSKAYA I, DEMIN A, et al. Povrezhdaemost niobiya impulsnymi potokami ionov geliya igelievoj plazmoj [Damage of niobium by pulse ows of helium ions and helium plasma][J]. Fizika i khimiya obrabotki materialov [Physics and Chemistry Of Materials Treatment], 2021, 6: 5.

[9] PIMENOV V N, MASLYAEV S A, DYOMINA E V, et al. Vzaimodejstvie moshchnyh impul'snyh potokov energiis poverhnost'yu vol'frama v ustanovke Plazmennyj Fokus [Interaction of powerful pulsed energy flows with the tungsten surface in the Plasma Focus installation][J]. Fizika i khimiya obrabotki materialov [Physics and Chemistry Of Materials Treatment], 2008, 3: 5-14.

[10] LUTTEROTTI L, Bortolotti. Algorithms for solving crystal structure using texture[J]. Acta Crystallographica, 2005, 61: 158-159.

[11] CEPELEV A B, PERLOVICH JU A, ISAENKOVA M G, et al. Strukturno-fazovye izmenenija v austenitnoj stali pristacionarnom i ciklicheskom jelektronnom obluchenii [Structural and phase changes in austenitic steel under stationary and cyclic electron irradiation], Fizika i himija obrabotki materialov [Physics and Chemistry of Materials Treatment], 2008, 1: 9-19.

[12] GRIGOREVA I S, MEJLIHOVA E Z. FIZICHESKIE V. Spravochnik [Physical Quantities Guide][M]. Jenergoatomizdat, 1991: 1232.

Numerical Simulation's Possibility of Materials' Damping Properties

Robert BAITEMIROV[1], Iurii DUBINOV[2], Alexander PRYGAEV[2],
Alexander KUZNETSOV[2], Dmitriy VISHNIVETSKIY[2]

1　LLC "Gazpromdobycha Noyabrsk", Chayandinskoye oil and gas field, Noyabrsk, 678144, Russian
2　Gubkin University, Moscow, 119991, Russian

Abstract: The process of testing materials to determine their mechanical and technological properties is a very long process. In recent years, digital materials science has been actively developing, which makes it possible to simulate the properties of materials without conducting empirical tests. In addition, this direction is relevant in connection with a new approach to testing, when not just the material is tested, but the entire structure. These approaches, combining the use of empirical data and computer simulation, are useful at the stage of creating new parts and assemblies of machines, which will operate under conditions of increased vibration loads. This makes it possible to improve and optimize the geometry of mechanism nodes and makes it possible predict the damping properties of materials according to the capabilities of digital materials science.

Keywords: Damping materials; Digital materials science; Numerical simulation; Titanium nickelide

This article discusses the process of modeling the damping properties of materials used for the manufacture of oil and gas equipment.

The damping parameters were evaluated using numerical models in various add-ons to the Solidworks program.

As constraints, we accept that the behavior of the system is linear, i.e. stresses and deformations are described by linear laws (after the load is removed, the system returns to its original position without residual deformations)[1].

Estimation of damping parameters in real time allows you to track the change in parameters, which can be used later to calculate damping characteristics; this allows you to bring it closer to the conditions of a real experiment[2, 3].

The calculation model itself differs from the real model, that is, it does not imitate the entire stand as a whole, but only those parts of it that are directly involved in the simulation (Figure 1)[4].

Since damping propertiesare being investigated, a test stand[5] was chosen as an analogue.

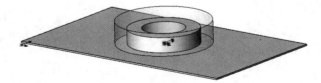

Figure 1　Model used in Solidworks: Washer 20 mm high,
90 mm outer diameter and 54 mm inner diameter

In order to subject this design to vibration, was used an option that allows you to set the nature of vibrations with a constant amplitude and adjustable frequency of vibrations. For example, the following driving force functionis taken:

$$y(x) = 0.75\sin^2(16x) \qquad (1)$$

As a driving force, youcan take either an expression similar to (1), or proposed by Solidworks.

During modeling in Solidworks, one of the characteristics obtained is the vibration displacement of a conditional "table" on which the tested structure made of the material under study is located, in this case it is a washer made of titanium nickelide.

The resulting characteristicis imported into Excel.

The debugging of the numerical model is carried out according to [6] by writing 2 Newton's law in differential form for this system:

$$\sum F_x = 0.75\sin^2(16x) - mg - k(x-U) - r\dot{x} = m\ddot{x} \tag{2}$$

where: k, U, r are the coefficients of vibration stiffness, displacement of the washer made of titanium nickelide and friction coefficient, respectively.

This model imitates the law of motion of a spring pendulum under the action of an elastic force.

After solving the second order inhomogeneous differential equation and finding the coefficients C_1 and C_2, the equation for the vibration displacement of the simulated "table" has the following form

$$y(x) = C_1 e^{-0.028}\sin(31.94x) + C_2 e^{-0.028}\cos(31.94x) - 0.007\sin(32x) + 0.022\cos(32x) + 0.74 \tag{3}$$

where: C_1 and C_2 are 0.007 and -0.7621944 respectively.

During the debugging of the numerical model, it was noted that the greatest influence on the convergence is the vibration stiffness coefficient k.

With a properly selected vibration stiffness coefficient k, the initial curve and the curve whose equation is described by expression (4) have a small average error relative to each other in amplitude (in this case, no more than 0.73%).

After verification of the numerical model, the vibration acceleration curveswere obtained for the frequency range of the excitation force from 10 Hz to 200 Hz (Figure 2).

To test the numerical model, a physical experimentwas carried out, the results of which are shown in Figure 3.

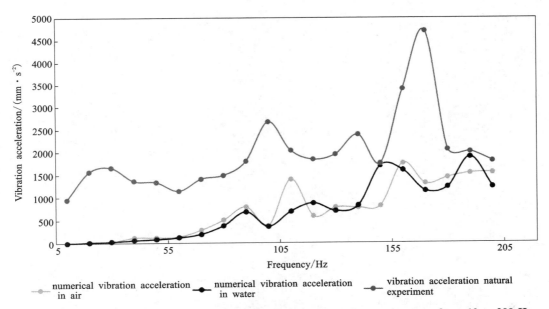

Figure 2 Vibration acceleration curves for the excitation force frequency range from 10 to 200 Hz

The results show that the vibration peaks of oscillations in different media are localized in the same ranges and have the same curvature, but when the washer vibrates inwater medium that simulates operation in real conditions, the value of the internal energy of the considered oscillatory system is higher, as evidenced by an increase in the width of resonance peaks, which corresponds to theoretical and practical data.

Differences in amplitudes between numerical and physical experiments can be explained by the fact that despite the fact that it is possible to set elastic and viscous properties in the Solidworks settings, during the solution their influence on the simulation is not fully taken into account by the program, it should also be taken into account that there are inelastic damping properties of titanium nickelide.

Based on the research, the following conclusions can be drawn:

(1) A method is proposed for analytically obtaining the vibration stiffness coefficient in the process of numerical simulation.

(2) The amplitude-frequency characteristic of a full-scale sample of titanium nickelide, made in the form

of a washer 20 mm high, with an outer diameter of 90 mm and an inner diameter of 54 mm, is described.

(3) The values obtained during the numerical experiment are compared with the values from the physicalexperiment; the average error is no more than 34%.

Acknowledgments: this research became possible thanks to the well-coordinated work of the team of the Department of physical metallurgy and non-metal materials, Gubkin University

References

[1] GOLOVIN I S. Internal friction and mechanical spectroscopy of metallic materials[J]. Moscow: National University of Science and Technology MISIS Publishing House, 2012-247.

[2] PRYGAEV A K, DUBINOV I S, ELAGINA O I, et al. Damping properties of titanium nickelide as one of the key features in the creation of oil and gas equipment[J]. Equipment and Technologies for the Oil and Gas Complex, 2020, 3(117): 8-13.

[3] DUBINOV Y S, ELAGINA O Y, DUBINOVA O B, et al. Influence of the vibration parameters on the destruction of abrasives under sliding friction[J]. Inorganic Materials: Applied Research, 2021, 12(2): 576-580.

[4] GOST 27242-87 "Vibration isolators. General requirements for testing. Introduction date 1988-01-01.

[5] DUBINOV I S. The use of damping properties of materials for the manufacture of oil and gas equipment [EB/OL]: Educational manual. Iu. DUBINOV S, PRYGAEV A K, DUBINOVA O B, et al. [M]. National University of Oil and Gas (Gubkin University), 2021.

[6] CHANG K H. Motion simulation and mechanism (Design with COSMOS Motion 2007)[M]. Oklahoma: KHC Norman, 2008.

Amorphous Ferromagnetic Wires for Structural Health Monitoring

Andrey ALPATOV, Vyacheslav MOLOKANOV, Andrey KRUTILIN, Natalia PALII

Baikov Institute of Metallurgy and Materials Science, Russian Academy of Sciences (IMET RAS), Moscow, 119334, Russia

Abstract: The paper presents the results of the labs' efforts to master the technology of amorphous wire production using the Ulitovsky-Taylor process. Studies of the structure and physical properties of amorphous cobalt-base wires have revealed the presence of tensoresistive effects. The paper presents working models of stress control in different products (pipes, cylinders, panels). It is shown that amorphous wires can be used as long-length stress and strain sensors, as well as pillage sensors in pipe-line health monitoring. Application prospects of different types of long-length sensors on the basis of amorphous wires for structural health monitoring are outlined.

Keywords: Amorphous wire; Strain sensor; Ulitovsky-Taylor method; Structural health monitoring

1 Introduction

The idea of using amorphous microwires as strain sensors has been widely discussed in the literature[1-4]. There is a suggestion for using the electromagnetic properties of thin ferromagnetic glass-coated amorphous microwires with a diameter of 5-20 μm. Anyhow application of thin amorphous microwires as stress sensors is troublesome because of their poor mechanical properties, small diameter, glass sheath and the lack of suitable instrumentation for signal acquisition.

IMET RAS has developed an original technology and mastered production of "thick" Co-Fe amorphous wires (AWs) with diameter of 40-120 μm[4-7].

2 Materials and methods

The modified Ulitovsky-Taylor technique (U-TT) was used to made glass-coated Co-Fe AWs (Figure 1). After glass removal AWs had the diameter of 40-80 μm.

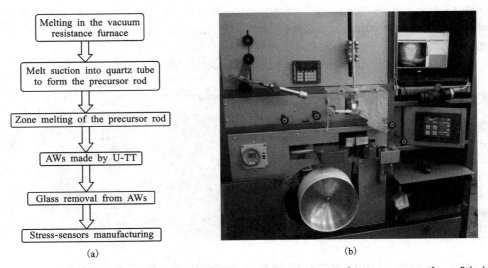

Figure 1 Process flowchart for the production of amorphous wires and stress sensors thereof (a) and general view of the laboratory set up used to produce AWs (b)

The use of zone melting of the rod precursors ensures an increase and stability of the mechanical properties AWs produced by U-TT[8].

An Instron 5848 universal testing machine was used to study the uniaxial tension of AWs. The gage length of wire sample was 50 mm; the tension rate was 1 mm/min.

The electrical resistance of AW samples was measured on the laboratory bench, simultaneously with tension of the samples. The moving gripper with micrometric screw produced a tensile load up to 7 N, registered by a strain gauge. Mastech MS8239C multimeter was used to measure the electrical resistance of the AWs attached to the laboratory stand clamps (wire length l = 328 mm).

3 Results and discussion

AWs obtained by U-TT possess unique complex of mechanical, physical, magnetic and corrosion properties; they exhibit super-elasticity effect, elastic elongation up to 3%, analogue of "metallic rubber" [Figure 2(a)]. They have high electrical resistance (1.2×10^{-6} $\Omega \cdot$ m), high tensile strength (3000 MPa) is combined with high ductility under bending. It is important that the electrical resistance varies linearly over the entire range of applied elastic stress [Figure 2(b)].

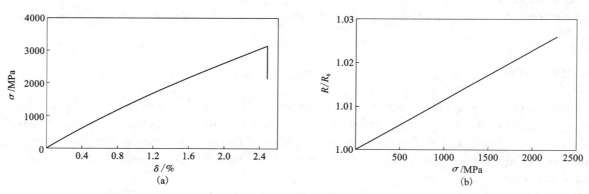

Figure 2　Tensile diagram (a) and reduced electrical resistance R/R_0 versus applied load (b) for $Co_{69}Fe_4Cr_4Si_{12}B_{11}$ AW with diameter d = 71 μm

High sensitivity of electrical resistance of AWs to applied loads, combined with high strength and corrosion resistance, offers good prospects for using such AWs as long-length sensors and stress-sensitive elements of critical structures. In contrast to conventional strain gauges that monitor point mechanical stresses in an object, an AW sensor can monitor the stress-strain state (SSS) along the length of tens of meters. In this case, one or more interconnected long strain gauges made of AWs can be attached to the surface of the object or inserted into the composite structural component (pipe, aircraft skin, working beam, etc.) during the manufacturing process.

Figure 3 shows the different types of strain sensors made of AWs.

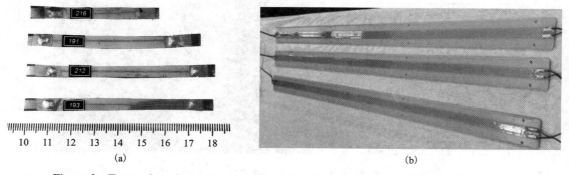

Figure 3　Types of strain sensors based on AWs: short, l<0.1 m (a), long, l>1 m (b)

Some typical applications of amorphous wire strain sensors are shown in Figure 4 and Figure 5. Figure 4(a) shows AW strain sensors on the surface of the 3 m polymer pipe: four sensors mounted longitudinally to monitor bending and thermal expansion, two sensors mounted spirally to monitor damage and leakage. As can be seen in Figure 4(b), a strain sensor from AW is spirally placed on the surface of the cylinder in order to check the state of the permissible dimensions of the cylinder in the process of operation. The crosswise attachment of the AW strain sensors on the panel allows monitoring of the tensile and compressive loads applied to the free edges of the panel (Figure 5).

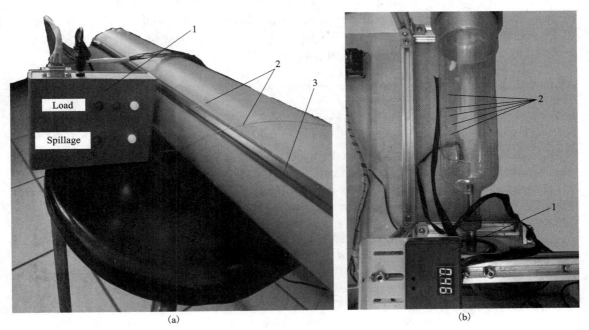

1—Recorder, 2—AW spillage sensor, 3—AW strain sensor

Figure 4 Long-length strain gauges from AWs to record strain-stress state on polymer pipe surface (a) and gas cylinder (b)

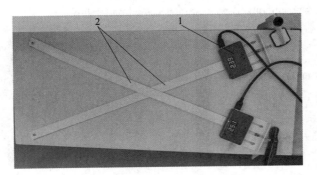

1—Recorder, 2—AW strain sensor

Figure 5 Strain gauge placement on the flexible panel, subjected to free bending (the right site is fixed)

The closest competitor to AW strain gauges in terms of functionality is fibre optic sensors[9, 10]. However, in comparison with them, the main advantages of AW sensors are the ease of placement and switching of wires on the controlled objects, high mechanical properties, the possibility of AW implementation inside polymer and composite products, and the possibility of using a simple existing device base for output signal registration and monitoring in "online" mode. Other advantages include the use of self-contained power supplies (battery and rechargeable), the light weight of the cables, their resistance to corrosion and the ability to use grid-based swathing techniques to monitor large areas. At the same time, the sensor's readings are not affected by strong electromagnetic fields or by the operating temperature (the operating range of the AW sensors is from −50 ℃ to +150 ℃).

4　Conclusions

Simple, reliable and economical variant of signal registration from AW sensor allows to develop new systems of SSS monitoring of large and extended objects such as gas and oil pipelines, objects and constructions of railway and road transport (bridges, tunnels, roads), products of aircraft and shipbuilding equipment, gas cylinders, tanks, building objects, etc.

Currently IMET RAS is working on introduction of tensoresistive sensors based on amorphous ferromagnetic microwires into different branches of economy. Taking into account specific tasks, not only new types of amorphous sensor elements are being developed, but also equipment for processing and transmission of information for real-time monitoring of critical structures.

Acknowledgments: The work was carried out under government assignment № 075-01176-23-00.

References

[1] PANINA L, DZHUMAZODA A, NEMATOV M, et al. Soft magnetic amorphous microwires for stress and temperature sensory applications[J]. Sensors, 2019, 19(23): 5089.

[2] ZHUKOV A, IPATOV M, CORTE-LEON P, et al. Advanced functional magnetic microwires for magnetic sensors suitable for biomedical applications//Magnetic materials and technologies for medical applications[M]. Woodhead Publishing, 2022: 527-579.

[3] CHURYUKANOVA M, STEPASHKIN A, SARAKUEVA A, et al. Application of ferromagnetic microwires as temperature sensors in measurements of thermal conductivity[J]. Metals, 2023, 13(1): 109.

[4] MOLOKANOV V V, MOROZ O V, KRUTILIN A V, et al. Fabrication and physicomechanical properties of amorphous microwires and microspirals[J]. Russian Metallurgy (Metally), 2022, 2022(4): 300-308.

[5] MOLOKANOV V V, KRUTILIN A V, PALII N A, et al. Mechanical, electromagnetic, and tribological properties of microspirals made from amorphous and crystalline metallic materials[J]. Russian Metallurgy (Metally), 2021, 2021(4): 363-366.

[6] MOLOKANOV V V, SHALYGIN A N, UMNOV P P, et al. Conditions for obtaining "thick" amorphous wires by the ulitovsky-taylor method[J]. Inorganic Materials: Applied Research, 10: 463-466.

[7] Ferromagnetic amorphous wires A/A/Baikov Institute of Metallurgy and Materials Science (IMET RAS) Laboratory No. 24. URL: https://www.amorphous-wires.ru/(accessed on 05 July 2023).

[8] RU 2796511 C1. 24. 05. 2023 Bull. № 15.

[9] RUMANOVSKY I G. The use of fiber-optic sensors for the main gas pipelines monitoring[J]. Far East: Problems of Development of the Architectural and Construction Complex, 2020, 1(1): 382-385.

[10] WU T, LIU G, FU S, et al. Recent progress of fiber-optic sensors for the structural health monitoring of civil infrastructure[J]. Sensors, 2020, 20(16): 4517.

Changing the Technological Properties of Materials by Conducting Thermomagnetic Treatment

Yu. S. DUBINOV, A. K. PRYGAEV, G. T. BOKOYEV, A. D. KOTOV, M. A. DUBROVIN, O. B. DUBINOVA

Gubkin University, Moscow, 119991, Russia

Abstract: The influence of thermomagnetic treatment modes on the technological properties of steels of various classes is considered: 30Cr13Mn8P, 12Cr18Ni9Ti, Cr11Ni8Cu20, 0Ni9. It is proved that treatment in a high inductance magnetic field at a temperature of 1000 ℃ increases the resistance to abrasive wear. The proposed method and steel treatment technology can be used to create wear-resistant surfaces on details in oil and gas mechanical engineering.

Keywords: Thermomagnetic treatment; Alloy steel; Abrasive wear

1 Introduction

To improve the mechanical properties of steels and alloys, various processing methods are used, both volumetric and surface, but the development of innovative treatment methods remains an urgent direction. One of these methods is thermomagnetic treatment, which is based on the use of electromagnetic fields combined with high temperatures. Thermomagnetic treatment is an effective method to change the microstructure and properties of materials, providing an increase in their strength, hardness, corrosion resistance and other important indicators[1,2].

There are two types of treatment in a magnetic field:

1-magnetic/thermal processing (MagTO) that is used to change the technological properties of materials. Magnetic treatment is carried out in a furnace at high temperature. This technology remains unclaimed due to the use of magnets at high temperatures.

2-thermomagnetic treatment (TMO) that is used for soft magnetic materials/magnetically soft materials. The treatment is carried out by cooling the samples in a magnetic field.

The authors propose to change the known technologies.

Therefore, the aim of the research is to explore the influence of thermomagnetic treatment (TMO) modes on the technological properties of alloy steels.

The effect is achieved due to the combined thermomagnetic treatment of steel products, including thermal treatment with heating 30-50 ℃ above the critical points of Ac_3 for steels that have polymorphic transformations and up to 900-1050 ℃ for steels that do not have polymorphic transformations, exposure at these temperatures and air cooling under the influence of a constant magnetic field with a magnetic induction value of 19.6×10^6 A/m in the transverse or longitudinal direction relative to the central axis of the treated products (work piece) to a temperature of 500 ℃ and subsequent air cooling without a magnetic field to full cooling down.

Improving the operational properties of products made of ferritic, ferrite-perlite, austenitic and martensitic steels due to the action of a magnetic field directly during the course of structural-phase transformations or redistribution of alloying elements during the implementation of diffusion processes during thermal treatment. The result of such thermomagnetic treatment is expressed in a change in impact strength and wear resistance.

The results of changing the properties of materials by the proposed method are shown in Figure 1 and Figure 2.

Based on the test results, the wear resistance of 30Cr13Mn8P steel after thermomagnetic treatment increased by 60%, and the impact strength increased by 3%, the wear resistance of 12Cr18Ni9Ti steel after thermomagnetic treatment increased by 16%, and the impact strength remained approximately the same. The

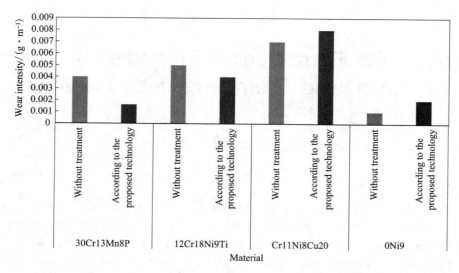

Figure 1　Changing wear intensity on samples with and without TMO

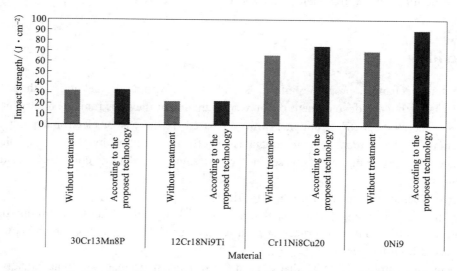

Figure 2　Changing the impact strength on samples with and without TMO

impact strength of steels 45 and 0Ni9 increased by 12% and 30% respectively, although at the same time the wear intensity decreased by 11% and 90% for steel 45 and 0Ni9 respectively.

According to the results of the performed research, the following conclusions can be drawn:

(1) A technology for thermomagnetic treatment of steels of various classes has been developed.

(2) It has been found that the use of the proposed thermomagnetic treatment makes it possible to increase the wear resistance of the material in some cases by 60% for martensitic grade steels.

(3) It has been found that the use of the proposed thermomagnetic treatment makes it possible to increasethe impact strength in some cases by 90% for ferritic grade steels.

(4) The influence of chemical elements on the effectiveness of the proposed thermomagnetic treatment has been revealed.

References

[1] MALYGIN B V. Magnetic hardening of tools and machine [M]. Moscow: Mashinostroenie, 1989: 112.
[2] BERNSTEIN M L, PUSTOVOIT V N. Heat treatment of steel products in a magnetic field [M]. Mechanical Engineering, 1987, 232.
[3] Application 036600 Russian Federation. Method of thermomagnetic processing of steel products; applicant Gubkin Russian State University of Oil and Gas (NIU)-No. 2023117137; application 29.06.2023, 33.

Application of Data from Photometric Analysis of Structural Images in the Choice of Heat Treatment Modes for Heat-resistant Alloys

V A ERMISHKIN, N A MININA, D L MIKHAILOV, N A PALII

Baikov's Institute of Metallurgy and Materials Science, Russian Academy of Sciences,
Leninskii Prospect 49, Moscow, 119334, Russia

Abstract: The article considers the results of a study of the kinetics of structural transformations under given heat treatment modes based on the data of photometric analysis of structural images (PHASI) of a new corrosion-resistant alloy Kh28M9N. The research program included testing samples of this alloy with two types of heat treatment: 1) single-stage short-term heat treatment in the temperature range 1380-1395 K; 2) two-stage heat treatment at temperatures from the same interval, followed by annealing of the samples at temperatures of 1073 K. The kinetic coefficients were calculated in the kinetic equations describing structural transformations under given heat treatment modes. It has been established that the temperature-time dependences of the kinetic coefficients in both groups of samples are not the same. It is shown that PHASI makes it possible to determine the generalized structural index, which is suitable for the analysis of heat treatment kinetics.

Keywords: Structural state; Photometric analysis; Chromium-nickel alloys; Heat treatment

1 Introduction

It is known that when creating new alloys, researchers usually do not go beyond the classical triad "composition-structure-properties". The choice of chemical composition largely determines the range of operating conditions of the developed alloy, while the level of its properties is mainly determined by the technology of its thermal and mechanical processing. An important role of heat treatment follows from its ability to form the required phase composition and grain structure of the alloy. This work aims to experimentally verify the use of photometric analysis of structural images on the example of the formation of a new corrosion-resistant heat-resistant alloy Kh28M9N of high resistance to intergranular corrosion in the medium of molten chlorides of metals of group IV of the D. I. Mendeleev Periodic Table. This alloy belongs to the Ni-Cr-Mo system and has a chromium equivalent of 43.94 wt. % and nickel equivalent of 61.63 wt. %., which provides the austenitic composition of the matrix. Attempts to describe[1] the structural state of alloys as a function of temperature-time parameters have not been successful so far. The most successful is the complex representation proposed by Dorn[2] in the form of effective time:

$$t_{ef} = t \cdot \exp\left(-\frac{U}{RT}\right) \tag{1}$$

where: t is the chronometric duration of the structure transformation, T is the Kelvin temperature at which the structure transformation process is carried out, U is the activation energy of the structural transformation process. The complexity of the analytical description of structural transformations lies in the absence of generalized structural parameters that express various aspects of the structural state of the material-phase and defect structure (defects in the crystal lattice and technological defects). The use of activation energy requires sufficiently reliable methods for its estimation. Previously, we made an attempt to use J. A. Brinell hardness values as a generalized structural parameter[1]. However, it cannot be considered completely successful, since the hardness values depend on the crystallography of the grains and their sizes. In this work, as a generalized parameter of the area under the spectral brightness curve of the reflection of visible light from fragments of the surface of the object under study[3-6].

2 Experiment

To solve the problem, a series of samples of the investigated alloy was prepared, processed according to various heat treatment modes. These data are presented in the Table 1.

Table 1 Modes of heat treatment of Kh28M9N alloy

No.	Heat treatment mode	Cooling	Annealing mode	Cooling	φ
1	1380 K/20 min	water			0.5677
2	1380 K/15 min	water	1073 K/30 min	air	0.4755
3	1380 K/15 min	water			0.3672
4	1380 K/25 min	water	1073 K/30 min	air	0.3671
5	1394 K/30 min	water			0.4865
6	1394 K/20 min	water	1073 K/30 min	air	0.4967
7	1380 K/25 min	water	1073 K/30 min	air	0.5388
8	1380 K/30 min	water	1073 K/30 min	air	0.5020
9	1380 K/25 min	water			0.4715
10	1380 K/30 min	water			0.5852

Table 2 System of equations describing the kinetics of structural transformations using φ

No.	System of equations
1	$155663 = C_1 \cdot 20 \cdot \exp\left(-\dfrac{U_1}{RT}\right)$
2	$130382 = C_2 \cdot 15 \cdot \exp\left(-\dfrac{U_1}{RT}\right) + C_o \cdot 30 \cdot \exp\left(-\dfrac{U_0}{RT}\right)$
3	$100686 = C_3 \cdot 15 \cdot \exp\left(-\dfrac{U_1}{RT}\right)$
4	$100659 = C_4 \cdot 25 \cdot \exp\left(-\dfrac{U_1}{RT}\right) + C_o \cdot 30 \cdot \exp\left(-\dfrac{U_0}{RT}\right)$
5	$133398 = C_5 \cdot 30 \cdot \exp\left(-\dfrac{U_1}{RT}\right)$
6	$136195 = C_6 \, 20 \cdot \exp\left(-\dfrac{U_1}{RT}\right) + C_o \cdot 30 \cdot \exp\left(-\dfrac{U_0}{RT}\right)$
7	$147739 = C_7 \cdot 25 \cdot \exp\left(-\dfrac{U_1}{RT}\right) + C_o \cdot 30 \cdot \exp\left(-\dfrac{U_0}{RT}\right)$
8	$137648 = C_8 \cdot 30 \cdot \exp\left(-\dfrac{U_1}{RT}\right) + C_o \cdot 30 \cdot \exp\left(-\dfrac{U_0}{RT}\right)$
9	$129285 = C_9 \, 25 \cdot \exp\left(-\dfrac{U_1}{RT}\right)$
10	$160462 = C_{10} \cdot 30 \cdot \exp\left(-\dfrac{U_1}{RT}\right) + C_o \cdot 30 \cdot \exp\left(-\dfrac{U_0}{RT}\right)$

The heat treatment modes in Table 1 were selected taking into account the range of expected operating temperatures (873-923 K) of a chemical reactor for processing chloride melts of chemical compounds of metals of group IV in the periodic system. The data in Table 1 made it possible to write down a system of equations describing the kinetics of structural transformations using the generalized structural criterion φ. The driving force behind these transformations, which were carried out by the mechanism of thermally activated diffusion, was the activation energy of self-diffusion U_{sd}. According to [2], its value is numerically equal to the activation energy of high-temperature creep. For the lower limit of high-temperature creep is considered to be $(0, -0.6) \cdot T_m K$. Assuming $U_1 = U_{cd}$, taking this fact into account, the values of the structural index were measured after cooling to room temperature, the values of the internal specific energy accumulated in the alloy were estimated by the expression: φU, the kinetic equations for all modes of heat treatments used are given in Table 2. Kinetic equation relating annealing parameters $-0.02852 = 30 C_0 \cdot \exp(-U/8921)$.

3 Results and discussion

10 equations contain 10 unknowns (C_1-C_{10}), which indicates the possibility of finding its solution. T_0 reduce the number of variables, $U_0/8921.459$ was expressed in terms of the U_1 value as $U_0/8921.459 = U_1 8.71511 \times 10^{-5}$. Under the assumptions made, from the solution of system (2) it is possible to obtain the values of the kinetic coefficients C_i. The resulting C_i values can be written as follows: $C_1 = 8.299144$; $C_2 = 4.639569$; $C_3 = 7.157349$; $C_4 = 4.741394$; $C_5 = 7.261201$; $C_6 = 4.892455$; $C_7 = 6.3013$; $C_8 = 6.310328$; $C_9 = 5.514$; $C_{10} = 5.7043$. The dependence of the kinetic coefficients of structural transformations on the effective time of the process is shown in Figure 1.

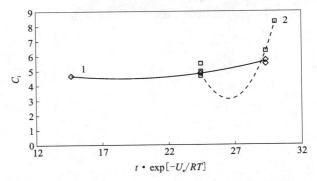

Figure 1 Dependence of the kinetic coefficients of structural transformations on the effective time of the process. 1) dependence for samples with heat treatment without additional annealing at T-1073 K; 2) also, but with additional annealing at 1073 K.

Obtained analytical descriptions of dependencies C_i for the case without additional annealing ($C_i = 0.009 \, t \cdot \exp(-U_1/RT)^2 - 0.368 \cdot \exp(-U_1/RT) + 7.91$, $R^2 = 0.968$) and with additional annealing ($C_i = 0.428 \, t \exp(-U_1/RT)^2 23.68 \cdot \exp(-U_1/RT) + 309.8$, $R^2 = 0.998$). This decision became possible due to the use as an integral characteristic of the structural state of the alloy, the value of the area under the spectral curve of reflection brightness, determined by photometric measurements of the spectral density of the reflection intensity of visible light reflected from the surface of the selected fragment of the object under study before and after physical or chemical impact on it. The effect of this exposure depends on the reflectivity. In our case, the intensity of thermal action is understood as the degree of structural change caused by the light flux incident on the fragment. Structural imaging impacts make it possible to evaluate the effect of structure change in relative energy units. To determine this effect in specific dimensional physical units, it is necessary to multiply the conditional energy effect by the activation energy of the process of self-diffusion of the alloying element with the highest concentration. In our case, nickel with $U_{ak} = 274.2$ kJ/mol[7].

A visual representation of the distribution of the structural components of the alloy under known temperature-time conditions is given in Figure 2-Figure 4. Under the images of the structure, the spectral curves of the reflection brightness are shown.

4 Conclusions

(1) An analytical description of the kinetic equations describing the energy of structural transformations of the new Kh28M9N alloy during heat treatments has been developed.

(2) Based on the use of photometric analysis of structural images (PHASI), a generalized criterion for the structural state of materials is proposed.

(3) Analytical descriptions of the dependences of the kinetic coefficients of the equations describing structural transformations in the investigated alloy under the selected heat treatment modes are obtained. It is shown that these dependences are different for heat-treated alloys with and without subsequent annealing.

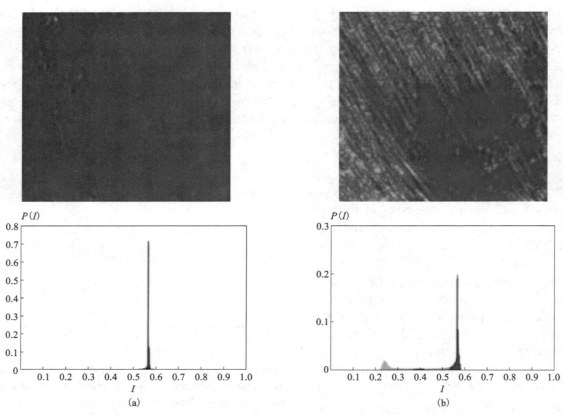

Figure 2 Image of fragment 1 and spectrum from it (left side of the figure) (a),
Fragment 3 and its brightness spectrum of visible light reflection from it (b)

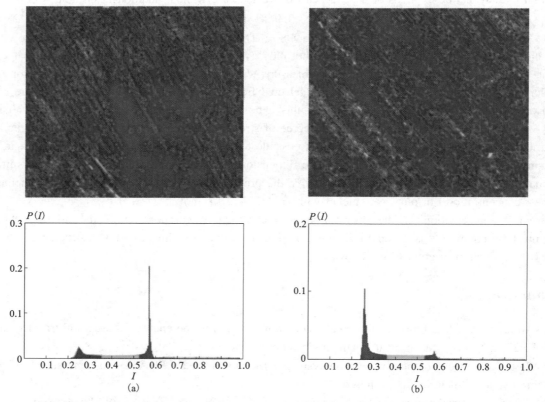

Figure 3 Image of fragment 2 and spectrum from it (left side of the figure) (a),
Fragment 4 and its brightness spectrum of visible light reflection from it (b)

Acknowledgments: The work was carried out according to the state task No. 075-01176-23-00.

References

[1] PECKNER D, BERNSHTEIN I M. Handbook of stainless steels[M]. V. C. Graw-Hill Book Co N. Y, 1977.
[2] DORN H. In the collection of creep and return. Per. from English[M]. Metallurgizdat, 1961, 124-160.
[3] ERMISHKIN V A, LEPESHKIN Y D, MURAT D P, et al. A method for photometric diagnostics of the structural state of materials, according to the analysis of a digital encoded image of their surface[P]. Patent No. 2387974, 2010, 12.
[4] ERMISHKIN V A, MININA N A, FEDOTOVA N L. Method for photometric diagnostics of phase transformations in solids according to the analysis of the brightness spectra of light reflection from their surface[P]. Patent No. 2387978, 2010, 12.
[5] MIKHAILOV D L, ERMISHKIN V A, MININA N A. Thermal stability of heat-resistant chromium-nickel alloys according to differential thermal analysis [C]. Proceedings of the XXIX International Scientific and Technical Conference "Engineering and Technosphere of the XXI Century", September 12-18, 2022, 206-208.
[6] ERMISHKIN V A, MIKHAILOV D L, KULAGIN S P, et al. Influence of heat treatment on the structure and properties of chromium-nickel alloy G-35//the Fourth interdisciplinary scientific forum with international participation "New materials and promising technologies"[C]. BukiVedi LLC, 2018(1): 648-653.
[7] KABLOV D E, SIDOROV V V, PUCHKOV Y A. Features of the diffusion behavior of impurities and refining additives in nickel and single-crystal heat-resistant alloys[J]. Aviation Materials and Technologies, 2016(1): 24-31.

Fiber Optical Current Sensor — A Method for Online Monitoring of Electrode Current in Electrolytic Metallurgy Process

Yi MENG[1], Jun TIE[1], Chun LI[1], Rentao ZHAO[1],
Hongwei JIANG[1], Xingzu PENG[1], Hao XIAO[2], Dongwei LIU[2]

1 School of Mechanical and Materials Engineering, North China University of Technology, Beijing, 100144, China
2 Beijing SIO Technology Co., Ltd., Beijing, 100085, China

Abstract: Fiber optical current sensor (FOCS), designed on the basis of Faraday Magneto-optic Effect principle, has the ability of being installed in the narrow space to measure current due to its characteristics of small size, light weight, softness, anti-magnetic field interference, good insulation and wide measurement range (from several to thousands of ampere). In this manuscript, the measurement principle of FOCS was introduced firstly, and then the electrode current measurement results of the aluminum electrolysis cell and copper electrorefining cell carried out by FOCS were reported, respectively. As a result, the methods for the online monitoring of electrode current on the electrolysis cell by using FOCS were provided, which might designate the development direction of the electrolysis cell digitization.

Keywords: Current measurement; Electrolytic metallurgy; Aluminum electrolysis; Copper electrorefining

1 Introduction

Electrolytic metallurgy is one of the important methods to extract metals. For some metals with lower than or similar to hydrogen element in chemical activity can be electrolytic deposited or refined in the aqueous solution, such as gold, copper, lead, zinc, nickel, etc. Conversely, the metals with much higher chemical activity than hydrogen will be extracted or refined by means of fused-salt electrolysis, such as aluminum, magnesium, rare earth, etc. The cryolite-alumina molten salt electrolytic process to produce the aluminum and the electrorefining in the aqueous solution of Cu_2SO_4 to produce the refined copper can be considered as the typical applications of electrolytic metallurgy of molten salt and aqueous solution, respectively.

The electrolyte systems and operating conditions of aluminum electrolysis are quite different from those of copper electrolysis, but the electrolysis cells between aluminum electrolysis and copper electrolysis have a similar characteristic: multiple electrode. For example, 48 anodes are placed in a 500 kA aluminum electrolysis cell, and they jointly carry the electrolysis series/line current by means of parallel connection, resulting in the 10 kA current passing through each anode. Similarly, there are 54 anodes connected in parallel and 53 cathodes with the parallel connection in a 35kA copper electrorefining cell, leading to the current being as high as 660 A passing through each cathode. The reaction space of the electrolysis cell mentioned above is filled with dozens of electrodes. The abnormal electrochemical reaction on the electrode surface, the concentration variation of active substances near the electrode, the circulation difference and space change between electrodes will be revealed directed by the change of electrode current. Therefore, the electrode in the electrolysis cell can be thought as a good sensor, and the electrode current can be considered as the information source for the local state of electrolytic production.

However, in the actual production process, the current values passing through the dozens of anodes or cathodes connected in parallel with each other do not show the even distribution but the normal distribution instead. Particularly, the change of electrode during the producing process will lead to a large change of current distribution. The failure occurring on the individual electrode in an electrolysis cell can not be discovered and excluded until it transforms into a serious malfunction, leading to the abnormal state in the entire cell. That is, a small illness could not be discovered and treated until it changed to a serious illness. At this moment, the production conditions might be destroyed largely, such as the reduce of current efficiency and product quality.

Unfortunately, due to the existence of strong background magnetic field, strong corrosive medium and narrow space, the online measurement of electrode current has always been a worldwide difficult problem in the industry and has not been resolved yet. It will significantly hinder the digital and intelligent development of electrolysis cells.

Fiber Optical Current Sensor (FOCS), designed on the basis of the Faraday Magneto-optic Effect principle, has been successfully used in the measurement of series current in electrolytic metallurgy processes since it has the characteristics of small size, light weight, softness, anti-magnetic field interference, good insulation, wide measurement range (from several to thousands of ampere) and high measurement accuracy (0.1%). Our research team has pioneered the use of FOCS to measure the anode current of the aluminum electrolysis cell and the cathode current of the copper electrorefining cell, and obtained some meaningful discoveries. With the significant progress of semiconductor technology and the significant decrease in the cost, a method for the online monitoring of electrode current in the electrolysis cell by using FOCS was proposed to optimize the control and management of electrolysis metallurgy process.

2 Measurement principle of FOCS

The structure of the FOCS, as shown in Figure 1, is composed of a sensing optical fiber ring, a photo-electric signal module, and the optical fiber between them. The sensing optical fiber ring consists of a $\lambda/4$ wave plate, sensing optical fiber, and reflector. The photo-electric signal module includes light sources, optical components, and electronic components. During the measurement process, the photo-electric signal module will emit a linearly polarized light beam, which will in consequence become circularly polarized through the $\lambda/4$ wave plate. The circularly polarized light will reach the reflector along the sensing optical fiber and turn back, during which the Faraday Effect will occur twice in total. Then, the light will pass through the $\lambda/4$ wave plate again, changing back to linearly polarized light and transmitting to the photoelectric signal module. When the electric current passes through the sensing optical fiber ring, the change of the light wave phase can be measured in the photoelectric signal module. Thus, the current intensity can be detected by coherent detection and digital closed-loop feedback technology.

According to Faraday Effect, the magnetic declination θ of the polarization plane corresponding to the polarized light transmitting in the optical fiber is as shown in Equation (1):

$$\theta = V \oint_l H dl \tag{1}$$

where: V is the Verdet constant of the optical medium, l is the propagation distance of the light in the medium, H is the magnetic field intensity.

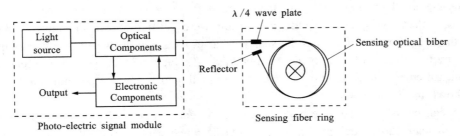

Figure 1 Principle diagram of FOCS

In the sensing optical fiber ring, when the light encircles the current carrying conductor whose current intensity is I, the light path around the conductor is closed. According to Ampere's circuital law, Equation (1) can be rewritten as Equation (2):

$$\theta = VnI \tag{2}$$

where: n is the coil of the closed light path around a current carrying conductor.

According to the measurement principle, the light passes through the polarizer, the optical fiber and the incoming signal detection system, so the current is obtained by measuring the polarization plane rotation angle, which is only dependent on the value of current surrounded by the closed coil and independent on the background magnetic field outside the sensing optical fiber ring. Therefore, this method has a good ability of resistance to electromagnetic interference. Ampere loop integration only requires infinite proximity between the reflector and the wave plate shown in Figure 1 to form a closed loop, and there is no requirement for the shape of the integrated path, i.e. the shape of optical fiber. Therefore, the sensing optical fiber ring can be produced into complicated shape based on the measurement environment to realize the current measurement in different narrow spaces under the condition of without affecting the optical properties of the fiber.

3 Electrode current measurement (routing inspection) by FOCS

3.1 Anode current of aluminum electrolysis cell

ABB reported the fiber-optic current sensor being used to measure the line currents of aluminum electrolysis rectifier unit for the first time[1]. Compared to the traditional current transformer measurement technology based on the hall sensor, the current value can be obtained easily by wrapping the sensing optical fiber around the large conductive busbar to form a closed fiber ring with the help of the epoxy resin bracket, as shown in Figure 2. The measurement error of series current value was within ±0.1%, and fiber-optic current sensor took on incomparable advantages in space, weight, transport and installation at the same time. So far, fiber-optic current sensor is used at almost all the newly built electrolytic aluminum plant to measure the series current value. However, the

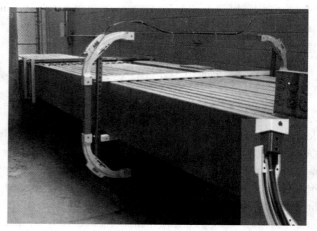

Figure 2 Fiber-optic current sensor installed at an aluminum smelter with a rated current of 260 kA. The width of sensing head housing is 2.25 m[1]

high price limits the application of the fiber-optic current sensor, compared to the traditional measurement methods on the basis of voltage-drop and hall sensor[2].

Based on the structure and measurement principle of fiber optical current sensor mentioned above shown in Figure 1, it can be indicated that the reflector and wave plate can be separated before the measurement, and then closed with each other during the current measurement process. The separated sensing optical fiber ring can be straightened into a thin and flexible optical fiber and then pass through various narrow spaces conveniently to measure the current. According to this characteristic, our team designed a fiber optical current sensor with a disassembled sensing ring, and then used it to successively measure the current values of 48 anodes in a 400 kA electrolysis cell. A preferable results were got and shown in Figure 3. Further, current distributions of the anode, the cathode and the shaft busbar in the electrolysis cell, as well as the zone current variation caused by the change of electrode were obtained[4-6]. Figure 3(a) shows the scene diagram of the anode current measurement in an aluminum electrolysis cell and the corresponding typical anode current distribution is shown in Figure 3(b).

Since FOCS can be used to measure both the anode current and cathode current of aluminum electrolysis cell accurately, our team further designed a portable FOCS which was convenient to the routing inspection in the production site as shown in Figure 4(a), considering the on-site requirements of aluminum electrolysis[7]. Figure 4 illustrates the newest measuring fork design, which realizes the close of both the wave plate and reflector by the two fork heads on the measuring fork [Figure 4(b)]. Its measurement accuracy is within ±1%, resulting in the accurate data support for the fault inspection on the production site.

3.2 Cathode current of copper electrorefining cell

The current values of both anode and cathode in copper electrorefining cells is as small as nearly 650A, but the distance between the anode conductive rod and cathode conductive rod is quite small, that is, 5-25 mm. So far, there is still not effective method to measure the electrode current. The optical fiber was used by our team to pass through the gap between the conductive rods firstly and then wrapped around the cathode conductive rod to form a closed loop with the help of the softness of the optical fiber. As a result, the

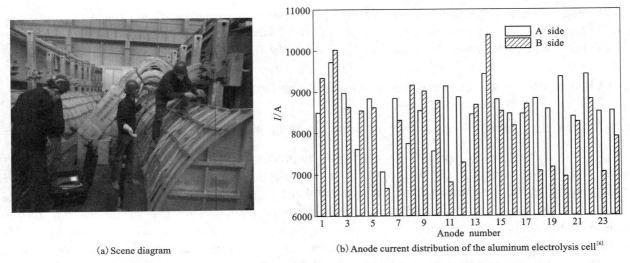

(a) Scene diagram (b) Anode current distribution of the aluminum electrolysis cell[6]

Figure 3 The measurement of anode current in a aluminum electrolysis cell by FOCS

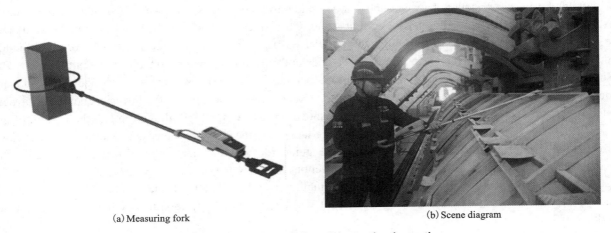

(a) Measuring fork (b) Scene diagram

Figure 4 Portable FOCS used in routing inspection

change of electrode current value was obtained. Figure 5(a) shows the scene diagram of cathode current measurement in a copper electrorefining cells. Figure 5(b) indicates the change of cathode current values in the process of the short-circuit occurrence between the anode and cathode, caused by the growth of nodulation on the cathode surface[8]. Consequently, the short-circuit can be found timely according to the change of cathode current value to avoid the reduce of current efficiency and product quality.

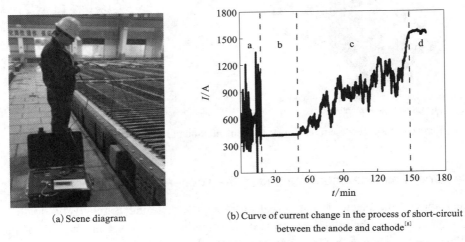

(a) Scene diagram (b) Curve of current change in the process of short-circuit between the anode and cathode[8]

Figure 5 Cathode current measurement in a copper electrorefining cell

During the electrode current measurement of copper electrorefining cell, our team found that the closed gap between the wave plate and the reflector would affect the measurement results. Based on our research[9], the shielding rings could reduce the impact of closing gap effectively.

4 Online electrode current measurement method of electrolysis cell by FOCS

4.1 Electrode current measurement methods of aluminum electrolysis cell

It is not ultimate goal to use FOCS to measure the electrode current point by point and routing inspection, and it is far from meeting the actual production need either. The way to meet the actual needs of the production and realize the function of FOCS is to carry out the real-time on-line measurement of electrode current, and the subsequent process control, management and optimization. The key to realizing on-line measurement of current is to avoid the interference to the electrolysis operation, such as the change of electrode operation. The solution for the on-line measurement of electrode current in the aluminum electrolysis cell is provided by our team and shown in Figure 6(a)[10]. Its characteristic is that, a sensing optical fiber ring on the parallel busbar is installed between two adjacent anode guide rods and another sensing optical fiber ring on the parallel busbar is installed between the anode guide rod and the shaft busbar (including the jumper busbar). Thus, the corresponding anode current value will be measured accurately by measuring the value and the direction of the current passing through the parallel busbar located on left and right sides of anode guide rod. However, the method shown in Figure 6(a) does not fully utilize the deformability of optical fiber. The results obtained by FOCS are only just the current passing through the sensing optical fiber ring and are independent on the shape and location of electric conductor inside the optical fiber ring. Therefore, a new method was subsequently proposed by our team and shown in Figure 6(b). Both the top and the bottom of the optical fiber rings are folded backward. Then they are lapped on the horizontal busbars. The optical fiber rings surround the connection position between the anode guide rod and the horizontal busbars, resulting in the measurement of current passing through the connection between the anode guide rod and the horizontal busbars, without the influence on both the disassembling and installation of the anode at the same time. Finally, the measurement of individual anode current will be carried out successfully by the application of the single FOCS.

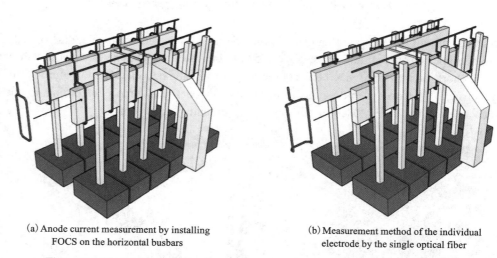

(a) Anode current measurement by installing FOCS on the horizontal busbars

(b) Measurement method of the individual electrode by the single optical fiber

Figure 6 Electrode current measurement methods of aluminum electrolysis cell

4.2 Electrode current measurement method of the copper electrorefining cell

As mentioned above, the electrode current of copper electrorefining is relatively small. The very tight space between the anode and cathode also prevents the installation of FOCS with the long length. So a mini-sensing optical fiber ring, as shown in Figure 7, was designed by our team according to Reference [11]. It is twisted by dozens of turns of optical fiber, with the measurement range of 0.1-2000 A and the measurement error of <±0.2%. It can be easily fastened around the specially designed inter-slot conductive plate[12] shown in Figure 8 for the on-line precise measurement of electrode current. Further design of conductive plate was made to protect both the sensing ring and the extend line of FOCS and simultaneously cover the electroconductive strip. Therefore, the current change of anode or cathode will be real-time tracking, and the occurrence of both the short circuit and broken circuit between the anode and cathode will be monitored without the impact on the works on the cell surface, as a result of the increase of current efficiency and cathode product quality.

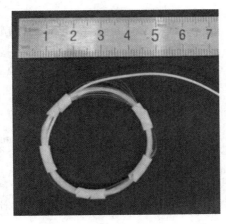

Figure 7   Mini-sensing optical fiber ring

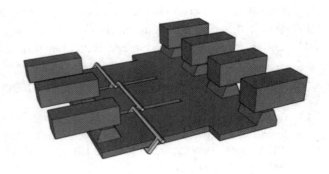

Figure 8   Cathode current measurement method in copper electrorefining cell by use of mini-FOCS

4.3 Cost Reduction Program

The current FOCS system, which is used for the aluminum electrolysis inspection, has been gradually accepted by various aluminum electrolysis companies. However, the number of FOCS is quite large if they are used to carry out the on-line measurement of electrode current of the electrolysis cell. So, further reduction in manufacturing cost is needed. We found that the main cost of FOCS system lay in the photoelectric signal module. In fact, the photoelectric signal module is entirely composed of semiconductor devices. If the number of photoelectric signal module needed is quite large,

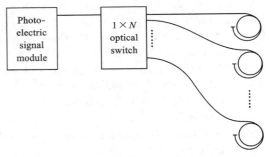

Figure 9   Using optical switch networking technology to achieve photoelectric signal modules sharing

it might also be produced by the way of high integration of various devices, such as transforming into a chip. Taking an aluminum electrolysis cell as an example: The world's aluminum production is nearly 70 million tons per year. If there are 32 anodes in the 300 kA aluminum electrolysis cell, nearly 2.8 million sensors will be needed. Similarly, the world's refined copper production is about 25 million tons in 2022. If there are 53 anodes in the 35 kA copper electrorefining cell, with the production of 350 tons per year for the single cell, 3.8 million sensors will be needed. Due to the huge demand, it should be worth transforming the photoelectric signal module of FOCS into the chip. Once it is produced into the chip, its cost should reduce significantly.

On the other hand, our team proposed a networking technology by using optical switches, as shown in Figure 9, which could also reduce the cost of FOCS largely.

5 Conclusion

The electrode current is the key information for digitizing metallurgical electrolysis cells. FOCS is an ideal technology for achieving accurate online measurement of electrode current. By utilizing the flexibility and deformability of fiber optic current sensor, electrode current measurement can be technically achieved through the following methods:

(1) By designing the sensing optical fiber rings with different shapes according to the measuring objects, the online measurement of electrode current without affecting production operations can be realized.

(2) The huge demand will promote the transformation of photoelectric signal module into integrated chip and the application of optical switch networking, thereby the significantly reduced costs.

(3) The application of FOCS provides an accurate electrode current information for electrolysis cell, promotes technological upgrading in traditional industries, and helps to save energy, reduce emissions, and improve product quality further.

Acknowledgments: This work was supported by the national natural science foundation of CHINA (No. 21978004), the national key research and development program of CHINA (No. 2022YFB3304900) and the 2023 College Student Innovation and Entrepreneurship Training Program (No. 10805136023XN262-114).

References

[1] BOHNERT K, NEHRING J, BRÄNDLE H, et al. Fiber-optic current sensor for electrowinning of metals[J]. Journal of Lightwave Technology, 2007, 25(11): 3602-3609.

[2] ZIEGLER S, WOODWARD R C, IU H H-C, et al. Current sensing techniques: A review[J]. Ieee Sensors Journal, 2009, 9(4): 354-376.

[3] WANG Y, TIE J, SUN S, et al. Testing and characterization of anode current in aluminum reduction cells[J]. Metallurgy and Material Transaction B, 2016, 47: 1986-1998.

[4] TIE J, ZHAO R T, ZHANG Z F. Accurate measurement and its application of anodic current in aluminium electrolysis (end)[J]. Metallurgical Industry Automation, 2017, 41(6): 49-54+81.

[5] TIE J, ZHAO R T, ZHANG Z F. Accurate measurement and its application of anodic current in aluminium electrolysis (to be continued)[J]. Metallurgical Industry Automation, 2018, 42(1): 49-53.

[6] FAN H T, TIE J, ZENG Q Y, et al. Measurement and analysis of anode current in 400 kA aluminum reduction pots[J]. Light Metal, 2019(4): 26-30.

[7] LI J G, XIAO H, LIU D W, et al. Research on the handheld fiber-optic current sensor for aluminum electrolysis current measurement[J]. Chinese Journal of Scientific Instrument, 2022, 43(12): 39-48.

[8] ZENG Q, LI C, MENG Y, et al. Analysis of interelectrode short-circuit current in industrial copper electrorefining cells[J]. Measurement, 2020, 164: 108015.

[9] LI C, ZHANG S, MENG Y, et al. Research of magnetic shielding on the closing gap of optical fiber current sensor[J]. Journal of Sensors, 2022: 3591818.

[10] TIE J, ZHAO R T, ZHANG Z F, et al. System and method for measuring anode current of aluminum electrolytic cell[J]. US20200032408A1, 2020.

[11] TIE J, LI C, MENG Y. A current measurement device based on fiber ring current sensor[J]. CN115639388A, 2022.

[12] TIE J, ZHAO R T, ZHANG Z F, et al. A system for measuring cathode current[J]. CN108411339A, 2018.

Mathematical Models and Software for Dynamic Simulation of Ladle Treatment Technology

O. A. KOMOLOVA, K. V. GRIGOROVICH

Baikov Institute of Metallurgy and Material Science RAS, Moscow, Russia

Abstract: The original software for dynamic modelling of ladle treatment processing of steel has been developed. We used physical and chemical models based on the law of conservation of mass and energy, as well as on the principles of nonequilibrium thermodynamics. In this software, all stages of the process (zones) were taken into account. This software takes into account such initial data as: power of ladle equipment, initial temperature, chemical composition of slag and metal, injection time and mass of additives, blowdown modes, electrical and thermal modes, thermodynamic database, thermal, physicochemical database of additives and inert gas, production database (for statistics). This software allows you to calculate the main characteristics of ladle processing, such as temperature and chemical composition of the slag and molten steel. To validate the program, the results of ladle processing of real steel melts and the results of selective control were used. This developed software and digital twins of the units can be used for online calculations and control of technological parameters in ladle processing, for modelling and optimizing the technology of out-of-furnace processing, for teaching and training steelmaking personnel.

Keywords: Mathematical models; Ladle treatment; Nonequilibrium thermodynamics

1 Introduction

The production technology of modern steel grades is based on the production of metal with narrow intervals of chemical composition, of alloying elements, modifiers, reducing the content of harmful impurities and non-metallic inclusions. Achieving these parameters requires fine-tuning of steelmaking technologies at each stage, taking into account changes in the temperature and composition of the steel melt and slag and the additives introduction mode. Modern metallurgical technologies of the XXI century provide various methods of ladle treatment processing to control the quality of steels and alloys. All industrial experiments on technology optimization are complex and extensive. The best way of this optimization is a computational modeling of metallurgical technologies.

Modelling of metallurgical processes is a complex task that requires the development of physicochemical models and mathematical algorithms that allow one to adequately describe high-temperature processes occurring in open nonequilibrium systems. Most computer software that simulates a real metallurgical process is based on approximation and statistical models that require a huge amount of experimental data[1-3]. This fact significantly limits the capabilities of the software, which is not able to adequately respond to various perturbations and random processes in a wide range of parameter changes. The use of calculation models that adequately describe the processes in ladle processing of steel makes it possible to calculate the optimal technology for the production of certain steel grades, simulating this on a computer without conducting a series of expensive industrial experiments, develop new steel production technologies and identify new factors affecting product quality.

The purpose of this study was to develop mathematical models, algorithms and software for dynamic modeling of steel processing technology in a ladle furnace and RH vacuum degasser.

2 Methods and models

The software for dynamic modeling of ladle processing technology was based on physicochemical models and thermodynamic models[4-5]. The task of this program was modeling and operational control of steel temperature and the chemical composition of slag and steel melt in steelmaking processes (ladle furnace and RH vacuum degasser). Physicochemical models based on the law of conservation of mass and energy and the principles of nonequilibrium thermodynamics were used. This program took into account all stages (zones) of the process. It was assumed that metallurgical systems do not reach equilibrium and are in non-equilibrium stationary states. In accordance with the thermodynamics of L. Onsager, it was assumed that the reaction rate is proportional to the chemical potential gradient according to the formulas:

$$V_i = - SL\mathrm{grad}\mu_i$$

where: V_i is reaction speed of i-component, mol/s; S is interaction surface, m^2; L is Onsager's coefficient, mol^2/(J·s·m); gradμ_i is gradient of the chemical potential of i component, J/(mol·m).

All components in the interaction zone in the slag-metal system are equal to the turbulent mass transfer conditions. Therefore, it was assumed that the surface area of interaction, Onsager coefficients, temperature and boundary layer thickness-δ are the same for all reactions. If the coefficient β is:

$$\beta^* = SL\frac{1}{\delta}$$

Then reaction speed of i-component is:

$$V_i = \beta^* RT\ln\frac{K_r}{K_e}$$

Where K_e and K_r are the equilibrium and real reaction constants. An iterative algorithm was developed to calculate the reaction rates of interaction between the components of the slag-metal system. The model determines the direction of chemical reactions for the metal-slag system and is presented as a matrix of k reactions and takes into account the mass and energy balance equations. All interaction zones are described by deterministic rather than statistical dependencies, the models are stable in a wide range of variables and are stable even after technology changes. This software takes into account input data such as: temperature, mass and composition of slag and metal, injection time and mass of additives, purge modes, electrical and time modes. Additional data to be used in the calculations are ladle equipment parameters (ladle geometry, transformer parameters, electrode consumption, number of lances, type of refractory materials), thermodynamic database, thermal, physico-chemical database on additives and inert gas, production database (for statistics). It is shown that the software works stably even after changing the technological scheme. Mathematical model consists of the following blocks:

—Calculation the speed of interaction between the components in the slag-metal system;

—Calculating the amount of metal and slag in the interaction zone depending on the power of stirring of the bath;

—Calculation of the mass of metal and slag;

—Calculation of the chemical composition and temperature of the slag and metal bath.

Calculations of energy balance for metal-slag system and in all areas including arc heating and takes into consideration heats of chemical reactions;

Calculations of heat of metal and slag melts, alloying elements and fluxes;

Heat loss calculations through the lining by radiation, for heating the inert gas and the reacting components at the boundary of the slag-metallining;

Calculations of nonmetallic inclusions formation and removal.

Figure 1 represents of the calculating scheme of the Ladle-Furnace software package.

Figure 2 shows a diagram of the interaction zones of the mathematical model of the RH-degasser. The following zones of interaction are considered:

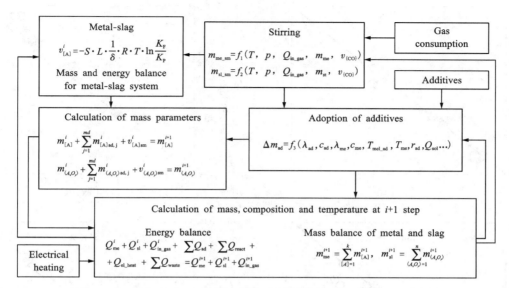

Figure 1 Calculating scheme of the Ladle Furnace software package

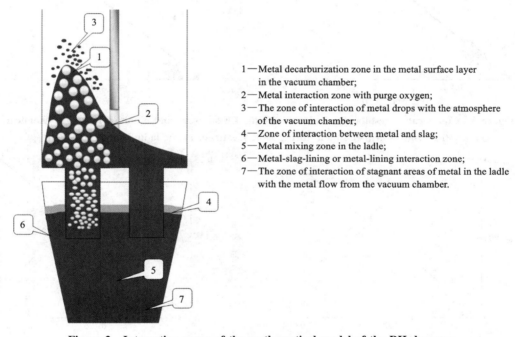

1 — Metal decarburization zone in the metal surface layer in the vacuum chamber;
2 — Metal interaction zone with purge oxygen;
3 — The zone of interaction of metal drops with the atmosphere of the vacuum chamber;
4 — Zone of interaction between metal and slag;
5 — Metal mixing zone in the ladle;
6 — Metal-slag-lining or metal-lining interaction zone;
7 — The zone of interaction of stagnant areas of metal in the ladle with the metal flow from the vacuum chamber.

Figure 2 Interaction zones of the mathematical model of the RH-degasser

3 Results and discussion

To test the software and validate the model, the results of out-of-furnace processing of 25 real steel melts for the pipeline and the results of selective control of the metal composition were used. Comparison of the results of the calculated values obtained using the developed software and the results of monitoring the chemical composition of the metal melt during processing in a 165-ton ladle furnace is shown in Figure 3.

Figure 4 presents a comparison of the calculated data and the results of monitoring the chemical composition of samples of molten metal during processing in a ladle furnace with a volume of 355 t. It is shown that the developed software allows dynamic modeling and optimization of the out-of-furnace processing technology. It has been established that the software adequately describes the dynamic changes in the main characteristics of the metal, slag and the response of the system to control actions in the technological process. It is shown that the developed digital twins for dynamic modelling of ladle processing of steel make it possible

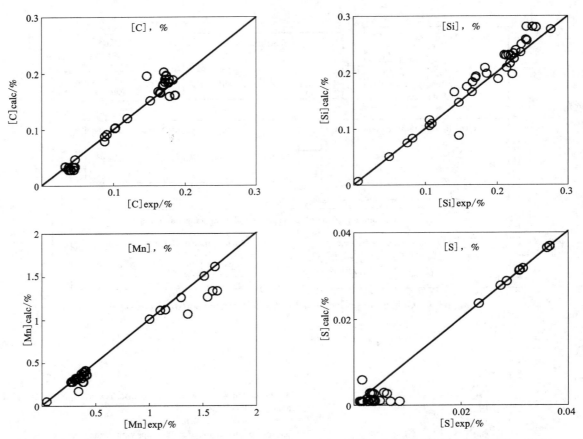

Figure 3 Comparative results of calculations by the LF software and obtained results of chemical composition control of metal melt during ladle treatment at the ladle furnace 165 tonn[2]

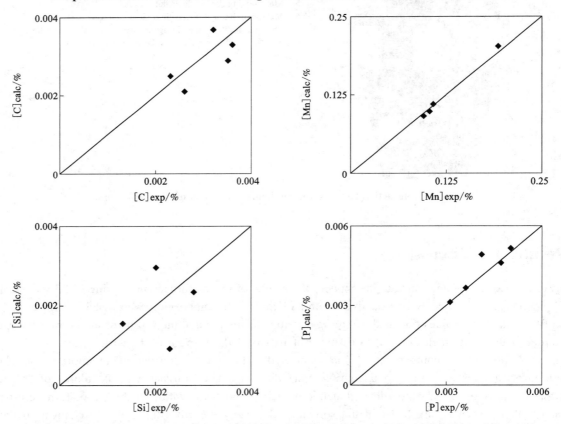

Figure 4 Comparative results of calculations by the LF software and obtained results of chemical composition control of metal melt during ladle treatment at the ladle furnace 355 t

to optimize the technology. The developed software can be used for online calculations and control of technological parameters of ladle processing, modeling and optimization of refining processing technology, education and training of steelmaking personnel.

The adequacy of the developed mathematical model of the RH-vacuum degasser was verified by comparison with the results of industrial melting on a 150-tonne pH degasser. In the process of evacuation, samples of molten metal were taken every 2-3 minutes using special samplers. The results of determining the carbon content in metal samples and comparison with the results of dynamic simulation using the developed software are shown in Figure 5. A good agreement between the calculated and experimental data has been obtained.

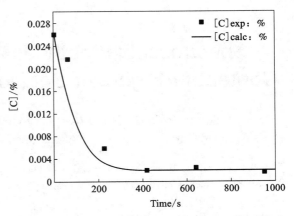

Figure 5　The results of determining the carbon content in metal samples and the simulation results of the RH-degasser

4　Conclusions

Original digital twins and software were developed for dynamic simulation of ladle steel processing installations: ladle furnace and RH degasser. For the development, physical and chemical models based on the laws of conservation of mass and energy, as well as on the principles of non-equilibrium thermodynamics, were used. In this software, all stages of the process (zones) were taken into account. The software for dynamic modelling of ladle processing technologies was used to simulate real industrial melts. It was shown that the developed software allows dynamic simulation of all stages of steel ladle processing technology and conducts the process along the optimal trajectory.

References

［1］ TSYMBAL V P. Mathematical modeling of complex systems in metallurgy: A textbook for high schools［M］. Publishing Association "Russian Universities": Kuzbassvuzuzdat-ASTSh, 2006: 431.
［2］ D'YACHKO A G. Mathematical and simulation modeling of production systems［M］. Moscow: MISiS, 2007: 538.
［3］ SOVETOV B Y A, YAKOVLEV S A. Modeling systems［M］. Higher School, 2001: 343.
［4］ KOMOLOVA O A. Modeling of the components interaction of slag and metal phases in the production of steel, development of algorithms and software for the processes description, Extended Abstract of Cand. Sci. (Tech.) Dissertation［M］. Moscow: Moscow Inst. Steel Alloys, 2014.
［5］ GRIGOROVICH K, KOMOLOVA O, TEREBIKINA D. Analysis and optimization of ladle treatment technology of steels processing［J］. Journal of Chemical Technology and Metallurgy, 2015, 50, 6, 574-580.

Spectroscopic Methods for Determination of Gold Content in High-salt Solutions of Complex Composition

Valentina VOLCHENKOVA[1,2], Evgeny KAZENAS[1], Nadezhda ANDREEVA[1],
Boris TAGIROV[2], Irina ZLIVKO[2], Alexander ZOTOV[2], Vladimir REUKOV[2], Irina NIKOLAEVA[3]

1 Baikov Institute of Metallurgy and Materials Science, Russian Academy of Sciences, Moscow, 119334, Russia
2 Institute of Geology of Ore Deposits, Petrography, Mineralogy and Geochemistry, Russian Academy of Sciences, Moscow, 119017, Russia
3 Department of Geology, Moscow State University, Moscow, 119991, Russia

Abstract: Methods for quantitative determination of gold concentrations by spectroscopic methods of analysis: AAS, AES-ICP and MS-ICP in high-salt solutions of complex composition have been developed. Analytical parameters of the element determination are defined, operating conditions of spectrometers are optimized. The methods of calibration are determined. Matrix interferences and methods of their accounting are revealed. The possibilities of application of AES-ICP, AAS and MS-ICP methods for the analysis of high-salt solutions have been compared. It is shown that the methods of AES-ICP and flame AAS are dominant for determination of element contents from 0.1 μg/mL to 500 μg/mL. The use of the mentioned methods of analysis with complementary analytical capabilities ensured the control of Au content in a wide range of concentrations from 0.0005 μg/mL to 500 μg/mL without preliminary separation of the matrix with good metrological characteristics. The relative standard deviation (S_r) is 0.06-0.005 at Au content from 1 μg/mL to 500 μg/mL and does not exceed 0.18 at element content from 0.0005 μg/mL to 1 μg/mL. The proposed methods were used to analyze gold content in the experimental determination of Au solubility in sulfur-containing hydrothermal solutions at high temperatures and pressures (450 ℃, 1 kbar). The obtained experimental data allowed to calculate thermodynamic constants of stability of hydrosulfide complexes of Au and to estimate the Au content in metal-bearing sulfide fluids and brines. These data will be used to build models of hydrothermal ore formation.

Keywords: Spectroscopic methods of analysis; Gold; High-salt solutions

To carry out studies on the forms of Au transfer and determination of dissolved gold content in hydrothermal fluids in the range of temperatures and pressures at which natural hydrothermal systems exist[1-2], analytical support is necessary: determination of gold content in high-salt solutions of complex composition. In the present work, gold was determined in solutions obtained from Au solubility experiments at 450 ℃, and pressure 1 kbar. The high-temperature experimental fluids contained on average up to 0.2 mol/L H_2S, 0.2 mol/L H_2SO_3 or H_2SO_4, and variable concentrations of KOH (0-1.25 mol/L). The experiments were performed in titanium autoclaves (VT-8 alloy), which complicated the composition of the solutions due to dissolution of the walls. After quenching the experiments, quenching condensates were combined with flushes of the autoclave walls performed with aqua regia. The obtained initial samples for analysis contain H_2SO_4, KCl (up to 5 wt.%), Ti, Ni, Fe on the background of 50% aqua regia. The studied solutions are rather difficult objects of research, both in terms of the influence of the matrix and the wide range of concentrations of the determined element (from $n \times 10^{-3}$ to $n \times 10^2$ μg/mL).

A variety of analysis methods have been proposed for the quantitative determination of gold content: gravimetry, titrimetry, spectrophotometry, atomic absorption spectrometry (AAS); inductively coupled plasma atomic emission spectrometry (ICP-IES); inductively coupled plasma mass spectrometry (ICP-MS). A detailed review of these methods is presented in various monographs[3-7]. The requirements of analytical control for the determination of gold content in high-salt solutions of complex composition are best satisfied by: AAS, AES-ICP and MS-ICP[8-12]. The published works[13-19] show the possibility of determining Au in various compounds by the above methods. The information presented in the papers is not complete, sometimes contradictory and does not reflect some specific problems. The issues of the influence of the matrix composition of complex systems used in experiments on the AAS, AES-ICP and MS-ICP determination of gold were either not considered or are insufficiently covered. Table 1 shows the comparative metrological characteristics of gold determination by various spectroscopic methods of analysis. The main problem is to take into account the influence of the matrix. This problem is solved by separating the determined and matrix elements, which is associated with significant difficulties (each new stage of analysis introduces additional errors by increasing the value of the control experiment)[14, 17-19]. The aim of the present studies was to investigate the influence of the salt background and concentration of various acids on the results of gold determination, to select the optimal operating modes of spectrometers and to evaluate the possibility of direct (without matrix separation) determination of the analyte in complex salt systems.

Table 1 Comparative metrological characteristics of gold determination by various spectroscopic methods of analysis

Analysis method	Method of determination	Detection limits/10^{-6}		Linear concentration interval /($\mu g \cdot mL^{-1}$)	Relative standard deviation, S_r
		1 mol/L HCl	In the background KCl		
AES-ICP	without matrix separation	2.5	5.5	0.006-100	0.15-0.003
	after matrix separation	0.1	0.5	0.0002-20	0.12-0.002
AAS, flame version	without matrix separation	5.0	10	0.005-5	0.18-0.004
	after matrix separation	0.2	1.4	0.00002-2	0.10-0.003
AAS, thermal atomisation	without matrix separation	1		0.001-0.02	0.18-0.004
	after matrix separation	0.1	0.8	0.0005-0.01	0.15-0.003
MS-ICP	without matrix separation	0.05	0.5	0.0002-20	0.15-0.005
	after matrix separation	0.001	0.01	0.00001-20	0.10-0.003

1 The experimental part

1.1 Equipment and reagents

Studies were carried out on atomic absorption spectrometer Thermo Fisher Scientific (USA), model iCE 3000; on sequential atomic emission spectrometer with inductively coupled plasma HORIBA JOBIN YVON (France-Japan), model ULTIMA 2; mass spectrometer with inductively coupled plasma Thermo Scientific (USA), model Element -2. All reagents used in the work corresponded to the qualification "p. a." and above. Standard solutions of gold were prepared from metal of high purity[20] or from Merk fixanal with Au concentration of 1.00 mg/mL.

1.2 Optimization of the analytical parameters

The optimal analytical parameters for AES-ICP and AAS determination of gold were found. The following optimal modes of the plasma spectrometer for determining gold in potassium chloride solution were experimentally established: discharge power 1.2 kW; argon cooling flow 14, transporting 0.70, plasma-forming 0.5 L/min; observation height 14 mm above the upper turn of the induction coil; sample feed rate 1.0 mL/min. The analytical Au wavelength of 242.795 nm was used.

Gold determination by the AAS method was carried out in an air-acetylene flame with a deuterium background corrector. A lamp with a hollow cathode served as a source of resonant radiation. Gold absorption was measured at a monochromator slit width of 0.7 nm at an analytical wavelength of 242.8 nm. The detection limits for gold on other resonant lines are 2-10 times worse[10].

It has been established that the maximum absorption of gold occurs during atomization in an air-acetylene flame depleted in fuel (Figure 1), in which the sensitivity of determining the element is 2 times higher.

In a nitrous oxide-acetylene flame, the gold absorption is several times worse[16]. To increase the analytical signals of gold and eliminate the influence of the sample composition, buffering is used: introduction of an excess of inorganic salts into solutions: lanthanum chloride and others[11]. Interferences can be eliminated or significantly reduced in the analysis of solutions with a low salt background (concentration of related elements from 0.1 mg/mL to 1 mg/mL each)[12]. For the studied solutions, this technique is not always suitable, since the concentration of salts in some is significantly higher.

Table 2 presents the metrological characteristics of the AES-ICP and flame AAS determination of gold in experimental solutions of various compositions.

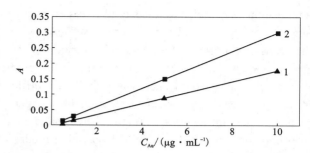

Figure 1 AAC calibration curves for the determination of gold in an air-acetylene flame: stoichiometric (1) and depleted fuel (2)

Table 2 Metrological characteristics of the AES-ICP (λ 242.795 nm) and flame AAS (λ 242.8 nm) determination of gold in experimental solutions of various compositions

Composition of solutions	Gold concentration/($\mu g \cdot mL^{-1}$)						S_r for the optimal range	
	Detection limit		Graph linearity		Optimal range			
	AAS	AES-ICP	AAS	AES-ICP	AAS	AES-ICP	AAS	AES-ICP
1 mol/L HCl	0.005	0.002	0.005-5	0.002-100	0.1-5	0.02-20	0.008-0.004	0.008-0.004
1 mol/L HNO$_3$+HCl	0.005	0.002	0.005-5	0.002-100	0.1-5	0.02-20	0.01-0.004	0.009-0.004
1 mol/L HNO$_3$+3 mol/L HCl+KCl	0.01	0.008	0.01-5	0.01-100	0.2-5	0.05-20	0.02-0.01	0.02-0.009
1 mol/L HNO$_3$+3 mol/L HCl+1 mol/L H$_2$SO$_4$+KCl	0.02	0.015	0.02-5	0.02-100	0.3-5	0.1-20	0.05-0.02	0.04-0.02

The concentration of the matrix element in the solution, at which the stability of the burner-atomizer system operation is not disturbed, has been established. For potassium the favourable concentration is not more than 10 g/L in the flame variant of AAS, not more than 0.5 g/L in AES-ICP and not more than 0.05 g/L in MS-ICP.

1.3 Dependence of gold signal intensity on the nature and concentration of acid

The effect of the nature and concentration of acids on the determination of the element was studied. Figure 2 and Figure 3 show the concentration dependences of Au analytical signals for HCl, HNO$_3$, and H$_2$SO$_4$.

As follows from the experimentally obtained data, a change in the acid concentration in the analyzed solution leads to a change in the analytical signals, and the degree of change depends on the nature and concentration of the acid. Solutions with high acid concentrations have a depressing effect on the analytical signals of gold. High acid concentrations decreased spray efficiency and, as a result, sensitivity. Changes in the analytical signals of the element are associated both with changes in the spray system and with processes occurring in the flame. However, practical error can be avoided by maintaining adequate acid levels in samples and standards. Uncontrolled fluctuations in the concentration of acids, especially sulfuric acid, can lead to errors that significantly exceed the instrumental ones.

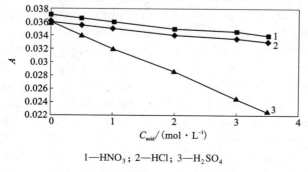

1—HNO$_3$; 2—HCl; 3—H$_2$SO$_4$

Figure 2 Dependence of AAS determination of Au (1 μg/mL) on nature and concentration of acids

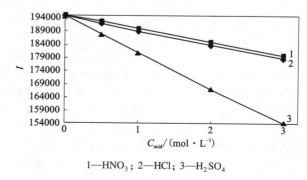

1—HNO$_3$; 2—HCl; 3—H$_2$SO$_4$

Figure 3 Dependence of AES-ICP determination of Au (1 μg/mL) on nature and concentration of acids

1.4 Influence of solution temperature on gold determination

Since the analyzed solutions contained high concentrations of various acids (up to 50%), they were diluted before imaging. The degree of dilution of the final sample solution should ensure that the analytical signals are measured in the optimal concentration range, and that matrix interferences from the sample base components and acids are minimized. A commonly used concentration of the analyzed solution for matrix components (without base separation) is 0.1-5 mg/mL. To avoid gold loss due to acidity reduction, dilution was performed immediately prior to measurement. When diluting test solutions, especially sulphuric acid solutions, they should be cooled and equalized with the temperature of calibration standards. Figure 4 shows the dependences of atomic absorption of gold solutions (1 μg/mL) containing: 1—3 mol/L HCl; 2—3 mol/L HNO_3 +1 mol/L HCl; 3—3 mol/L H_2SO_4 +1 mol/L HCl from temperature. At a temperature change of 5 degrees the atomic absorption of 3 mol/L H_2SO_4 solutions changes by 8%, which is more significant than the instrumental error of measurements not exceeding 1%-2%.

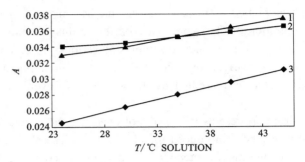

1—3 mol/L HCl; 2—3 mol/L HNO_3 +1 mol/L HCl;
3—3 mol/L H_2SO_4 +1 mol/L HCl

Figure 4 Dependence of atomic absorption (A) of gold solutions (1 μg/mL) on temperature of solutions containing

1.5 Influence of solution compositions on the values of gold analytical signals

The effects of the compositions of the solutions used in the experiments and their concentrations on the values of gold analytical signals were studied. It was found that in the flame variant of AAS the gold absorption is not affected by the content of up to 100 g/L KCl in the solution when using deuterium background corrector. But in order to avoid soiling of the spray system and the burner it is necessary to rinse the system for a very long time after each sample with high salt content (more than 10 g/L). In addition, the metrological characteristics of the analyte determination deteriorate. When using AES-ICP it is allowed no more than 0.5 g/L KCl and no more than 0.05 g/L KCl in gold determination by MS-ICP method. The influence of Ti, Ni and Fe on the analytical signals of gold was studied. At the content of elements at the level of 1-10 μg/mL no influence is shown. The determination of low Au concentrations (less than 0.01 μg/mL) by AAC and AES-ICP methods is slightly influenced by Fe[18].

The detection limits of gold in the presence of matrix were found. The evaluation of the lower limits of the analyte contents to be determined was carried out using a series of comparison solutions against the background of potassium chloride by approximating the power or exponential function of the experimental dependences of the relative standard deviation S_r from the concentration of the element being determined[8]. Table 2 shows the values of the lower limits of analyte concentrations calculated at S_r = 0.33. The detection limits on the matrix background differ from those in pure solutions (1 mol/L HCl) and not for the better (Table 2). The detection limit for Au depends on the composition of the solution, the method of analysis, and varies from 0.0002 μg/mL to 0.50 μg/mL.

As the experimental solutions contain gold in a wide range of concentrations, the linearity of the graphs was studied. It was found that in 0.5 wt.% KCl solution, the calibration graphs are linear in the concentration range from 0.01 μg/mL to 10 μg/mL when determining Au by the AAS method and from 0.01 μg/mL to 100 μg/mL using AES-ICP method.

In order to reduce the influence of the salt background, the solutions were pre-diluted whenever possible. To achieve greater accuracy of the analysis, the determination of gold is preferably carried out in the optimal concentration range (Table 2) of each method of analysis. To reduce the error of determination of low content of the element (less than 0.01 μg/mL), the calibration solutions were prepared identical in composition and acid content with the solutions of the analyzed samples. In the areas of calibration curves with a slight violation of linear dependence, the concentration of gold in the analyzed solution was calculated by measuring the absorbance of the analyte and two nearby standards.

To eliminate matrix interferences in the determination of gold, caused by the changing concentration of the same cationic and anionic composition of experimental solutions, the method of interactive matrix matching was used[6]. All tested solutions were diluted to the same background concentration, on which the calibration graph is plotted. The use of this method made it possible to eliminate the influence of changes in the concentration of the sample base on the results of analysis and to speed up the determination.

Series of experimental samples of various salt composition were analyzed. The measurement of analytical signals was carried out under the found optimal conditions indicated above. The result of the element determination is presented as the average of several values (at least three) measurements of the analyzed solution.

Table 3 presents the results of checking the correctness of the AES-ICP and flame AAS determination of gold using the "introduced-found" method. The absence of a systematic error between the entered and found values of the mass contents of the determined component is shown.

Table 3 Results of checking the correctness of the AES-ICP and flame AAS determination of gold using the "introduced-found" method ($n=10$, $P=0.95$)

№ sample	AES-ICP			Flame AAS		
	Au content found/($\mu g \cdot mL^{-1}$)			Au content found/($\mu g \cdot mL^{-1}$)		
	Initial	Introduced	Initial plus additive	Initial	Introduced	Initial plus additive
1	0.058	0.060	0.012	0.06	0.06	0.012
2	0.172	0.15	0.33	0.164	0.15	0.30
3	0.38	0.40	0.76	0.37	0.40	0.80
4	0.76	0.75	1.52	0.75	0.75	1.45
5	1.02	1.00	2.06	1.02	1.00	1.96

Table 4 presents the results of gold determination in experimental solutions with different potassium chloride content. Due to the lack of standard solutions of the composition of the studied materials to confirm the correctness of the results of the determination of the analyte were compared the data obtained by different methods of analysis: AES-ICP, AAS and MS-ICP. Good convergence of the results of gold determination by different methods within the errors of the methods was obtained.

Table 4 Results of gold determination in experimental solutions obtained by various methods ($n=10$, $P=95$)

№ sample	Sample composition		Au content/($\mu g \cdot mL^{-1}$)		
		+KCl g/L	AES-ICP	Flame AAS	MS-ICP
1		0	0.0022±0.0005		0.0022±0.0005
2		0	0.062±0.004	0.060±0.006	0.056±0.002
3		1.0	0.170±0.006	0.172±0.007	0.164±0.004
4		2.1	0.357±0.006	0.371±0.008	0.368±0.005
5		3.0	0.754±0.016	0.740±0.020	0.762±0.012
6		3.3	1.04±0.02	1.02±0.03	1.02±0.02
7	Ti, Ni, Fe	3.9	1.90±0.09	1.86±0.09	1.96±0.08
8	50% aqua regia	4.8	2.97±0.12	3.14±0.11	2.87±0.10
9	(HCl : HNO$_3$ 3 : 1)+	5.8	5.92±0.12	5.90±0.12	5.80±0.11
10	1 mol/L H$_2$SO$_4$	7.9	14.7±0.24	14.1±0.26	14.9±0.19
11		37	14.0±0.36	14.7±0.36	14.1±0.26
12		10	35.7±0.21	35.8±0.33	35.2±0.32
13		9.7	43.6±0.41	44.0±0.46	43.2±0.45
14		13	175±2.2	180±2.6	172±5.6
15		28	199±2.0	202±2.3	196±5.7
16		20	440±2.3	437±3.6	448±8.6

2 Discussion of the results

The possibilities of using AES-ICP, AAS and MS-ICP methods for the analysis of high-salt solutions are compared (Table 1, Table 4). ICP-MS method is the most sensitive and versatile, but expensive and does not allow the analysis of high-salt solutions without strong dilution. When determining high analyte contents (more than 50 μg/mL), the determination error increases due to the huge dilution. The method is preferable when determining the concentration of gold in the systems under study from 0.0005 μg/mL to 50 μg/mL. AES-ICP method is optimal and dominant in the determination of gold from 10^{-2} μg/mL to $n \cdot 10^2$ μg/mL. The linear section of the calibration dependence of 4-5 orders is an advantage of the method, which allows analysis using a sample solution with a reasonable dilution for solutions with unknown element content, which reduces the time of analytical work. The AAC method has a small dynamic range (2 orders of magnitude), does not allow determining a large analyte concentration interval, but is not burdened by the influence of high concentrations of potassium chloride, and is accessible and quite competitive with AES-ICP in terms of metrological characteristics.

3 Conclusion

The use of AES-ICP, AAS and MS-ICP analysis methods with complementary analytical capabilities, provided the control of gold content in complex solutions by composition and salt content, and determine the element in them in a wide range of concentrations from 0.0005 μg/mL to 500 μg/mL without preliminary separation of the matrix with good metrological characteristics. The relative standard deviation (S_r) is 0.06-0.005 at Au content from 1 μg/mL to 500 μg/mL and does not exceed 0.15 at element content from 0.0005 μg/mL to 1 μg/mL. It is shown that the methods of AES-ICP and flame variant of AAS are dominant at determination of element contents from 0.01 μg/mL to 500 μg/mL. MS-ICP allows to determine gold in the studied systems from 0.0005 μg/mL to 50 μg/mL. The proposed methods were used in experiments to study the forms of gold transfer in high-temperature hydrothermal fluids containing sulfur in various oxidation degrees. The key values of stability constants of Au aqueous complexes have been determined, which is necessary in calculating the solubility of gold under the parameters of fluid-magmatic interaction and in the formation of hydrothermal deposits of this metal.

Acknowledgments: The methodological and research part of the reported study was funded by the Russian Science Foundation grant No. 23-17-00090. Access to the electronic database of the scientific publication was obtained under the state assignment No. 075-01176-23-00.

References

[1] AKINFIEV N N, ZOTOV A V. Thermodynamic description of chloride, hydrosulfide, and hydroxo complexes of Ag(I), Cu(I), and Au(I) at temperature of 25-500 ℃ and pressure of 1-2000 bar[J]. Geochemistry International, 2001,39(10): 990-1006.

[2] ZOTOV A V, TAGIROV B R, KOROLEVA L A, et al. Experimental modeling of Au and Pt Co-transport by chloride hydrothermal fluids (350-450 ℃, 500-1000 bar)[J]. Geology of Ore Deposits, 2017, 59(5): 434-442.

[3] GINZBURG S I, EZERSKAIA N A, PROKOFEVA I V, et al. Analytical chemistry of platinum metals[M]. Moscow: Nauka, 1972: 616c.

[4] BIMISH F. Analytical chemistry of precious metals[M]. Moscow: Mir, 1969, 1: 297.

[5] ZOLOTOVIU A, VARSHAL G M, IVANOV V M. Analytical chemistry of platinum group metals[M]. Moscow: Editorial URSS, 2003: 591.

[6] KARPOVA I A, BARANOVSKOI V B, ZHITENKO L P. Analytical Control of Noble Metals[M]. Moscow: Technosphere, 2019: 400.

[7] BUSEV A I, IVANOV V M. Analytical chemistry of gold[M]. Moscow: Nauka, 1973: 264.

[8] Atomic-emission analysis with inductive plasma. The results of Science and Technology, Ser [J]. Analytical Chemistry, 1990(2): 255.

[9] THOMPSON M P, WALSH J, N. Handbook of inductively coupled plasma spectrometry[M]. New York: Blackie, 1989: 316.
[10] PRAIS V. Analytical atomic absorption spectroscopy[M]. Moscow: Mir, 1976: 358.
[11] PUPYSHEV A A. Atomic absorption spectral analysis[M]. Moscow: Technosphere, 2009: 784.
[12] BRITSKE M E. Atomic absorption spectrochemical analysis[M]. Moscow: Chemistry, 1982: 223.
[13] IUDELEVICH I G, STARTSEVA E A. Atomic absorption determination of noble metals[M]. Novosibirsk: Nauka, 1981: 160.
[14] TIUTIUNNIK O A, NABIULLINA S N, ANOSOVA M O, et al. Determination of trace contents of elements of the platinum group and gold in ultramafic rocks using sorbents AG-X8 И LN-resin by inductively coupled plasma mass spectrometry[J]. J. Analytical Chemistry, 2020, 17(6): 527-536.
[15] SEREGINA I F, BUKHBINDER G L, SHABANOVA L N, et al. Determination of platinum metals by extraction and inductively coupled plasma atomic emission-spectrometry[J]. J. Analytical Chemistry, 1986, 41(5): 681-688.
[16] STOLIAROVA I A, FILATOVA M P. Atomic absorption spectrometry in the analysis of mineral raw materials[M]. Leningrad: Nedra, 1981: 12.
[17] LOSEV V N, MAZNIAK N V, TROFIMCHUK A K, et al. Sorption-photometric determination of gold after its isolation with silica, chemically modified with thiourea derivatives[M]. Diagnostics of Materials, 1998, 64(6): 11-13.
[18] GUSKOVA E A. Extraction-reextraction concentration of platinum metals and gold in AAS-ETA and AES-ICP methods of analysis of technological and geological objects. Abstract of the dissertation for the degree of candidate of chemical sciences[M]. Novosibirsk, 2013: 23.
[19] TORGOV V G, KORDA T M, DEMIDOVA M G., et al. Extraction-reextraction preconcentration in a system based on i-alkylaniline and petroleum sulfides for the determination of platinum metals and gold by inductively coupled plasma atomic emission spectrometry[J]. J. Analytical Chemistry, 2009, 64(9): 901-909.
[20] LAZAREV A I, KHARLAMOV I P, IAKOVLEV P I. Hhandbook of chemist-analyst[M]. Metallurgy, 1976: 121-125.

Effect of Growth Rate on Dendritic Morphology and Secondary Dendrite Arm Spacing of Pd-20W Alloy

Jiming ZHANG, Youcai YANG, Ming XIE, Li CHEN,
Yongtai CHEN, Jiheng FANG, Saibei WANG, Aikun LI

Yunnan Precious Metals Laboratory, Sino-platinum Metals Co. Ltd., Kunming, 650106, China

Abstract: The Pd-20W alloy is a kind of excellent precision resistance alloy with comparatively good comprehensive performances. In view of the wide range of crystallization temperature and the thick dendritic structures, its processing performance drops sharply. The effect of growth rates on Dendrite Morphology and secondary dendrite arm spacing (SDAS) was studied by three different solidification methods. The SDAS was measured by the method of the longitudinal section. The results show that with the increase of growth rate, the primary dendrite is slenderer and denser while the secondary dendrite becomes shorter and less obvious. There is the dendrite segregation of Pd-20W alloy at the different growth rate. As the growth rate increases, the degree of dendritic segregation in grain and in grain boundary decreases. By regression analysis of the curve between the growth rate and the SDAS, the fitting function between the SDAS and growth rate was obtained, It will provide the theory evidence for optimizing melting process, reducing casting segregation and avoiding crack.

Keywords: Pd-20W alloy; Growth rate; Dendritic morphology; Secondary dendrite arm spacing (SDAS)

1 Introduction

The Pd-20W alloy has a high and stable resistivity, a low temperature coefficient of resistance, a low thermoelectric power versus copper and a good corrosion resistance, which is mainly used for high resistance or small precision potentiometer winding materials. There is a growing tendency to use precious metal alloys with precision resistance at present. The melting points of Pd and W of Pd-20W alloy are considerably different, and the temperature range of solidification crystallization is up to 200 ℃[1-2]. The solidification phase is a more developed dendritic framework and a radial growth of the ingot from center to periphery, which cuts off the feeding channel. The casting is easy to produce defects such as the dispersed shrinkage, porosity, segregation, etc., which results in a sharp decline in the processing performance of the alloy. Concerning the main forms of dentritic growth, the characteristics of dendritic structure and the solidification interface have directly influenced the properties of the solidification structure and materials in the process of metal solidification. The dendrite arm spacing in the solidification structure is an important physical parameter, which directly affects the component segregation, the second phase and the distribution of the microvoid. The mechanical properties of the alloy is directly influenced by the secondary dendrite arm spacing (SDAS). In this paper the effect of growth rate on the solidification structure and SDAS of Pd-20W alloy is studied to obtain the relationship between SDAS and growth rate. It was of great significance to provide the oretical basis for the optimization of cast structure of Pd-20W alloy[3-5].

2 Experiment

Using 99.99% Pd and 99.95% W as raw materials, the Pd-20W (mass fraction) alloy was prepared by the nominal alloy composition of the alloy. Firstly, the configured Pd-20W alloy was smelted with the use of a high frequency furnace. The alloy repeatedly melted three times in order to make the ingot fully homogenized. The components of samples were analyzed by inductively coupled plasma atomic emission spectrometry (ICP-AES, PerkinElmer Optima 8000), and the mass fraction of W was 20.85%. Experiments were conducted in a self-made solidification device, the methods of solidification modes are used such as the graphite insulation, the quartz sand insulation and the water-cooled copper mold, the corresponding solidification times are 25.01 s, 7.91 s, 1.35 s respectively. The maximum solidification temperature is 1800 ℃, which is monitored by laser infrared thermometer. The average growth rate were (a) 50 μm/s, (b) 3.2×10^2 μm/s and (c) 1.8×10^3 μm/s respectively. The samples were cut along the axial direction, the section of which were grinded and polished. The corrosion of samples were wiped off by concentrated hydrochloric acid with the solution ratio of 10 g and Chromium anhydride with the solution ratio of 1 g. Then the dendritic structure was obtained. The morphology was observed by the scanning electron microscope (HITACHI S-3400N+EDS) and the optical microscope (LEICA DM4000M).

3 Experiment results and discussion

3.1 Dendritic morphology

The microstructures of Pd-20W alloy with different growth rates are shown in Figure 1 and Figure 2, respectively. Figure 1 is in longitudinal section and Figure 2 is in transverse section. When the growth rate is low [Figure 1(a) and Figure 2(a)], the primary dendrite is thick and big with irregular distribution and staggered growth. The secondary dendrite is relatively developed with a short rod distribution in the primary dendrite. When the growth rate is high [Figure 1(c) and Figure 2(c)], the primary dendrite is relatively small, straight and inter-parallel. The growth of secondary dendrite is strongly inhibited with an dense point distribution in the primary dendrite[6]. The effect of the dendrite refinement is very significant.

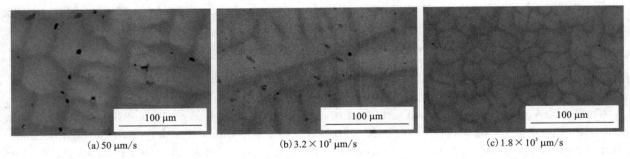

(a) 50 μm/s　　(b) 3.2×10^2 μm/s　　(c) 1.8×10^3 μm/s

Figure 1　Microstructures of Pd-20W alloy in longitudinal section at growth rates

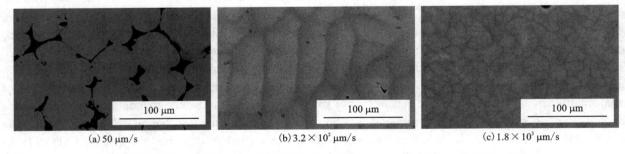

(a) 50 μm/s　　(b) 3.2×10^2 μm/s　　(c) 1.8×10^3 μm/s

Figure 2　Microstructures of Pd-20W alloy in transverse section at growth rates

3.2 Dendritic segregation and Microvoid

It is seen from the Table 1 that there is the dendrite segregation of Pd-20W alloy at different growth rates and there are significant componental differences in the grain and in the grain boundary. As the growth rate increases, the componental differences in the grain boundary and in the grain can be reduced. Moreover, the high growth rate has a certain effect on the inhibition of the dendritic segregation.

Table 1 The composition distribution of Pd-20W alloy

The cooling rates /(μm·s^{-1})	W content/%		Difference in W content/%
	In grain	In grain boundary	
50	22.62	14.56	8.06
3.2×10^2	22.00	15.35	6.65
1.8×10^3	20.89	15.94	4.95

Figure 3 illustrates a typical microstructure of Pd-20W alloy at the growth rate of 50 ℃/s. The circular cavities are distributed along the grain boundaries, and there are nothing in the grain. The sizes of cavities are about 4.0 μm and the point A belongs to the micro-hole which is located in the inner side of the cavity. The size is 1.52 μm. According to the characteristics of the solidification of peritectic alloy, it can be known that the peritectic reaction occurs after the primary phase and the liquid phase. The surface of the primary phase is covered with a layer of peritectic phase which hinders the direct contact of the primary phase and the liquid phase. The peritectic reaction is terminated, subsequently the growth of the peritectic phase can only be spread by solid phase diffusion (known as peritectic transformation) to the primary phase and the direct solidification of a subcooled liquid. Since it is difficult to find the primary presence in metallographic tissue, the large numbers of circular rod Pd-20W samples were for destructive fracture analysis to find the primary presence, Finally, the initial phase is found from the fracture morphology, as shown in Figure 4. A half-empty cardiac sphere was observed, about 8 μm in diameter. After the energy spectrum analysis, the mass score of W in the small circular sphere is 79.3%, The W content is much higher than the nominal component in the Pd-20W alloy, but is very close to the lowest W content in the primary phase which is 97%. On the basis of the above theory, it is deduced that the microhole is the primary phase (W) and the cavity is the peritectic phase. The cause of the above results is that the reaction is not completed under the non-equilibrium solidification. Once the W phase is separated from the liquid, the peritectic reaction is terminated. Since the growth rate is 50 μm/s, the W is not completed in the process of peritectic reaction at a lower growth rate, which causes the phenomenon that part of the W remains.

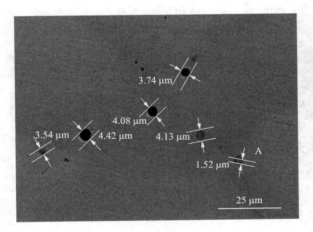

Figure 3 The typical microstructure of Pd-20W alloy at 50 μm/s

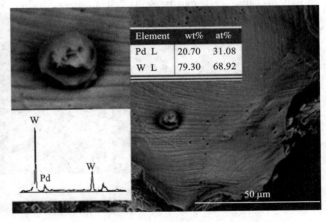

Figure 4 The leading-phase of Pd-20W alloy

3.3 The relationship between the growth rate and the SDAS

As for the parallel arrangement of the dendrite, the spacing between the dendrites is defined as a primary dendrite arm spacing (λ_1), and the spacing between each dendrite branch is defined as the SDAS(λ_2). λ_2 and its measuring method of Pd-20W alloy are shown in Figure 5(a). It can be clearly seen from Figure 5(b) that the fracture morphology of Pd-20W alloy presents very obvious dendrite structure at 50 μm/s, and it is obvious of the distribution of primary dendrite arm spacing and the SDAS. λ_2 is chosen as a measurement on the longitudinal section. In the experiment, the dendrite arm spacing is determined by using a calibrated eyepiece to observe the axial profile of the sample. The number (N_r) of the dendrite is measured with the

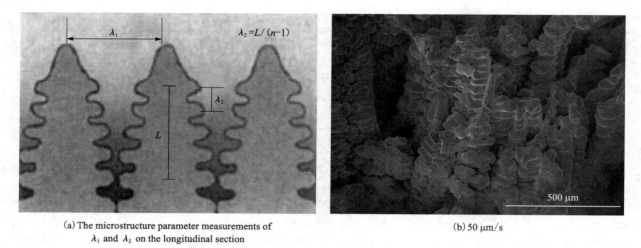

Figure 5 The schematic illustration of the SDAS and the fracture microstructures of Pd-20W alloy

intersection of the scale line, and λ_2 is calculated by using formula (1)[7][8]:

$$\lambda = \frac{L_r}{N_r - 1} \tag{1}$$

Where L_r is the total scale of the eyepiece. The average value of the dendrite arm is measured by the number of different regions in the same sample.

The microstructure of alloy at the half position from the center of the ingot under three growth conditions was selected as the basis for statistical SDAS parameters, as shown in Figure 6. As can be seen from Figure 6, in order to count the number of parameter groups and facilitate measurement, the scale of the microscopic morphology picture was finally selected as 200 μm. With the increase of growth rate, dendrites of the alloy not only become very fine and dense, but also have certain directionality. In general, the SDAS has a great influence on mechanical properties. λ_2 is determined by the growth rate and the characteristics of the alloy. Regarding a certain composition of the alloy, the main factor is the growth rate. The statistical results of λ_2 are shown in Table 2. It can be seen that λ_2 gradually increases with the decrease of growth rates, which indicates a trend of coarsening.

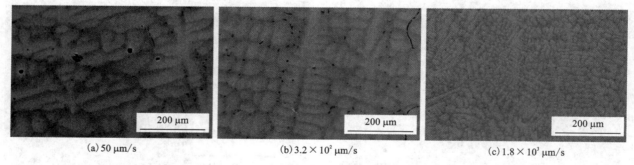

Figure 6 Microstructures of Pd-20W alloy in longitudinal section at growth rates

Table 2 The statistical sizes of the SDAS

The growth rate /(μm·s^{-1})	The point	L_r/μm	N_r/piece	λ_2/μm	The average value of λ_2/μm
50	1	259.45	6	51.89	61.68
	2	271.28	5	67.82	
	3	262.38	5	65.60	
	4	126.4	3	63.20	
	5	179.81	4	59.94	

The growth rate /(μm·s^{-1})	The point	L_r/μm	N_r/piece	λ_2/μm	The average value of λ_2/μm
3.2×10^2	1	283.05	11	28.31	39.59
	2	179.77	7	29.96	
	3	221.13	6	44.23	
	4	173.58	4	57.86	
	5	150.33	5	37.58	
1.8×10^3	1	80.8	8	11.54	14.00
	2	113.45	9	14.18	
	3	66.91	7	11.15	
	4	53.4	4	17.80	
	5	61.39	5	15.35	

The cause of the coarsening maybe that some secondary dendrite arms formed at the beginning become unsteady at the later stage of solidification so that this dendrite arm begins to dissolve when other dendritic arms are still growing. In the process of the dendrite growth, different curvatures of the SDAS lead to the difference in the liquid phase solute concentration near each dendrite arm. The curvature radius of the dendrite becomes shorter with the increase in growth rate. Furthermore, due to the presence of a solute concentration gradient, the solute will be redistributed and it will diffuse from the coarse dendrite arm to the fine dendrite arm along the concentration gradient, while the solvent will diffuse from the fine dendrite arm to the coarse dendrite arm. This will cause the fine dendrite arm to dissolve and the thick dendrite arm to become coarser. As a result, the longer it remains in the solid-liquid phases, the more sufficient the process is and the bigger SDAS becomes[9].

The relationship between SDAS and the growth rate is consistent with the exponential relationship represented in formula(2)[10][11]:

$$\lambda_2 = AV^{-n} \quad (2)$$

Where λ_2 is the SDAS(μm), A and n indicate a constant respectively, V is the growth rate.

The fitting curve is obtained by data fitting and regression processing as shown in Figure 7. According to its slope and intercept, the values of material constants n and A can be obtained as follows: $n = -0.412$, $A = 5.837$. The fitting function between the SDAS and the growth rate of Pd-20W alloy was obtained from the fitting curve: $\lambda_2 = 15.867 V^{-0.412}$. As shown in Figure 7, the growth rate has a significant effect on the dendrite spacing. The greater the growth rate and the crystallization degree of super-cooling are, the higher the dendritic nucleation rate is and the less the time of solidification is. Consequently SDAS is smaller. In addition, the

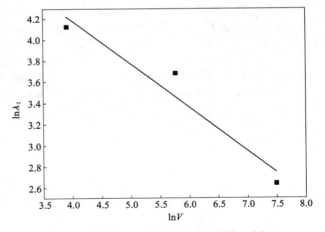

Figure 7 Relationships between ln$V$ and lnλ_2

high growth rate has a certain effect on the inhibition of SDAS. He[11] et al. showed that with the decrease of growth rate, the secondary dendrite arm coarsened, and the SDAS increased. Hosseini[12] et al. showed that the spacing of secondary dendrite arms decreased with the increase of growth rate, and the ultimate shear stress, yield shear stress and normalized displacement were improved at higher growth rate. Sun[10] et al. obtained the empirical formula of growth rate and secondary dendrite arm spacing, which provided theoretical basis for judging the cooling temperature of castings, establishing the prediction mechanism of high temperature diffusion annealing, and reducing the segregation of castings.

4 Conclusion

(1) With the increase of growth rates, the longitudinal primary dendrite is slenderer and denser while the secondary dendrite becomes shorter and less obvious. In a certain range, the higher growth rate can significantly refine the dendrite.

(2) There is the dendrite segregation of Pd-20W alloy at different growth rates. As the growth rate increases, the degree of dendritic segregation in the grain boundary and in the grain decreases. The high growth rate has a certain effect on the inhibition of the dendritic segregation.

(3) The relationship between the SDAS and the growth rate is consistent with the exponential relationship shown below: $\lambda_2 = 15.867 V^{-0.412}$. It is found that the fine dendritic structure can be obtained by the reasonable growth rate, which can improve the comprehensive properties of Pd-20W alloy.

References

[1] YIN J M, GUO X M, SHEN L Q, et al. Research on microstructure and properties of Pd-20W alloys[J]. Precious Metals, 2013, 34(2): 20-25.

[2] ZHANG J M, XIE M, YANG Y C. Phase selection during solidification of Pd-W20 peritectic alloys[J]. Precious Metals, 2014, 35(S1): 72-76.

[3] WU Y H, LI Y T, JIA L, et al. Effect of cooling rate on secondary dendrite arm spacing and crack defects of 42Cr Mo casting[J]. Journal of Mechanical Engineering, 2014, 50(16): 104-111.

[4] JABBARI BEHNAM M M, DAVAMI P, VARAHRAM N, et al. Effect of cooling rate on microstructure and mechanical properties of gray cast iron[J]. Materials Science and Engineering: A, 2010, 528(2): 583-588.

[5] TURHAL M S, SAVASKAN T. Relationships between secondary dendrite arm spacing and mechanical properties of Zn-40Al-Cu alloys[J]. Journal of Materials Science, 2003, 38: 2639-2646.

[6] RAMESH G, NARAYAN K. Themml analysis and microstructure of ZA8 alloy solidifying against chills[J]. Trans Indian Inst Met, 2012, 65(6): 719-723.

[7] CHE J B, LIAO D M, SUN F, et al. Simulation calculation of secondary dendritic arm spacing during solidification of aluminum alloy casting[J]. Foundry, 2020, 69(4): 382-387.

[8] VANDERSLUIS E, RAVINDRAN C. Comparison of measurement methods for secondary dendrite arm spacing[J]. Metallography, Microstructure and Analysis, 2017, 6: 89-94.

[9] LIU J F, LI R D. Research progress on dendrite arm spacing of ZA alloy[J]. Foundry, 2013, 62(10): 958-963.

[10] SUN D R, ZHANG H, MEN Z X, et al. Simulation research on the effect of cooling rate on secondary dendrite arm pacing in casting[J]. Heavy Casting and Forging, 2014, 04: 1-4.

[11] HE J, ZOU Y Z, HUANG W M, et al. Effect of cooling rate on secondary dendrite arm spacing of ZL114A alloy[J]. Foundry Technology, 2008, 29(90): 1213-1216.

[12] HOSSEINI V A, SHABESTARI S G, GHOLIZADEH R, et al. Study on the effect of cooling rate on the solidification parameters, microstructure, and mechanical properties of LM13 alloy using cooling curve thermal analysis technique[J]. Materials & Design, 2013, 50: 7-11.

Investigation into the Effect of Precursor Solution pH on the Phase Structure of the YBCO Superconducting Target

Wenyu ZHANG[1], Benshuang SUN[1,2], Xiaokai LIU[1], Huiyu ZHANG[1], Hetao ZHAO[1], Yongchun SHU[1,2], Yang LIU[1,2], Jilin HE[1,2]

1 School of Materials Science and Engineering, Zhengzhou University, Zhengzhou, 450001, China
2 Zhongyuan Critical Metals Laboratory, Zhengzhou, 450001, China

Abstract: In this work, $YBa_2Cu_3O_{7-x}$ (YBCO) compound powders were prepared using the oxalic acid coprecipitation method under a pH value range of 3-7, and the YBCO superconducting targets were obtained by further calcination and sintering. The effects of pH on the microstructure and phase composition of the YBCO superconducting targets were investigated using scanning electron microscopy (SEM), X-ray diffraction (XRD), thermogravimetric-differential scanning calorimetry (TG-DSC), and energy dispersive spectrometry (EDS). The results showed that the composition and morphology of the compound powders were significantly affected by the pH of the precursor solution, and the three compound powders with different pH values were composed of two phases consisting of lamellar and spherical particles. The number and size of the lamellar nanoparticles increased with an increase in pH. The XRD and EDS results indicated the Y_2BaCuO_5 green phase and CuO secondary phase in the YBCO superconducting target, prepared at pH values of 3 and 5, in addition to the superconducting phase $YBa_2Cu_3O_7$. The YBCO superconducting target with a pH value of 7 did not contain the $YBa_2Cu_3O_7$ phase, and only contained the Y_2BaCuO_5 and CuO phases.

Keywords: YBCO compound powders; Precursor solution; pH; Superconducting phase; Targets

1 Introduction

The discovery of Hg superconductivity in 1911 initiated the study of superconductivity[1], and subsequently, metal element superconductors such as Pb[2,3], Sn, Nb[4], and Th were found. Binary compounds with superconductivity were shown to further improve the critical transition temperature (T_c). Superconductors such as Nb_3Sn[5], Nb_3Ge[6], and La[7] have successively increased T_c to a maximum value of 23 K. Early superconducting materials, known as low-temperature superconducting materials, required the use of liquid helium as a refrigerant, due to a low critical transition temperature. The application of low-temperature superconducting materials has been subjected to the high cost of liquid helium. In 1987, the discovery of $YBa_2Cu_3O_{7-x}$[8] (YBCO or Y-123 for short) changed this situation, and its critical transition temperature T_c reached 92 K. Since then, high-temperature superconductors such as Bi-based[9-11], Tl-based[12-14], Hg-based[15-17], and MgB_2[18-20] have been discovered. Due to the advantages of a high critical transition temperature T_c, high critical current density J_c, high critical magnetic field H_c, low anisotropy, low alternating current loss rate, and high mechanical strength, Y-based superconductors have attracted widespread attention in the industry. These superconductors have been widely used in microelectronics, transmission, superconducting devices, magnetic energy storage, and medical testing[21-24].

Theoretically, YBCO has a typical perovskite structure (space group Pmmm) with a lattice constant of $a = 0.3817$ nm, $b = 0.3883$ nm, and $c = 1.1633$ nm. Changes in the x value in the molecular formula between 0 and 1 (i.e., the change of oxygen content) will lead to the formation of two different structural phases of YBCO, namely, the orthorhombic phase $YBa_2Cu_3O_7$ and tetragonal phase $YBa_2Cu_3O_6$. When the x value is close to 0, the material will have an orthogonal phase with superconducting properties, which can be regarded as the orderly accumulation of three anoxic perovskite-type structural units along the Z axis. When the x value is oriented to 1, the crystal structure will change from the orthorhombic phase to the tetragonal phase, with the elimination of superconductivity[25-29].

For mechanism research and market application of superconducting properties, high-temperature superconducting targets with excellent performance need to be prepared, requiring the advantages of fine particle size, uniform distribution, and high purity of the target. The preparation process of the powder will directly determine the performance of the target. Therefore, finding the most suitable preparation and synthesis route is particularly important. Current preparation methods of YBCO high-temperature oxide superconducting powder can be generally divided into two categories: the solid phase method[30-32] and the chemical wet method[32-35]. The solid phase method will adopt the traditional sintering process, with the directly mixed powder by calcination to obtain the final product. This has the significant advantage of simple operation. However, it also has obvious disadvantages, such as a long production cycle, uneven particle size, low purity, and large particle size. The chemical wet method can produce nano-sized powders with good uniformity and high purity.

Research on the preparation of the YBCO superconducting target has mainly focused on the optimization of powder treatment and the effect of the sintering process on the target[36-39]. The precursor of the YBCO superconducting target has been mainly prepared using the traditional solid phase method[30-32, 37]. Several studies have been reported on the preparation of pure and fine YBCO precursors using the oxalic acid co-precipitation method[33, 40, 41], and the process conditions of preparing YBCO precursors by oxalic acid coprecipitation have been discussed. However, analysis of the effect of precursor solution pH on the microstructure and composition of YBCO precursor is lacking, along with the composition and electrical properties of the final synthesized YBCO superconducting target.

In this study, YBCO precursor powder was prepared using the oxalic acid coprecipitation method, and the YBCO superconducting target was obtained by further calcination and sintering. The effects of pH value of the precursor solution on the growth, morphology, and superconducting properties of the YBCO superconducting targets were investigated, and the thermal characteristics of different temperature stages and phase identification were studied using thermogravimetric-differential scanning calorimetry (TG-DSC) and X-ray diffraction (XRD). Subsequently, the microstructure and chemical composition of the material were systematically analyzed using scanning electron microscopy (SEM) with energy dispersive spectrometry (EDS).

2 Experimental methodology

2.1 Preparation of YBCO compound powders and targets

High-grade pure Y_2O_3, $BaCO_3$, and CuO were weighed according to a molar ratio of Y:Ba:Cu = 1:2:3 and dissolved in concentrated nitric acid to produce an aqueous nitrate solution. The prepared oxalic acid solution (1.2 M) was slowly added to the above-mixed solution while continuously stirring to form complete precipitation. The oxalate coprecipitation solutions were adjusted to pH values of 3, 5, and 7 using ammonia solution (33wt%), and then thermally aged at 80 ℃ for 10 h. The compound powders of YBCO-3 (pH=3), YBCO-5 (pH=5), and YBCO-7 (pH=7) were centrifuged; dried at 100 ℃ for 24 h; and calcinated at 930 ℃ for 12 h. Subsequently, the calcined powders were pressed into green compacts at a pressure of 6 MPa and then cold and isostatically pressed at a pressure of 250 MPa for 10 min. The green compacts were sintered at 940 ℃ for 12 h in oxygen, and the heating and cooling rate was 3 ℃/min. The sintered YBCO superconducting targets were prepared after annealing at 500 ℃ for 2 h. The YBCO superconducting target preparation flowchart is shown in Figure 1.

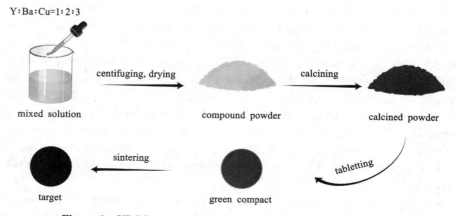

Figure 1　YBCO superconducting target preparation flowchart

2.2 Characterization of the YBCO sample

The thermal stability of the compound powders was tested using TG-DSC (SDT Q600, USA), and the test temperature was increased from 25 ℃ to 1300 ℃ in an air atmosphere at a heating rate of 10 ℃/min. In addition, the crystallinity and phase compositions of the compound powders and sintered targets were analyzed with XRD (D8, Bruker, Karlsruhe, Germany) using a CuK_α radiation source. SEM (Quanta 250 FEG, USA) was used to investigate the grain morphology and elemental composition of the YBCO powders and targets. After grinding and polishing, the sintered targets were analyzed with EDS (INCA x-act, Germany), and the relative density of the YBCO superconducting targets was calculated by the Archimedes method. A precision four-probe resistivity meter (HPS2663) was used to measure the resistivity at room temperature ($\rho_{300\,K}$) of the YBCO superconducting targets.

3 Results and discussion

3.1 Morphology and phase structure analysis of the original and compound YBCO powders

Figure 2 shows the morphology and phase structure analysis of the purchased original powders. Figure 2(a)-(c) presents the XRD diffraction patterns of the Y_2O_3, $BaCO_3$, and CuO particles, which were entirely consistent with the PDF#41-1105, PDF#05-0378, and PDF#48-1548 standard card diffraction peaks, respectively. The micromorphology and particle size distribution of three original powders are shown in Figure 2(d)-(g), indicating that the Y_2O_3 particles agglomerated into lamellar substances, and the particle size of D(50%) was 1051.00 nm. The $BaCO_3$ nanoparticles agglomerated and were rod-shaped, with anaverage size of 629 nm. The original CuO particles presented irregular blocks of different sizes, and the particle size distribution range was mainly 462-2818 nm. The particle size was mainly distributed at approximately 1152 nm [D(50%)].

A comparative experiment was performed in which we prepared compound powders with different pH values by the oxalic acid coprecipitation method. As shown in the SEM image in Figure 3, the particle size of the compound powders prepared using the oxalic acid coprecipitation method was relatively large, regardless of the pH value of the precursor solution. Two different morphologies were observed in the YBCO compound powders. One consisted of irregular spherical particles with a particle size between 0.2 and 2 μm, and the other had a larger-sized lamellar substance. With an increase in pH value, the number of larger-sized lamellar substances in the compound powders increased, the crystallite size also increased, and the size of the YBCO-7 compound powders could reach 20 μm. Therefore, the precursor solution of the pH significantly affected the morphology and size of the compound powders prepared using the oxalic acid coprecipitation method.

3.2 Thermal behavior and phase analysis of the YBCO

The TG-DSC method was used to thermally analyze the compound powders and study the main reaction temperature during the YBCO phase formation process, as shown in Figure 4, presenting the TG-DSC curve of the YBCO compound powders at temperatures of 25-1300 ℃. We observed four endothermic peaks at 85.88 ℃, 287.66 ℃, 424.98 ℃, and 889.23 ℃, and the phase formation process of the compound powders could be divided into four stages. The first stage occurred before 276 ℃, which corresponded to 10.14 wt% weight loss.

In this stage, the endothermic peak at 85.88 ℃ corresponded to the crystalline water leaving from the lattices of the oxalate salts. The temperature ranged from 276 ℃ to 344 ℃ in the second stage, and the mass loss was 22.72 wt%. The main reaction consisted of the decomposition of the oxalate salts to the corresponding carbonates (the initial reaction temperature was 287.66 ℃). The third stage occurred from 344 ℃ to 685 ℃, and the mass loss was 13.95 wt%. The peaks at 424.98 ℃ could be attributed to the decomposition of the carbonates and to their corresponding oxides. In the fourth stage, the mass loss of the compound powders was 5.35 wt% at temperatures of 685 ℃ to 962 ℃[33]. This reaction process mainly consisted of superconducting phase $YBa_2Cu_3O_7$ formation, according to an endothermic peak of 889.23 ℃.

Figure 5 shows the XRD patterns of the compound powders prepared by hydrothermal treatment at 90 ℃ for 10 h, the calcined powders at 930 ℃ for 12 h, and the superconducting targets sintered at 940 ℃ for 12 h.

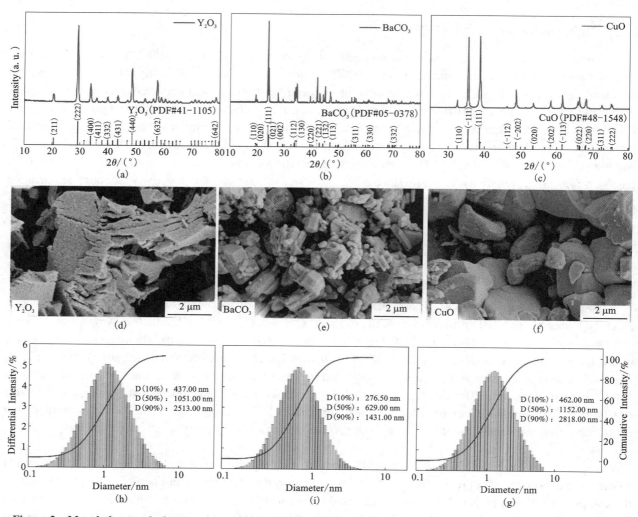

Figure 2 Morphology and phase structure analysis of the original powders, where (a), (b) and (c) correspond to XRD patterns of Y_2O_3, $BaCO_3$, and CuO, respectively, (d), (e), and (f) correspond to SEM morphologies of Y_2O_3, $BaCO_3$, and CuO, respectively, (h), (i) and (g) correspond to the particle size distribution of Y_2O_3, $BaCO_3$, and CuO, respectively.

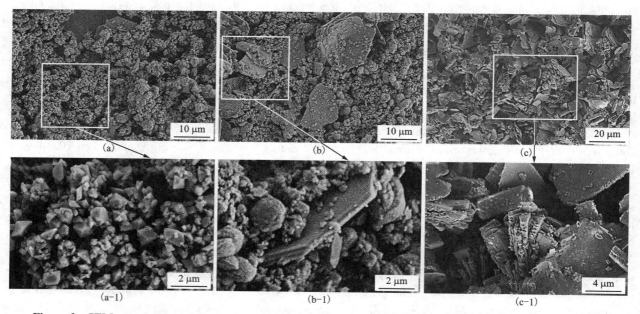

Figure 3 SEM morphology of the YBCO compound powders, where (a) and (a-1) correspond to YBCO-3, (b) and (b-1) correspond to YBCO-5, and (c) and (c-1) correspond to YBCO-7.

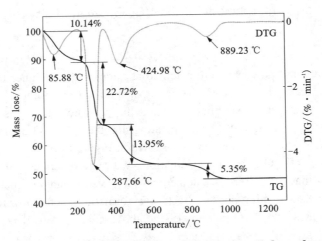

Figure 4 TG-DSC diagram of the YBCO compound powders.

Figure 5(a) shows that the superconducting phase $YBa_2Cu_3O_7$ with perovskite structure formed in the YBCO-3 and YBCO-5 calcined powders after calcination. The prominent diffraction peaks corresponded to the (013) and (103) crystal orientations of $2\theta = 32.54°$ and $2\theta = 32.83°$, which was consistent with the standard card PDF#78-1312. However, in addition to the superconducting phase $YBa_2Cu_3O_7$, we also observed the Y_2BaCuO_5 and CuO secondary phases, which corresponded to standard cards PDF#38-1413 and PDF#48-1548, respectively. Furthermore, no diffraction peak of the corresponding superconducting phase $YBa_2Cu_3O_7$ was detected in the YBCO-7 calcined powders, indicating that it was not conducive to the synthesis of the Y-123 phase under this pH condition.

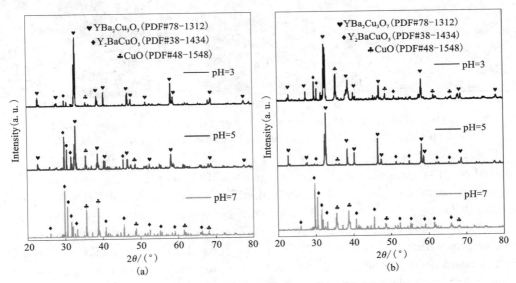

Figure 5 XRD pattern of YBCO with different pH values

Figure 5(b) shows that the sintered YBCO-3 and YBCO-5 superconducting targets did not obtain a single superconducting phase, and secondary phases remained. The diffraction peak of the superconducting phase $YBa_2Cu_3O_7$ of the sintered YBCO-5 superconducting target was enhanced, and the diffraction peak intensity of the secondary phases was reduced. This indicated that further sintering could promote the formation of superconductivity, while the YBCO-3 superconducting target showed the opposite phenomenon. YBCO-7 did not form the $YBa_2Cu_3O_7$ phase throughout the heat treatment process, and only the Y_2BaCuO_5 and CuO phases formed. This could be explained by the high pH of the precursor solution leading to incomplete precipitation of the Cu^{2+} and Ba^{2+} ions, which made the ratio of the three ions deviate from the strict stoichiometric ratio. The precipitation conditions at a pH of 5 could make the three ions precipitate to entirely obtain the $YBa_2Cu_3O_7$ phase.

3.3 Microstructure and grain morphology of the YBCO calcined powders and superconducting targets

According to the TG-DSC curve shown in Figure 4, the compound powders started to phase transition at 889 ℃. Therefore, the compound powders were calcined at a temperature higher than the above temperature, and the compound powders were calcined at 930 ℃ for 12 h in an air atmosphere. As shown in Figure 6, the size of the compound powder increased significantly, and the ceramic block was sintered. The solid blocks with different sizes are presented in Figure 6(a), indicating that the YBCO-3 compound powders were almost entirely sintered into solid blocks after calcination. As shown in Figure 6(b), the YBCO-5 compound powders formed partial solid blocks of $YBa_2Cu_3O_7$ after calcination, with agglomerated particles that were not entirely sintered. Figure 6(c) shows that the YBCO-7 calcined powders still contained many un-sintered flake phases filled with voids, and no superconducting phase $YBa_2Cu_3O_7$ formed, which confirmed that no superconducting phase $YBa_2Cu_3O_7$ was present in the XRD patterns of the above-calcined powders.

Figure 6 SEM morphology of the YBCO calcined powders, where (a) and (a-1) correspond to YBCO-3, (b) and (b-1) correspond to YBCO-5, and (c) and (c-1) correspond to YBCO-7.

The calcined powders were used to prepare the green YBCO compacts. The compacts were sintered at 940 ℃ in an oxygen atmosphere for 12 h and annealed at 500 ℃ for 2 h. Table 1 shows the relative density and ρ_{300K} of the YBCO superconducting targets. As shown in Figure 7(a) and (a-1), the grains grew with fewer holes, and the relative density reached 90.61%. The YBCO-5 superconducting target had almost no holes, and the density further improved to 92.58% [Figure 7(b) and (b-1)]. Figure 7(c) and (c-1) shows that the YBCO-7 superconducting target still exhibited a state of particle aggregation, the grains did not grow with many pores, and the relative density was only 85.45%.

Table 1 Relative density and $\rho_{300\ K}$ of YBCO superconducting targets

Sample	pH	Relative density/%	$\rho_{300\ K}/(\Omega \cdot m)$
YBCO-3	3	90.61	1.884×10^{-3}
YBCO-5	5	92.58	3.351×10^{-4}
YBCO-7	7	85.45	3.843×10^{2}

Figure 8 shows the surface element distribution map of the YBCO-3 superconducting target, and Figure 8(a) shows the three regions with different hues, which indicated the existence of three phases where the Y,

Figure 7 SEM cross-section of the YBCO superconducting targets, where (a) and (a-1) correspond to YBCO-3, (b) and (b-1) correspond to YBCO-5, and (c) and (c-1) correspond to YBCO-7

Ba, and Cu elements were unevenly distributed [Figures 8(c)-(e)]. The results of combining the SEM and EDS data distributions of the YBCO-3 superconducting target (Figure 9) indicated that there were indeed three phases with different compositions. They were the $YBa_2Cu_3O_7$ main phase corresponding to the light color region 1, the green phase of Y_2BaCuO_5 corresponding to the bright color region 2, and the CuO secondary phase corresponding to the dark color region 2.

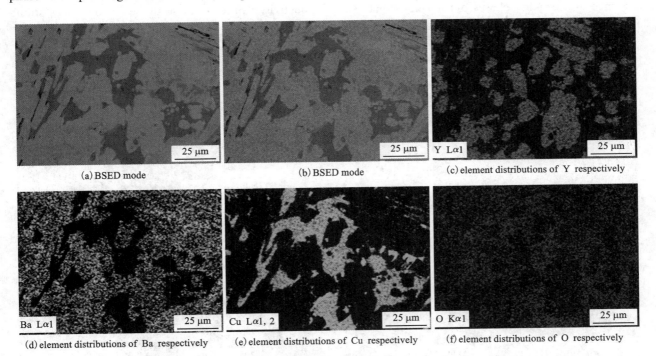

Figure 8 EDS diagrams of the YBCO-3 superconducting target

Figure 10 shows the surface element distribution map of the YBCO-5 superconducting target. Combining the SEM and EDS data distribution of the YBCO-5 superconducting target (Figure 11) indicated that three phases were observed with different components. The phase composition was the same as the YBCO-3 superconducting target, namely, $YBa_2Cu_3O_7$, Y_2BaCuO_5, and CuO, but the distributions of the secondary

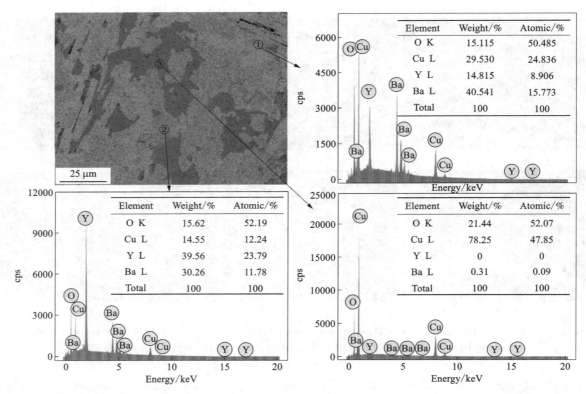

Figure 9 SEM image and EDS spectra and chemical compositions (tables) of the YBCO-5 superconducting target

phases decreased obviously. This was consistent with the conclusions from the above XRD patterns. We observed that the YBCO superconducting target was not composed of a single phase $YBa_2Cu_3O_7$, and the presence of secondary phases did not deteriorate superconductivity. Instead, these phases could act as strong pinning centers and increase the critical current density, J_c.

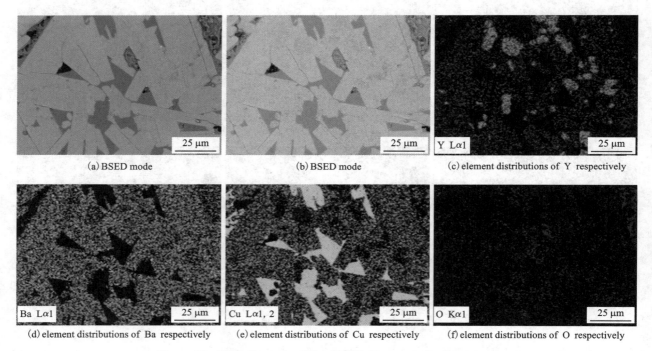

Figure 10 EDS diagrams of the YBCO-5 superconducting target

Figure 12 shows the surface elemental distribution of the YBCO-7 superconducting target, and Figure 12(a) shows the two areas with different shades of color, indicating two phase distributions. Combining the

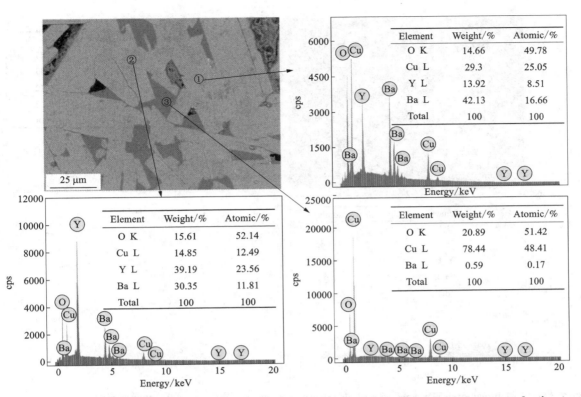

Figure 11 SEM image and EDS spectra and chemical compositions (tables) of the YBCO-5 superconducting target

SEM and EDS data distributions of the YBCO-7 superconducting target in Figure 13 indicated that the CuO secondary phase corresponded to dark area 1 and the Y_2BaCuO_5 green phase corresponded to light area 2.

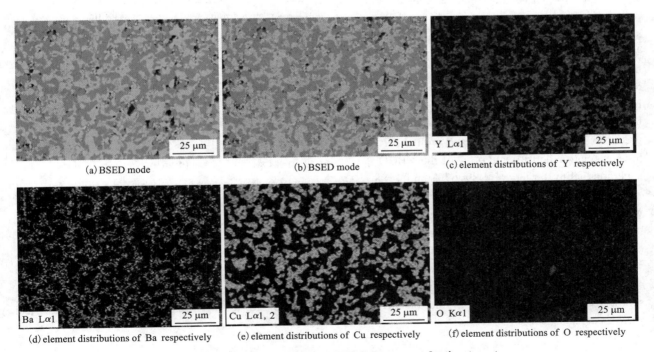

Figure 12 EDS diagrams of the YBCO-7 superconducting target

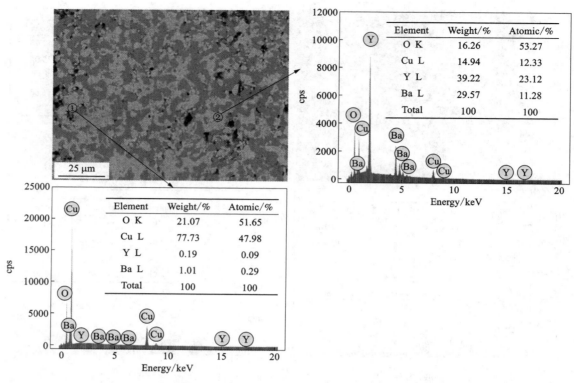

Figure 13 SEM image and EDS spectra and chemical compositions (tables) of the YBCO-7 superconducting target

3.4 Physical property analysis of the YBCO superconducting targets

Figure 14 shows the relative density and resistivity at room temperature of the YBCO superconducting targets. According to Figure 14, the YBCO-5 superconducting target had the highest relative density of 89.83% and the lowest $\rho_{300\,K}$ of $3.351 \times 10^{-4}\ \Omega \cdot m$. This indicated that the YBCO superconducting target had the best electrical properties when the precursor solution was synthesized with a pH of 5. The relative density and $\rho_{300\,K}$ of the YBCO-3 superconducting target were 89.67% and $1.884 \times 10^{-3}\ \Omega \cdot m$, respectively. All of the values were lower than the YBCO-5 superconducting target, which could be explained by an excessive number of non-superconducting second phases. YBCO-7 had the lowest density of 86.38% and the highest $\rho_{300\,K}$ of $3.843 \times 10^{2}\ \Omega \cdot m$. The $\rho_{300\,K}$ was six orders of magnitude higher than YBCO-5. The superconducting phase $YBa_2Cu_3O_7$ could not be synthesized when the precursor solution pH was 7, and this led to high resistivity and poor electrical performance.

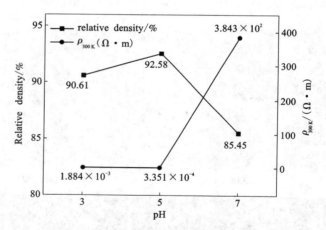

Figure 14 Relative density and resistivity $\rho_{300\,K}$ with pH of the YBCO superconducting targets

4 Conclusions

The YBCO compound powders were prepared using the oxalic acid coprecipitation method and used to prepare the YBCO superconducting target. The SEM results showed that the YBCO compound powders were composed of two phases of large-sized lamellar substances and mini-size spherical particles. The number and size of the lamellar material in the compound particles increased with increasing pH. After XRD analysis, the YBCO-3 and YBCO-5 superconducting targets were composed of $YBa_2Cu_3O_7$, Y_2BaCuO_5, and CuO phases. The second aryphase content was less in the YBCO-5 superconducting target than in YBCO-3. However, the YBCO-7 superconducting target did not contain the superconducting phase $YBa_2Cu_3O_7$ and only contained the Y_2BaCuO_5 and CuO phases. Following SEM and EDS analyses, the additional elemental analysis results of the YBCO superconducting targets were consistent with the XRD results. The YBCO-5 superconducting target had the highest density and the lowest ρ_{300K} of $3.351\times10^{-4}\Omega \cdot m$, and the optimal pH of the precursor solution for YBCO synthesis was 5. Superconducting performance was the decisive factor in measuring the prepared superconducting target. Subsequently, the superconducting properties of the prepared superconducting target will be tested and characterized, such as critical transition temperature T_c, critical current density J_c, and magnetization performance.

Acknowledgments: The authors would like to acknowledge the support from the National Natural Science Foundation of China (Grant No. 92062223), Joint Funds of the National Natural Science Foundation of China (Grant No. U21A2065), and Supported by Project of Zhongyuan Critical Metals Laboratory, Zhengzhou University, China.

References

[1] ONNES H K. The resistance of pure mercury at helium temperatures, Commun[J]. Phys. Lab. Univ. Leiden, b, 120 (1911).

[2] INDOVINA P, ONORI S, TABET E, Magnetic field modulation of the microwave impedance of Pb superconducting films[J]. Solid State Communications, 1970(8): 1721-1724.

[3] BRINKMAN W, TSUI D. Asymmetry effects in superconducting tunneling[J]. Physical Review B, 1972(6): 1063.

[4] WHETSTONE C, CHASE G, RAYMOND J, et al. Thermal stability for Ti-22 atomic percent Nb superconducting cables and solenoids[J]. IEEE Transactions on Magnetics, 1966(2): 307-310.

[5] GAVALER J, JANOCKO M, BRAGINSKI A, et al. Superconductivity in Nb_3Ge[J]. IEEE Transactions on Magnetics, 1975(11): 192-196.

[6] BACHNER F, GOODENOUGH J, GATOS H. Superconducting transition temperature and electronic structure in the pseudo binaries Nb_3Al-Nb_3Sn and Nb_3Sn-Nb_3Sb[J]. Journal of Physics and Chemistry of Solids, 1967(28): 889-895.

[7] BEDNORZ J G, MÜLLER K A. Possible high T_c superconductivity in the Ba-La-Cu-O system[J]. Zeitschrift für Physik B Condensed Matter, 1986(64): 189-193.

[8] CHU C, HOR P, MENG R, et al. Superconductivity at 52.5 K in the lanthanum-barium-copper-oxide system[J]. Science, 1987(235): 567-569.

[9] MAEDA H, TANAKA Y, FUKUTOMI M, et al. A new high-T_c oxide superconductor without a rare earth element[J]. Japanese Journal of Applied Physics, 1988(27): L209.

[10] MICHEL C, HERVIEU M, BOREL M, et al. Superconductivity in the Bi-Sr-Cu-O system[J]. Zeitschrift für Physik B Condensed Matter, 1987(68): 421-423.

[11] AKIMITSU J, YAMAZAKI A, SAWA H, et al. Superconductivity in the Bi-Sr-Cu-O system[J]. Japanese Journal of Applied Physics, 1987(26): L2080.

[12] SHENG Z, HERMANN A. Superconductivity in the rare-earth-free Tl-Ba-Cu-O system above liquid-nitrogen temperature[J]. Nature, 1988(332): 55-58.

[13] SHENG Z, HERMANN A, EL ALI A, et al. Superconductivity at 90 K in the Tl-Ba-Cu-O system[J]. Physical Review Letters, 1988(60): 937.

[14] SHENG Z, HERMANN A. Bulk superconductivity at 120 K in the Tl-Ca/Ba-Cu-O system[J]. Nature, 1988(332): 138-139.

[15] PUTILIN S, ANTIPOV E, CHMAISSEM O, et al. Superconductivity at 94 K in $HgBa_2CuO_{4+\delta}$[J]. Nature, 1993(362): 226-228.

[16] SCHILLING A, CANTONI M, GUO J, et al. Superconductivity above 130 K in the Hg-Ba-Ca-Cu-O system[J]. Nature, 1993(363): 56-58.

[17] GAO L, XUE Y, CHEN F, et al. Superconductivity up to 164 K in $HgBa_2Ca_{m-1}Cu_mO_{2m+2+\delta}$ ($m=1, 2,$ and 3) under quasihydrostatic pressures[J]. Physical Review B, 1994(50): 4260.

[18] NAGAMATSU J, NAKAGAWA N, MURANAKA T, et al. Superconductivity at 39 K in magnesium diboride[J]. Nature, 2001(410): 63-64.

[19] KIM T H, MOODERA J S. Magnesium diboride superconductor thin film tunnel junctions for superconductive electronics[J]. Journal of Applied Physics, 2006(100): 113904.

[20] ISIKAKU-IRONKWE O P. Possible high-T_c superconductivity in LiMgN: a MgB_2-like material[J] Physics, 2012.

[21] TOMITA M, MURAKAMI M. High-temperature superconductor bulk magnets that can trap magnetic fields of over 17 tesla at 29 K[J]. Nature, 2003(421): 517-520.

[22] GUREVICH A. To use or not to use cool superconductors[J]. Nature Materials, 2011(10): 255-259.

[23] ZHIJIAN J, ZHIYONG H, YUE Z, et al. Review of technology and development in the power applications based on second-generation high-temperature superconductors[J]. Journal of Shanghai Jiaotong University, 2018(52): 1155.

[24] ZHANG Y M, GUO J, ZHAO H X, et al. The research of preparing high temperature superconductive materials for YBCO by the traditional solid-state reaction technique[J]. Physical Experiment of College, 2011.

[25] WANG L, LI T, GU H. Effect of humidity on microstructure and properties of YBCO film prepared by TFA-MOD method[J]. Journal of Rare Earths, 2009(27): 486-489.

[26] JORGENSEN J, BENO M A, HINKS D G, et al. Oxygen ordering and the orthorhombic-to-tetragonal phase transition in $YBa_2Cu_3O_{7-x}$[J]. Physical Review B, 1987(36): 3608.

[27] KWOK W, CRABTREE G, UMEZAWA A, et al. Electronic behavior of oxygen-deficient $YBa_2Cu_3O_{7-\delta}$[J]. Physical Review B, 1988(37): 106.

[28] JORGENSEN J, VEAL B, KWOK W, et al. Structural and superconducting properties of orthorhombic and tetragonal $YBa_2Cu_3O_{7-x}$: The effect of oxygen stoichiometry and ordering on superconductivity[J]. Physical Review B, 1987(36): 5731.

[29] HARABOR A, ROTARU P, HARABOR N A, et al. Orthorhombic YBCO-123 ceramic oxide superconductor: Structural, resistive and thermal properties[J]. Ceramics International, 2019(45): 2899-2907.

[30] ALIABADI A, FARSHCHI Y A, AKHAVAN M. A new Y-based HTSC with T_c above 100 K[J]. Physica C: Superconductivity and its Applications, 2009(469): 2012-2014.

[31] PęCZKOWSKI P, KOWALIK M, JAEGERMANN Z, et al. Synthesis and physicochemical properties of $Er_{0.5}Dy_{0.5}Ba_2Cu_3O_{6.83}$ cuprate high-temperature superconductor[J]. Acta Physica Polonica A, 2019(135): 28-35.

[32] SAFRAN S, BULUT F, NEFROW A R A, et al. Characterization of the $CoFe_2O_4$/Cu displacement effect in the Y123 superconductor matrix on critical properties[J]. Journal of Materials Science: Materials in Electronics, 2020(31): 20578-20588.

[33] OCHSENKÜHN-PETROPOULOU M, ARGYROPOULOU R, TARANTILIS P, et al. Comparison of the oxalate co-precipitation and the solid state reaction methods for the production of high temperature superconducting powders and coatings[J]. Journal of Materials Processing Technology, 2002(127): 122-128.

[34] PAN H, XU X, GUO J. TEM characterization of nanoscale YBCO particles[J]. Materials Letters, 2003(57): 3869-3873.

[35] YANG Y, OUT P, ZHAO B, et al. Characterization of $YBa_2Cu_3O_{7-x}$ bulk samples prepared by citrate synthesis and solid-state reaction[J]. Journal of Applied Physics, 1989(66): 312-315.

[36] OHMUKAI M. The Effect of the pressure for the formation of $YBa_2Cu_3O_{7-d}$ bulk ceramics with domestic microwave oven[J]. Engineering, 2011(3): 1095-1097.

[37] GONZALEZ J L, PIUMBINI C K, SCOPEL W L, et al. Pore structure dependence with the sintering time for dense ceramic bulk $YBa_2Cu_3O_y$[J]. Ceramics International, 2013(39): 3001-3006.

[38] SUASMORO S, KHALFI M F, KHALFI A, et al. Microstructural and electrical characterization of bulk $YBa_2Cu_3O_{7-\delta}$ ceramics[J]. Ceramics International, 2012(38): 29-38.

[39] DAHIYA M, KUMAR R, KUMAR D, et al. Flux pinning characteristics of YBCO: $NaNbO_3$ by introducing artificial pinning centers with different morphology[J]. Ceramics International, 2021(47): 34189-34198.

[40] WANG A, CUI L, XIE J, et al. Thermodynamic analysis of YBCO powder by oxalate co-precipitation preparation[J]. Journal of the Chinese Society of Rare Earths, 2014.

[41] MUJAINI M, YAHYA S Y, HAMADNEH I, et al. Synthesis of $YBa_2Cu_3O_{7-\delta}$ high temperature superconductor by co-precipitation method[J]. AIP Conference Proceedings, 2008(1017): 119-123.

A Multi-scale Study of the Thermal Transport Properties of Graphene-reinforced Copper/diamond Composites

Jiarui ZHU[1,2,3], Hui YANG[1,2,3,*], Shuhui HUANG[1,2,3], Hong GUO[1,2,3], Zhongnan XIE[1,2,3]

1. State Key Laboratory of Nonferrous Metals and Processes, GRINM Group Co., Ltd., Beijing, 100088, China
2. GRIMAT Engineering Institute Co., Ltd., Beijing, 101407, China
3. General Research Institute for Nonferrous Metals, Beijing, 100088, China

Abstract: As a potential modifying interlayer for copper/diamond composites, graphene has higher intrinsic thermal conductivity (TC) than the carbide layer generated by widely applied strategies of surface metallization and matrix alloying. However, the mechanism of graphene's influence on the heat transport process is not clear, and it is not known what the enhancement potential of graphene modification is. The thermal boundary conductance (TBC) is calculated using the non-equilibrium molecular dynamics (NEMD) method and the heat transport mechanism is analyzed through the phonon spectra beyond the interface. The TBC achieves the highest value after modifying with single-layer graphene, decreases with the increment of the number of graphene layers and converges at 3-layer of graphene. The TC of composites are also influenced by crystal planes and performs the best with Cu(110)/dia(111) interface. By applying the TBC values obtained through NEMD method into numerical simulation, we verifies that graphene can effectively improve the TC of copper/diamond composites and single-layer graphene enhances the heat transport process better. The TC of graphene-modified copper/diamond composite are influenced by the volume ratio of diamond and reaches 1054.51 W/(m·K) with 70 vol.% diamond. This work provides theoretical guidance and support for future experimental work in optimizing the copper/diamond composites.

Keywords: Graphene; Copper/diamond composites; Interfacial modification; Molecular dynamics; Numerical simulation

1 Introduction

With the rapid development of electronic technology, electronic components are gradually miniaturized, integrated and functionalized and are more and more widely used. The significant increase in power density makes the rapid transfer of heat generated by components the key to improving their life and reliability, creating a great demand for high-performance thermal management materials[1,2]. Copper/diamond composites combine the higher thermal conductivity of diamond with a lower coefficient of thermal expansion on top of the high thermal conductivity of copper[3,4], making it one of the most mature thermal management materials in use today. For the performance enhancement of copper/diamond composites, common strategies are mainly focused on copper matrix alloying and surface metallization to enhance the interfacial bonding between copper and diamond. However, the interface of the composites obtained in those methods consists of various types of metal carbides with low TC, and the TC of the corresponding copper/diamond composites is difficult to break through the bottleneck of 800 W/(m·K)[5,6].

Graphene, as a two-dimensional carbon material with excellent TC, is an allotropic isomer with diamond, becoming one of the options to potentially enhance the TC of copper/diamond composites[7,8]. Currently, some scholars have produced graphene-modified copper/diamond composites with a TC of 572.9 W/(m·K) by in situ growth method, which proves that graphene can improve the TC of composites[9]. However, the mechanism of graphene's influence on the heat transport effect is not clear, and it is not known what the enhancement potential of graphene modification is. In this paper, the interfacial thermal transport mechanism of graphene-modified copper/diamond composites and the effects of the number of graphene layers and crystal planes on the thermal transport properties are investigated by molecular dynamics (MD) methods. The TBC data obtained at the atomic scale are imported into a finite element model to explore the thermal conductivity changes of the composites with different TBC values and different diamond volume fractions.

2 Models and methods

MD software LAMMPS[10, 11] is used to study the interfacial heat transport mechanisms at atomic scale. The model is shown in Figure 1(a), with the normal z-direction of Cu(001)//dia(001), x and y direction of Cu(100)//dia(100) and Cu(010)//dia(010). To avoid introducing large mismatch strains at the interface, the period of Cu/dia in the in-plane direction is chosen to be 13 : 13, and the mismatch strain is 1.2% (Other interface models are built in a similar way). Copper and diamond are larger than 5 nm in the z-direction to avoid surface effects. Periodic boundary conditions are applied in both x, y and z directions, and a vacuum layer of 5 nm is introduced along the top and bottom of the z-direction to simulate a free surface. Two layers of atoms are fixed at both ends in the z-direction, and the two layers of atoms adjacent to the fixed atomic layer are set as the cold and heat sources. The force fields of Cu and diamond are EAM[12] and Tersoff[13] respectively, the LJ force fields is used between copper and diamond atoms. The TBC is calculated using the NEMD method[14, 15, 16] and the heat transport mechanism is analyzed by calculating the phonon spectra of the atoms on both sides of the interface, details of which can be found in our previous work[17]. The graphene modification layer is introduced as shown in Figure 1(b), the TC of the three-phase, two-interface structure is calculated by Equation (1)[18].

$$\frac{1}{h_c} = \frac{1}{G_1} + \frac{a}{K_d} + \frac{1}{G_2} \tag{1}$$

Where h_c denotes the interfacial thermal conductivity (ITC). G_1 and G_2 denote the TBC of the boundary on both sides of the inter-layer. K_d denotes the TC of the interfacial material. The a denotes the thickness of the interface layer.

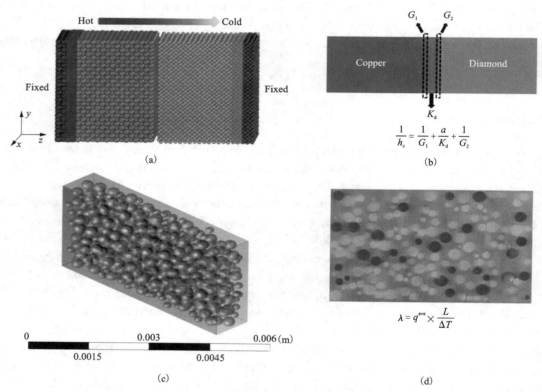

Figure 1 Schematic of copper/diamond model(a), interfacial thermal conductivity of copper/graphene/diamond model(b), diamond particle random distribution model(c) and heat flux of copper/diamond composites(d)

The thermal conductivity of macro-scale copper/diamond composites is calculated using the Mechanical module of ANSYS V19.0 software[19]. The model is shown in Figure 1(c), where diamond particles of 0.2 mm and 0.3 mm (volume ratio of 1 : 1) have been randomly distributed within the copper matrix. TC of diamond is set to be 2000 W/m·K. TC of copper is set to be 400 W/(m·K). TBC at the Cu/dia interface needs to be set according to the results of MD simulations. Then a temperature difference is applied on both sides of the model, and the heat flow is obtained by steady state heat transfer simulation, as shown in Figure 1 (d). The TC of the composite material is obtained using Equation (2),

$$\lambda = q^{ave} \times \frac{L}{\Delta T} \tag{2}$$

Where: λ denotes the TC of the composite material; L denotes the length of the model heat transfer direction; ΔT denotes the temperature difference.

3 Results and discussions

3.1 Interfacial thermal conductivity of Cu/dia interface

The ITC of Cu/dia interface was calculated using the NEMD methods. Figure 2(a) illustrates the temperature distribution in the direction perpendicular to the interface and shows a significant temperature difference (ΔT) of 148.697 K. The heat flux (J) obtained is 1.08×10^{-7} J/s. The TBC calculated using formula $G = J/A\Delta T$ is 33.165 MW/(m²·K). Figure 2(b) shows the in-plane and out-of-plane vibrational phonon spectra of copper and diamond atoms at the interface. It can be seen that the phonon overlap is very low in both the low frequency (0-10 THZ) and high frequency (10-30 THZ) zones. The overlap concentration of phonon spectrum S_{out} of out-of-plane mode is 0.03609, and the S_{in} is 0.03932. Such a low S means that the phonon vibrations of copper and diamond cannot well-matched over the interface. Therefore it is difficult for copper/diamond composites to show good heat transfer performance.

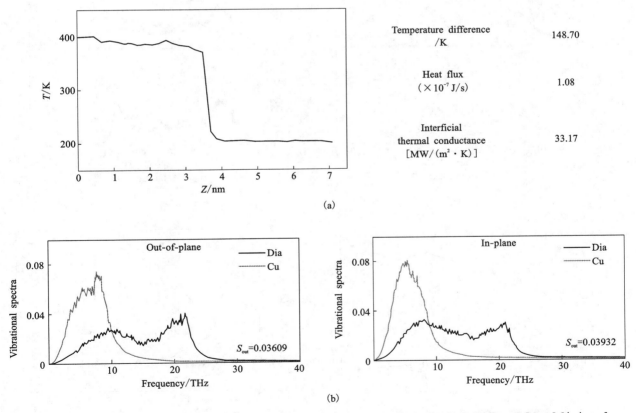

Figure 2 Temperature distribution and parameters of heat transfer process in copper/diamond model(a) and phonon spectrum of copper/diamond model(b)

3.2 Interfacial thermal conductivity of Cu/gra/dia interface

As can be seen from Figure 1(b), the interfacial structure after graphene modification consists of three phases and two interfaces. The TBC of such structure can be calculated through the TBC of Cu/gra and dia/gra. By simulating the dia(001)/multilayer graphene model, laws of TBC with the number of graphene layers were obtained. With single graphene, the TBC of dia(001)/gra^s has the highest value of 1.96 GW/(m^2·K). When the number of graphene layers increases to 3, the TBC of dia(001)/gra^t(t denotes triple) converges to about 1.53 GW/(m^2·K). The phonon spectrum can be seen in Figure 3(a). The S_{out} is always larger than S_{in}, which means the out-of-plane phonon transport dominates interfacial thermal transport processes. With the increment of graphene layers, S_{out} gradually decreases, the interfacial heat transport becomes weaker, therefore the TBC becomes smaller.

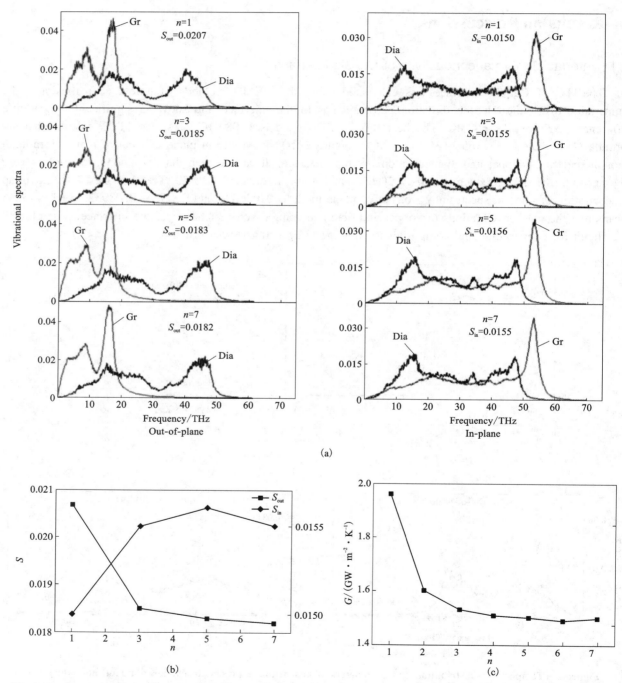

Figure 3 Out-of-plane and in-plane phonon spectra of diamond and graphene atoms at the interfaces of different graphene layers(a), variation of phonon spectral overlap in different graphene layer models(b) and variation of phonon spectral overlap in different graphene layer models(c)

To investigate the effect of different crystal planes on the TBC, we built dia(001)/gra^s and dia(111)/

gra^s models and got the corresponding values of 1.96 GW/(m^2·K) and 2.23 GW/(m^2·K). As shown in Figure 4, the dominant S_{out} of dia(111)/gra^s is higher than dia(001)/gra^s (0.0263>0.0207). Therefore combining graphene with dia(111) is beneficial for obtaining better interfacial heat transfer properties.

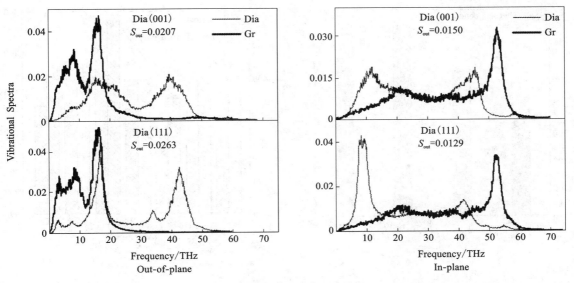

Figure 4 Phonon spectra of diamond(001) and diamond(111) surfaces with graphene

Our former work[17] has concluded similar rules of how the number of graphene layers and different crystal plane of copper influence the TBC of Cu/gra interface. With single graphene, the TBC of Cu(001)/gra has the highest value of 640.19 MW/(m^2·K). When the number of graphene layers increases to 3, the value of TBC converges to about 185.58 MW/(m^2·K). Cu(011)/gra^s possesses a higher TBC than Cu(001)/gra^s and Cu(111)/gra^s. The TBC values of different dia/gra and Cu/gra interfaces are summarized in Table 1. According to the formula in Figure 1(b), the TBC of distinct Cu/dia interfaces with different modification methods are calculated in Table 2. There exist multiple crystal planes in forming Cu/dia interface. However, in this paper, we assume there only exists equal Cu(001)/dia(001) and Cu(110)/dia(111) in actual composites and derive the equivalent values of TBC in Table 2.

Table 1 Thermal boundary conductance of different Cu/gra and dia/gra interfaces

Interface Model	Cu(001)/Dia(001)		Cu(011)/Dia(111)	
Single-layer graphene	Cu(001)/Gra^s 640.19 MW/(m^2·K)	Dia(001)/Gra^s 1.96 GW/(m^2·K)	Cu(011)/Gra^s 673.61 MW/(m^2·K)	Dia(111)/Gra^s 2.23 GW/(m^2·K)
Three-layer graphene	Cu(001)/Gra^t 185.58 MW/(m^2·K)	Dia(001)/Gra^t 1.53 GW/(m^2·K)	Cu(011)/Gra^t 246.52 MW/(m^2·K)	Dia(111)/Gra^t 1.67 GW/(m^2·K)

Table 2 The interfacial thermal conductivities of distinct Cu/dia interfaces with different modification methods

Interface	Without modification /(MW·m^{-2}·K^{-1})	Single-layer graphene /(MW·m^{-2}·K^{-1})	Three-layer graphene /(MW·m^{-2}·K^{-1})
(100)/(100)	33.17	467.68	161.10
(110)/(111)	39.65	505.20	210.48
Average	36.41	486.44	185.79

3.3 Thermal conductivity of copper/diamond composites

Some experimental works have obtained graphene modified copper/diamond composites by introducing in-situ grown graphene as a highly effective interlayer[9]. As shown in Figure 5(b), TC of the composites increases and then decreases with increasing graphene content, meanwhile the value can reach 572.9 W/(m·K). It is not clear how the graphene content affects the TC of the composites and how much TC can be achieved with graphene-modified interlayer.

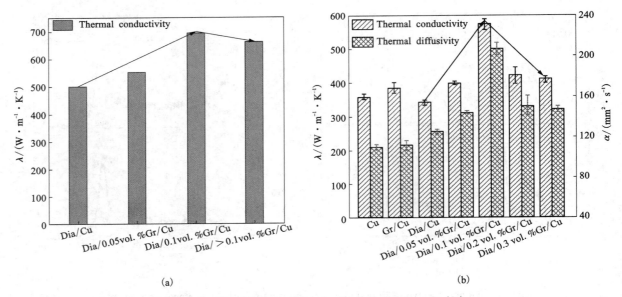

Figure 5 Simulation results(a), Experimental results(b)

We conducted some numerical simulations of steady-state heat transfer with a copper/diamond model (diamond volume fraction of 30%). The TC of copper/diamond composites without interlayer is 502.17 W/m·K. Changing the TBC of half Cu/dia interface in the model to 486.44 MW/(m²·K) means half of the interface has been modified by single-layer graphene (volume fraction of graphene is 0.05%). The TC of copper/diamond composite with 50% modified single-layer graphene, full modified single-layer graphene and full modified 3-layers graphene are calculated to be 549.26 W/(m·K), 694.37 W/(m·K) and 663.82 W/(m·K) respectively. Shown in Figure 5(a), the values display a similar pattern with the experimental results, which verifies that graphene can effectively improve the TC of copper/diamond composite and single-layer graphene enhance the heat transport process better.

The volume ratio of diamond has a strong influence on the TC of composites. In this paper, we constructed models for 5%-30% diamond volume fraction and applied the TBC shown in Table 2 for the TC calculation of the composites. Besides, the TBC of TiC [21 W/(m·K)][20] has also been used to make a comparison. The TC increases monotonically with the increment of diamond as Figure 6 displays. The value of single-layer graphene modified copper/diamond with 30 vol.% diamond reaches 694.37 W/(m·K), which is higher than the value of 3-layers graphene and TiC modified composites. Due to highly computation cost, we haven't built models with over 30 vol.% diamond. However, the values are extrapolated through existing results to 70 vol.% diamond. The TC of single-layer graphene modified copper/diamond

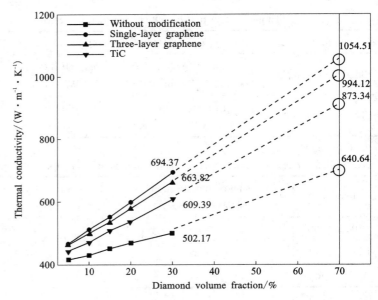

Figure 6 Data fitting and prediction

composite with 70 vol. % diamond reaches 1054.51 W/(m·K). Such a result means that graphene has high potential as a modification interlayer in the optimization of copper/diamond composites.

4 Conclusions

In this work, a systematic theoretical research work was carried out to address the inadequacy of the existing experimental work on graphene-modified copper/diamond composites. The mechanism of how graphene enhances the TC of copper/diamond composites is explained, the influence played by different factors is analyzed, and the microscopic TBC results are transferred to the macroscopic model to perform a multi-scale study. The main findings are as follows:

(1) Single-layer graphene has the best modification effect, while the TBC value converges after the number of graphene layer reaches 3.

(2) The Cu(011)/dia(111) interface possesses prominent TBC value.

(3) With the increase of graphene content, the percentage of graphene-modified interfaces increases, which can effectively enhance the TC of composites. However, the TC of the composite material decreases after graphene exceeds a certain content.

(4) TC of single-layer graphene-modified composites reaches 1054.51 W/(m·K) at 70% diamond volume fraction, which shows great application prospects.

This work explains the mechanism and law of enhancing the TC of copper/diamond composites by graphene modification through a systematic study, proposes the upper limit of the composites under this modification method, and provides theoretical guidance and support for future experimental work in optimizing the copper/diamond composites.

References

[1] MOORE A L, SHI L. Emerging challenges and materials for thermal management of electronics[J]. Materials Today, 2014, 17(4): 163-174.

[2] ZIDAN M A, STRACHAN J P, LU W D. The future of electronics based on memristive systems[J]. Nature Electronics, 2018, 1(1): 22-29.

[3] LIAO M, WANG Y, WANG F, et al. Unexpected low thermal expansion coefficients of pentadiamond[J]. Physical Chemistry Chemical Physics, 2022, 24(38): 23561-23569.

[4] WANG Z, TANG Z, XU L, et al. Thermal properties and thermal cycling stability of graphite/copper composite fabricated by microwave sintering[J]. Journal of Materials Research and Technology, 2022, 20: 1352-1363.

[5] PAN Y, HE X, REN S, et al. High thermal conductivity of diamond/copper composites produced with Cu-ZrC double-layer coated diamond particles[J]. Journal of Materials Science, 2018, 53: 8978-8988.

[6] ZHANG J, ROSENKRANZ A, ZHANG J, et al. Modified wettability of micro-structured steel surfaces fabricated by elliptical vibration diamond cutting[J]. International Journal of Precision Engineering and Manufacturing-Green Technology, 2022, 9(5): 1387-1397.

[7] MOAYEDI H, ALIAKBARLOU H, JEBELI M, et al. Thermal buckling responses of a graphene reinforced composite micropanel structure[J]. International Journal of Applied Mechanics, 2020, 12(01): 2050010.

[8] NASEER A, AHMAD F, ASLAM M, et al. A review of processing techniques for graphene-reinforced metal matrix composites[J]. Materials and Manufacturing Processes, 2019, 34(9): 957-985.

[9] CAO H J, TIAN Z Q, LU M H, et al. Graphene interlayer for enhanced interface thermal conductance in metal matrix composites: An approach beyond surface metallization and matrix alloying[J]. Carbon, 2019, 150: 60-68.

[10] NOSÉ S. A molecular dynamics method for simulations in the canonical ensemble[J]. Molecular Physics, 1984, 52: 255-268.

[11] HUANG Z X, TANG Z A. Evaluation of momentum conservation influence in non-equilibrium molecular dynamics methods to compute thermal conductivity[J]. Physica B: Condensed Matter, 2006, 373(2): 291-296.

[12] CHERNE F J, BASKES M I, DEYMIER P A. Properties of liquid nickel: A critical comparison of EAM and MEAM calculations[J]. Physical Review B, 2001, 65(2): 024209.

[13] TERSOFF J. Modeling solid-state chemistry: ineratomic potentials for multicomponent systems[J]. Phys. Rev. B, 1989, 395(56): 6-8.

[14] WEI N, ZHOU C, LI Z, et al. Thermal conductivity of aluminum/graphene metal-matrix composites: From the

thermal boundary conductance to thermal regulation[J]. Mater. Today Commun, 2022, 30: 103-117.

[15] OU B, YAN J, WANG Q, et al. Thermal conductance of graphene-titanium interface: A molecular simulation[J]. Molecules, 2022, 905: 76-80.

[16] GHOSH S, BAO W, NIKA D L, et al. Dimensional crossover of thermal transport in few-layer graphene[J]. Nat. Mater, 2010, 9: 555-558.

[17] ZHU J, HUANG S, XIE Z, et al. Thermal conductance of copper-graphene interface: a molecular simulation[J]. Materials, 2022, 15: 7588.

[18] LI J, ZHANG H, ZHANG Y, et al. Microstructure and thermal conductivity of Cu/diamond composites with Ti-coated diamond particles produced by gas pressure infiltration[J]. Journal of Alloys and Compounds, 2015, 647: 941-946.

[19] ZHANG Y J, DONG Y H, ZHANG R Q. Numerical simulation of thermal conductivity of diamond/copper composites[J]. Cailiao Rechuli Xuebao/Transactions of Materials and Heat Treatment, 2018, 39(6): 110-117.

[20] Guo C, Fangyuan S, Jialiang D, et al. Effect of Ti interlayer on interfacial thermal conductance between Cu and diamond[J]. Acta Materialia, 2018: S1359645418306980.

Application of Rare Earth Metal Ferroalloys in Special Steels

Lifeng ZHANG[1], Ying REN[2], Qiang REN[3]

1 School of Mechanical and Materials Engineering, North China University of Technology, Beijing, 100144, China
2 School of Metallurgical and Ecological Engineering, University of Science and Technology Beijing (USTB), Beijing, 100083, China
3 School of Mechanical Engineering, Yanshan University, Qinhuangdao, 066004, China

Abstract: Two types of rare earth metal ferroalloys including lanthanum and cerium ferroalloys were characterized. Intermetallic compounds of $CeFe_2$ and Ce_2Fe_{17} were observed in the cerium ferroalloy leading to a high chemical stability of the alloy while not in the lanthanum ferroalloy. From the viewpoint of practical application, the cerium ferroalloy was easier to preserve while the lanthanum ferroalloy should be preserved carefully to prevent oxidation. After the addition of lanthanum ferroalloy in non-oriented silicon steels, inclusions were modified to lanthanum-containing oxysulfides, leading to a remarkable decrease in the number of fine MnS inclusions. The core loss of non-oriented silicon steels was decreased after the addition of lanthanum ferroalloy. For the stainless steel, when the content of cerium in the steel increased from 0 to 250 ppm, the modification sequence of inclusions was: Si-Mn(-Al)-O and MnS → Ce-Si-Mn-O-S → Ce(-Si)-O-S → CeS and CeC_2, which agreed well with thermodynamic calculations. In-situ observation of the localized corrosion induced by different kinds of inclusions in a stainless steel was performed to analyze the correlation between different kinds of inclusions and the localized corrosion. The order of the volume expansion rate of pits induced by inclusions was CeS>Si-Mn(-Al)-O>MnS>Ce-C-O-S>Ce-Si-Mn(-Al)-O>Ce-O-S.

Keywords: Rare earth element; Inclusion; Non-oriented silicon steel; Stainless steel

1 Introduction

Rare earth metals (REMs) include the 15 lanthanide elements with the atomic number of 57 to 71 in the periodic table and the two elements of scandium (Sc) and yttrium (Y) which have similar chemical properties to lanthanide elements. Due to their unique outer shell electronic structure, REMs have a high chemical activity and have been widely used as additives to steels. The dispersion, size distribution, composition and morphology control of inclusions in steels has significant impact on properties of the steel[1-4], and inclusions could also be effectively modified by rare earth elements[5-8]. Besides, the microstructure and property of steels could be significantly improved by an appropriate addition of rare earth elements[9-12]. The main functions of REMs in the steel include the following three aspects: (a) deeply purify molten steel by removing residual elements including oxygen, sulfur, arsenic, etc., (b) modify inclusions to reduce their harm to the steel performance, (c) act as micro-alloying elements. However, due to the high chemical activities of REMs, they are easy to be oxidized by the air, the slag and refractories, causing a low yield ratio in the actual production of REMs-treated steels. In the current study, rare earth ferroalloys were prepared and their morphology and composition were characterized. The apply effect of rare earth ferroalloys in typical special steels including a non-oriented silicon steels and a stainless steel was reported.

2 Morphology and Composition of Rare earth Ferroalloy

The rare earth ferroalloy was grand using sandpapers and polished using diamond polishing paste. During polishing, alcohol was used to prevent the reaction with water. The morphology and composition of a 30 wt.% cerium ferroalloy and a 30 wt.% lanthanum ferroalloy is shown in Figure 1. As shown in Figure 1(a), there were three phases in 30 wt.% CeFe. Two intermetallic compounds including $CeFe_2$ and Ce_2Fe_{17} were formed except for a less pure Fe phase. As shown in Figure 1(b), there was no intermetallic compound between lanthanum and iron. Pure lanthanum phase existed between the dendritic shape iron. It is also shown that some pure lanthanum have fell off from the alloy matrix which was attributed to the oxidation and powderization of pure lanthanum. The difference in the morphology between LaFe and CeFe indicated cerium ferroalloy was more conductive to long-term preservation. The binary phase diagram of Ce-Fe and La-Fe was calculated using FactSage^{TM}, as shown in Figure 2. It was consistent with the result in Figure 1 that two types of intermetallic compounds were formed in CeFe but not in lanthanum ferroalloy.

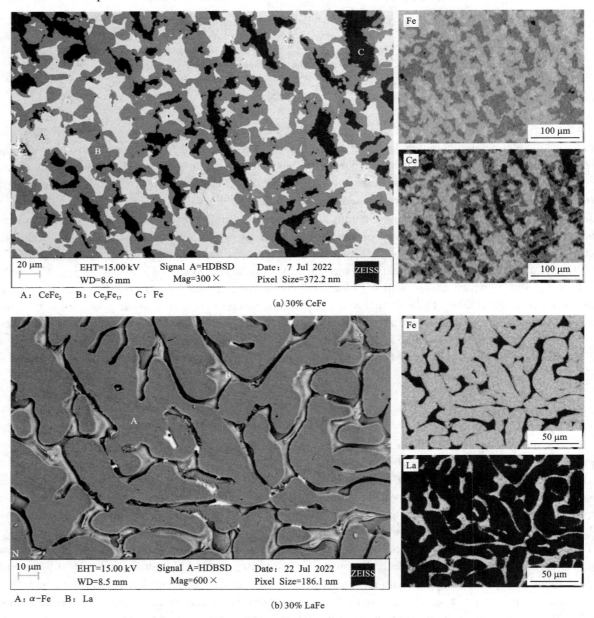

Figure 1　Morphology and EDS elemental mapping of cerium ferroalloy

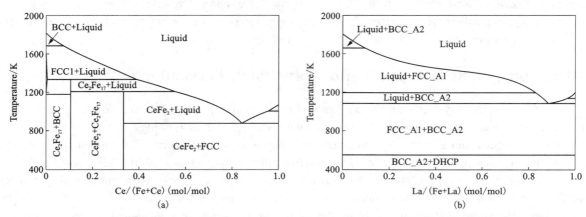

Figure 2　Binary phase diagrams of Ce-Fe (a) and La-Fe (b) calculated using FactSage^{TM}

3 Application in non-oriented silicon steels

After the addition of lanthanum ferroalloy in non-oriented silicon steel, the original MgO · Al_2O_3 inclusions were modified to lantahnum-containing inclusions which were composed of main La_2O_2S and less $LaAlO_3$ ones. The elemental mapping of a composite MgO-CaS-La_2O_2S inclusion is shown in Figure 3[13]. Since the dissolved sulfur was mainly combined with lanthanum, the precipitation of fine MnS in the solid steel was significant reduced. As shown in Figure 4(a)[13], the number density of fine MnS decreased from 1.58×10^5 to 2.23×10^4. In non-oriented silicon steels, fine inclusions lead to an increase in the hysteresis loss by directly pinning the motion of domain walls and indirectly retarding the grain growth during recrystallization annealing. Thus, as shown in Figure 4(b)[13], the addition of lanthanum ferroalloy contributed to reduce the core loss of non-oriented silicon steels.

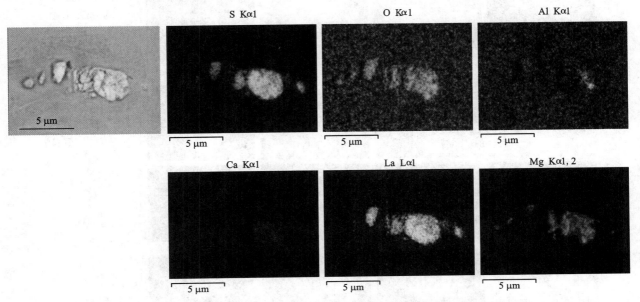

Figure 3 Elemental mapping of a typical lanthanum-containing inclusion in LaFe treated non-oriented silicon steel[13]

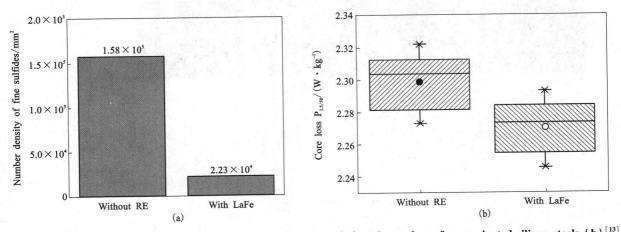

Figure 4 Effect of lanthanum ferroalloy on fine MnS inclusions (a) and core loss of non-oriented silicon steels (b)[13]

4 Application in stainless steel

The variation of inclusions compositions with the T. Ce content in the steel is shown in Figure 5[14]. When the content of cerium in the SS increased from 0 to 250 ppm, the modification sequence of inclusions was: Si-Mn(-Al)-O and MnS→Ce-Si-Mn-O-S→Ce(-Si)-O-S→CeS and CeC_2. When the T. Ce content in the steel increased from 0 to 250 ppm, the number density and area fraction of non-CeC_2 inclusions decreased from 77.7 mm² to 0.8 mm² and from 544.7×10⁻⁶ to 1.8×10⁻⁶, respectively. The average diameter of non-CeC₂ inclusions first decreased then gradually increased, and when the T. Ce content in the steel was 150×10⁻⁶, the average diameter of non-CeC₂ inclusions reached the minimum value (1.4 μm). When the content of T. Ce in the steel increased to 150×10⁻⁶, CeC₂ inclusions were formed, and when T. Ce content in the steel increased from 150×10⁻⁶ to 250×10⁻⁶, the number density, area fraction, and average of CeC₂ inclusions increased.

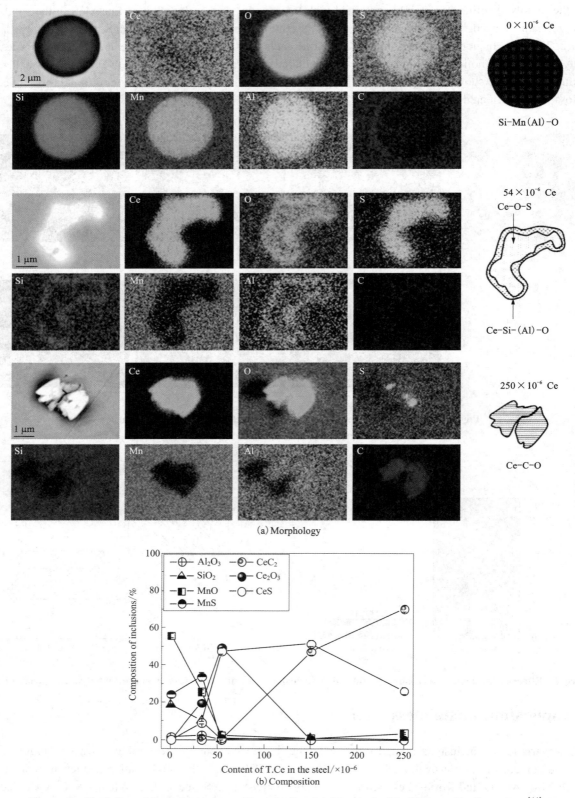

Figure 5 Variation of inclusions compositions in the stainless steel with different T. Ce content[14]

In-situ observation of the localized corrosion induced by different kinds of inclusions in a stainless steel was performed to analyze the correlation between different kinds of inclusions and the localized corrosion, as shown in Figure 6[15]. Inclusions of MnS, CeS and Ce-O-S dissolved prior to the steel matrix and cerium-containing oxides dissolved behind the steel matrix due to the difference between electron work functions of inclusions and the steel matrix. The order of the volume expansion rate of pits induced by inclusions was CeS>Si-Mn(-Al)-O>MnS>Ce-C-O-S>Ce-Si-Mn(-Al)-O>Ce-O-S, which is the effect of O/Ce ratio and S/Ce ratio on electron work function of cerium-containing inclusions. The electron work function of MnS, CeS and Ce-O-S was smaller than that of the steel matrix, thus, these inclusions first dissolved as the anode. However, the electron work function of cerium-containing oxides was bigger than that of the steel matrix, and the steel matrix dissolved prior to cerium-containing oxides. For cerium-containing inclusions, the electron work function of inclusions increased with the increase of the O/Ce ratio and the S/Ce ratio of inclusions. The order of the electron work function of cerium-containing inclusions was cerium oxides>cerium sulfides>cerium oxy-sulfide.

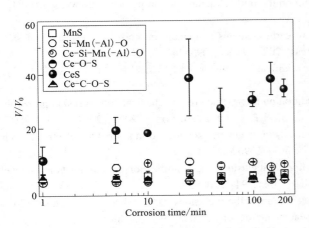

Figure 6 Index of the volume expansion of pits induced by different kinds of inclusions[15]

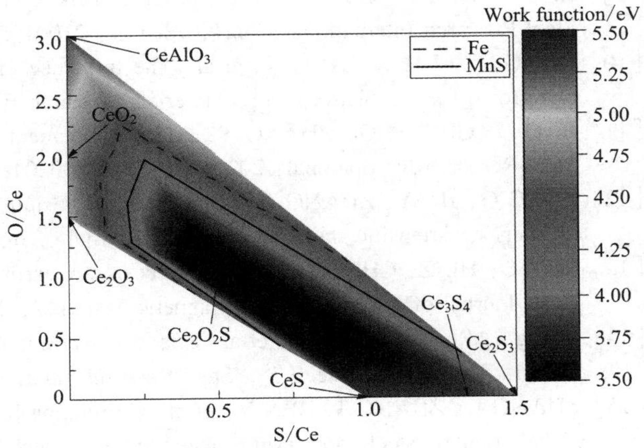

Figure 7 Effect of O/Ce ratio and S/Ce ratio on electron work function of cerium-containing inclusions[15]

4 Conclusions

(1) Two types of intermetallic compounds including $CeFe_2$ and Ce_2Fe_{17} were observed in cerium ferroalloy leading to a high chemical stability of the alloy while not in the lanthanum ferroalloy. From the view point of practical application, cerium ferroalloy was easier to preserve while the lanthanum ferroalloy should be preserved carefully to prevent oxidation.

(2) After the addition of lanthanum ferroalloy in non-oriented silicon steels, inclusions were modified to lanthanum-containing oxysulfides, leading to a remarkable decrease in the number of fine MnS inclusions. The core loss of non-oriented silicon steels was decreased after the addition of lanthanum ferroalloy.

(3) When the T. Ce content in the stainless steel increased from 0 to 250 ppm, the modification sequence of inclusions was: Si-Mn(-Al)-O and MnS→Ce-Si-Mn-O-S→Ce(-Si)-O-S→CeS and CeC_2. The order of the volume expansion rate of pits induced by inclusions was CeS>Si-Mn(-Al)-O>MnS>Ce-C-O-S>Ce-Si-Mn(-Al)-O>Ce-O-S.

Acknowledgments: The authors are grateful for support from the National Natural Science Foundation of China (grant No. U22A20171), the High Steel Center (HSC) at North China University of Technology, University of Science and Technology Beijing and Yanshan University.

References

[1] ZHANG L. Non-metallic inclusions in steels: Industrial practice (in Chinese)[M]. Beijing: Metallurgical Industry Press, 2019: 1.

[2] REN Y, ZHANG L, PISTORIUS P C. Transformation of oxide inclusions in type 304 stainless steels during heat treatment[J]. Metallurgical and Materials Transactions B, 2017, 48(5): 2281-2292.
[3] ZHANG L. Nucleation, growth, transport, and entrapment of inclusions during steel casting[J]. JOM, 2013, 65(9): 1138-1144.
[4] ZHANG L, THOMAS B G. State of the art in evaluation and control of steel cleanliness[J]. ISIJ International, 2003, 43(3): 271-291.
[5] ZHANG J, ZHANG L, REN Y. Effect of yttrium content on the transformation of inclusions in a Si-Mn-killed stainless steel[J]. Metallurgical and Materials Transactions B, 2021, 52(4): 2659-2675.
[6] WANG X, LI G, LIU Y, et al. Cerium addition effect on modification of inclusions, primary carbides and microstructure refinement of h13 die steel[J]. ISIJ International, 2021, 61(6): 1850-1859.
[7] LI B, ZHU H, ZHAO J, et al. Effect of rare-earth La on inclusion evolution in high-Al steel[J]. Steel Research International, 2021, 92(1): 2100347.
[8] YU Z, LIU C. Modification mechanism of spinel inclusions in medium manganese steel with rare earth treatment[J]. Metals, 2019, 7(9): 804-816.
[9] ZHONG L, WANG Z, CHEN R, et al. Effects of yttrium on the microstructure and properties of 20 MnSi steel[J]. Steel Research International, 2021, 92(11): 2100198.
[10] DONG R, LI H, ZHANG X, et al. The influence of rare earth elements lanthanum on corrosion resistance of steel plate for off shore platform[J]. Materials Research Express, 2021, 8(9): 096526.
[11] BAO D, CHENG G, HUANG Y, et al. Refinement of solidification structure of h13 steel by rare earth sulfide[J]. Steel Research International, 2021, 92(1): 2100304.
[12] ZANG Q, JIN Y, ZHANG T, et al. Effect of yttrium addition on microstructure, mechanical and corrosion properties of 20cr13 martensitic stainless steel[J]. Journal of Iron and Steel Research International, 2020, 27(12): 451-460.
[13] REN Q, HU Z, CHENG L, et al. Effect of rare earth elements on magnetic properties of non-oriented electrical steels[J]. Journal of Magnetism and Magnetic Materials, 2022, 560: 169624.
[14] Zhang J, Ren Y, Huang R, et al. Effects of cerium addition on inclusions and solidification structure of a low-nickel Si-Mn-killed stainless steel[J]. Steel Research International, 2023, 94: 2200956.
[15] ZHANG L, ZHANG J, REN Y, et al. Correlation between nonmetallic inclusions and localized corrosion of a low-nickel stainless steel with cerium treatment[J]. Steel Research International, 2023, 94: 2300041.

Effect of Ca Treatment on Sulfide Inclusion of High-strength Low-alloy Steel

Lifeng ZHANG[1], Xiaoyong GAO[2]

1 School of Mechanical and Materials Engineering, North China University of Technology, Beijing, 100144, China
2 School of Mechanical Engineering, Yanshan University, Qinhuangdao, 066004, China

Abstract: The effect of Ca treatment on sulfide inclusions in a high-strength low-alloy steel was investigated. Scanning electron microscope was used to characterize sulfide inclusions. Before Ca treatment, sulfide inclusions in the steel were mainly irregular Al_2O_3-MnS composite sulfide and pure MnS. Pure MnS exhibited severe aggregation and uneven distribution and the proportion of pure MnS was 74.6%. The average sizes of Al_2O_3-MnS and pure MnS were 3.72 μm and 3.54 μm, respectively. After Ca treatment, sulfide inclusions were mainly spherical Al_2O_3-CaO-CaS-MnS composite sulfides and pure MnS. The proportion of pure MnS decreased to 40.1%. The average sizes of Al_2O_3-CaO-CaS-MnS composite sulfides and pure MnS were 4.21 μm and 3.54 μm, respectively. Ca treatment can effectively control sulfide inclusions in high-strength low-alloy steel. The pure MnS was changed to composite sulfides including CaS and MnS, and the number, area and maximum size of pure MnS decreased. The formation mechanism of sulfide inclusions after Ca treatment was explained by thermodynamic calculation. The original pre-existing Al_2O_3 inclusions had positive effect on MnS control because they offered nucleation sites for MnS after Ca treatment. The industrial trial demonstrated that calcium treatment had a good effect on the modification of large MnS inclusions and prevented the generation of large-sized Type Ⅱ MnS inclusions at the top of the ingot.

Keywords: High-strength low-alloy steel; Inclusion; Sulfide; Calcium treatment; Thermodynamic calculation

1 Introduction

Inclusions such as MnS are usually defects of ultrasonic testing for high-strength low-alloy steel with heavy weight[1]. MnS inclusions are softer than the steel matrix and they are elongated during hot forging[2]. It is essential to control MnS inclusions of heavy steel ingots. Ca treatment is a common method to modify inclusions[3-6]. When adding Ca in Al-killed steels, oxide inclusions transformed from Al_2O_3 to calcium aluminate which had a lower melting point[7]. Modifying such Al_2O_3 inclusions by Ca treatment has a positive impact on casting by reducing nozzle clogging caused by Al_2O_3 inclusions[8-11]. In addition, Ca treatment results in (Ca, Mn)S sulfides which are harder than MnS and hense remain nearly unchanged during hot processing such as hot forging and hot rolling. Ca treatment on inclusions has been extensively studied and successfully applied in continuous casting[12-14]. However, the study of Ca treatment in heavy ingot casting is rarely reported. Unlike continuous casting in a water-cooled copper mold, ingot casting takes place in a big mold without water cooling. The objective of this study is to study the effect of Ca treatment on MnS inclusions in a high-strength low-alloy steel. Experimental studies were first conducted to simulate the Ca treatment process in industry. Then the application of Ca treatment in industry was conducted.

2 Experimental procedure

The raw material in the experimental trials was 16Mn billet. The chemical composition of the billet is shown in Table 1. Two heats were carried out in a vertical heat resistant furnace. The experimental melting temperature was 1873 K and the melting time was 20 minutes. After melting, the liquid steel was cooled in the furnace. The cooling rate was set as low as 2 K/min in order to simulate the slow cooling of heavy steel ingot with hundreds of tons of weight in industry. In one heat, no Si-Ca alloy was added and the sample was named as 16Mn. In the other heat, Si-Ca alloy was added and the sample was named as 16Mn-Ca. Except Ca addition amount, other parameters in two heats were the same.

Table 1 Chemical composition of 16Mn steel (wt%)

C	Mn	Si	P	S	Ni	Cr	Al	Fe
0.15	1.25	0.32	0.01	0.005	0.31	0.19	0.02	Bal.

After the experiments, the characteristic (composition, size, number and area) of sulfide inclusions in the steels were quantitatively and automatically analyzed by scanning electron microscope (SEM) with an energy dispersive spectrometer (EDS). The minimum sulfide inclusion size was set at 1.0 μm. The scanning area of inclusions analyzing for each sample was more than 10 mm^2, in order to obtain enough inclusions for statistics.

3 Features of sulfide inclusions

Figures 1 and 2 show the morphology and element distribution of typical sulfides in both steels. Without Ca treatment, sulfide inclusions in 16Mn steel were mainly Al_2O_3-MnS composite sulfides and pure MnS, as shown in Figure 1. The fraction of Al_2O_3-MnS composite sulfides was about 25.4%, as shown in Figure 3. Pure MnS exhibited severe aggregation and uneven distribution and the proportion of pure MnS was 74.6%. However, sulfide inclusions in 16Mn-Ca steel were mainly spherical Al_2O_3-CaO-CaS-MnS and pure MnS, as shown in Figure 2. In Al_2O_3-CaO-CaS-MnS composite sulfides, Al_2O_3-CaO was in the core and CaS-MnS was in the outer layer. The fraction of pure MnS decreased to 40.1% after Ca treatment, as shown in Figure 3.

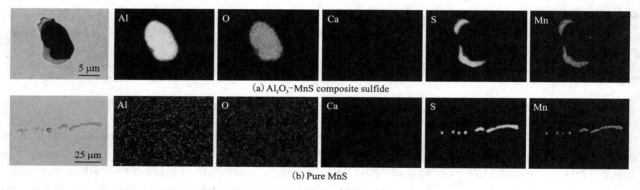

Figure 1 Typical sulfide inclusions in 16Mn steel

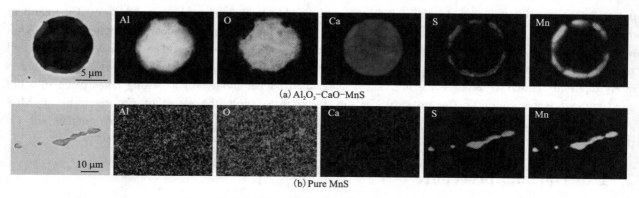

Figure 2 Typical sulfide inclusions in 16Mn-Ca steel

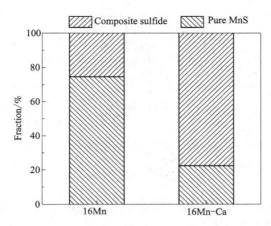

Figure 3　Fraction of different sulfides in the steels

Table 2 shows the number, area and size of different sulfide inclusions in the steels. For composite sulfides, the number and area in 16Mn-Ca steel increased, and the average and maximum size increased. For pure MnS, however, the number and area in 16Mn-Ca steel decreased, and the maximum size decreased. It is clear that Ca treatment has an effective effect on MnS control. The pure MnS was changed to composite sulfides including CaS and MnS.

Table 2　Characteristic of sulfide inclusions in the steels

Steel No.	Scanning area /mm^2	Type of sulfide	Number	Area /μm^2	Average size /μm	Maximum size /μm
16Mn	12.74	Composite sulfide	78	924.86	3.72	7.89
		Pure MnS	229	2318.10	3.54	32.15
16Mn-Ca	12.74	Composite sulfide	194	2882.83	4.21	13.42
		Pure MnS	130	1052.28	3.54	16.35

4　Thermodynamic calculation

In order to understand the effect of Ca treatment on MnS inclusions in the steel, FactSage software with FactPS, FToxid, and Fsteel databases was applied for thermodynamic analysis. The phase transformation of inclusions during cooling is shown in Figure 4. In 16Mn steel, Al_2O_3 was the only inclusion in the liquid steel. This is because the steel was Al-killed and the deoxidation product was Al_2O_3. When the temperature decreased to about 1677 K (1404 ℃), MnS started to precipitate. The content of MnS at 1473 K was 0.012 wt.%. It should be noted that some MnS precipitated on Al_2O_3 according to experimental results above. Therefore, the content of pure MnS was lower than 0.012 wt.%. It is difficult to calculate the contents of Al_2O_3-MnS composite sulfide and pure MnS under the current technology condition.

As shown in Figure 4(b), after Ca treatment, Al_2O_3 in the molten steel was transformed to Al_2O_3-CaO-CaS, which hence decreased the content of dissolved sulfur in the molten steel. The content of Al_2O_3-CaO-CaS was higher than Al_2O_3. When the temperature decreased to about 1685 K (1412 ℃), MnS started to precipitate. The content of MnS at 1473 K decreased to 0.0069 wt.% due to lower content of dissolved sulfur. Similarly, since some MnS precipitated on Al_2O_3-CaO-CaS according to experimental results above, the content of pure MnS was lower. It is also difficult to calculate the contents of Al_2O_3-CaO-CaS-MnS composite sulfide and pure MnS under the current technology condition.

Based on the preceding analysis, the formation mechanism of composite sulfide inclusions after Ca treatment was proposed, as shown in Figure 5. There are two stages: melting stage and solidification stage. In the melting stage, irregular Al_2O_3 was transformed to spherical Al_2O_3-CaO. Then Al_2O_3-CaO reacted with

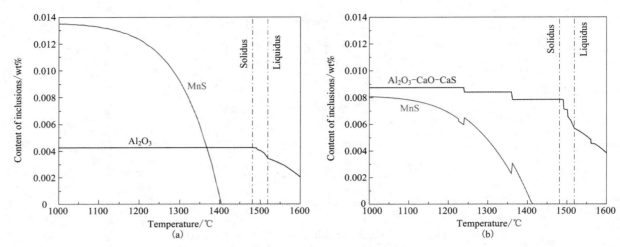

Figure 4　Phase transformation of inclusions during cooling of 16Mn steel (a) and 16Mn-Ca steel (b)

dissolved sulfur atom to form CaS in the outer layer of the inclusion. During solidification stage, CaS reacted with Mn and S atoms to form CaS-MnS because of excellent miscibility between CaS and MnS. It can be seen that the original pre-existing Al_2O_3 inclusions had positive effect on MnS control because they offered nucleation sites for MnS.

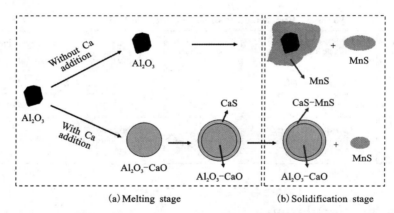

Figure 5　Schematic diagram for the formation and evolution of sulfide inclusions in the steel

5　Application of Ca treatment in industry

Based on the laboratory scale experimental results, an industrial trial was performed to study the effect of calcium treatment on sulfide inclusions in 16Mn heavy steel ingot. The production route was "electric arc furnace→ladle furnace→vacuum degassing→tundish→vacuum ingot casting". Calcium treatment was carried out by inserting calcium wire into molten steel after the vacuum break during vacuum degassing. Figure 6 shows the evolution of inclusions in the steel with effective calcium treatment[1]. The solid CaO-Al_2O_3 type inclusions were changed to liquid CaO-Al_2O_3 type with a large amount of CaS after calcium treatment. During the solidification stage, the liquid inclusions containing CaS and MnS were fully floated and the small-sized homogeneous and heterogeneous MnS inclusions were precipitated at the top of the ingot. In addition, the aspect ratio of the sulfide inclusions was about 1.6 (Figure 7), which indicated that calcium treatment had a good effect on the modification of large MnS inclusions. It is demonstrated that the calcium treatment prevented the generation of large-sized Type Ⅱ MnS inclusions at the top of the ingot.

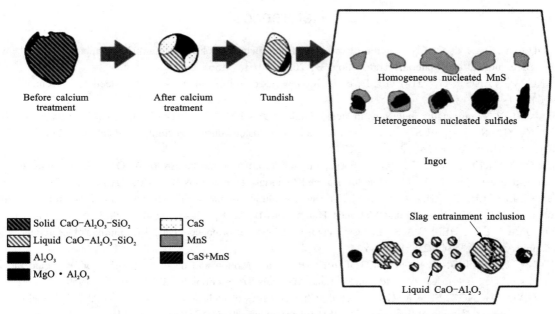

Figure 6 Evolution diagram of inclusions in heavy 303-ton ingots after Ca treatment[1]

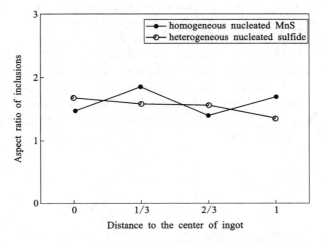

Figure 7 Aspect ratio of sulfide inclusions at the top of the ingot[1]

6 Conclusions

In the present study, laboratory scale experiments and thermodynamic analysis were performed to investigate the effect of calcium treatment on inclusions in a high-strength low-alloy steel. Ca treatment has an effective effect on MnS control. Without Ca treatment, sulfide inclusions were mainly Al_2O_3-MnS composite sulfide inclusions and pure MnS. After calcium treatment, sulfide inclusions were mainly spherical Al_2O_3-CaO-CaS-MnS and pure MnS. The proportion of pure MnS decreased from 74.6% to 40.1%. The number, area and maximum size of pure MnS decreased. The industrial trial demonstrated that calcium treatment had a good effect on the modification of large MnS inclusions and prevented the generation of large-sized Type II MnS inclusions at the top of the ingot.

Acknowledgments: The authors are grateful for support from the National Natural Science Foundation of China (grant No. U22A20171), the High Steel Center (HSC) at North China University of Technology, University of Science and Technology Beijing and Yanshan University.

References

[1] ZHOU Q Y, BA J T, CHEN W, et al. Evolution of non-metallic inclusions in a 303-ton calcium-treated heavy ingot [J]. Metallurgical and Materials Transactions B, 2023, 54(3): 1565-1581.

[2] YANG W, YANG X G, ZHANG L F, et al. Review of control of mns inclusions in steel[J]. Steelmaking, 2013, 29(6): 71-78.

[3] ZHANG L F. Non-metallic inclusions in steel: Fundamentals [M]. Beijing: Metallurgical Industry Press, 2019.

[4] REN Y, ZHANG L F, LI S S. Transient evolution of inclusions during calcium modification in linepipe steels[J]. ISIJ International, 2014, 54(12): 2772-2779.

[5] REN Q, YANG W, CHENG L, et al. Formation and deformation mechanism of Al_2O_3-CaS inclusions in Ca-treated non-oriented electrical steels[J]. Metallurgical and Materials Transactions B, 2020, 51(1): 200-212.

[6] ZHANG L F, LIU Y, ZHANG Y, et al. Transient evolution of nonmetallic inclusions during calcium treatment of molten steel[J]. Metallurgical and Materials Transactions B, 2018, 49(4): 1841-1859.

[7] HOLAPPA L E K, HELLE A S. Inclusion control in high-performance steels[J]. Journal of Materials Processing Technology, 1995, 53(1-2): 177-186.

[8] XUAN C J, PERSSON E S, SEVASTOPOLEV R, et al. Motion and detachment behaviors of liquid inclusion at molten steel-slag interfaces[J]. Metallurgical and Materials Transactions B, 2019, 50(4): 1957-1973.

[9] YANG W, ZHANG L F, WANG X H, et al. Characteristics of inclusions in low carbon Al-killed steel during ladle furnace refining and calcium treatment[J]. ISIJ International, 2013, 53(8): 1401-1410.

[10] TANAKA Y, PAHLEVANI F, KITAMURA S, et al. Behaviour of sulphide and non-Alumina-based Oxide inclusions in Ca-treated high-carbon steel[J]. Metallurgical and Materials Transactions B, 2020, 51(4): 1384-1394.

[11] GUO Y T, HE S P, CHEN G J, et al. Thermodynamics of complex sulfide inclusion formation in Ca-treated Al-killed structural steel[J]. Metallurgical and Materials Transactions B, 2016, 47(4): 2549-2557.

[12] XU J F, HUANG F X, WANG X H. Formation mechanism of CaS-Al_2O_3 inclusions in low sulfur Al-killed steel after calcium treatment[J]. Metallurgical and Materials Transactions B, 2016, 47(2): 1217-1227.

[13] ZHANG X W, ZHANG L F, YANG W, et al. Characterization of the three-dimensional morphology and formation mechanism of inclusions in linepipe steels[J]. Metallurgical and Materials Transactions B, 2017, 48(1): 701-712.

[14] CHEN X R, CHENG G G, HOU Y Y, et al. Oxide-inclusion evolution in the steelmaking process of 304L stainless steel for nuclear power[J]. Metals, 2019, 9(2): 257.

Numerical Simulation on Entrapment of Inclusions During Electroslag Remelting Process

Wei CHEN[1], Tianjie WEN[2], Lifeng ZHANG[3, *]

1 School of Mechanical Engineering, Yanshan University, Qinhuangdao, 066004, China
2 Hunan Aerospace Magnetoelectric Company, Limited, China Aerospace Science and Industry Corporation (CASIC), Changsha, 410200, China
3 School of Mechanical and Materials Engineering, North China University of Technology, Beijing, 100144, China

Abstract: In the current study, the movement, removal, dissolution, and entrapment of inclusions during the electroslag remelting (ESR) process were numerical simulated using a coupled multiphase flow model, solidification model, electromagnetic field model, and discrete phase model (DPM). The results show that the entrapment of 1 μm, 10 μm, and 100 μm inclusions were 4.24%, 3.49%, and 0.01%, respectively. Two inclusion accumulation zones in the ESR ingot were found, including the droplet-influenced zone and the ESR edge zone. The removal fraction of inclusions in the liquid film stage first increased and then decreased. As the inclusion diameter increased from 1 μm to 100 μm, the dissolution fraction of inclusions in the liquid film stage increased sharply from 1% to 88%. The removal fraction of inclusions first decreased and then increased as the inclusion diameter increased from 1 μm to 10 μm and 100 μm.

Keywords: Electroslag remelting; Inclusions; Entrapment; Mathematical simulation

1 Introduction

During the electroslag remelting (ESR) process, various complex phenomena such as the electrode melting, droplet dripping, directional solidification of the molten steel, steel-slag multiphase flow, chemical reaction, and removal of inclusions occur in the mold. Since the 1970s, many researchers have established and improved the numerical model for the ESR process and realized the simulation of the electromagnetic field, velocity field, and temperature field[1-3]. Complex processes such as the vibrating electrodes[4], three-phase electrodes[5], and rapid electroslag remelting[6] have also been realized. However, the movement, removal, and entrapment of inclusions in the ESR process still need to be further studied. Therefore, based on the complex multiphase flow and electromagnetic field distribution in the ESR process, the current study further coupled the inclusion transport model to study the movement, removal, dissolution, and entrapment of inclusions.

2 Mathematical model

A three-dimensional mathematical model was developed to predict the entrapment of inclusions during the ESR process. The melting current and frequency were 9 kA and 50 Hz, and the melting rate was 0.15 kg/s. The calculation model and distribution of the multiphase flow and electromagnetic field during the ESR process can be found in the author's previous study[7-9].

As shown in Figure 1, there are mainly four final destinations of the inclusions, including the removal of inclusions at the liquid film-slag interface, the removal of inclusions at the droplet-slag interface, the removal of inclusions at the metal pool-slag interface, and the entrapment of inclusions at the solidification front. In the current study, the removal, dissolution, and entrapment of inclusions during the ESR process were investigated and the discrete phase model (DPM) was employed to solve the transport of inclusions[10].

It is assumed that the random walk model was not considered when the inclusions move in the liquid film and droplet, but the random walk model was added when the inclusions move in the metal pool. The inclusions were assumed to be removed where the volume fraction of the molten steel was less than 0.1, and it is considered that the inclusions were entrapped by the solidification front where the solid fraction was greater than 0.6. The inhibition of the steel-slag interface on the movement of inclusions was achieved by introducing an interface resistance[11]. In addition, it was assumed that the inclusions staying in the liquid film at the electrode bottom and the metal pool were all dissolved through the steel-slag interface[12]. The initial position of the inclusions was at the bottom of the electrode, and the injection speed was 0.676 m/s. Three groups of inclusions with a diameter of 1 μm, 10 μm, and 100 μm were injected, and the number of inclusions in each group was 13025.

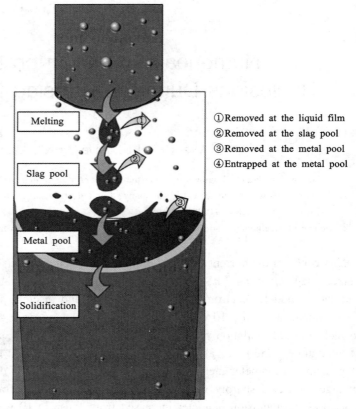

Figure 1 Four destinations of inclusions in the ESR process

3 Removal, dissolution, and entrapment of inclusions

3.1 Motion of inclusions during ESR process

Figure 2 shows the spatial distribution of 1 μm inclusions in the steel liquid film, droplet, and metal pool during the ESR process. The inclusions aggregated on the upper part of the liquid film due to the action of buoyancy. Therefore, some inclusions stayed in the liquid film and do not enter the molten pool with the droplet or remove at the steel-slag interface, especially the large size inclusions. In the dripping droplet, the inclusions were affected by the flow field and aggregated at the lower part of the droplet. When the inclusions enter the metal pool with the molten droplets, they were initially distributed in a conical shape at the upper center of the metal pool due to the impact of the droplets. Subsequently, the inclusions moved toward the side wall of the mold with the action of buoyancy and flow field. Finally, some inclusions floated to the upper surface of the metal pool to be removed or dissolved, and some inclusions moved with the molten steel to the solidification front and were entrapped.

3.2 Distribution of inclusions in slag pool

Figure 3 shows the effect of the inclusion diameter on the dissolution fraction and removal fraction at the liquid film stage. As the inclusion size increases from 1 μm to 100 μm, the dissolution fraction of inclusions increased sharply from 1% to 88%, while the removal fraction of inclusions first increased and then decreased. The proportion of 1 μm, 10 μm, and 100 μm inclusions entering the metal pool were 73.98%, 72.95%, and 0.31%, respectively. With the effect of the buoyancy, it is difficult for large size inclusions to move down into the metal pool with the droplet, and it is also difficult to move down into the slag pool through the liquid film-slag interface at the electrode bottom. Therefore, most of the large size inclusions stayed in the liquid film at the electrode bottom and finally dissolved into the slag phase through the liquid film-slag interface.

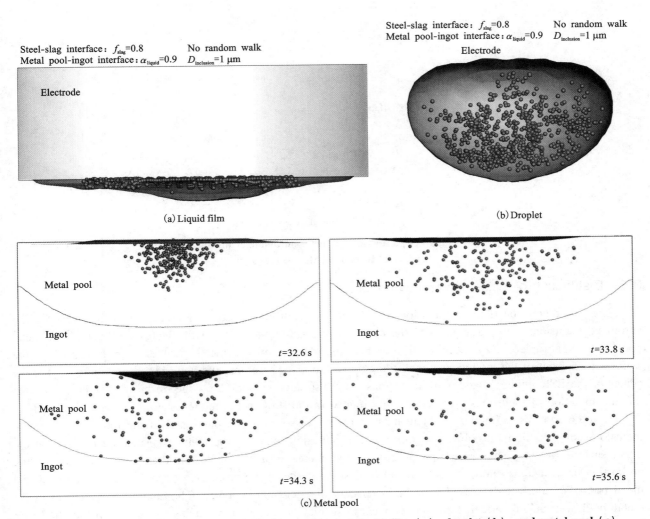

Figure 2 Spatial distribution of 1 μm inclusions in the steel liquid film (a), droplet (b), and metal pool (c)

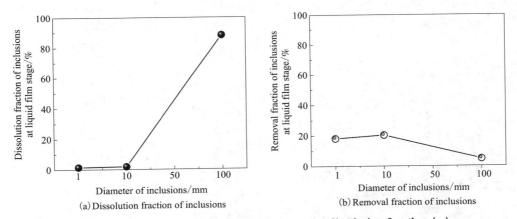

Figure 3 Effect of the inclusion diameter on the dissolution fraction (a) and removal fraction (b) at the liquid film stage

Figure 4 shows the effect of the inclusion diameter on the removal fraction at the droplet stage. The removal fraction of inclusions in the droplet stage was less affected by the inclusion diameter. When the inclusion diameter increased from 1 μm to 10 μm and 100 μm, the removal fraction of inclusions first decreased and then increased.

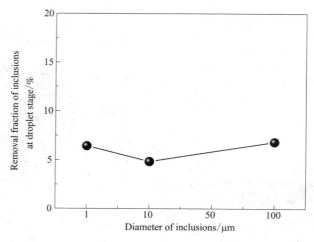

Figure 4 Effect of the inclusion diameter on the removal fraction at the droplet stage

3.3 Distribution of inclusions in metal pool

The entrapment locations of inclusions on the cross section of the ESR ingot are shown in Figure 5. The entrapped inclusions were aggregated at the center of the ESR ingot. When the inclusions entered the metal pool with the droplet, they were easily reached the bottom of the metal pool and be captured by the solidification front of the molten steel due to the impact force. Therefore, the inclusions were aggregated at the center of the ESR ingot. In addition, inclusions were also aggregated at the edge of the ingot, indicating that the accumulation of inclusions here was mainly related to the flow of the molten steel.

Figure 6 shows the fraction of inclusions removed, dissolved, and entrapped in the molten pool to the inclusions entering the melt pool. The proportion of inclusions entering the metal pool that are removed at the upper surface of the metal pool increased with the increase of the inclusion diameter. Contrary to the proportion of inclusions removed at the metal pool-slag interface, the proportion of inclusions dissolved at the metal pool-slag interface decreased with the increase of the inclusion diameter. The floating of inclusions in the metal pool was mainly affected by the buoyancy, while the removal of inclusions at the metal pool-slag interface was affected by the flow of the molten steel and the interface resistance. As shown in Figure 6, the proportion of inclusions entrapped by the solidification front in the metal pool decreased with the increase of the inclusion diameter.

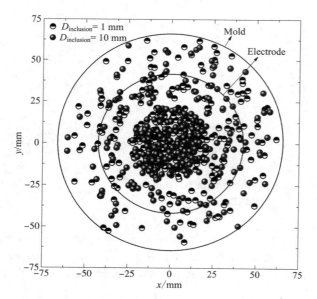

Figure 5 Entrapment locations of inclusions on the cross section of the ESR ingot

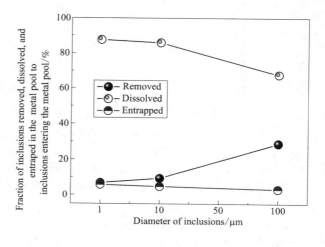

Figure 6 The fraction of inclusions removed, dissolved, and entrapped in the metal pool to the inclusions entering the metal pool

The effect of the inclusion diameter on the dissolution fraction, removal fraction, and entrapment fraction at the metal pool stage is compared in Figure 7. It can be seen that the dissolution fraction, removal fraction, and entrapment fraction of inclusions generally decreased with the increase of the inclusion diameter. Although the 100 μm inclusions were more likely to float up to the metal pool-slag interface, only 0.31% of the 100 μm inclusions entered the metal pool. Thus, the dissolution fraction, removal fraction, and entrapment fraction of 100 μm inclusions in the metal pool stage were significantly smaller than those of 1 μm and 10 μm inclusions. Table 1 summarizes the destinations of 1 μm, 10 μm, and 100 μm inclusions during the ESR process.

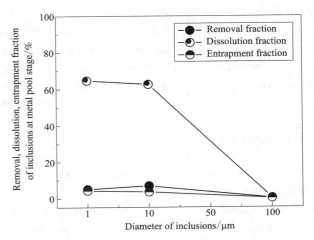

Figure 7 Effect of the inclusion diameter on the dissolution fraction, removal fraction, and entrapment fraction at the metal pool stage

Table 1 Effect of the inclusion diameter on the destination of inclusions

	Diameter/μm	1	10	100
Liquid film stage	Removal fraction/%	18.14	20.28	4.79
	Dissolution fraction/%	1.45	1.92	88.06
Droplet stage	Removal fraction/%	6.43	4.85	6.84
Metal pool stage	Removal fraction/%	5.04	6.84	0.09
	Dissolution fraction/%	64.7	62.61	0.21
	Entrapment fraction/%	4.24	3.49	0.01

Figure 8 shows the distribution of inclusions on the cross section of the ESR ingot in mathematical simulation and industrial trial. The accumulation zones of inclusions in the center and below the electrode edge were defined as droplet impact zone and electrode side zone, respectively. For the mathematical simulation, a droplet was formed and falled at the center of the mold, thus the droplet impact zone was located at the center of the electroslag ingot with a radius of about 20 mm. For the industrial trial, the electrode was welded by two billets with a cross-section of 150 mm×150 mm. The position where the droplet was formed during remelting

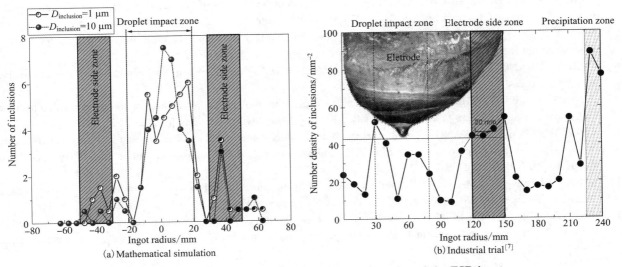

Figure 8 Distribution of inclusions on the cross section of the ESR ingot

was not located at the geometric center of the mold, but slightly off from the geometric center, as shown in Figure 8(b). Therefore, the droplet impact zone obtained from the industrial trial was located at the position where the radius of the electroslag ingot was 60 mm. The radius of the droplet impact zone was about 22 mm. For 1 μm inclusions, the entrapment fraction obtained using mathematical simulation was closer to the industrial trial results. For 10 μm inclusions, the simulated entrapment fraction was quite different from the industrial trial results due to ignore of the precipitation of inclusions.

4 Conclusions

(1) As the inclusion diameter increased from 1 μm to 100 μm, the dissolution fraction of inclusions in the liquid film stage increased sharply from 1% to 88%, while the removal fraction of inclusions in the liquid film stage first increased and then decreased. Most of the large size inclusions stayed in the liquid film at the bottom of the electrode and finally dissolved into the slag phase through the steel-slag interface.

(2) The removal fraction of inclusions in the droplet stage was less affected by the inclusion diameter. The interfacial resistance reduced the removal fraction of 1 μm and 10 μm inclusions in the droplet stage, but the removal fraction of 100 μm inclusions increased slightly.

(3) The entrapment fraction of 1 μm, 10 μm, and 100 μm inclusions were 4.24%, 3.49%, and 0.01%. There were two inclusion accumulation zones in the ESR ingot, including the droplet-influenced zone and the ESR edge zone.

Acknowledgments: The authors are grateful for the support from the National Science Foundation China (grant No. U22A20171, No. 52104343), the High Steel Center (HSC) at Yanshan University and North China University of Technology, China.

References

[1] DILAWARI A, SZEKELY J. A mathematical model of slag and metal flow in the ESR process[J]. Metallurgical Transactions B, 1977, 8: 227-236.
[2] KHARICHA A, LUDWIG A, WU M. Shape and stability of the slag/melt interface in a small DC ESR process[J]. Materials Science and Engineering A, 2005, 413: 129-134.
[3] YU J, LIU F, LI H, et al. Numerical simulation and experimental investigation of nitrogen transfer mechanism from gas to liquid steel during pressurized electroslag remelting process[J]. Metallurgical and Materials Transactions B, 2019, 50: 3112-3124.
[4] WANG F, LOU Y, CHEN R, et al. Effect of vibrating electrode on temperature profiles, fluid flow, and pool shape in ESR system based on a comprehensive coupled model[J]. China Foundry, 2015, 12(4): 285-292.
[5] WANG Q, LI G, HE Z, et al. A three-phase comprehensive mathematical model of desulfurization in electroslag remelting process[J]. Applied Thermal Engineering, 2017, 114: 874-886.
[6] KARIMI SIBAKI E, KHARICHA A, WU M, et al. A multiphysics model of the electroslag rapid remelting (ESRR) process[J]. Applied Thermal Engineering, 2018, 130: 1062-1069.
[7] WEN T, REN Q, ZHANG L, et al. Evolution of nonmetallic inclusions during the electroslag remelting process[J]. Steel Research International, 2021, 92(6): 2000629.
[8] WEN T, ZHANG H, LI X, et al. Numerical simulation on the oxidation of lanthanum during the electroslag remelting process[J]. JOM, 2018, 70: 2157-2168.
[9] ZHANG L, WEN T, CHEN W, et al. Mathematical modeling on the initial melting of the consumable electrode during electroslag remelting process[J]. Metallurgical and Materials Transactions B, 2021, 52: 4033-4045.
[10] CHEN W, ZHANG L, WANG Y, et al. Prediction on the three-dimensional spatial distribution of the number density of inclusions on the entire cross section of a steel continuous casting slab[J]. International Journal of Heat and Mass Transfer, 2022, 190: 122789.
[11] STRANDH J, NAKAJIMA K, ERIKSSON R, et al. A mathematical model to study liquid inclusion behavior at the steel-slag interface[J]. ISIJ International, 2005, 45(12): 1838-1847.
[12] FEICHTINGER S, MICHELIC S K, KANG Y B, et al. In situ observation of the dissolution of SiO_2 particles in $CaO-Al_2O_3-SiO_2$ slags and mathematical analysis of its dissolution pattern[J]. Journal of the American Ceramic Society, 2014, 97(1): 316-325.

Influence of Microstructure on the Flow Boiling Heat Transfer Characteristics of Diamond/Cu Heat Sink

Mingmei SUN[1,2,3], Nan WU[1,2,3], Hong GUO[1,2,3,*], Zhongnan XIE[1,2,3], Shuhui HUANG[1,2,3]

1. State Key Laboratory of Nonferrous Metals and Processes, GRINM Group Co., Ltd, Beijing, 100088, China
2. GRIMAT Engineering Institute Co., Ltd., Beijing, 101407, China
3. General Research Institute for Nonferrous Metals, Beijing, 100088, China

Abstract: Micro heat sink is an effective means of solving high-power chip heat dissipation. Diamond/Cu composites exhibit high thermal conductivity and a thermal expansion coefficient that is compatible with semiconductor materials, rendering them ideal heat sink materials. This study aims to incorporate secondary diamond particles into the gaps of the main diamonds, thus constructing a three-dimensional heat conduction network within the composite material. By incorporating a Cr_3C_2 interface layer, the interface bonding strength can be enhanced, and the interfacial mismatch can be reduced, thereby decreasing the interfacial thermal resistance and ensuring high thermal conductivity at the connections within the heat conduction network. The diamond/Cu composite material was fabricated into a heat sink and subjected to a flow boiling heat transfer experiment. Results indicate that the diamond/Cu heat sink displayed a decrease in wall superheat of 10.2-14.5 ℃ and an improvement in heat transfer coefficient of 37.5%-51.1% compared with a Cu heat sink under identical heat fluxes. The micro heat sink based on diamond/Cu shows great potential in high-power chip packaging.

Keywords: Diamond/Cu; Microstructure; Heat transfer; Flow boiling

1 Introduction

With the development of electronic information technology, electronic devices are demonstrating a trend toward high frequency, high speed, high integration, and miniaturization[1-3]. The heat generated by electronic devices per unit area is rapidly increasing, and the large amount of heat produced severely affects the stability and reliability of electronic devices. More than 55% of electronic device failures are caused by heat dissipation issues related to temperature[4].

In the face of increasingly severe heat dissipation issues, the main solution is to develop efficient cooling methods[5]. The closer the heat sink is to the chip, the higher its heat dissipation efficiency. Materials with a low coefficient of thermal expansion, such as W-Cu, Mo-Cu, SiC/Al, C/Cu, diamond/Cu, and diamond, have been extensively investigated for their compatibility with semiconductor materials in terms of the thermal conductivity near the chip area[6,7]. In addition to enhancing thermal conductivity, introducing thermal convection to improve the heat dissipation efficiency is another important approach[8]. This approach involves increasing the heat transfer through convection to augment the overall cooling effect. Indeed, techniques such as cold plates and microchannel heat sinks (MCHS) utilize heat convection to enhance heat transfer capability[9]. One particular area of research focuses on heat sinks that utilize the latent heat of phase change to dissipate large amounts of heat[10]. These heat sinks aim to take advantage of the high heat transfer capacity associated with phase change processes, such as boiling or evaporation, to achieve efficient cooling[11,12]. When facing the demands of miniaturization and high heat flux for heat dissipation, the optimal solution is to combine heat sink with phase change cooling. However, previous research efforts have primarily focused on the design of structures[13], with limited studies on the influence of materials on phase change micro heat sinks. Researchers commonly use metal materials (copper and aluminum) to fabricate heat sinks[14]. However, the high thermal expansion coefficient of metal materials presents a significant disparity with the thermal expansion coefficient of semiconductor materials, making them unsuitable for heat transfer applications near the chip. Qi[15] and Yang[16] proposed the use of diamond thin films as a material for heat sinks. Compared with aluminum MCHS, diamond MCHS exhibited an improved heat transfer coefficient of 37% to

73% and higher stability. However, diamond thin film as a material presents challenges in processing, a limited thickness, and an inability to cater to various usage scenarios. Diamond/Cu composites have high thermal conductivity [650-850 W/(m·K)] and a thermal expansion coefficient (3.8×10^{-6}-5.8×10^{-6}/K) that matches that of semiconductor materials. Moreover, diamond/Cu composites possess better processability and can be manufactured in larger volumes than diamond thin films. These characteristics make diamond/Cu composite heat sinks promising for addressing high thermal loads. Given that the thermal expansion coefficient closely matched that of semiconductor materials, diamond/Cu composites can theoretically be applied in the field of heat transfer for semiconductor chip packaging[17]. Heat sinks fabricated by utilizing the high thermal conductivity of diamond/Cu composites also exhibit excellent thermal dissipation performance near the chip. However, few have studied the heat transfer performance of diamond/Cu composites in relation to their inherent composite material properties.

In this study, a complex heat conduction network was constructed within the composite material by incorporating diamond particles of two size ranges. This approach ensures that the composites exhibit a high thermal conductivity. The composite was then processed into a micro heat sink for the flow boiling heat transfer experiments. Through interdisciplinary research, this experiment combined the unique material microstructure of diamond/Cu composites with a theoretical model designed specifically for flow boiling heat transfer experiments, providing guidance for the application of diamond/Cu composites in the field of flow boiling heat transfer.

2 Experimental setup

Figure 1 shows the schematic of the experimental flow loop system. All components of the entire setup are connected using stainless steel pipes, forming a closed-loop system. Deionized water flows out of the liquid reservoir and then pumped into a flow meter by a gear pump before entering the preheater. The preheater is responsible for heating the deionized water to the desired temperature for experimental purposes. After passing through the preheater, the deionized water enters the testing section. The temperature and pressure of the fluid at the inlet and outlet of the test section are monitored using two K-type thermocouples and two pressure transducers. The heating power of the cartridge heaters in the testing section is controlled by a DC power supply unit, and the heating temperature is monitored by 15 T-type thermocouples, which are distributed across three layers. The fluid then flows back to the liquid reservoir after passing through the condenser.

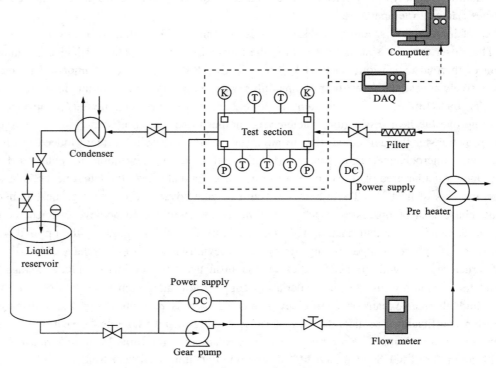

Figure 1　Schematic of the flow loop

Figure 2 illustrates the decomposition diagram of the test section and presents a schematic representation of the heat sink in the present study. The test section comprises seven main components: a cover plate, an upper aluminum plate, a polytetrafluoroethylene (PTFE) housing, a copper substrate with a heat sink, a thermal resistant insulation board, and a lower aluminum plate. The cover plate is made of polysulfone and secured to the upper aluminum plate using bolts. The cover plate is designed with liquid inlet and outlet ports and spaces for liquid flow. Fifteen holes are distributed on both the PTFE housing and the copper substrate to accommodate the insertion of thermocouples. The copper substrate is made of oxygen-free copper and connected to the micro heat sink through welding. Additionally, the copper substrate has four holes for inserting cartridge heaters. The bottom thermal insulation board is used to reduce heat dissipation. The lower aluminum plate supports the entire structure and is securely fixed to the upper aluminum plate using bolts.

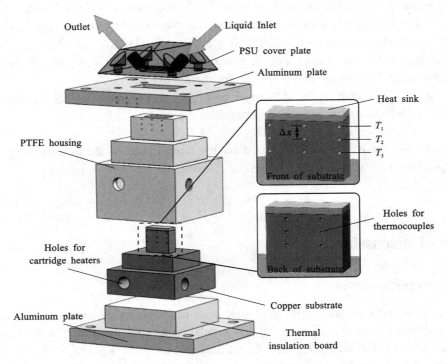

Figure 2　Schematic of the test section

3　Heat transfer data reduction

Mass flux G is fixed at 507 kg/(m² · s) and obtained using the following equation:

$$G = \frac{v\rho}{A} \tag{1}$$

where: v is the velocity of the liquid, ρ is the density of the liquid, and A is the cross-sectional area for flow (10 mm×0.5 mm).

Effective heat flux q_{eff} can be represented using the one-dimensional Fourier's law of heat conduction:

$$q_{\text{eff}} = -\lambda_i \frac{dT}{dx} \tag{2}$$

where λ_i is the thermal conductivity of the material. For copper, the thermal conductivity is 389 W/(m · K), and that for diamond/Cu composites is 780 W/(m · K). dT/dx represents the temperature gradient in the longitudinal direction of the copper substrate when it is being heated. This temperature gradient can be estimated using the temperature data from the three thermocouples embedded in the substrate, using a three-point backward Taylor series approximation as follows:

$$\frac{dT}{dx} \cong \frac{3T_{1.5} - 4T_{2.5} + T_{3.5}}{2\Delta x} \tag{3}$$

where: T_1, T_2, and T_3 represent the temperatures of the thermocouples on the substrate from top to bottom, as shown in Figure 3. Δx is the distance between them and is equal to 4 mm. This approach was used by Buchling and Kandlikar[18] and Gupta and Misra[19].

The temperature of the heat-exchange surface (T_w) refers to the temperature of the heat sink surface that is in contact with the fluid. T_w can be calculated using the one-dimensional Fourier's law of heat conduction based on the temperature values obtained from the top thermocouples and q_{eff}. In this study, a soldered connection of tin between the substrate and the heat sink was used. Therefore, when calculating the thermal conductivity, the thermal resistances of the substrate, the solder layer, and the heat sink must be considered. T_w is calculated using the following equation:

$$T_w = T_1 - q_{eff}\left(\frac{L_1}{\lambda_{Cu}} + \frac{L_T}{\lambda_T} + \frac{L_{HE}}{\lambda_{HE}}\right) \quad (4)$$

where: L_1 is the distance between the top thermocouple and the top surface of the substrate, which is 1.5 mm. L_T and λ_T represent the thickness and thermal conductivity of the tin solder layer, respectively. According to reference[20-22], the following can be assumed: $L_T = 0.1$ mm and $\lambda_T = 58$ W/(m·K). L_{HE} is the thickness of the heat sink, which is 2.0 mm. Depending on the material of the heat sink, the thermal conductivity is denoted as λ_{Cu} for oxygen-free copper and $\lambda_{diamond/Cu}$ for diamond/Cu.

Wall superheat ΔT_{sat}, defined as the difference between heat-exchange surface T_w and liquid saturation temperature T_{sat}, is calculated using the following formula:

$$\Delta T_{sat} = T_w - T_{sat} \quad (5)$$

Heat transfer coefficient h is calculated based on the temperature of the heat-exchange surface (T_w), outlet liquid temperature T_f, and effective heat flux q_{eff}, using the following equation:

$$h = \frac{q_{eff}}{T_w - T_f} \quad (6)$$

4 Results and discussion

4.1 Heat transfer performance

The primary indicator for evaluating the heat sink is its heat transfer performance. Figure 3(a) presents the boiling curves obtained from the experimental and simulation data for diamond/Cu and Cu heat sinks at the following condition: inlet subcooling $\Delta T_{sub} = 20$ ℃ and mass flux $G = 507$ kg/(m²·s). The x-axis represents the wall superheat (ΔT_{sat}), while the y-axis represents the effective heat flux (q_{eff}) of the heat sink. The slope and position of the boiling curve reflect the intensity of boiling of the working fluid. The boiling curve of the diamond/Cu heat sink is shifted to the right and has a steeper slope than the boiling curve of the Cu heat sink. This result indicates that the diamond/Cu heat sink exhibits a greater boiling intensity than the Cu heat sink at the same q_{eff} and effectively reduces the heat-exchange surface temperature, which is due to the higher thermal conductivity of the diamond/Cu composite and the heat conduction network created by the diamond particles within it. To study their effects, we used the finite volume method in the computational fluid dynamics software ANSYS Fluent to simulate the heat sink and the boiling heat transfer process with fluid flow. In this simulation, the multiphase flow was modeled using the volume of fluid model, and the viscous model utilized was RNG k-epsilon. The boundary conditions were set based on the ΔT_{sub} and G of the fluid in the experiment. Only the thermal conductivity of the heat sink material was varied. After considering the heat loss due to environmental and design factors, the simulated results of the Cu heat sink were fitted to the experimental data. The simulation curve of the Cu heat sink closely matches the experimental curve. The simulated data of the diamond/Cu heat sink with a thermal conductivity of 780 W/(m·K) were fitted using the same method used for the Cu heat sink. However, the experimental boiling curve of the diamond/Cu heat sink is shifted even higher than the simulation curve, with a significant difference. This result suggests that merely altering the thermal conductivity of the material is insufficient to simulate the flow boiling state of a diamond/Cu heat sink under realistic conditions. Figure 3(b) displays the experimental and simulated heat transfer coefficient curves of the two materials in the heat sink. The x-axis represents the effective heat flux (q_{eff}) of the heat sink, while the y-axis represents heat transfer coefficient h. The heat transfer coefficient curve visually illustrates the heat sink's heat transfer capacity. At a low heat flux ($q_{eff} < 480$ kW/m²), the heat transfer coefficients of the two materials of the heat sink are relatively similar. However, the h of the diamond/Cu heat sink rapidly increases as the q_{eff} increases and becomes significantly higher than that of the Cu heat sink. Compared with the Cu heat sink at the same q_{eff}, the diamond/Cu heat sink exhibits a reduction in T_w ranging from 10.2 ℃ to 14.5 ℃. Furthermore, h was enhanced by 37.5% to 51.1%.

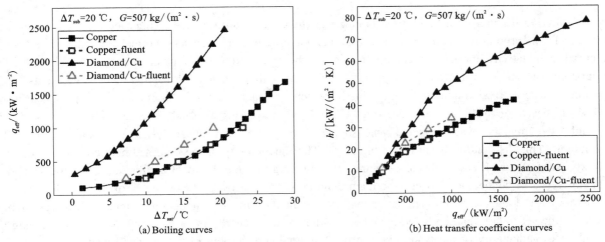

Figure 3 Heat transfer performance comparison between diamond/Cu and Cu heat sink

4.2 Thermal conduction model and refinement of heat transfer performance

The experiment employed Cu-Cr alloy as the matrix and utilized Cr for interface control. Under high-temperature preparation conditions, the C element on the surface of the diamond combined with the Cr element in the matrix, leading to the formation of a Cr_3C_2 interface layer. To further investigate the interfacial thermal conductivity, the experiment adopted the diffusion mismatch model (DMM). By considering the phonon scattering at the interface as elastic scattering rather than specular reflection, the DMM model calculates the phonon interface transmission probability (η) from material 1 to material 2 as follows:

$$\eta = \frac{\sum_j v_{j,2}^{-2}}{\sum_j v_{j,1}^{-2} + \sum_j v_{j,2}^{-2}} \tag{7}$$

The interface thermal conductivity (h) in the DMM model is expressed as follows:

$$h = \frac{1}{4}\rho_1 c_1 v_1 \eta = \frac{\rho_1 c_1 v_1^3}{4(v_1^2 + v_2^2)} \tag{8}$$

where ρ represents the density of the material, c denotes the specific heat capacity of the material, and v signifies the Debye phonon velocity of the material.

Based on the material parameters provided in Table 1, the corresponding values of h for different materials can be determined. The thermal conductivity coefficients at both the incident and outgoing interfaces must be considered when calculating the interfacial thermal conductivity because phonons pass through Cr_3C_2 before reaching the diamond from the copper substrate and then exit. According to the calculations using the DMM method, the obtained interface h values are shown in Table 2. After obtaining the interfacial thermal conductivity coefficients, we can incorporate them into the thermal conduction model and further adjust the simulation results by setting the boundary conditions at the interface.

Table 1 Material parameters for the calculation of interface thermal conductivity by DMM

Material	Density $\rho/(\text{kg}\cdot\text{m}^{-3})$	Thermal conductivity $\lambda/[\text{W}\cdot(\text{m}\cdot\text{K})^{-1}]$	Specific heat $c/[\text{J}\cdot(\text{kg}\cdot\text{K})^{-1}]$	Phonon velocity $v/(\text{m}\cdot\text{s}^{-1})$
Cu	8900	398	386	2881
Cr_3C_2	6680	19	456	5628
Diamond	3520	1500	512	12775

Table 2 Interfacial thermal conductivity coefficients calculated by DMM

Interface	Cu-Cr_3C_2	Cr_3C_2-Diamond	Diamond-Cr_3C_2	Cr_3C_2-Cu
$h/(W \cdot m^{-2} \cdot K^{-1})$	5.138×10^8	6.966×10^8	4.820×10^9	1.955×10^9

Considering the uniformity and good bonding of the Cr_3C_2 interface layer, it can be regarded as an ideal interface. By incorporating the interfacial thermal conductivity into the transient thermal simulation in ANSYS, a thermal conductivity model that is applicable to the diamond/Cu composite can be established. The temperature was set according to the temperature gradient within the material of the same dimensions in the experiment. The thermal conductivity at the interface between each diamond particle and the copper matrix was determined based on Table 2. By utilizing a graphical processing method, the real cross-section of the diamond/Cu composite was extracted and subjected to finite element analysis to obtain the heat flux distribution. This distribution is shown in Figure 4. The functionality of the diamond heat conduction network was experimentally verified, with the main heat flux flowing through the main diamonds, while the secondary diamonds serve as connectors between the main diamonds. The heat flow tended to prioritize the shortest heat-conducting path along the cooling direction within the main diamond conduction channels. In cases of high heat flux, the secondary diamonds redirected excess heat toward nearby main diamonds, further enhancing the internal heat conductivity of the material. The high thermal conductivity of the diamond neatwork is bound to impact the heat transfer performance of the composite heat sink.

After incorporating the microstructure and the interfacial thermal conductivity, further refinements were made to the heat transfer data. Under the same wall superheat conditions, the heat flux from the transient heat transfer model in ANSYS was included. The modified boiling and heat transfer coefficient curves are shown in Figure 5. At the same wall superheat, the introduction of the heat conduction network significantly increased the heat flux in the diamond/Cu composite. The modified boiling and heat transfer coefficient curves after the refinement closely resemble the experimental data. The coupling of the fluent and transient thermal models formed a new model that can accurately simulate the flow boiling heat transfer performance of the diamond/Cu heat sink. The excellent heat transfer performance in the flow boiling of the diamond/Cu heat sink is mainly attributed to the high heat conduction network created by the stacking of diamond particles.

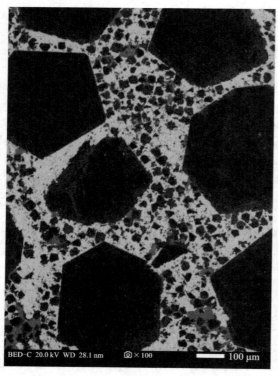 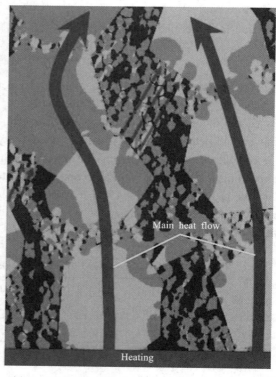

(a) Cross-sectional BED of the diamond/Cu composite   (b) Simulation of heat flux based on transient thermal analysis

Figure 4

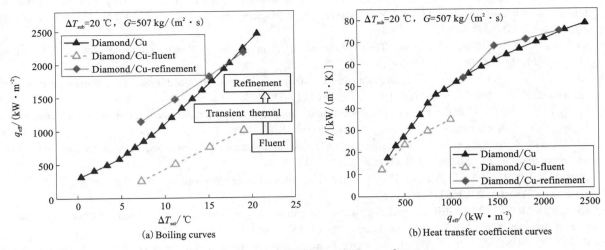

Figure 5 Refinement of heat transfer performance

5 Conclusions

The heat dissipation capability and temperature uniformity of the diamond/Cu heat sink are superior to those of traditional Cu heat sink. The diamond/Cu heat sink exhibits a low wall superheat when subjected to high heat fluxes, greatly decreasing the surface temperatures on chips and helping suppress the occurrence of local overheating, thereby protecting the heating elements:

(1) Under the same heat load, the diamond/Cu heat sink exhibits higher boiling intensity and an improved heat transfer coefficient h of 37.5%-51.1% compared to the Cu heat sink. At a high heat flux (q_{eff}) of approximately 1600 kW/m^2, the diamond/Cu heat sink has a temperature difference that is 29.06% lower than that of the Cu heat sink.

(2) The heat conduction network constructed using diamonds is the main factor enhancing the heat transfer performance of the composite. The low thermal resistance of the diamond channels significantly increases the internal heat flux of the material. The utilization of a uniform, complete, and continuous transition layer enhances the interfacial bonding and compatibility while ensuring a high interfacial thermal conductivity.

(3) Combining the transient thermal analysis model, which considers the thermal conduction network and the interfacial layer thermal conductivity, with the fluent model allows for a highly accurate simulation of the flow boiling heat transfer performance of the diamond/Cu composite heat sink.

Acknowledgments: This work was supported by Youth Fund Project of Grinm and State Key Laboratory of Nonferrous Metals and Processes.

References

[1] KANG J S, LI M, WU H, et al. Experimental observation of high thermal conductivity in boron arsenide[J]. Science, 2018(361): 575-578.

[2] Ghani I A, Kamaruzaman N, Sidik N. Heat transfer augmentation in a microchannel heat sink with sinusoidal cavities and rectangular ribs[J]. Int. J. Heat Mass Transf. 2017(108): 1969-1981.

[3] LEE E, MENUMEROV E, HUGHES R A, et al. Low-cost nanostructures from nanoparticle-assisted large-scale lithography significantly enhance thermal energy transport across solid interfaces[J]. ACS Appl. Mater. Interfaces, 2018(10): 34690-34698.

[4] MALLIK S, EKERE N, BEST C, et al. Investigation of thermal management materials for automotive electronic control units[J]. Appl. Therm. Eng. 2011(31): 355-362

[5] KUDDUSI L, DENTON J C. Analytical solution for heat conduction problem in composite slab and its implementation in constructal solution for cooling of electronics[J]. Energ. Convers. Manage. 2007(48): 1089-1105.

[6] MOLINA J M, NARCISO J, WEBER L, et al. Thermal conductivity of Al-SiC composites with monomodal and bimodal particle size distribution[J]. Materials Science & Engineering A, 2008, 480(1-2): 483-488. DOI: 10.1016/j.msea.2007.07.026

[7] CHEN P, LUO G, SHEN Q, et al. Thermal and electrical properties of W-Cu composite produced by activated sintering[J]. Mater. Des. 2013(46): 101-105.

[8] LIN Y, LI J, LUO Y, et al. Conjugate heat transfer analysis of bubble growth during flow boiling in a rectangular microchannel[J]. Int. J. Heat Mass Transf., 2021(181): 121828.

[9] WANG G, HAO L, CHENG P. An experimental and numerical study of forced convection in a microchannel with negligible axial heat conduction[J]. Int. J. Heat Mass Transf., 2009(52): 1070-1074

[10] QIAN J, LI X, WU Z, et al. A comprehensive review on liquid-liquid two-phase flow in microchannel: flow pattern and mass transfer[J]. Microfluid Nanofluid., 2019(23): 1-30.

[11] MCHALE J P, GARIMELLA S V. Heat transfer in trapezoidal microchannels of various aspect ratios[J]. Int. J. Heat Mass Transf., 2010(53): 365-375.

[12] KANDLIKAR S G, WIDGER T, KALANI A, et al. Enhanced flow boiling over open microchannels with uniform and tapered gap manifolds[J]. J. Heat Transf., 2013(135): 061401.

[13] DENG D, ZENG L, SUN W, et al. Experimental study of flow boiling performance of open-ring pin fin microchannels[J]. Int. J. Heat Mass Transf., 2021(167): 120829.

[14] YIN L, JIANG P, XU R, et al. Heat transfer and pressure drop characteristics of water flow boiling in open microchannels[J]. Int. J. Heat Mass Transf., 2019(137): 204-215.

[15] QI Z, ZHENG Y, ZHU X, et al. An ultra-thick all-diamond microchannel heat sink for single-phase heat transmission efficiency enhancement[J]. Vacuum, 2020, 177: 109377-. DOI: 10.1016/j.vacuum.2020.109377.

[16] YANG Q, ZHAO J Q, HUANG Y P, et al. A diamond made microchannel heat sink for high-density heat flux dissipation[J]. Applied Thermal Engineering, 158[2023-07-15]. DOI: 10.1016/j.applthermaleng.2019.113804.

[17] LI Y, ZHOU H, WU C, et al. The interface and fabrication process of diamond/Cu composites with nanocoated diamond for heat sink applications[J]. Metals-Open Access Metallurgy Journal, 2021, 11(2): 196. DOI: 10.3390/met11020196.

[18] BUCHLING P, KANDLIKAR S G. Enhanced flow boiling of ethanol in open microchannels with tapered manifolds in a gravity-driven flow[J]. J. Heat Transf., 2016(138): 031503.

[19] GUPTA S K, MISRA R D. Experimental study of pool boiling heat transfer on copper surfaces with $Cu-Al_2O_3$ nanocomposite coatings[J]. Int. Commun. Heat Mass Transf., 2018(97): 47-55.

[20] WEIBEL J A, GARIMELLA S V, NORTH M T. Characterization of evaporation and boiling from sintered powder wicks fed by capillary action[J]. Int. J. Heat Mass Transf., 2010(53): 4204-4215.

[21] DENG D, TANG Y, LIANG D, et al. Flow boiling characteristics in porous heat sink with reentrant microchannels [J]. Int. J. Heat Mass Transf., 2014(70): 463-477.

[22] YIN L, SUN M, JIANG P, et al. Heat transfer coefficient and pressure drop of water flow boiling in porous open microchannels heat sink[J]. Appl. Therm. Eng., 2023(218): 119361.

Basic Research on Preparation of Al-Si Alloy from Oxides of Al and Si by Molten Salt Electrolysis

Jiaxin YANG[1,2], Zhaowen WANG[1,2,*], Wenju TAO[1,2,*], Youjian YANG[1,2]

1 Key Laboratory of Ecological Metallurgy of Polymetallic Minerals, School of Metallurgy, Northeastern University, Shenyang, 110819, China
2 School of Metallurgy, Northeastern University, Shenyang, 110819, China

Abstract: In view of the problems in the comprehensive utilization of solid waste with high content of Al and low Si, the basic problems of preparing Al-Si alloy by electrolysis of oxides mainly containing Al and Si in cryolite molten salt have been studied. The influence of the mass fraction of silicon in the electrolyte on the physical and chemical properties of molten salt was studied. Besides, the dissolution rate of mullite and kaolinite in molten salt was measured. The reduction mechanism of silicon was explored by the electrochemical method to further optimize the electrolytic process. Under the optimum process conditions, Al-Si alloy with a silicon content of 0-22wt% was obtained.

Keywords: Al-Si alloy; Molten salt electrolysis; Electrochemical behavior

1 Introduction

Considering the huge demand for energy in the world today, thermal power generation is still an irreplaceable mode of production capacity, accounting for about 40% of the world's total power generation[1-3]. The by-product of coal combustion at high temperatures is an oxide mainly composed of alumina and silica (named OAS)[4]. It is reported that 600 million to 1000 million tons of OAS will be produced in the world every year[5-6]. Moreover, further efficient recycling is limited due to the extremely complex components[7].

Relevant reports show that only 10% of the mixture will have the opportunity to be used in the fields of high value-added applications, such as nanomaterials and energy storage. Most of the mixtures are still mainly used in poor-value fields such as fertilizers and roads. Some even forced to pile up or landfill[8] leading to serious environmental problems.[9-10] Therefore, finding a green, efficient and high value-added recycling method has become an urgent necessity.

In this study, molten salt electrolysis was used to electrochemical reduce OAS to prepare high-value Al-Si alloys. The dissolution rate of OAS in cryolite molten salt was tested. The electrochemical reduction mechanism of Si was analyzed by cyclic voltammetry to optimize the electrolytic process. High-value Al-Si alloy is prepared by molten salt electrolysis.

2 Experiment

2.1 Materials

The oxides of Al and Si (named OAS) used in this experiment was provided by an aluminum plant in China. From Table 1, it can be seen that the main compositions in OAS are Al, Si and O. Besides, the XRD results [Figure 1(c)] show that Al and Si in OAS mainly exist in the form of mullite, alumina and silica. The particle size analysis results [Figure 1(a)] are consistent with the SEM [Figure 1(b)], indicating that the OAS is mainly composed of spherical particles with a size of 10 μm. Meanwhile, all the reagents were purchased from Aladdin Reagent Co., Ltd., (Shanghai, China).

Table 1 Chemical composition (mass fraction) of OAS used in the experiment

Composition	Al	Si	Ti	Fe	P	Ca	O	Others
Mass fraction/wt%	28.51	20.67	0.66	0.22	0.06	0.09	49.62	0.16

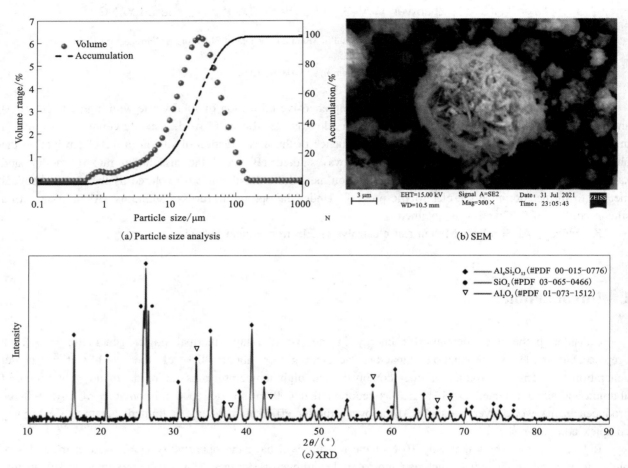

Figure 1 Characterization of OAS

2.2 Methods

Before the experiment, all experimental reagents were placed in a muffle furnace and kept at 673 K for 4 h to remove attached water and volatile impurities.

2.3 Experimental devices

The self-made comprehensive physicochemical properties tester of molten salt is applied to explore the influence of the change of Si mass fraction on the properties of molten salt. The assembly structure is shown in Figure 2(a). Tungsten and platinum are used as the working electrode (WE) and reference electrode (RE) in electrochemical experiments due to their high stability. And the electrolytic experimental device is shown in Figure 2(c).

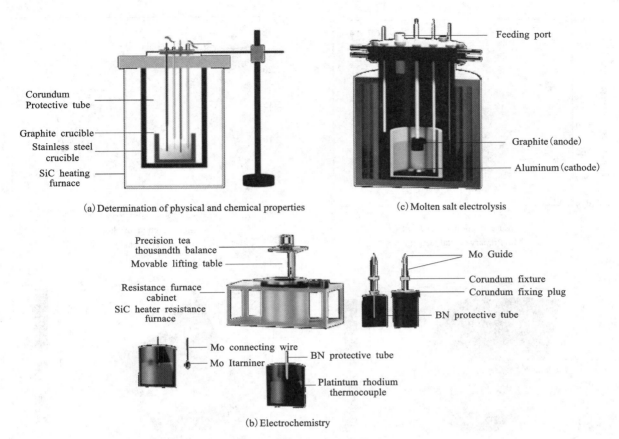

Figure 2 Experimental device

3 Results and discussion

3.1 Effect of SiO_2 mass fraction on physicochemical properties and dissolution behavior

Different mass fractions of SiO_2 (0-8 wt%) were introduced to (CR=2.2 and 2.4) NaF·AlF_3-4.8 wt% CaF_2-2.6 wt% LiF molten salt to explore the effect lead by the content of Si on the physical and chemical properties of molten salt. The total mass of molten salt was chosen to be 100 g. The results obtained are shown in Figure 3. It can be seen that when 4 wt% SiO_2 was introduced, both of the molten salt systems with different molecular ratios performed the lowest initial crystallization temperature. In addition, the density of the electrolyte gradually increased with the improvement of the mass fraction of SiO_2 and the molecular ratio of the molten salt. Furthermore, electrolytes with higher molecular ratios and silica concentrations showed more sensitive fluctuations. And under the same temperature and molecular ratio, the conductivity of the electrolyte decreased with the addition of SiO_2 in the electrolyte.

The reason might be that the complex ion group would be formed through the combination of Si^{4+}, F^- and O^{2-} ions in the electrolyte, thus impeding the directional movement of Na^+, which played the major role in current migration. Meanwhile, from the results of the average dissolution time of kaolinite and mullite obtained in the molten salt system with different molecular ratios, it can be seen that the high molecular ratio would be beneficial to improve the dissolution rate of OAS in cryolite molten salt.

3.2 Electrochemical reduction mechanism of Si^{4+}

The electrochemical reduction mechanism of Si during molten salt electrolysis was studied by cyclic voltammetry. The theoretical decomposition voltage was shown in Table 2, which proved that the reduction peak at -1.3 V in the cyclic voltammetric curve of the blank system should be the precipitation process of Al. With the addition of 3 wt% SiO_2, two new redox peaks at -0.8 V and -1.0 V were found in the cyclic

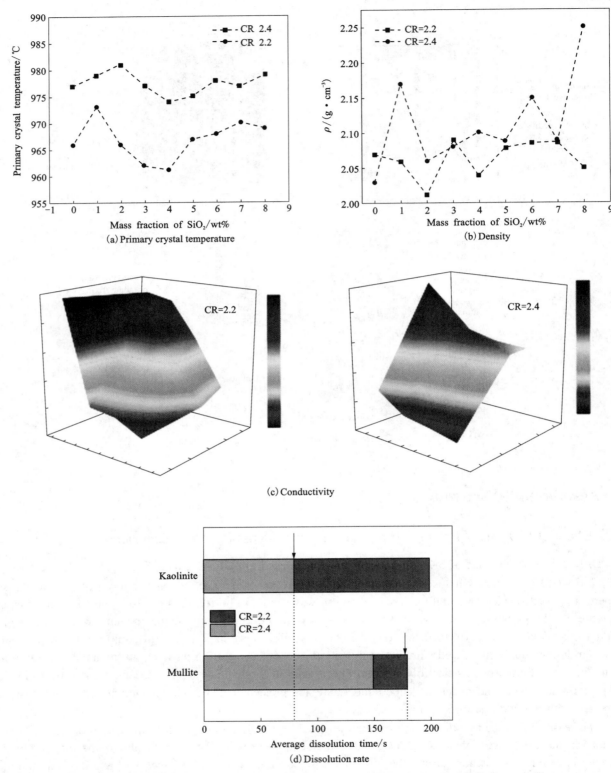

Figure 3 Effect of different mass fractions of SiO₂ on the physical and chemical properties of (CR=2.2 or 2.4) NaF-AlF₃-4.8 wt%CaF₂-2.6 wt%LiF molten salt system

voltammetric results, whose potentials shifted slightly with the increase of scanning rate, which proved that it was a quasi-reversible process. And the number of electrons transferred according to the quasi-reversible system could be calculated as follows.

$$|Ep\text{-}Ep/2| = \frac{1.875RT}{n\alpha F} \tag{1}$$

where R is the gas constant, T is the temperature in K, F is the Faraday constant, and α is the transfer coefficient, which is assumed to be 0.5.

Table 2 Theoretical decomposition voltage of each component in molten salt system

Reaction	$\Delta G_T^0 / (J \cdot mol^{-1})$	E_T^0 / V
$AlF_3 = Al(1) + 3/2F_2(g)$	1189497.290	-4.109
$NaF = Na(g) + 1/2F_2(g)$	441709.625	-4.577
$KF = K(1) + 1/2F_2(g)$	446444.584	-4.626
$CaF_2 = Ca(1) + F_2(g)$	993229.335	-5.146
$LiF = Li(1) + 1/2F_2(g)$	500880.158	-5.190

Based on Equation 1, when the scanning speed is 0.1 V/s, the half-peak width potential difference of R_2 is 0.1661, the number of electrons transferred (n) was calculated to be 2.3 or ~2. Therefore, it was proved that Si should be reduced in two steps: S1: $Si^{4+} + 2e^- = Si^{2+}$ S2: $Si^{2+} + 2e^- = Si$

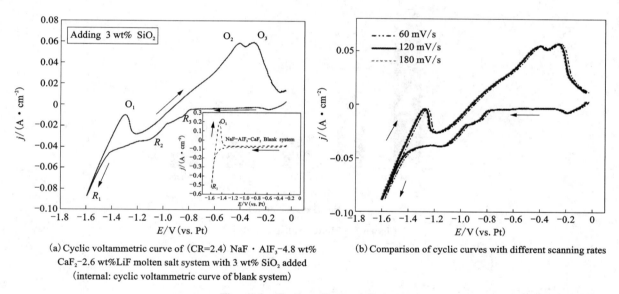

(a) Cyclic voltammetric curve of (CR=2.4) NaF · AlF_3-4.8 wt% CaF_2-2.6 wt%LiF molten salt system with 3 wt% SiO_2 added (internal: cyclic voltammetric curve of blank system)

(b) Comparison of cyclic curves with different scanning rates

Figure 4 Cyclic voltammetric test

3.3 Preparation of Al-Si alloy

14.5wt% OAS was added into the system of (CR = 2.4) NaF · AlF_3-4.8 wt% CaF_2-2.6 wt% LiF for molten salt electrolysis to prepare Al-Si alloy. The total mass of molten salt was 400 g. In addition, the current density of the anode was selected as 1.0 A/cm^2, the bottom area of the anode was 12.56 cm^2, the electrolysis time was 10 h, and the electrode distance was 4 cm.

During the electrolysis process, the cell voltage and the back electromotive force (BEMF) measured by the power-off method were shown in Figure 5. It can be found that the cell voltage slightly increased due to the continuous consumption of oxides. The BEMF measured by power-off method was consistent with the decomposition voltage of alumina and silica on graphite anode. After electrolysis, 52.03 g of alloy was collected, in which the mass fraction of Si was 22.0 wt%. At the same time, the alloy also contained a small amount of Ti and Fe. The results of phase characterization were consistent with those of chemical analysis.

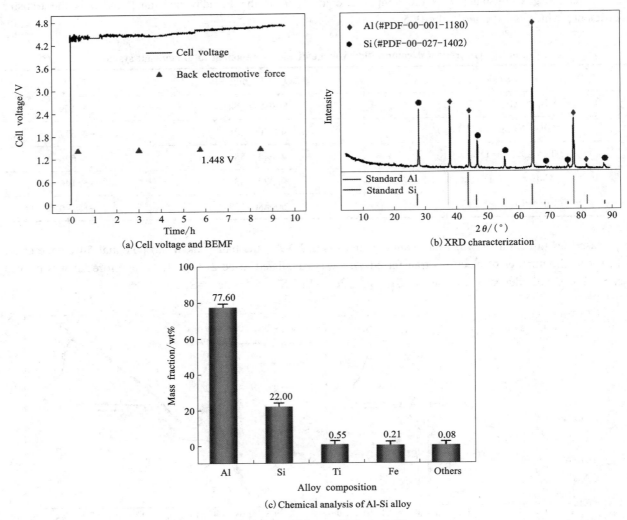

Figure 5　Molten salt electrolysis

3　Conclusion

A method for preparing Al-Si alloy by recovering by-products of thermal power generation through molten salt electrolysis was proposed. The initial crystal temperature of molten salt changed with the content of SiO_2. And the lowest initial crystal temperature of molten salt can be obtained when 4 wt% SiO_2 is added. The conductivity of molten salt decreases with the increase of SiO_2 content in the system. The addition of high content of SiO_2 will be often accompanied by a more obvious increase in density. At higher temperature, the molten salt with a higher molecular ratio was more conducive to the dissolution of OAS. The electrochemical reduction of Si^{4+} in cryolite molten salt was a multi-step: Si^{4+}-Si^{2+}-Si. Under the optimized electrolytic process conditions, Al-Si alloy with 22wt% Si was obtained.

References

[1] MURAKAMI T, OTSUKA K, FUKASAWA T, et al. Hierarchical porous zeolite synthesis from coal fly ash via microwave heating[J]. Colloids and Surfaces A: Physicochemical and Engineering Aspects, 2023, 661: 130941.

[2] YI T, XIAOTING Z, SHUOLIN Z, et al. Efficient synthesis of alkyl levulinates fuel additives using sulfonic acid functionalized polystyrene coated coal fly ash cat[J]. Journal of Bioresources and Bioproducts, 2023(2), 8: 198-213.

[3] SIQI H, CONGCONG C, ZHIBO Z, et al. Highly efficient separation of uranium from wastewater by in situ

synthesized hydroxyapatite modified coal fly ash composite aerogel[J]. Journal of Industrial and Engineering Chemistry, 2023, 118: 418-431.

[4] MOKOENA B K, MOKHAHLANE L S, CLARKE S. Effects of acid concentration on the recovery of rare earth elements from coal fly ash[J]. International Journal of Coal Geology, 2022, 259: 104037.

[5] CHAO W, GUOGUANG X, XINYUE G, et al. High value-added applications of coal fly ash in the form of porous materials: A review[J]. Ceramics International, 2021 (16), 47: 22302-22315.

[6] GADORE V, AHMARUZZAMAN M D. Tailored fly ash materials: A recent progress of their properties and applications for remediation of organic and inorganic contaminants from water[J]. Journal of Water Process Engineering, 2021, 41: 101910.

[7] MUSHTAQ F, ZAHID M, BHATTI I A, et al. Possible applications of coal fly ash in wastewater treatment[J]. Journal of Environmental Management, 2019, 240: 27-46.

[8] SLEAP S B, TURNER B D, SLOAN S W. Kinetics of fluoride removal from spent pot liner leachate (SPLL) contaminated groundwater[J]. Journal of Environmental Chemical Engineering, 2015(4), 3: 2580-2587.

[9] PALMIERI M J, ANDRADE-VIEIRA L F, CAMPOS J. Cytotoxicity of spent pot liner on allium cepa root tip cells: A comparative analysis in meristematic cell type on toxicity bioassays[J]. Ecotoxicology and Environmental Safety, 2016, 133: 442-447.

[10] XIUJUAN S, ZHICONG L, JINPENG F, et al. High-efficiency and rapid cyanide removal in SPL derived from aluminum electrolysis production under microwave and sensitizer synergistic action[J]. Journal of Cleaner Production, 2022, 365: 132691.

Study on the Plasma Sphero Process of Rare and Precious Metal Powders

Kang YUAN[1,2,3], Xiaoxiao PANG[1,2,3], Yubai HOU[1,2,3]

1 BGRIMM Advanced Materials Science & Technology Co., Ltd., Beijing, 102206, China
2 BGRIMM Technology Group, Beijing, 100160, China
3 Beijing Engineering Technology Research Center of Surface Strengthening and Repairing of Industry Parts, Beijing, 102206, China

Abstract: In this paper, two plasma sphero systems were used to manufacture sphero rare and precious powders, including tantalum and rhenium powders. The influence of powder feeding rate on the powder sphero effect and manufacture efficiency was particularly analyzed. By using a slower feeding rate, the sphericity of the powders can be improved. The elemental gasification and impurity cleaning in the sphero process were also discussed in this study.

Keywords: Plasma sphero; Powder; Rare and precious material

1 Introduction

With the rapid development of 3D printing additive manufacturing, thermal spraying, and other industries, the market demand for high-quality metal, alloy, and ceramic spherical powder is strong, and the application fields cover aerospace, automobile, medical, electronics, and other industries[1-5]. In recent years, with the rapid development of aerospace and high-end electronics industries, the manufacture of high-purity spherical powders for 3D printing and thermal spraying has become one of the research and application hotspots[6-9]. Rhenium and tantalum belong to rare and precious refractory-metal systems with high melting points (about 3000 ℃) and excellent chemical properties. In orthopedic medicine, tantalum orthopedic implants have also become a research hot spot in the medical field[6-8, 10, 11] because they can rapidly induce bone tissue growth and has good histocompatibility with the human body. In the electronics industry, tantalum is also widely used in integrated circuits and capacitor equipment[9, 12, 13]. Rhenium is an important additive element in high-temperature alloys, widely used in aviation engines and aerospace vehicles[14-16].

In industry, metallurgical tantalum or rhenium powders are prepared by reducing their oxides or salts[17, 18]. Due to irregular shape and poor fluidity, these metallurgical powders cannot be directly applied to fields such as 3D printing and thermal spraying. To improve the fluidity of the powder, Ren Ping et al. used the hydrogenation crushing technology to prepare spherical tantalum powder. The particle size distribution of this tantalum powder is concentrated, but the sphericity is still poor[19]. In recent years, with the rise of radio frequency induction plasma spheroidizing technology, the spheroidization of refractory metals has developed rapidly[20-22]. Yang Kun et al. used radio frequency induction plasmasphero technology to study the spheronization treatment of irregular metallurgical grade tantalum powder, and obtained high spherical tantalum powder. Liang Dong, Shang Fujun, and others prepared nano tantalum powder for capacitors using induction plasma technology, and the obtained nano tantalum powder was spherical[23, 24]. There are few studies on the spheroidization of rhenium powder. Spherical tantalum or rhenium powder can be prepared using induction plasma sphero technology, and its quality and powder production efficiency are related to equipment power and process control parameters. At present, there are few research reports on this aspect. This article will focus on studying the sphero behavior of tantalum and rhenium powders, as well as the mechanism of material loss and impurity removal, to ultimately obtain high-quality spherical tantalum and rhenium powders.

2 Experiment

This study conducted tantalum and rhenium powder spheroidization manufacture using induction plasma spheroidization equipment (Tekna, Canada). Figure 1 shows the schematic diagram of Tekna device structure (the design of the device is introduced in [25]). The 15 kw Tekna induction plasma spheroidization system uses a PFV100 vibration powder feeder. Its working principle is to input a mechanical vibration wave of about 100 hertz to the powder feeder, causing the powder to disperse and move under a certain amplitude of energy continuously. This powder dispersion movement is limited to the spiral track designed in the powder feeder, so the powder will spiral from bottom to top, driven by vibration energy, and ultimately be transported to the plasma torch by the carrier gas at the outlet.

Figure 1 Structure drawings of 15 kW Tekna inductive plasma sphero equipments

The tantalum and rhenium powders used in this study are metallurgical reduced powders with irregular shapes and poor flowability. After spheroidizing, the powders were observed using Hitachi SU5000 scanning electron microscopy (SEM) for the microscopic morphology of the sample. This study also applied Thermo Calc software for thermodynamic calculations of the carbon-hydrogen binary system. The TCFE9 database was used to calculate the phase composition at different molar ratios of hydrogen content at 3500 ℃ (above the melting point of tantalum and rhenium) to explain the effect of hydrogen on removing carbon impurities.

3 Results

The transfer speed of powders in the PFV100 vibrating powder feeder used in the 15 kW Tekna system is influenced by various factors, mainly including vibration frequency, amplitude, total powder amount, powder fluidity, etc. Generally, the vibration frequency is relatively fixed because it is related to the overall quality of the powder feeder. The amplitude is an important parameter that can be used to adjust the powder feeding rate. The larger the amplitude, the faster the powder feeding. As the powder is added to the powder feeder, it will also contribute to the overall quality of the feeder, thus affecting the powder

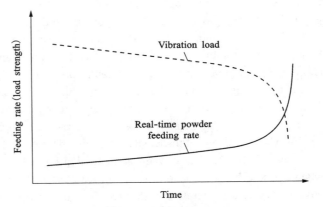

Figure 2 The powder feeding rate development in 15 kW equipment

feeding rate. For the PFV100 vibrating powder feeder used for 15 kW equipment, when the amount of powder added is less than 100 g, the total amount of powder has little effect on the uniformity of the powder feeding rate. But when several hundred grams are added (for high-density powders such as tantalum and rhenium, the powder feeder can accommodate up to 1 kg of powder), as shown in Figure 2, the powder feeding rate will show a significant dynamic change pattern. During the initial powder feeding process, due to the large total amount of powder and high vibration load, the powder feeding rate is relatively low. The powder feeding rate will slowly increase over time, as the total amount of powder continues to decrease and the vibration load continues to decrease. When the powder feeding is approaching its end, the vibration load of the powder feeder significantly decreases, and the powder feeding rate will significantly accelerate at this time. For the control of the powder spheronization process, special attention should be paid to the powder at the end of the feeding section, as this phenomenon can cause a decrease in the overall spheroidization rate of the powder. Generally, it is necessary to lower the amplitude value at the end of the powder feeding to reduce the excessively fast powder feeding rate and improve the stability of powder spheroidization. The spheroidization process also requires the powder to have good flowability, otherwise, it will affect the stability of powder transportation.

Figure 3 shows the typical powder feeding rates of powders with different melting points under extreme parameter conditions. The so-called "extreme parameter" refers to the maximum power recommended by the equipment (i.e., 15 kW) while keeping the carrier gas flow rate as low as possible (with no powder blockage as the lower limit). For tantalum and rhenium powder, if 200 mesh raw materials are used, the maximum powder feeding rate is tens of g/min to ensure sphericity higher than 90%, and the hourly yield is at the kilogram level. However, applying limit parameters will significantly increase the difficulty of on-site production, such as the possibility of discontinuous or even blocked powder feeding. Therefore, the actual spheronization process will refer to the limit parameters. Still, relatively conservative process parameters will be used to

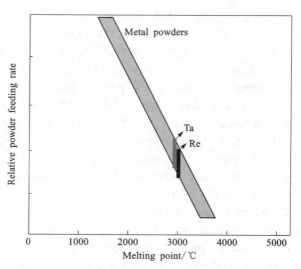

Figure 3　The relative powder feeding rates of materials with different melting points in 15 kW equipment

ensure the stability of the production process and powder quality. The spheroidization effect of powder is also related to many factors, such as the state of the raw material powder. Generally, internally dense powders are more prone to spheroidization, such as sintered and crushed powders, which are easier to spheroid than agglomerated or metallurgical irregular powders. This is because internal pores, to some extent, block heat transfer and powder melting.

Figure 4 shows the appearance morphology of spherical tantalum powder under different powder feeding rates on 15 kW equipment. At a lower powder feeding rate, the spheroidization rate of the powder is very high (even up to 99%), and the powder basically presents a regular spherical shape. When the powder feeding rate is high, some raw tantalum powder is not completely melted in the plasma, so the spheroidized powder also contains many irregular particles. If tantalum powder with a low spheroidization rate is used for 3D printing, it may cause printing defects and affect the product's performance. But for thermal spraying technologies (such as thermal spraying and laser cladding), even if the spheroidization rate is not high because the fluidity of the powder has been dramatically improved, these spheroidized tantalum powders can show good spraying adaptability and prepare high-density coatings. The typical physical properties of spheroidized tantalum powder are shown in Table 1. The oxygen content of the powder depends on the oxygen content of the raw tantalum powder. Generally, after spheroidization treatment, the oxygen content will decrease by more than 40%, mainly related to the high-temperature hydrogen reduction effect. The purity of the powder is also associated with the quality of the raw tantalum powder. Still, induction plasma spheroidization can remove low melting point metals or volatile non-metallic elements, so the purity of the powder after spheroidization will also be improved.

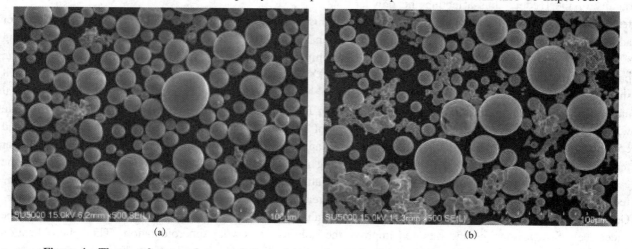

Figure 4　The tantalum powder morphology at (a) a lower feeding rate and (b) a quicker feeding rate

Table 1 Basic physical properties of sphero idized tantalum powder

Projects	Properties
Purity	>99.99%
Oxygen content	Decrease>40% than raw material, usually <1000×10^{-6}
Spherical ratio	>95%
Flowability	<10 s/50 g
Apparent density	>8.5 g/cm^3

Similar sphere behavior was observed in rhenium powders. In Figure 5, a lower feeding rate made high sphericity of the powders. The spherical rhenium powders produced by the plasma sphere process got excellent flowability. The oxygen content in the powders was shallow, about 200×10^{-6}.

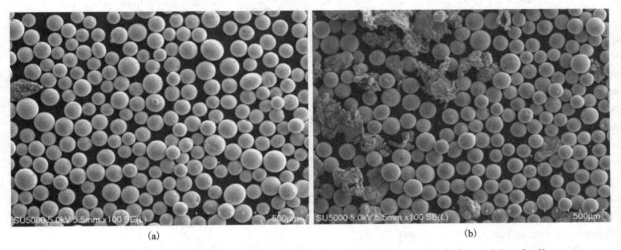

(a) (b)

Figure 5 The rhenium powder morphology at (a) a lower feeding rate and (b) a quicker feeding rate

Since plasma is a high-heat medium with tens of thousands of degrees Celsius, it can theoretically melt and vaporize any known element. However, due to the limited residence time of the powder in the plasma, reasonable control of the process (such as powder feeding rate, carrier gas flow rate, and powder feeding needle position) can achieve melting and shaping of the powder while minimizing gasification losses as much as possible. However, pure metal element powders may risk significant loss during shaping, even with optimized processes. The risk of loss is closely related to the difference in boiling point of the element. Figure 6 shows the melting and boiling points of different factors (arranged in descending order of melting point). Tantalum and rhenium, which are the focus of this paper, belong to the system of refractory metals, and their boiling points are also very high, so they are relatively not volatile loss elements. For aspects with lower melting points (<1000 ℃), the difference between melting and boiling points is mostly <1000 ℃. In general, induction plasma technology is rarely used to spheroid materials with a melting point below 1000 ℃. For refractory metals such as tantalum and rhenium, when the powder melts at more than 3000 ℃, it also reaches the boiling point of some low melting point metals. At this time, these soft melting point elements can be vaporized, thus achieving the effect of powder purification. In addition to common melting point metal impurities, some non-metallic components, such as phosphorus and sulfur, are also removed by gasification. After gasification, these impurities will enter into the micro nano particles. After subsequent cleaning, the impurity content in the final product powder will be reduced.

Among non-metallic elements, the effective removal of oxygen and carbon is more related to the hydrogen content in the plasma. In conventional atmospheres, metal oxides exist stably at high temperatures. However, in a hydrogen atmosphere, hydrogen can replace oxygen from the powder to reduce oxygen content. For tantalum and rhenium powders, the reduction in oxygen content of the powders after induction plasma spheroidization can reach over 40% compared to the raw material. The removal of carbon is also

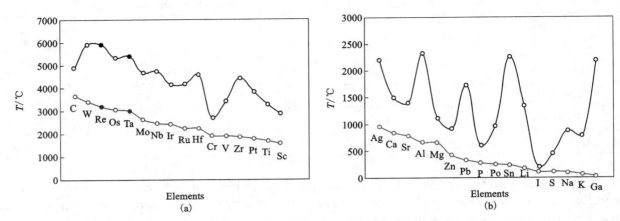

Figure 6 Statistical analyzing of the elements' melting and boiling points (the easily lost elements are green marked whose melting and boiling points' mismatch is less than 1000 ℃)

related to the plasma atmosphere. The results in Figure 6(a) show that the melting point of carbon itself (taking graphite as an example) exceeds that of tungsten (the highest melting point element in the metal), and the boiling point is also close to 5000 ℃. Therefore, the removal of carbon does not solely rely on high temperatures. Figure 7 shows a possible explanation for the transition of gas and solid phases in the carbon-hydrogen binary system with increasing hydrogen content. When the hydrogen content is low, carbon mainly exists in a solid state (graphite). But as the hydrogen ratio increases, carbon will enter the gas phase in a linear pattern. From a thermodynamic perspective, when the hydrogen content reaches a specific value, solid carbon is no longer stable, and therefore all carbon will enter the gas phase. Thus, at high temperatures, the gas-phase reaction between hydrogen and carbon may become a meaningful way to reduce carbon content.

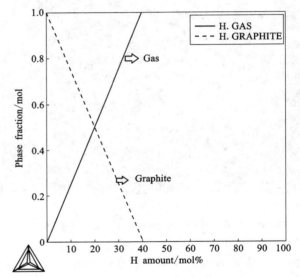

Figure 7 C-H binary phase constitute under H amount changing (temperature: 3500 ℃, pressure: 10^5 Pa)

4 Conclusion

By applying plasmasphero technique, tantalum and rhenium powders were spheroidized with controlling their feeding rates. It was found that the powder feeding rate was not a constant by using the PVF100 feeding system. The elemental gasification and impurity cleaning in the sphero process were found. The metallic impurities can be removed due to their lower boiling points. Oxygen and carbon were removed by reacted by hydrogen.

Acknowledgments: This work was funded by Yunnan Province Scientific Project (NO. 202302AB080021).

References

[1] HAN S B, ZHANG Y W, TIAN X J, et al. Research and application of high quality 3D printing metal powders for aerospace use[J]. Powder Metallurgy Industry, 2017, 27(6): 44-51.

[2] QIN S S, YU Y, ZENG G Y, et al. Research on the preparation of metal powder for 3D printing[J]. Powder Metallurgy Industry, 2016, 26(5): 21-24.
[3] GAO C F, YU W Y, HU Q L, et al. Performance characteristics and research progress of metal powders for 3D printing[J]. Powder Metallurgy Industry, 2017(5): 59-64.
[4] ZHANG Y H, DONG B b. Study on the method of gas atomization for production of 3D printing metal powders[J]. Mechanical Research & Application, 2016, 29(2): 203-205.
[5] ZHANG X J, TANG S Y, ZHAO H Y, et al. Research status and key technologies of 3D printing[J]. Journal of Materials Engineering, 2016, 44(2): 122-128.
[6] YANG L, WANG F Y. Progress of 3D printed porous tantalum in orthopedics[J]. Journal of Third Military Medical University, 2016, 44(2): 122-128.
[7] LI Y. Study on the process and microstructure of porous biomedical metal preparation by laser addition manufacturing (3D printing)[D]. Suzhou: Suzhou University, 2015.
[8] YE L. One kind of using 3D printing preparation of porous tantalum method of medical implant material? CN201210022122.1.
[9] ZHANG L, TIAN D B, WU Q, et al. Fabrication of gradient electronlytic capacitors anode pellet based on multi-material printing[J]. Rare Metal Materials and Engineering, 2020, 49(11): 3909-3913.
[10] WU H, GUO Z. Comparison of orthopedic implants 3D printed with porous titanium versus porous tantalum and prospects of their applications[J]. Chinese Journal of Orthopaedic Trauma, 2020, 22(10): 916-920.
[11] GUO M, ZHENG Y F. Manufacture technique and clinical application of porous tantalum implant in orthopaedic surgery[J]. Chinese Journal of Clinical and Basic Orthopaedic Research, 2013, 000(1): 47-55.
[12] GAO F L. Development of tantalum powder forultra high reliability capacitor[J]. Rare Metal Materials and Engineering, 1985(5): 22-26.
[13] DOU J M. Production of tantalum powder for capacitor[J]. Rare Metals, 1987(6): 40.
[14] MOTTURA A, REED R C, GUÉDOU J Y, et al. What is the role of rhenium in single crystal superalloys[J]. MATEC Web of Conferences, 2014, 14: 01001. DOI: 10.1051/matecconf/20141401001.
[15] JIN T, WANG W Z, SUN X F, et al. Role of rhenium in single crystal Ni-based superalloys[J]. Materials Science Forum, 2010, 638: 2257-2262. DOI: 10.4028/www.scientific.net/MSF.638-642.2257.
[16] LUO Y S, LIU S Z, SON F L. Status of study on strengthening mechanism of rhenium in single crystal superalloys [J]. Materials Review, 2005.
[17] HU X F, XU Q, WU Y. Research progress tantalum powder production technology[J]. Materials Review, 2005 (10): 105-107.
[18] REN P, YANG G, CHEN Y W, et al. Manufacturing technique and performance of metallurgical grade tantalum powder with low oxygen[J]. Nonferrous Metals(Extractive Metallurgy), 2018(8): 60-63, 73.
[19] REN P, ZHOU H Q, CHEN X Q, et al. Preparation of spherical or near spherical tantalum powder[J]. Nonferrous Metals(Extractive Metallurgy), 2018, 000(12): 43-46.
[20] GU Z T, YE G Y, LIU C D, et al. Study on the titanium powders spheroidization by using RF induction plasma[J]. Powder Metallurgy Technology, 2010(2): 120-124.
[21] YE G Y, GU Z T, LIU C D. RF plasma system for spheroidization of powder materials[J]. 2009 Annual Meeting of China Nuclear Society, 2009: 3833-3836.
[22] SHENG Y W, GUO Z M, HAO J J, et al. Characterization of spherical molybdenum powders prepared by RF plasma processing[J]. Powder Metallurgy Industry, 2011(6): 6-10.
[23] LIANG D, HOU J B, YIN T, et al. Effect of induction plasma process parameters on the prepared capacitor grade tantalum powder performance[J]. Ordnance Material Science and Engineering, 2010(3): 46-50.
[24] SHANG F J, SHI H G, WANG Y Q, et al. Effect of raw powder on characteristic of Ta powder prepared by induction plasma technology[J]. Ordnance Material Science and Engineering, 2009, 32(005): 64-68.
[25] Teksphero Spheroidization Systems[EB/OL]. https://www.tekna.com/spheroidization-systems.

Effects of Co Doping on Martensite Transformation and Thermal Hysteresis in NiTi Alloys

Baolin PANG, Zhenqiang WANG, Baoxiang ZHANG, Zan LIAO, Jiangbo WANG

GRIMED Medical Equipment (Beijing) Co., Ltd., Beijing, 102200, China

Abstract: The martensite transformation and thermal hysteresis mechanism of $Ni_{50-x}Ti_{50}Co_x$ alloys with different Co doping concentration are studied by density functional theory and plane-wave pseudopotential technique with generalized gradient approximation (GGA). The calculation results show that the phase transformation type before and after Co doping is B2 → B19 → B19′, and B19′ is the stable martensite structure of the alloy. After Co doping, the d-state electrons of Co and the d-state electrons of Ti form hybridization, which reduces the density of electronic states near the Fermi level. With the increase of Co doping, the thermal stability of martensite increases gradually, and the martensite transformation temperature decreases. Furthermore, the lattice of B2 phase and B19′ martensite phase tends to match, and the nucleation barrier of martensite phase decreases, which reduces the transformation hysteresis of the alloy. The research results provide reliable theoretical support for further optimization and design of NiTi alloys.

Keywords: First principles calculation; Co doping; $Ni_{50-x}Ti_{50}Co_x$; Martensite transformation; Thermal hysteresis

1 Introduction

NiTi alloy has excellent shape memory effect, super elasticity, corrosion resistance and other properties, and is widely used in aerospace, biomedicine and other fields[1-3]. However, NiTi alloy still has the problem of poor functional stability. Due to the wide thermal hysteresis of phase transformation, the shape memory effect, martensite transformation temperature and other functional characteristics of NiTi alloy attenuate after multiple thermal cycles or stress cycles, which leads to the failure of alloy devices after multiple cycles[4]. The results show that the shape memory effect and superelastic of NiTi alloys are related to martensite transformation[5], and the thermal hysteresis is higher when the transformation temperature is lower than room temperature[6,7]. In order to improve the phase transition temperature of NiTi alloy and broaden its application prospect in many fields, a lot of researches have been carried out by different scholars. Among them, the addition of alloying elements can effectively improve the phase transition temperature of NiTi alloy. Nam T H et al.[8] found that the addition of Cu can change the phase transition temperature of NiTi alloy, and the trend of alloy phase transition temperature varied with the addition amount of Cu. When the content of Cu was less than 10 at%, the phase transition temperature decreased, and when the content of Cu was higher than 10 at%, the phase transition temperature increased slightly with the increase of Cu content. Cui J et al.[9] found that the phase transition hysteresis of NiTi alloy gradually decreased with the increase of Cu, and the phase transition hysteresis was minimal when Cu content was near 12 at%. However, the addition of Cu element is not significant in improving the thermal stability of phase transition temperature. Meng X L et al.[10] found that as the number of thermal cycles increased, the phase transition temperature of TiNiHfCu alloy decreased and then stabilized, and the decrease of phase transition temperature was only 20-30 ℃. The addition of appropriate Pd can also reduce the thermal hysteresis of NiTi alloy[11,12]. Ma J et al.[13] found that the addition of different amounts of Pd to NiTi alloy had different effects on the phase transition temperature. When the Pd content was lower than 10 at%, the phase transition temperature decreased with the increase of Pd content; when the Pd content was higher than 10 at%, the phase transition temperature increased rapidly with the increase of Pd content. Similar to Pd, the addition of an appropriate amount of Pt can also reduce the phase transition temperature of NiTi alloy to a certain extent, thereby reducing the thermal hysteresis[14]. However, the introduction of precious metals such as Pd and Pt led to a sharp increase in the cost of NiTi alloys. The element Co in the fourth cycle has similar mechanical properties to Ni and exhibits good corrosion resistance and chemical stability. Research shows that Co doped NiTi alloy tends to replace the Ni element[15], thus reducing the content of the sensitizing element Ni. In addition, the appropriate amount of Co element can stimulate the human skeletal hematopoietic system, promote hemoglobin synthesis and increase the number of red blood cells, which makes its application in the medical field possible. In order to study the influence of Co doping on NiTi alloy and its internal mechanism, the martensite transformation and thermal hysteresis mechanism of $Ni_{50-x}Ti_{50}Co_x$ alloy with different Co doping concentration are studied by first principles calculation to explain its microscopic mechanism.

2 Experiment

Based on the first principles method of density functional theory, the first principles calculation is carried out through the quantum mechanics module Cambridge serial total energy package (Castep) module in Materials Studio (MS) software. The exchange correlation energy function of electron-electron interaction is calculated using the Perdew-Burke-Ernzerhof (PBE) functional under the general gradient approximation (GGA). Ultrasoft pseudopotential is used to describe the interaction between electrons. NiTi alloy has three cell models of phase structure, namely, high-temperature austenite phase (B2) with body-centered cubic structure, low-temperature martensite phase (B19′) with monoclinic structure and intermediate phase (B19) with rhombic structure, as shown in Figure 1. Considering the reliability of the calculated data and the requirement of the doping concentration, the crystal structure model of $Ni_{50-x}Ti_{50}Co_x$ (x = 0, 3.125, 6.25, 9.375, 12.5) with different doping concentrations is obtained by establishing a supercell. In order to ensure the consistency of doping concentration, 2×2×4 and 2×2×2 supercells are constructed for B2, B19, and B19′, and k points in Brillouin region are selected as 4×4×2 and 5×3×3, respectively. The plane wave cutoff energy used in the calculation is 450 eV, the self-consistent convergence accuracy is 1.0×10^{-5} eV/atom, the tolerance offset is 1.0×10^{-4} nm, and the force on each atom is 0.05 eV/nm. Firstly, the cell models with different phase structures are geometrically optimized, and then their lattice constant, density of states, and elastic constants are calculated respectively.

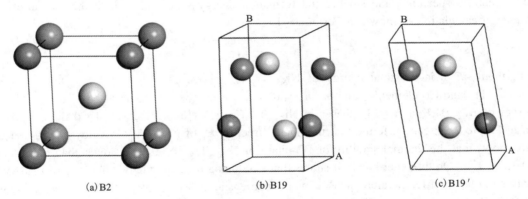

(a) B2 (b) B19 (c) B19′

Figure 1 Cell model of NiTi alloy with different phase structures

3 Results and discussion

3.1 Effects of Co doping on structural stability

$Ni_{50-x}Ti_{50}Co_x$ alloys with different Co doping contents are obtained by replacing Ni atoms in NiTi alloy with Co atoms. Structure optimization is carried out, and lattice parameters of different phase structures are obtained as shown in Table 1. It can be seen that for the same phase structure, as the Co doping amount increases, the cell parameters of different alloys slightly increase, which is the same as the effect when Cu and Pb atoms are doped[16]. The change of the crystal cell parameters is related to the atomic radius. According to the atomic radius table of elements, the atomic radius of Co (1.26 Å) is higher than that of Ni (1.24 Å).

When the elements with large atomic radius are replaced with Ni atoms, serious lattice distortion occurs in the alloy, thus increasing the cell parameters.

Table 1 Lattice parameters of B2, B19 and B19′ phases in $Ni_{50-x}Ti_{50}Co_x$ alloys

Alloy	Phase	a/Å	b/Å	c/Å	β/(°)
$Ni_{50}Ti_{50}$	B2	5.850	-	-	90
	B19	5.241	9.333	8.073	90
	B19′	5.494	9.062	8.400	100.18
$Ni_{46.875}Ti_{50}Co_{3.125}$	B2	5.902	-	-	90
	B19	5.283	9.318	8.054	90
	B19′	5.514	9.035	8.365	100.15
$Ni_{43.75}Ti_{50}Co_{6.25}$	B2	5.945	-	-	90
	B19	5.312	9.294	8.050	90
	B19′	5.525	9.013	8.340	99.96
$Ni_{40.625}Ti_{50}Co_{9.375}$	B2	5.976	-	-	90
	B19	5.334	9.251	8.062	90
	B19′	5.524	8.990	8.327	99.42
$Ni_{37.5}Ti_{50}Co_{12.5}$	B2	5.985	-	-	90
	B19	5.350	9.228	8.070	90
	B19′	5.525	8.967	8.317	99.20

In order to analyze the effect of Co doping on the stability of $Ni_{50-x}Ti_{50}Co_x$, the formation energy (E_F) is calculated. Generally speaking, the smaller the formation energy, the more stable the structure[17, 18]. The formation energy calculation is shown in Formula (1):

$$E_F = \frac{[E_{total} - (mE_{Ni} + nE_{Ti} + lE_{Co})]}{m+n+l} \quad (1)$$

Among them, E_{total} is the total energy of $Ni_{50-x}Ti_{50}Co_x$ alloy, E_{Ni}, E_{Ti}, and E_{Co} are the single atom energies of Ni, Ti, and Co elements, and m, n, and l are the number of atoms of Ni, Ti, and Co elements. The E_F curves of B2, B19, and B19′ phases in $Ni_{50-x}Ti_{50}Co_x$ alloy as a function of Co doping concentration are shown in Figure 2. It can be found that the phase structure E_F of $Ni_{50-x}Ti_{50}Co_x$ alloy is satisfied B2>B19>B19′. This shows that the thermal stability of B2 phase of $Ni_{50-x}Ti_{50}Co_x$ alloy is lower than that of martensite phase (B19, B19′). On the other hand, it shows that Co doping does not change the type of phase transition. B2 phase is still the high-temperature phase, while martensite phase (B19, B19′) is the low-temperature phase and B19 exists as an intermediate phase. The martensite transformation path is B2→B19→B19′. In addition, the E_F of phase structures in $Ni_{50-x}Ti_{50}Co_x$ alloys gradually decreases with the increase of Co doping concentration, indicating that the addition of Co increases the thermal stability of B2 phase and martensite phase (B19, B19′).

The transformation temperature of martensite can be analyzed by the energy difference (ΔE) between martensite phase and parent phase. The larger ΔE, the higher the phase transition temperature[19]. Figure 3 shows the relationship curve of ΔE between B2 phase and the martensitic phase (B19, B19′) in $Ni_{50-x}Ti_{50}Co_x$ alloys with the doping concentration of Co. It can be seen that both $\Delta E_{B19′}$ and ΔE_{B19} decrease with the increase of Co doping, indicating that Co doping leads to the decrease of martensite transformation temperature of $Ni_{50-x}Ti_{50}Co_x$ alloys. The energy difference between martensite and the parent phase of $Ni_{50-x}Ti_{50}Co_x$ alloys satisfies $\Delta E_{B19} < \Delta E_{B19′}$, indicating that the phase transition temperature of B2→B19 is lower than B2→B19′, and the stable martensite phase of alloys is B19′, which is consistent with the results shown in Figure 2.

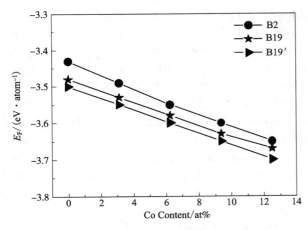

Figure 2 E_F of B2, B19 and B19' phases in $Ni_{50-x}Ti_{50}Co_x$ alloys as a function of Co doping concentration

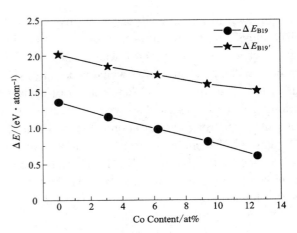

Figure 3 Changes in ΔE_{B19} and $\Delta E_{B19'}$ of $Ni_{50-x}Ti_{50}Co_x$ alloys with Co content

3.2 Effects of Co doping on martensite transformation

Elastic constant is a basic parameter of structural materials, which can indirectly judge the stability of materials. The B2 phase of NiTi alloy has three independent elastic tensors: C_{11}, C_{12} and C_{44}, in which C_{44} represents the basal plane shear modulus along the $<1\bar{1}0>$ direction, $C' = (C_{11}-C_{12})/2$ represents the basal plane shear modulus along the $<1\bar{1}0>$ direction, and the anisotropy factor $A = C_{44}/C'$ represents the degree of correlation between C_{44} and C'[16]. When A is large and C_{44} is much larger than C', the {001} plane corresponding to C_{44} is not easy to shear along the $<1\bar{1}0>$ direction, while the {110} plane corresponding to C' is easy to shear along the $<1\bar{1}0>$ direction[3]. In order to study the influence mechanism of Co doping on the martensite transformation behavior from the perspective of elastic properties, the elastic constants of B2 phase of $Ni_{50-x}Ti_{50}Co_x$ alloys are calculated, and the results are shown in Table 2. The B2 phases of different alloys satisfy the stability conditions: $C_{11}>0$, $C_{44}>0$, $C_{11}-C_{12}>0$, $C_{11}+2C_{12}>0$, indicating that the B2 phase can still exist stably after Co doping. With the increase of Co doping, both C_{44} and C' values gradually increase. The increase (decrease) of C_{44} means that the shear difficulty of B2 phase in the {001} plane along the direction of $<1\bar{1}0>$ increases (decreases), which decreases (increases) the transformation temperature of B2→B19' martensite. The increase (decrease) of C' means that the difficulty of shear of B2 phase in the {110} plane along the $<1\bar{1}0>$ direction increases (decreases), which decreases (increases) the martensite transition temperature of B2→B19[20]. This further indicates that Co doping results in the decrease of martensitic transition temperature. In addition, it can also be found that the A value of NiTi alloy is 2.3, which is relatively close to the experimental value[23], indicating that the martensite transformation product of NiTi alloy is B19'. Compared with NiTi alloy, the A value of $Ni_{50-x}Ti_{50}Co_x$ alloy increases with the increase of Co doping amount, but it is still much smaller than that of NiAl ($A = 9$)[21,22], CuZn ($A = 11$)[23] and other martensite transformation alloys, indicating that the martensite transformation product of the alloy after Co doping is still B19' phase.

Table 2 Elastic constants of B2 phase in $Ni_{50-x}Ti_{50}Co_x$ alloys

Alloy	C_{11}	C_{12}	C_{44}	C'	A
$Ni_{50}Ti_{50}$	251.2	209.2	48.3	21	2.3
$Ni_{46.875}Ti_{50}Co_{3.125}$	255.9	208.9	56.2	23.5	2.4
$Ni_{43.75}Ti_{50}Co_{6.25}$	256.8	206.2	65.8	25.3	2.6
$Ni_{40.625}Ti_{50}Co_{9.375}$	259.6	204.4	74.5	27.6	2.7
$Ni_{37.5}Ti_{50}Co_{12.5}$	254.6	198.4	84.4	29.1	2.9

In order to further analyze the mechanism of the effect of Co doping on the phase transition behavior of NiTi alloys, the electronic density of states (DOS) of $Ni_{50-x}Ti_{50}Co_x$ alloys are calculated. Figure 4 shows the total energy state density and fractal density of B2 phase in NiTi alloy. It can be observed that TiNi alloy has an obvious bimodal structure, which is because NiTi alloy has a cubic crystal structure and bimodal is a typical characteristic of the electron state density of cubic crystal structure[24]. The total energy state density is mainly determined by the d-state density, and there is a pseudo forbidden band at the Fermi level -1.25 eV, which is related to the strong hybridization between Ni and Ti atoms. In the low-energy region below the Fermi level of B2 phase of NiTi alloy, there is a strong peak in the d-state of Ni, and the total energy state density is mainly determined by the d-state of Ni. In the high-energy region above the Fermi level of B2 phase of NiTi alloy, there is a strong peak in the d state of Ti, and the total energy state density is mainly determined by the d state of Ti (Figure 5).

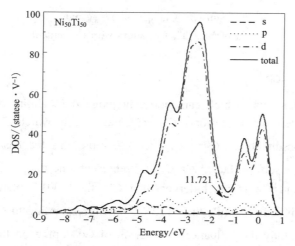
Figure 4 Total density of state and partial density of state of B2 phase in NiTi alloy

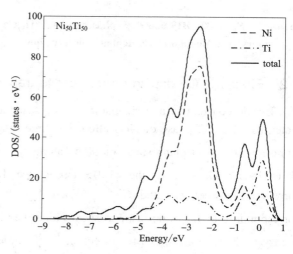
Figure 5 Total energy density of state and Ni d, Ti d partial density of state of B2 phase in NiTi alloy

Figure 6 shows the total energy density of states, partial density of states, and d-partial density of states for different atoms of the B2 phase in the $Ni_{50-x}Ti_{50}Co_x$ alloys. It can be found that the total energy density of states of different alloys is a typical bimodal structure, indicating that Co doping does not change the crystal structure, and the $Ni_{50-x}Ti_{50}Co_x$ alloy is still the cubic crystal structure. In addition, similar to NiTi alloy, the total energy state density of $Ni_{50-x}Ti_{50}Co_x$ alloys is mainly determined by the density of d state, and there is obvious pseudo forbidden bands. As the Co doping amount increases, the pseudo forbidden band decreases from 11.721 states/eV of $Ni_{50}Ti_{50}$ alloy to 10.696 states/eV of $Ni_{37.5}Ti_{50}Co_{12.5}$ alloy. The level of pseudo forbidden band can be used to measure the stability of intermetallic compounds. The lower the pseudo forbidden band, the more stable the structure[25]. Therefore, Co doping makes the structure more stable, which is consistent with the results shown in Figure 2. Different from NiTi alloy, a new group of peaks appears near -2.0 eV below the Fermi level in the density of d partial states of $Ni_{50-x}Ti_{50}Co_x$ alloy. According to the d-fractal state density diagram of atoms in Figure 6, it can be seen that the new peak near -2.0 eV is mainly determined by the d-state electron of Co. The total energy state density in the high-energy region above the Fermi level is mainly determined by the d-fractal state density of Ti, and the total energy state density in the low-energy region below the Fermi level is mainly determined by the d-fractal state density of Ni and Co, indicating that there is strong hybridization between Ni and Ti, and there is also significant hybridization between Ti and Co. In addition, the d-state density of Co gradually increases with the increase of Co doping, while the d-state density of Ni gradually decreases, which is due to the replacement of Co by Ni atoms and the reduction of electrons involved in Ni atoms, resulting in a reduction of the hybrid effect between Ni and Ti, and the enhancement of the hybrid effect between Ti and Co. The more the density of charge states overlap, the stronger the hybridization and the stronger the structural stability[20]. Comparing the hybridization strength between Ti and Co in different $Ni_{50-x}Ti_{50}Co_x$ alloys, it can be seen that Co doping improves the structural stability of B2 phase and decreases the phase transition temperature.

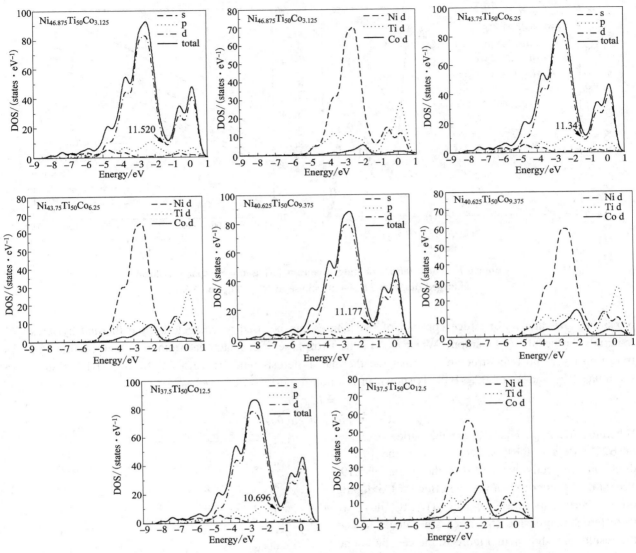

Figure 6 Total energy density of states and partial density of states of B2 phase in $Ni_{50-x}Ti_{50}Co_x$ alloys, as well as the d-partial density of states of each atom

Figure 7 shows the total state density of B2 phase of $Ni_{50-x}Ti_{50}Co_x$ alloy and the corresponding local magnifications diagram. It can be found that the peaks dominated by Ti in the high-energy region above Fermi level shift to the high-energy region with the increase of Co doping, while the peaks dominated by Ni in the low-energy region below Fermi level shift to the low-energy region. This shows that Co doping makes Ti lose fewer electrons, while Ni gains more electrons, resulting in increased bonding strength between Ti and Ni, thus increasing the stability of the alloy. It can be seen from Figure 7 (b) that the total energy density of states near the Fermi level gradually decreases with the increase of Co doping amount. The higher the total energy state density near the Fermi level, the more unstable the system, and the smaller the total energy state density, the more stable the system[25]. Therefore, $Ni_{50-x}Ti_{50}Co_x$ alloy becomes more and more stable with the increase of Co doping, and the phase transition temperature gradually decreases.

3.3 Effects of Co doping on phase transition hysteresis

Shape memory alloys with narrow hysteresis have good thermal/stress cycle stability[26, 27]. The alloy with narrow hysteresis satisfies two conditions: one is that there is no volume change during phase transformation, and the other is that there is a stress-free and non-twin perfect interface between martensite and the parent phase, which is also a necessary and sufficient condition for the alloy with narrow hysteresis[28-30]. In order to reflect the relationship between the change of lattice volume and thermal hysteresis during phase

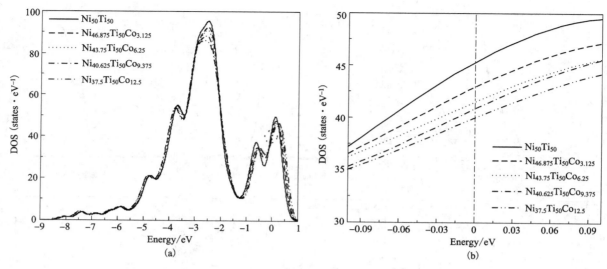

Figure 7 Total density of states diagram (a) and corresponding local enlarged diagram (b) of B2 phase in $Ni_{50-x}Ti_{50}Co_x$ Alloys

transition, the ratio λ_2 of lattice constant of two phases can be used to characterize the thermal hysteresis of materials during phase transition. When $\lambda_2 = 1$, the two phases can match each other without producing stress transition layer and twin structure, so that the thermal hysteresis tends to zero. λ_2 of B2→B19 is shown in Formula (2), and λ_2 of B2→B19' is shown in Formula (3):

$$\lambda_2^{B19} = b_{B19}/b_0 \qquad (2)$$
$$\lambda_2^{B19'} = (b_{B19'} \times \sin\theta)/b_0 \qquad (3)$$

Wherein, b_0, b_{B19}, and $b_{B19'}$ are the lattice constants of B2, B19 and B19' structures along the [010] direction, respectively, and θ is the angle of B19' structure. In order to analyze the effect of Co doping on the thermal hysteresis of $Ni_{50-x}Ti_{50}Co_x$ alloy, λ_2 of different martensitic structures are calculated and the results are shown in Figure 8. It can be found that whether the martensite structure of $Ni_{50-x}Ti_{50}Co_x$ is B19 or B19', λ_2 decreases with the increase of Co doping, indicating that Co doping improves the structural compatibility. The λ_2 of B19 structure is higher than that of B19' structure, which shows that martensite of B19' structure has higher structural compatibility than B19. In addition, the stable martensite structure is B19', and λ_2 tends to 1 with

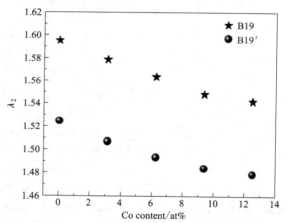

Figure 8 Relationship between different martensitic structures and λ_2 of $Ni_{50-x}Ti_{50}Co_x$ alloys

the increase of Co doping, indicating that the thermal hysteresis of phase transition decreases with the increase of Co doping concentration. It is worth noting that when the Co doping amount is 12.5%, λ_2 is closest to 1, which has the narrowest phase transition hysteresis.

In order to explain the effect of Co doping on the hysteresis phase transformation of alloys from the perspective of energy, the energy difference between stable martensite B19' and B2 phase and λ_2 of different alloys are analyzed, and the results are shown in Figure 11. With the increase of Co doping, ΔE and λ_2 values decrease. The formation of martensite is related to the lower nucleation barrier. Co doping reduces the energy difference between B2 phase and martensite B19' phase, thus reducing the energy barrier that needs to overcome when B2 → B19', reducing the driving force required for martensitic nucleation. The structural compatibility of B19' and B2 phase increases, leading to a reduction in the number of dislocations due to structural mismatch during phase transformation and a reduction in the hysteresis phase transformation.

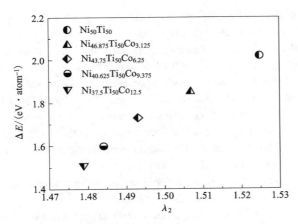

Figure 9 Relationship between ΔE and λ_2 of $Ni_{50-x}Ti_{50}Co_x$ alloys about stable martensitic phase

4 Conclusions

The effects of Co doping on martensitic phase stability, phase transition type, phase transition temperature and phase transition hysteresis of $Ni_{50-x}Ti_{50}Co_x(x=0, 3.125, 6.25, 9.375, 12.5)$ shape memory alloys are investigated by first principles method of density functional theory. The following conclusions are drawn:

(1) Co replaces the position of Ni in NiTi alloy, causing severe lattice distortion and increasing the lattice parameters. Co doping does not change the transformation type of the alloy, and the transformation path of martensite is still B2→B19→B19'. The formation energy phase structure decreases gradually and the thermal stability of martensite increases with the increase of Co doping.

(2) The d-state electrons of Co and the d-state electrons of Ti form hybridization after doping, while the hybridization intensity between the d-state of Ti and the d-state of Co is weaker than that between the d-state of Ti and the d-state of Ni. The density of electronic states near the Fermi level decreases, which improves the stability of B2 phase and leads to the reduction of martensite transformation temperature.

(3) The structural compatibility of B19' phase and B2 phase in the $Ni_{50-x}Ti_{50}Co_x$ alloys is higher than that of B19 phase. The increase of Co content makes the λ_2 value of B19' phase gradually decrease and tends to 1, thus making the stability of B19' martensite phase gradually increase.

(4) With the increase of Co doping, the lattice parameter b of the martensite phase shows a good proportional relationship with the value of λ_2, indicating that the increase of b makes the lattice of B2 phase and martensite phase gradually match, thus decreasing the phase transformation hysteresis. The energy difference between B2 phase and stable martensite phase decreases with the increase of Co doping, which reduces the nucleation barrier of martensite phase, improves the compatibility of B2 phase and martensite phase, and then decreases the phase transformation hysteresis. The calculation explains the influence mechanism of Co doping from the microscopic perspective, providing reliable theoretical support for further optimization and design of NiTi alloy.

Acknowledgments: This work was financially supported by Youth Fund Project of China Grimed Technology Group Co., Ltd (G126202231229049).

References

[1] CHENG W J, QIANG Y F, LIN Y J, et al. Positron study of defects in TiNi alloys[J]. Chinese Physics, 2000, 9(3): 216.
[2] CHEN Z G, XIE Z, LI Y C, et al. Stability of small Ni-Ti bimetallic clusters studied by density functional theory [J]. Chinese Physics B, 2010, 19(4): 043102.
[3] OTSUKA K, REN X. Physical metallurgy of Ti-Ni-based shape memory alloys[J]. Progress in Materials Science,

2005, 50(5): 511-678.
- [4] TADAKI T, SHIMIZU K I. Thermal cycling effects in an aged Ni-rich Ti-Ni shape memory alloy[J]. Transactions of the Japan Institute of Metals, 1987, 28(11): 883-890.
- [5] LIU H C, SUN G A, WANG Y D, et al. Study on strain rate correlation of deformation mechanism of NiTi shape memory alloy[J]. Acta Physica Sinica, 2013, 62(18): 357-365.
- [6] TAN C L, HUANG Y W, TIAN X U, et al. Origin of magnetic properties and martensitic transformation of Ni-Mn- in magnetic shape memory alloys[J]. Applied Physics Letters, 2012, 100(13): 132402-132405.
- [7] ZHANG K G, WANG F, YE J J, et al. Comparison of phase transformation behaviors of annealed Ti-50Ni and Ti-45Ni-5Cu shape memory alloy wire[J]. Journal of Functional Materials, 2020, 51(9): 09109-09113.
- [8] NAM T H, SABURI T, SHIMIZU K, et al. Cu-content dependance of shape memory characteristics in Ti-Ni-Cu alloys[J]. Materials Transactions, 1990, 31(11): 959-967.
- [9] CUI J, CHU Y S, FAMODU O O, et al. Combinatorial search of thermoelastic shape-memory alloys with extremely small hysteresis width[J]. Nature Mater, 2006, 5(4): 286-290.
- [10] MENG X L, CAI W, LAU K T, et al. Phase transformation and microstructure of quaternary TiNiHfCu high temperature shape memory alloy[J]. Intermetallics, 2005, 13(2): 197-201.
- [11] ATLI K C, KARAMAN I, NOEBE R D, et al. Improvement in the shape memory response of Ti50.5Ni24.5Pd25 high-temperature shape memory alloy with scandium microalloying[J]. Metallurgical and Materials Transactions A, 2010, 41(10): 2485-2497.
- [12] LI H, MENG X L, CAI W. Martensitic transformation and microstructure of (Ni, Cu, Pd)-rich Ti49.5Ni39-xCu11.5Pdx alloys with near-zero hysteresis and excellent thermal stability[J]. Intermetallics, 2020, 126: 106927.
- [13] MA J, NOEBE R D. High temperature shape memory alloys[J]. International Materials Reviews, 2013, 55(5): 257-315.
- [14] BRIAN L, MAIER H J, et al, Structure and thermomechanical behavior of NiTiPt shape memory alloy wires[J]. Acta Biomaterialia, 2009, 5(1): 257-267.
- [15] SHANG X P. First-principles study of transition element doped NiTi alloy[J]. Rare Metal Materials and Engineering, 2016, 45(8): 2041-05.
- [16] SUN S Y. First-principles study on the influence of alloying elements on the phase transition of martensitic in TiNi-based shape memory alloys[D]. Harbin: Harbin Engineering University, 2020.
- [17] HU Q M, YANG R, LU J M, et al. Effect of Zr on the properties of (TiZr)Ni alloys from first-principles calculations[J]. Physical Review B, 2007, 76(22): 224201-224207.
- [18] LI G F, LU S Q, DONG X J, et al. Microcosmic mechanism of carbon influencing on NiTiNb9 alloy[J]. Journal of Alloys and Compounds, 2012, 542: 170-176.
- [19] CHEN J, LI Y, SHANG J X, et al. First principles calculations on martensitic transformation and phase instability of Ni-Mn-Ga high temperature shape memory alloys[J]. Applied Physics Letters, 2006, 89(23): 231921-231923.
- [20] TAN C L. Study on the alloying mechanism of Ti-Ni and Nb(Ta)-Ru and Ni-Mn-Ga memory alloys[D]. Harbin: Harbin Institute of Technology, 2008.
- [21] NIU X, HUANG Z, WANG B, et al. Effects of point defects on properties of B2 NiAl: A first-principles study[J]. Rare Metal Materials and Engineering, 2018, 47(9): 2687-2692.
- [22] CHAO J, GLEESON B. Effects of Cr on the elastic properties of B2 NiAl: A first-principles study[J]. Scripta Materialia, 2006, 55(9): 759-762.
- [23] ENAMI K, NAGASAWA A, et al, Elastic softening and electron-diffraction anomalies prior to the martensitic transformation in a Ni-Al β_1 alloy[J]. Scripta Metallurgica, 1976, 10(10): 879-884.
- [24] KUMAR P, WAGHMARE U V. First-principles phonon-based model and theory of martensitic phase transformation in NiTi shape memory alloy[J]. Materialia, 2020, 9: 100602.
- [25] XU J H, OGUCHI T, FREEMAN A J. Crystal structure, phase stability, and magnetism in Ni3V[J]. Physical Review B, 1987, 35(13): 6940.
- [26] ZARNETTA R, TAKAHASHI R, YONG M L, et al. Identification of quaternary shape memory alloys with near-Zera thermal hysteresis and unprecedented functional stability[J]. Advanced Functional Materials, 2010, 20(12): 1917-1923.
- [27] GU H, BUMKE L, CHLUBA C, et al. Phase engineering and supercompatibility of shape memory alloys[J]. Materialia Today, 2018, 21(3): 265-277.
- [28] DELVILLE R, KASINATHAN S, ZHANG Z Y, et al. Transmission electron microscopy study of phase compatibility in low hysteresis shape memory alloys[J]. Philosophical Magazine, 2010, 90(1-4): 177-195.
- [29] ZARNETTA R, TAKAHASHI R, YOUNG M L, et al. Shape memory materials: Identification of quaternary shape memory alloys with near-zero thermal hysteresis and unprecedented functional stability[J]. Advanced Functional Materials, 2010, 20(12).

[30] KONIG D, BUENCONSEJO J S P, GROCHLA D, et al. Thickness-dependence of the B2-B19 martensitic transformation in nanoscale shape memory alloy thin films: Zero-hysteresis in 75 nm thick Ti51Ni38Cu11 thin films [J]. Acta Materialia, 2012, 60(1): 306-313.

[31] BUENCONSEJO P J S, ZARNETTA R, YOUNG M, et al. On the mechanism that leads to vanishing thermal hysteresis of the B2-R phase transformation in multilayered (TiNi)/(W) shape memory alloy thin films [J]. Thin Solid Films, 2014, 564: 79-85.

Stabilization of the β'-Cu_4Ti Phase in Cu-Ti Alloys by Micro-alloying with Gd

Yumin LIAO, Chengjun GUO, Chenyang ZHOU, Weibin XIE, Bin YANG, Hang WANG*

Faculty of Materials Metallurgy and Chemistry, Jiangxi University of Science and Technology, Ganzhou, 341000, China

Abstract: The stability of metastable β'-Cu_4Ti is a crucial factor that affects the properties of Cu-Ti alloys. Previous studies have found that a Ti content in the β'-Cu_4Ti phase equal to or greater than 19 at% results in a lower stability of β'-Cu_4Ti relative to the stable β-Cu_4Ti phase, causing a re-dissolution of Ti atoms so that the Ti content in the Cu matrix is maintained at a high level. In this study, the effects of different alloying elements (X = W, Re, and Gd etc.) on the stability of the β'-Cu_4Ti and the β-Cu_4Ti phases have been examined using first-principles calculation. The results have revealed that the inclusion of Gd improves the stability of the β'-Cu_4Ti phase at a Ti content of 19 at%. In addition, the experimental results have shown that Gd significantly delayed the precipitation of the β-Cu_4Ti phase, indicating an enhanced stability of the β'-Cu_4Ti phase.

Keywords: Cu-Ti alloy; Metastable phase; First-principles calculation; Stability; Gd element

1 Introduction

The mechanical properties and electrical conductivity of elastic copper alloys make them key components in electronic and electrical products. The Cu-Ti system is considered an elastic copper alloy. Due to the stress relaxation properties and elasticity of Cu-Ti alloys (1 at%-6 at% Ti)[1-3], they are widely used in the production of connectors and lead frames in electrical equipment. The Cu-Ti alloy is recognized as a potential replacement for Cu-Be alloys[4-6]. However, the electrical conductivity of Cu-Ti alloys is insufficient, which has restricted the range of applications. In order to adapt to the current trend of miniaturization and multi-functionalization of electronic and electrical equipment, it is essential to improve the mechanical properties and electrical conductivity of Cu-Ti alloys.

During the early stage of aging, spinodal decomposition of a supersaturated solid solution in the Cu-Ti alloy occurs, forming a Ti-rich zone. The metastable β'-Cu_4Ti phase precipitates and grows in the Ti-rich zone[7]. As the phase is coherent with the Cu matrix, the finely dispersed β'-Cu_4Ti in the Cu matrix serves as the main strengthening phase and guarantees the high elastic properties of the Cu-Ti alloys. In the later stage of aging, the discontinuous β-Cu_4Ti phase nucleates and grows[8]. As a result of the incoherent relationship between the β-Cu_4Ti phase and the Cu matrix and the coarse lamellar structure of the β-Cu_4Ti phase, the mechanical properties of the Cu-Ti alloy deteriorate after the nucleation and growth of the stable β-Cu_4Ti phase[9,10]. Controlling the nucleation and growth of β-Cu_4Ti is key in the manufacture and processing of Cu-Ti alloys[11]. Alloying has been used in many studies to promote the precipitation of dissolved Ti atoms in the Cu matrix, delaying or inhibiting the precipitation of the β-Cu_4Ti phase[12,13]. This is an effective method to improve electrical conductivity and maintain the high strength and high elasticity of Cu-Ti alloys[14,15]. Wang[16] studied the Cu-Ti-Sn alloy, and found that the inclusion of Sn resulted in a $CuSn_3Ti_5$ phase that reduced the solid solubility of Ti in the Cu matrix, and the electrical conductivity of the alloy increased after aging. Liu[17] found that Ni addition improved the electrical conductivity of the alloy by precipitating NiTi and β'-Ni_3Ti phases. The higher the Ni content, the higher the electrical conductivity of the alloy. Konno[18] analyzed the Cu-Ti-Al alloy, and observed that the inclusion of Al reduced the content of Ti in the Cu matrix by forming an $AlCu_2Ti$ phase, thereby improving the electrical conductivity of Cu-Ti alloys. Semboshi[19] reported that the addition of H delayed the precipitation of the β-Cu_4Ti phase by H segregation at the grain boundary of Cu-Ti alloys, preferentially forming TiH_2. Semboshi[20] also found that B as an additive segregated at the grain boundary, inhibiting nucleation and growth of the β-Cu_4Ti phase. Although many studies have considered improvements in the mechanical properties and electrical conductivity of the Cu-Ti alloy by alloying, there are few reports concerning the mechanism of alloying in the formation and dissolution of the metastable β'-Cu_4Ti phase, and the inhibition of nucleation and growth of the stable β-Cu_4Ti phase.

In previous work, the relationship between the formation and dissolution of the β'-Cu_4Ti phase and the stability of precipitates has been systematically studied[21]. It was found that a Ti content of the β'-Cu_4Ti phase greater than or equal to 19 at% serves to lower the stability of the β'-Cu_4Ti phase relative to β-Cu_4Ti, and the Ti atoms re-dissolved in the matrix. This translates into a low electrical conductivity of the Cu-Ti alloy when the β'-Cu_4Ti phase is the main precipitate. Based on the metastable characteristics of the β'-Cu_4Ti phase, it is expected that the stability at a high Ti content (greater than or equal to 19 at%) should be improved by alloying, which hinders the re-dissolution of Ti atoms. The electrical conductivity of the alloy may be enhanced by decreasing the Ti atoms in the Cu matrix. In addition, a lower Ti content in the matrix will inhibit the nucleation and growth of the β-Cu_4Ti phase, ensuring the requisite mechanical properties of the alloy with high elasticity and high strength.

This study builds on prior work that established a decrease in the stability of the β'-Cu_4Ti phase in the Cu-Ti alloy caused by a segregation of Ti atoms, resulting in the re-dissolution of Ti atoms which hampers improvements in the electrical conductivity of the alloy. From a micro-alloying perspective, first-principles calculation[22-25] was employed to explore possible alloying elements that can improve the stability of the β'-Cu_4Ti phase. Combined with experimental characterization, the influence of alloying elements on the stability of the β'-Cu_4Ti and β-Cu_4Ti phases was evaluated to facilitate a mechanistic understanding of the Cu-Ti alloy, and inform the development of high-performance materials.

2 Calculations and experimental methods

2.1 Calculation method

The Vienna Ab-initio Simulation Package (VASP) with the Projector Augmented Wave (PAW) method was used to perform density functional theory (DFT) calculations[26, 27]. The Perdew-Burke-Ernzerhof (PBE) of generalized gradient approximation (GGA) was selected as the exchange correlation functional[28]. The cutoff energy for plane wave expansion was set at 500 eV. In the case of the β'-Cu_4Ti phase calculations, the Brillouin zone was sampled using a 5×1×4 k-point mesh. A 6×1×2 k-point mesh was used for the β-Cu_4Ti phase calculations. The convergence accuracy for the energy was set at 1.0×10^{-5} eV/atom.

The calculation model in this study was constructed based on the unit cell of the β'-Cu_4Ti and the β-Cu_4Ti phases. The prototype of β'-Cu_4Ti is Ni_4Mo, the space group is I4/m, with lattice parameters $a=0.584$ nm and $c=0.362$ nm[3]. The prototype of β-Cu_4Ti is Au_4Zr, the space group is Pnma, with lattice parameters $a=0.4528$ nm, $b=0.4345$ nm, and $c=1.2932$ nm[29]. Based on the unit cells of β'-Cu_4Ti and β-Cu_4Ti, β'-$Cu_{80}Ti_{20}$ and β-$Cu_{80}Ti_{20}$ supercells with 100 atoms were constructed. In order to investigate the influence of alloying element X (X = W, Re, Gd etc.) on β'-Cu_4Ti and β-Cu_4Ti phase stability, Cu-Ti-X ternary alloy models were constructed based on the β'-$Cu_{80}Ti_{20}$ and β-$Cu_{80}Ti_{20}$ supercells. Models for β'-$Cu_{80}(Ti_{19}X)$ and β-$Cu_{80}(Ti_{19}X)$ with 19 at% Ti content were constructed, as shown in Figure 1. All the models were calculated using DFT, and the calculations met the convergence accuracy requirements.

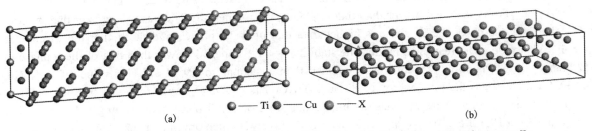

Figure 1 Calculation models for (a) β'-$Cu_{80}(Ti_{19}X)$, and (b) β-$Cu_{80}(Ti_{19}X)$ supercells

The formation energy ΔH (eV/atom)[30, 31] of the $A_x B_y C_z$ phase is calculated by:

$$\Delta H = \frac{1}{x+y+z}(E_0 - xE_{solid}^A - yE_{solid}^B - zE_{solid}^C) \quad (1)$$

where E_0 is the total energy of the $A_x B_y C_z$ phase, E_{solid}^A, E_{solid}^B, and E_{solid}^C are the energies of atoms A, B and C in the crystal, respectively, and x, y, z are the number of A, B and C atoms in the supercell, respectively. The formation energy is associated with alloying capability, and it reflects the difficulty of phase formation. A lower ΔH indicates a greater alloying capability and the phase is more readily formed.

In order to investigate the effect of different alloying elements (X) on the stability of the β'-Cu_4Ti and β-Cu_4Ti phases, the total energy difference $\Delta E_0(X)$ and formation energy difference $\Delta H^{\beta'\text{-}\beta}(X)$ were calculated using Eqs. (2) and (3), respectively:

$$\Delta E_0(X) = E_0(X)_{\beta'} - E_0(X)_{\beta} \quad (2)$$
$$\Delta H^{\beta'\text{-}\beta}(X) = \Delta H(X)_{\beta'} - \Delta H(X)_{\beta} \quad (3)$$

where $E_0(X)_{\beta'}$ and $E_0(X)_{\beta}$ are the total energy of the β'-$Cu_{80}(Ti_{19}X)$ and β-$Cu_{80}(Ti_{19}X)$ supercells, respectively; $\Delta H(X)_{\beta'}$ and $\Delta H(X)_{\beta}$ represent the formation energy of the β'-$Cu_{80}(Ti_{19}X)$ and β-$Cu_{80}(Ti_{19}X)$ supercells, respectively.

2.2 Experimental methods

The Cu-4.3Ti and Cu-4.3Ti-0.72Gd alloys were prepared using pure copper sticks (99.95%), pure titanium sticks (99.99%) and pure gadolinium blocks (99.95%) in a vacuum induction furnace (VGL-400) under argon atmosphere. The melt was poured into a pre-heated pure graphite 25 mm×75 mm×120 mm mold. The ingots were homogenized and the solution treated at 850 ℃ for 4 h, then quenched in cold water at room temperature. The quenched samples were aged at 500 ℃ for 0.5, 1, 2, 4, 6, 8, 12, 18 and 24 h. The homogenization treatment, solid solution treatment and aging were conducted in a box-type resistance furnace (SX2-18-18). The aged samples were ground using SiC paper to 2000 mesh and etched in a solution of 10 vol% nitric acid at room temperature for 10.

Sample phase was determined using an X-ray diffractometer (XRD, Bruker D8 Focus) with a Cu radiation target. Optical micrographs of the samples were obtained using an optical microscopy (OM, AxioskoFe, P2). Sample microstructure was assessed by transmission electron microscopy (TEM, Thermofisher Talos F200X) with an acceleration voltage of 200 kV; the elemental distribution was detected by energy dispersive spectroscopy (EDS, Thermofisher Talos F200X). The TEM samples were double sprayed using an electrolytic double sprayer device; the double spray solution was a mixed solution of 20% nitric acid and 80% methanol. The electrical conductivity of the samples was measured by an eddy current conductivity tester (SIGMASCOPE SMP350). The hardness of the samples was determined using a Vickers hardness tester (Future-Tech) with a 1 kg load for 10 s.

3 Results

3.1 Calculated stability of the precipitated phase

The total energy differences and formation energy differences of the β'-Cu_4Ti and β-Cu_4Ti phases are shown in Figure 2. The value of $\Delta E_0(Cu)$ is 0.12 eV [Figure 2(a)], which means the total energy of β'-$Cu_{80}(Ti_{19}Cu)$ is 0.12 eV higher than that of β-$Cu_{80}(Ti_{19}Cu)$. As the value of $\Delta E_0(Ti)$ is 1.34 eV, the total energy of β'-$Cu_{80}Ti_{20}$ is 1.34 eV higher than that of β-$Cu_{80}Ti_{20}$. When the Ti content is greater than or equal to 19 at%, the total energy of the β' phase is higher than that of the β phase, and the higher the Ti content the greater the ΔE_0 value. The results show that the stability of the β' phase is lower than that of the β phase in the Cu-Ti binary alloy when the Ti content is greater than or equal to 19 at%. The value of $\Delta E_0(Gd)$ is −2.02 eV, which means the total energy of β'-$Cu_{80}(Ti_{19}Gd)$ is −2.02 eV lower than that of β-$Cu_{80}(Ti_{19}Gd)$, indicating that the stability of the β' phase is higher than that of the β phase after adding Gd. The values of $\Delta E_0(W)$ and $\Delta E_0(Re)$ are 1.43 eV and 1.39 eV, respectively, which are higher than the value of $\Delta E_0(Ti)$. The total energy of β'-$Cu_{80}(Ti_{19}X)$ ($X=W$ or Re) is higher than that of the β-$Cu_{80}(Ti_{19}X)$, indicating that W and Re reduce the stability of the β' phase but serve to stabilize the β phase. The $\Delta E_0(X)$ values of other elements, with the exception of W, Re and Gd, in Figure 2(a) are between 0.12 and 1.34 eV, which means that elements other than W, Re and Gd have little effect on the stability of the β and β' phases.

As shown in Figure 2(b), the value of $\Delta H^{\beta'\text{-}\beta}(\text{Cu})$ is 0.0012 eV/atom, establishing that the formation energy of $\beta'\text{-}\text{Cu}_{80}(\text{Ti}_{19}\text{Cu})$ is 0.0012 eV/atom higher than that of $\beta\text{-}\text{Cu}_{80}(\text{Ti}_{19}\text{Cu})$. As the value of $\Delta H^{\beta'\text{-}\beta}(\text{Ti})$ is 0.0134 eV/atom, the formation energy of the $\beta'\text{-}\text{Cu}_{80}\text{Ti}_{20}$ is 0.0134 eV/atom higher than that of $\beta\text{-}\text{Cu}_{80}\text{Ti}_{20}$. When the Ti content is greater than or equal to 19 at%, the formation energy of the β' phase exceeds that of the β phase, and the higher the Ti content the greater is the value of $\Delta H^{\beta'\text{-}\beta}(X)$. The results demonstrate that in the Cu-Ti binary alloy (without the addition of a third element), when the Ti content is greater than or equal to 19 at% the possibility of forming the β' phase is less than that of β phase formation. As shown in Figure 2(b), the value of $\Delta H^{\beta'\text{-}\beta}(\text{Gd})$ is -0.0202 eV/atom, and the formation energy of $\beta'\text{-}\text{Cu}_{80}(\text{Ti}_{19}\text{Gd})$ is lower than that of $\beta\text{-}\text{Cu}_{80}(\text{Ti}_{19}\text{Gd})$, indicating a greater possibility of β' phase formation relative to the β phase. The values of $\Delta H^{\beta'\text{-}\beta}(\text{W})$ and $\Delta H^{\beta'\text{-}\beta}(\text{Re})$ are 0.0138 eV/atom and 0.0143 eV/atom, respectively, which are both higher than the value of $\Delta H^{\beta'\text{-}\beta}(\text{Ti})$. The formation energy of $\beta'\text{-}\text{Cu}_{80}(\text{Ti}_{19}X)$ is higher than that of $\beta\text{-}\text{Cu}_{80}(\text{Ti}_{19}X)$, indicating that W and Re can reduce the possibility of β' phase formation but promote the formation of the β phase. The $\Delta H^{\beta'\text{-}\beta}(X)$ values of the other elements included in Figure 2(b) fall between 0.0012 and 0.0139 eV/atom, suggesting that the elements, with the exception of W, Re and Gd, have little effect on the formation of the β and β' phases.

In brief, the inclusion of W and Re improves the formation and stability of β phase, while reducing the formation and stability of β' phase. The addition of Gd enhances the formation and stability of the β' phase, and has the opposite effect in the case of the β phase. In contrast, other elements with the exception of W, Re and Gd (in Figure 2) have little effect on the formation or stability of the β and β' phases.

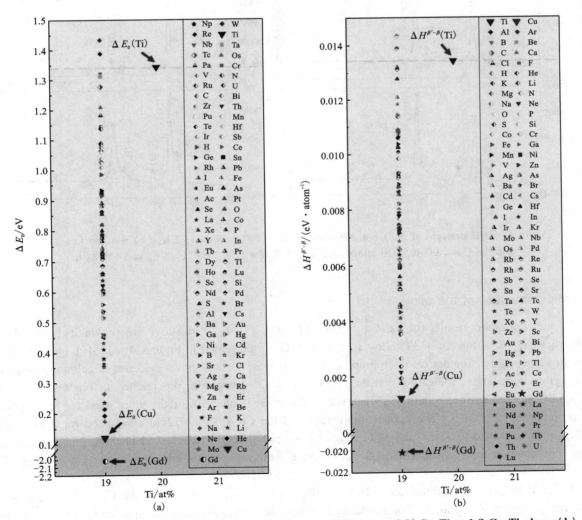

Figure 2　Total energy differences (a) and formation energy differences of $\beta'\text{-}\text{Cu}_4\text{Ti}$ and $\beta\text{-}\text{Cu}_4\text{Ti}$ phases (b)

3.2 Analysis of the alloy phase composition

The XRD patterns of the Cu-4.3Ti and Cu-4.3Ti-0.72Gd alloys aged at 500 ℃ are shown in Figure 3. In the patterns of the Cu-4.3Ti alloy [Figure 3(a)], only the peaks due to the fcc Cu matrix of the alloy aged for 2 h were detected. As shown in Figure 3(b), peaks attributed to the Cu matrix and peaks indexed to the β'-Cu_4Ti and β-Cu_4Ti phases were observed for the sample aged for 8 h. The XRD pattern for the alloy aged for 24 h is shown in Figure 3(c), and includes peaks due to the Cu matrix and β-Cu_4Ti phase. The XRD patterns for the Cu-4.3Ti-0.72Gd alloy are presented in Figures 3(d), 3(e) and 3(f). In the case of the samples aged for 2 h [Figure 3(d)] and 8 h [Figure 3(e)], there are no detectable peaks due to the β-Cu_4Ti phase, and only the peaks attributed to the Cu matrix and β'-Cu_4Ti phase are evident in the patterns. Peaks due to β-Cu_4Ti and β'-Cu_4Ti phases and the Cu matrix are detected for the sample aged for 24 h [Figure 3(f)].

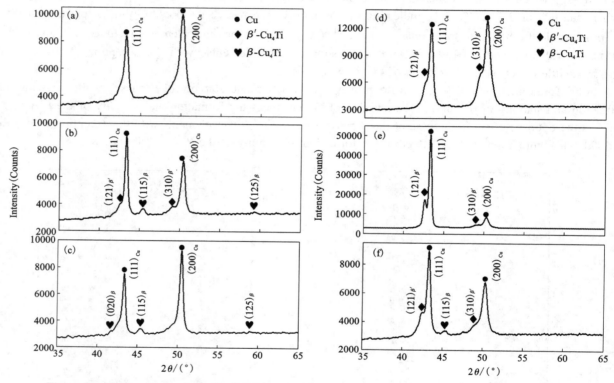

Figure 3 XRD patterns of the Cu-4.3Ti alloys aged at 500 ℃ for (a) 2 h, (b) 8 h and (c) 24 h, and the Cu-4.3Ti-0.72Gd alloys aged at 500 ℃ for (d) 2 h, (e) 8 h, and (f) 24 h.

3.3 Microstructure of the alloy

The metallographic results for the Cu-4.3Ti and Cu-4.3Ti-0.72Gd alloys are shown in Figure 4. The optical micrographs of the Cu-4.3Ti alloy aged at 500 ℃ are presented in Figures 4(a), 4(b) and 4(c). As shown in Figure 4(a), the optical micrograph of the alloy aged for 2 h shows evidence of cellular precipitation at some grain boundaries, indicating that the β-Cu_4Ti phase had begun to precipitate. In Figures 4(b) and 4(c), significant cellular precipitation is observed at the grain boundaries of the alloy aged for 8 h, which is more pronounced after 24 h. The optical micrographs of the Cu-4.3Ti-0.72Gd alloy aged for 2 to 24 h are included in Figures 4(d), 4(e) and 4(f). No cellular precipitation is observed in these images, indicating that the precipitation and growth of the β-Cu_4Ti phase were delayed or inhibited in the Cu-4.3Ti-0.72Gd alloy.

The TEM images and selected-area electron diffraction (SAED) patterns of the Cu-4.3Ti alloy aged at 500 ℃ are presented in Figure 5. The modulated contrast of finely dispersed precipitates in the copper matrix of the alloy aged for 2 h can be seen in Figure 5(a). The precipitates were determined to be β'-Cu_4Ti from the

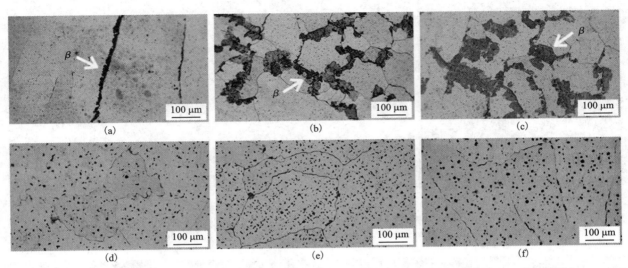

Figure 4 Optical micrographs of Cu-4.3Ti alloys aged at 500 ℃ for (a) 2 h, (b) 8 h and (c) 24 h, and Cu-4.3Ti-0.72Gd alloys aged at 500 ℃ for (d) 2 h, (e) 8 h, and (f) 24 h

SAED patterns. As shown in Figure 5 (b), there is evidence of lamellar cellular precipitates, which were identified as the β-Cu_4Ti phase and copper matrix according to the SAED patterns. A large area of cellular precipitation was observed for the alloy aged for 24 h [Figure 5(c)], which were composed of the β-Cu_4Ti phase and copper matrix as determined by SAED analysis.

Figure 5 TEM images and SAED patterns (inset) of Cu-4.3Ti alloy aged at 500 ℃ for (a) 2 h, (b) 8 h, and (c) 24 h

The TEM images and SAED patterns of the Cu-4.3Ti-0.72Gd alloy aged at 500 ℃ are shown in Figure 6. Only nano-scale precipitates dispersed in the matrix were observed for the alloys aged for 2 h [Figure 6(a)] and 8 h [Figure 6(b)]. Based on the SAED patterns, the precipitates were determined to be the β'-Cu_4Ti phase. As shown in Figure 6(c), the TEM image of the alloy aged for 24 h reveals a small amount of cellular precipitation, which was identified by SAED as the β-Cu_4Ti phase and copper matrix.

A scanning TEM high-angle annular dark field (HAADF) image and the associated elemental energy-dispersive X-ray spectroscopy (EDS) of the Cu-4.3Ti-0.72Gd alloy aged at 500 ℃ for 8 h are presented in Figure 7. The STEM-HAADF image [Figure 7(a)] shows the β'-Cu_4Ti phase as black square particles of ca. 20-30 nm size. The EDS results for Cu, Ti and Gd are given in Figures 7(b), 7(c) and 7(d), respectively. The Cu element is mainly distributed in the matrix, Ti is mainly concentrated in the β'-Cu_4Ti phase, and Gd evenly distributed in the matrix and the β'-Cu_4Ti phase.

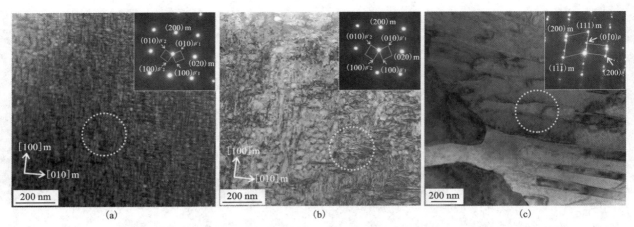

Figure 6 TEM images and SAED patterns (inset) of the Cu-4.3Ti-0.72Gd alloy aged at 500 ℃ for (a) 2 h, (b) 8 h, and (c) 24 h

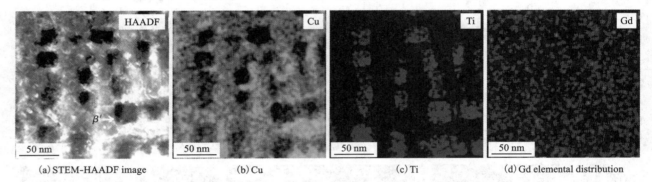

(a) STEM-HAADF image   (b) Cu   (c) Ti   (d) Gd elemental distribution

Figure 7 STEM-HAADF images and EDS maps of the Cu-4.3Ti-0.72Gd alloy aged at 500 ℃ for 8 h

3.4 Conductivity and hardness of the alloy

The changes in the electrical conductivity of the Cu-4.3Ti and Cu-4.3Ti-0.72Gd alloys aged at 500 ℃ are shown in Figure 8(a). The electrical conductivity of the quenched Cu-4.3Ti alloy was 10.5%IACS, which increased to 19.1%IACS after aging for 0.5 h. Following aging for 1 h to 12 h, the electrical conductivity fluctuated between 19.7% and 20.7%IACS, and was essentially unchanged. After aging for 18 h, the electrical conductivity increased to 22.7%IACS, and was slightly higher after aging for 24 h, reaching 23.4%IACS. The electrical conductivity of the quenched Cu-4.3Ti-0.72Gd alloy was 4%IACS, increasing to 10.5% and 16.9%IACS after aging for 0.5 h and 4 h, respectively. The electrical conductivity of the alloy decreased slightly after extending the aging time to 24 h, varying between 13.8% and 16.2%IACS.

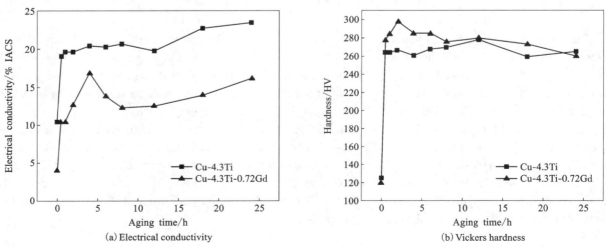

(a) Electrical conductivity   (b) Vickers hardness

Figure 8 Changes in the mechanical and electrical properties of the Cu-4.3Ti and Cu-4.3Ti-0.72Gd alloys aged at 500 ℃

The changes in the Vickers hardness of the Cu-4.3Ti and Cu-4.3Ti-0.72Gd alloys aged at 500 ℃ were illustrated in Figure 8(b). The Vickers hardness of the quenched Cu-4.3Ti alloy was 125 HV, increasing to 264 HV after aging for 0.5 h, and maintained between 261 and 270 HV after aging from 1 h to 8 h. The maximum hardness of the alloy was 279 HV after aging for 12 h, decreasing to 265 HV after aging for 24 h. The Vickers hardness of the quenched Cu-4.3Ti-0.72Gd alloy was 120 HV, and increased to 278 HV after aging for 0.5 h, reached a maximum of 298 HV following aging for 2 h, and decreased to 260 HV after aging for 24 h.

4 Discussion

4.1 Effect of Gd on the stability of the β'-Cu_4Ti phase

Previous studies have shown that when the Ti content of the β'-Cu_4Ti phase is greater than or equal to 19%, the stability of the β-Cu_4Ti phase is higher than that of the β'-Cu_4Ti phase. The resultant instability of β'-Cu_4Ti causes a re-dissolution of Ti atoms into the matrix, which has a deleterious effect on electrical conductivity that is difficult to correct. Consequently, it is difficult to attain a 19 at% Ti content in the β'-Cu_4Ti phase. The first-principles calculations have shown that the inclusion of Gd improves the formation and stability of the β'-Cu_4Ti phase when the Ti content is 19 at% (Figure 2), inhibiting the dissolution of the β'-Cu_4Ti phase and the Ti atoms during aging. With a reduction of Ti content in the matrix, the nucleation and growth of the β-Cu_4Ti phase which requires Ti atoms from the matrix are inhibited (Figures 3, 4, and 6). Allowing for the influence of calculation errors, other elements may exhibit similar effects, but this study has demonstrated firstly the role of Gd in the Cu-Ti alloy system.

Analysis of alloy microstructure has revealed a significant β-Cu_4Ti phase in the Cu-Ti alloy after aging at 500 ℃ for 8 h, which is more pronounced following aging for 24 h (Figure 5). After adding Gd to the Cu-Ti alloy, only a small amount of the β-Cu_4Ti phase was observed in the Cu-Ti-Gd alloy after aging for 24 h. It is confirmed that Gd has a strong effect in inhibiting the nucleation and growth of the β-Cu_4Ti phase, and can improve the formation and stability of the β'-Cu_4Ti phase.

The role of Gd in the segregation of Ti atoms and stability of the β'-Cu_4Ti phase is illustrated by the schematic representation of precipitate formation and Ti atom diffusion shown in Figure 9. The Ti atom segregation and re-dissolution during aging of the Cu-Ti alloy without Gd is presented in Figures 9(a)-(e). The initial stage of Ti atom segregation in the aging process is addressed in Figure 9(a). Ti atoms are enriched, due to spinodal decomposition, in Ti-rich region where the Ti atoms are randomly distributed. With further Ti segregation, the arrangement of Ti atoms becomes gradually ordered (D1a), as shown in Figure 9(b). As the aging time is extended, the ordered structure proceeds to form the metastable β'-Cu_4Ti phase. With further segregation of Ti atoms, the stability of the β-Cu_4Ti phase is higher than that of β'-Cu_4Ti when the Ti content is greater than or equal to 19 at%. The Ti atoms tend to re-dissolve from the β'-Cu_4Ti phase into the matrix, as shown in Figure 9(d). In the later stage of aging, Ti atoms diffuse into the β-Cu_4Ti phase, so the β-Cu_4Ti phase grows at a fast rate [Figure 9(e)].

The diffusion of Ti atoms during the aging of the Cu-Ti-Gd alloy is illustrated in Figure 9(f)-(j). The segregation of Ti atoms, short-range ordering of Ti atoms, nucleation and growth of the β'-Cu_4Ti phase is similar to that observed for the Cu-Ti alloy without the addition of Gd [Figure 9(f)-(h)]. However, the inclusion of Gd at a Ti content of 19 at% serves to promote the formation and stability of the β'-Cu_4Ti phase as opposed to β-Cu_4Ti, and the Ti atoms are retained in β'-Cu_4Ti [Figure 9(i)]. The precipitation and growth of the β-Cu_4Ti phase is inhibited due to the low content of Ti atoms in the matrix [Figure 9(j)].

4.2 Effect of Gd on alloy properties

As the aging time is extended, the electrical conductivity of Cu-Ti alloy first increases, and then remains largely unchanged before ultimately increasing again (Figure 8). Moreover, the Vickers hardness first increases and then decreases. At the initial stage of the aging process, a significant component of finely dispersed β'-Cu_4Ti phase precipitates in the matrix [Figure 5(a)] with an accompanying increase in the hardness of the alloy. The precipitation of the β'-Cu_4Ti phase consumes a large number of Ti atoms in the

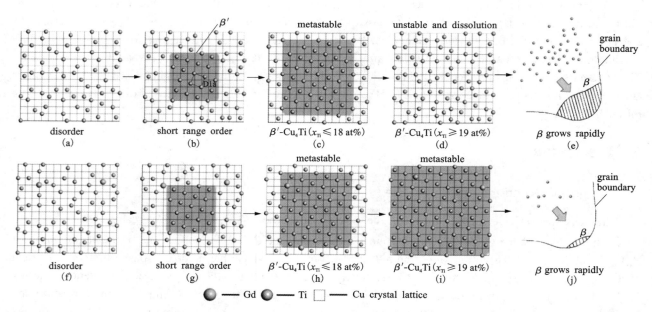

Figure 9 Schematic representation of precipitate formation and Ti atom diffusion with and without the inclusion of Gd

matrix, resulting in an increase in electrical conductivity over the 0-1 h period [Figure 8(a)]. In the intermediate stage of the aging process, the β-Cu_4Ti phase begins to nucleate at the grain boundary (Figure 4 and Figure 5), and the β'-Cu_4Ti and β-Cu_4Ti phases coexist. When the Ti content in the β'-Cu_4Ti phase is greater than or equal to 19 at%, the Ti atoms re-dissolve in the matrix and diffuse to the β-Cu_4Ti phase. The Ti content in the matrix remains essentially unchanged, and the electrical conductivity of the alloy is maintained between 19.7% and 20.7% IACS [Figure 8(a)]. At this point, the Vickers hardness of the alloy is slightly increased to 279 HV. In the later stage of aging process, an appreciable component of the β-Cu_4Ti phase has precipitated and grown [Figure 5(c)]. The Ti content in the matrix no longer maintains a dynamic equilibrium, and begins to decrease with a consequent increase in electrical conductivity [Figure 8(a)]. However, the Vickers hardness of the alloy gradually decreases due to the dissolution of the β'-Cu_4Ti phase into the matrix [Figure 8(b)].

With the addition of Gd to the Cu-Ti alloy, the change in mechanical properties essentially follow that exhibited by the Cu-Ti alloy without Gd. Inclusion of Gd increases the maximum Vickers hardness to 298 HV [Figure 8(b)], which is higher than that of the binary Cu-Ti alloy. The increase in hardness of the Cu-Ti-Gd alloy is due to the precipitation of the finely dispersed β'-Cu_4Ti phase (Figure 6). The β'-Cu_4Ti phase effectively hinders dislocation movement to achieve the effect of precipitation strengthening. In addition, Gd atoms are dissolved in the Cu matrix [Figure 7(d)], which results in a lattice distortion of the Cu matrix and a consequent solid solution strengthening of the Cu-Ti-Gd alloy. Therefore, the mechanical properties of the alloy are improved with the addition of Gd.

In order to characterize the role of Gd in this study, the amount of Gd added is relatively large, which can deteriorate the electrical conductivity of the Cu-Ti-Gd alloy. In the alloy preparation process, it is necessary to properly control the level of Gd addition to tune the ultimate properties of the alloy.

5 Conclusions

(1) The addition of W and Re to the Cu-Ti alloy reduces the formation and stability of the β'-Cu_4Ti phase, but improves formation and stability of the β-Cu_4Ti phase. The inclusion of Gd improves the formation and stability of the β'-Cu_4Ti phase, but has the opposite effect in the case of the β-Cu_4Ti phase. The other elements considered in this study have little effect on the formation and stability of the β'-Cu_4Ti and β-Cu_4Ti phases, and will be the subject of future work.

(2) The addition of Gd serves to reduce the total energy and formation energy of the β'-Cu_4Ti phase. It has been established that Gd can improve the formation and stability of the β'-Cu_4Ti phase, thereby reducing

the re-dissolution of β'-Cu_4Ti and Ti atoms, inhibiting nucleation and growth of the β-Cu_4Ti phase. In addition, Gd inclusion increases the Ti content in the metastable β'-Cu_4Ti phase to 19 at%, reducing the Ti content in the Cu matrix.

(3) Experimental results have demonstrated that Gd delayed and inhibited the nucleation and growth of the β-Cu_4Ti phase, confirming the effect of Gd element in improving the stability of the β'-Cu_4Ti phase.

(4) The Vickers hardness of the Cu-Ti-Gd alloy is higher than that of the Cu-Ti alloy, which is attributed to a precipitation strengthening effect of the finely dispersed β'-Cu_4Ti phase and a solid solution strengthening effect promoted by the micro-alloying element Gd.

Acknowledgments: This work was supported by The Youth Jinggang Scholars Program In Jiangxi Province, and The Program for Innovative Youth Talents of Jiangxi Province.

References

[1] HUANG L, CUI Z S, MENG X P, et al. Effect of trace alloying elements on the stress relaxation properties of high strength Cu-Ti alloys[J]. Materials Science and Engineering: A, 2022, 846: 143281.

[2] WANG X, XIAO Z, QIU W T, et al. The evolution of microstructure and properties of a Cu-Ti-Cr-Mg-Si alloy with high strength during the multi-stage thermomechanical treatment[J]. Materials Science and Engineering A, 2021, 803: 140510.

[3] DATTA A, SOFFA W A. The structure and properties of age hardened Cu-Ti alloys[J]. Acta Metallurgica, 1976, 24: 987-1001.

[4] SEMBOSHI S, KANENO Y, TAKASUGI T, et al. High strength and high electrical conductivity Cu-Ti alloy wires fabricated by aging and severe drawing[J]. Metallurgical and Materials Transactions A, 2018, 49(10): 4956-4965.

[5] NAGARJUNA S, SRINIVAS M. Grain refinement during high temperature tensile testing of prior cold worked and peak aged Cu-Ti alloys: Evidence of superplasticity[J]. Materials Science and Engineering A, 2008, 498: 468-474.

[6] TANG Y C, KANG Y L, YUE L J, et al. The effect of aging process on the microstructure and mechanical properties of a Cu-Be-Co-Ni alloy[J]. Materials and Design, 2015, 85(15): 332-341.

[7] LAUGHLIN D E, CAHN J W. Spinodal decomposition in age hardening copper-titanium alloys[J]. Acta Metallurgica, 1975, 23: 329-339.

[8] ZHANG J P, YE H Q, KUO K H, et al. A hrem study of the crystal structure of Cu_4Ti[J]. Physica Status Solidi A, 1985, 88: 475-482.

[9] THOMPSON A W, WILLIAMS J C. Age hardening in Cu-2.5 Wt Pct Ti[J]. Metallurgical Transactions A, 1984, 15A: 931-937.

[10] SEMBOSHI S, AMANO S, FU J, et al. Kinetics and equilibrium of age-induced precipitation in Cu-4 at. Pct Ti binary alloy[J]. Metallurgical and Materials Transactions A, 2017, 48: 1501-1511.

[11] SOFFA W A, LAUGHLIN D E. High-strength age hardening copper-titanium alloys: Redivivus, progress in materials science, 2004, 49: 347-366.

[12] BOZIC D, STASIC J, RUZIC J, et al. Synthesis and properties of a Cu-Ti-TiB_2 composite hardened by multiple mechanisms[J]. Materials Science and Engineering A, 2011, 528(28): 8139-8144.

[13] XU Y X, QIU X M, WANG S Y, et al. Microstructures, mechanical properties and formation mechanisms of tungsten heavy alloy brazed joints using Cu-Ti-Ni-Zr amorphous filler[J]. Materials & Design, 2022, 223: 111181.

[14] LI S, LI Z, XIAO Z, et al. Microstructure and property of Cu-2.7Ti-0.15Mg-0.1Ce-0.1Zr alloy treated with a combined aging process[J]. Materials Science and Engineering A, 2016, 650: 345-353.

[15] LI C, WANG X H, LI B, et al. Effect of cold rolling and aging treatment on the microstructure and properties of Cu-3Ti-2Mg alloy[J]. Journal of Alloys and Compounds, 2020, 818: 152915.

[16] WANG X H, CHEN C Y, GUO T T, et al. Microstructure and properties of ternary Cu-Ti-Sn alloy[J]. Journal of Materials Engineering and Performance, 2015, 24(7): 2738-2743.

[17] LIU J, WANG X H, GUO T T, et al. Microstructure and properties of Cu-Ti-Ni alloys[J]. International Journal of Minerals Metallurgy and Materials, 2015, 22(11): 1199-1204.

[18] KONNO T J, NISHIO R, SEMBOSHI S, et al. Aging behavior of Cu-Ti-Al alloy observed by transmission electron microscopy[J]. Journal of Materials Science, 2008, 43(11): 3761-3768.

[19] SEMBOSHI S, KANENO Y, TAKASUGI T, etal. Suppression of discontinuous precipitation in Cu-Ti alloys by aging in a hydrogen atmosphere[J]. Metallurgical and Materials Transactions A, 2020, 51(7): 3704-3712.

[20] SEMBOSHI S, IKEDA J, IWASE A, et al. Effect of boron doping on cellular discontinuous precipitation for age-hardenable Cu-Ti alloys[J]. Materials, 2015, 8(6): 3467-3478.

[21] LIAO Y M, GUO C J, ZHOU C Y, et al. Stability of the metastable β'-Cu_4Ti phase in Cu-Ti alloys: Role of the Ti content[J]. Materials Characterization, 2023, 203: 113164.

[22] XU Y, TIAN M L, HU C Y, et al. Structural, electronic, mechanical, and thermodynamic properties of Cu-Ti intermetallic compounds: First-principles calculations[J]. Solid State Communications, 2022, 352: 114814.

[23] ZHANG D L, WANG J, KONG Y, et al. First-principles investigation on stability and electronic structure of sc-doped θ'/Al interface in Al-Cu alloys[J]. Transactions of Nonferrous Metals Society of China, 2021, 31(11): 3342-3355.

[24] ZANG W A. A half century of density functional theory[J]. Physics Today, 2015, 68(7): 34-39.

[25] ZHU Y D, YAN M F, ZHANG Y X, et al. First-principles investigation of structural, mechanical and electronic properties for Cu-Ti intermetallics[J]. Computational Materials Science, 2016, 123: 70-78.

[26] KRESSE G, FURTHMULLER J. Efficient iterative schemes for Ab initio total-energy calculations using a plane-wave basis set[J]. Physical Review B, 1996, 54: 11169-11186.

[27] RANGEL T, CALISTE D, GENOVESE L, et al. A wavelet-based projector augmented-wave (Paw) method: Reaching frozen-core all-electron precision with a systematic, adaptive and localized wavelet basis set[J]. Computer Physics Communications, 2016, 208: 1-8.

[28] PERDEW J P, BURKE K, ERNZERHOF M. Generalized gradient approximation made simple[J]. Physical Review Letters, 1996, 77: 3865-3868.

[29] SEMBOSHI S, ISHIKURO M, SATO S, et al. Extraction of precipitates from age-hardenable Cu-Ti alloys[J]. Materials Characterization, 2013, 82: 23-31.

[30] LI C M, ZENG S M, CHEN Z Q, et al. First-principles calculations of elastic and thermodynamic properties of the four main intermetallic phases in Al-Zn-Mg-Cu alloys[J]. Computational Materials Science, 2014, 93: 210-220.

[31] SUN D Q, WANG Y X, ZHANG X Y, et al. First-principles calculation on the thermodynamic and elastic properties of precipitations in Al-Cu alloys[J]. Superlattices and Microstructures, 2016, 100: 112-119.

Effect of Process Parameters and Alloy Composition on Residual Stress of Martensitic Stainless Steel Cladding Layer

Kaiping DU*, Ziqiang PI, Xing CHEN, Zhaoran ZHENG, Jie SHEN

1　BGRIMM Technology Group, Beijing, 100160, China
2　Beijing Key Laboratory of Special Coating Materials and Technology, Beijing, 102206, China
3　Beijing Industrial Parts Surface Hardening and Repair Engineering Technology Research Center, Beijing, 102206, China

Abstract: Laser cladding is a technology that uses high-energy-density lasers to quickly melt and solidify alloy powder on the surface of the metal substrate to form a cladding layer with good performance. In this work, the martensitic stainless steel layers were fabricated on the C45 steel substrate by the laser cladding with different process parameters. The results show that holes in the cladding layer is unavoidable. The laser cladding process parameters have the important influence on the residual stress in the cladding layer. Under the action of residual stresses, the holes in the cladding layer will be the source of cracks, which will cause cracks in the cladding layer. By optimizing the composition of the alloy powder, the increase in the content of Ni from 1% to 4% makes residual stress in the cladding layer further reduce to 93 MPa

Keywords: Laser cladding; Process parameters; Powder composition; Martensitic stainless steel cladding layer

1 Introduction

Laser cladding technology is a surface modification technology with broad application prospects. Its advantages include metallurgical bonding between the cladding layer and the substrate, small thermal deformation, narrow heat-affected zone, low dilution rate of the cladding layer, high powder utilization rate and convenient for industrial production. [1-4] Martensitic stainless steel is one of the main material systems of laser cladding, which is widely used in the surface protection of wear-resistant parts in mining, machinery, automobile, petrochemical and other fields. [5-6]

Laser cladding is a very complex physical and chemical change, and its process parameters, including laser power, scanning speed, layer thickness and overlap ratio, have a great influence on the quality and performance of the cladding layer[7]. Laser power is an important parameter in laser cladding process. With the increase of laser power, the increasing surface temperature of cladding material is easy to produce holes and cracks, which affect the shape, microstructure, microhardness and other properties of the cladding layer[8]. Scanning speed is also an important parameter in laser cladding. Lalas[9] obtained the relationship between the scanning speed, powder feeding rate and the morphology of the cladding layer during laser cladding, by measuring the depth, height and width of the cladding layer. The powder feeding rate determines the thickness of the cladding layer, which further affects the laser energy absorption of the powder. It will affect the microstructure and properties of the cladding layer. As a process control parameter, the overlap ratio refers to the ratio of the superposition distance of two adjacent cladding layers to the beam size. In order to obtain the cladding layer with uniform structure and stable performance as far as possible, reasonable overlap ratio has become an inevitable choice[10].

The alloy powder composition has a significant impact on the cladding layer performance. In order to reduce the cracking tendency of the cladding layer, it is necessary to reduce the residual stress of the cladding layer. As an austenite forming element, Ni element can promote the formation of austenite, which is beneficial for reducing the residual stress of the cladding layer and reducing the cracking tendency. [11-14]

Based on the above research results, the effect of laser cladding process parameters and powder composition on the residual stress of martensitic stainless steel cladding layer would be systematically investigated, and the causes of defects in cladding layer would be also analyzed.

2 Experiments

The martensitic stainless steel powder was prepared by gas atomization process. The chemical composition and morphology of the powders were shown in Table 1 and Figure 1. The cladding layer was prepared on the C45 steel substrate, whose chemical composition were shown in Table 1. No. 1 alloy powder with Ni content of 1% was used for the study on the optimization of process parameters. No. 2, 3 and 4 alloy powder were used for the study on the composition optimization.

Table 1 Chemical composition of martensitic stainless steel powder and c45 steel substrate   %

Element	Martensitic stainless steel				C45 steel
	No. 1	No. 2	No. 3	No. 4	
Fe	Bal.				Bal.
Cr	16.00				≤0.25
Ni	1.00	2.00	4.00	6.00	≤0.25
Mn	<0.5				0.50
Si	1.00				0.17
C	<0.20				0.42

The laser cladding experiment was carried out by using the MF-LC 2000 equipment of GTV Company. Process parameters are as follows: laser power from 2200 W to 3000 W, scanning speed from 6 mm/s to 10 mm/s, layer thickness from 1 mm to 3 mm and overlap ratio from 30% to 70%. The microstructure of the cladding layer was observed with a Hitachi SU 5000 scanning electron microscope (SEM). The residual stress test of the cladding layer was carried out by using D8 ADVANCE X-ray diffractometer (XRD) produced by BRUKER Company. The friction test of the cladding layer was carried out by using the UMT friction and wear tester of BRUKER Company. After the test, the samples were cleaned and dried respectively, and then weighed by an analytical balance to calculate the wear loss. The electrochemical test was carried out on the sample using an electrochemical workstation with a 3.5 wt% NaCl aqueous solution.

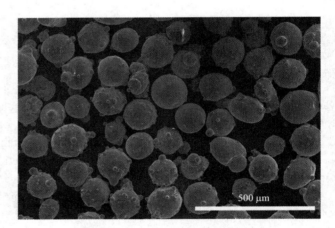

Figure 1 Morphology of martensitic stainless steel powder

3 Results and discussion

3.1 Effect of process parameters on microstructure and residual stress of cladding layer

The porosity in the cladding layer is closely related to the process parameters. Through the observation of SEM, the porosity of the cladding layer under different process parameters are measured. By using the method of multiple linear regression analysis, it can be seen that the influence degree of process parameters on porosity from high to low is thickness of cladding layer, laser power, scanning speed and overlap ratio, which is shown in Equation (1).

$$\omega = -2.625\times10^{-2}P + 42.28D - 3.315\times10^{-1}\delta + 1.309\times10^{-5}P^2 - 2.178\times10^{-2}PD +$$
$$1.794\times10^{-1}V^2 - 4.332V\delta + 4.22D^2 + 58.89\delta^2 \qquad (1)$$

where ω: porosity P: laser power, W; D: thickness of cladding layer, mm; δ: overlap ratio V: scanning speed, mm/s.

The research shows that although the porosity in laser cladding could be reduced by adjusting the process parameters, it is difficult to avoid. There are mainly three types of holes. The first type is lap hole. A large amount of powder mixed with carrier gas directly enters the molten pool, forming voids between powders and unmelted spherical particles, as shown in Figure 2(a). The second type is solidification shrinkage. At the later stage of solidification, the proportion of solid phase increases, the dendrite grows and connects to form a skeleton, dividing the molten metal into small isolated molten pools. These molten metals are difficult to be supplemented during solidification, thus forming many small and dispersed holes, as shown in Figure 2(b). The third type is that the carbon in the cladding powder reacts with oxygen to produce gas. Once the bubble nucleates and grows, it is easy to form diffuse holes, namely, precipitated holes, as shown in Figure 2(c). The formation schematic diagram of the above three types of holes is shown in Figure 3.

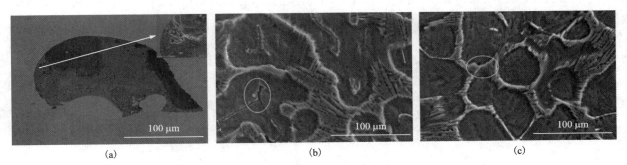

Figure 2 Microstructure of three types of holes

In addition to holes, cracks often appear in the laser cladding layer. These cracks occur at the junction (a) of the cladding layer and the heat affected zone and inside the cladding layer (b), as shown in Figure 4.

The stress in the cladding layer is the most important factor leading to cracks. The high heating and cooling rates in laser cladding process make the temperature distribution in the cladding layer seriously uneven, which restricts the deformation, and forms large macro residual stress in the cladding layer and the heat affected zone. At the same time, due to the large temperature gradient in the molten pool, a series of crystal form transformations of planar crystal, columnar crystal and fine equiaxed crystal occur in a very small range during the crystallization of the molten pool metal, resulting in the formation of large structural stress in the cladding layer, namely, micro residual stress. Under the combined action of above two kinds of residual stresses, the holes in the cladding layer will be the source of cracks, which will cause cracks in the cladding layer.

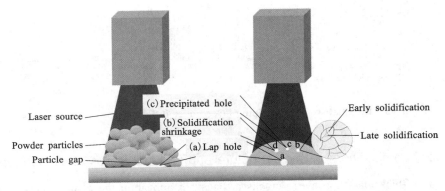

Figure 3 Formation schematic diagram of three types of holes

Residual stress of cladding layer under different process parameters is shown in Figure 5. Under the

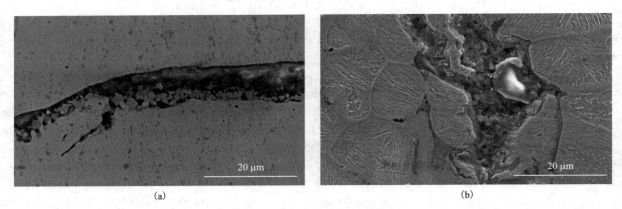

Figure 4　Crack at junction (a) of cladding layer and heat affected zone and inside cladding layer (b)

optimum process conditions in this study, the minimum residual stress of the cladding layer is only 245 MPa, which is 30.6% lower than the maximum residual stress of 353 MPa.

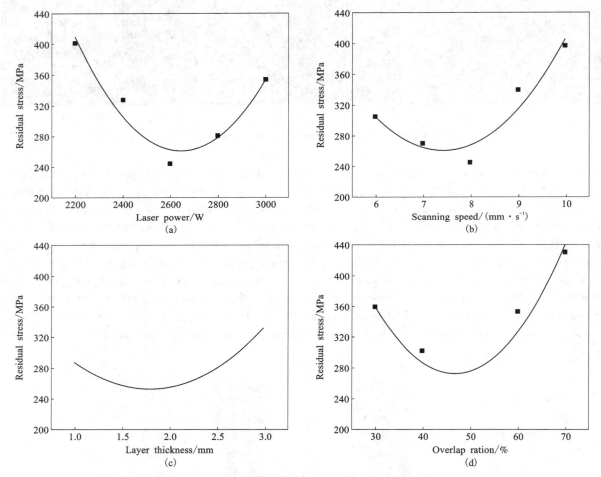

Figure 5　Residual stress of cladding layer under different process parameters

3.2　Effect of alloy composition on microstructure and residual stress of cladding layer

When preparing a thick cladding layer, cracks are easy to appear in the cladding layer, due to the high content of Martensite in the alloy material. In order to reduce the possibility of cracks in the cladding layer, it is necessary to increase the austenite content in the cladding layer, leading to the reducing of the cladding layer residual stress. Ni element, as an austenite forming element, could promote the formation of austenite, reduce residual stress, and prevent cracking of the cladding layer. Therefore, the thermodynamic phase diagram calculation software Thermo-Calc was used in this work to calculate the phase precipitation of cladding layer in solidification process under different Ni contents.

The variations of proportions of retained austenite and Sigma phase in the cladding layer under different Ni contents are shown in Figure 6. The residual austenite only appears in the cladding layer when the Ni content is 3%, 4%, and 5%. And the proportion of residual austenite in the cladding layer is the highest when the Ni content is 4%. Therefore, it could be inferred that the toughness of the cladding layer is the best when the Ni content is 4%. The proportion of Sigma phase decreases with the increase of Ni content in the cladding layer. When the Ni content is 6%, the Sigma phase basically disappears. Considering that the higher the retained austenite content leading to the better the toughness and lower crack sensitivity of the cladding layer, and the more brittle phases represented by Sigma phase leading to the higher the probability of crack occurrence, it is recommended to optimize the Ni content from 1% to 4% to reduce the number of cracks in the cladding layer.

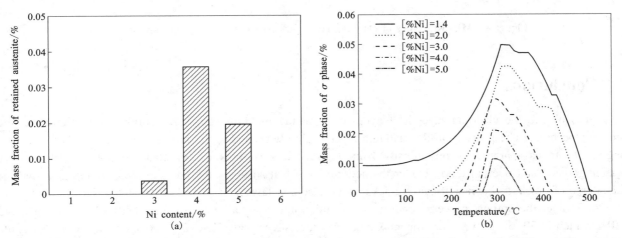

Figure 6　Variations of proportions of retained austenite and sigma phase in cladding layer under different Ni contents

The residual stress of cladding layer under different Ni contents are shown in Figure 7. The Plus-minus sign in the Figure 6 indicates the direction of residual stress, independent of the value. When the Ni element content is 4.0 wt%, the residual stress of the cladding layer is the lowest, with a value of 93 MPa. This is consistent with the calculated results.

The microstructure of the cladding layer with different Ni contents are described in Figure 8. When the Ni element content is 1%, cracks, extending along grain boundaries, could be observed inside the cladding layer. When the Ni element content is 4%, the microstructure of the cladding layer is dense and there are no obvious cracks. A large amount of retained austenite distributed near grain boundaries appears in the microstructure of the

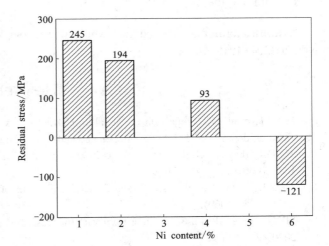

Figure 7　Residual stress of cladding layer under different Ni contents

cladding layer, which not only does not significantly reduce the hardness and strength of the cladding layer, but also reduces the stress of the cladding layer and reduces the probability of crack occurrence. On the basis of ensuring the hardness and strength of the cladding layer, this will reduce the stress in the cladding layer and the probability of crack occurrence.

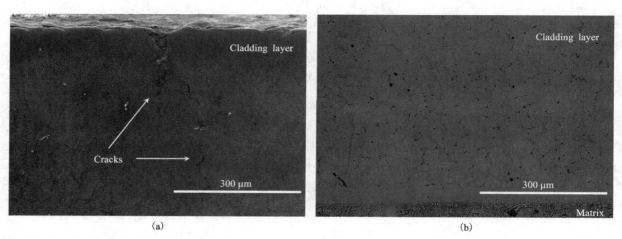

Figure 8 Microstructure of cladding layer with Ni content of 1% (a) and 4% (b)

4 Conclusions

(1) Holes in the cladding layer is unavoidable, mainly including lap hole, solidification shrinkage and precipitated hole. The porosity in the cladding layer is closely related to the process parameters. The influence degree of process parameters on porosity from high to low is thickness of cladding layer, laser power, scanning speed and overlap ratio. Under the action of residual stresses, the holes in the cladding layer will be the source of cracks, which will cause cracks in the cladding layer.

(2) Under the optimal process conditions, the minimum residual stress of the cladding layer is only 245 MPa, which is 30.6% lower than the maximum residual stress with 353 MPa. Ni element, as an austenite forming element, could promote the formation of austenite, reduce residual stress, and prevent cracking of the cladding layer. When the Ni element content is 4.0 wt%, the residual stress of the cladding layer is the lowest, with a value of 93 MPa.

Acknowledgments: The authors would like to thank the sponsorship by Beijing Nova Program of China (NO. 20220484019).

References

[1] MA M, et al. A comparison on metallurgical behaviors of 316L stainless steel by selective laser melting and laser cladding deposition[J]. Materials Science and Engineering A, 2017, 685: 265-273.

[2] SIDDIQUI A A, et al. Recent trends in laser cladding and surface alloying[J]. Optics & Laser Technology, 2021, 134: 106619.

[3] ZHU L, et al. Recent research and development status of laser cladding: A review[J]. Optics & Laser Technology, 2021, 138: 106915.

[4] SINGH S, et al. Laser cladding technique for erosive wear applications: A review[J]. Materials Research Express, 2020, 7(1): 012007.

[5] GAO W, et al. Study on the laser cladding of FeCrNi coating[J]. Optik, 2019, 178: 950-957.

[6] ZHANG Y, et al. Effect of Nb content on microstructure and properties of laser cladding FeNiCoCrTi$_{0.5}$Nb$_x$ high-entropy alloy coating[J]. Optik, 2019, 198: 163316.

[7] WANG X, et al. Laser melting deposition of duplex stainless-steel coating on high strength low alloy pipeline steels for improving wear and corrosion resistance[J]. Materials Express, 2019, 9(9): 1009-1016.

[8] ANJOS M A, et al. Fe-Cr-Ni-Mo-C alloys produced by laser surface alloying[J]. Surface & Coatings Technology, 1995, 70(3-5): 235-242.

[9] LALAS C, et al. An analytical model of the laser clad geometry[J]. International Journal of Advanced Manufacturing Technology, 2007, 32(1-2): 34-41.

[10] WATKINS K G, et al. Influence of the overlapped area on the corrosion behaviour of laser treated aluminium alloys

[J]. Materials Science and Engineering A, 1998, 252(2): 292-300.
[11] CHIU K Y, et al. Laser cladding of austenitic stainless steel using NiTi strips for resisting cavitation erosion[J]. Materials Science and Engineering A, 2005, 402(1-2): 126-134.
[12] LIU X B, et al. Synthesis of a nickel silicide-base composite coating on austenitic steel by laser cladding[J]. Journal of Materials Science Letters, 2001, 20(16): 1489-1492.
[13] KAUL R, et al. Laser cladding of austenitic stainless steel with hardfacing alloy nickel base[J]. Surface Engineering, 2013, 19(4): 269-273.
[14] BHADURI A K, et al. Hardfacing of austenitic stainless steel with nickel-base NiCr alloy[J]. International Journal of Microstructure & Materials Properties, 2011, 6(1/2): 40-42.

Nano-$Li_{1.3}Al_{0.3}Ti_{1.7}(PO_4)_3$ Modified Ni-Rich Cathode Materials for High Energy Density Lithium-ion Batteries

Zongpu SHAO[1,2,3], Yafei LIU[2,3], Xuequan ZHANG[2,3], Yanbin CHEN[2,3], Yueguang YU[1,4]

1　Northeastern University, Shenyang, 110167, China
2　BGRIMM Technology Group Co., Ltd., Beijing, 100160, China
3　Beijing Easpring Material Technology Co., Ltd., Beijing, 102628, China
4　China Iron & Steel Research Institute Group Co., Ltd., Beijing, 100053, China

Abstract: Nano scale $Li_{1.3}Al_{0.3}Ti_{1.7}(PO_4)_3$ (LATP) modified $LiNi_{0.90}Co_{0.05}Mn_{0.05}O_2$ (NCM9) was synthesized by a solid-state method. Transmission electron microscopy (TEM) revealed that the nano-layer is about 100 nm thick. The discharge specific capacities and cycle performance of 0.5% LATP modified NCM9 based on lithium ion battery and sulfide solid state battery are both much higher than those of the pristine. Meanwhile, the modified NCM9 exhibited 3.5 ℃ higher exothermal peak temperature and 281 J/g lower overall heat generation. The results demonstrate that LATP coating layer can prevent the direct contact between the NCM particle and the electrolyte, stabilize the crystal structure and improve the thermal stability of NCM9.

Keywords: $LiNi_{0.90}Co_{0.05}Mn_{0.05}O_2$; $Li_{1.3}Al_{0.3}Ti_{1.7}(PO_4)_3$; Lithium ion battery; Surface modification; Sulfide solid state battery

1　Introduction

Rechargeable lithium-ion batteries (LIBs) are widely used in electric vehicles and portable electronic devices.[1-2] However, the energy density and safety performance of lithium-ion batteries have been put forward higher requirements. In terms of energy density, it can be achieved by increasing the Ni contends of $LiNi_xCo_yMn_{1-x-y}O_2$ (NCM). However, NCM reacts with the electrolyte (liquid electrolyte or sulfide electrolyte)[3,4] during the cycle and high temperature storage, and the reaction becomes more intense as the nickel content increases. This means that the safety performance of nickle-rich cathode materials faces greater challenges. In order to further improve battery safety performance, it is urgent to enhance the crystal structural and electrochemical stability of NCM and restrain its strong oxidizing property at discharged state. Surface modification, mainly including metal oxides coating and lithium compound coating[5], was confirmed an effective method. Metal oxides coating (i.e. Al_2O_3[6], ZrO_2[7], etc.), first reported by Cho et al.[8], considerably enhances the capacity retention and structural stability of NCM. Nevertheless, metal oxides were lithium ionic insulators, resulting in ionic resistance between the cathode and electrolyte. Surface modification with a lithium ion conductor can solve this issue. NASICON-type $Li_{1+x}Al_xTi_{2-x}(PO_4)_3$ (LATP) with high ionic conductivity, attracted much attention among various solid state electrolyte. Morimoto et al.[9] prepared the composite by mechanical milling LATP and NCM, which exhibited high capacity (around 180 mA·h/g) and excellent cycle performance. However, most of the related coating research mainly uses sol-gel method, which is difficult to ensure good crystallinity of coating layer and hard to achieve large-scale production[10,11]. Here, we report a high-performance NCM cathode coated with nano LATP ($x = 0.3$ in this work) by solid-state method, then characterize the layer by scanning electron microscopy (SEM) and transmission electron microscopy (TEM). The cycling stability of the modified NCM is evaluated based on not only conventional lithium-ion battery but also sulfide solid state lithium battery.

2　Experiment

Single crystal $LiNi_{0.90}Co_{0.05}Mn_{0.05}O_2$ (NCM9) powder (Beijing Easpring Material Technology Co. Ltd., China) was used for preparation. The coating layer was formed from a nanoscale LATP powder. LATP was synthesized by a solid-state method. Li_2CO_3, Al_2O_3, TiO_2 and $NH_4H_2PO_4$ were used as raw materials, which were mixed by ball milling with a rotation speed of 900 r/min at room temperature, sintered 850 ℃ for 10 h in air. The obtained LATP powder was fine ground by a pulverizer (VGREEN, China) with deionized water as the dispersing agent, then dried and pulverized to get the nanoscale powder. 0.5wt% LATP nanopowders were mixed with NCM9 by super mixer (KAWATA, Japan), followed by heating treatment at 500 ℃ for 8 h.

2025-type-coin-cells were fabricated to exam the cycling performance with lithium metal as the anode and 1M $LiPF_6$ in ethylene carbonate/dimethly carbonate (EC/DMC) (1∶1 vol%) as the electrolyte. The cathode was prepared by mixing the modified NCM9 (or pristine), carbon black and polyvinylidene fluoride (PVDF) binder with a weight ratio of 95∶2.5∶2.5 and blade coating on Al foil. Charge-discharge measurements were carried out at the cut-off potential range from 3.0 to 4.3V (vs Li^+/Li) with a current density of 0.1C, 0.33C and 1C(200 mA/g) rate at 25 ℃ and 45 ℃ respectively.

Sulfide solid state batteries were fabricated with the Li_6PS_5Cl-NCM9 as the cathode, Li_6PS_5Cl as the solid electrolyte, and the Li foil as the anode. The electrolyte layer was obtained by compressing Li_6PS_5Cl powder (China Automotive Battery Research Institute Co., Ltd., China) at 80 MPa in a stainless steel mold with a diameter of 12 mm. Subsequently, the composite cathode(pristine or modified NCM9 mix sulfide electrolyte with a weight ratio of 70∶30) powder was pressed on the top of the pellet at 80 MPa and the loading was around 10 mg/cm^2. After that, the Li foil was attached onto the other side of the pellet as the anode at 80 MPa. Finally, the formed three-layered pellet was pressed under 80 MPa with stainless steel as the current collector. The electrochemical performance of the solid state battery was tested in a Swagelok cell at room temperature under a cut-off voltage of 2.5-4.3V (vs Li^+/Li). The first charge-discharge and cycling performance were evaluated at 0.02C and 0.1C rate respectively for solid state batteries.

3 Results and discussion

Figure 1(a) shows the nano-sized LATP aqueous sol and the Tyndall effect is distinct when a laser beam irradiates it, and Figure 1(b) shows the SEM of LATP powder, which reveals that the most particle size of LATP is below 300 nm. The XRD patterns [Figure 1(c)] show all of the diffraction peaks can be indexed to a well-defined $LiTi_2(PO_4)_3$ structure with R-3c(167) space group, and no impurity phase was detected.

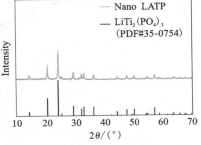

(a) The photograph of the nano-sized LATP aqueous sol and its Tyndall effect   (b) SEM images of LATP powder   (c) XRD patterns of nano LATP powder

Figure 1

The SEM images of pristine and modified NCM9 are shown in Figure 2. NCM9 [Figure 2(a)] powder is of smooth surface and modified NCM9 powders [Figure 2(b)] exist membrane-like layer with some dots on it. The dots are larger size particles of LATP. The TEM images of modified NCM9 [Figure 2(d)] revealed the LATP layer is less than 100 nm thick and exists in the form of nanocrystalline. The XRD patterns show all of the diffraction peaks can be indexed to a well-defined a-$NaFeO_2$-type layered structure with R-3m space group, and no impurity phase was detected, and this revealed that LATP coating cannot change the phase structure of NCM.

Charge and discharge curves at 0.1C, 0.33C, 1.0C current density of pristine and modified NCM9 with the cut-off potential of 4.3 V at room temperature, are shown in Figure 3(a). The discharge specific capacities of modified NCM9 at 0.1C, 0.33C and 1C are 222.7, 211.4 and 203.6 mA·h/g respectively,

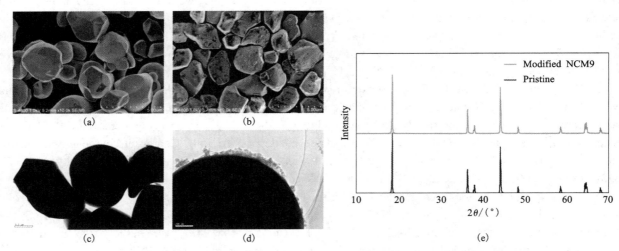

Figure 2　The SEM images (a)(b) and TEM images (c)(d) of pristine and modified NCM9 powders

which is much higher than those of pristine (218.2, 208.3 and 203.5 mA·h/g at 0.1C, 0.33C and 1C, respectively). The voltage platform of modified NCM9 during charging is lower than that of pristine, while its discharge voltage is slightly higher than that of pristine, indicating that the internal resistance polarization of NCM9 is reduced by the coating of LATP. The cycling performance of cathode materials is shown in Figure 3 (b). The cells with LATP modified NCM9 exhibit excellent electrochemical stability with 198.7 mA·h/g discharge capacity after 80 cycles tests. While the cell of pristine demonstrates a rapid discharge capacity degeneration with 194.3 mA·h/g discharge capacity after 80 cycles.

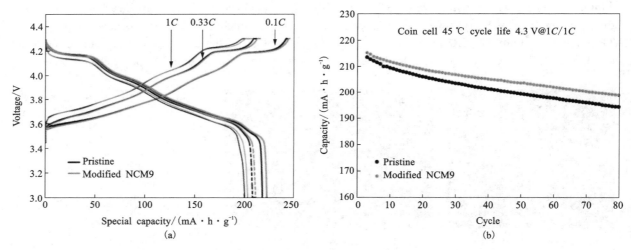

Figure 3　Charge and discharge curves at 0.1C, 0.33C, 1C (a) and cycling performance (b) of pristine and modified NCM9

Apart from the rate and cycling performance, the thermal stability of cathode materials is of significant importance for their applications in commercial electrical vehicles and energy storage systems. Figure 4 shows the DSC curves of the pristine and modified NCM9 electrodes charged to 4.3 V. The pristine exhibited an exothermal peak at 201.8 ℃ with the overall heat generation of 1173 J/g. Conversely, the initial exothermic reaction temperature for the LATP modified NCM9 reached a higher value of 205.3 ℃ with greatly reduced heat generation (892 J/g). This good thermal stability of the modified NCM9 materials can be ascribed to the highly thermal stability of the LATP coating layer.

The electrochemical performance of NCM9 for sulfide solid state battery tested at room temperature is shown in Figure 5. The discharge specific capacities of modified NCM9 at 0.02C is 132.0 mA·h/g, which is much higher than the pristine (83.1 mA·h/g). The voltage platform of modified NCM9 during charging is much lower than that of pristine, while its discharge voltage is significantly higher than that of pristine,

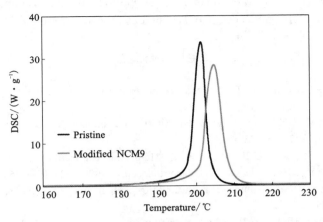

Figure 4　DSC curves of pristine and modified NCM9

indicating that the solid-solid interface resistance in solid state batteries are reduced effectively by the coating of LATP. The capacity of the modified NCM9 is 93.2% of the initial discharge capacity after 10 cycles (76.9% capacity retention after 40 cycles), while the capacity retention of the pristine exhibits an accelerated deterioration after the 10 cycles at 4.3 V, resulting in only 66.1% of the initial capacity. The results demonstrate that the side reaction between sulfide and cathode can be restrained by the LATP nanosized layer.

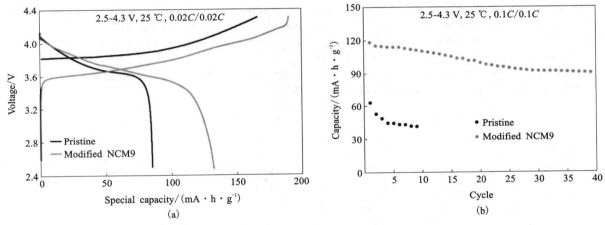

Figure 5　Charge and discharge curves at first cycle (a) and
cycle performance of pristine and modified NCM9 for sulfide solid state cells (b)

4　Conclusion

In this study, nano $Li_{1.3}Al_{0.3}Ti_{1.7}(PO_4)_3$ (LATP) modified $LiNi_{0.9}Co_{0.05}Mn_{0.05}O_2$ (NCM9) was synthesized by a solid-state method. SEM images show that the surface of modified NCM9 powder are a membrane-like layer with some dots on it. Transmission electron microscopy (TEM) revealed that the nano-layer is about 100 nm thick. The discharge specific capacities of 0.5% LATP modified NCM9 in liquid state cells at $0.1C$, $0.33C$ and $1C$ are 222.7, 211.4 and 203.6 mA·h/g respectively, which is much higher than the pristine. Meanwhile, the modified NCM9 based on sulfide solid state batteries (without additional pressure) exhibited 132.0 mA·h/g discharge capacity and 93.2% discharge retention of the initial discharge capacity after 10 cycles at the cut-off potential of 4.3 V, which is much higher than the pristine. The modified NCM9 exhibited 3.5 ℃ higher exothermal peak temperature and 281 J/g lower overall heat generation than the pristine. The results demonstrate that the LATP coating layer can prevent the direct contact between the NCM particle and the electrolyte, stabilize the structure of NCM, improve the thermal stability of NCM, and be widely used in high safety and high energy density batteries.

Acknowledgments: This work was financially supported by the Beijing Municipal Science and Technology Commission and Beijing Science and Technology Project (Grant No. Z191100004719001).

References

[1] GOODENOUGH J B, PARK K S. The Li-ion rechargeable battery: A perspective[J]. J Am Chem Soc, 2013, 135: 1167-1176.

[2] XIAO J, SHI F, GLOSSMANN T, et al. From laboratory innovations to materials manufacturing for lithium-based batteries[J]. Nat Energy, 2023, 8: 329-339.

[3] NISHIMOTO A, AGEHARA K, FURUYA K, et al. High ionic conductivity of polyether-based network polymer electrolytes with hyperbranched side chains[J]. Macromolecules, 1999, 32: 1541-1548.

[4] DONG Y H, LI J. Oxide cathodes: Functions, instabilities, self healing, and degradation mitigations[J]. Chem. Rev, 2023, 123(2): 811-833.

[5] SHAO, Z P, LIU, Y F, CHEN, Y B, et al. Significantly improving energy density of cathode for lithium ion batteries: The effect of Li-Zr composite oxides coating on $LiNi_{0.6}Co_{0.2}Mn_{0.2}O_2$[J]. Ionics, 2020, 26: 1173-1180.

[6] CHO J, KIM Y J, PARK B. Novel $LiCoO_2$ cathode material with Al_2O_3 coating for a Li ion cell[J]. Chem Mater, 2000, 12: 3788-3791.

[7] CHO J, KIM T J, KIM Y J, et al, High-performance ZrO_2-coated $LiNiO_2$ cathode material[J]. Electrochem, Solid-State Lett, 2001, 4: A159-A161.

[8] CHO J, KIM Y J, PARK B. $LiCoO_2$ cathode material that does not show a phase transition from hexagonal to monoclinic phase[J]. J Electrochem Soc, 2001, 148: A1110-A1115.

[9] MORIMOTO H, AWANO H, TERASHIMA J, et al. Preparation of lithium ion conducting solid electrolyte of NASICON-type $Li_{1+x}Al_xTi_{2-x}(PO_4)_3$ ($x=0.3$) obtained by using the mechanochemical method and its application as surface modification materials of $LiCoO_2$ cathode for lithium cell[J]. J Power Sources, 2013, 240: 636-643.

[10] WANG Y, LI H, et al. An in situ formed surface coating layer enabling $LiCoO_2$ with stable 4.6 V high-voltage cycle performances[J]. Adv Energy Mater, 2020, 10: 2001413.

[11] KIM M Y, SONG Y W, HAN J H, et al. LATP-coated NCM for high-temperature operation of all-solid lithium battery[J]. Materials Chemistry and Physics, 2022, 290: 126644.

The Influence of Feed Rate on the Abradability of AlSi/PHB Seal Coatings

Xinwo ZHAO[1,2,3], Jiangang SUN[1,2,3], Deming ZHANG[1,2,3]

1. Beijing General Research Institute of Mining & Metallurgy Group Co., Ltd, Beijing, 100160, China
2. Beijing Engineering Technology Research Centre of Surface Strengthening and Repairing of Industry Parts, Beijing, 102206, China
3. Beijing Key Laboratory of Special Coating Material and Technology, Beijing, 102206, China

Abstract: Abradable seal coating is a material used in aero engines to improve its efficiency. In this paper, using the method of simulated working conditions, the self-developed BGRIMM-ATR high-temperature and ultra-high-speed abradability test machine is used to conduct abradability tests. In this paper, using the method of simulated working conditions, the self-developed BGRIMM-ATR high-temperature and ultra-high-speed abradability test machine is used to conduct abradability experiments on the AlSi/PHB seal coating at different feed rates to study the effect of feed rate changes on AlSi/the influence of PHB seal coating on abradability, the macroscopic and microscopic morphology of the coating were analysed by stereo microscope and scanning electron microscope (SEM), and a comprehensive evaluation of the abradability was carried out by a variety of characterization methods. The results show that at room temperature, with a linear velocity of 350 m/s and a feed depth of 500 μm, when the feed rate increases from low to high, the abradability of the AlSi/PHB coating shows a trend that first increases and then decreases. The change of the feed rate has a significant effect on the relative abradability of AlSi/PHB. The abradability of the coating is excellent at the medium feed rate, which is mainly manifested by the cutting mechanism; at the low feed rate and the high feed rate, the abradability of the coating is excellent. rate, the abradability of the coating is reduced, which is mainly manifested as deep furrows, cracks, and chipping of the coating. Chips, the main Chips, the main mechanisms are ploughing, melting, and collision tearing.

Keywords: Sealing coating; Abrasion behaviour; Abradability; Abradability testing machine

1 Introduction

With the rapid development of the global economy, energy saving and consumption reduction has become a serious task facing mankind in the 21st century. The field of aviation industry is no exception, in the design of aviation vehicles, safer, more efficient, more comfortable has become the development direction of the new generation of aircraft, which also brings new and higher requirements to the design of aviation engines, low fuel consumption, large thrust, long life has become the overall goal of the new generation of aviation engine design[1]. It has been found that the sealing of the engine's air path, i.e., the size of the radial clearance between the rotor and the receiver, has a great influence on the power, efficiency and fuel consumption rate of the compressor and turbine. According to the data report, the increase of 0.076 mm radial clearance of the compressor of a typical engine will increase the corresponding unit fuel consumption by about 1%; the increase of 0.00127 mm rotor and static sub-clearance of a high-pressure turbine will correspond to the increase of the unit fuel consumption by about 0.5%; and the efficiency can be improved by about 1% when the average reduction of the rotor and static sub-clearance is 0.245 mm[2].

Although the reduction of the gap can improve the performance of the aero-engine, excessive reduction of the gap between the rotor and the static sub is likely to cause the rotating part and the magazine of the abrasion damage, which is extremely unfavourable to the engine. Sealant coating technology has become a common method of controlling the rotor and stator gap. When the blades scrape against the coating, the sealant coating will "absorb" most of the abrasion energy, thus ensuring that the rotor blades do not wear out or stick, and maintaining a minimum airway gap to improve engine performance.

Aluminium silicon polyphenylene ester (AlSi/PHB) wearable sealing coatings are widely adopted by major engine manufacturers at home and abroad due to their easy production process, low cost, easy rework and adjustment performance, and good sealing effect[3-4], and are used in a large number of applications, such as fan, compressor and other parts manufacturing and repair, which can significantly increase engine power and reduce fuel consumption. In the aluminium-silicon polyphenylene ester material, AlSi nominal content of 60%, polyphenylene ester nominal content of 40%. AlSi as the metal skeleton phase, the presence of Al reduces the shear strength[5], Si can improve the alloy's fluidity, reduce the tendency to thermal cracking, reduce the loosening, improve the airtightness of the alloy, so that the alloy has a good corrosion resistance and medium machinability, with a medium strength and hardness, but the plasticity of the lower, can be good abrasive properties. Polyphenylene ester has good self-lubricating properties, making it very suitable for use as an abradable phase in abradable sealing coatings. AlSi/PHB is generally used at temperatures of less than 325 ℃, and is mainly used in low-pressure pressurised parts of the magazine sealing.

In this paper, the abrasion test of AlSi/PHB sealing coating material was carried out under different simulated working conditions using the self-developed BGRIMM-ATR high-temperature and ultra-high-speed abrasion tester, and the abrasionability was investigated under different feed rate conditions. By analysing the wear quality, wear height and wear morphology of the tested coatings and blades, the effects of different test parameters on the abrasiveness of the sealing coatings were investigated, and a preliminary evaluation of the abrasiveness of the AlSi/PHB sealing coating materials was carried out.

2 Experimental process and methodology

2.1 Experimental materials

The simulated blade used is GH4169, the main chemical composition and performance parameters are shown in Table 1 and Table 2. It is processed by EDM wire cutting, with a tip height of 1.5 mm, a thickness of 0.5 mm, a width of 25 mm, and is ultrasonically cleaned with acetone and then blown dry for spare parts.

Table 1 Chemical composition of GH4169(wt/%)

Ni	Cr	Fe	Mo	Nb	Co	Ti
55	21	tolerance (i.e. allowed error)	3.3	5.5	1.0	1.15

Table 2 Mechanical properties of GH4169

Density $\rho/(g \cdot cm^{-3})$	Brinell hardness (HBS)	Yield strength $\sigma_{p0.2}$/MPa	Modulus of elasticity E/GPa	Thermal conductivity $W/(m \cdot K)$	Tensile strength σ_b/MPa	Elongation δ/%
8.24	≥363	550	199.9	14.7	965	30

AlSi/PHB coatings were prepared using agglomerated aluminium-silicon polyphenylene ester (PHB) sealing coating material (grade KF-120) developed by Mining and Metallurgy Science and Technology Group Ltd. The material is nearly spherical powder particles, and the basic properties of the powder material are shown in Table 3.

Table 3 Typical performance of AlSi/PHB

Grades	Composition/%	Particle size/μm	Loose packing density/(g·cm^{-3})	Flowability /[s·(50 g)$^{-1}$]
KF-120	40PHB+60AlSi	45-297	0.60	190

The substrate is sandblasted and roughened and degreased to form a clean rough surface to improve the bonding strength between the substrate and the coating. The powder was dried before spraying. The AlSi/PHB coating was prepared on the surface of the substrate using a German GTV spraying system with an F6 plasma spray gun, in which NiAl95/5 composite powder was selected for the bonding layer, the thickness of the bonding layer was about 120 μm, and the thickness of the AlSi/PHB composite surface layer was 2.5 mm. The optimized spraying process parameters were: main gas (Ar) flow rate 60 L/min, auxiliary gas (H_2) flow rate 7 L/min, powder feeding rate of 26 g/min, arc current of 400 A, spray distance of 90 mm.

2.2 High-temperature ultra-high-speed abradable testing machine and experimental parameters

The evaluation of abradable performance of sealing coatings and the study of abrasion mechanism are of great significance in guiding the preparation and composition optimisation of sealing coating materials. High-temperature and ultra-high-speed abrasion test is the evaluation method closest to the actual working conditions of aero-engine adopted at home and abroad. The method is to evaluate the coating and study the abrasion mechanism by simulating the mutual scraping process between moving and static components. In order to simulate the friction and wear phenomenon under the service conditions of aero-turbine engine airway sealing mating, study the abrasion mechanism of the sealing coating, and evaluate the abrasion performance of the coating, Mining and Metallurgy Science and Technology Group has developed a full-size BGRIMM-ATR high-temperature and ultra-high-speed abrasion test rig[6]. The parameters of the equipment are as follows: maximum rotational speed of the disc 15500 r/min, maximum linear speed 450 m/s, maximum test temperature 1200 ℃, micro-feed rate range 2-2000 μm/s. The abradable testing machine is able to accurately record key data in the test, such as the rotational speed of the disc, the linear speed of the tip of the blade, the feed speed of the specimen, the depth of the feed, the heating temperature of the specimen, the scraping force, etc. The test machine is also able to accurately record the abrasion performance of the coating.

Abrasion tests were carried out on AlSi/PHB sealcoat specimens using a BGRIMM-ATR abrasion tester with the following dimensions: 100 mm×40 mm×8 mm. Since AlSi/PHB materials are generally operated at temperatures lower than 325 ℃, the test temperature was selected to be 25 ℃. The experimental parameters are shown in Table 4.

Table 4 Parameters of AlSi/PHB sealer coating abradable tests

Test No.	Linear velocity/(m·s^{-1})	Feed rate/(μm·s^{-1})	Feed depth/μm	Temperature/℃
1#	350	5	500	25
2#	350	50	500	25
3#	350	100	500	25

2.3 Sample characterisation

An electronic balance with a measurement accuracy of 0.001 g was selected to measure the mass of the AlSi/PHB tightly coated specimens and blades before and after the test, and the wear ratio of the two was calculated. Macroscopic morphology analysis of the coating and blade wear marks after the test was carried out using an optical microscope, and the microstructure morphology of the coating specimens was observed and analysed using Hitachi's Hitachi SU-5000 scanning electron microscope (SEM).

2.4 Blade/coating wear quality, IDR definition

Define the blade mass wear ratio as: blade wear mass÷blade mass before experiment×100%. Define the coating mass wear ratio as: coating-to-substrate wear mass÷coating-to-substrate mass before the experiment×100%. A positive blade mass wear ratio means that the blade is worn, and a negative ratio means that there is adhesion of the coating on the blade. The lower the blade mass wear ratio, the less damage to the blade when the coating and blade are grinding against each other. Coating mass wear ratio represents the loss of coating mass before and after the test. The smaller the change of the wear mass ratio under different working conditions, the more stable the abrasion performance of the coating is, and there is no extreme situation such as large-scale flaking and falling off.

Due to the irregular surface shape of the scratch, it is impossible to obtain accurate scratch depth data by direct measurement. According to the geometric relationship between rotor radius and scratch length, more accurate scratch depth data is calculated, and the geometric relationship between scratch depth and scratch length is shown in Figure 1. Before and after the abrasion test, vernier calipers were used to measure and record the length of the coating scratches, which meets the provisions of GB/T 1214. 3 with an accuracy of 0. 02 mm.

Formula for calculation of coating scraping depth:

$$D = R_b - \sqrt{R_b^2 - \frac{L^2}{4}} \qquad (1)$$

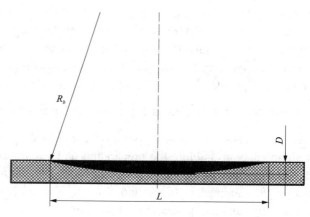

Figure 1 Geometric relationship between scratch depth and scratch length

Where D is the calculated value of coating scraping depth; R_b is the radius of the wheel disc + blade height; L is the coating scraping length.

The ratio of the blade height change to the total feed depth before and after the scraping test was defined as the feed depth ratio (IDR), which was calculated by the following equation:

$$IDR = \begin{cases} \dfrac{\Delta h}{D} (\Delta h < 0) \\ \dfrac{\Delta h}{D + \Delta h} (\Delta h > 0) \end{cases} \qquad (2)$$

Where Δh is the change value of blade height, Δh = height before scraping-height after scraping; when the blade height increases after scraping test ($\Delta h < 0$), i. e., there is the adhesion of coating material to the tip of the blade height increase, the total feed depth value = the calculated value of the depth of the scrape mark; when the blade height decreases after scraping test ($\Delta h > 0$), the total feed depth = the calculated value of the depth of the scrape mark + when the blade height decreases ($\Delta h > 0$), the total feed depth = the calculated value of scratch depth + the value of blade height change, i. e., the total feed depth is the sum of the depth of the tip wear and the depth of the coating being scraped.

The feed depth ratio (IDR) is a quantitative indicator of theabradability of a coating, and the smaller the absolute value of the IDR, the better the abradability. IDR is positive when blade wear is dominant, and negative when the coating material adheres to the blade. In general, the absolute value of IDR less than 10 per cent of abradability is excellent, 10%-20% of abradability is good, and 20%-30% of abradability is acceptable.

3 Results and analyses

3.1 Macroscopic morphology of coating and blade wear

The macroscopic morphology of the coatings and simulated blades after the abradable test was observed, as shown in Figure 2. As can be seen from Figure 2, all the coatings are abraded to different degrees, and there are obvious scraping arc grooves on their surfaces, the abrasion morphology presented by the coating specimens is basically similar under different feed rates, and there are obvious ploughing grooves and cutting traces on the scraping area of the coating surfaces, and the abrasion mechanism is dominated by ploughing and cutting, and the coatings have not been flaked off and fallen off under all the experimental conditions, and the scraping direction of the blade is from left to right, and there is no obvious difference within the whole scraping range, the wear morphology of the coating surfaces is basically the same. The direction of the blade scraping is from left to right, in the whole range of scraping; the wear morphology of the coating surface is basically the same, there is no obvious difference, and it can be assumed that although the feed rate has changed, the overall abrasion mechanism has not changed, for the ploughing wear. From the figure can also be seen, the coating surface are unevenly distributed pits, and with the increase of the feed rate, the coating scraping pits become more, which indicates that when the feed rate increases, because of the increase in the intensity of the collision the blade and the coating wear and tear, and with the increase of the feed rate, the number of times to scrape the same depth of scraping becomes less, each time to touch the grinding of the feed depth becomes larger, the blade and the coating of the collision wear effect is more intense! The number of craters produced by the collision is larger and more, and the number and size of the craters affects the sealing of the coating.

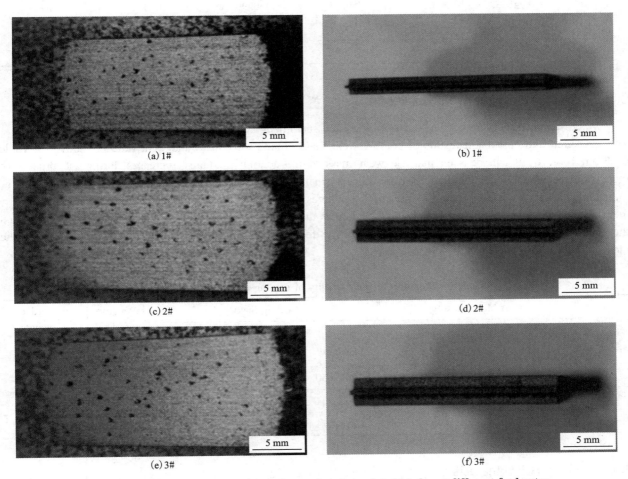

Figure 2 Macroscopic morphology of coated and simulated blades at different feed rates

From Figures 2(a)-(d), it can be seen that the blade tips of 1# and 2# are basically free of wear, and only some positions have scratches due to scraping and touching, while the blade tips of 3# [Figures 2(e) and 2(f)] have relatively obvious scraping furrow scratches and some smearing marks, but the wear is relatively flat and even. From the analysis of the macroscopic images of the coating material and the blades after being scraped, it can be seen that compared with the coating wear, the blade wear is relatively light, indicating that the AlSi/PHB coating material has good abradable properties and can withstand some extreme conditions. According to the scraping state of the coating, the larger the feed rate under the same conditions, the worse the sealing effect of the AlSi/PHB coating will be after scraping.

3.2 Coating scraping depth, blade wear height and total wear depth

Macroscopic tests were carried out on the coated specimens and simulated blades after the abradable test. Measurement and calculation results of coating scraping depth, blade wear height and total wear depth are shown in Table 5. As can be seen from Table 5, under the conditions of a line speed of 350 m/s, a feed rate of 5-100 μm/s, and a feed depth of 500 μm: the scraping depth of the AlSi/PHB coatings at different feed rates is greater than the feed depth, and the reduction in the height of the counter-abrasive blades is lower than the feed depth. Comparing the three sets of experimental data, it can be seen that the blade wear height and total wear depth have a tendency to decrease first and then increase during the process of increasing feed rate.

Table 5 Coating scraping depth, blade wear height and total wear depth

Test No.	Coating scraping length/mm	Calculated depth of coating scraping /mm	Blade height before experiment /mm	Blade height after experiment /mm	Blade wear height/mm	Total depth of abrasion/mm
1#	40.32	0.597	50.91	50.90	0.01	0.597
2#	41.92	0.645	50.98	50.97	0.01	0.655
3#	41.58	0.635	50.96	50.91	0.05	0.685

3.3 Coating and blade wear quality analysis

Macroscopic tests were carried out on the coated specimens and simulated blades after the abradable test. The measured and calculated results of coating wear mass, blade wear mass and wear ratio are shown in Table 6. From Table 6, it can be seen that the wear mass of the coating gradually increases during the process of increasing feed rate, while the blade has only a very small mass increase at feed rates of 5 μm/s and 100 μm/s, indicating that a very small portion of the coating adheres to the tip of the blade. When the feed rate is 50 μm/s, there is no mass change of the blade, which indicates that the AlSi/PHB coating has the best wearability and the smallest mass-to-wear ratio of the blade when the feed rate is 50 μm/s, and the coating can provide the most effective protection for the blade.

Table 6 Statistics of wear quality of coating and blade specimens

Test No.	Leaf blade mass/g			Coating mass/g			Proportion of blade mass worn /%	Proportion of coating quality worn /%
	Initial quantity	Final volume	Wear and tear	Initial quantity	Final volume	Wear and tear		
1#	28.942	28.943	-0.001	206.050	205.630	0.420	-0.003	0.204
2#	28.861	28.861	0	206.324	205.886	0.438	0	0.212
3#	28.863	28.866	-0.003	204.474	203.999	0.475	-0.010	0.232

3.4 Coating SEM analysis

The SEM morphology of the scraped area of the coating after the abradable test was observed, as shown in Figure 3. As can be seen in Figure 3, the coating specimens all have obvious scraping and furrowing morphology under 200× electron microscope, and the light-coloured part in the figure is the polyphenylene ester solid lubrication phase, and the dark-coloured part is the AlSi metal phase, and there are also some holes. And in 1# [Figure 3(a)], 3# [Figure 3(c)] specimens appeared some deeper and larger furrows, and these furrows also produced a lot of adhesion around the metal phase. As can be seen from specimen #1 [Figure 3(a)], when the coating is touch-ground at a feed rate of 5 μm/s, the depth of a single scrape is small and the furrows are more obvious. The metal phase around the solid lubrication phase will be torn off and some holes will be formed, and at the same time, some scale-like structures will be formed on the metal phase along the direction of the plough furrow, there are a lot of fine cracks and crushing around the plough furrow, and there are some traces of melt adhesion on both sides of the furrow. From 2# [Figure 3(b)] specimens can be seen, the coating in the 50 μm/s feed rate conditions touching the grinding, a single scraping depth increase in furrow uniform and solid lubrication phase around the holes compared to the 1# a lot lower, and there is no obvious scaly structure, the overall scraping morphology is relatively homogeneous, indicating that the conditions of the main wear mechanism for cutting wear. From 3# [Figure 3(c)] specimens can be seen, the coating in the 100 μm/s feed rate conditions touching the grinding, the single scraping depth is the largest, the furrow becomes obvious and deep, in the furrow around the rupture of the metal phase and touching the grinding debris can also be observed in the solid lubrication phase around the

tearing effect of the largest in the three groups, the holes are the deepest and the most obvious, and at the same time, it can be seen along the direction of the furrow of the coating that there are a number of larger pieces of scaly structure.

Figure 3　SEM morphology of coatings under different feed rates

3.5　*IDR* analysis

The *IDR* values were calculated for the coated specimens after the abradability test and the results are shown in Table 7. From Table 7, it can be seen that the abradability of 1#, 2# and 3# coatings are all excellent, but in comparison, the *IDR* value of 2# coating is 0, and the abradability is the best among these three groups. Meanwhile, when the feed rate is increased from 5 μm/s to 50 μm/s, the change of *IDR* value is small, but when the feed rate is increased from 50 μm/s to 100 μm/s, the change of *IDR* value is greatly increased compared with the previous section, which indicates that too high feed rate will cause the blade and the coating to touch the abrasion process more intensely.

Table 7　Calculation results of coating *IDR* values

Test No.	Calculated depth of coating scraping D/mm	Blade wear height Δh/mm	Total abrasion depth/mm	IDR
1#	0.597	0.01	0.607	1.65
2#	0.645	0.01	0.655	1.53
3#	0.635	0.05	0.685	7.30

In summary, the results of the experiment can be seen, the experimental touch-grinding mode for intermittent touch-grinding mode, the blade and the coating of the touch-grinding process time is very short, in each high-speed touch-grinding after the surface of the coating will form the friction heat, in the same conditions of the depth of feed, the low feed rate of the blade and the coating of the touch-grinding number of times, long time, a single scraping amount is small, so the blade each time nearly with the surface of the previous touch-grinding, which will result in the gathering of heat on the surface of the coating. This will cause the heat gathered in the coating of the scraping surface is not easy to diffuse, when the heat reaches a certain level, it will form a molten form in the region. At the same time, the radial impact of each touching of the blade will locally compact and plastically damage the area, causing the furrows to deform and break, so that at a feed rate of 5 μm/s, there are many small cracks and layers of broken material around the furrows. The ploughing and cutting morphology of the coating surface indicates that the main deformation mechanism is ploughing and cutting deformation, and the more uniform the scraping grooves are, the more stable the touching and grinding process is. At the high feed rate of 100 μm/s, because of the low number of touch-grinding, the single feed is very large, and the ploughing and impact tearing effect of the blades on the coating is also very large, which results in the deep furrows and holes around the lubrication phase increasing, the cracks in the metal phase increasing, the abrasive debris increasing, and the abrasive wearability becoming poor.

4 Conclusions

In this paper, the BGRIMM-ATR high-temperature ultra-high-speed abrasive testing machine was used to conduct experiments on AlSi/PHB abrasive coating materials with different feed rates at a linear speed of 350 m/s, with feed rates of 5 μm/s, 50 μm/s, 100 μm/s. The results of the coating and blade wear quality, surface macro and micro wear morphology, wear height, etc. were obtained at different feed rates. Height and other results were obtained, and conclusions were drawn from the study:

(1) With the increase of feed rate, the *IDR* value shows a tendency of decreasing and then increasing, and when the feed rate is 50 μm/s, the abradability of the coating is relatively optimal.

(2) At low feed rates, the wear mechanisms of AlSi/PHB coatings are cutting, ploughing and melting; at medium feed rates the wear mechanisms are mainly cutting; at high feed rates the wear mechanisms are mainly ploughing and tearing.

(3) The change in feed rate has a significant effect on the relative wearability of AlSi/PHB coatings under ambient service conditions. However, the AlSi/PHB coatings still have excellent abradability under ambient service conditions, micro, conventional and extreme feed conditions, making them an excellent abradable sealing coating.

References

[1] YIN C L, CHEN M Y, ZHAN J, et al. Research progress of wearable sealing coatings[J]. Aerospace Manufacturing Technology, 2008, 20: 92-94.

[2] WU X. A review of laser fabrication of metallic engineering components and of materials[J]. Materials Science and Technology, 2007, 23(6): 631-640.

[3] NOVINSKI E R. The design thermal sprayed abradable seal coatings for gas turbine engines[C]//Proceedings of 4th National Thermal Spray Conference USA, 1991: 451-454.

[4] DEMASI J T. Protective coating in the gas turbine engine[J]. Surface and Coatings Technology, 1994, 68: 1-9.

[5] LIU T, YU Y, SHEN J, et al. Influence of feed rate on abradability of AlSi/hBN sealing coatings[J]. Thermal Spraying Technology, 2014, 6(3): 43-49.

[6] SUN J G, MA C C, WANG H. Research progress of high-temperature and ultra-high-speed test system and sealing coating evaluation system[J]. Thermal Spraying Technology, 2019(2): 5-11.

Single Crystalline LiNi$_{0.65}$Co$_{0.15}$Mn$_{0.20}$O$_2$ for High-energy Li-ion Batteries with Outstanding Cycling Stability

Jun Wang, Xuequan ZHANG, Shunlin SONG, Yafei LIU, Yanbin CHEN

BGRMM Technology Group, Beijing Easpring Materials Technology Co., Ltd., Beijing, 100029, China

Abstract: Nickel-rich cathode materials have been considered as the most promising candidate because of their high energy density, but the polycrystalline nickel-rich cathode materials suffer from severe problems such as poor cycle stability and thermal stability. Herein, a single crystalline LiNi$_{0.65}$Co$_{0.15}$Mn$_{0.20}$O$_2$ cathode material were synthesized by a coprecipitation method, and its electrochemical performance were compared with polycrystalline cathodes. SEM and XRD results show that single crystalline materials have increased particle strength and structural stability during cycling. Full cells test results show that single crystalline cathode materials have an outstanding cycling stability of 93.5% and 91.3% capacity retention after 1000 cycles at 25 ℃ and 45 ℃, respectively.

Keywords: Nickel-rich cathode materials; Single crystalline; Cathode; Cycling stability

1 Introduction

To meet the demand of electric vehicles and electrical energy storage systems, lithium-ion batteries with high energy density, long-term cycling stability, high rate capability, and thermal stability have been required.[1] Among the many cathode materials, nickel-rich cathode materials have been considered as the most promising candidate because of their high capacity, low cost, and environmental benignity.[2] However, the polycrystalline nickel-rich cathode materials are facing two serious problems: (1) The micro-crack and inter-grain separation generated at the cathode active materials during electrochemical cycling process allows the electrolyte permeate into the particle interior through the crack network and impede the lithium ion diffusion. This result in a loss of connectivity within the particles and give rise to increased polarization, which lead to the performance degradation and capacity fade. (2) The polycrystalline nickel-rich cathode materials show serious gas evolution and poor thermal stability.[3]

To solve or ameliorate the above-mentioned disadvantages of polycrystalline nickel-rich cathode materials, one of the best strategies is to build a single crystalline structure with a sub-micron single particle.[4,5] This structure has the following advantages: (1) The single crystalline morphology can be stably maintained even after electrochemical cycling process, which could inhibit the generation of microcrack and restrain the continuous formation of the resistance layer. It is helpful to improve the cycle stability and extend the cycle life of battery. (2) The single crystalline cathode materials have less gas evolution during high temperature storage and higher thermal stability compared with polycrystalline cathode materials. Therefore, the single crystalline characteristics could resolve the long-term cycling stability and poor thermal stability issues, providing a possibility for the wide application of nickel-rich cathode materials.

Herein, we report a single crystalline LiNi$_{0.65}$Co$_{0.15}$Mn$_{0.20}$O$_2$ cathode material (named as 65SC), which exhibits higher cycling stability and rate capability in half-cells compared with polycrystalline LiNi$_{0.65}$Co$_{0.15}$Mn$_{0.20}$O$_2$ cathode materials. The 65SC cathode achieved a capacity retention of 89.7% with a capacity of 187 mA·h·g^{-1} at 25 ℃ and a capacity retention of 94.6% at 45 ℃. Moreover, in full-cell configuration, the 65SC also shows ultrahigh cycling stability.

2 Results and discussion

To obtain the 65SC cathode materials, the Ni$_{0.65}$Co$_{0.15}$Mn$_{0.20}$(OH)$_2$ precursor was first prepared by a coprecipitation method. ICP results indicate the stoichiometric ratio of Ni/Co/Mn in as-prepared precursor is

65.15/15.07/19.78, which is consistent with the target values. Then the 65SC were prepared by calcining the mixture of precursor and LiOH·H_2O at 900 ℃. For comparison, the polycrystalline materials (named as 65PC) were also prepared by the similar method, but the optimized sintering temperature is 850 ℃. It is clearly seen that the 65SC sample shows good single crystal morphology [Figures 1(a), (b)], while 65PC exhibits obvious polycrystalline spherical morphology composed of many primary particles [Figures 1(c), (d)]. The XRD patterns show all of the diffraction peaks can be indexed to a well-defined a-$NaFeO_2$-type layered structure with R-3m space group, and no impurity phase was detected (Figure 2). The obvious splitting of (018)/(110) peaks for both samples implies that a well-defined layered structure without disordered atomic arrangements was formed. Moreover, the intensity ratio of the (003)/(104) peaks for 65SC and 65PC is 1.22 and 1.24, respectively, which suggests that no obvious cation mixing occurred for both samples.

Figure 1 SEM images of (a), (b) 65SC and (c), (d) 65PC

Figure 2 The XRD patterns of 65SC and 65PC

The SEM images of 65SC and 65PC electrodes with different electrode density (3.0, 3.5, and 4.0 g/cm^3) show that the morphology of 65SC cathode at 4.0 g/cm^3 was still retained (Figure 3). However, the spherical morphology of 65PC cathode at 4.0 g/cm^3 have been destroyed due to the high pressure. Therefore, the electrode density of 3.5 g/cm^3 and 3.0 g/cm^3 for 65SC and 65PC were used to evaluate their electrochemical properties in this paper.

The first charge-discharge curves in Figure 4(a) show the 65SC with capacity of 187 mA·h/g, which is higher than that of 65PC of 184 mA·h/g at 0.1 C and at 25 ℃ in a voltage ranged from 3.0 to 4.3 V, indicating that the single crystalline cathodes can deliver higher reversible capacity than polycrystalline cathodes. To investigate the rate capability, the cells were subjected to galvanostatic cycling at various C rates ranged from 0.1C to 2C [Figure 4(b)]. At 2C, the 65SC exhibits higher discharge capacity retention of 89.8% compared with

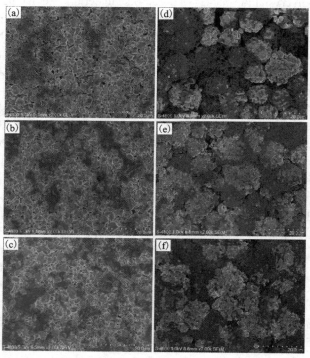

Figure 3 The SEM images of (a)-(c) 65SC and (d)-(e) 65PC electrodes with different electrode density of (a), (d) 3.0, (b), (e) 3.5 and (c), (f) 4.0 g·cm^{-3}

65PC of 88.4%. Then, the both cathodes were cycled at 1C and the operating voltage is 3.0-4.5 V and 3.0-4.4 V for 25 ℃ and 45 ℃, respectively. After 80 cycles at 25 ℃, 89.7% of initial discharge capacity was retained for 65SC cathodes, whereas 87.4% was retained for 65PC cathode [Figure 4(c)]. A similar trend was also observed at 45 ℃ [Figure 4(d)], the capacity retention still following the order 65SC (94.6%) > 65PC (91.2%), indicating the single crystalline structure can significantly enhance the cycling stability.

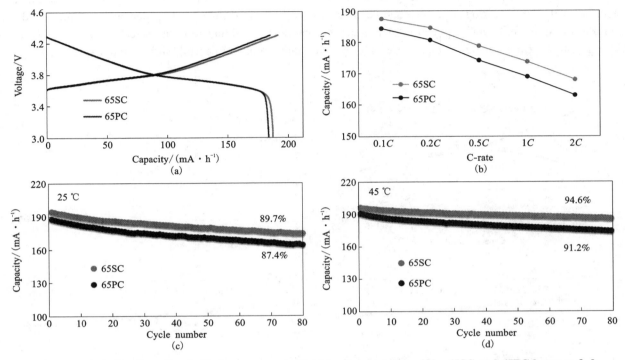

Figure 4　(a) The first charge-discharge curves and (b) rate capability of the 65SC and 65PC between 3.0 and 4.3 V and at 0.1C; The cycling performances of the 65SC and 65PC at (c) 25 ℃ and (d) 45 ℃

To get an insight into the high specific capacity and superior cycling stability of 65SC, the structural changes of cycled 65SC and 65PC electrodes have been investigated by SEM and XRD (Figure 5). The SEM images of 65PC electrode after 80 cycles have some broken and some primary particles have fallen off, but the 65SC electrode is still well retained, indicating good structural stability. In addition, the XRD patterns indicating that the layered structure is still retained for 65SC and 65PC electrodes after 80 cycles, however, the intensity ratio of the (003)/(104) peaks for 65SC and 65PC electrodes after 80 cycles is 3.67 and 0.86, respectively, which suggests that the single crystalline structure could significantly inhibit cation mixing occurred during cycling and improve structural stability.

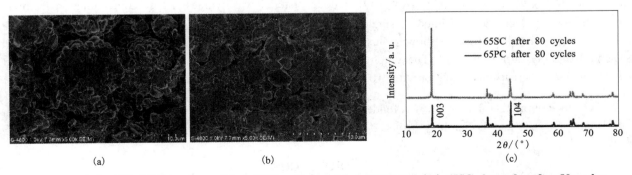

Figure 5　(a), (b) SEM images and (c) XRD patterns of (a) 65PC and (b) 65SC electrodes after 80 cycles

Finally, we combined the 65SC cathodes with graphite-based anodes to evaluate their performance in lithium-ion full cells at 25 ℃ and 45 ℃ between 2.8 V and 4.2 V. At 25 ℃, based on the cathode active material, the full-cell delivered an initial charge and discharge capacity of 200.8 and 176.2 mA·h/g with a

Coulombic efficiency of 87.7% at 0.2C [Figure 6(a)]. Increasing the dis-/charge rate subsequently to 0.33C, 0.5C, and 1C, results in discharge specific capacities of 175.8, 172.1, and 167.4 mA·h/g, respectively [Figure 6(b)]. After the formation cycles, the full cells have been subjected to a long-term cycling at different temperatures of 25 ℃ and 45 ℃ at 1 C, the graphite/65SC full cells at 25 ℃ delivered a capacity of 167.1 mA·h/g with a capacity retention of 93.5% after 1000 cycles [Figure 6(c)]. The graphite/65SC full cells at 45 ℃ exhibited an outstanding cycling stability for 1000 cycles with a capacity retention of 91.3%, leading to an average capacity loss of only 0.0156 mA·h/g per cycle [Figure 6(d)]. Considering the commercial application of this active materials as lithium-ion battery cathodes, this advanced cycling stability even at high temperature is very important, thus further highlighting this single crystalline material.

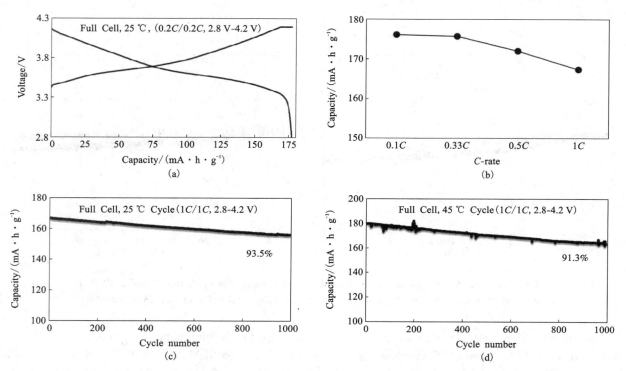

Figure 6 Performance evaluation of graphite/65SC lithium-ion full cells between 2.8 V and 4.2 V: (a) the full cell charge-discharge curves; (b) rate capability test; long-term galvanostatic cycling for 1000 cycles at (c) 25 ℃ and (d) 45 ℃

3　Conclusion

A single crystalline $LiNi_{0.65}Co_{0.15}Mn_{0.20}O_2$ cathode material was synthesized by coprecipitation method. Compared with polycrystalline cathodes, single crystalline cathodes deliver higher reversible capacity and cycle stability. SEM and XRD results show that single crystalline materials have increased particle strength and structural stability during cycling. The graphite/65SC full cells exhibited an outstanding cycling stability of 93.5% and 91.3% capacity retention after 1000 cycles at 25 ℃ and 45 ℃, respectively.

Acknowledgments: This work was funded by the Beijing Municipal Science and Technology Commission and Beijing Science and Technology Project (Grant No. Z191100004719001)

References

[1] KIM Y, SEONG W M, MANTHIRAM A. Cobalt-free, high-nickel layered oxide cathodes for lithium-ion batteries: Progress, challenges, and perspectives[J]. Energy Storage Materials, 2021, 34: 250-259.

[2] TIAN F, BEN L, YU H, et al. Understanding high-temperature cycling-induced crack evolution and associated atomic-scale structure in a Ni-rich $LiNi_{0.8}Co_{0.1}Mn_{0.1}O_2$ layered cathode material[J]. Nano Energy, 2022, 98:

107222.
[3] SHI C G, PENG X, DAI P, et al. Investigation and suppression of oxygen release by $LiNi_{0.8}Co_{0.1}Mn_{0.1}O_2$ cathode under overcharge conditions[J]. Advanced Energy Materials, 2022: 2200569.
[4] FAN X, HU G, ZHANG B, et al. Crack-free single-crystalline Ni-rich layered NCM cathode enable superior cycling performance of lithium-ion batteries[J]. Nano Energy, 2020, 70: 104450.
[5] TREVISANELLO E, RUESS R, CONFORTO G, et al. Polycrystalline and single crystalline NCM cathode materials: Quantifying particle cracking, active surface area, and lithium diffusion[J]. Advanced Energy Materials, 2021: 2003400.

A Bulk Oxygen Vacancy Dominating WO₃ Photocatalyst for Carbamazepine Degradation

Weiqing GUO[1], Qianhui WEI[1], Gangrong LI[1], Feng WEI[1,*], Zhuofeng HU[2,*]

1 GRINM (Guangdong) Institute for Advanced Materials and Technology, Foshan, 528000, China
2 School of Environmental Science and Engineering, Guangdong Provincial Key Laboratory of Environmental Pollution Control and Remediation Technology, Sun Yat-sen University, Guangzhou, 510006, China

Abstract: Creating oxygen vacancy in tungsten trioxide (WO_3) has been considered as an effective strategy to improve the photocatalytic performance for degrading organic pollutants. In this study, a series of WO_3 samples with both surface and bulk oxygen vacancies were prepared and their catalytic degradation performance for carbamazepine (CBZ) was evaluated. They exhibit higher photocurrent densities, charges transfer efficiency and photocatalytic degradation efficiency than oxygen vacancy free samples. Moreover, it was observed that the surface oxygen vacancy could be recovered upon being exposed to air for a period of time, resulting in a bulk oxygen vacancy dominating WO_3 (bulk-OV-WO_3) with even much higher degradation efficiency. The mechanism study shows bulk-OV-WO_3 mainly degrade the CBZ by producing OH radicals and superoxide radicals, while oxygen vacancy free sample that mainly oxidize the CBZ by the photoexcited hole which requires the CBZ to be adsorbed on the surface for degradation. The radical generated by bulk-OV-WO_3 exhibits stronger oxidizing capacity by migrating to the solution for CBZ degradation. The results of this study could potentially broaden our understanding of the role of oxygen vacancies and shed some lights for us to develop more advanced photocatalyst in the future.

Keywords: Tungsten trioxide; Oxygen vacancy; Photocatalytic degradation

1 Introduction

Along with the industrialization and urbanization, organic pollutants have caused severely environmental problems and resulted in increasingly severe shortage of water resources[1]. Therefore, there is urgent demand of developing new technique for resolving the problem of organic pollutants. In the past decades, research attention has been focused on developing and cost-effective photocatalytic technique as it is a green and efficient approach for degradation of organic pollutants[2-5]. In a photocatalytic degradation system, photocatalyst plays a vital role in the degradation efficiency of pollutants. Among all investigated photocatalysts, tungsten trioxide (WO_3) exhibits competitive advantage for the degradation of organic pollutants. Owing to its relatively narrow band gap (2.4-2.8 eV) and positive valence band (VB), WO_3 can utilize visible light energy and have stronger oxidation ability[6,7]. They are both beneficial to oxidize organic pollutants. However, WO_3, as photocatalysts, also has the shortcomings of low photogenerated carries density, poor charges transfer efficiency and electron-hole separation[7].

Previous studies have suggested that introducing defect structure into WO_3 could be an effective way to overcome above shortcomings, and oxygen vacancy is a simple but effective defect structure for improving the photocatalytic performance of WO_3[8]. According to previous literature, the promotion from oxygen vacancy can be reflected in several aspects. Some researches demonstrated that oxygen vacancy could improve the utilization of light energy due to extending light absorption of WO_3 in the visible regions[9]. Some studies demonstrated that oxygen vacancy could effectively promote the separation of photogenerated electron-hole so that it could maximize utilization rate of photogenerated carriers[10]. Other reports verified that the existence of oxygen vacancy could increase the charges transfer efficiency[11]. However, some reports showed that oxygen vacancies could cause negative influence on the photocatalytic performance as they may become recombination centers of electron and hole and weaken crystal structure[12].

Oxygen vacancies can be introduced to WO_3 by thermal treatment in inert or reducing atmosphere[13-16]. It is investigated that the thermal treatment with different atmosphere, temperature or duration can result in different impacts on the distribution of oxygen vacancies and thereby the photocatalytic performance. It is considered that surface and bulk oxygen vacancies generated by thermal treatment play different roles on photocatalytic performance of samples[7]. Bulk oxygen vacancies could format intermediate electronic bands[17, 18] which can inhibit the recombination of photogenerated electron-hole pairs[19, 20]. Meanwhile, surface oxygen vacancies could result in decrease of VB due to the local band caused by surface defect[21, 22].

In this study, we prepare WO_3 with oxygen vacancies ($OV-WO_3$) samples by thermal treatment with different duration under Ar atmosphere. Then the influence on their physical properties, photocatalytic performance, photocatalytic degradation mechanism, and the distribution of oxygen vacancies are investigated. Herein, we discover that the surface oxygen vacancies will be suppressed and recovered when the sample is placed in air. Importantly, when the surface oxygen vacancies are recovered, the sample with mainly bulk oxygen vacancies exhibit much higher than sample with both surface and bulk oxygen vacancies. This bulk oxygen vacancy dominating sample exhibits very high activity in the photocatalytic degradation of Carbamazepine (CBZ). 5×10^{-6} of CBZ can be totally degraded within 120 min. Compared with oxygen vacancy free sample, this bulk oxygen vacancy dominating WO_3 exhibits more effective degradation mechanism. The oxygen vacancy free sample mainly oxidize the CBZ by the photoexcited holes and requires the CBZ to be adsorbed on the surface. However, the bulk oxygen vacancy dominating WO_3 mainly degrade the CBZ by producing OH radicals and superoxide radicals, which can migrate to the solution and have more effective degradation efficiency.

2 Experiment

2.1 Preparation of pristine WO_3 and WO_3 with oxygen vacancies

Pristine WO_3 sample is prepared by a hydrothermal method as reported before.

$OV-WO_3$ samples: The pristine WO_3 nanoparticles was thermally treated in pure Ar flow at 550 ℃ for 0.5 h, 3 h, 6 h or 10 h to prepare four $OV-WO_3$ samples which were denoted as WO_3-0.5 h, WO_3-3 h, WO_3-6 h, or WO_3-10 h samples, respectively.

2.2 Characterizations

The scanning electron microscopy (SEM) images were taken on a Quanta 400 Thermal FE environmental scanning electron microscope (FEI, Netherlands). The diffuse reflectance absorption spectra of the samples were obtained by a UV-visible spectrophotometer equipped with an integrated sphere attachment (UV-3600, SHIMADZU, Japan). The X-ray diffraction (XRD) data were recorded on a Rigaku diffraction instrument (CuK_α radiation source with $\lambda = 0.15418$ nm). Photoelectrochemistry (PEC) measurements were performed on an electrochemical workstation (CHI760E, Chenhua, China) equipped with a three-electrode cell. The working electrode was made as follows: 2 mg of samples was fixed on FTO conducting glass mixed with 0.05 mL 0.05 wt% Nafion, using graphite rod as counter electrodes and Ag/AgCl as the reference electrodes, and 0.1 M Na_2SO_4 was used as electrolyte. The EPR spectra were detected by a CW/Pulse EPR system (A300, Bruker Co., Germany) with a microwave frequency of 9.64 GHz, a microwave power of 0.94 mW, a modulation frequency of 100 kHz, and a modulation amplitude of 2.0 G. X-Ray photoelectron spectroscopy (XPS) was performed using a Sengyang SKL-12 spectrometer equipped with a VG CLAM 4 MCD electron energy analyzer and twin anode MgK_α radiation (1253.6 eV) or AlK_α radiation (1496.3 eV) X-ray sources.

2.3 Photocatalytic performance experiments

All photocatalytic degradation experiments were carried out in a quartz tube, with 0.25 g/L photocatalyst and 5 ppm carbamazepine (CBZ) under ultraviolet-visible irradiation from a 350 W xenon lamp. At an interval of 20 min, 0.5 mL of the solution was extracted. The residual concentrations of CBZ in reaction solution were detected every 20 min by HPLC (Shimadzu LC-20AD) equipped with Poroshell 120 EC-C18 column (4.6 mm×100 mm, 2.7 μm, Agilent Technology, USA). The mobile phase for HPLC detection was a mixture of methanol and water with a ratio of 6 : 4 and at a 0.6 mL/min flow. The photocatalytic activity was evaluated using a time profile of C_t/C_0, where C_t is the concentration of the pollutants at the irradiation time t, and C_0 is the concentration at the equilibrium before irradiation, respectively.

2.4 Detection and measurements of photogenerated radicals

The formed active oxygen radicals including $\cdot OH$, $\cdot O_2^-$, and $^1O_2$ in photocatalytic degradation systems were all identified by a Bruker A300 electron paramagnetic resonance (EPR) spectrometer using 5, 5-dimethyl-1-pyrroline-N-oxide (DMPO), DMPO with methanol, and TEMP as the rapping agent. Typically, 100 mM of DMPO was mixed with 20 mL solution (including 0.25 g/L catalyst, 5×10^{-6} CBZ). All samples had been illuminated for 5 min before ERP measurements.

3 Conference programme and proceedings

3.1 Structures and morphologies properties

The SEM images shows that the WO_3 powder particle is a cluster of two-dimensional nanosheets with the size of $ca.$ 1 μm [Figure 1(a) & (b)]. After thermally treated for 0.5 h to 10 h, there is no remarkably visual change in particle surface morphology and size of the samples [Figure 1(c)-(f)], suggesting the morphology of the WO_3 is stable.

Figure 2 shows the optical properties of all samples. As compared with untreated WO_3, the color of thermally treated WO_3 samples changed from yellow to olive-green (0.5 h and 3 h), and then to indigo after 6 h and 10 h thermal treatment [Figure 2(a)]. According to literature, the change of color could be attributed due to the formation of bulk oxygen vacancies[7]. Correspondingly, their light absorption in the visible regions is extented with the increase of oxygen vacancy [Figure 2(b)]. It can be found that more light can be absorbed when the WO_3 samples is thermally treated for longer time. Figure 2(c) shows the band gap of the

Figure 1 SEM images of (a) & (b) WO_3; and OV-WO_3 samples: (c) WO_3-0.5 h; (d) WO_3-3 h; (e) WO_3-6 h; (f) WO_3-10 h

samples calculated by the linear approximation of $(\alpha h\nu)^2$ vs. photo energy[23]. Visibly, all OV-WO_3 samples possess narrower band gap than pristine WO_3 due to oxygen vacancy[7, 18, 24]. The calculated value of pristine WO_3 is around 2.77 eV, whereas the optical absorption of the OV-WO_3 samples gradually shifts to the longer wavelengths with the band gap value (E_g) ranging from 2.53 to 2.71 eV. Among all OV-WO_3 samples, WO_3-10 h shows the most intense light absorption and the narrowest band gap.

The XRD pattern [Figure 3(a)] of all samples could be indexed to the monoclinic phase of WO_3 (JCPDF20-1324). There is no observable shift for all characteristic diffraction peaks of the WO_3 after thermal treatment. It suggests that the main crystal structure is preserved even though oxygen vacancy is increased[25]. Nevertheless, the intensity of the diffraction peaks slightly weakens with the increase of their thermal-treated time, indicating the degradation of the WO_3 crystallinity[26]. The crystalline degradation was mainly reflected

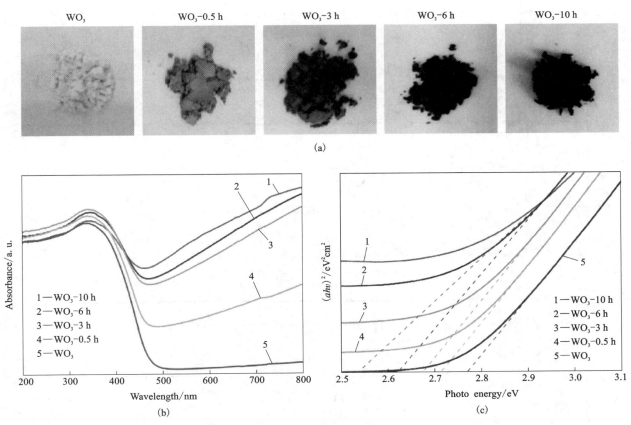

Figure 2 The photographs (a), UV-vis diffuse reflectance spectra (b) and Band gaps calculated from Kubellka-Munk plots of the samples (c)

in the orientations on (020), (022) and (202) planes [Figure 3(b) and (c)].

Based on the results of SEM, UV-vis diffuse reflectance spectra and XRD, it can conclude that the introduction of oxygen vacancy has little influence on the surface morphology and size of WO_3, except narrowing its band gap and weaken its crystal structures.

3.2 Photoelectrochemistry (PEC) properties

PEC measurements were conducted in a three-electrode electrochemical cell in 0.1 M Na_2SO_4 solution (pH: $ca.$ 7.0). The photocurrent density is measured at 0.6 V and the result is shown in Figure 4(a). It illustrates that OV-WO_3 samples can produce higher photocurrent density than pristine WO_3, as reflected in the i-t patterns with on-off recycles of intermittent irradiation. This suggests that the introduction of oxygen vacancy, to some extent, can improve the photo-irradiated charges transfer efficiency of WO_3. Among all OV-WO_3 samples, WO_3-3 h exhibits highest photocurrent density which is about triple of pristine WO_3 at 0.6 V vs. RHE. Nevertheless, the photocurrent density of OV-WO_3 samples decreases when the oxygen vacancies exceed the optimal proportion, indicating that the presence of more oxygen vacancies in WO_3 nanosheets does not necessarily further increase.

Similar trend has been observed in EIS, which is the common analytical tools to estimate the charge separation and migration. As displayed in Figure 4(b), four OV-WO_3 samples exhibit smaller arc radius than pristine WO_3 verifying that the existence of oxygen vacancy can reduce interfacial charge transport resistance, thus leading to more effective charge separation. The charge transport resistance reaches the minimum value when WO_3 had been thermally treated for 3 h. Trend from rising then declining generally means the charges transfer efficiency of OV-WO_3 samples suffers from two opposite (positive and negative) influence factors. Based upon the result of XRD (Figure 3), the negative influence may arise from the crystalline destroy when the samples have excessive oxygen vacancies. As a result, optimal proportion of oxygen vacancies appeared when WO_3 was thermal treated for 3 h, and thus WO_3-3 h presents the best electrical performance.

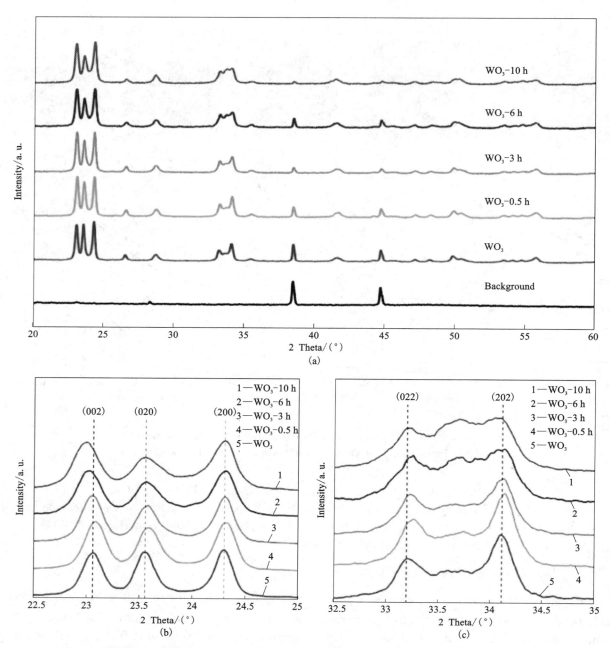

Figure 3 XRD patterns of WO_3, WO_3-0.5 h, WO_3-3 h, WO_3-6 h and WO_3-10 h (a); their XRD patterns on (002), (020) and (200) planes (b), and on (020) and (200) planes (c)

The Mott-Schottky plots [Figure 4(c)] of pristine and OV-WO_3 samples are plotted according to ref [27-29]. The flat-band potentials are calculated to be 0.03, 0.09, 0.13, 0.07 and 0.06 V vs NHE for WO_3, WO_3-0.5 h, WO_3-3 h, WO_3-6 h and WO_3-10 h, respectively. As their E_g were 2.77, 2.71, 2.67, 2.62 and 2.53 eV, their VB can be calculated according to the equation of $E_g = E_{VB} - E_{CB}$. The detail of band structures has been illustrated in Figure 4(d).

3.3 Photocatalytic degradation of CBZ

The photocatalytic degradation of 5 ppm CBZ by as-prepared samples is investigated, and results are shown in Figure 5 and Figure 6, and photocatalytic degradation efficiency (%) can be calculated by the following equation[30].

$$\text{Degradation efficiency} = (1 - C_0/C_t) \times 100\% \tag{1}$$

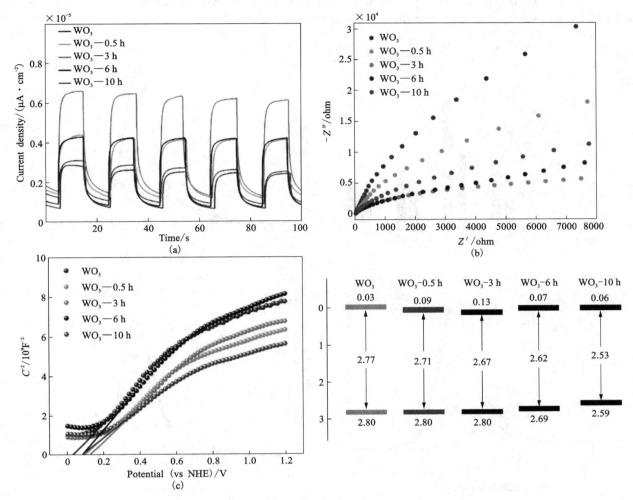

Figure 4

As showed in Figure 5, only 36% CBZ has been removed by pristine WO_3 after 100 min while those of the OV-WO_3 samples ranged from 49% to 94%. Obviously, the degradation efficiencies on OV-WO_3 samples are remarkably higher than that of pristine WO_3. Here, the kinetic constant (k) for CBZ is further estimated by the following equation[30].

$$k = \frac{\ln(C_0/C_t)}{t} \quad (2)$$

The plots of $\ln(C_0/C_t)$ vs. t are found to be a linear relationship, clarifying that the photocatalytic degradation reaction of CBZ can be described by using pseudo-first-order model (Figure S1). From the kinetic curves, the kinetic constant k values on WO_3 is calculated to

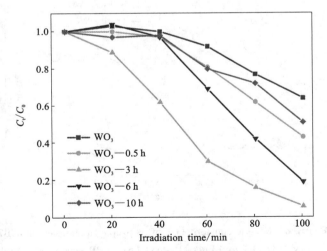

Figure 5 Photocatalytic degradation curve of as-prepared samples for CBZ under UV-vis

0.0060 min^{-1}, while those of WO_3-0.5 h, WO_3-3 h, WO_3-6 h and WO_3-10 h are increased to 0.0106, 0.0340, 0.0212 and 0.0081 min^{-1}, respectively [Figure 6(a)]. Hence, the k value on WO_3-3 h was the largest and 5.67 times than that of on WO_3, suggesting that the introduction of oxygen vacancy can significantly improve the photocatalytic degradation performance of WO_3. However, the photocatalytic performance of OV-WO_3 samples decreased when the oxygen vacancies exceed the optimal proportion (WO_3-

3 h). The improved photocatalytic performance is considered to be originated from the promotion of the dissociation of excitons and accelerate the transfer of charges.

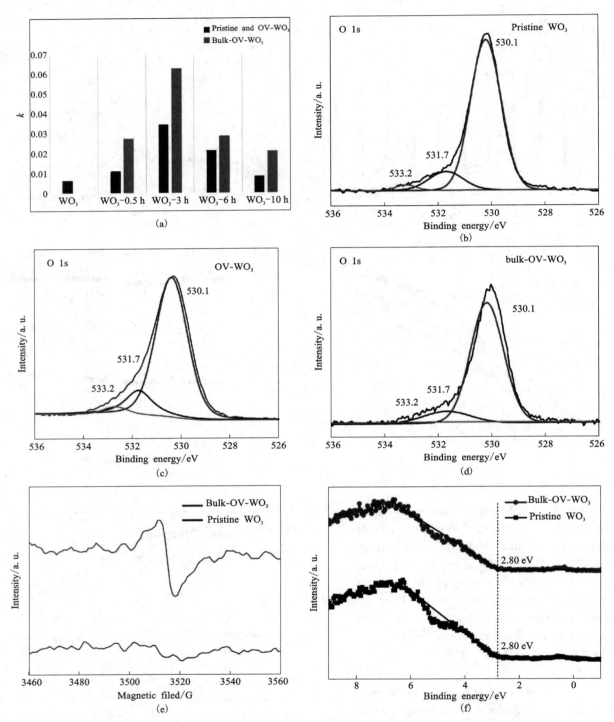

Figure 6 Kinetic constant (k) of as-prepared samples for CBZ under UV-vis (a); O 1s XPS spectra of WO_3(b), OV-WO_3(b) and bulk-OV-WO_3(c); EPR spectra (e) and VB XPS spectra (f) of WO_3 and bulk-OV-WO_3

3.4 Optimization of oxygen vacancy

The activity of OV-WO_3 can also be enhanced by optimizing the oxygen vacancy distribution. As mentioned above, the thermal treatment causes oxygen vacancies in both bulk and on the surface, and leads to higher activity. Interestingly, we discovered that the sample would exhibit even higher activity when it was placed in the air for a period of time. The degradation efficiency of bulk-OV-WO$_3$ is higher and their kinetic constant increase to 0.0271, 0.0628, 0.0284, 0.0208 min^{-1}. They are 1.34-2.57 times than that of OV-WO$_3$ just after thermal treatment. This indicates oxygen-exposure of the thermally treated sample will exhibit higher activity as compared.

This interesting phenomenon can presumably be attributed to the recovery of the surface oxygen vacancies. The surface oxygen vacancies are investigated by XPS. Figure 6(b), (c) and (d) show the O 1s XPS spectra of WO$_3$, OV-WO$_3$ (represented by OV-WO$_3$-3 h) and bulk-OV-WO$_3$ (represented by bulk-OV-WO$_3$-3 h) samples, which can be deconvoluted in three peaks. The peaks at about 530.2, 531.7 and 533.2 eV correspond to surface lattice oxygen species (O_L), hydroxyl species (O_{OH}) and adsorbed water species (O_{H_2O}), respectively. It is considered that the percentage of O_{OH} signal links to the quantity of surface oxygen vacancies[7, 20, 31]. The percentages of the three oxygen species of WO$_3$, OV-WO$_3$ and bulk-OV-WO$_3$ samples are listed in Table 1. Their percentage of O_{OH} increases obviously from 11.8% for WO$_3$ to 15.3% for WO$_3$-3 h-new, suggesting the thermal treatment cause surface oxygen vacancies.

Moreover, when the OV-WO$_3$ sample is placed in the air for a period of time, the density of surface oxygen vacancies decreases apparently. As shown in Figure 7(d), the percentage of O_{OH} decreases from 15.3% to 11.3%. This demonstrates that much surface oxygen vacancies are recovered. It is reported that the surface oxygen vacancies will cause the shift of VB position[7, 21, 22]. However, no drift is observed between WO$_3$ and bulk-OV-WO$_3$ samples, which demonstrates that the much surface oxygen vacancies are recovered and the density of surface oxygen vacancies in OV-WO$_3$-3 h sample becomes similarly to that of pristine WO$_3$. Meanwhile, EPR spectra indicates the formation of bulk oxygen vacancies [Figure 6(e)]. The EPR signal can still be observed on bulk-OV-WO$_3$-3 h sample, but no signal in pristine WO$_3$ sample. It indicates the existence of bulk oxygen vacancy bulk-OV-WO$_3$-3 h sample.

Therefore, it can conclude that the activity of sample increases when both bulk and surface oxygen vacancies are presented, while the activity is further increased when the surface oxygen vacancies are recovered.

Table 1 Detail results of O 1s XPS spectra of pristine WO$_3$, OV-WO$_3$-3 h and bulk-OV-WO$_3$

		Total	O_L	O_{OH}	O_{H_2O}
Area	WO$_3$	84376.662	72615.110	10022.600	1738.952
	OV-WO$_3$	174771.533	142919.300	26817.080	5035.153
	Bulk-OV-WO$_3$	78938.520	64014.520	8907.012	6016.988
Percentage/%	WO$_3$	100	86.1	11.8	2.1
	OV-WO$_3$	100	81.8	15.3	2.9
	Bulk-OV-WO$_3$	100	81.1	11.3	7.6

3.5 Mechanism of photocatalytic degradation

This interesting phenomenon inspires us that the bulk oxygen vacancy dominating sample should be the optimal sample for photocatalytic degradation. Subsequently, the mechanism for the degradation is studied.

To investigate the contribution of the various oxidative species generated in the photocatalytic system of WO$_3$ and bulk-OV-WO$_3$ samples for the degradation of CBZ, active species are tested by EPR. The generation of reactive oxygen species in the system is monitored through EPR technique after adding DMPO or TEMP as trapping agent under UV-vis radiation. As shown in Figure 7(a), a four-line EPR signal of DMPO-·OH adduct with an intensity ratio of 1∶2∶2∶1 is also detected in both of WO$_3$ and bulk-OV-WO$_3$ systems, and the intensity of DMPO-·OH adduct signal in the system with WO$_3$-3 h is much stronger than that of WO$_3$. It indicates oxygen vacancy is significantly conducive to the generation of ·OH. As the VB of WO$_3$ and WO$_3$-3 h are both 2.80 V, more positive than the standard redox potential of H$_2$O/·OH (2.37 V)[32], both samples can oxidize H$_2$O to ·OH under UV-vis thermodynamically. Therefore, ·OH can be generated from the oxidation of H$_2$O directly (h$^+$+H$_2$O→·OH+H$^+$). Then, methanol is added during the EPR measurement

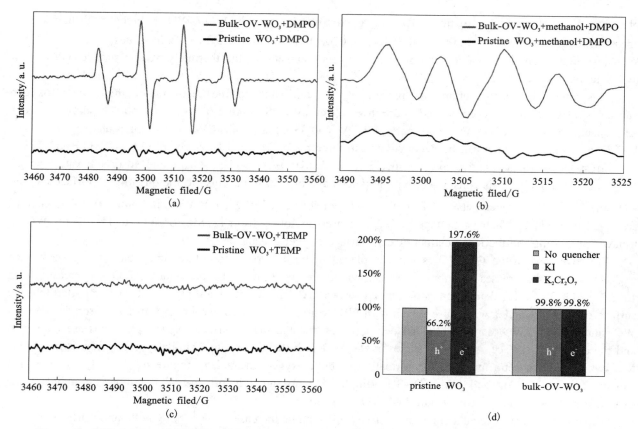

Figure 7 EPR spectra of DMPO-·OH adduct (a); DMPO-·O_2^- adduct (b) and TEMP-$^1O_2$(c) for WO_3 and bulk-OV-WO_3 under UV-vis irradiation; the changes of degradation efficiencies of CBZ in WO_3 and bulk-OV-WO_3 photocatalytic systems with KI or $K_2Cr_2O_7$ as quenchers (d)

to catch ·OH· Four very weak peaks with an intensity ratio of 1:1:1:1 are detected in bulk-OV-WO_3 with oxygen vacancies [Figure 7(b)]. It suggests that ·O_2^- is generated in bulk-OV-WO_3 systems, as they are the characteristic quartet signal of DMPO-·O_2^- adduct. Here, ·O_2^- may be generated by the reduction of O_2(e^-+O_2→·O_2^-). However, these characteristic signals cannot be observed in WO_3 systems without oxygen vacancies. Therefore, this strongly indicates that the introduction of oxygen vacancies is favorable for the formation of ·O_2^-, which is beneficial to the degradation efficiency. Besides, there is no signal of TEMP-$^1O_2$ adduct detected in WO_3 nor bulk-OV-WO_3 systems [Figure 7(c)].

In addition, active species trapping tests are also executed by adding KI and $K_2Cr_2O_7$ as the quenchers of photoexcited hole (h^+) and e^-, respectively. When 10 mM KI is added into the WO_3 system, the degradation efficiency of CBZ remarkably decreases to 66.2% of pristine value [Figure 7(d)]. It suggests that photoexcited hole (h^+) is one of major reactive species contributed to the catalytic degradation of CBZ by WO_3. However, the addition of KI in the bulk-OV-WO_3 does not influence the degradation efficiency. This means that the hole does not directly cause the degradation of CBZ, instead, the hole mainly oxidize water into OH radicals for the degradation.

Meanwhile, the degradation efficiency of CBZ remarkably increases to 197.6% as compared with pristine value when 10 mM $K_2Cr_2O_7$ is added. The increase may be originated from the promotion of charge separation due to quenching of e^-. In comparision, the degradation efficiency of CBZ in bulk-OV-WO_3 photocatalytic system has little change by adding 10 mM $K_2Cr_2O_7$. This should be due to that the high charge transfer efficiency of WO_3-3 h resulted in the insignificant enhancement caused by the introduction of $K_2Cr_2O_7$.

In summary, the bulk oxygen vacancy dominating WO_3 optimize the photodegradation mechanism. The oxygen vacancy free sample mainly oxidizes the CBZ by the photoexcited hole (h^+), requiring the CBZ to be adsorbed and cause low activity. By contrast, the bulk oxygen vacancy dominating WO_3 mainly produce ·OH or ·O_2^- to degrade CBZ, the stronger oxidization power of radicals and their ability to migrate to the solution is beneficial to the degradation of CBZ.

4 Conclusion

In summary, introduction of oxygen vacancy can improve photoelectric properties and photocatalytic degradation performance of WO_3 samples. But the improvment would be diminished when vacancies exceed the optimal proportion. In this study, a series of WO_3 without OV, with both bulk and surface OV and with dominating bulk OV were prepared and evaluated for their catalytic degradation performance for CBZ. Results showed that the surface oxygen vacancy would be recovered in air, and then the bulk oxygen vacancy dominating WO_3 samples even exhibit better photocatalytic degradation performance. Therefore, both proportion and location of oxygen vacancy could exert different influence on the photocatalytic performance of WO_3. The recovered behavior of surface oxygen vacancy and characteristic is deserved to be further studied. The results of this study could potentially provide an effective method for developing advanced photocatalyst by effectively utilizing oxygen vacancy, as well as broaden our understanding of the role of oxygen vacancy in photocatalyst.

Acknowledgments: This work was supported by Key-Area Research and Development Program of Guangdong Province, China (2021B0909060001), the National Natural Science Foundation of China (Grant No. 51902357), the Natural Science Foundation of Guangdong Province, China (2019A1515012143), Guangdong High Level Innovation Research Institute, China (2021B0909050001).

References

[1] ZHAO T Y, XING Z, XIU Z, et al. [J]. J. Hazard. Mater., 2019, 364: 117-124.
[2] ZHAO H, LIU X, DONG Y M, et al. [J]. Appl. Catal. B: Environ., 2019, 256: 117872.
[3] WANG H, ZHANG L, CHEN Z, et al. [J]. Chem. Soc. Rev., 2014, 43: 5234-5244.
[4] ZHANG W, LI G, LIU H, et al. [J]. Environ. Sci. Nano., 2019, 6: 948-958.
[5] FENG Y, LI H, LING L, et al. [J]. Environ. Sci. Technol., 2018, 52: 7842-7848.
[6] KIM J, LEE C, CHOI W. [J]. Environ. Sci. Technol., 2010, 44: 6849-6854.
[7] WANG Y T, CAI J M, WU M Q, et al. [J]. Appl. Catal. B: Environ., 2018, 239: 398-407.
[8] LI Y S, TANG Z L, ZHANG J Y, et al. [J]. J. Phys. Chem. C., 2016, 120: 9750-9763.
[9] MI Q X, ZHANAIDAROVA A, BRUNSCHWIG B S, et al. [J]. Energy Environ. Sic., 2012, 5: 5694-5700.
[10] WU J X, QIAO P Z, LI H, et al. [J]. J. Colloid Interf. Sci., 2019, 557: 18-27.
[11] KOO B R, AHN H J. [J]. Nanoscale, 2017, 9: 17788-17793.
[12] SHI W N, LI H, CHEN J Y, et al. [J]. Electrochim. Acta, 2017, 225: 473-481.
[13] SOLARSKA R, KRÓLIKOWSKA A, AUGUSTYńSKI J. [J]. Angew. Chem. Int. Edit., 2010, 49(43): 7980-7983.
[14] LIU G, HAN J F, ZHOU X, et al. [J]. J. Catal., 2013, 307: 148-152.
[15] MOULZOLF S C, DING S A, LAD R J. [J]. Sensors & Actuat. B-Chem., 2001, 77: 375-382.
[16] SOLARSKA R, KRÓLIKOWSKA A, AUGUSTYńSKI J. [J]. Chem. Int. Edit., 2010, 122(43): 8152-8155.
[17] LONG R, ENGLISH N J. [J]. Phys. Rev. B, 2011, 83: 024416.
[18] ANSARI S, KHAN M, KALATHIL S, et al. [J]. Nanoscale, 2013, 5: 9238-9246.
[19] LINSEBIGLER A, LU G. [J]. J. Yates, Chem. Rev., 1995, 95: 735-758.
[20] CAI J M, ZHU Y M, LIU D S, et al. [J]. ACS Catal., 2015, 5: 1708-1716.
[21] SCHMID M, SHISHKIN M, KRESSE G, et al. [J]. Phys. Rev. Lett., 2006, 97: 046101.
[22] WEINHARDT L, BLUM M, BÄR M, et al. [J]. J. Phys. Chem. C., 2008, 112: 3078-3082.
[23] ZHANG T T, ZHAO K, YU J G, et al. [J]. Nanoscale, 2013, 5: 8375-8383.
[24] PAN X Y, YANG M Q, FU X Z, et al. [J]. Nanoscale, 2013, 5: 3601-3614.
[25] LI W J, DA P M, ZHANG Y Y, et al. [J]. ACS Nano, 2014, 8: 11770-11777.
[26] ZHAO C Z, KONG X G, LIU X M, et al. [J]. Nanoscale, 2013, 5: 8084-8089.
[27] HU Z F, XU M K, SHEN Z R, et al. [J]. J. Meter. Chem. A, 2015, 3: 14046-14053.
[28] YANG X Y, WOLCOTT A, WANG G M, et al. [J]. Nano Lett., 2009, 9: 2331-2336.
[29] WOLCOTT A, SMITH W A, KUYKENDALL T R, et al. [J]. Small, 2009, 5: 104-111.
[30] YANG Y, ZHANG C, HUANG D L, et al. [J]. Appl. Catal. B: Environ., 2019, 245: 87-99.
[31] RAHIMNEJAD S, HE J H, PAN F, et al. [J]. Mater. Res. Express, 2014, 1: 045044.
[32] MA J Z, WANG C X, HE H. [J]. Appl. Catal. B: Environ. 2016, 184: 28-34.

Application of Acidophilic Bacteria in the Metal Enrichment of Electroplating Sludge Use the External Circulation Bioreactor

Bingyang TIAN[1,2], He SHANG[1,2], Wencheng GAO[1,2], Jiankang WEN[1,2]

1 National Engineering Research Center for Environment-friendly Metallurgy in Producing Premium Non-ferrous Metals, China GRINM Group Corporation Limited, Beijing, 101407, China
2 Beijing Engineering Research Center of Strategic Nonferrous Metals Green Manufacturing Technology, GRINM Resources and Environment Tech. Co., Ltd., Beijing, 101407, China

Abstract: A microbial consortium of Acidithiobacillus thiooxidans, Acidithiobacillus ferrooxidans, and Leptospirillum ferriphilum was applied in external circulation bioreactor to bioleaching Ni, Cu, Zn and Cr from electroplating sludge. By way of cascade expansion, the microbial consortium can maintain higher-density cell growth (10^9 cells/mL) in external circulation bioreactor after 9 days. Bacteria and biological acids were quickly separated by external circulation bioreactor. The electroplating sludge is leached with biological acid according to the solid-liquid ratio of 8%. The leachate is returned to the external circulation bioreactor and regenerated to a highly active state under the action of bacteria. The highly active regeneration liquid can be used to leach the target metal from the second batch of electroplating sludge. In this way, on the one hand, it maintains the highest activity of bacteria, on the other hand, it enriches the concentration of target metals in the leaching solution. Additionally, at the end of the cycle enrichment process, the concentrations of the target metals Ni, Cu, Zn and Cr were 5.52 g/L, 6.60 g/L, 15.60 g/L and 25.68 g/L, respectively. In addition, the pilot production of electroplating sludge disposal by external circulation bioreactor has been completed.

Keywords: Biotechnology; External circulation bioreactor; Metal enrichment; Electroplating sludge

1 Introduction

The rapid development of the electroplating industry has generated a large amount of electroplating wastewater (plating part cleaning wastewater and waste bathsolution)[1]. Its composition is complex, containing not only a variety of metal ions (Ni, Cu, Zn, Cr, etc.), but also inorganic acids, cyanide, organic additives and other harmful substances[2]. Chemical precipitation accounts for about 41% of the electroplating wastewater treatment in my country, but it also brings a lot of secondary pollutants-electroplating sludge[3]. Statistically, nearly 10^5 tonnes of electroplating sludge is produced annually in China[4]. How to properly treat and dispose of electroplating sludge is an urgent problem to be solved at present. On the one hand, if the electroplating sludge is only disposed of in a harmless and reduced amount, a large amount of metal resources will be lost; on the other hand, if the electroplating sludge is disposed of as a resource, a certain economic benefit is required, and the production process will cause secondary pollution to the environment[5].

At present, electroplating sludge is mainly disposed of in three ways: incineration, cement kilncoordination or safe landfill[6]. The advantages of these methods mainly lie in large batch capacity and simple process operation[7]. However, these disposal processes still have potential environmental risks (flue gas diffusion during incineration, excessive carbon emissions, leachate leakage from landfills, etc.), and also cause the loss of a large amount of valuable metals[8]. Therefore, the selection of treatment and disposal methods for electroplating sludge should first evaluate its economic value and environmental risk, and then choose the best resource recovery or harmless disposal[9].

Under the above mention of both environmental protection and efficient resource reuse, finding an effective and economical way to maximize the metals extraction and recycling from electroplating sludge will become increasingly important to industries[10]. Bioleaching, as an emerging alternative to the hydrometallurgical technologies, has attracted more attention owing to its environmental friendliness, low-cost and mild reaction conditions[11]. Cultivate the bacteria to the logarithmic phase of high activity, and then separate the bacteria from the biological acid[12]. Then put the bacteria into the new culture medium to continue the cultivation, and use biological acid to leach the valuable metals in the electroplating sludge, without the participation of the bacteria in the leaching process[13]. Obviously, indirect bioleaching avoids the direct poisoning of electroplating sludge to bacteria[14]. In addition, by intervening in the leaching process, the solid-liquid ratio of leaching can be increased, the leaching time can be shortened, and the efficiency of bioleaching can be improved[15]. However, indirect bioleaching is the same as direct bioleaching, and its leaching process is also affected by different factors such as initial pH (biological acid pH 0.8-1.0), leaching temperature (30-40 ℃) and Fe^{3+} concentration (1-2 g/L, too low Fe^{3+} will reduce the leaching rate of valuable metals, too high will form jarosite precipitation and hinder the contact reaction between active substances and materials)[16]. Therefore, more research is still needed on the process of bioleaching electroplating sludge recycling valuable technology to promote its application in industrial production[17].

In this research, a series of detailed leaching experiments were designed to study the metals leaching efficiency and enrichment concentration from electroplating sludge inexternal circulation bioreactor. Results of the metal leaching and enrichment analysis during bioleaching were presented. The cell growth and pH change under different operational conditions were evaluated. In addition, the toxic leaching experiments were performed for the bioleaching residues to check the metal residue.

2 Materials and methods

2.1 Electroplating sludge sample and microbial media

The electroplating sludge sample came from an electroplating factory in Zhongshan city, Guangdong Province. The samples were dried and crushed for analysis. The contents of Ni, Cu, Zn, Cr, Al, and Fe in sludge are 1.38%, 1.65%, 3.90%, 6.42%, 1.14%, and 3.31%, respectively. Three typical bioleaching strains in the form of mixed flora were used in this study. The strains include one sulfur-oxidizing strains (*Acidithiobacillus thiooxidans* (*A. t*, ATCC, 19377)), one ferrous/sulfuroxidizing strains (*Acidithiobacillus ferrooxidans* (*A. f*, ATCC 23270)) and one ferrous-oxidizing strains (*Leptospirillum ferriphilum* (*L. f*, ATCC 49881)). *At*, *Af* and *Lf* could resist 51.2 g/L Cu^{2+}, 70 g/L Zn^{2+}, 36.4 g/L Cr^{3+}, 56 g/L Cd^{2+}, and 59 g/L Ni^{2+}. An inorganic salt medium used for cultivating bacteria is composed of $(NH_4)_2SO_4$ (2.0 g/L), K_2HPO_4 (1.0 g/L), $MgSO_4 \cdot 7H_2O$ (0.5 g/L) and $CaCl_2$ (0.25 g/L), in which elemental sulfur (10.0 g/L) and FeS_2 (2.0 g/L) were supplemented with mixed flora as energy substrates[18].

2.2 External circulation bioreactor

In this study, the external circulation bioreactor was operated under a couple of major configurations: (a) Bacterial culture and high activity biological acid regeneration unit (Effective volume 1 m^3). Submerged continuous separation of biological acid and bacteria device. (b) Electroplating sludge leaching tanks. (c) Solid-liquid separation device. (d) Inlet/Outlet pumps. The schematic diagram of the reactor was given in Figure 1.

2.3 Bioleaching tests

Before starting the experimental period, the external circulation bioreactor was seeded 100 L of mixed bacteria. The temperature of outer loop control in the MBR was 35 ℃, and the aeration rate was 10 L/min, stirring speed was 150 r/min. After running 9 days, the pH value declined to 0.8. The cell density was controlled at 10^9 cells/mL. Turn on the inlet pumps (Supplement of inorganic salt medium) and outlet pumps (Extract highly active biological acid into the leaching tank) of the external circulation bioreactor at the same time. The speed of circulation was 41.67 L/h.

After the amount of biological acid reaches the set demand, it is stirred and leached by adding electroplating sludge according to the solid-liquid ratio. The leaching tank is provided with an overflow port, the hydraulic retention time is 2 h, and the slurry is collected into the solid-liquid separator. After the slurry is separated by solid-liquid, the mud cake is discharged after on-line cleaning. The filtrate is regenerated through the inlet pump into the external circulation reactor. The regenerated highly active biological acid enters the leaching of the next batch of electroplating sludge. When the pH in the external circulation bioreactor is greater

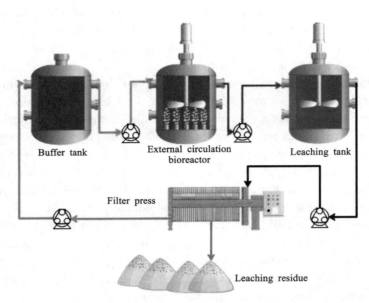

Figure 1 Schematic diagram of external circulation bioreactor

than 0.8, or the leaching rate of the target metal in the electroplating sludge is less than 100%, the recycling of filtrate should be stopped. The filtrate of metal enrichment can enter the extraction section to recover the target metal.

2.4 Analytical methods

The pH value of solution was determined using a pH meter (Hana HI2221, Italy). The released concentrations of heavy metals were determined by inductively coupled plasma atomic emission spectrometry (ICP-OES, Perkin Elmer Optima 8300). The morphology change of materials was analyzed with a scanning electron microscope (SEM, Hitachi S-4800, Japan) at an accelerating voltage of 20 kV. Cell numbers were determined using a microscope (Olympus IX71, Japan). The toxicity characteristic leaching procedure (TCLP) of the bioleaching residues were conducted according to the Chinese environmental standard HJ/T 300 2007.

3 Results and discussion

3.1 Operation of external circulation bioreactor

After adding mixed bacteria, inorganic salt medium, sulfur and pyrite to the external circulation bioreactor, the pH value of the system was 0.8 and the number of bacteria was 10^9 cells/mL. Turn on the effluent pump and separate the highly active biological acid from the bacteria through the immersion cell separator. Highly active biological acid is put into the leaching tank and set aside. The bacteria are left in the external circulation bioreactor. At the same time, the feed pump of the external circulation bioreactor is turned on to supplement the inorganic salt medium to keep the liquid level in the bioreactor unchanged. Electroplating sludge is continuously added to the leaching tank according to the set solid-liquid ratio. The hydraulic retention time is guaranteed for 2 h. The slurry flows out through the overflow port of the leaching tank and is collected into the solid-liquid separator. The mud cake is discharged after on-line cleaning in the solid-liquid separator. After the filtrate was collected, the inorganic salt culture medium was replaced by the inlet water pump and entered into the external circulation bioreactor for regeneration. The regenerated highly active biological acid is used to leach the next batch of electroplating sludge. This cycle enriches the concentration of valuable metals in the solution. When the pH in the external circulation bioreactor is greater than 0.8, or the leaching rate of electroplating sludge is less than 100%, the cycle should be stopped. At this time, the filtrate enters the next extraction and separation process to separate and recover the valuable metals in the solution.

3.2 Disposal capacity of electroplating sludge

The growth curve of bacteria is shown in Figure 2. The strains density grew steadily from the 1st to the 9th day, and reached a stable state of 10^9 cells/mL in the subsequent leaching time. The bacteria grew rapidly to the logarithmic growth phase, and no obvious lag phase could be found. The strains with high activity, which ensure the application of bioleaching technology is extremely important. In the regeneration-circulation process, the pH of external circulation bioreactor was 0.8. After 5 times of regeneration-circulation process, Ni, Cu, Zn and Cr extraction concentration were 5.52 g/L, 6.60 g/L, 15.60 g/L and 25.68 g/L, respectively. At the same time, the external circulation

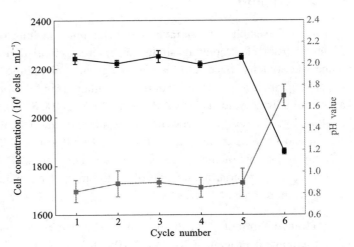

Figure 2 Variations of cell concentration and pH with the cycle number during cycle enrichment process

bioreactor had a sharp increase of pH from 0.8 to 1.7, and the number of cells had a sharp decline. The results showed that the high concentration of heavy metals in the leaching solution could inhibit the growth of the strain, and the batch experiment was over. The handling capacity of sludge was 400 kg.

3.3 The evaluation of electroplating sludge detoxification

The TCLP test was carried out to evaluate the potential risk of electroplating sludge and the bioleached residue. From Figure 3, the concentration of Ni, Cu, Zn and Cr in test of electroplating sludge were well above the limit value of Chinese National Standard (GB5085.3-2007 and GB16889-2008). However, the mix metals in the bioleached residue leachate were well below the limits. It could be concluded that the Cr remaining in the solid residue have a non-soluble form (residual fraction) and could not be dissolution. Hence, the bioleached residue was well considered as non-hazardous waste and could be used as a raw material for the construction industry.

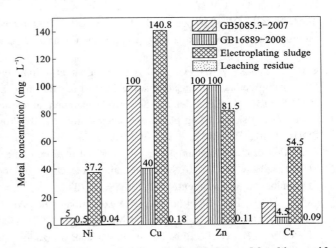

Figure 3 TCLP test of electroplating sludge and leaching residue

Suggesting that the bioleached residue had a low risk of environmental pollution and can be reused in construction material safely. Therefore, the bioleaching can reclaim metals of the electroplating sludge effectively, and realize resourceful utilization and harmlessness.

3.4 Analysis on the advantages of external circulation bioreactor

The external circulation reactor increased the concentration of bacteria through culture and enrichment, and kept the activity of the bacteria in the logarithmic phase. It solves the problems of slow growth of bacteria and low efficiency of biological oxidation. The external circulation bioreactor greatly improves the bioleaching efficiency and significantly shortens the bioleaching cycle, and the heavy metals concentrated by reflux of bioleaching solution provide favorable conditions for subsequent recovery, purification and solidification stability. The equipment operates under normal temperature and pressure with low investment and simple operation. It can be used in many kinds of hazardous wastes (waste batteries, waste catalysts, electronic circuit boards, electroplating sludge, pickling sludge, smelting residue) to realize the leaching and enrichment of valuable metals.

4 Conclusion

In this research, we unravel the behaviour of improving metal leaching concentration and leaching efficiency from electroplating sludge in external circulation bioreactor. The method of indirect contact can reduce the toxicity of electroplating sludge to acidophilic bacteria, therefore, allowing us to probe the bioleaching process in 1 m^3 external circulation bioreactor. Under the conditions of 8% pulp density, 150 r/min stirring velocity and 2 h leaching time, 100% leaching rates of Ni, Cu, Zn and Cr were achieved. Additionally, the design of enrichment experiment dramatically improved the target metal concentration (Ni, Cu, Zn and Cr were 5.52 g/L, 6.60 g/L, 15.60 g/L and 25.68 g/L, respectively), followed by a broadly reducing in the cost of metal recovery at the later stages. Overall, this research can provide a demonstration project for the industrial application of bioleaching electroplating sludge in MBR, with a view to applying this method to more types of hazardous wastes.

Acknowledgments: We highly appreciate and thank financial support from the National Key Research and Development Program (2022YFC2105304).

References

[1] TIAN B, CUI Y, ZHAO J, et al. Stepwise recovery of Ni, Cu, Zn, and Cr: A green route to resourceful disposal of electroplating sludge[J]. Journal of Environmental Chemical Engineering, 2023, 11(3): 109767.

[2] YANG Y, LIU X, WANG J, et al. Screening bioleaching systems and operational conditions for optimal Ni recovery from dry electroplating sludge and exploration of the leaching mechanisms involved[J]. Geomicrobiology Journal, 2016, 33(3-4): 179-184.

[3] TIAN B, CUI Y, QIN Z, et al. Indirect bioleaching recovery of valuable metals from electroplating sludge and optimization of various parameters using response surface methodology (RSM)[J]. J Environ Manage, 2022, 312: 114927.

[4] CAO C, XU X, WANG G, et al. Characterization of ionic liquids removing heavy metals from electroplating sludge: Influencing factors, optimisation strategies and reaction mechanisms[J]. Chemosphere, 2023, 324: 138309.

[5] YAN K, LIU Z, LI Z, et al. Selective separation of chromium from sulphuric acid leaching solutions of mixed electroplating sludge using phosphate precipitation[J]. Hydrometallurgy, 2019, 186: 42-49.

[6] PINTO F M, PEREIRA R A, SOUZA T M, et al. Treatment, reuse, leaching characteristics and genotoxicity evaluation of electroplating sludge[J]. Journal of Environmental Management, 2021, 280: 111706.

[7] SUN J, ZHOU W, ZHANG L, et al. Bioleaching of copper-containing electroplating sludge[J]. Journal of Environmental Management, 2021, 285: 112133.

[8] QU Z, SU T, ZHU S, et al. Stepwise extraction of Fe, Al, Ca, and Zn: A green route to recycle raw electroplating sludge[J]. Journal of Environmental Management, 2021, 300: 113700.

[9] ZHOU W, ZHANG L, PENG J, et al. Cleaner utilization of electroplating sludge by bioleaching with a moderately thermophilic consortium: A pilot study[J]. Chemosphere (Oxford), 2019, 232: 345-355.

[10] ZHANG L, ZHOU W, LIU Y, et al. Bioleaching of dewatered electroplating sludge for the extraction of base metals using an adapted microbial consortium: Process optimization and kinetics[J]. Hydrometallurgy, 2020, 191: 105227.

[11] NIKFAR S, PARSA A, BAHALOO-HOREH N, et al. Enhanced bioleaching of Cr and Ni from a chromium-rich electroplating sludge using the filtrated culture of Aspergillus niger[J]. Journal of Cleaner Production, 2020, 264: 121622.

[12] SU R, LIANG B, GUAN J. Leaching effects of metal from electroplating sludge under phosphate participation in hydrochloric acid medium[J]. Procedia Environmental Sciences, 2016, 31: 361-365.

[13] WU P, ZHANG L, LIN C, et al. Extracting heavy metals from electroplating sludge by acid and bioelectrical leaching using *Acidithiobacillus ferrooxidans*[J]. Hydrometallurgy, 2020, 191: 105225.

[14] GUNARATHNE V, RAJAPAKSHA A U, VITHANAGE M, et al. Hydrometallurgical processes for heavy metals recovery from industrial sludges[J]. Critical Reviews in Environmental Science and Technology, 2020: 1-41.

[15] KALOLA V, DESAI C. Biosorption of Cr(VI) by halomonas sp. DK4, a halotolerant bacterium isolated from chrome electroplating sludge[J]. Environ Sci Pollut Res Int, 2020, 27(22): 27330-27344.

[16] SUNDRAMURTHY V P, BASKAR R. Detoxification of electroplating sludge by bioleaching: Process and kinetic aspects[J]. Polish Journal of Environmental Studies, 2015, 24: 1249-1257.

[17] RASTEGAR S O, MOUSAVI S M, SHOJAOSADATI S A. Cr and Ni recovery during bioleaching of dewatered metal-plating sludge using *Acidithiobacillus ferrooxidans*[J]. Bioresource Technology, 2014, 167: 61-68.
[18] WANG J, CUI Y, CHU H, et al. Enhanced metal bioleaching mechanisms of extracellular polymeric substance for obsolete $LiNi_xCo_yMn_{1-x-y}O_2$ at high pulp density[J]. Journal of Environmental Management, 2022, 318: 115429.

Comparative Analysis of Domestic and Foreign Indentation Test Standards

Qinli Lü*, Sitong YE, Shuai XU, Hong JI

China United Test & Certification Co., Ltd., China Grinm Group Co., Ltd., Beijing, 101407, China

Abstract: Instrumented indentation test method is an important method to study the hardness of coatings, and the formulation and revision of the standards for the instrumented indentation test have been made by both domestic and foreign relevant institutions. Comparison and analysis of the standards ASTM E2546, GB/T 21838, GB/T 22458 and GB/T 25898 have been carried out from the aspects such as the scope of application, test method, calibration and verification of test machines, and calibration of reference blocks. ASTM E2546 has relatively simple regulations for the test method and the calibration and verification of test machines. GB/T 21838 has a wide range of application and high accuracy requirements for calibration and verification of test machines. GB/T 22458 has more accurate and detailed requirements on the test environment, test pieces and test surface as well as the test operation for micro-nano range test. GB/T 25898 is targeted at micro-nano scale films and provides a test method for the indentation hardness and modulus of thin films that exclude the influence of substrate as much as possible. Compared with foreign indentation test standards, domestic standards have more detailed requirements and are facilitating the practical use, which provide a strong basis and support for the standardization of instrumented indentation tests in China.

Keywords: Instrumented indentation test; ASTM E2546; GB/T 21838; GB/T 22458; GB/T 25898

1 Introduction

With the rapid development of preparation technology for micro-nano coatings, the coating thickness gradually develops to the micro-nano level, and the traditional hardness and modulus measurement methods can not meet the needs. Because of its high accuracy and easy operation, instrumented indentation test has gradually become an important test method in the field of nanomechanics characterization. Instrumented indentation technology, also known as depth sensitive technology, is to continuously control and record the force and displacement data of indentation, and analyze these data to permit the determination of hardness and material properties. However, the instrumented indentation test is susceptible to many factors such as method, instrument, sample and test personnel during the test, which often leads to poor comparability of test results. Therefore, it is very important to maintain the consistency of domestic and foreign instrumented indentation test standards.

The International Organization for Standardization (ISO) began to standardize the instrumented indentation test in 1997, and officially issued ISO 14577-2002 "Metallic Materials-Instrumented Indentation Test for Hardness and Material Parameters" in 2002[1-4]. In 2007, American Society of Testing Materials (ASTM) issued ASTM E2546-2007 "Standard Practice for Instrumented Indentation Testing", and revised ASTM E2546-15 in 2015[5]. In 2008, in reference to ISO and ASTM standards and considering the future trend of technological development, China issued national standards: GB/T 21838-2008 "Metallic materials-Instrumented indentation test for hardness and material parameters" and GB/T 22458-2008 "General rules of instrumented nanoindentation test"[6]. GB/T 21838 is equivalent to ISO 14577. After revised in 2019, 2020 and 2022, the current standards GB/T 21838.1-2019, GB/T 21838.2-2022, GB/T 21838.3-2022 and GB/T 21838.4-2020 contain four parts, which are test method, verification and calibration of testing machines, calibration of reference blocks and test method for metallic and non-metal coatings[7-10]. In 2010, the standard GB/T 25898-2010 "Instrumented nanoindention test-indentation hardness and modulus of thin film" was issued in China for the hardness and modulus of micro-nano thin films[11].

Domestic and foreign standards tend to be consistent in terms of test method and data analysis method, but there are still many differences in other details such as the scope of application, test method, calibration and verification of test machines, calibration of reference blocks, etc. This paper summarizes and sorts out the latest versions of these standards, and compares their similarities and differences.

2 Comparison of standards

2.1 Scope of application

The standard ASTM E2546-15 specifies the requirements for the instrument capability and the basic test method, and gives the data analysis method required to determine the hardness and material parameters. Parts 1, 2 and 3 of the standard GB/T 21838 specify the methods and requirements for instrumented indentation test of block materials, and its scope of application is relatively broad. In the fourth part of GB/T 21838, detailed test methods are given for the covering layer in the micro-nano range. The standard GB/T 21838 applies not only to metal block materials, but also to thin metallic or non-metallic coatings and non-metallic materials. The standard GB/T 22458-2008 is suitable for the indentation depth in nano range, so that the measurement of material parameters for both blocks and films is more accurate and reliable in the nano range. The standard GB/T 25898-2010 applies to thin films attached to the solid surface, and to determine the indentation hardness and modulus of thin film with the indentation depth from nanometers to several microns.

On the whole, compared with foreign instrumented indentation test standards, our national standards are relatively more targeted.

2.2 Test principle and method

Domestic and foreign instrumented indentation test standards are consistent in the test principle, and the basic idea is to obtain the indentation hardness and material properties by continuously recording the force and displacement of indentation (the F-h curve)[12]. Figure 1 and Figure 2 respectively give the typical loading and unloading F-h curve and the schematic cross section of an indentation, where h_{max} is the maximum indentation depth, h_p is the permanent indentation depth after unloading, and h_c is the contact indentation depth, which is used to analyze and calculate the projected area of contact of the indenter.

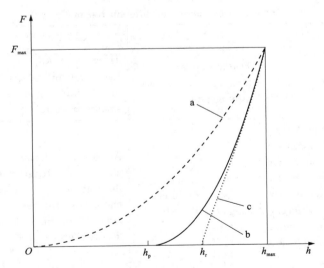

a—application of the test force; b—removal of the test force; c—tangent to curve b at F_{max}.

Figure 1 Schematic representation of the indentation test procedure

The indentation hardness H_{IT} and modulus E_{IT} of the tested materials can be obtained from the F-h curve. This method has clear physical concept and simple calculation, so it is widely used in commercial nanoindentation instruments. E_{IT} and H_{IT} are as given in the following formulas.

$$H_{IT} = \frac{F_{max}}{A_p} \tag{1}$$

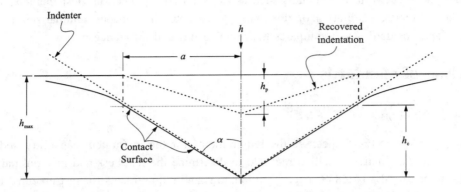

Figure 2 Schematic representation of the cross section of indentation

$$E_r = \frac{S\sqrt{\pi}}{2\sqrt{A_p}} \tag{2}$$

$$E^* = \frac{1}{\dfrac{1}{E_r} - \dfrac{1-\nu_i^2}{E_i}} = \frac{E_{IT}}{(1-\nu_s^2)} \tag{3}$$

$$E_{IT} = \frac{(1-\nu_s^2)}{\dfrac{1}{E_r} - \dfrac{1-\nu_i^2}{E_i}} \tag{4}$$

Here F_{max} is the maximum applied force, A_p is the projected (cross-section) area of contact between the indenter and the test piece, S is the contact stiffness determined by fitting the first 50%-90% of the test force removal curve, E_r is the reduced modulus of the indentation contact, E_i is the modulus of the indenter (for diamond 1140 GPa), ν_s is the Poisson ratio of the test piece, ν_i is the Poisson ratio of the indenter (for diamond 0.07), E^* is the plane strain modulus.

Table 1 Comparison of different test methods

Compared items	ASTM E2546	GB/T 21838	GB/T 22458	GB/T 25898
Test control mode	/	Two test cycles: force control, displacement control	Three test cycles: force control, displacement control, compressive strain rate control	Three test cycles: force control, displacement control, compressive strain rate control
Environment	Prior to performing any tests, the instrument and the test piece shall be stabilized to the temperature of the environment. Temperature change during each test should be less than 1.0 ℃	Ambient temperature: 23(±5)℃. Relative humidity: 45%±10%	Ambient temperature: 23(±5)℃. Temperature fluctuation: <1 ℃. Thermal drift rate during each test: ≤0.05 nm/s. Relative humidity: <50%	Ambient temperature: 23(±5)℃. Temperature fluctuation: <1 ℃. Thermal drift rate during each test: ≤0.05 nm/s. Relative humidity: <50%
Test piece thickness	The test piece thickness should be at least 10 times the indentation depth or 6 times greater than the indentation radius, whichever is greater	The test piece thickness should be at least 10 times the indentation depth or 3 times the indentation diameter, whichever is greater	The test piece thickness should be at least 10 times the indentation depth or 6 times the indentation diameter, whichever is greater	/

Continue the Table 1

Compared items	ASTM E2546	GB/T 21838	GB/T 22458	GB/T 25898
Test surface	/	The test surface tilt is less than 1°	The indentation depth should be at least 20 times the surface roughness R_a, which is difficult to meet at depths less than or equal to 0.2 μm and should be stated in the test report. The test surface tilt is less than 1°	The surface roughness R_a should be less than 5% of the maximum indentation depth. The test surface tilt is less than 1°
Loading process	The test environment shall be clean and free from vibrations, electromagnetic interference, or other variations that could adversely affect the instrument performance	The test force shall be applied without shock or vibration that can significantly affect the test results	The test force shall be applied without the influence of shock and vibration	/
Test position	Locate the test at lease six indent radii away from free surfaces such as edges, voids and other indentations	For ceramic materials and metals: indentations shall be at least 3 R away from interfaces or free surfaces and the minimum distance between indentations shall be at least 5 R. For other materials: separations are at least 10 R. R: the indentation diameter	For ceramic materials and metals: indentations shall be at least 6 R away from interfaces or free surfaces and the minimum distance between indentations shall be at least 6 R. For other materials: separations are at least 20 R. R: the indentation diameter	Indentations shall be at least 30 h_{max} away from interfaces or free surfaces. The distance between indentations shall be at least 20 h_{max}. h_{max}: the maximum indentation depth
Approach speed of the indenter	/	For micro range indentations, the approach speed should not exceed 2 μm/s. Typical micro/nano range approach speeds are 10 nm/s to 20 nm/s or less during final approach	During the process of determining the contact zero point between the indenter and the piece, the indenter approach speed should not exceed 2 μm/s, and the typical approach speeds are 10 nm/s to 20 nm/s or less during final approach	The approach speed should not exceed 2 μm/s. Typical approach speeds are 10 nm/s to 20 nm/s or less during final approach

In this paper, the test methods of the latest standards ASTM E2546, GB/T 21838, GB/T 22458 and GB/T 25898 compared, as shown in Table 1.

Compared with the standard ASTM E2546, domestic standards not only specify the ambient temperature and relative humidity of test, but also specify the temperature change range and the thermal drift rate, which is very important. If the material is thermally stable without time dependence, the indentation depth measurement results can be corrected by thermal drift calibration technology. Wang Chunliang et al.[13] found that temperature fluctuations must be strictly controlled for materials with time dependence. Because the influence of thermal drift and creep cannot be distinguished, otherwise large errors will be brought to the indentation depth measurement. Both domestic and foreign standards emphasize that the instrument and the test piece shall be stabilized to the ambient temperature, and that the stability of the temperature is more important than the actual test temperature during the test.

For requirements of the test piece thickness, the standards ASTM E2546, GB/T 21838 and GB/T 22458 require the piece thickness to be at least 10 times the indentation depth, and when testing coatings, the coating thickness is considered as the test piece thickness, but the standard GB/T 25898-2010 has no such requirements. In order to prevent the test results from being affected by the test piece support, it is necessary to consider the relationship between the indentation depth and the piece thickness, which is recommended to supplement to the standard GB/T 25898-2010. In addition, for ultra-high hardness films ($\geqslant 40$ GPa) the indentation depth should not exceed 5% of the coating thickness to ensure that the matrix does not produce the plastic deformation[14].

As the inclination, roughness and cleanliness of the test surface have a great impact on the test results, domestic standards require that the test surface tilt is less than 1°, the surface roughness is less than 5% of the indentation depth, and the test surface is dry and clean, while the standard ASTM E2546 has no relevant requirements. Shen Lin et al.[15] found that the results of indentation hardness, modulus and force-depth curve are more dispersed with the rougher test surface in the actual test. For test pieces with the same surface roughness, the influence of surface roughness on the test results is smaller with the greater indentation depth, which verifies the great influence of surface roughness on the test results, thus it is recommended that the standard ASTM E2546 supplements this. In addition, all standards did not analyze and specify the bumps and depressions on the test surface, but Zhang Zhongli et al.[16] found that the surface concave defects would cause the size effect of thin film materials, which had a great impact on the yield load of thin films, therefore all standards should be supplemented at this point.

In addition to the standard GB/T 25898, the other standards have no specific provisions for loading, holding and unloading time, but these standards clearly specify the need to record the time of each stage in the test report. The standard GB/T 25898 clearly points out that the loading, holding and unloading time should be set to 30 s in the test cycle based on a single stiffness measurement method, and the load holding time should be set to 10-30 s in the test cycle based on a continuous stiffness measurement method. In fact, Chen Yajun et al.[17] conducted nanoindentation tests on the aluminum coating on the surface of TC4 titanium alloy and studied the effects of loading and holding time on the mechanical behavior of coatings, and they found that different loading and holding times would have a certain impact on the test results.

For the test process, the standards ASTM E2546 and GB/T 21838 both require that there should be no vibration that seriously affects the test results during the load application, while the standard GB/T 22458 requires no vibration and the standard GB/T 25898 does not specify this. At this point, the requirement of GB/T 22458 is too strict and idealized, and the provisions of ASTM E2546 and GB/T 21838 are more reasonable. In addition, if the indenter moves too fast during the test, the indenter may cause large vibration or damage during contacting with the test surface, which will affect the test accuracy. The standard ASTM E2546 does not explain the indenter approach speed, while all domestic standards specify the maximum speed of the indenter during the entire test process and the approach speed of the final approach. This regulation is helpful to improve the accuracy of test results.

For the test process, the standards ASTM E2546 and GB/T 21838 both require that there should be no vibration that seriously affects the test results during the load application, while the standard GB/T 22458 requires no vibration and the standard GB/T 25898 does not specify this. At this point, the requirement of GB/T 22458 is too strict and idealized, and the provisions of ASTM E2546 and GB/T 21838 are more reasonable. In addition, if the indenter moves too fast during the test, the indenter may cause large vibration or damage during contacting with the test surface, which will affect the test accuracy. The standard ASTM E2546 does not explain the indenter approach speed, while all domestic standards specify the maximum speed of the indenter during the entire test process and the approach speed of the final approach. This regulation is helpful to improve the accuracy of test results.

2.3 Verification and calibration of test machines

In order to ensure the accuracy and usability of the test results, it is necessary to verify and calibrate test machines. The methods for verification and calibration of test machines are a direct verification method and an indirect verification method. The standards ASTM E2546-15, GB/T 21838.2-2022 and GB/T 22458-2008 are compared in terms of the direct verification method, as shown in Table 2.

For methods of the force and displacement verification, both the standard GB/T 21838.2-2022 and GB/T 22458-2008 require the indentation number. Because if the indentation number in the calibration test is too small, the dispersion degree of test results may be large. The standard GB/T 21838.2-2022 also rigorously requires a force calibration device for tests in nano range, which is several times more accurate than GB/T 22458-2008 and ASTM E2546-15. As the accuracy of the calibration device is the basis for the accuracy of the entire instrument, thus the high requirements of GB/T 21838.2-2022 should be followed at this point. For the allowable error of calibration, the standard GB/T 21838.2-2022 is also more stringent, but the error of ±1% is difficult to achieve with the maximum test force F_{max} less than 2 μN. The requirements of ASTM E2546-15 are relatively more reasonable, which should be explained in the test report. In addition, the standard GB/T 21838.2-2022 does not mention the time verification, and there will be a large deviation in the determination of time-related parameters when the time is not recorded accurately in the test. Therefore, it is necessary for GB/T 22458-2008 and ASTM E2546-15 to require that the difference between the time reported by the test equipment and that measured by an independent timing device must be less than 1 s.

Table 2 Verification and calibration of test machines-the direct verification

Contrasted item	ASTM E2546-15	GB/T 21838.2-2022	GB/T 22458-2008
Method of force verification	/	A minimum of 16 evenly distributed points in the test force range shall be calibrated	A minimum of 16 evenly distributed points in the test force range shall be calibrated
Device of force verification	The device used to verify forces shall be accurate to within 0.25% or 1 μN of each verification force, whichever is greater	Electronic balance with a suitable accuracy of 0.1% of the maximum verification force or 0.1 μN for the nano range	The device used to verify forces shall be accurate to within 0.25% or 1 μN of each verification force, whichever is greater
Tolerance of test forces	Every measurement shall be within 1% or 2 μN of its nominal value, whichever is greater. When the 2 μN tolerance is used, the F_{max} shall be accurate to within 5% of the stated value		
Method of displacement verification	At least ten verification lengths shall be chosen so as to span evenly the defined displacement range	The device shall be calibrated at a minimum of 16 points in each direction evenly distributed throughout its travel	The device shall be calibrated at a minimum of 16 points in each direction evenly distributed throughout its travel
Time verification	The time required for a test segment at least 10 s in duration shall be verified by an independent timing device. The difference between the time reported by the test equipment and that measured by an independent timing device must be less than 1 s	/	The time required for a test segment at least 10 s in duration shall be verified by an independent timing device. The difference between the time reported by the test equipment and that measured by an independent timing device must be less than 1 s

Both domestic and foreign standards stipulate that an indirect verification is required after a direct verification, and the indirect verification of test machines is to verify the overall performance of test machines. All standards have basically the same requirements for indirect verification methods, but the standard ASTM E2546-15 has more stringent requirements for the indirect verification cycle. In addition, for the evaluation of indirect verification results, the standard ASTM E2546-15 is relatively simple, which requires only the matching degree of the indentation modulus value reported by the test machine and the nominal Young's modulus, while the standards GB/T 21838.2-2022 and GB/T 22458-2008 are relatively complex and accurate, which need to analyze the repeatability of test machines through the coefficient of variation and analyze whether its performance meets the requirements through the error of test machines, and these analyses are greatly helpful to improve the accuracy of the test machine.

2.4　Calibration of reference blocks

Table 3　Requirements of reference blocks

Compared items	ASTM E2546-15	GB/T 21838.3-2022	GB/T 22458-2008
Material selection	Materials for reference blocks should have the follow characteristics: The accuracy of both dynamic Young's modulus and Poisson's ratio for reference blocks is better than 1.0%; a well-known, uniform composition; an amorphous or single crystal structure or known grain size distribution; isotropic elastic properties; a chemically stable surface; a melting or glass transition temperature well above room temperature; little or no pile-up of material about the perimeter of the indentation site	Materials for reference blocks should have the follow characteristics: values of dynamic Young's modulus and Poisson's ratio, each determined to an accuracy better than 1.0%; nonmagnetic	Materials for reference blocks should have the follow characteristics: uniform composition and properties; a chemically stable surface; little or no pile-up of material about the perimeter of the indentation site; no obvious time correlation; nonmagnetic
Thickness	/	For the nano range: ≥2 mm. For the micro range: ≥5 mm. For the macro range: ≥16 mm	≥2 mm of 20 times the depth, whichever is greater
Test surface orientation	For blocks placed with their bottom surface on a specimen mounting plate, the parallelism between top and bottom surfaces shall be within 0.5 degrees. For blocks mounted by their sides, the side surfaces shall be perpendicular to the test surface to within 0.5 degrees	If the reference block is mounted on its bottom, this condition is valid if the maximum deviation in flatness of the test and support faces does not exceed 5 μm in 50 mm and the maximum error in parallelism does not exceed 10 μm in 50 mm	For blocks placed with their bottom surface on a specimen mounting plate, the parallelism between top and bottom surfaces shall be within 0.5 degrees. For blocks mounted by their sides, the side surfaces shall be perpendicular to the test surface to within 0.5 degrees
Test surface finish	The test surface should be as smooth as possible. R_a is required to be no more than 10 nm for many applications measured over a 10 μm trace. Blocks intended specifically for very-low-force verification will require lower roughness levels	The test surface shall be free from scratches that interfere with the measurement of the indentations. Indentations between scratches are permitted. For the macro and micro range, R_a shall not exceed 50 nm and 10 nm respectively for the test surface. For the nano range, R_a shall not exceed 10 nm. It is recommended that the R_a be less than 1 nm to be practical use for calibration purpose	The test surface should be as smooth as possible. R_a is required to be no more than 10nm for many applications measured over a 10 μm trace. Blocks intended specifically for very-low-force verification will require lower roughness levels

Continue the Table 3

Compared items	ASTM E2546-15	GB/T 21838.3-2022	GB/T 22458-2008
Uniformity	The elastic properties of the block at its test surface do not deviate from the bulk values by more than 5%	For the macro range, the maximum permissible coefficient of variation for the indirect verification purpose for HM, H_{IT} and E_{IT} is 2%. For the micro range, the maximum permissible coefficient of variation for the indirect verification purpose for E_{IT} is 2%. For the nano range, the maximum permissible coefficient of variation for the indirect verification purpose for E_{IT} is 5%	Ensure the performance uniformity of each part of the bulk material, and the deviation between the measured value of each part and the average value of all parts should be within 5%

The requirements of reference blocks for the standards ASTM E2546-15, GB/T 21838.3-2022, and GB/T 22458-2008 are compared, as shown in Table 3.

In Table 3, it can be seen that the standard GB/T 21838.3-2022 pays more attention to the calibration procedure of reference blocks, and the requirements involve the test machine, the indenter verification, the calibration method, the indentation numbers, etc., which are more detailed on the basis of GB/T 21838.1-2019 for some test machines. The standards ASTM E2546-15 and GB/T 22458-2008 focus more on the qualities and requirements of reference blocks, such as the material selection, installation requirements and test surfaces. For the material and manufacture of reference blocks, the standard ASTM E2546-15 does not mention material magnetism, the standard GB/T 21838.3-2022 does not mention bumps and depressions around the indentation, and the standard GB/T 22458-2008 requires that materials have no significant temporal correlation, while these three points have a great impact on the accuracy of calibration results. Therefore, it is suggested that all standards complement each other in this respect.

For the validity of reference blocks, the standard GB/T 21838.3-2022 stipulates that the validity period of the calibration of reference blocks is 5 years, and the calibration value is only valid for the test conditions at the time of calibration. In addition, Zhou Liang et al.[18] studied the size effect of indentation hardness in micro-nano scale tests and found that the plastic deformation of materials would produce the size effect of nanoindentation hardness through the dislocation mechanism. For tests in the micro-nano range, this effect had a highly significant impact on the hardness measurement, and only the hardness values of the same indentation size could be compared. Therefore, the standard GB/T 21838.3-2022 specifies that only the indentation modulus can be used as a material parameter for the calibration and verification of the flexibility and the indenter area function in the micro-nano range. The standard GB/T 22458-2008 only requires that the stability of reference blocks should be checked regularly. Thus both the standards ASTM E2546-15 and GB/T 22458-2008 should supplement the validity of reference blocks by reference to the requirements of GB/T 21838.3-2022.

3 Conclusions

Through the comparison and analysis of domestic and foreign instrumented indentation test standards, the following conclusions are obtained.

(1) For the scope of application, the standards GB/T 21838 and ASTM E2456 stipulate a wide range of test forces and depths from the conventional range to the nano range, with GB/T 21838 having specific requirements for indentation tests of micro-nano coatings, while the standards GB/T 22458 and GB/T 25898 are mainly suitable for nanoscale tests, and their regulations are more targeted.

(2) For the test method, the standard ASTM E2456 has relatively simple requirements for the test environment, test pieces, test surfaces and the test operation, and ignores some requirements for factors that have a greater impact on test results, while the domestic standards have more detailed requirements for the

above aspects, which is conducive to quantitative practical operation.

(3) For the test method, the standard ASTM E2456 has relatively simple requirements for the test environment, test pieces, test surfaces and the test operation, and ignores some requirements for factors that have a greater impact on test results, while the domestic standards have more detailed requirements for the above aspects, which is conducive to quantitative practical operation.

(4) For the calibration of reference blocks, the standard GB/T 21838 pays more attention to the calibration procedure of reference blocks, while the standards ASTM E2456 and GB/T 22458 pay more attention to the nature and requirements of reference blocks.

In summary, compared with the foreign indentation test standard, the domestic standards have more detailed requirements for test methods, calibration and verification of test machines and calibration of reference blocks, which are more conducive to practical operation. These standards provide a strong basis and support for the standardization of instrumented indentation tests in China.

Acknowledgments: This work was supported by The Self-Funded Technology Investment Project of China United Test & Certification Co., Ltd.

References

[1] Metallic materials-instrumented indentation test for hardness and materials parameters-Part 1: Test method, ISO 14577-1: 2015[S]. 2015.

[2] Metallic materials-instrumented indentation test for hardness and materials parameters-Part 2: Verification and calibration of testing machines, ISO 14577-2: 2015, 2015.

[3] Metallic materials-instrumented indentation test for hardness and materials parameters-Part 3: Calibration of reference blocks, ISO 14577-3: 2015, 2015.

[4] Metallic materials-instrumented indentation test for hardness and materials parameters-Part 4: Test method for metallic and non-metallic coatings, ISO 14577-4: 2016, 2016.

[5] Standard practice for instrumented indentation testing, ASTM E2546-15, 2015.

[6] General rules of instrumented nanoindentation test, GB/T 22458-2008, 2008.

[7] Metallic materials-instrumented indentation test for hardness and materials parameters-Part 1: Test method, GB/T 21838.1-2019, 2019.

[8] Metallic materials-instrumented indentation test for hardness and materials parameters-Part 2: Verification and calibration of testing machines, GB/T 21838.2-2022, 2022.

[9] Metallic materials-instrumented indentation test for hardness and materials parameters-Part 3: Calibration of reference blocks, GB/T 21838.3-2022, 2022.

[10] Metallic materials-instrumented indentation test for hardness and materials parameters-Part 4: Test method for metallic and non-metallic coatings, GB/T 21838.4-2020, 2020.

[11] Instrumented nanoindentation test-indentation hardness and modulus of thin film, GB/T 25898-2010, 2010.

[12] OLIVER W, PHARR G M J. An improved technique for determining hardness and elastic modulus using load and displacement sensing indentation experiments[J]. Materials Research Society, 1992, 7(6), 1564-1583.

[13] WANG C L. Study on the methods of nanoindentation testing[J]. Shanghai Research Institute of Materials, 2007.

[14] DU J, ZHANG P, ZHAO J J, et al. Comparison of micro-hardness with nano-indentation hardness in hardness measuring for micron level hard coating[J]. Physical Testing and Chemical Analysis Part A: Physical Testing, 2008, 44(4), 189-192.

[15] LIN S. Research on the influences of surface roughness on the nanoindentation results of sample material[D]. Jilin: Jilin University, 2012.

[16] ZHANG Z L. Multi-scale study of nanoindentation based on surface pit defects[D]. Shanghai: Fudan University, 2012.

[17] CHEN Y J, YU J Q, WANG F S, et al. Comparative study on instrumented nano indentation test standards[J]. China Measurement & Test, 2018, 44(1): 9-15.

[18] ZHOU L, YAO Y X. Research development of hardness indentation size effect at micro/nano scale[J]. Journal of Harbin Institute of Technology, 2008, 40(4): 597-602.

Evolution of S(Al$_2$CuMg) Phase During Fabrication Process and Its Influence on Mechanical Property in a Commercial Al-Zn-Mg-Cu Alloy

Kai WEN[1,2,3,*], Hongwei YAN[1,2,3], Lizhen YAN[1,2,3], Hongwei LIU[1,2,3],
Wei XIAO[1,2,3], Ying LI[1,2,3], Guanjun GAO[1,2,3], Rui LIU[1,2,3], Weicai REN[1,3,4]

1. State Key Laboratory of Non-ferrous Metals and Processes, GRINM Group Co., Ltd., Beijing, 100088, China
2. GRIMAT Engineering Institute Co., Ltd., Beijing, 101407, China
3. General Research Institute for Nonferrous Metals, Beijing, 100088, China
4. Northeast Light Alloy Co., Ltd., Harbin, 150060, China

Abstract: In the present study, the evolution of S(Al$_2$CuMg) phase in a commercial Al-Zn-Mg-Cu alloy during homogenization, deformation and solution treatments was investigated and the effect of residual S phase on mechanical properties in finished plates was revealed. The results showed that S phase formed during solidifications and existed in eutectic structure. Parts of AlZnMgCu phase transformed to S phase after the first homogenization treatment with a regime of 470 ℃/24 h. During the second homogenization treatment at 480 ℃, most of the S phase had gradually dissolved into the matrix. However, a small amount of large-sized S phase remained even after extending the second stage homogenization time to 48 h. During deformation these remaining S phase was broken, resulting in size reduction and apparent distribution along the rolling direction, which could effectively dissolved into the matrix by a two-stage solution treatment with a regime of 470 ℃/1.5 h + 480 ℃/2 h. Comparing the microstructure and properties of final plates treated by homogenization processes, it was evident that the alloy subjected to a single-stage homogenization treatment exhibited a significantly higher amount of residual S phase, which directly compromised fracture toughness but had a relatively minor impact on strength.

Keywords: S(Al$_2$CuMg) phase; Microstructure; Property; Al-Zn-Mg-Cu alloy

1 Introduction

Since the aging hardening phenomenon is accidentally found by Alfred Wilm, the initial Al-Zn-Mg-Cu alloy is furtherly developed by trial and error, and a series of new alloys are used in aerospace and military fields due to good match of high strength, preferential fracture toughness, satisfying fatigue crack propagation property and stress corrosion cracking resistance[1]. Considering the development of Al-Zn-Mg-Cu alloys, the control of phase dissolution and precipitation is a key point for final performance[2]. Thereinto, a favorable implement of the former one makes a good foundation for the latter one, which raises enough demand to investigate phase dissolution during fabrication processes[3,4]. In as-cast microstructure of Al-Zn-Mg-Cu alloys, typical eutectic phases include Mg(Zn, Cu, Al)$_2$ phase, MgZn$_2$ phase, S(Al$_2$CuMg) phase and θ(Al$_2$Cu) phase[5]. It needs to be pointed out that the former two usually exist in all alloys while part or none of the latter two can be detected, which are close to chemical compositions. Thereinto, S(Al$_2$CuMg) phase is a unique one, which can be formed not only in solidification processes, but also by phase transformation from Mg(Zn, Cu, Al)$_2$ phase[6]. As a dissolvable phase with relative high melting temperature, the dissolution of it is harder than Mg(Zn, Cu, Al)$_2$ phase, which means higher heat treatment temperature and longer preserving time. Besides, S(Al$_2$CuMg) phase is not coherent with the matrix, which may serve as crack initiation source due to the deformation inconformity with the matrix and furtherly damage dynamic property. Hence, the dissolution of S(Al$_2$CuMg) phase is quite important for Al-Zn-Mg-Cu alloys.

Previous literatures have mentioned the dissolution of S(Al$_2$CuMg) phase during homogenization. Fan[7] investigated the microstructure evolution during homogenization of an Al-6.31Zn-2.33Mg-1.7Cu alloy, who proposed the formation of S(Al$_2$CuMg) phase and the phase transformation of primary particle from Mg(Zn, Cu, Al)$_2$ phase to S(Al$_2$CuMg) phase. He[8] also found the phase transformation in 7085 alloy and asserted that the amount of S(Al$_2$CuMg) phase decreased by a higher homogenization treatment with a regime of 480 ℃/8 h. Similarly, Deng[9] raised a homogenization regime of 400 ℃/10 h+470 ℃/24 h+485 ℃/4 h to diminish S(Al$_2$CuMg) phase in 7050 alloy, so do the investigations of Liu[10] and Li[11]. Though their investigations are detailed and valuable, however, in industrial conditions, the dissolution of S(Al$_2$CuMg) phase should be considered under a whole fabrication process, and the influence of remaining ones on final property for commercial alloy products lacks enough analysis.

In present study, a commercial Al-Zn-Mg-Cu alloy is employed and the evolution of S(Al$_2$CuMg) phase during homogenization, deformation and solution treatments is thoroughly investigated and the influence of residual S(Al$_2$CuMg) phase on mechanical property is analyzed and discussed.

2 Experimental procedure

A commercial Al-Zn-Mg-Cu alloy fabricated under industrial conditions was used in present study. The chemical composition of it was Al-6.5Zn-2.4Mg-2.2Cu-0.12Zr-0.05Fe-0.03Si (wt. %). The size of initial ingots was 420 mm×1320 mm in cross section while that for rolled plates was 76 mm×1210 mm. Samples for as-cast microstructure observation and homogenization investigations were cut from the one quarter of thickness at the width center. Similarly, samples for deformation microstructure observation and solution treatment investigations were from the center of thickness within a width range from a quarter to half of the plate width. The heat treatments were carried out in an industrial induction heating furnace with a temperature range of 3 ℃. A single stage homogenization regime of 470 ℃/24 h and a double stage one of 470 ℃/24 h+480 ℃/24 h were performed on the ingots and the homogenized ingots were surface-milled and then rolled to plates by multi-passes. Metallographic structure was observed by a Zeiss Axiovert 200MAT optical microscope, phase types were identified by a JEOL JSM-7900 field emission scanning electron microscope. Differential scanning calorimetry (DSC) analysis were performed on a NETZSCH DSC 404F3 machine. The area fraction of residual S(Al$_2$CuMg) phase was quantitatively calculated with Image Plus Pro (IPP) software on the basis of SEM images. Samples for mechanical property tests were also cut from the center of thickness within a width range from a quarter to half of the plate width, tensile property along longitudinal and long transverse directions was measured while fracture toughness along L-T and T-L directions was measured. Tensile strength was measured on a CMT4304 tensile test machine attached by an extensometer with a max loading of 3 t. The measurement processes met the standard of GB/T 228.1-2010 and the tensile speed was 2 mm/min. The fracture toughness was measured on an MTS 370.10 hydraulic servo fatigue testing machine following a standard of GB/T 4161-2007.

3 Results and discussion

3.1 Evolution of S(Al$_2$CuMg) phase in homogenization-rolling deformation-solid solution heat treatment

(1) Second phase analysis in as-cast microstructure

Figure 1 shows the as-cast microstructure of the alloy ingot, revealing a typical dendritic structure [Figures 1(a) and (c)] which predominantly distributed at grain boundaries. Within the grains, there are fine and densely dispersed MgZn$_2$ phases [circled in Figures 1(b) and (d)]. The dendritic structure consists of two distinct phases with noticeable contrast in brightness. Comparing OM and SEM images, it can be observed that the prominent brownish-yellow phase in OM corresponds to the bright white phase in SEM, while the less abundant brown phase in OM corresponds to the dark gray phase in SEM. Table 1 presents the energy-dispersive X-ray spectroscopy (EDS) analysis of the second phase in the as-cast microstructure, revealing that the dendritic structure mainly consists of Mg(Zn, Cu, Al)$_2$ phase, with a relatively small amount of the S(Al$_2$CuMg) phase.

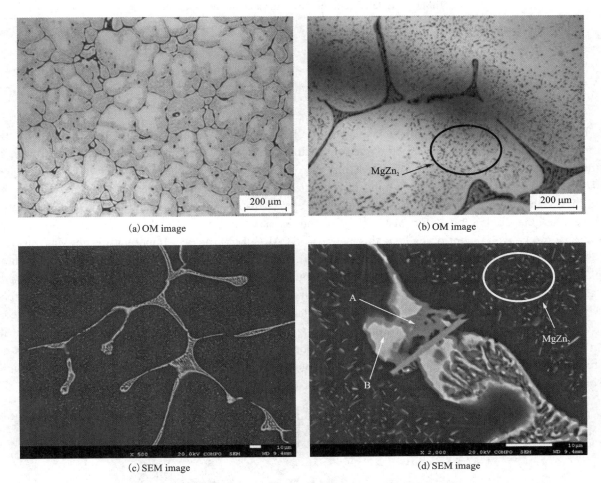

Figure 1　Microstructure images of as-cast Al-Zn-Mg-Cu alloy

Table 1　Chemical composition of second phase of the as-cast alloy (at%)

Label	Mg	Al	Cu	Zn	Phase type
A	22.28	53.98	21.59	2.15	S(Al_2CuMg) phase
B	32.69	19.83	14.15	33.33	Mg(Zn, Cu, Al)$_2$ phase

Figure 2 illustrates the corresponding DSC curve while Table 2 provides the corresponding statistical results. It can be observed that the lower-melting quaternary phases dominate the entire melting absorption peak, with an initial melting temperature of approximately 480 ℃ and a peak melting temperature of approximately 482 ℃. On the other hand, the higher-melting S(Al_2CuMg) phase exhibits a significantly smaller peak area compared to the Mg(Zn, Cu, Al)$_2$ phase, with a peak melting temperature of approximately 489 ℃. This further supports the observations made through microstructural analysis.

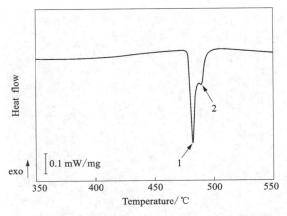

Figure 2　DSC curve of as-cast Al-Zn-Mg-Cu alloy

Table 2 Related data of DSC curve of the as-cast alloy

Initial melting temperature/℃	Peak melting temperature/℃		Endothermic peak enthalpy/(J/g)
479.0	1	482.3	18.02
	2	489.4	

(2) Second phase analysis after homogenization treatments

Figure 3 presents the microstructure images of the alloy after being subjected to a homogenization treatment at 470 ℃ for 24 h. It can be observed that the $MgZn_2$ phases within the grain boundaries in the as-cast microstructure have completely dissolved, and the reticular eutectic structures at the grain boundaries have partially dissolved, resulting in rounded and disconnected edges. The remaining second phase appears brown color in OM and dark gray color in SEM, and the energy-dispersive X-ray spectroscopy (EDS) analysis indicates that the compositions for Mg, Al, Cu and Zn are 22.55%, 53.45%, 21.83% and 2.17% (in at%), respectively, confirming that this phase is $S(Al_2CuMg)$ phase. Importantly, no $Mg(Zn, Cu, Al)_2$ phase is found in the homogenized microstructure, indicating the complete dissolution of it. Figure 4 displays the DSC curve of the homogenized microstructure. It can be observed that only one endothermic peak corresponding to $S(Al_2CuMg)$ phase is presented in the homogenized microstructure, confirming the absence of $Mg(Zn, Cu, Al)_2$ phase. Besides, the onset and peak melting temperatures of the homogenized $S(Al_2CuMg)$ phase exhibit a slight rightward shift compared to the as-cast microstructure. Table 3 indicates that the peak area of the heat absorption for the alloy approaches 5 J/g, which is larger than that of the as-cast microstructure. This is because during the primary homogenization treatment, apart from the direct dissolution of some $Mg(Zn, Cu, Al)_2$ phases, another portion of the phase is transformed to $S(Al_2CuMg)$ phase.

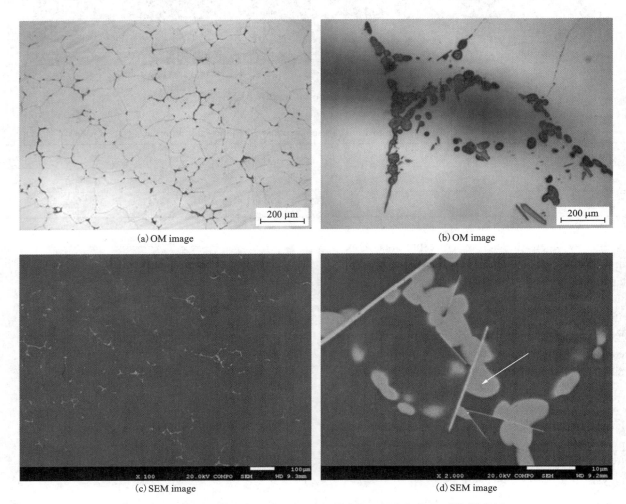

Figure 3 Microstructure images of the alloy homogenized by 470 ℃/24 h

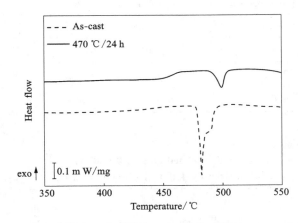

Figure 4 DSC curve of the alloy homogenized by 470 ℃/24 h

Table 3 Related data of the alloy homogenized by 470 ℃/24 h

Initial melting temperature/℃	Peak melting temperature/℃	Endothermic peak enthalpy/(J·g^{-1})
491.5	499.2	4.97

Figure 5 illustrates OM and SEM images of the alloy after undergoing a two-stage homogenization treatment at 470 ℃/24 h and 480 ℃ for various durations (x = 12, 24, 48 h). Table 4 presents the corresponding EDS analysis of remaining second phase. It can be observed that when the second stage homogenization time is 12 h, there is a significant presence of coarse S(Al$_2$CuMg) phase in the alloy. When the time extends to 24 h, the quantity of coarse S(Al$_2$CuMg) phase decreases, indicating that prolonging second stage homogenization time can effectively promote the dissolution of S(Al$_2$CuMg) phase. However, even with an extending of 48 h, there are still localized regions with a small amount of aggregated coarse S(Al$_2$CuMg) phase, suggesting that S(Al$_2$CuMg) phase can partially dissolve during homogenized at 480 ℃, but extending the time to 48 h does not guarantee a complete dissolution.

Figure 6 presents corresponding DSC curves. It can be observed that the sample treated for 48 h still exhibits a small endothermic peak, indicating the presence of undissolved S(Al$_2$CuMg) phase. Table 5 provides related initial melting temperature, peak melting temperature and enthalpy values. It can be observed that these values align with the microstructural observations, confirming the presence of coarse S(Al$_2$CuMg) phase even after an extended time of 48 h.

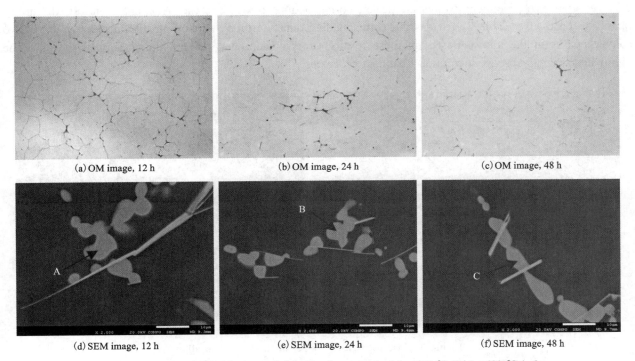

Figure 5 OM and SEM images of the alloy homogenized by 470 ℃/24 h+480 ℃/x h

Table 4 Chemical composition of the second phase of the alloy homogenized by 470 ℃/24 h+480 ℃/x h (at%)

Label	Mg	Al	Cu	Zn	Phase type
A	23.52	52.56	21.83	2.10	S(Al$_2$CuMg) phase
B	21.61	55.91	20.30	2.17	S(Al$_2$CuMg) phase
C	22.40	54.40	21.18	2.02	S(Al$_2$CuMg) phase

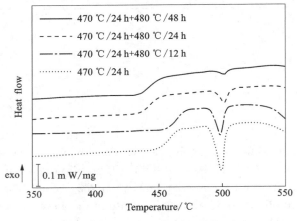

Figure 6 DSC curve of the alloy homogenized by 470 ℃/24 h+480 ℃/x h

Table 5 Related data of the alloy homogenized by 470 ℃/24 h+480 ℃/x h

Second stage homogenization time/h	Initial melting temperature /℃	Peak melting temperature /℃	Endothermic peak enthalpy /(J·g^{-1})
12	491.7	498.2	2.699
24	494.0	501.7	1.429
48	495.2	501.8	0.388

(3) Second phase analysis of hot-rolled microstructure

After the deformation process, the microstructure of the hot-rolled alloy was analyzed. OM and SEM microstructure images of the alloy at long-transverse plane are shown in Figure 7 and Table 6 presents the corresponding EDS results for the second phase. It can be observed that the coarse S(Al$_2$CuMg) phase remaining from the homogenization process undergoes fragmentation during deformation and disperses in a streamline pattern along the rolling direction. Furthermore, the second phase exhibits smaller dimensions and densely distributed characteristics. Figure 8 displays corresponding DSC curve. It can be observed that the curve exhibits two prominent exothermic peaks, corresponding to Mg(Zn, Cu, Al)$_2$ phase and S(Al$_2$CuMg) phase, respectively. Table 7 provides the corresponding data, furtherly supporting the presence of these two soluble phases in the alloy. Based on these results, it is evident that the remaining S phase needs to be eliminated during subsequent solution treatment processes.

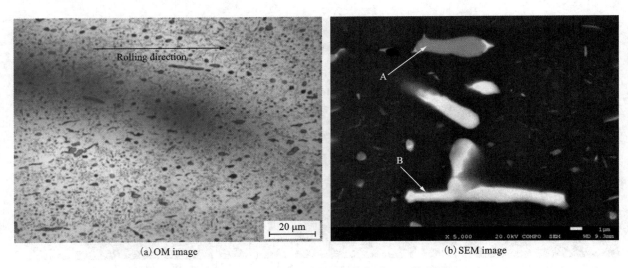

Figure 7 Microstructure images of the hot-rolled alloy

Table 6 Chemical composition of the second phase of the hot-rolled alloy (at%)

Label	Mg	Al	Cu	Zn	Phase type
A	17.91	64.67	14.88	2.55	S(Al_2CuMg) phase
B	19.26	53.42	8.66	18.66	Mg(Zn, Cu, Al)$_2$ phase

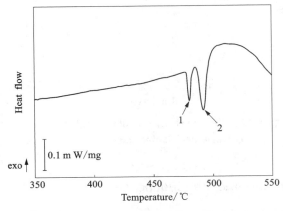

Table 7 Related data of DSC curve of the hot-rolled alloy

Initial melting temperature/℃	Peak melting temperature/℃		Endothermic peak enthalpy/(J·g^{-1})
478.2	1	480.4	0.813
	2	492.1	2.974

Figure 8 DSC curve of the hot-rolled alloy

(4) Second phase analysis after solution treatments

Figure 9 shows SEM images of the alloy plate after solution treatment and Table 8 presents EDS analysis of the second phase. It can be observed that after the solution treatment at 470 ℃ for 1.5 h, Mg(Zn, Cu, Al)$_2$ phase in the hot-rolled microstructure has completely dissolved into the matrix, leaving only the S phase. Furthermore, adding a higher temperature at 480 ℃ for 2 h, it is found that the S phase completely diminished, leaving behind only the Fe-rich phase. The DSC curves for both cases are shown in Figure 10. Comparing the DSC curves of single and double stage solution treatment, it can be observed that the curve for double stage solution treatment is smooth and devoid of any exothermic peaks. This indicates the complete dissolution of the S phase can be obtained by a double stage solution treatment.

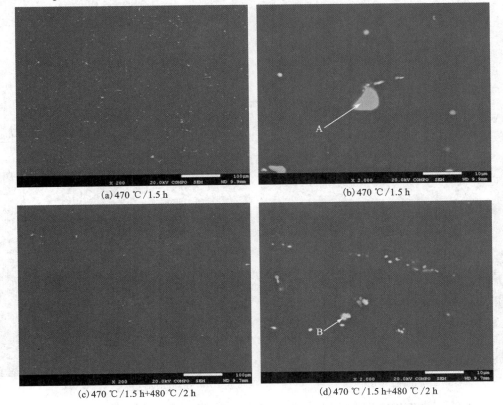

Figure 9 SEM images of the alloy solution treated by single and double stage regimes

Table 8 Chemical composition of the second phase of the solution treated alloy

Label	Mg	Al	Cu	Zn	Fe	Phase type
A	24.26	51.02	22.76	1.96	-	S(Al_2CuMg) phase
B	1.44	78.36	6.03	13.26	0.91	Fe-rich phase

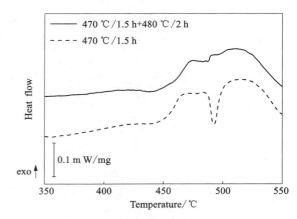

Figure 10 DSC curves of the alloy solution treated by single and double stage regimes

3.2 Effect of residual S(Al_2CuMg) phase on strength and fracture toughness

Figure 11 shows the microstructure images of the finished plate for the alloy with single-stage (470 ℃/24 h) and double-stage (470 ℃/24 h+480 ℃/24 h) homogenization heat treatments. It can be observed that the single-stage homogenization treatment results in a microstructure with a higher amount of residual second phase compared to the microstructure after the double-stage homogenization treatment. Additionally, the residual second phase exhibits a streamline distribution along the rolling direction. EDS analysis in Table 9 confirms the presence of residual S phase in the microstructure after the single-stage homogenization treatment, while the microstructure after the double-stage homogenization treatment mainly consists of noticeable Fe-rich phase.

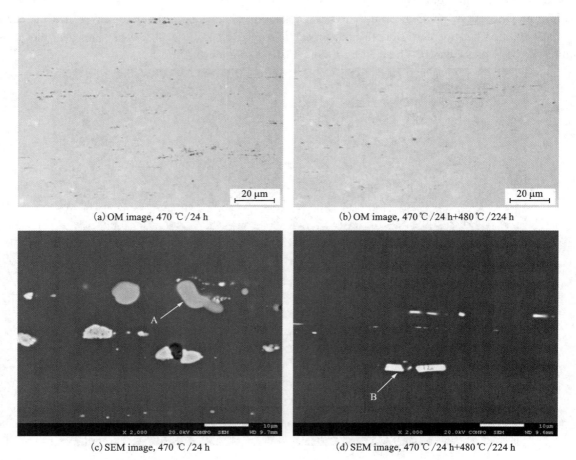

(a) OM image, 470 ℃/24 h   (b) OM image, 470 ℃/24 h+480 ℃/224 h

(c) SEM image, 470 ℃/24 h   (d) SEM image, 470 ℃/24 h+480 ℃/224 h

Figure 11 Microstructure images of the finished plate with different homogenization regimes

Table 9 Chemical composition of second phase in the finished plate (at%)

Label	Mg	Al	Cu	Zn	Fe	Phase type
A	24.28	50.96	22.84	1.92	-	S(Al$_2$CuMg) phase
B	1.44	78.36	13.26	0.91	6.03	Fe-rich phase

The area fractions of the residual S(Al$_2$CuMg) phase in the aforementioned finished plate were quantitatively analyzed, the values for single (470 ℃/24 h) and double stage (470 ℃/24 h+480 ℃/24 h) homogenization ones were 0.11‰ and 0.04‰, respectively, which indicates that the residual S phase in the finished sheet after single stage homogenization treatment is higher than that for the double stage one. Related tensile strength and fracture toughness along longitudinal and long-transverse directions are shown in Figure 12. The strength for both does not differ significantly, with a difference of approximately 10 MPa. However, the alloy with a higher fraction of residual S(Al$_2$CuMg) phase resulting from the single-stage homogenization treatment exhibits lower fracture toughness values in the L-T and T-L orientations, with reductions of 7.5 MPa·m$^{1/2}$ and 5 MPa·m$^{1/2}$, respectively, compared to the finished plate with double-stage homogenization treatment.

It can be concluded that the presence of residual S phase in the alloy plate has minimal influence on strength but significantly reduces fracture toughness. Therefore, it is important to achieve complete dissolution of the S phase during the heat treatment process to minimize its impact on the fracture toughness of the alloy.

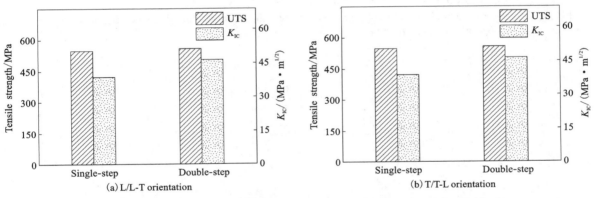

Figure 12 Comparison of tensile strength and fracture toughness for finished plate products with single and double stage homogenization treatments

4 Conclusion

In the present study, as-cast microstructure of a commercial Al-Zn-Mg-Cu alloy fabricated under industrial conditions was investigated, the evolution of S(Al$_2$CuMg) phase after homogenization, hot-rolling and solution treatment was detailedly observed and mechanical property of finished plate with different homogenization regimes was evaluated. S(Al$_2$CuMg) phase initially formed during the solidification process, after the first-stage homogenization treatment (470 ℃/24 h), a portion of the Mg(Zn, Cu, Al)$_2$ phase transformed into the S(Al$_2$CuMg) phase. Through the second-stage homogenization treatment (480 ℃/x h), most of the S(Al$_2$CuMg) phase gradually dissolved into the matrix. However, even after extending the second-stage time to 48 h, a small amount of S(Al$_2$CuMg) phase remained. The residual S(Al$_2$CuMg) phase underwent fragmentation and reduction in size after deformation, after a double-stage solution treatment (470 ℃/1.5 h+480 ℃/2 h), the residual S(Al$_2$CuMg) phase can be essentially diminished. The alloy plate subjected to single stage homogenization treatment possessed a significantly higher amount of residual S(Al$_2$CuMg) phase, which directly compromised fracture toughness while had a minor impact on its strength.

References

[1] AZARNIYA A, TAHERI A K, TAHERI K K. Recent advances in ageing of 7XXX series aluminum alloys: A physical metallurgy perspective[J]. J. Alloy. Compd., 2019, 11, 781: 945-983.

[2] ROMETSCH P A, ZHANG Y, KNIGHT S. Heat treatment of 7XXX series aluminium alloys — Some recent developments. T. Nonferr. Metal. Soc., 2014, 7, 24(7): 2003-2017.

[3] ZHOU B, LIU B, ZHANG S. The advancement of 7XXX series aluminum alloys for aircraft structures: A Review[J]. Metals., 2021, 4, 11(5): 718.

[4] DURSUN T, SOUTIS C. Recent developments in advanced aircraft aluminium alloys[J]. Mater. Des., 2014, 4, 56: 862-871.

[5] JARRY P, RAPPAZ M. Recent advances in the metallurgy of aluminium alloys. Part I: Solidification and casting[J]. CR Phys., 2018, 12, 19(8): 672-687.

[6] JIA P, CAO Y, GENG Y, et al. Studies on the microstructures and properties in phase transformation of homogenized 7050 alloy[J]. Mater. Sci. Eng. A, 2014, 8, 612: 335-342.

[7] FAN X, JIANG D, MENG Q, et al. The microstructural evolution of an Al-Zn-Mg-Cu alloy during homogenization[J]. Mater. Lett., 2006, 6, 60(12): 1475-1479.

[8] HE L, LI X, ZHU P, et al. Effects of high magnetic field on the evolutions of constituent phases in 7085 aluminum alloy during homogenization[J]. Mater. Charact., 2012, 9, 71: 19-23.

[9] DENG Y, YIN Z, CONG F. Intermetallic phase evolution of 7050 aluminum alloy during homogenization[J]. Intermetallics, 2012, 7, 26: 114-121.

[10] LIU Y, JIANG D, XIE W, et al. Solidification phases and their evolution during homogenization of a DC cast Al-8.35 Zn-2.5 Mg-2.25 Cu alloy[J]. Mater. Charact., 2014, 7, 93: 173-183.

[11] LI N, CUI J. Microstructural evolution of high strength 7B04 ingot during homogenization treatment[J]. T. Nonferr. Metal. Soc., 2008, 8, 18(4): 769-773.

Tuning the Mechanical and Corrosion Properties of Selective Laser Melted Al-Mg-Sc-Zr Alloy Through Heat Treatment

Jinglin SHI[1,2], Qiang HU[3,*], Xinming ZHAO[3], Yonghui WANG[3], Jinhui ZHANG[3], Yingjie LIU[3]

1　Beijing COMPO Advanced Technology Co. , Ltd, Beijing, 101407, China
2　Industrial Research Institute for Metal Powder Material, Beijing, 101407, China
3　GRINM Additive Manufacturing Technology Co. , Ltd, Beijing, 101407, China

Abstract: The current interest in Al-Mg-Sc-Zr alloy fabricated by selective laser melting (SLM) is centred around excellent mechanical property without taking into account its corrosion behavior. In this work, gas-atomized Al-Mg-Sc-Zr alloy powder was manufactured by SLM. The influence of heat treatment on microstructure, phase evolution, mechanical properties and electrochemical corrosion behaviors were investigated and microstructure-property relationship was established to obtain high-performance alloy. The result suggests that the sample heat treated at 350 ℃ for 4 h possesses superior mechanical properties, but is more susceptible to corrosion. When heat treatment temperature reaches 400 ℃, Oswald ripening of Al_3(Sc, Zr) and Al_6Mn particles declined the strength and the Al_6Mn phase is corroded as cathode instead of the α-Al matrix, thereby inhibiting the occurrence of pitting corrosion. The specimen heat treated at 300 ℃ for 4 h reconciled desired strength and high toughness as well as superior corrosion resistance.

Keywords: Selective laser melting; Al-Mg-Sc-Zr alloy; Heat treatment; High strength; Corrosion resistance

1　Introduction

Additive manufacturing (AM) is a state-of-the-art manufacturing technology, which has grown from the field of rapid prototyping. Benefiting from the layer-by-layer building, AM is suitable for the complex-shaped components and the customized products in small runs[1]. Selective laser melting (SLM), one of the mainstream AM processes, can effectively form dense metallic components by laser beam melting[2]. During the SLM process, a focused laser source moves on the basis of the CAD file data, making materials melt and metallurgical bond to fabricate the near net shaped parts[3]. Comparing with the traditional processing method, SLM have advantages in unlimited shape design for topology optimization, which attracts great attentions in the field of aerospace, marine and automotive.

Aluminum (Al) alloy is regarded as a better candidate for lightweight structural materials. However, most of the conventional high-strength Al alloys are not suitable for SLM process owing to undesirable columnar grains and periodic cracks[4,5]. Near-eutectic Al-Si alloys have been extensively produced by SLM, but mechanical properties of the annealed alloy hinder their applications[6,7]. In recent years, Scalmalloy © alloy system (Sc-and Zr-modified Al-Mg alloys) has promoted the development of heat treatable and SLM-processable Al alloys with maximal tensile strength surpassing 520 MPa[8,9,10]. The excellent mechanical properties not only come from the solid solution strengthening of Mg but also the grain boundary strengthening of fine grains and the precipitation strengthening of coherent Al_3(Sc, Zr)[11]. Meanwhile, Al-Mg alloy is the traditional antirust Al alloy, so the corrosion behavior of SLM-processed Al-Mg-Sc-Zr alloys has drawn much attention. Gu et al.[12] investigated the microstructures and corrosion behaviors on different planes, showing that the XY-plane had a better corrosion resistance than that of XZ-plane in 3.5 wt.% NaCl solution. Gu et al.[13] also compared the corrosion behaviors of SLM-fabricated with heat-treated (325 ℃ for 4 h) Al-Mg-Sc-Zr specimens. However, the effect of different heat treatments on the corrosion behavior has not been reported, which is essential to manipulate the mechanical and corrosion properties.

In our work, SLM-processed Al-Mg-Sc-Zr alloys were heat treated at different temperatures. The relationship between microstructure and properties was investigated by the electrical conductivity, microhardness, room-temperature tensile and electrochemical corrosion tests, so as to develop the high-strength and corrosion-resistant SLMed Al alloys by regulating heat treatment.

2 Experimental materials and methods

2.1 Raw material and specimen preparation

Gas-atomized powder raw material with chemical composition of Al-4.66Mg-0.54Mn-0.76Sc-0.37Zr (wt.%) was used for this study. Figure 1(a) shows the secondary electron image of the sieved powder. The illustration presents a cellular dendritic structure. Particle size distribution of the powder was analyzed by Beckman Coulter LS 13 320, with a mean particle diameter d_{50} = 35.28 μm [Figure 1(b)]. SLM process was performed on CASIC ASA-260M machine equipped with a 500 W fiber laser under argon atmosphere, and oxygen content was kept less than 0.5%. Cuboid samples sized of 10 mm×10 mm×8 mm and 10 mm×10 mm ×70 mm were deposited parallel to the building platform using stripe hatch strategy with 67° laser beam rotation between adjacent layers. Optimized working parameters are a laser power P = 380 W, a scan speed v = 1100 mm/s, layer thickness t = 0.03 mm, and a hatch spacing h = 0.1 mm. The volumetric energy densities E, according to the relationship $E = P \cdot (vht)^{-1}$, is 115.15 J/mm^3[4]. Heat treatment temperature ranged from 300 ℃ to 400 ℃ (with intervals of 25 ℃) was performed for 4 hours in air atmosphere, and samples then cooled in furnace to room temperature.

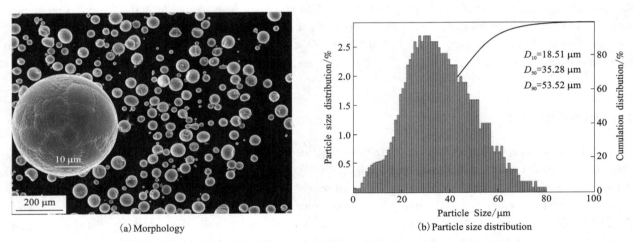

(a) Morphology (b) Particle size distribution

Figure 1 Characterization of the gas-atomized Al-Mg-Sc-Zr alloy powder

2.2 Characterization of properties

The Vickers microhardness was performed on a Buehler Wilson HV1150 hardness tester (America) with a load of 5 kg for 15 s. The electrical conductivity was measured at room temperature by a Sigmatest 2.069 eddy current instrument. Electrochemical tests containing the electrochemical impedance spectroscopy (EIS) and potentiodynamic polarization were conducted on a Princeton PARSTAT 3000 A electrochemical workstation. The tested surface of every sample was mechanically ground and polished. Corrosion process was operated in 3.5 wt.% NaCl solution at 25 ℃ using the conventional sealed traditional three-electrode system. A clean specimen surface, a platinum foil and a saturated calomel electrode (SCE) acted as the working electrode, the counter electrode and the reference electrode, respectively. EIS tests were carried out under a stable open circuit potential value and the frequency ranged from 100 kHz to 10 mHz with an amplitude of 10 mV. After EIS tests, potentiodynamic polarization curves were scanned from −1.7 V to −0.4 V with a scanning rate of 2 mV/s. All the electrochemical measurements were repeated three times to ensure data accuracy.

2.3 Microstructure observation

Phases composition were characterized by a Rigaku Smart Lab X-ray diffractometer with CuK$_\alpha$ radiation at 40 kV, 200 mA. Surface morphologies of specimens after potentiodynamic polarization tests were observed by a Zeiss Axio Vert. Al optical microscopy. Microstructure was examined by a Talos F200X emission transmission electron microscope (TEM). The TEM foils were mechanically ground down to about 50 μm, followed by punching to 3 mm discs and then the twin-jet polishing was conducted in an electrolyte solution (30% nitric acid and 70% methanol) cooled to -30 ℃ at a voltage of about 20 V.

3 Results and discussion

3.1 Microstructure features

Figure 2 depicts the HAADF-TEM images of heat-treated samples at 300 ℃ (HT300), 350 ℃ (HT350) and 400 ℃ (HT400) for 4 h. Figures 2(a)-(c) represented the typical known bi-modal grain size distribution[10]. The fine-grained (FG) region is attribute to the heterogeneous nucleation of high melting point phases and the coarse-grain (CG) region consists of epitaxial columnar crystal[14, 15]. Details of the FG region in Figures 2(e)-(f) show that nearly spherical, gray square or needle-like particles are located at the grain boundaries and there are more needle-like phases in HT400 specimen.

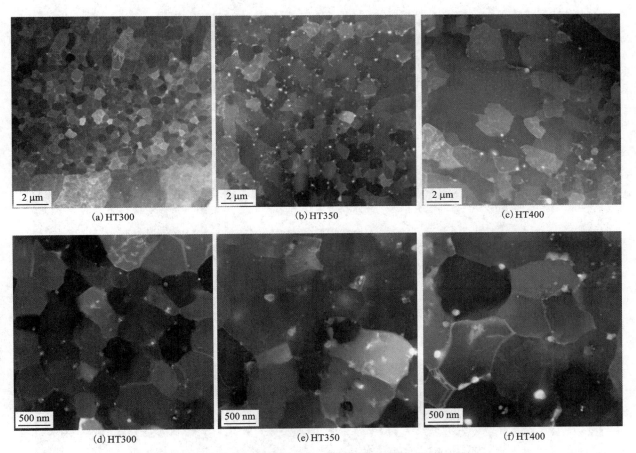

Figure 2 HAADF-TEM images of SLMed Al-Mg-Sc-Zr alloys

3.2 Phase analysis

XRD spectra of SLMed Al-Mg-Sc-Zr alloys with different heat treatment temperatures are depicted in Figure 3. Variations of precipitate intensity can be clearly seen from the enlarged drawing. The peak of Al$_6$Mn gradually highlights with temperature rising, so the needle-like phases in Figure 2 are determined as Al$_6$Mn and Nikulin et al[16] also reported the similar phenomenon in Al-Mg alloy. With ordered L1$_2$ structure, Al$_3$(Sc, Zr) phase is coherent with the Al matrix[17] and can be distinguished at diffraction peaks in high angle (see Figure 3).

Figure 4 describes the high-resolution electron microscopy (HRTEM) image of $Al_3(Sc, Zr)$ phase, which is validated by the corresponding Fast Fourier Transform (FFT) diffraction patterns. Those coherent nano-sized precipitations uniformly distributed in matrix and gradually coarsened with the increase of heat treatment temperature. A large number of Al_6Mn particles precipitated inside the grain of HT400 specimen are shown in Figure 5(a) and its energy dispersive spectrometry (EDS) result is presented in Figures 5(b)-(f).

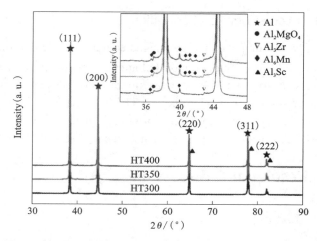

Figure 3 X-ray diffraction patterns of heat-treated samples

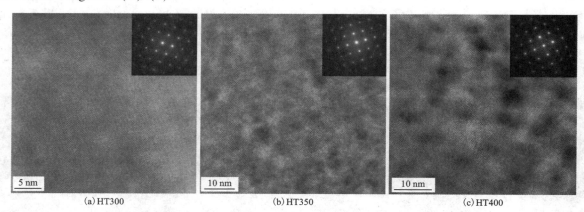

Figure 4 TEM images of $Al_3(Sc, Zr)$ precipitates for different samples

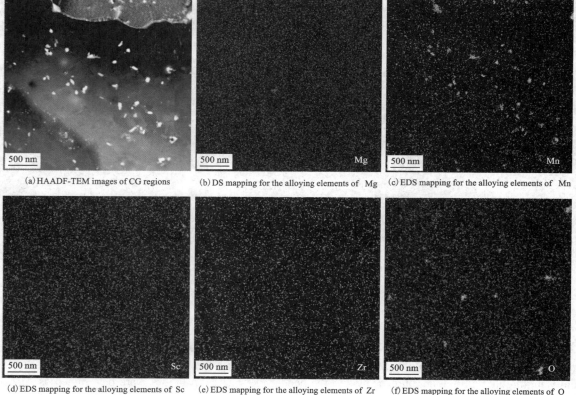

Figure 5 STEM of HT400 specimen

3.3 Electrical conductivity and mechanical properties

The effect of different heat treatment temperatures on Vickers microhardness and electrical conductivity are depicted in Figure 6(a). When heat treatment temperatures range from 300 ℃ to 400 ℃, a continuous ascent of electrical conductivity can be seen. The coarsening of Al_3(Sc, Zr) precipitates and the gradual precipitation of Al_6Mn particles remit lattice distortion energy, thus reducing the electron scattering coefficient. For the temperature-hardness curve, there is a turning point at 350 ℃. The microhardness increases significantly before 350 ℃ and then decreases, which results from the evolution of Al_3(Sc, Zr) and Al_6Mn particles (see Figure 4 and Figure 5). The influence of heat treatment temperature on tensile strength is the same as that on microhardness as described in Figure 6(b). When the heat treatment temperature reaches 400 ℃, Ostwald ripening results in the loss of coherent relationship between the precipitated phase and matrix, so strength and plasticity of the alloy are declined. Dislocations and grain boundaries in the HT300 specimen are effectively hindered, thus obtaining the highest elongation compared with others.

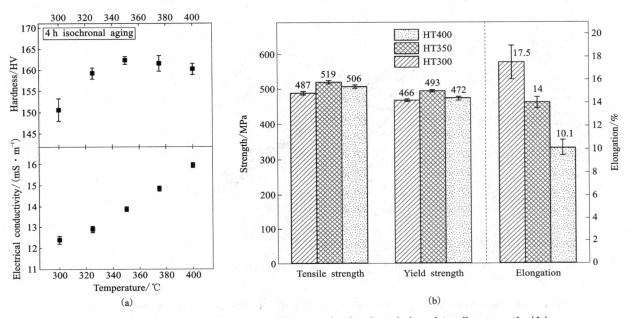

Figure 6 Electrical conductivity and Vickers microhardness(a) and tensile properties(b)

3.4 Corrosion properties

Potentiodynamic polarization and EIS measurements were performed to research the corrosion resistance of heat-treated Al-Mg-Sc-Zr alloys. The polarization curves in Figure 7(a) exhibit apparent passivated behavior. The passive film of HT300, HT350 and HT400 samples formed in the range of −1050 mV to −680 mV, −1180 mV to −920 mV and −1250 mV to −950 mV, respectively. Pitting potential (E_{pit}) is an important indication to evaluate the pitting resistance of aluminum alloy[18] and HT300 specimen shows the highest E_{pit} (−0.752 V). On the basis of the polarization curve, the values of corrosion current density (I_{corr}), anodic Tafel slope (β_a) and cathodic Tafel slope (β_c) can be obtained by linear fitting and relevant corrosion parameter values are listed in Table 1. To ensure the reliability of the data, three samples were tested for each heat treatment condition. The polarization resistance (R_p) was calculated by[19]:

$$R_p = \frac{\beta_a \beta_b}{2.3(\beta_a + \beta_c) i_{corr}} \tag{1}$$

The corrosion current is directly proportional to the electrochemical corrosion rate, which can directly determine the level of corrosion resistance. The value of I_{corr} ranked in the following order: HT350 (1.841 μA/cm²) > HT400 (0.631 μA/cm²) > HT300 (0.104 μA/cm²). In addition, the HT300 specimen possessed the maximum R_p value of 5.053×10⁵ Ω·cm², so it signifies a superior corrosion resistance.

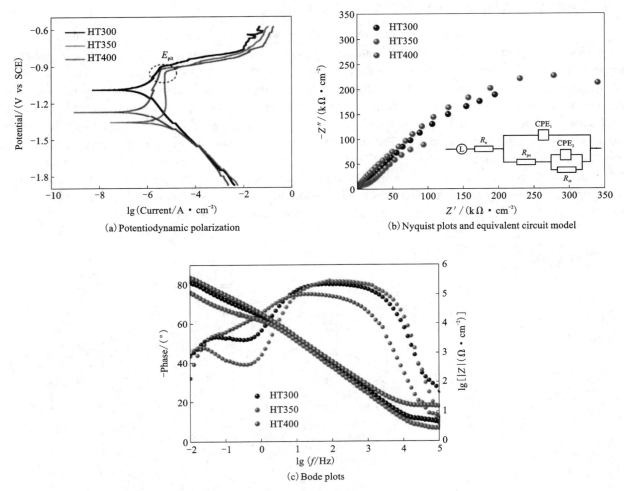

Figure 7 Electrochemical corrosion results

Table 1 Fitting results of polarization curves for HT300, HT350 and HT400 samples in 3.5 wt. % NaCl solution (average value for three parallel specimens)

	$i_{corr}/(\mu A \cdot cm^{-2})$	$\beta_a/(V \cdot dec^{-1})$	$-\beta_c/(V \cdot dec^{-1})$	$R_p/(\Omega \cdot cm^{-2})$	E_{pit}/V
HT300	0.104	0.285	0.201	5.053×10^5	-0.752
HT350	1.841	0.239	0.122	2.264×10^4	-0.914
HT400	0.631	0.262	0.183	7.199×10^4	-0.906

EIS analysis can evaluate the condition of electrode surface. Nyquist plots in Figure 7(b) consist of two capacitive loops deviating from semicircle in first quadrant and the corresponding phase angle and impedance bode plots are presented in Figure 7(c). There are also two peaks in the phase angle bode plot. One peak at about 80° is originated from the formation of oxide film on the surface. The other one is at about 50°, relating to the blocking effect of $Al(OH)_3$ on Cl^- migration[20]. The impedance moduli with a wide linear part at middle frequency reveals that the property of oxide film dominants the impedance behavior. Based on the EIS characteristic, an equivalent circuit model of Figure 7(a) was established. R_{ct} and CPE_2 describe the charge transfer resistance and capacitance of the electric double layer. R_{po} and CPE_1 are the pore resistance and capacitance of the anodic film. R_s represents solution resistance. Fitting results are shown with solid lines in Figure 7 and the values of relevant parameters are summarized in Table 2. With the highest average R_{po} value, HT300 specimen possesses a superior corrosion resistance, and the film resistance of HT400 specimen is higher than that of HT350. Therefore, the results of EIS tests are consistent with the conclusion of the potentiodynamic polarization measurements.

Table 2 Fitting results of EIS measurements for HT300, HT350 and HT400 samples in 3.5 wt.% NaCl solution (average value for three parallel specimens)

	$R_s/(\Omega \cdot cm^2)$	$R_{po}/(k\Omega \cdot cm^2)$	$CPE_1/(F \cdot cm^{-2})$	n_1	$R_{ct}/(k\Omega \cdot cm^2)$	$CPE_2/(F \cdot cm^{-2})$	n_2	$\chi^2 \times 10^{-3}$ Chi-squared values
HT300	3.77	40.91	7.58×10^{-6}	0.89	59.80	3.62×10^{-5}	0.75	1.44
HT350	3.45	26.03	5.52×10^{-6}	0.91	41.23	3.74×10^{-5}	0.78	1.22
HT400	4.63	41.61	7.45×10^{-6}	0.90	36.45	3.28×10^{-5}	0.79	1.07

Surface morphologies of the heat-treated samples after polarization tests are presented in Figure 8. Some large holes originating from the SLM processing can be observed and they tend to be large corrosion pits by enrichment of Cl^-. In addition, molten pool boundaries seem to be corroded preferably. This is because primary $Al_3(Sc, Zr)$ phase mostly distribute in the FG regions along the molten pool boundary and it has been confirmed to act as the anode to accelerate the corrosion of the surrounding α-Al matrix[21]. The unavoidable Oswald ripening of $Al_3(Sc, Zr)$ phases can accelerate the corrosion of samples. In contrast, the corrosion resistance of the HT400 specimen is improved [see Figure 8(c)], which lies in the the higher potential of Al_6Mn phase[22]. According to the Figure 5, Al_6Mn particles appear at the grain boundary or nearby $Al_3(Sc, Zr)$ phase, so that Al_6Mn phase was corroded as the cathode instead of the surrounding α-Al matrix.

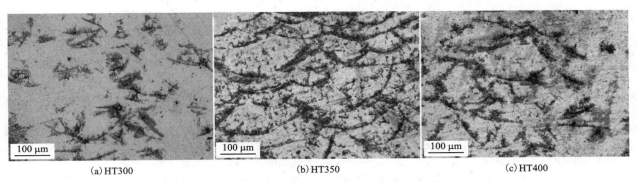

(a) HT300   (b) HT350   (c) HT400

Figure 8 Optical micrographs of samples after polarization tests in 3.5 wt.% NaCl solution

4 Conclusion

In this work we studied the effect of different heat treatment temperatures on corrosion behavior of Al-Mg-Sc-Zr alloy processed by selective laser melting. The following conclusions can be drawn from this study:

(1) The mechanical and corrosion properties of samples are closely related to the evolution of $Al_3(Sc, Zr)$ and Al_6Mn particles, which can be tuned through heat treatment.

(2) Although 350 ℃ heat-treated specimen possesses the highest tensile strength, it is more susceptible to corrosion due to the Oswald ripening of $Al_3(Sc, Zr)$ and Al_6Mn particles.

(3) With higher potential, Al_6Mn particles preferentially form a corrosive galvanic cell with $Al_3(Sc, Zr)$ particles rather than the surrounding α-Al matrix, so the specimen heat treated at 400 ℃ shows superior corrosion resistance compared with the 350 ℃ heat-treated specimen.

References

[1] SAMES W J, LIST F A, PANNALA S, et al. The metallurgy and processing science of metal additive manufacturing [J]. Int Mater Rev, 2016, 61: 1-46.

[2] HERZOG D, SEYDA V, WYCISK E, et al. Additive manufacturing of metals[J]. Acta Mater, 2017, 117: 371-392.

[3] YAP C Y, CHUA C K, DONG Z L, et al. Review of selective laser melting: Materials and applications[J]. Appl Phys Rev, 2015, 2: 041101.
[4] KAUFMANN N, IMRAN M, WISCHEROPP T M, et al. Influence of process parameters on the quality of aluminium alloy EN AW 7075 using selective laser melting (SLM)[J]. Phys Procedia, 2016, 83: 918-926.
[5] ZHANG H, ZHU HH, QI T, et al. Selective laser melting of high strength Al-Cu-Mg alloys: Processing, microstructure and mechanical properties[J]. J Mater Sci Eng A, 2016, 656: 47-54.
[6] PRASHANTH K G, SCUDINO S, KLAUSS H J, et al. Microstructure and mechanical properties of Al-12Si produced by selective laser melting: Effect of heat treatment[J]. J Mater Sci Eng A, 2014, 590: 153-60.
[7] ABOULKHAIR N T, MASKERY I, TUCK C, et al. The microstructure and mechanical properties of selectively laser melted AlSi10Mg: The effect of a conventional T6-like heat treatment[J]. J Mater Sci Eng A, 2016, 667: 139-146.
[8] SPIERINGS A B, DAWSON K, VOEGTLIN M, et al. Microstructure and mechanical properties of as-processed scandium-modified aluminium using selective laser melting[J]. CIRP Ann Manuf Technol, 2016, 65: 213-216.
[9] SPIERINGS A B, DAWSON K, HEELING T, et al. Microstructural features of Sc-and Zr-modified Al-Mg alloys processed by selective laser melting[J]. Mater Des, 2017, 115: 52-63.
[10] SPIERINGS A B, DAWSON K, KERN K, et al. SLM-processed Sc-and Zr-modified Al-Mg alloy: Mechanical properties and microstructural effects of heat treatment[J]. J Mater Sci Eng A, 2017, 701: 264-273.
[11] FULLER C B, KRAUSE A R, DUNAND D C, et al. Microstructure and mechanical properties of a 5754 aluminum alloy modified by Sc and Zr additions[J]. J Mater Sci Eng A, 2016, 338: 8-16.
[12] GU D D, ZHANG H, DAI D H, et al. Anisotropic corrosion behavior of Sc and Zr modified Al-Mg alloy produced by selective laser melting[J]. Corros Sci, 2020, 170: 108657.
[13] ZHANG H, GU DD, DAI D H, et al. Influence of heat treatment on corrosion behavior of rare earth element Sc modified Al-Mg alloy processed by selective laser melting[J]. Appl Surf Sci, 2020, 509: 145330.
[14] SPIERINGS A B, DAWSON K, DUMITRASCHKEWITZ P, et al. Microstructure characterization of SLM-processed Al-Mg-Sc-Zr alloy in the heat treated and HIPed condition[J]. Addit Manuf, 2018, 20: 173-181.
[15] MA R L, PENG C Q, CAI Z Y, et al. Effect of bimodal microstructure on the tensile properties of selective laser melt Al-Mg-Sc-Zr alloy[J]. J Alloys Compd, 2020, 815: 152422.
[16] NIKULIN I, KIPELOVA A, MALOPHEYEV S, et al. Effect of second phase particles on grain refinement during equal-channel angular pressing of an Al-Mg-Mn alloy[J]. Acta Mater, 2012, 60: 487-497.
[17] TAENDL J, NAMBU S, ORTHACKER A, et al. In-situ observation of recrystallization in an Al-Mg-Sc-Zr alloy using confocal laser scanning microscopy[J]. Mater Charact, 2015, 108: 137-144.
[18] SZKLARSKA-SMIALOWSKA Z. Pitting corrosion of aluminum[J]. Corros Sci, 1999, 41: 1743-1767.
[19] LI C, PAN Q L, SHI Y J, et al. Influence of aging temperature on corrosion behavior of Al-Zn-Mg-Sc-Zr alloy[J]. Mater Des, 2014, 55: 551-559.
[20] YANG Y, CHEN Y, ZHANG J X, et al. Improved corrosion behavior of ultrafine-grained eutectic Al-12Si alloy produced by selective laser melting[J]. Mater Des, 2018, 146: 239-248.
[21] WLOKA J, VIRTANEN S. Influence of scandium on the pitting behaviour of Al-Zn-Mg-Cu alloys[J]. Acta Mater, 2007, 55: 6666-6672.
[22] TAN L, ALLEN T R. Effect of thermomechanical treatment on the corrosion of AA5083[J]. Corros Sci, 2010, 52: 548-554.

Effect of Grain Boundary Diffusion of Nano and Micron $Tb_{70}Cu_{15}Al_{15}$ on the Coercivity of Hydrogenation Disproportionation Desorption Recombination $Nd_2Fe_{14}B$ Powders

Xuhua WANG[1,2], Dunbo YU[1,2,*], Yang LUO[1,2,*], Zilong WANG[1], Yuanfei YANG[1],
Ningtao QUAN[1], Zhongkai WANG[1], Weikang SHAN[1], Wenjian YAN[1], Wenlong YAN[1,2]

1　National Engineering Research Center for Rare Earth, Grirem Advanced Materials Co., Ltd., Beijing, 100088, China
2　General Research Institute for Nonferrous Metals, Beijing, 100088, China

Abstract: The effect of diffusion source particle size on the coercivity of hydrogenation-disproportionation-desorption-composite (HDDR) $Nd_2Fe_{14}B$ powders was investigated. In order to optimize the magnetic properties and microstructure of $Nd_2Fe_{14}B$ powders, the change of coercivity after diffusion of nano-$Tb_{70}Cu_{15}Al_{15}$ (n-TCA) and micro-$Tb_{70}Cu_{15}Al_{15}$ (m-TCA) was investigated. The low melting point of the nanoscale diffusion source facilitates the substitution of terbium into neodymium by grain boundary diffusion process (GBDP) and promotes the optimization of Cu and Al element on the grain boundary. The coercivity of the sample treated with n-TCA increased by 5.2 kOe, whereas that of the sample treated with m-TCA increased by only 4.4 kOe. In addition, the loss of remanent magnetization was small for both powders. The microstructure indicates that the nano-diffusion source has a well-distributed microstructure, which improves the utilization of heavy rare earth elements. Therefore, the treatment of HDDR magnetic powders with nano-sized diffusion sources can further improve the coercivity and heat resistance of the materials.

Keywords: Grain boundary diffusion; $Nd_2Fe_{14}B$ powders; Magnetic properties; Coercivity

1　Introduction

$Nd_2Fe_{14}B$ magnets find widespread application in the traction motors of hybrid and electric vehicles due to their exceptional maximum energy density product, denoted as $(BH)_{max}$. As a pivotal component of these motors, thin-walled magnetic rings, fabricated using anisotropic bonded $Nd_2Fe_{14}B$ magnetic powders, exhibit noteworthy magnetic and technological properties. This renders them a promising solution for fulfilling the high-efficiency and energy-saving requirements of motors in the domain of new energy automobiles and active heat dissipation systems for high-performance computers.

At present, the hydrogenation disproportionation desorption recombination method (HDDR)[1-4] stands as the predominant approach for producing highly anisotropic $Nd_2Fe_{14}B$ magnetic powders. The HDDR method offers an appealing route to obtain such powders, owing to its significant merits, including magnet uniformity, efficiency, and cost-effectiveness.

Nevertheless, the magnetic powders produced through the HDDR process exhibit high remanence. Regrettably, the presence of irregular coarse grains in the $Nd_2Fe_{14}B$ hard magnetic phase leads to a coupling effect, resulting in the magnetic particles suffering from inadequate temperature stability and low coercivity[5]. In fact, the unsatisfactory coercivity stands as the primary factor significantly impeding the practical application of these magnets[6]. Therefore, for real-world uses, anisotropic $Nd_2Fe_{14}B$ magnetic powders synthesized via the HDDR method require enhanced coercivity (H_{ci}) at room temperature to maintain high magnetization levels even at elevated temperatures.

It is widely acknowledged that incorporating Heavy Rare Earth (HRE) elements such as Dy or Tb in place of Nd to form compounds with significantly higher magnetocrystalline anisotropy field (HA) can effectively elevate coercivity. One general approach for enhancing H_{ci} involves the formation of the $(Nd, HRE)_2Fe_{14}B$ phase. However, it should be noted that the magnetic moments of HRE elements align antiparallel with $Nd_2Fe_{14}B$, leading to a reduction in magnetization saturation. Furthermore, the scarcity of heavy rare earth reserves poses a potential long-term impediment to their widespread application[7-10]. To overcome these challenges, the grain boundary diffusion (GBD) method has emerged as an excellent approach to optimize properties[11]. Through HRE GBD, the grains can be magnetically hardened with a minimal amount of HRE, while maintaining almost no change in magnetic remanence[12].

Currently, the research on grain boundary diffusion in HDDR magnetic powders primarily centers around the impact of different diffusing sources on coercivity. However, there is a relative scarcity of studies regarding the particle size and diffusion mechanism of these sources[13-16]. This becomes particularly significant when dealing with HDDR magnetic powders with nanoscale grain sizes, as the use of traditional micron-sized diffusing sources from bulk $Nd_2Fe_{14}B$ materials hampers the uniform distribution of these sources within the material.

To address this issue, our study utilized nano-sized particles of the heavy rare earth alloy $Tb_{70}Cu_{15}Al_{15}$ to conduct grain boundary diffusion on HDDR $Nd_2Fe_{14}B$ magnetic powders. Through magnetic performance testing and microscopic observations, we made a compelling discovery. The samples treated with nano-sized diffusing sources displayed a more uniform and continuous distribution of grain boundaries in comparison to those treated with micron-sized diffusing sources. The nano-sized diffusing sources exhibited enhanced migration capabilities between grains, facilitating the formation of continuous Tb and Cu shell structures that enveloped the grains. In contrast, micron-sized diffusing sources led to the depletion of most Tb and Cu elements in the coarse triangular grain boundary regions.

2 Experiments and discussions

The magnetic powder comprises a nominal composition of 12.5%Nd, 6.4%B, 0.3Ga%, 0.2%Nb, balance of Fe, consisting of a 2∶14∶1 phase and a Nd-rich phase. TbCuAl was prepared through arc melting and subsequent melt spinning under a high-purity argon atmosphere. The resulting thin TbCuAl ribbons were fragmented into micrometer-sized diffuse sources (m-TCA) using the HD process. A portion of the m-TCA was ball-milled in an organic medium for 20 hours to produce a nanoscale diffusion source (n-TCA). Both diffusion sources were thoroughly mixed with the HDDR magnetic powder and subjected to a heat treatment at 790 ℃ for two hours, followed by air-cooling for testing purposes. The magnetic properties were evaluated using Vibrating Sample Magnetometer (VSM) measurements. To study the detailed microstructure and compositional distribution, Transmission Electron Microscopy (TEM) with its ancillary High Angle Annular Dark Field (HAADF) measurements were employed.

2.1 Magnetic properties

As shown in Figure 1, for the addition of the same mass of anisotropic magnetic powder, the coercivity of the magnetic powder using $Tb_{70}Cu_{15}Al_{15}$ nano-alloy diffusion source particles is higher, and the coercivity of the sample is further enhanced by 0.8 kOe compared to that of the sample using the micron-sized diffusion source particles of $Tb_{70}Cu_{15}Al_{15}$. The coercivity enhancement of magnetic powder is observed as the amount of the diffusion source added increases. However, the efficiency of coercivity enhancement per unit mass of heavy rare earth elements decreases. The best overall magnetic properties of the magnetic powder were achieved when 1.2 wt.% of $Tb_{70}Cu_{15}Al_{15}$ nano-alloy diffusion source was added to the sample. The coercivity of the

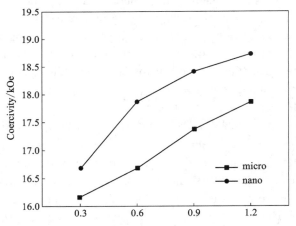

Figure 1 Coercivity of two types of diffusion sources and diffusion source content

magnetic powder was 18.66 kOe, while the maximum magnetic energy product was also enhanced to 40.93 MGOe.

It is worth noting that the coercivity optimization from the diffusion source particle size effect was the weakest when the diffusion source addition was 0.3 wt.%. When the diffusion source addition is lower, the diffusion source is less likely to have excessive localized concentration, which corroborates the effect of the concentration distribution of the diffusion source on the magnetic powder performance from another perspective. Comparing the nanoalloy diffusion source-treated samples to the original magnetic powders with heat treatment only, although the remanent magnetization appears to be reduced, the squareness and the maximum magnetic energy product of the samples are more than that of the heat-treated original magnetic powders.

Taking the average of the magnetic properties of the samples with the addition of the nano-diffusion source, the remanent magnetization decreases from 13.45 kGs to 13.07 kGs, while the squareness improves from 0.46 to 0.51, and the maximum magnetic energy product improves from 38.46 to 39.5 MGOe. This indicates that the diffusion source of $Tb_{70}Cu_{15}Al_{15}$ nanoalloys, in addition to increasing the coercive force of the samples, improves the microstructure and orientation of the samples.

2.2 Microstructure

Figure 2(a) shows the TEM picture of the magnetic powder with diffused m-TCA. The size of the main phase of NdFeB grains is between 200 and 500 nm similar to the size of single domains of NdFeB materials. From the figure, it can be seen that the grain boundary diffusion has little effect on the main phase of the grains, but there are many coarse massive rare-earth-rich grain boundary phases at the grain boundaries, which are mostly distributed in the triangular grain boundaries and the boundary between them and the main phase is relatively blurred. Figure 2(b) shows the TEM picture of magnetic powder with grain boundary diffusion of n-TCA. From the figure, it can be seen that the heat treatment process did not significantly change the size of the main phase of $Nd_2Fe_{14}B$ grains, but the edges of the main phase look smoother and the morphology of the grains is more regular. The boundary between the grain boundary phase and the main phase is clearer, and the distribution of the grain boundary phase in the magnet is more uniform.

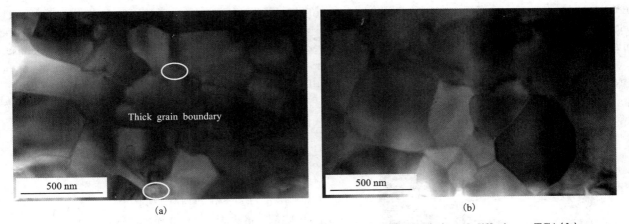

Figure 2 TEM picture of grain boundaries of samples diffusing m-TCA(a) and diffusing n-TCA(b)

The elemental distribution at the grain boundaries of the samples was analyzed by TEM-EDS, and the results are shown in Figures 3 and 4. Figure 3 shows that the wide grain size contains a large amount of Cu and Tb elements, resulting in a large waste of heavy rare earth elements. And the Cu in the grain boundary phase does not appear to be significantly elevated on the surface of the grains, indicating that the Cu element fails to wet the surface of the grains well. This further indicates that the low melting point alloy fails to effectively broaden the grains of nanocrystals.

Figure 3 shows that the concentration of Cu and Tb elements is abnormally elevated on both sides of the grain boundaries, indicating that Cu-rich microstructures have been generated at the junction of the grain boundaries and the main phases, and that the Cu bias not only reduces the ferromagnetism of the grain boundary phases, but also reduces the melting point of the boundary phases, promotes the migration of Tb elements within the grain boundaries and improves the wettability between the grain boundary phases and the main phases. The Cu element bias not only helps to reduce the ferromagnetism in the grain boundary phase, but also lowers the melting point of the grain boundary phase, promotes the migration of Tb element in the grain boundary, and improves the wettability between the grain boundary phase and the main phase.

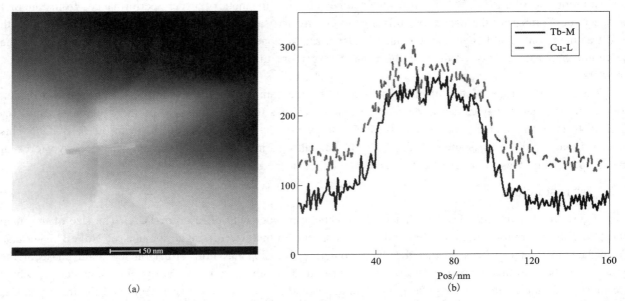

Figure 3 TEM picture of grain boundaries of samples diffusing n-TCA(a) and distribution of Tb and Cu in the grain boundary phase(b)

Figure 4 shows that the concentration of Cu and Tb elements is abnormally elevated on both sides of the grain boundaries, indicating that Cu-rich microstructures have been generated at the junction of the grain boundaries and the main phases, and that the Cu bias not only reduces the ferromagnetism of the grain boundary phases, but also reduces the melting point of the boundary phases, promotes the migration of Tb elements within the grain boundaries and improves the wettability between the grain boundary phases and the main phases. The Cu element bias not only helps to reduce the ferromagnetism in the grain boundary phase, but also lowers the melting point of the grain boundary phase, promotes the migration of Tb element in the grain boundary, and improves the wettability between the grain boundary phase and the main phase.

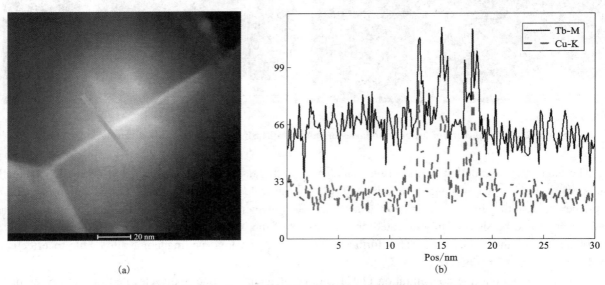

Figure 4 TEM picture of grain boundaries of samples diffusing n-TCA(a) and distribution of Tb and Cu in the grain boundary phase(b)

3 Conclusion

The grain boundary diffusion of HDDR magnetic powders by nanosizing the diffusion source can further improve the coercivity and heat resistance of the material by about 20%. The nano-diffusion source not only has the potential to prepare ultra-high coercivity magnetic powders, but also can be used as a means to reduce the amount of heavy rare earth elements.

Acknowledgments: This work was sponsored by national engineering research center for rare earth, GRINM Group Co., Ltd., Beijing.

References

[1] ZHANG X D, HAN J Z, YANG Y C, et al. Turing pattern formation in $Pr_2Fe_{14}B$-based alloy and its role in anisotropy inducement at the early stage of disproportionation[J]. Applied Physics Letters, 2008, 92(17): 5040.

[2] WAN F M, HAN J Z, ZHANG Y F, et al. Coercivity enhancement in HDDR near-stoichiometric ternary Nd-Fe-B powders[J]. Journal of Magnetism & Magnetic Materials, 2014, 360: 48.

[3] NAKAYAMA R, TAKESHITA T, ITAKURA M, et al. Magnetic properties and microstructures of the Nd-Fe-B magnet powder produced by hydrogen treatment-(II) (abstract)[J]. Journal of Applied Physics, 1990, 67: 4665.

[4] SONG T, TANG X, YIN W, et al. Coercivity enhancement of hot-pressed magnet prepared by HDDR Nd-Fe-B powders using Pr-Cu eutectic alloys diffusion[J]. Journal of Magnetism and Magnetic Materials, 2019, 471: 105.

[5] CHOI M, CHO S, SONG Y, et al. Simultaneous enhancement in coercivity and remanence of $Nd_2Fe_{14}B$ permanent magnet by grain boundary diffusion process using NdH_x[J]. Current Applied Physics, 2015, 15(4): 461.

[6] SONG T, WANG H, TANG X, et al. The effects of Nd-rich phase distribution on deformation ability of hydrogenation-disproportionation-desorption-recombination powders and magnetic properties of the final die-upset Nd-Fe-B magnets[J]. Journal of Magnetism & Magnetic Materials, 2019, 476(APR.): 194.

[7] LIU W Q, CHANG C, YUE M, et al. Coercivity, microstructure, and thermal stability of sintered Nd-Fe-B magnets by grain boundary diffusion with TbH_3 nanoparticles[J]. Rare Metals, 2017, 36(9): 718.

[8] YANG X, GUO S, DING G, et al. Effect of diffusing TbF_3 powder on magnetic properties and microstructure transformation of sintered Nd-Fe-Cu-B magnets[J]. Journal of Magnetism and Magnetic Materials, 2017, 443: 179.

[9] ZHANG Q, GUO S, YANG X, et al. Coercivity enhancement of sintered Nd-Fe-B magnets by chemical bath deposition[J]. AIP Advances, 2018, 8(5): 056220.

[10] JINGHUI D, GUANGFEI D, XU T, et al. Highly efficient Tb-utilization in sintered Nd-Fe-B magnets by Al aided TbH_2 grain boundary diffusion[J]. Scripta Materialia, 2018, 155: 50.

[11] HIROTA K, NAKAMURA H, MINOWA T, et al. Coercivity enhancement by grain boundary diffusion process to Nd-Fe-B sintered magnets[J]. IEEE Transactions on Magnetics, 2006, 42: 2909.

[12] SEPEHRI-AMIN H, OHKUBO T, HONO K, et al. Grain boundary structure and chemistry of Dy-diffusion processed Nd-Fe-B sintered magnets[J]. Journal of Applied Physics, 2010, 107(9pt.2): 873.

[13] KIM T H, LEE S R, BAE K H, et al. Effects of Al/Cu co-doping on crystal structure and chemical composition of Nd-rich phases in Nd-Fe-B sintered magnet[J]. Acta Materialia, 2017.

[14] GOTO, RYOTA, NISHIO, et al. Wettability and interfacial microstructure between $Nd_2Fe_{14}B$ and Nd-Rich phases in Nd-Fe-B alloys[J]. IEEE Transactions on Magnetics, 2008, 44(11): 4232.

[15] SEPEHRI-AMIN H, OHKUBO T, NISHIUCHI T, et al. Coercivity enhancement of hydrogenation-disproportionation-desorption-recombination processed Nd-Fe-B powders by the diffusion of Nd-Cu eutectic alloys[J]. Scripta Materialia, 2010, 63(11): 1124.

[16] LU K, BAO X, CHEN G, et al. Modification of boundary structure and magnetic properties of Nd-Fe-B sintered magnets by diffusing Al/Cu co-added alloy ribbons[J]. Scripta Materialia, 2019, 160: 86.

Recent Developments of Purification of Rare Earth Metals

Xiaowei ZHANG[1,2,*], Hongbo YANG[1,2], Zhiqiang WANG[1,2], Zongan LI[1,2], Chuang YU[1,2], Xinyu GUO[1,2], Jiamin ZHONG[1,2], Penghong HU[1,2], Fan YANG[1,2], Wenli LU[1,2], Chenchen XU[1,2]

1 National Engineering Research Center for Rare Earth Materials, GRINM Group Co., Ltd., Beijing, 100088, China
2 GRIREM Advanced Materials Co. Ltd., Beijing, 100088, China

Abstract: The recent development of purification methods of rare earth metal, including vacuum refining, vacuum distillation, zone refining, solid state electrotransport, electrorefining, hydrogen plasma melting, solid state external getting and electron beam melting, was discussed. This work can help to provide some advises and directions for further study on purification of rare earth metal.

Keywords: Rare earth metals; Purification methods; Development trend

High purity rare earth metals are the basis material of high-performance magnetic materials, optical materials and electrical functional materials, which are widely used in new energy automobiles, integrated circuits, display screen, 5G communications and other strategic emerging industries.

In recent years, the preparation and application of high purity rare earth metals have been paid more attention around the world, and there are a great deal of research works on the purification theory of rare earth metals, new technology development, technology and process optimization. At present, the common purification methods include vacuum refining[1], vacuum distillation[2], zone melting[3], solid state electrotransport[4], etc. In recent years, plasma melting, solid state external getting and other purification methods were introduced into the purification of rare earth metals. The present research status of rare earth purification is reviewed in present paper, aiming to provide reference for further development of rare earth metal purification.

1 Common purification methods

1.1 Vacuum melting

In the vacuum melting process, rare earth metal is heated to a temperature above its melting points under argon or vacuum, with impurities separated by volatilization. The impurities mainly consist of excess reductants, metals, non-metals, and interstitial impurities with high vapor pressures. The excess reductants (i.e. Ca or Li), residual fluoride (i.e. REF_3), partial fluoride products (i.e. CaF_2 or LiF), and other interstitial impurities (i.e. O, C, and H) are removed from rare earth metal prepared by metal thermal reduction. Moreover, the electrolyte, such as REF_3, and LiF can be also removed from rare earth metals prepared by molten salt electrolysis.

The vacuum melting method is suitable for the initial purification of low purity rare earth metals except Sm, Eu, Tm and Yb. Tungsten and Tantalum are generally used as crucible materials during the medium frequency vacuum induction melting. However, the significant amounts of W and Ta elements can be dissolved into the rare earth melt. Due to the higher density of Ta compared to light rare earth such as La, Ce, Pr, and Nd, the impurity of Ta usually precipitates and settles to the bottom of crucible during solidification. Subsequently, the Ta contamination in the light rare earth metals can be removed by mechanical method[5] or by zone refining[6]. Additionally, the impurities of W and Ta in the heavy rare earth metals such as Y, Gd, Dy, and Lu, can be removed by vacuum distillation.

To avoid crucible contamination, the water-cooled copper crucible is used instead of tungsten and tantalum crucibles during vacuum arc melting, electron beam melting, and vacuum levitation melting processes. Water-cooled crucible can be heated to higher temperatures than tungsten and tantalum crucibles for more efficient removal of gaseous impurities. The purity of rare earth metal generally reaches 99.5% to 99.9%

by weight after initial purification by the vacuum melting process.

1.2 Vacuum distillation

The impurity removal process of vacuum distillation is opposite to that of vacuum melting, and is suitable for rare earth metals with high saturation vapor pressure. It is a purification method that uses the significant difference between the vapor pressure of the base metal and the impurities to separate the base metal and impurities by heating in high temperature and vacuum. Impurities with lower saturated vapor pressure remain in the residue, and impurities with higher saturated vapor pressure than the base metal condense in the exhaust gas or places with lower temperature.

Vacuum distillation is generally considered as molecular distillation. Metal vapor is composed of molecules with various atomic structures, such as monoatomic gas molecules, diatomic gas molecules, polyatomic gas molecules, etc. For rare earth metals, Pr, Sm, Dy, Er, Tm and Yb only have monoatomic gas molecules, and other rare earth elements have monoatomic gas molecules and diatomic gas molecules at the same time[7]. With the decrease of pressure and the increase of temperature, diatomic gas molecules tend to decompose into monoatomic gas molecules.

Vacuum distillation purification can be completed by sublimation or evaporation, distillation purification of other rare earth metals except La and Ce can be divided into three categories: (1) Sm, Eu, Tm and Yb: sublimation purification is generally adopted; (2) Sc, Dy, Ho and Er: distillation purification is generally used, and sublimation or multiple distillation purification can be used to prepare higher purity metals; (3) Y, Pr, Nd, Gd, Tb and Lu: generally purified by distillation. For the selection of distillation temperature, it is necessary to comprehensively consider the properties of rare earth metals (melting point, boiling point and saturated vapor pressure), the types of impurities, target purity and other factors. If the distillation temperature is too low, the metal impurities with slightly lower saturated vapor pressure and most crucible impurities can be effectively removed. However, due to the low evaporation rate of rare earth metals, the collision probability between rare earth metals and residual gas in the furnace will increase, and the contents of gas impurities such as O and N may be high, and the purification period will be prolonged. If the distillation temperature is too high, the evaporation rate of rare earth metals is too fast, which can easily cause impurity entrainment. When the distillation temperature exceeds 1650 ℃, it will cause the volatilization of low valent rare earth oxide, limit the effectiveness of oxygen removal[8], and deteriorate the purification effect.

Vacuum distillation is the mainstream technology for massive preparation of rare earth metals above 3N at present. For high purity rare earth metals above 4N, multiple distillation and purification are required. In order to improve the purification efficiency of 4N rare earth metals and strengthen the distillation purification effect, scholars at home and abroad have explored the following measures. (1) Placing a baffle above the metal evaporation surface: According to the characteristic that impurities with higher saturated vapor pressure are preferentially condensed in the low temperature zone, a multilayer condensation baffle is arranged above the rare earth metal, so that the rare earth metal can be condensed on baffles with different temperatures, and high-purity rare earth metal can be obtained on the baffle with high temperature[9]. (2) Adding W in the distillation process: In the process of distillation and purification of rare earth metals, using W filter screen or adding W powder can remove impurities such as Fe, F, Ti, Zr, etc[10, 11]. It is considered that impurities such as Fe, F and Ti in rare earth metals may combine with W, so that these impurities cannot be co-distilled with rare earth metals[12]. Domestic scholars use "the theory of alloy phase formation" to explain that the interaction strength between W and impurities is much smaller than that between Sc and impurities, and the solubility of impurities in W is much greater than that in W to explain why adding W element improves the impurity removal effect[13]. However, the above two viewpoints are contradictory, and the reasons for strengthening purification need to be further studied. (3) Rare earth metal vapor passes through liquid metal La or Ce: The saturated vapor pressure of metal Yb and Sm is very high. When its metal vapor passes through the extremely low saturated vapor pressure of La and Ce melt, the impurities in Yb and Sm enter into liquid La and Ce, playing the role of "washing", while not causing La and Ce pollution[14].

In the distillation and purification experiments of rare earth metals, it was found that the impurities in the distillation products showed an uneven distribution, such as Al and Fe in metal Dy showed an exponential decrease trend[2], and Cu in metal Tb showed a linear decrease trend[15]. In view of the above phenomenon, the movement behavior of impurities in the distillation and purification of rare earth metals was studied.

According to the physical properties of impurities, the impurities in rare earth metals were divided into five categories. (1) Extremely high volatile impurities: the saturated vapor pressure of impurities is much higher than the saturated vapor pressure of the main metal. (2) Highly volatile impurities: The k_0 of impurities is within the range of 10 to 10^3 ($k_0 = C_g^*/C_l^*$, where C_g^* represents the concentration of impurities in the gas phase at the interface and C_l^* represents the impurity concentration in the liquid phase at the interface). (3) Medium volatile impurity: The k_0 of the impurity is within the range of $1 \sim 10$. (4) Lowly volatile impurities: The k_0 of the impurities is within the range of 10^{-3} to 1. (5) Extremely low volatile impurities: the saturated vapor pressure of impurities is much lower than the saturated vapor pressure of the main metal. For category $2 \sim 4$ impurities, through kinetic analysis of impurity volatilization process, the restrictive links in the volatilization process of three types of impurities were identified, that is, category 2 impurities were controlled by liquid boundary layer diffusion mass transfer, and category 3 and 4 impurities were controlled by liquid boundary layer diffusion mass transfer and volatile mass transfer. According to the rate equation of mass transfer in the liquid boundary layer and the conservation of mass of solute[16, 17], the distribution models of high, medium and low volatile impurities in the distillation products were established, and the experiments of purification of Tb by vacuum distillation were carried out. The theoretical calculation results were consistent with the experimental results (Figure 1).

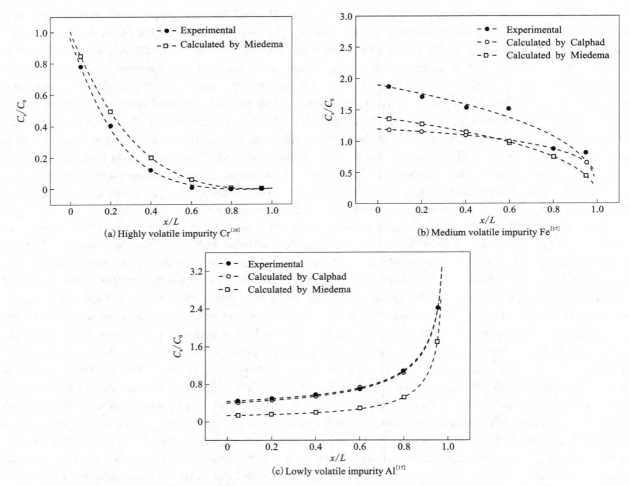

Figure 1   Comparison between the calculated and experimental results of impurities in Tb

Vacuum distillation is the main purification method for preparing high purity rare earth metals. At present, more than ten kinds of rare earth metals with purity above 99.99%, such as Sm, Yb, Tm, Er, Ho, have been prepared by this method (relatively 60 kinds of impurities, including C, N, O and S)[18], and the purity of some metals has reached 99.99% ~ 99.995%.

1.3 Solid state electrontransport

Solid state electrontransport (SSE), also known as solid state electrolysis, is a method to purify metal by orderly migration of impurity atoms to both ends of metal rod under a DC electric field. The principle is that when direct current passes through metal conductor, the kinetic energy of carrier (electron or hole) is transmitted to solute ion through collision. Therefore, in addition to electrostatic force applied to solute ion, there is also a migration force generated from momentum exchange with the carrier. Under action of its resultant force, solute ion will eventually move in conductor at a stable migration rate.

The phenomenon of SSE was first discovered in 1861. In the experiment, it was found that Pb-Sn alloy became soft at one end and brittle at the other end after direct current was applied[19]. The study of element migration in metals began in 1935, and it is found that C migrated in Fe[20]. Since then, this method was gradually extended to purify other metals.

SSE was first applied in rare earth metals in 1961. J. M. Williams[21] studied migration of interstitial impurity in Y metal by SSE, and then, a lot of researches on purification of rare earth metals, the results shows that interstitial impurities such as C, N, O, and H migrate to the anode in rare earth metals such as La[4], Pr[22], Gd[23], Lu[22] and Y[20]. Some scholars considered that migration direction of interstitial impurity is related to the radius of matrix metal at 0.9 times melting point temperature. At this temperature, interstitial impurity migrates to anode in Y, Gd, Lu, and Sc, with atomic radius greater than Ti, and migrates to cathode in others rare earth metals with the atomic radius less than Ti[24].

SSE has a significant effect on removal of interstitial impurities such as C, N, O and H. The O content in high melting point metals such as Y and Tb can be reduced to less than 20 μg/g. In addition to interstitial impurities, some metallic impurities also migrate obviously, but the migration directions of different impurities in rare earth metals are not the same. For example, Fe impurity migrates to anode, Al and Cu migrate to cathode in Pr metal[22], Cu migrates to cathode in Gd metal[22], and Fe, Ni, Al and Cu migrate to anode in La metal. Considering migration direction of most impurities, the highest purity of rod after SSE purification appears in the middle part of the rod.

In order to overcome the SSE's disadvantages of long purification period, low yield and low purification efficiency, some research work has been carried out. (1) SSE purification operation at above crystal transition temperature: When the temperature increases to the crystal transition temperature, the crystal structure of most rare earth metals transform from dense hexagonal structure to body-centered cubic structure, the migration rate of interstitial impurities increase significantly. For example, migration rate of impurity such as O and N in Sc metal can be increased by an order of magnitude[25]. (2) Application of pulse current: Pulse current can significantly increase current intensity under condition of the same input power, and migration rate of impurities, that

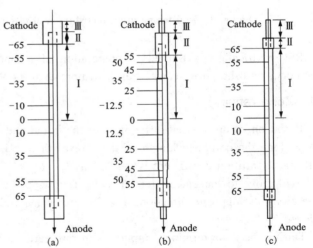

Figure 2 Electrode connection with different structures and their metal rods[27]

were not obviously migrated under steady DC current, is significantly improved under pulse current, such as impurity elements Al, Cu and O in La; and impurities that do not migrate under DC current are also migrated under pulse current, such as Mo and O in Ce[26]. (3) Application of static magnetic field: The impurity removal ratio of O, N in Tb metal increases 20% with 1T steady magnetic field than without magnetic field; (4) Optimizing electrode connection structure: Through redesign and optimize the electrode structure, the temperature difference of metal La rod in SSE purification is reduced from 150 ℃ of traditional structure to less than 30 ℃, the residual rate of Fe impurity in La metal is reduced from 60% to about 20%, and the purification efficiency is greatly improved[27] (Figure 2 and Figure 3).

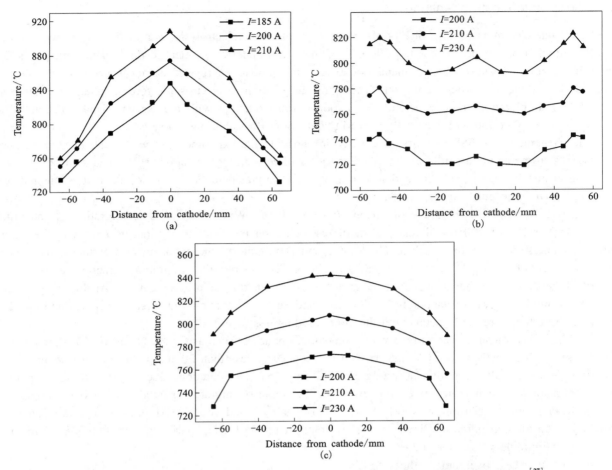

Figure 3 Rod temperature distribution of electrode connection with different structures[27]

SSE purification is limited to preparation of a small amount of high purity rare earth metals in the laboratory, and the purity can be reached more than 99.99%.

1.4 Zone refining

Based on compositional differences of the liquid and solid in equilibrium, a narrow molten zone is moved slowly along the complete length of the rare earth metal specimen to bring about impurity segregation. The equilibrium partition coefficient is defined as $k_0 = C_S/C_L$, during the zone refining process, the impurities of $k_0 < 1$ will move to the end region of metal following the zone refining direction but impurities of $k_0 > 1$ will move along the opposite direction. For the impurities with a value of $k \approx 1$, the purification effect cannot be achieved.

During the zone refining, impurities in rare earth metals can be classified into four categories.

(1) Highly saturated vapor pressure impurities, such as Ca, Mg, Mn, etc., will be preferentially evaporated and removed due to the high temperature in the melting zone[28].

(2) Lower saturation vapor pressures, which moving direction determined by the k_0. For instance, Co, Fe, Ni in Y exhibit the same moving direction as the molten zone, whereas Al and Si show no significant moving effect[28]; Ta, Cu, Co in La metal[29,30], Al, Fe, Si, Co in Ce metal[30], and Fe, W, Ni in Gd metal[29,31], exhibit the same moving direction as the molten zone also.

(3) Interstitial impurity, including O, C, N, H, and S, exhibit distinct moving direction within various rare earth metals. Significant moving of O and C is absent in Y[29], while notable moving of O, H, and N occurs in La, Ce, and Gd[29-31]. Similarly, S exhibits prominent moving in La and Ce[30], with a moving direction opposite to the molten zone. C, on the other hand, demonstrates moving in La, Ce, and Gd, aligning with the movement of the molten zone[30,31].

(4) Rare earth impurity: Due to the similarity properties of neighboring rare earth elements, no

significant moving of impurity Ce was observed during the zone melting of La metal[3].

After zone refining, impurities tend to accumulate at both ends of the specimen. By cutting the specimen ends, high purity rare earth metals can be obtained. The purity of the refined rare earth metals can reach above 99.99%.

1.5 Molten salt (Electrochemical) deoxidation

Based on the partition coefficient differences of oxygen in the molten metal and the slag, oxygen can diffuse from rare earth metal to the slag, and the rare earth metal is purified. In the molten salt, rare earth metals to be purified act as anode, high purity graphite or carbon rod act as cathode, the O impurity in rare earth metal will react with metal precipitated on the cathode surface and solved into molten salt, and then transport to anode surface to form CO or CO_2.

The molten salt deoxidation was began to study in $CaCl_2$ molten salt system[32], appropriate amount Ca was added in the molten salt, O can be removed by the formation of CaO in the molten salt. Due to the high statured vapor pressure of chloride, the deoxidation temperature is lower than 1000 ℃ usually; after 7 to 10 days' deoxidation, O in Tb and Y metal decrease from 540 μg/g and 9700 μg/g to 100 μg/g and 1150 μg/g respectively[33]. Considered that the deoxidation temperature lower than 1000 ℃, most rare earth metal is in solid state, the diffusion rate of O impurity in solid metal is very slow, resulting that the deoxidation period is very long; in the other hand, after a long purification time, the deoxidation reaction reaches equilibrium, so that there is a limit of O content for this method.

In order to improve deoxidation efficiency and overcome deoxidation limit, there are two modification measures: (1) Fluoride molten salt with lower statured vapor pressure instead of chloride molten salt, rare earth metal and the fluoride salt are heated to liquid state, and O migrates from liquid metal to the slag, this is known as molten salt extraction deoxygenation. The molten salt is usually individual REF_3 or REF_3-LiF; in the $W_{REF_3}/W_{RE} = 80\%$ molten salt

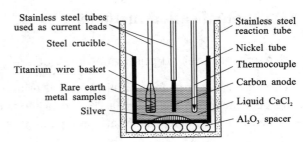

Figure 4 Schematic diagram of electrochemical deoxidation experimental device[36]

system, O in Dy and Y metal decrease from 2900 μg/g and 3500 μg/g to 340 μg/g and 1100 μg/g respectively[34]. Due to the high melting points of rare earth fluoride, slag is entrapped in solidified metal easily, and the F content in rare earth metal is high; in order to decrease melting point of the fluoride slag, LiF was added, the O content of rare earth metal can further decreased, and the F impurity can be partly remove[35]. (2) A DC current applied on the two electrodes (rare earth metal to be purified acting as cathode, and high purity graphite or carbon rod act as cathode), Oxygen diffuses into $CaCl_2$-Ca molten salt, and transports to anode surface to generate CO or CO_2; therefore, the O content of molten salt decreases gradually, and it is beneficial to remove O in the metal. After electrochemical deoxidation for 10 hours, O content in La, Pr, Nd, Gd and Tb metal decreases less than 30 μg/g[36, 37] (Figure 4).

1.6 Electrolytic refining

In the molten salt system, the coarse rare earth metals act as the anode, while pure rare earth metals or inert metals serve as the cathode. The purification of rare earth metals is accomplished through an electrochemical process involving anode dissolution and cathode deposition. During the electrolytic process, highly electrically active rare earth metals dissolve and deposit on the cathode, while less electrically reactive metallic impurities such as Ta, W, Fe, and others do not undergo anode dissolution, thus achieving the separation of impurities.

To prevent the reaction of metal products and electrolyte with air, moisture, electrodes, and crucible materials at high temperatures during the electrolytic refining process, purification must be conducted in a closed and inert atmosphere environment. The electrolyte system primarily comprises chloride[38, 39], fluoride[40], or chloride-fluoride hybrid systems[41, 42].

Currently, only a limited number of rare earth metals have been publicly reported in research studies regarding electrolytic refining purification, such as Y, Gd, La, Ce. When utilizing the YCl_3-LiCl system for electrolytic refining of metal Y, excellent removal efficiency is observed for impurities such as Al, Ca, C, Cu, Fe, Ni, Ta, Ti. However, the removal of oxygen (O) is only effective in electrolytic refining when the YCl_3 concentration is below 5.4% (molar fraction), while variations in current density do not impact the purification efficiency[38].

In the electrolytic refining experiments of Gd, it has been observed that the electrolyte system and current density have a certain influence on the shape of the rare earth metals deposited at the anode[42], specifically, the 75% LiF-25% GdF_3 system exhibits a needle-like morphology, while the 80% LiCl-18% LiF-2% GdF_3 system exhibits a flake-like morphology. Electrolytic refining enables the removal of impurities such as Ta and Fe from Gd, while reducing the O content from 810 μg/g to below 100 μg/g[41, 42]. Regarding the metal La, electrolytic refining is an effective method to remove crucible metal impurities such as Ta and W, as well as interstitial impurities like C and O. This refining process can enhance the purity of La from 99% to 99.867%[40].

Electrolytic refining demonstrates remarkable effectiveness in removing low-electronegativity impurities and interstitial impurities such as O and C from rare earth metals. Moreover, it offers a short purification cycle, making it widely applicable for the high-purity preparation of low-saturation vapor pressure rare earth metals like La, Ce, Pr, Nd, Gd, Tb, Y, among others. It is poised to become the primary purification technique for the production of high-purity rare earth metals with low-saturation vapor pressure.

2 Other purification methods

2.1 Plasma arc melting purification method

Plasma arc melting is a purification method using plasma arc as a heat source. Whereinto, an inert gas (such as Ar), reducing gas (such as H_2) or a mixture of the two gases are used as the plasma generating gases. Due to the high temperature of plasma arc, it is commonly used to melt refractory metals as shown in Figure 5.

In the process of traditional Ar plasma melting and purification of rare earth metals, Ca, Mg and other high saturated vapor pressures are preferred to volatilize, but the removal effect of O in rare earth metal Gd is not obvious[44]. However, by combining the technical advantages of plasma melting and zone melting, it is not only beneficial for the metal impurities migrate with the melting zone, the directional movement of the gas impurities O, N, C, S also takes place in the purification process. But some gas impurities move in opposite directions in different rare earth metals[30], and the reasons are not yet clear.

Under high temperature conditions, H_2 can dissociate into H atoms, and H in the atomic state has very high reducibility, which is conducive to removing gas impurities in metals. By using Ar+H_2 as a plasma source, impurities such as O, N, C in rare earth metals react with H atoms to generate H_2O, NH_3, and CH_4. Finally, these gas impurities are removed effectively. Compared with argon plasma melting purification, the removal rate of O, N and C impurities in hydrogen plasma melting purification is significantly improved. The O content of Gd metal drops down to 55 μg/g[45]. The O, N content of Tb metal drops below 10 μg/g[46].

Figure 5 Schematic diagram of impurity removal in hydrogen plasma melting process[43]

2.2 Electron beam melting

Under vacuum condition, electron beam melting is a purification method of using electron beam to heat the metal material in the water-cooled copper crucible to purify rare earth metals. This method can effectively remove the high saturated vapor pressure impurities and some interstitial impurities in rare earth metals.

Because of its high temperature and high vacuum, electron beam melting is often used to purify high melting point metals.

In the electron beam melting purification of La metal, it was found that electron beam melting can effectively remove impurities such as Mg, Li, Mn, Cr, Fe, Ti, etc., and the content of the above impurities can be less than 1 μg/g after purification[1]. For the impurity of Ni, Si impurity, its saturated vapor pressure is much higher than that of La metal, however, they are almost unchanged after electron beam melting, the main reason is that the activity coefficient of these impurities in liquid La metal is very small, and its separation coefficient is less than 1, so that it cannot be removed[47].

2.3 Slid state external getter

In high temperature sealing condition, based on the stronger binding force between the getter and interstitial impurity, the interstitial impurities can be diffuse from rare earth metal to be purified to the getter, and the rare earth metal can be purified[48].

The common used getters for rare earth metals are Ca and Y, in this purified process, the getter and rare earth metal are wrapped in a Ta sheet and heat to a setting temperature, for the Ca getter, O impurity in Gd metal decreases from 1600 μg/g to 800 μg/g, and N impurity almost no decrease; for the Ca getter, the impurities of O, N and H in Gd, Tb and Dy metal decrease less than 20 μg/g, 45 μg/g and 10 μg/g respectively[49]. In order to improve contact area, the Y film was sputtered on the rare earth metal surface, O and N in Gd metal decrease from 187 μg/g and 97 μg/g to 55 μg/g and 33 μg/g respectively[50], O in Tb metal decreases from 490 μg/g to 62 μg/g[49].

3 Future development direction

In conclusion, the following aspects on the basic research and key technology development about rare earth metal purification need to be strengthened in the future:

(1) Vacuum distillation is the main massive purification method for high purity rare earth metals at present, but the basic theoretical research is still not systematic, especially the thermodynamic equilibrium and removal limit of trace impurities, which need to be further studied.

(2) Electrolytic refining has a very large application potential in the massive purification of rare earth metals with low saturated vapor pressure, and it is necessary to increase the research and development of basic theories, key technologies and equipment of electrolytic refining.

(3) High-purity rare earth metals need to use multifarious purification methods, resulting in a long preparation process, low production efficiency, high industrial costs. Developing new high-efficiency purification technology of rare earth metals, improving the purification efficiency of existing purification technology and optimizing the process flow are the key development directions in the future.

References

[1] PANG S, LU W, YANG Z, et al. Mechanism of removing ferrum impurity in lanthanum refined by electron beam melting[J]. Journal of Rare Earths, Accepted

[2] ZHANG X, MIAO R, LI C, et al. Impurity distribution in metallic dysprosium during distillation purification[J]. Journal of Rare Earths, 2016, 34(9): 924-930.

[3] XU K. Distribution of impurities of lanthanum during zone refining process[D]. Beijing: General Research Institute for Nonferrous Metals, 2020.

[4] SCHMIDT F A, MARTSCHING G A, CARLSON O N. Electrotransport of carbon, nitrogen and oxygen in lanthanum[J]. Journal of the Less-common Metals, 1979, 68(1): 75-83.

[5] GSCHNEIDNER J K A, EYRING L. Handbook on the Physics and Chemistry of rare earths[M]. North-Holland Publishing Company, 1978.

[6] FORT D, JONES D W, BEAUDRY B J, et al. Zone refining of rare earth metals: Lanthanum, cerium and gadolinium[J]. Journal of the Less-common Metals, 1981, 81(2): 273-292.

[7] DAI Y N, YANG B. Vacuum metallurgy of nonferrous metal materials[M]. Beijing: Metallurgical Industry Press, 2000: 30.

[8] BUSCH G, KALDIS E, MUHEIM J, et al. The purification of europium[J]. Journal of Less-common Metals, 1971, 24(4): 453-457.
[9] HABERMANN C E, DAANE A H. Vapor pressures of the rare-earth metals[J]. The Journal of Chemical Physics, 1964, 41(9): 2818-2827.
[10] HABERMANN C E, DAANE A H. The preparation and properties of distilled yttrium[J]. Journal of Less-common Metals, 1963, 5: 134-139.
[11] KARL A, GSCHNEIDNER J. The application of vacuum metallurgy in the purification of rare-earth metals[R]. 1965.
[12] LI G D, LIU Y L. Study on technology and optimization of purifying scandium metal by vacuum distillation method [J]. J. Chin. Soc. Rare Earths, 2000, 18(2): 183.
[13] IONOV A M, NIKIFOROVA T V, RYTUS N N. Aspects of the purification of volatile rare earth metals by UHV sublimation: Sm, Eu, Tm, Yb[J]. Vacuum, 1996, 47(6-8): 879-883.
[14] ZHANG X W, MIAO R Y, WU D G, et al. Impurity distribution in distillate of terbium metal during vacuum distillation purification[J]. Transactions of Nonferrous Metals Society of China, 2017, 27(6): 1411-1416.
[15] ZHANG L, ZHANG, X W, LI Z A, et al. Distribution model of highly volatile impurity in distillate of rare earth metal[J]. Vacuum, 2020, 176: 1-8.
[16] ZHANG L, ZHANG, X W, LI Z A, et al. Distribution model of lowly volatile impurity in rare earth metal purified by vacuum distillation[J]. Separation and Purification Technology, 2021, 262: 118314.
[17] ZHANG X, WANG Z, CHEN D, et al. Preparation of high purity rare earth metals of samarium, ytterbium and thulium[J]. Rare Metal Materials and Engineering, 2016, 45(11): 2793-2797.
[18] VERHOEVEN J. Electrotransport in metals[J]. Metallurgical Reviews, 1963, 8(1): 311-368.
[19] CARLSON O N, SCHMIDT F A, PETERSON D T. Electrotransport of interstitial atoms in yttrium[J]. Journal of the Less-common Metals, 1966, 10(1): 1-11.
[20] WILLIAMS J M, HUFFINE C L. Solid state electrolysis in yttrium metal[J]. Nuclear Science and Engineering, 1961, 9(4): 500-506.
[21] FU S, LI Z A, ZHANG Z Q, et al. Research on the purification of praseodymium metal by solid-state electromigration[J]. Rare Metals, 2015, 39(11): 1018-1023.
[22] PETERSON D T, SCHMIDT F A. Electrotransport of carbon, nitrogen and oxygen in gadolinium[J]. Journal of the Less-common Metals, 1972, 29(3): 321-327.
[23] SCHMIDT F A, CARLSON O N. Electrotransport of carbon in molybdenum and uranium[J]. Metallurgical Transactions A (Physical Metallurgy and Materials Science), 1976, 7A(1): 127-132.
[24] MUIRHEAD C M, JONES D W. Electrotransport of carbon, nitrogen and oxygen in scandium[J]. Journal of the Less-Common Metals, 1976, 50(2): 237-244.
[25] MARCHANT J D, SHEDD E S, HENRIE T A, et al. Electrotransport of impurities in rare-earth metals, using a pulsed current[R]. 1971.
[26] ZHONG J M, LI Z A, ZHANG X W, et al. Electrode connection optimization for both temperature difference and purification of lanthanum rod during solid-state electrotransport[J]. Rare Metals, 2016, 42: 713-718.
[27] MURPHY J E, WONG M M. Purification of yttrium metal by zone refining and field freezing[J]. Journal of the Less-common Metals, 1975, 40(1): 65-77.
[28] MIMURA K, SATO T, ISSHIKI M. Purification of lanthanum and cerium by plasma arc zone melting[J]. Journal of Materials Science, 2008, 43(8): 2721-2730.
[29] FORT D, BEAUDRY B J, GSCHNEIDNER JR K A. Ultrapurification of rare earth metals: Gadolinium and neodymium[J]. Journal of the Less-common Metals, 1987, 134(1): 27-44.
[30] OKABE T H, DEURA T N, OISHI T, et al. Thermodynamic properties of oxygen in yttrium-oxygen solid solutions [J]. Metallurgical and Materials Transactions B, 1996, 27(5): 841-847.
[31] OKABE T H, HIROTA K, KASAI E, et al. Thermodynamic properties of oxygen in RE-O (RE5Gd, Tb, Dy, Er) solid solutions[J]. Journal of Alloys and Compounds, 1998, 279(2): 184-191.
[32] LIU Y L, LI G Y, LI G D. Study on the preparation of dysprosium hypoxic metal by molten salt extraction[J]. J. Chin. Soc. Rare Earths, 2002, 20(supplement): 33-35.
[33] LIU Y L, YUN Y H. Study on the extraction and purification of yttrium metal from LiF-YF$_3$ molten salt[J]. Rare Earth, 2010, 31(1): 92-94.
[34] OKABE T H, HIROTA K, WASEDA Y, et al. Thermodynamic properties of Ln-O (Ln = La, Pr, Nd) solid solutions and their deoxidation by molten salt electrolysis[J]. Shigen-to-Sozai, 1998, 114(11): 813-818.
[35] HIROTA K, OKABE T H, SAITO F, et al. Electrochemical deoxidation of RE-O (RE = Gd, Tb, Dy, Er) solid solutions[J]. Journal of Alloys and Compounds, 1999, 282(1): 101-108.
[36] MERRILL C C, WONG M M. Electrorefining yttrium[R]. 1967.

[37] WU Y K, LI H Y, CHEN Y W, et al. Study on electrolytic refining of cerium in NaCl-KCl-CeCl$_3$ molten salt system [J]. J. Chin. Soc. Rare Earths, 2020, 38(1): 76-82.

[38] WANG W, LI Z A, WANG Z Q, et al. Research on the electrolytic refining of lanthanum metal by molten salts [J]. Rare Metals, 2013, 35(5): 770-777.

[39] ZWILLING G, JR, K. A. G. Fused salt electrorefining of gadolinium: An evaluation of three electrolytes [J]. Journal of the Less Common Metals, 1978, 60(2): 221-230.

[40] ZWILLING G, BAILEY D M, JR., K. A. G. Fused salt electrorefining of gadolinium: An evaluation of three electrolytes II: Deposition characteristics in LiCl-LiF-GdF$_3$, LiF-BaF$_2$-GdF$_3$ and LiF-GdF$_3$ [J]. Journal of the Less Common Metals, 1981, 78(1): 109-118.

[41] LI G L. Scientific basic research on the high purification process of gadolinium and terbium in rare earth metals [D]. Beijing: Beijing University of Science and Technology, 2016.

[42] LI L, LI G L, TIAN F, et al. Research on the preparation of high-purity gadolinium of new functional materials [J]. Material Protection, 2013, 46(S2): 28-29.

[43] LI G L, TIAN F, LI L, et al. Hydrogen plasma arc melting technology to prepare high-purity rare earth functional material gadolinium [J]. Material Protection, 2013, 46(S2): 25-27.

[44] LI G L, LI L, GUO H, et al. Effect of plasma arc melting method on carbon content in terbium metal [J]. High Voltage Technology, 2015, 44(1): 2925-2929.

[45] YANG Z F. Research on electron beam melting to purify metal lanthanum [D]. Beijing: Beijing Nonferrous Metals Research Institute, 2019.

[46] TIAN F. Study on the removal of oxygen_nitrogen_hydrogen impurities in rare earth metal gadolinium, terbium, dysprosium by solid-phase external suction method [D]. Beijing Nonferrous Metals Research Institute, 2014.

[47] LI G L, LI L, FU K, et al. Effect of active metal film method on oxygen content in rare earth metal terbium [J]. Functional Materials, 2015, 46(23): 23061-23063.

[48] LI L, LI G L, HAN L H, et al. Effect of magnetron sputtering Y film on N and O content in gadolinium [J]. Functional Materials, 2014, 45(17): 17034-17-39.

Effect of Pre-deformation Heat Treatment Process on Microstructure and Mechanical Properties of TB3 Titanium Alloy

Baohui ZHU[1,2,3], Feng DU[3], Xiaofei LI[3], Lin CHEN[1,2], Dongxin WANG[1,2], Jingming ZHONG[1]

1 State Key Laboratory of Special Rare Metal Materials, Northwest Rare Metal Materials Research Institute, Shizuishan, 753000, China
2 CNMC Ningxia Orient Group Co. Ltd, Shizuishan, 753000, China
3 Ningxia Horizontal Titanium Industry Co., Ltd, Shizuishan, 753000, China

Abstract: In order to strengthening the TB3 alloy wire, different heat treatment processes (solid solution, solid solution + aging, solid solution + cold deformation, solid solution + cold deformation + polygonization+aging) were preformed to analyze the properties and strengthening mechanism of TB3 wire. The OM, EBSD, TEM and universal testing machine were used to characterize the microstructure and properties. The result indicated that TB3 grains were flattened due to the deformation increased, the defects increased and the grain boundaries were distorted as well. The relatively characters showed obvious work hardening. After the polygonization+aging process, the small angle grain boundaries increased, the α phase precipitated decreased and priority to growth near the slip or high-density dislocation zone as the cold deformation increased. Moreover, a saturated metastable β phase formed through the solid solution process, and the polygonization process promoted the formation of dislocation cell. As a conclusion, the dislocation cell and the precipitated α phase comprehensive strengthened the TB3 wire. The optimal heat treatment process for TB3 titanium alloy wire is: 790 ℃×20 min, AC solid solution+cold deformation 50%+instantaneous heating to 630 ℃×15 min, AC +550 ℃×16 h, AC.

Keywords: Titanium alloy; TB3; Wire; Pre-deformation heat treatment; Mechanical

1 Introduction

TB3 alloy (Ti-10Mo-8V-1Fe-3.5Al) is a kind of metastable B-type titanium alloy that can be strengthened by heat treatment. The main advantage of the alloy is that it has excellent cold forming performance in the solid solution treatment state, and then can obtain high strength after aging treatment. It is mainly used for manufacturing high-strength aerospace fasteners above 1100 MPa grade[1-5]. However, with the development of aerospace vehicles, the demand for material properties has further increased.

Deformation and solution aging are the main means of strengthening TB3 titanium alloy. The strengthening mechanism of the deformation is that the movement of the dislocation is hindered, resulting in an increase in the strength of the alloy. The strengthening method of solution aging is to precipitate the strengthened phase and form dislocation distortion field, which leads to dislocation obstruction[6-7]. However, the limited strengthening effect of deformation or solution aging on TB3 alloy, the combination of the above strengthening method may achieve a better strengthening effect. Among various combinations of deformation and heat treatment systems, the suitable heat treatment is high temperature heat treatment and low temperature heat treatment for titanium alloys. High temperature deformation heat treatment conducts the plastic deformation under the phase transformation temperature, and enhanced the mechanical properties due to the crystallization process suppressed. However, the high temperature heat treatment accuracy is poor, and hardly to apply to the manufacturing of aviation fasteners such as rivets. The low temperature deformation heat treatment is the deformation of titanium alloy at room temperature, and the subsequent aging effect is strengthened by increasing the dislocation density of titanium alloy. In order to further improve the strength of TB3 titanium alloy, it is an effective and feasible scheme to introduce cold deformation (that's the hard state, H) in the heat treatment process. Its purpose is to couple the dislocation and precipitation to strengthen the ideal mechanical properties, that is, the pre-deformation heat treatment scheme[8-9].

In this paper, TB3 titanium alloy wire was treated with different heat treatment processes: solid solution (ST); Solution+aging (STA); Solution+x% cold deformation+polygonization+aging (STHxPA) to study the effects of pre-deformation heat treatment on the microstructure and properties of titanium alloy, and to provide a suitable heat treatment process reference for the strengthening of the alloy.

2 Experiment

2.1 Preparation

The Φ7.88 mm TB3 wires were prepared by three times melting of a Φ510 mm ingot in 3t vacuum arc remelting furnace, forged in the single-phase zone of 1150 ℃ firstly, whereafter 3 rounds of forging were preformed above the β phase zone, and finally rolled in the range of 50-80 ℃ above the phase transition point. The solution process was used by a chamber electric furnace (the temperature error was ±5 ℃). Several processes were driven during the experiments, that is, a solution treatment of 790 ℃×20 min (air cooling, AC) was performed, and then different pre-treatment of 30%-50% cold deformation were implemented on the solid solution treated wire (i.e Φ7.88→Φ6.60 mm (30%), Φ7.88→Φ6.10 mm (40%), Φ7.88→Φ5.57 mm (50%). Further more, a polygonization heat and aging treatment (i.e. instantaneous heating to 630 ℃×15min, AC, 550 ℃×960 min, AC) were followed. The specific process parameters were shown in Table 1.

Table 1 Experimental Process Parameters

Sample	Solid solution		Pre-treatment	Polygonization heat		Age	
	Temperature /℃	Time /min	Cold deformation /%	Temperature /℃	Time /min	Temperature /℃	Time /min
TB3	790	20	30	630	15	550	960
	790	20	40	630	15	550	960
	790	20	50	630	15	550	960

2.2 Characterization

The TB3 longitudinally-samples prepared by different processes were tested for tensile test and optical metallographic (OM) examination. The solution aging samples were further subjected to transmission electron microscopy (TEM) testing. The variations of grain morphology, texture, and KAM map of as-received sample obtained by Electron Backscattered Diffraction (EBSD) technique. The tensile properties were measured using an electronic universal testing machine, which based on the standard of GB/T 228—2010 Metallic Materials-Tensile testing-Part 1: Room Temperature Test Methods on an electronic universal testing machine, with a gauge length of 25 mm. Aims to ensure the accuracy of the experiment, two samples shall be tested for each group. A OLYMPUS GX51 inverted metallographic microscope was used to observe the surface microscope morphology. In order to further observe the multilateral structure and the morphology of aging precipitates, the solution aging samples was thinned by electrolytic double spraying, and was observed by FEI TECNAI F30 transmission electron microscopye.

3 Results and Discussion

3.1 Mechanical properties

Figure 1 shows the effect of cold deformation on the mechanical properties of TB3 alloy wires at room temperature, and the tensile properties of different treatment processes are shown in Figure 2.

The strength of the sample increases as the cold deformation increases, while the elongation decreases, as shown in Figure 1. It exhibits the work hardening characteristics obviously. A extremely continuous increase occurred in tensile strength, the yield strength initially significant increases, and then dramatic declines. The elongation shows decrease originally, and then gradually stabilizes. Figure 2(a) shows the strengthening level of polygonization treatment. Compared to the solid solution aging process, a higher strength of 200 MPa of polygonization treatment was obtained, and less reduction in plasticity. Speciffcally, the strength of the polygonization treatment increased as the cold deformation increased, moreover, the plasticity slightly increased. In a word, the polygonization treatment indicated a significant strengthening and toughening effect. Comparing the comprehensive properties of strength and plasticity, the mechanical

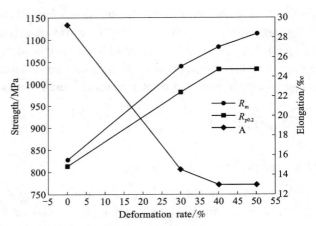

Figure 1 Effect of cold deformation on tensile properties of TB3 alloy

comprehensive properties of STH50PA state were considered to be optimal, that is, the optimal polygonization treatment process was 790 ℃×20 min, AC solid solution+50% cold deformation+instantaneous heating to 630 ℃×15 min, AC+550 ℃×16 h.

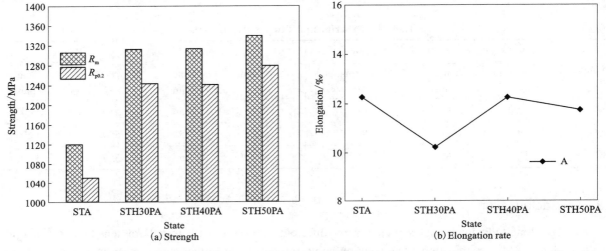

Figure 2 Tensile properties of TB3 alloy in different states

3.2 Microstructure

The morphologies of different cold deformation TB3 samples are shown in Figure 3. The solution heat treatment [Figure 3(a)] was equiaxed metastable β single phase morphology with an average grain size of approximately 60 μm. As the deformation increased, the shape of equiaxed grains elongated and flattened along the rolling direction, and gradually becoming fibrous structures. The elongation equiaxed grains was not significant in the 30% deformation [Figure 3(b)]. As the deformation was 50%, the flattening of the grains was very obvious [Figure 3(d)]. As a result, the deformation level and structural defects of grains gradually increased along the longitudinal direction due to the deformation increased, furthermore, some grain boundaries undergo distorted deformation.

The morphology of TB3 solution aging alloy wires presented in Figure 3(e), and the Figure 3(f), (g), (h) show all four different polygonization processes. Compared the optical morphology, un-polygonization treatment TB3 sample had more precipitates, with a large amount of precipitates distributed in areas of α phase. The content of precipitates decreased as the deformation increased. Additionally, most of the polygonization treatment samples' precipitates were distributed near the slip bands within the crystal.

The variations of grain morphology, texture, and KAM map of as-received sample obtained by EBSD technique are shown in Figure 4. It is clear that the microsturcture evolution is closely related to deformation.

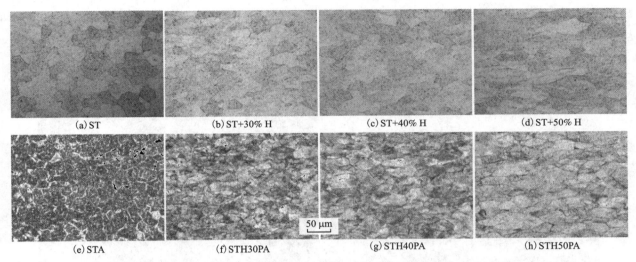

Figure 3 OM of cold deformation and treated with STA and different polygonization processes

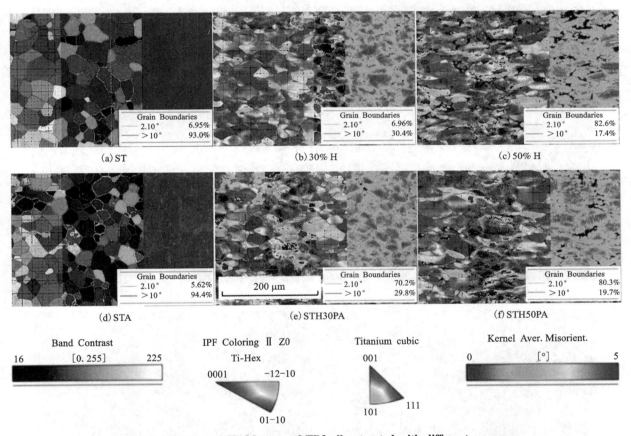

Figure 4 EBSD and KAM maps of TB3 alloy treated with different processes

It can be seen that a multi-directional selection presented as the deformation increased, which indicated the distortion in the grain. The small angle grain boundary increased substantially as the deformation, and the polygonization samples converted to the inside of the grain obviously. The various of deformation samples' small angle grain boundaries were due to the new dislocation clusters. After the aging+polygonization process, the conversion of small angle grain boundaries were contributed to the increased of subcrystal after polygonization. In the KAM maps, the orientations difference increased with the increase of deformation, and there were multiple dark areas. The appearance of dark areas indicated no preferred surface of the crystal. It is noted that the dark areas were repaired by the multilateralization treatment, which can be interpreted as the appearance of subcrystals that causes the crystal to grow in the preferred direction.

Figures 5 (a) and (b) show the TEM image and diffraction pattern spots of the solid solution aging TB3 alloy. It was observed that the precipitates α phase were distributed in a needle-like form uniformly. In addition, the precipitates and matrix β presented a certain coherent relationship, namely $\beta[111]//\alpha[2\bar{1}\bar{1}0]$.

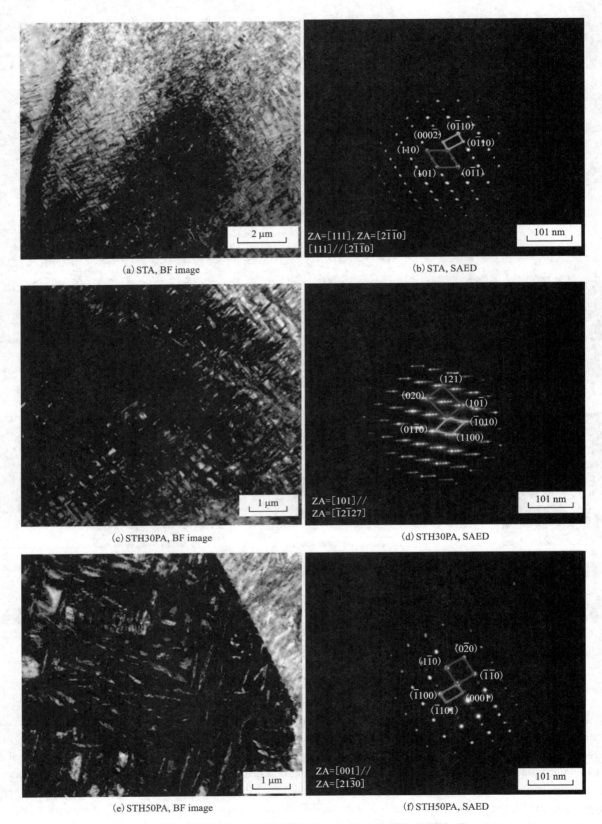

Figure 5 TEM images and diffraction pattern SAED of TB3 alloy

The TEM bright field image (BF) and selected area electron diffraction pattern (SAED) of solid solution aging TB3 alloy are shown in Figure 5. In pace with the deformation increased, the precipitates of needle-like α phases were getting thicker and interwoven. Precipitated α phase and matrix β phase took on the status of coherent relationship, and the diffraction pattern calibration for the three types of deformation were: β [101]//α [$\bar{1}2\bar{1}27$], β [111]//α [$17\bar{7}\,\bar{1}00$] and β [001]//α[$21\bar{3}0$].

3.3 Formation and strengthening mechanism of polygonization

TEM morphology of polygonization treatment TB3 alloy wires are shown in Figure 6. It is observed that the solution aging precipitated α phase distributed in a needle-like manner near the grain boundary (GB, Grain boundry), the aging precipitated α phase were formed near the slip band after 30% cold deformation and polygonization treatment. Furthermore, the dislocation morphology within the grain presented many uneven and high-density areas of local dislocations, forming dislocation entanglement (DT, Distribution angle, arrow), dislocation walls, and dislocation cells (DC, Distribution cell, arrow head). Specially, the dislocation cell was a subgrain or polygonized structure. The deformation increased, the number of polygonized structures increased, therefore, the needle-like precipitation α phase tended to become thicker, and it preferentially grew at the edges of dislocation entanglement or multilateral organizations.

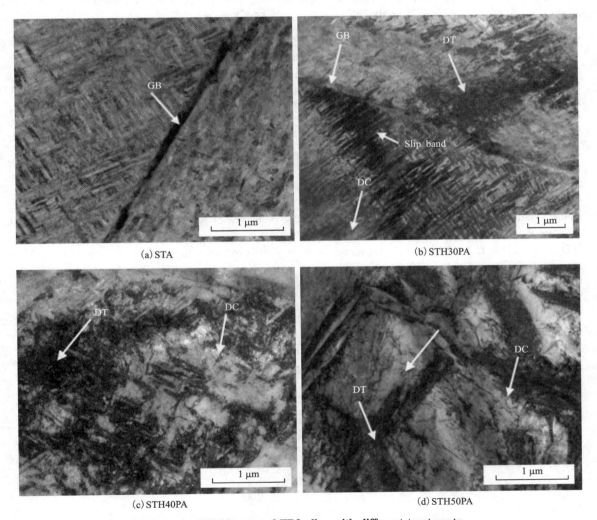

Figure 6 TEM images of TB3 alloy with different treatments

The polygonized structures were cellular in shape. It was due to dislocations' gather and wind in higher density local dislocations area as the deformation increased. It is to be noted that the dislocation aggregation in the cell wall was much higher than inside the cell. Especially, this type of dislocation cell was easy to be formed in high level auxiliary metals.

In general, the strengthening mechanism of the polygonization is that the saturated metastable state β phase formed after solid solution treatment, and a large number of dislocations were accumulated to form work hardening as the cold deformation. Additionally, the dislocation cells were formed by the polygonization process, and then precipitated needle-like α strengthening phase in high-density dislocation areas after aging. The deformation and precipitation phases worked together to obtain a comprehensive strengthening effect finally.

4 Conclusion

The TB3 wires were subjected to solid solution, solid solution+aging, solid solution+cold deformation and solid solution+cold deformation+polygonization+aging heat treatment processes. The TB3 grains got flatten and the boundaries twisted as the cold deformation increased, in addition, the precipitation α phase reduction. The cold deformation results macroscopically appeared as work hardening characteristics. Polygonization treatment transforms the work hardening generated by cold deformation into dislocation cells, and it leads to the small angle grain boundaries increased. The significant strengthening and toughening effect were observed due to the dislocation cells and the aging precipitation α phase. As a conclusion, the optimal heat treatment process for high hardness TB3 wire samples was 790 ℃×20 min, AC solid solution+50% cold deformation+instantaneous heating to 630 ℃×15 min, AC+550 ℃×16 h.

References

[1] HE C Y, YE H C, QU H L, et al. Effect of heat treatment on microstructure and mechanical properties of TB3 titanium alloy[J]. Hot Working Technology, 2011, 40(20): 181-182+185.

[2] SHA A X, WANG Q R, LI X W. Process analysis of BT16 titanium alloy fastener[J]. Rare Metal Materials and Engineering, 2006, 35(3): 455-458.

[3] ZHANG Y N, LI M Q, ZHANG S Q. Effects of hot-processing on microstructure and tensile properties for TB3 Alloy[J]. Heat Treatment of Metals, 2000, 9: 14-15.

[4] KARASEVSKAYA O P, IVASISHIN O M, SEMIATIN S L, et al. Deformation behavior of beta-titanium alloys [J]. Materials Science & Engineering A, 2003, 354(1-2): 121-132.

[5] China Aviation Materials Manual editorial board. Chinese Handbook of Aeronautical Material[M]. Beijing: China Standard Press, 2002.

[6] WU X D, YANG G J, GE P, et al. Inductions of β titanium alloy and solid state phase transition[J]. Titanium, Progress of Titanium Industry, 2008, 25(5): 1-6.

[7] GE P, ZHAO Y Q, ZHOU L. Strengthening mechanism of Beta titanium alloys[J]. Materials Reports, 2005, 19 (12): 52-56.

[8] SONG Z Y, SUN Q Y, XIAO L, et al. Effect of prestrain and aging treatment on morphologys and tensile properties of Ti-10Mo-8V-1Fe-3.5Al alloy[J]. Materials Science and Engineering A, 2010, 527: 691-698.

[9] SONG Z Y, SUN Q Y, XIAO L, et al. Precipitation behavior and tensile property of the stress-aged Ti-10Mo-8V-1Fe-3.5Al alloy[J]. Materials Science and Engineering A, 2011, 528: 4111-4114.

Prediction Techniques for Shrinkage and Porosity During Solidification of Be-Al Alloy

Yao XIE[1,2], Shenmin LI[2], Junyi LI[1], Dongxin WANG[1,*], Yajun YIN[2,*], Jingmin ZHONG[1]

1. State Key Laboratory of Special Rare Metal Materials, Northwest Rare Metal Materials Research Institute Ningxia Co., Ltd., Shizuishan, 753000, China
2. State Key Laboratory of Materials Processing and Die & Mould Technology, Huazhong University of Science & Technology, Wuhan, 430074, China

Abstract: As a new type of high performance material, Be-Al alloy has excellent properties such as light weight, high strength and high stiffness. The solidification temperature range of Be-Al alloy is very large, and the solidification process is mainly mushy solidification, which is easy to produce shrinkage and porosity defects. Numerous previous research on shrinkage and porosity defects of Be-Al castings rarely consider the resistance of mushy region to feeding, which can not be used to predict porosity of Be-Al castings. Therefore, this paper starts with the pressure loss in the paste region, carries out numerical simulation and porosity prediction of the solidification process of Be-Al alloy, furthermore, puts forward the prediction technology of porosity and porosity during the solidification process of Be-Al alloy. In the end, we simulate the step test piece and a Be-Al alloy casting using the InteCAST software. Compared to the actual pouring results, the model used in the experiment is able to accurately predict the solidification defects of the Be-Al alloy castings.

Keywords: Be-Al alloy; Shrinkage and porosity defects; Prediction of casting defect

1 Introduction

Beryllium (Be, melting point, 1287 ℃) has very low density, high strength and stiffness, and high thermal conductivity, this is the main reason that Be is of interest in aerospace, specialized commercial applications and military applications[1]. However, Be metal is a brittle material with an elongation of less than 3% at room temperature. It is apparent that an alloy is needed which would closely approach the desirable characteristics of Be but alleviate or eliminate the undesirable ones by using a very ductile material as the second component. The aluminium (Al, Face-centered cubic) is selected as the soft ductile envelope material for the hard Be particles, because good plasticity results from the multiple-slip systems of Al, and it has a relatively low density[2]. Be-Al alloy is also a kind of light metal[3] and been developed in the 1960s[2]. Combining the advantages of Be and Al, Be-Al alloys have excellent properties such as low density, high specific stiffness, specific strength, elevated thermal conductivity, and low expansion as well, and are widely used in aerospace and weapons under normal manufacturing conditions. When the temperature reaches (644± 1)℃, the eutectic point of Be-Al binary alloy is located at (2.4±0.5) at% Be. At this temperature, the solid solubility of Be in Al is only 0.3 at%, while Al is virtually insoluble in Be, moreover, the solubility is extremely limited. Therefore, in previous studies[4-6], it is considered that Be-Al alloy is a kind of composite material with discontinuous granular Be phase reinforcing Al matrix.

Be-Al alloys with expensive and toxic characters limit the experimental research and widely applications. In recent years, with the advancements of computer technology, numerical modeling and simulation have been widely used as a powerful tool to make a description of the solidification process of Be-Al alloys[7-10]. The extremely wide solidification temperature range leads to the solidification process of Be-Al alloy is mainly paste solidification. During the solidification process of Be-Al alloy castings, complex dendrite structure will be formed, which will produce greater resistance to the flow of metal liquid, resulting in the difficulty of feeding. As results, shrinkage holes and porosity defects are easily formed in the casting process of Be-Al alloy. Up to now, it has not been able to develop a highly complete preparation technology of Be-Al alloy castings, facing the urgent demand for large size, thin wall and complex Be-Al alloy components in the aerospace field. It is urgent to develop the corresponding preparation process.

Therefore, in this paper, an isolated liquid region search method considering the loss of feeding pressure is proposed, and a prediction model of shrinkage and porosity during solidification of Be-Al alloy is established. Meanwhile, the casting experiment was designed to verify the accuracy of the prediction model and numerical simulation of the shrinkage and porosity of Be-Al alloy during solidification.

2 Experimental procedures

During the solidification process of castings, due to the different cooling rates at different positions of castings, multiple isolated liquid phase zones will be formed, and the metal liquid feeding in each isolated liquid phase zone will not affect each other. The isolated liquid phase region will undergo shrinkage upon cooling, and these shrinkage quantities will remain in the isolated liquid phase region and eventually form porosity defects.

The solidification process of Be-Al alloy is mainly mushy solidification, during which a large number of complex dendrite structures will be formed, and these dendritic structures will increase the resistance on the feeding path. When the resistance reaches a certain value, the metallic liquid can no longer flow. Therefore, the feeding resistance of the paste region affects the formation of the isolated liquid region, thus, the contractile porosity formed. In this paper, we propose an isolated liquid region search algorithm that takes the paste region feeding resistance into account, then we construct a predictive model for the shrinkage and porosity during the solidification of Be-Al alloy based on this algorithm.

2.1 Pressure loss model of feeding path in paste region

The pressure loss caused by liquid metal flowing through a cell is related to numerous factors, such as solid-liquid ratio (the higher the solid phase ratio of the cell, the greater the pressure loss), liquid metal viscosity (the higher the viscosity, the greater the pressure loss), secondary dendrite arm spacing (the smaller the secondary dendrite arm spacing, the greater the pressure loss), etc. In order to calculate the pressure loss accurately, it is necessary to construct an accurate pressure loss model for the feeding path in the mushy region.

We treat the mushy region as a porous medium and derive the pressure loss formula for liquid metals flowing through the paste region from Darcy's law and the Carman-Kozeny equation (1)[11].

$$dP = \frac{\rho_s - \rho_l}{\rho_l} \frac{180\mu}{C_T^2} \frac{(1-f_l)^2}{f_l^2} \frac{\dot{T}^{\frac{5}{3}}}{G} dx \tag{1}$$

where μ is the viscosity of the melt, ρ_s and ρ_l are the density of the solid and liquid phase, and G is the modulus of the temperature gradient vector, $\dot{T}\left(=\left|\frac{\partial T}{\partial t}\right|=-\frac{\partial T}{\partial t}\right)$ is the mode of solidification rate, C_T is the secondary dendrite spacing coefficient, f_l is the liquid phase rate, then the pressure loss of metal liquid through a cell is

$$\Delta P = \int_{x_1}^{x_2} \frac{\rho_s - \rho_l}{\rho_l} \frac{180\mu}{C_T^2} \frac{(1-f_l)^2}{f_l^2} \frac{\dot{T}^{\frac{5}{3}}}{G} dx \tag{2}$$

Since the liquid phase rate has a linear relationship with temperature, the pressure loss calculation formula can be converted to:

$$\Delta P = \int_{T_1}^{T_2} \frac{\rho_s - \rho_l}{\rho_l} \frac{180\mu}{C_T^2} \frac{\dot{T}^{\frac{5}{3}}}{G} \frac{\left(\frac{T_1-T}{T_1-T_s}\right)^2}{\left(\frac{T-T_s}{T_1-T_s}\right)^2} \frac{1}{G} dT \tag{3}$$

In the solidification simulation process, the casting mesh is approximately millimeter level, the temperature gradient and solidification rate of the metal liquid within a cell length range can be regarded as a constant value, and the pressure loss of the metal liquid through a cell can be obtained by integrating:

$$\Delta P = \frac{\rho_s - \rho_l}{\rho_l} \frac{180\mu}{C_T^2} \frac{\dot{T}^{\frac{5}{3}}}{G^2} (A - 2B - C) \tag{4}$$

In which,

$$A = T_2 - T_1 \tag{5}$$

$$B = (T_1 - T_s) \ln \frac{T_2 - T_s}{T_1 - T_s} \tag{6}$$

$$C = (T_1 - T_s)^2 \left(\frac{1}{T_2 - T_s} - \frac{1}{T_1 - T_s} \right) \tag{7}$$

2.2 Isolated liquid region search algorithm based on pressure the loss model

In this paper, the Floyd algorithm is used to calculate the minimum pressure loss value of the casting cell and compare it with its water head value. If the minimum pressure loss value of a cell is greater than its water head value, the changed cell is regarded as a non-complementary cell, otherwise, it is regarded as a complementary cell. All adjacent complementary cells are treated as belonging to the same isolated liquid region.

The Floyd algorithm is an algorithm that finds the shortest path in a weighted graph with positive or negative edge weights (but no negative period). In the simulation of casting solidification process, we can regard casting cells as vertices in the graph, and the pressure loss between the center points of adjacent cells as the weight between two vertices, so as to build a weighted graph, and use the Floyd algorithm to calculate the minimum pressure loss between any two cells. The physical meaning of this value is the value of the minimum pressure loss when the metal liquid flows from one cell to another. The value obtained by adding the minimum pressure loss from one cell (cell A) to another cell (cell B) plus the maximum water head of the casting minus the water head of these cells A is called the minimum water head loss from cell A to cell B. The smallest of the minimum head loss values of all cells to a cell is called the minimum pressure loss value of the cell.

2.3 Isolation liquid phase region shrinkage distribution method

In time of the solidification process of the casting, since the density of the metal liquid is commonly inferior to that of the solid metal, the solidification process will produce shrinkage. As mentioned above, the shrinkage generated in the isolated liquid phase region will remain inside the isolated liquid phase region, so it is necessary to reasonably distribute these shrinkage amounts to the appropriate cells in the isolated liquid phase region. In the shrinkage distribution of the isolated liquid phase region of the paste, the larger the head of a cell and the smaller the pressure loss value, the stronger the shrinkage feeding capability.

Therefore, in determining the priority of the shrinkage distribution in the isolated liquid phase region of the paste, the algorithm determines the priority of the shrinkage distribution based on the difference between the head and the pressure loss value of the cell, with higher priority for larger difference. A parameter, Critical Delta P, needs to be introduced here. When the difference between the pressure value of the first cell of the previous priority and the pressure value of a cell of the next priority is less than this parameter, the cell can be added to the previous priority for shrinkage allocation calculation. For gravity casting, this parameter value is equal to Rou×9.8×dx, representing the increase in pressure value brought by gravity at 1 cell height.

As shown in Figure 1, although B1 and B2 are located in the same horizontal layer as A1-A4, because B1 and B2 are in the mushy region with greater feeding resistance, the pressure loss value is larger, and the feeding priority is lower than that of cell A1-A4. The feeding priority of B1 and B2 is the same as cell B3-B6 located in the layer below the horizontal. Similarly, C1-C6, D1-D4, and E1-E2 reside in the same priority. The pressure loss values of F1-F6 are too large to be replenished and therefore belong to the non-replenished cells.

When the amount of contraction generated from time t to time $t + \Delta t$ is distributed: If the amount of current contractions is small, all cells with the same highest priority are equally distributed. If the amount of fresh shrinkage is moderate, some of the highest priority cells will not be able to carry the shrinkage amount during the average allocation, and the excess cells that cannot carry the shrinkage amount will be reallocated to other cells of the same priority that can still distribute the shrinkage amount. If the amount of shrinkage is large and the amount of new shrinkage exceeds the sum of the amount of shrinkage that can be carried by all the cells of the highest priority, the amount of shrinkage that the sum of the amount of shrinkage that can be

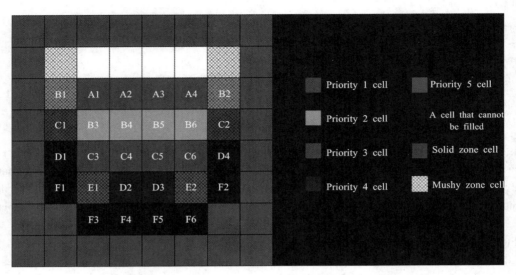

Figure 1　Priority of the isolated liquid phase region feeding cells

carried by all the cells of the highest priority is first allocated to these cells, and then the excess amount of shrinkage is allocated to the cells of the next priority. When the cells in the next priority are still unable to bear the amount of shrinkage, it continues until all shrinkage has been allocated.

For non-complementary cells, the shrinkage amount is treated separately, because it is regarded as an isolated liquid phase region with only one cell, the shrinkage amount generated during the solidification process of the cell itself cannot be supplemented by other cells, and it does not need to "consume" liquid metal to supplement other cells. Therefore, for a non-complementary cell, the gas phase ratio can be directly set as the product of the gas phase ratio before the cell becomes "non-complementary cell" plus the liquid phase ratio after the cell becomes "non-complementary cell" and the solidification shrinkage rate.

During the casting solidification process, the amount of shrinkage will be continuously generated, and the resulting shrinkage will be continuously distributed in the isolated liquid phase region according to the method described above to model the porosity generation. When all elements of the Gwynne degree are below the solid phase line, the casting solidification is complete and the entire simulation calculation is complete.

3　Prediction and analysis of porosity of Be-Al alloy

In order to verify the accuracy of the prediction model of shrinkage porosity and porosity during the solidification process of Be-Al alloy, a step piece of Be-Al alloy was designed in this paper, and the model was imported into InteCAST to simulate the solidification process of the casting, and the result was compared with the actual pouring result.

3.1　Experimental design of Be-Al alloy step casting

The current casting method for Be-Al alloy castings is mainly investment casting, and the castings are designed to be as compact as possible, with the step direction being downward. As shown in Figure 2, there are six levels of the ladder. The thinnest part is 5 mm, and the thickness of the ladder increases with 5 mm. The thickness of the maximum part is 30 mm, the height of the ladder is 15 mm, and the length is 80 mm.

The design of overall gating system is relatively large, and the casting process diagram is shown in Figure 3.

3.2　Solidification simulation of Be-Al alloy step parts

In this experiment, the temperature field is simulated using the InteCAST and the solidification process of the cast is simulated using the InteCAST gravity cast simulation system. First, the InteCAST pre-processed mesh molecular profile system was used for segmentation. In order to ensure the accuracy of the temperature field simulation results, at least two mesh cells are required in the thinnest part of the castings, which is the one with the lowest step size. The thickness is 5 mm, so the grid division step length is set to 2.5 mm×2.5

mm×2.5 mm. After division, the number of cast meshes was 8346, the number of poured mesh was 109339 and the total number of meshes was 117685. According to the actual casting process, the casting temperature is set at 1400 ℃, the ambient gas temperature is set at 25 ℃, and the mold shell temperature is set at 600 ℃. The heat transfer coefficient of the material in contact with the air is defined as: The heat transfer coefficient of casting/shell, shell/gas environment was set as 1000 W/(m^2·K), 50 W/(m^2·K), respectively. More thermophysical parameters are shown in Table 1.

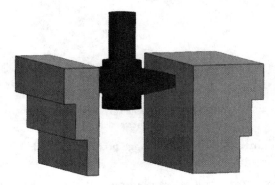
Figure 2 Three-dimensional diagram of Be-Al alloy castings

Figure 3 Three-dimensional diagram of gating system

Table 1 Thermophysical parameters of alloy and mold material[12-16]

Material property	Be-38 wt% Al	Ceramic mold	Air
Conductivity/(W·m^{-1}·K^{-1})	185	0.95	2.595
Density/(kg·m^{-3})	2150	966.7	1210
Specific heat/(kJ·kg^{-1}·K^{-1})	1.55	-	-
Liquidus/K	1355	-	-
Solidus/K	919	-	-
Viscosity/(cm^2·s^{-1})	0.06	-	-
Latent heat/(kJ·kg^{-1})	236.81	-	-

Figure 4 shows the porosity distribution of the stepped castings predicted by the shrinkage prediction model for the Be-Al alloy. Figure 4(a) shows the location distribution of pore pines with porosity greater than 0.1%. It can be seen that in addition to the position of the step near the edge of the thin wall side due to the relatively minor effect of the cooling speed of the mushy zone, additional positions are more or less affected by the mushy zone and produce a certain shrinkage. Figure 4(b) shows the location distribution of pore pines when the porosity is greater than 0.2%. At this time, the pines at the thinner step positions are virtually invisible, while the pines at the thicker step positions are still widely distributed. This indicates that the probability of porosity formation is higher and that contractile porosity defects are more pronounced in thicker staircases than in smaller staircases. These results are consistent with the actual CT results. In particular, the porosity of a cell does not represent the gas-phase ratio determined in the cell, but rather the probability of the formation of a pore. The larger the porosity, the higher the probability that the cell will form a pore.

Figure 5(a) shows cells with porosity greater than 0.4%. As the solidification shrinkage rate of the Be-Al alloy is 0.04, the conversion of a single liquid-phase cell to a solid-phase cell results in a shrinkage of 0.04 cell volume. In this paper, we define 10% of this quantity as a critical quantity, and cells with porosity greater than this value will have a strong tendency to produce constrictive holes. Figure 5(b) shows the area with the greatest probability of producing shrinkage holes at the step with a thickness of 20 mm (along the X direction) predicted by the shrinkage prediction model of Be-Al alloy. Figure 5(c) shows the enlarged image of the predicted shrinkage area at the 20 mm thickness step. This is consistent with the location of shrinkage holes in the actual pouring results of Be-Al alloy as shown in Figure 5(d). Furthermore, Figure 5(e) shows the comparison between the shrinkage area of the actual pouring result and the predicted result. By comparing the

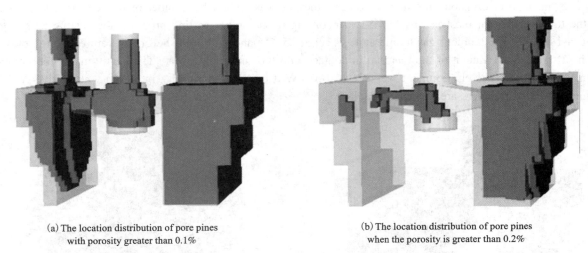

(a) The location distribution of pore pines with porosity greater than 0.1%

(b) The location distribution of pore pines when the porosity is greater than 0.2%

Figure 4　The porosity distribution of the stepped castings predicted by the shrinkage prediction model

shrinkage area of the actual pouring result with that of the predicted result, it can be found that the area of the overlap area between the predicted result and the actual result is 37783 pixels, and the area of the actual result is 43258 pixels, thus the prediction accuracy rate can be calculated as 87.34%.

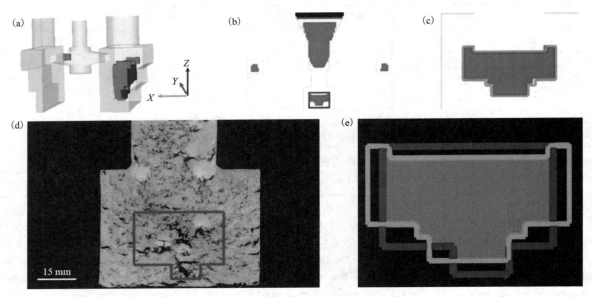

Figure 5　The distribution of defects, (a) the defects predicted by simulation, (b) the defects in thickness of 20 mm (along the X direction) predicted by simulation, (c) the enlarged image of the predicted shrinkage area at the 20 mm thickness step, (d) the defects in thickness of 20 mm (along the X direction) of Be-Al alloy, (e) the comparison between the shrinkage area of the actual pouring result and the predicted result.

4　Conclusion

(1) In this paper, we study the pressure loss of liquid metals flowing through a paste region and develop a pressure loss model for filling paths in the paste region. A search algorithm for the minimum pressure loss in the paste region and a search algorithm for the isolated liquid in the paste region have been proposed. Moreover, a predictive model for the shrinkage and porosity during the solidification of the Be-Al alloy castings has been developed.

(2) The prediction accuracy rate can be calculated as 87.34% by comparing the shrinkage area of the actual pouring result of Be-Al alloy with that of the predicted result.

Acknowledgments: This work was fully sponsored by NATIONAL KEY TECHNOLOGIES R & D PROGRAM, CHINA (GRANT NUMBER2021YFC2902304).

References

[1] TRUEMAN D L, SABEY P, GUNN G. Beryllium, in: Critical metals handbook, British Geological Survey[M]. Nottingham, England, 2013: 99-121.

[2] FENN R W, GLASS R A, Needham R A, et al. Beryllium-aluminum alloys[J]. J. Spacecr. Rockets., 1965, 2(1): 87-93.

[3] MOLCHANOVA L V, ILYUSHIN V N. Alloying of aluminum-beryllium alloys[J]. Russ. Metall. Met., 2013: 71-73.

[4] KUANG Z Y, YANG W S, JU B Y, et al. Achieving ultrahigh strength in Be/Al composites by self-exhaust pressure infiltration and hot extrusion process[J]. Mater. Sci. Eng. A, 2023, 862: 144473.

[5] KUANG Z, XIA Y, CHEN G, et al. Effect of interfacial strength on mechanical behavior of Be/2024Al composites by pressure infiltration[J]. Materials, 2023, 16(2): 752.

[6] YU L, WANG J, QU F, et al. Effects of scandium addition on microstructure, mechanical and thermal properties of cast Be-Al alloy[J]. J. Alloys. Compd., 2018, 737: 655-664.

[7] RAPPAZ M, GANDIN C A. Probabilistic modelling of microstructure formation in solidification processes[J]. Acta Metall. Mater., 1993, 41(2): 345-360.

[8] NASTAC L. Numerical modeling of solidification morphologies and segregation patterns in cast dendritic alloys[J]. Acta Mater., 1999, 47(17): 4253-4262.

[9] SHEHATA F, ABD-ELHAMID M. Computer aided foundry die-design[J]. Mater. Des., 2003, 24(8): 577-583.

[10] BELTRAN-SANCHEZ L, STEFANESCU D M. A quantitative dendrite growth model and analysis of stability concepts[J]. Mater. Sci. Eng. A, 2004, 35: 2471-2485.

[11] CARMAN P C. Fluid flow through granular beds[J]. Chem. Eng. Res. Des., 1997, 75: S32-S48.

[12] LIU X N, MA S G. Study and application of Al-Be alloy[J]. Chin. J. Rare Metals, 2003, 27(1): 62-65.

[13] WILLIAM S, OMAR S, SAID E S. Application of aluminum-beryllium composite for structural aerospace component[J]. Eng. Fail. Anal., 2004, 11(6): 895.

[14] ELMER J W, AZIZ M J, TANNER L E, et al. Formation of bands of ultrafine beryllium particles during rapid solidification of Al-Be alloys: Modeling and direct observations[J]. Acta Metall. Mater., 1994, 42(4): 1065-1080.

[15] CARTER D H, BOURKE M A M. Neutron diffraction study of the deformation behavior of beryllium-aluminum composites[J]. Acta Mater., 2000, 48(11): 2885-2900.

[16] WANG Z H, WANG J, YU L B, et al. Numerical simulation and process optimization of vacuum investment casting for Be-Al alloys[J]. Int. J. Metalcast., 2019, 13: 74-81.

Study on Synthesis of Submicron Polycrystalline Diamond Composite Using Cobalt Acetate as Cobalt Source

Jiarong CHEN[1,2], Jinghe ZHU[2], Qiaofan HU[2], Haiqing QIN[1,2],
Peicheng MO[1,2], Xiaoyi PAN[1,2], Jun ZHANG[1,2], Kai LI[1,2], Chen CHAO[1,2,*]

1 Guangxi Key Laboratory of Superhard Material, China Nonferrous Metal (Guilin) Geology and Mining Co., Ltd., Guilin, 541004, China
2 National Engineering Research Center for Special Mineral Material, China Nonferrous Metal (Guilin) Geology and Mining Co., Ltd., Guilin, 541004, China

Abstract: Acetic acid as a cobalt source was used as a catalyst on the surface of submicron diamond as a catalyst in the synthesis of polycrystalline diamond compact (PDC). The PDC is synthesized in a hinged Hexa-orientation Press. The catalytic mechanism of synthesis PDC is discussed. The polycrystalline diamond (PCD) phases in the PDC samples are confirmed by X-ray diffraction method (XRD). Scanning electron microscopy (SEM) images reveal that the PCD layer surface consists of a homogeneous diamond, cobalt and tungsten carbide alloy. Raman and abrasion ratio methods are used to analyze the performance of PDC. Experimental results show that the residual stress of PCD layer surface is compressive stress and cobalt acetate as a cobalt source can reduce the cobalt content in PDC.

Keywords: Polycrystalline diamond; Cobalt acetate; Cobalt source

1 Introduction

Polycrystalline diamond composite (PDC) has high hardness, wear resistance, thermal conductivity, strength and impact toughness.[1] Due to these excellent properties, PDC has been widely used in cutting tools,[2] wire drawing mould,[3] grinding drill bit and impact drill bit.[4,5] Moreover, PDC is considered as an excellent material for space drilling for its outstanding physical and chemical properties under vacuum condition[6]. Although the PDC has high hardness and good wear resistance, it cannot avoid damage after long-term use[7]. During use, when the polycrystalline diamond particles formed in the PDC are relatively coarse, the knife edge changes caused by the wear, cracking or shedding of the entire diamond during use will be relatively large. The wear, cracking or shedding of diamond particles like this is inevitable during the cutting process. The structural changes of this kind of cutting edge are difficult to observe with naked eye, but they will obviously appear on the surface of the workpiece to be processed, making the surface of the workpiece inconsistent. When the diamond particles in the synthetic PDC are relatively small, the damage structure changes are also relatively small, showing higher precision and smoothness. Therefore, the PDC synthesized by fine-grained diamond can realize the purpose of cutting instead of grinding. Submicron diamond is one of such fine particle diamonds. The basic structure of submicron PDC is the same as the traditional PDC. Both diamond powder and cemented carbide (WC-Co) substrate are sintered together. In this way, the PDC has both the high hardness and wear resistance of diamond, the strength and toughness of cemented carbide. The difference is that the submicron PDC uses the submicron diamond powder to synthesize the polycrystalline layer. Cobalt has been considered as one of the best catalysts for the synthesis of PDC. The traditional synthesis method of PDC uses cobalt powder as catalyst, cobalt content by volume accounts for more than 5%. High cobalt content is beneficial to the synthesis of PDC, while excessive cobalt content reduces the high temperature resistance of PDC. Therefore, in the application of high temperature environment, PDC decobalt treatment is needed. It is very important to reduce the catalyst content in PDC synthesis for sub-micron diamond composite.

In this work, we creatively introduce cobalt acetate as cobalt source to synthesize Submicron PDC. We also discuss the catalytic mechanism of cobalt acetate as cobalt source to synthesize PDC.

2 Materials and methods

2.1 Materials

Raw diamond powder was obtained from the Henan Yalong Superhard Material Co., Ltd (Zhengzhou, China), and has a size of 0.5-1 μm, and a purity of 99%. Sodium hydroxide, hydrochloric acid and nitric acid were obtained from the Sinopharm Chemical Reagent Co., Ltd (Shanghai, China). Cobalt powder was obtained from the Aladdin Bio-Chem Technology Co., Ltd (Shanghai, China), and has a size of 1-2 μm and has a purity of 99.9%. Absolute ethanol was obtained from Sinopharm Chemical Reagent Co., Ltd (Shanghai, China). Hydrogen was obtained from the Qixing District Hengyu Gas Wujiaohua Business Department (Guilin, China), and has a purity of 99.9%. All other materials were from the Tebang new materials Co., Ltd (Guilin, China).

2.2 Methods

The raw diamond powder was treated for 1.5 h in molten sodium hydroxide in a stainless steel crucible to remove pyrophyllite impurities followed by 3 h treatment in hot aqua regia to neutralize residual alkali. Cobalt acetate was dissolved in a beaker filled with deionized water, then 0.5-1.0 μm submicron diamond powder was added into the beaker and stirred for 1.5 h to obtain a suspension. The samples were described in Table 1.

Table 1 Description of the samples under study

Samples	a	b	c	d	e
Cobalt acetate/g	0.25	0.50	1.00	2.00	4.00
Diamond powder/g	20.00	20.00	20.00	20.00	20.00

The suspension was stirred and dried in 100 ℃ water bath to avoid solid-liquid stratification and obtain solid mixture. The solid mixture was processed for 1.5 h in a 150 ℃ oven and then placed into a stainless steel tank with stainless steel balls (200 g of 2 mm diameter balls & 500 g of 5 mm diameter balls), And a horizontal rotary mixing 120 r/min for 2 h to obtain submicron diamond powder with cobalt acetate coating. The powder and stainless steel balls were separated by 200-mesh stainless steel sieve. The powder was decomposed at 350 ℃ for 1 h, and then at 900 ℃ and vacuum degree 1×10^{-4} Pa in vacuum furnace for 1.5 h to remove impurity gas and decompose cobalt acetate, and the powder was restored and hydrogenated in a tube furnace at 900 ℃ for 1.5 h with hydrogen[8]. The hydrogenated powder and tungsten cobalt alloy substrate were loaded into a niobium cup with zirconium cover and charged into a high pressure block of pyrophyllite. PDC samples were prepared in hinge type hexahedral high pressure equipment, and have a temperature of 1600 ℃, a pressure of 5.5 GPa. All the PDC samples were polished to a mirror finish by a series of diamond wheels. The phase composition of polycrystalline compact diamond (PCD) layer surfaces was recorded on a X-ray diffraction (XRD: PANalytical Xpert PRO, D/max 2500 v/pc). The morphology of the PCD layer and the diamond powder were examined by scanning electron microscope (SEM: Gemini SEM 300). The element content was measured by X-ray energy dispersive spectroscopy (EDS: EDAX OXFORD INCAIE350). Micro-Raman spectroscopy was carried out using 514 nm excitation from an argon laser. All Raman data of PCD layer surfaces was taken using a×50 objective lens, 2400 line/mm grating, 20 mW laser power and 60 s collection time for consistency. There are three dots measured by Raman spectroscopy. The test of abrasion ratio is to rub the PCD layer with a high-speed silicon carbide grinding wheel which H value is 3.4 mm in wear tester. Figure 1 shows the schematic diagram of the cutter made of different positions of the PDC.

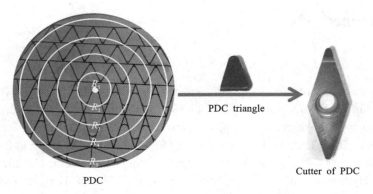

$R_1 = 0$ mm, $R_2 = 4.0$ mm, $R_3 = 8$ mm, $R_4 = 12$ mm, $R_5 = 16$ mm

Figure 1 The schematic diagram of the cutter made of different positions of the PDC

3 Results and discussion

The prepared samples (a, b, c, d & e) are analyzed by XRD and the related patterns are collected as shown in Figure 2. All of the samples [Figures 2(a), (b), (c), (d) and (e)] are indexed to cobalt (carbon: PDF05-0727). It can be found from the XRD pattern that the characteristic peak of cobalt becomes more and more obvious with the increase of cobalt content in samples (a, b, c, d & e). This is consistent with the weight rule of adding cobalt in the experiment. With the increase of cobalt content, the XRD characteristic peak of carbide tungsten (carbon tungsten: PDF 65-

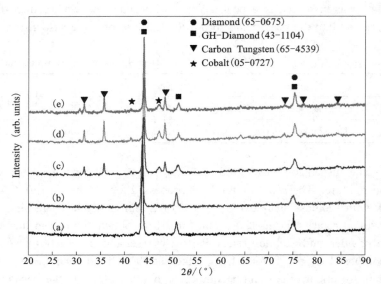

Figure 2 XRD patterns of PCD layer surfaces

4539) appears in the spectrum. This is the carbon tungsten penetrated into the polycrystalline layer by the high temperature wetting action of cobalt in tungsten-cobalt alloy. This result in accordance with most research results[9-10]. The main diffraction lines of all samples are related to the diamond crystal structure. The results show that the original phase of diamond is preserved during high temperature and high pressure synthesis. However, the formation of polycrystals is due to the formation of new D-D bonds by the graphitized carbon on the diamond surface under the action of high temperature and high pressure cobalt catalyst. Some diffraction lines are indexed to the GH-diamond carbon crystal structure and can be attributed to the (200) and (220) planes of the (111) cubic Gh-diamond carbon (PDF 43-1104). This is a new type of diamond phase and pressure synthesized at high temperatures, which joins raw diamonds together to form a polycrystalline diamond. The results show that a small amount of cobalt on the surface of diamond particles can catalyze the formation of D-D bonding between diamond particles.

The microstructure and the compositions on PCD layer surface and diamond powder with cobalt coating were analyzed by means of SEM and EDS. The SEM results indicate that the particle size of diamond raw materials ranges from 500 to 900 nm. The size of cobalt particles is less than 150 nm, and the layer formed by cobalt particles covers the surface of diamond particles. The PCD layer surface of all the samples were polished to a mirror finish before SEM test and diamond powder of all the samples were coated with cobalt acetate and transformed into a cobalt coating through a series of chemical reactions. The specific conversion mode is cobalt acetate decomposed at high temperature in air environment to obtain cobalt oxide, as shown in Eqs. (1 and 2).[11] Cobalt oxide is reduced by hydrogen to obtain elemental cobalt, as shown in Eq. (3). To prevent the oxidation of the fine cobalt element on the diamond surface, diamond particles loaded with cobalt were hydrogenated at high temperature[8].

$$Co(CHCOO)_2 4H_2O \xrightarrow{150\ ℃} Co(CHCOO)_2 + 4H_2O \quad (1)$$

$$6Co(CH_3COO)_2 + 13O_2 \xrightarrow{350\ ℃} 2Co_3O_4 + 24CO_2 + 18H_2O \quad (2)$$

$$Co_2O_3 + 3H_2 \xrightarrow{\geq 400\ ℃} 2Co + 3H_2O \quad (3)$$

All of the samples [Figures 3(a), (b), (c), (d) and (e)] show that the amount of cobalt added is relatively small, it is not easy for the discrete cobalt particles to appear outside the diamond. With the increase of the amount of cobalt added, the discrete cobalt particles gradually increase. After enlarging the SEM image of sample (a) from 20000 times to 100000 times, it can be clearly seen that the diamond surface is coated with granular cobalt, as shown in Figure 3(a_2). As shown in Figure 4, X-ray energy dispersive spectroscopy (EDS) analysis point can confirm that the composition of diamond surface is cobalt, in which the carbon peak is diamond. From the X-ray energy dispersive spectroscopy (EDS) analysis point shown in Figure 4, it can be determined that cobalt acetate has been converted to cobalt, as there is no distinct peak of the oxygen element. Figure 5 shows the surface morphology of polycrystalline layer of diamond composite sheet. In the figure, black is diamond and white is cobalt. Sample (a) has the lowest cobalt content and a large number of pores appear on the surface of its polycrystalline layer. Sample (e) has the highest cobalt content. The diamond particles are filled with too much cobalt, which prevents the formation of D-D bonds between diamonds.

Figure 3   Images (SEM) of diamond powder with cobalt coating

The practical application indicates that oversize and inhomogeneous distribution of residual stresses will induce PDC tools tipping and delamination, which would lead to great decrease in life performance of PDC[12]. So it is meaningful to compare the residual stress of the sample. According to the theory of solid molecules, the bond length of diamond molecules will become longer or shorter under the action of external forces. When diamond is under compressive stress, the bond length of diamond molecules will become shorter, the vibration frequency will increase, and the spectral band of Raman peak will move to high frequency. When diamond is under tensile stress, the bond length of diamond molecules will become longer, the spectral band will move to low frequency. The Raman shifts of the PCD layer surface is shown in Figure 6.

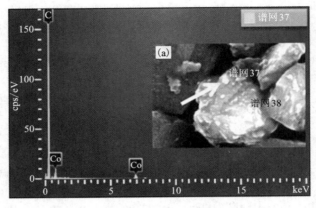

Figure 4 X-ray energy dispersive spectroscopy (EDS) analysis point for sample (a) of diamond powder with cobalt coating

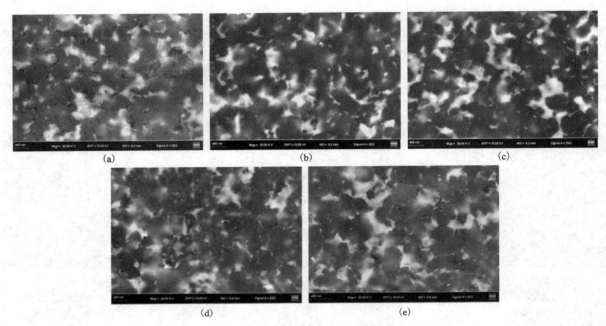

Figure 5 Images (SEM) of PCD layer surface

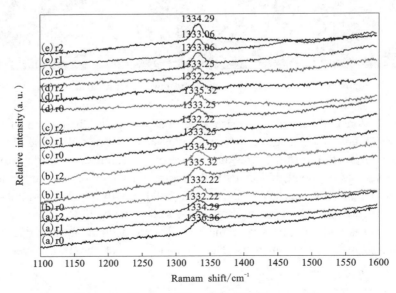

Figure 6 Raman shifts of the PCD layer surface

Table 2 Stress measurements for the PCD layer surface

Samples	Test point	Raman shift/cm^{-1}	$\Delta\gamma$/cm^{-1}	Stress/GPa
a	r0	1336.36	-4.36	-2.6914
a	r1	1334.29	-2.29	-1.4136
a	r2	1332.22	-0.22	-0.1358
b	r0	1332.22	-0.22	-0.1358
b	r1	1335.32	-3.32	-2.0494
b	r2	1334.29	-2.29	-1.4134
c	r0	1333.25	-1.25	-0.7716
c	r1	1332.22	-0.22	-0.1358
c	r2	1333.25	-1.25	-0.7716
d	r0	1335.32	-3.32	-2.0494
d	r1	1332.22	-0.22	-0.1358
d	r2	1333.25	-1.25	-0.7716
e	r0	1333.06	-1.06	-0.6543
e	r1	1333.06	-1.06	-0.6543
e	r2	1334.29	-2.29	-1.4136

According to the Raman test results and the stress calculation theory of diamond polycrystalline, the Raman absorption peaks of all samples are shifted to the wavenumber of diamond Raman characteristic peak 1332^{-1} cm in the state of no residual stress, and all the residual surface stresses are compressive stresses. The relationship between stress and frequency shift can be expressed by Eq. (4)[8].

$$\sigma_b = \frac{\gamma_0 - \gamma}{1.62} \quad (4)$$

As shown in Table 2, although there is a large difference in tensile stress on the surface of the samples, all samples show compressive stress from the center to the edge of the surface. The residual stress of diamond surface is related to the amount of metal catalyst, and the residual stress of diamond composite sheet can be controlled by controlling the content of catalyst, which is consistent with the literature reports[12]. The existence of compressive stress is due to the elastic deformation of the diamond particle structure of the submicron polycrystalline diamond composite sheet, which is not conducive to the stability of the diamond composite sheet, especially affecting the later application.

Table 3 PCD abrasion ratio for different samples

Samples & locations	a	b	c	d	e
R_1	1.8×10^4	1.8×10^4	2.3×10^4	2.3×10^4	2.3×10^4
R_2	1.8×10^4	1.9×10^4	2.3×10^4	2.4×10^4	2.3×10^4
R_3	1.8×10^4	2.0×10^4	2.4×10^4	2.4×10^4	2.3×10^4
R_4	1.8×10^4	2.0×10^4	2.4×10^4	2.3×10^4	2.3×10^4
R_5	1.9×10^4	2.1×10^4	2.4×10^4	2.4×10^4	2.4×10^4

Table 3 shows the wear ratio of diamond composite sheets. All samples show the same trend, the overall trend of wear ratio from center to edge of diamond composite sheets is that the wear ratio increases. With the increase of catalyst content, the wear ratio gradually tends to 2.4×10^4. Too much catalyst will also reduce the wear resistance of diamond composite sheet. The reason is that too much catalyst affects the formation of D-D bond between diamond particles.

4 Conclusion

In this paper, cobalt acetate can be used as cobalt source to catalyze the synthesis of polycrystalline diamond composite sheets. Polycrystalline diamond composite sheet can be synthesized by using cobalt acetate as cobalt source. Cobalt acetate as cobalt source can reduce the amount of catalyst in diamond composite. High cobalt content in high temperature and high pressure environment is conducive to tungsten carbide penetration into the PCD layer. Higher cobalt content can improve the wear resistance of the material, but too much cobalt can also reduce the wear resistance of the material. The residual stress of diamond composite sheet can be reduced with appropriate amount of catalyst.

Acknowledgments: This work was supported by the China Non-Ferrous Mining Group Co., LTD. Technology Planning Project (2022KJZX05), Guangxi Innovation Driven Development Special Fund Project (GUIKE AA17204098), development and industrialization of high performance superhard tool material technology, Guilin Scientific Research and Technology Development Project (20190204-1) and Science and Technology Innovation Fund Project (KDYCXJJ2020006) of China Nonferrous Metals (Guilin) Geology and Mining Co., Ltd.

References

[1] KANYANTA V, DORMER A, MURPHY N. Impact fatigue fracture of polycrystalline diamond compact (PDC) cutters and the effect of microstructure[J]. International Journal of Refractory Metals and Hard Materials, 2014, 46. DOI: 10.1016/j.ijrmhm.2014.06.003.

[2] YAHIAOUI M, PARIS J Y, DELBÉ K. Independent analyses of cutting and friction forces applied on a single polycrystalline diamond compact cutter[J]. International Journal of Rock Mechanics and Mining Sciences, 2016, 85: 20-26. DOI: 10.1016/j.ijrmms.2016.03.002.

[3] VEGA G, HADDI A, IMAD A. Investigation of process parameters effect on the copper-wire drawing[J]. Materials & Design, 2009, 30(8): 3308-3312. DIO: 10.1016/j.matdes.2008.12.006

[4] LI W, LING X, PU H. Development of a cutting force model for a single PDC cutter based on the rock stress state [J]. Rock Mechanics and Rock Engineering, 2019(8): 1-6. DOI: 10.1007/s00603-019-01893-7.

[5] JIA-LIN T, GANG L, SI-LU L. Geometric properties and three-dimensional rock-breaking mechanics of a new ring-embedded PDC drill bit[J]. Chinese Journal of Engineering Design, 2017, 24(1): 70-75. DOI: 10.3785/j.issn.1006-754X.2017.01.010.

[6] LI D F, LEI Y, XU S N. Particular coring bit for lunar soil drilling[J]. Journal of China University of Geosciences, 2013, 38: 167-173. DOI: 10.3799/dqkx.2013.S1.017.

[7] Quality and wear behavior of graded polycrystalline diamond compact cutters[J]. International Journal of Refractory Metals & Hard Materials, 2016, 56: 87-95.

[8] CHEN J R, MO P C, LIN F, et al. Effect of hydrogen protection on synthetic polycrystalline diamond composite[J]. Diamond and Related Materials, 2020, 1101-1105. DOI: 10.1016/j.diamond.2020.108118.

[9] QU G L, YAN L W, ZHANG WE B, et al. Research on solution of tungsten carbide and V anadium carbide in solid. phase of cobalt[J]. Cemented Carbide, 2014(1): 7. DOI: 10.3969/j.issn.1003-7292.2014.02.002.

[10] ZHANG L P. Research on preparation of gradient WC/Co cemented carbides by spark plasmSintering(SPS)[D]. Beijing: Beijing University of Technology, 2006.

[11] ZHANG K L, JIA M K, TANG H, et al. The thermal decomposition mechanism of cobaltous acetate[J]. Journal of Wuhan University (Natural Science Edition), 2002, 48(4): 4. DOI: 10.3321/j.issn:1671-8836.

[12] JIA H S, MA H G, GUO W, et al. HPHT preparation and micro-raman characterization of polycrystalline diamond compact with low residual stress[J]. Science China (Physics, Mechanics & Astronomy), 2010(8): 68-71. DOI: 10.1007/s11433-010-4045-7.

Wear-resistant CoCrNi Multi-principal Element Alloy at Cryogenic Temperature

Qing ZHOU*, Yue REN, Dongpeng HUA

State Key Laboratory of Solidification Processing, Center of Advanced Lubrication and Seal Materials, Northwestern Polytechnical University, Xi'an, 710072, China

Abstract: Traditional high strength engineering alloys suffer from serious surface brittleness and inferior wear performance when servicing under sliding contact at cryogenic temperature. Here, we report that the recently emerging CoCrNi multi-principal element alloy defies this trend and presents dramatically enhanced wear resistance when temperature decreases from 273 K to 153 K, surpassing those of cryogenic austenitic steels. The temperature-dependent structure characteristics and deformation mechanisms influencing the cryogenic wear resistance of CoCrNi are clarified through microscopic observation and atomistic simulation. It is found that sliding-induced subsurface structures show distinct scenarios at different deformation temperatures. At cryogenic condition, significant grain refinement and a deeper plastic zone give rise to an extended microstructural gradient below the surface, which can accommodate massive sliding deformation, in direct contrast to the strain localization and delamination at 273 K. In addition, the temperature-dependent cryogenic deformation mechanisms (stacking fault networks, twinning and phase transformation) are also revealed, which provide additional hardening and toughening of the subsurface material. The current work clarifies the superior wear resistance of the CoCrNi alloy, making it an excellent candidate for safety-critical applications involving cryogenic sliding conditions.

Keywords: Multi-principal element alloy; Wear resistance; Cryogenic temperature

Ce Substitution with La to Improve Magnetic Properties of RE-Fe-B Sintered Magnets

Ming YUE*, Hao CHEN, Weiqiang LIU

Faculty of Materials and Manufacturing, Beijing, 100124, China

Abstract: Nd-Fe-B sintered magnets play key role in many fields such as green energy technology. Nevertheless, the rapidly growing demand of the magnets arises the concern of overdue consumption of Nd metal. To balance the utilization of rare earth elements, Ce is commonly used to replace some Nd in Nd-Fe-B sintered magnet. Nevertheless, the magnetic properties are always undermined due to the appearance of $CeFe_2$ soft magnetic phase in the magnet. Moreover, the Ce^{4+} ratio, which is also harmful to the magnetic properties of the magnet, will increase with the Ce content in the magnet. In present study, we proposed a novel strategy to improve the magnetic properties via partial Ce substitution with La in the magnet. It is found that the La elements inhibit the generation of the $CeFe_2$ phase and tend to stay in the triple junctions, leading to microstructure modification that contributes directly to the enhancement of the coercivity of the magnet. Specifically, the La will promote the segregation of the Re/Cu/Ga elements and contribute to the formation of continuous thicker lamellar grain boundaries, which had a stronger magnetic isolation effect. On the other hand, partial La atoms entering the $Re_2Fe_{14}B$ phase are beneficial for promoting the Ce^{3+} ion ratio, which is benefit to the improvement of the remanence of the magnet. As a result, the coercivity, remanence, and maximum energy product are simultaneously enhanced for magnet with La substitution compared with those of undoped magnet. In detail, 10 wt.% substitution of Ce with La results in 2.8%, 0.5%, and 2.1% increment of coercivity, remanence, and maximum energy product of the magnet.

Keywords: Re-Fe-B sintered magnets; $CeFe_2$ phase; Ce-valence; Microstructure; Magnetic properties

A Novel Low-density ZrTiNbAl System Multi-principal Element Alloys with Unique Mechanical Properties

Xuehui YAN, Baohong ZHU*

GRIMAT Engineering Institute Co., Ltd., Beijing, 101400, China

Abstract: Next-generation high-performance structural materials are required to possess ultrahigh strength, excellent ductility, as well as low-density design. However, traditional alloy systems generally derive their properties from a dominant component, which brings great obstacles to breaking the existing performance limit. In recent years, many new alloys with promising properties are likely to be discovered near the center (as opposed to the corners) of phase diagrams, which no longer contain a single major component, but multiple major elements and form a concentrated solid-solution structure, as referred as multi-principal element alloys (MPEAs). The new concept provides the possibility to explore desirable performances in a wide composition space, and further achieve breakthroughs in performances. Here, a quaternary (Zr50Ti35Nb15)$_{100-x}$Al$_x$ (at%, x = 10, 20, 30, 40) alloys with a low density of less than 5.8 g/cm^3 were designed. The results show that these alloys have a remarkable tensile strength of 1.8 GPa and keep a tensile strain of 8%. Furthermore, the maximum strain of these alloys reached 25% and was accompanied by a tensile strength of 1.3 GPa. The current work does not only provide novel ultra-strong and tough structural materials with low density but also sheds new light on designing BCC-HEAs with attractive performances and strain-hardening ability. Moreover, the high-temperature deformation ability of these alloys was also studied. Alloys exhibited superior tensile ductility with a maximum elongation of up to 200%–a near-superplastic behavior–at elevated temperatures in the temperature range of 400-500 ℃, which provides the possibility for the subsequent preparation of complex structural components through thermal-mechanical processing.

Keywords: Multi-principal element alloys; Structural materials; Low-density; High strength and toughness

低密度且高强韧 ZrTiNbAl 系多主元合金的设计与性能研究

闫薛卉，朱宝宏*

1 有研工程技术研究院有限公司，北京，101400，中国

摘要：随着高精尖技术不断突破，各领域对结构材料的性能提出了更高要求，如低密度、更强韧等。然而，传统合金性能开发在空间逐渐趋于瓶颈，这为突破现有性能极限带来较大困难。多主元合金是一类基于"高熵"理念开发的复杂合金，旨在更广阔的成分空间中通过调控化学无序获得独特的理化性能。本工作基于"多主元"理念设计了一类低密度且综合力学性能优异的 ZrTiNbAl 系合金，该体系合金密度低于 5.6 g/cm^3，最高室温抗拉强度可达 1.8 GPa，最大拉伸塑性应变可达 25%，是一类具有发展潜力的新型高强韧结构材料。同时，为了满足该体系合金的工程应用需求，进一步开发了其超塑性变形能力，在中温区其变形量可达 200%，为后续通过热机械加工制备复杂结构件提供了可能性。

关键词：多主元合金；结构材料；低密度；高强韧

A Novel Nb-W-C Alloy with Special Microstructure and Excellent Mechanical Property

Xiaohong YUAN[1,2], Yan WEI[1,2], Li CHEN[1,2], Xian WANG[1], Changyi HU[1], Jialin CHEN[1]

1　State Key Laboratory of Advanced Technologies for Comprehensive Utilization of Platinum Metal, Kunming Institute of Precious Metals, Kunming, 650106, China
2　Yunnan Precious Metals Lab, Sino-Platinum Metals Co., Ltd., Kunming, 650106, China

Abstract: A new kind of Nb-W-C alloy with special microstructure and excellent mechanical property was designed and prepared by CVD subsequent with homogenizing heat treatment. Its microstructure is composed of columnar crystals with specific orientation, needle-like Nb_2C second phase and layered structure with Nb/W composition segregation. By analysing the effect of homogenizing heat treatment on the property and microstructure of the alloy, it was found that the as-deposited layered structure was retained in the microstructure after heat treatment at 1400 ℃, but disappeared by treated at 1600 ℃. The layered structure can effectively separate Nb_2C and inhibit the size coarsening of Nb_2C during carbide formation. During the homogenization process, needle-like carbides can maintain the strengthening effect under the condition of slight grain growth, but excessively high homogenization heat treatment temperature coarsens the grains and carbides, and ultimately weakens the strength of the alloy. The comprehensive strengthening effect of grain refinement and carbide pinning makes the room temperature strength of the sample reach 670 MPa after homogenizing treatment at 1400 ℃ for 6 hours, which is 35%-45% higher than the as-deposited Nb-W-C alloy and the existing widely used Nb521 alloy produced by melting-forging-annealing technology.

Keywords: Nb-W-C alloy; CVD; Homogenizing heat treatment; Mechanical property; Special microstructure

一种具有特殊组织和优异力学性能的新型 Nb-W-C 合金

袁晓虹[1,2]，魏燕[1,2,*]，陈力[1,2]，王献[1]，胡昌义[1]，陈家林[1]

1　稀贵金属综合利用新技术国家重点实验室，昆明贵金属研究所，昆明，650106，中国
2　云南贵金属实验室，贵研铂业股份有限公司，昆明，650106，中国

摘要：采用化学气相沉积(CVD)结合均匀化热处理工艺，设计并制备了一种具有特殊组织和优异力学性能的新型 Nb-W-C 合金。其微观组织由特定取向的柱状晶、针状 Nb_2C 第二相和具有 Nb/W 成分偏聚的层状结构组成。通过分析均匀化热处理对合金性能与组织的影响发现，沉积态的层状结构经 1400 ℃ 热处理后能够保留在组织中，而经 1600 ℃ 热处理后层状结构消失。层状结构的存在可以在热处理过程中有效分隔 Nb_2C，并抑制 Nb_2C 碳化物在加热过程中的尺寸粗化。在均匀化过程中，针状碳化物可以在晶粒略微长大的情况下保持强化效果，但过高的均匀化热处理温度会导致晶粒和碳化物粗化，最终弱化合金的强度。晶粒细化和碳化物钉扎的综合强化作用使得合金经 1400 ℃ 保温 6 小时均匀化处理后，其室温强度达到 670 MPa。与沉积态 Nb-W-C 合金和现有广泛使用的熔-锻-退火工艺生产的 Nb521 合金相比，均匀化处理后合金的室温强度提高了 35%~45%。

关键词：Nb-W-C 合金；CVD；均匀化热处理；力学性能；特殊微观结构

Effect of Compaction Pressure on the Sintering Activation Energy and Microstructure of IGZO Targets

Xiaokai LIU, Wenyu ZHANG, Benshuang SUN, Xueyun ZENG,
Huiyuyu ZHANG, Zhijun WANG, Chaofei LIU, Yongchun SHU, Yang LIU*, Jilin HE

School of Materials Science and Engineering, Zhengzhou University, Zhengzhou, 450001, China

Abstract: The effect of porosity and pore size on grain growth of IGZO target was studied by phase field simulation. The results show that when the number of pores is constant, the larger the pore size, the faster the grain growth rate and the coarser the grain. Under the condition of pores of the same size, the larger the number of pores, the stronger the pinning effect on grains, the slower the growth of grains, and the finer the grains. The porosity directly determines the density of the green compact. The influence of the green compact density on the densification of IGZO target was investigated by experiments. The IGZO targets with high density and uniform structure were obtained by molding-cold isostatic pressing (CIP) and two-step sintering (TSS) method. The effect of green density on the IGZO targets was studied under different pressures with CIP, and the critical relative density of the IGZO compact should reach 44% to obtain a higher density target. The densification activation energies of IGZO green compacts obtained by CIP at 180 MPa and 250 MPa were calculated to be 510 kJ/mol and 370 kJ/mol, respectively, through the main sintering curve (MSC). The smaller the grain size, the larger the resistivity can be obtained by the results of the resistivity test. The grain boundary area decreases with the increase of grain size, and the resistance of electrons passing through the grain boundary decreases, which reduces the resistivity of the target. Finally, the IGZO target with a relative density of 99.50%, a grain size of 8.59 μm and a resistivity of 4.3 mΩ·cm was obtained by optimizing the TSS process.

Keywords: IGZO; Target; Densification activation energy; Two-step sintering

Tribological Properties of MoS$_2$/a-C: Si Composite Films under High-temperature Atmosphere and Vacuum Environments

Yanjun CHEN[1,2], Songsheng LIN[1,*], Fenghua SU[2,*]

1　Institute of New Materials, Guangdong Academy of Sciences, National Engineering Laboratory of Modern Materials Surface Engineering Technology, Guangdong Provincial Key Laboratory of Modern Surface Engineering Technology, Guangzhou, 510651, China

2　School of Mechanical and Automotive Engineering, South China University of Technology, Guangzhou, 510640, China

Abstract: MoS$_2$/a-C: Si composite films with different contents of C and Si elements were prepared by a hybrid deposition system consisting of a superimposed high-power impulse magnetron sputtering (HiPIMS) - direct current magnetron sputtering (DCMS) and a mid-frequency magnetron sputtering (MFMS). The content of C and Si element in the films was controlled by varying the DC current of the graphite-silicon (C-Si) mosaic target. The effects of C and Si element doping contents on the microstructure and mechanical properties of the films were investigated. In addition, the tribological behavior of the films was evaluated in atmosphere environment at elevated temperature (up to 350 ℃) and vacuum environments were evaluated. The results showed that MoS$_2$/a-C: Si composite films exhibited excellent tribological properties in both high-temperature atmospheric and vacuum conditions. MoS$_2$/a-C: Si composite films exhibited low friction coefficient at elevated temperature due to the improved thermal stability. In particular, MoS$_2$/a-C: Si composite film with 40.37 at% C showed a low friction coefficient of 0.051 and a wear rate of 1.47×10^{-6} mm^3/(N·m) in ambient air at 350 ℃. MoS$_2$/a-C: Si composite film also exhibited excellent tribological properties due to the precipitation of C on the wear mark under vacuum environment. The average friction coefficient of MoS$_2$/a-C: Si composite film under vacuum environment was as low as 0.018, and the wear rate was as low as 1.99×10^{-7} mm^3/(N·m).

Keywords: MoS$_2$/a-C: Si composite film; High-power impulse magnetron sputtering; Mid-frequency; High-temperature friction; Vacuum friction

硅掺杂无氢类金刚石碳膜在大气环境中的热诱导超低摩擦

陈彦军[1,2]，林松盛[1,*]，苏峰华[2]

1　广东省科学院新材料研究所，现代材料表面工程技术国家工程实验室，广东省现代表面工程技术重点实验室，广州，510651，中国

2　华南理工大学机械与汽车工程学院，广州，510640，中国

摘要：通过由并联叠加的高功率脉冲磁控溅射-直流磁控溅射(HiPIMS-DCMS)和中频磁控溅射(MFMS)组成的混合沉积系统制备了不同含量的C和Si元素掺杂的MoS$_2$/a-C: Si复合薄膜。通过改变C-Si镶嵌靶的DC电流大小来控制薄膜中C和Si元素的含量。研究了C和Si元素掺杂量对复合薄膜微观结构和力学性能的影响。此外，研究了复合薄膜在高温大气环境(高达350 ℃)和真空环境下的摩擦学性能。结果表明：MoS$_2$/a-C: Si复合薄膜在高温大气和真空条件下均表现出优异的摩擦学性能。MoS$_2$/a-C: Si复合薄膜由于热稳定性的提高在高温下具有较低的摩擦系数。其中，C含量为40.37 at%的MoS$_2$/a-C: Si复合薄膜在350 ℃大气环境中的摩擦系数为0.051，磨损率为1.47×10^{-6} mm^3/(N·m)。在真空环境下，MoS$_2$/a-C: Si复合薄膜由于C在磨痕表面的析出也表现出优异的摩擦学性能。C含量为40.37 at.%的MoS$_2$/a-C: Si复合薄膜在真空环境下的平均摩擦系数低至0.018，磨损率低至1.99×10^{-7} mm^3/(N·m)。

关键词：MoS$_2$/a-C: Si复合薄膜；高功率脉冲磁控溅射；中频；高温摩擦；真空摩擦

Effect of Y Doping on Hot Corrosion Behavior of NiAlHf Coating

Xiaoya LI[1,2], Qian SHI[1,*], Songsheng LIN[1], Mingjiang DAI[1], Kesong ZHOU[1]

1 Institute of New Materials, Guangdong Academy of Sciences, National Engineering Laboratory of Modern Materials Surface Engineering Technology, Guangdong Provincial Key Laboratory of Modern Surface Engineering Technology, Guangzhou, 510651, China
2 State Key Laboratory of Powder Metallurgy, Central South University, Changsha, 410083, China

Abstract: The inlet temperature of aero engine turbine has gradually increased, and the environment has become more and more harsh, especially when working under marine atmosphere environment, the high-temperature components are subjected to the combined action of high-temperature oxidation and mixed molten salt, which needs higher requirements for high-temperature oxidation resistance and molten salt corrosion resistance of aero engine. NiAlHf coating is a candidate material for a new generation of high-temperature protective coatings because of its excellent oxidation resistance, however, the oxide scale is easy to fall off and has poor corrosion resistant to S at high temperature, leading to limiting its application. And Y element has a strong pinning effect on S, which can slow down the internal penetration of S and improve the corrosion resistance of the coating. Therefore, this study prepares Y-doping NiAlHf coating on nickel-based single crystal alloy by arc ion plating technology to obtain protective coating with both oxidation resistance and molten salt corrosion resistance at high temperature. The results show that a denser oxide scale can be formed on the surface of Y-doping NiAlHf coating during corrosion process, preventing the internal penetration of S to a certain extent. At the same time, compared with Ni, Al and Hf elements, Y has a stronger bonding effect on S, which further alleviates the catalytic corrosion effect of S, and then improves the corrosion resistance of NiAlHf coating.

Keywords: Hot corrosion; Y-doping; Mixed molten salt

Y 掺杂对 NiAlHf 涂层热腐蚀行为的影响

李小亚[1,2], 石倩[1,*], 林松盛[1], 代明江[1], 周克崧[1]

1 广东省科学院新材料研究所，现代材料表面工程技术国家工程实验室，广东省现代表面工程技术重点实验室，广州，510651，中国
2 中南大学粉末冶金研究院国家重点实验室，长沙，410083，中国

摘要：航空发动机涡轮进口温度逐步提升，服役环境越发苛刻，尤其在海洋大气环境下工作时，高温部件受到高温氧化与混合熔盐的共同作用，对航空发动机高温抗氧化与耐熔盐腐蚀性能提出了更高的要求。NiAlHf 涂层因具有优异的高温抗氧化性能成为新一代高温防护涂层的候选材料，但氧化皮易脱落，不耐 S 腐蚀，限制了其应用。而 Y 元素对 S 具有较强的钉扎作用，可减缓 S 的内渗透，提升涂层耐高温熔盐腐蚀性能。因而，本研究采用电弧离子镀技术，通过在镍基单晶合金上制备 Y 掺杂 NiAlHf 涂层以获得同时具有高温抗氧化与耐熔盐腐蚀性能的高温防护涂层。结果表明：Y 掺杂 NiAlHf 涂层在腐蚀过程中涂层表面可生成较致密的氧化层，可在一定程度上阻止 S 的内渗透；同时与 Ni，Al，Hf 元素相比，Y 对 S 具有更强的键合作用，进一步缓解了 S 对涂层的催化腐蚀，进而提升了 NiAlHf 涂层的耐熔盐腐蚀性。

关键词：热腐蚀；Y 掺杂；混合熔盐

The Correlation Mechanism Between the Grain Boundary Characteristics of Sintered Nd-Fe-B Base Magnets and the Heavy Rare Earth Element Diffusion Behaviors

Qingzheng JIANG[1,2,3,*], Qingfang HUANG[1,3], Zhenchen ZHONG[1,2], Hang WANG[1,2]

1. Jiangxi Key Laboratory for Rare Earth Magnetic Materials and Devices & College of Rare Earths, Jiangxi University of Science and Technology, Ganzhou, 341000, China
2. National Rare Earth Function Materials Innovation Center, Ganzhou, 341000, China
3. Fujian Key Laboratory for Rare Earth Functional Materials, Longyan, 366300, China

Abstract: Nd-Fe-B permanent magnetic material is widely used in many fields due to its outstanding hard magnetic properties. The application of heavy rare earths (HRE) grain boundary diffusion technology can effectively improve the coercivity of magnets with minor reduction in remanence. This method has already been used to fabricate the magnets with high coercivity and working temperature in permanent magnet motors. However, the HRE accumulate on the surface of the magnet. Anisotropic diffusion of HRE elements is observed in sintered Nd-Fe-B magnets. The deteriorated squareness of diffused magnets also limits the widely application of this technology. In this investigation, the addition of Nd-Al alloy is used to regulate the composition and distribution of the grain boundary for Nd-Fe-B base materials by dual alloy method. Thin and narrow grain boundary is formed in base materials and the diffusion channel is optimized. When HRE diffuse along hard magnetization direction with a depth of 10 mm, the increment of coercivity for original magnet is 1.3 kOe less than that for grain boundary modified magnets. The electronic probe micro-analysis results show that agglomerated Tb-rich phase is decreased and obvious Tb-rich shell can be observed in grain boundary modified magnets. The grain boundary modified magnet show more uniform microstructure and better squareness. For 10 mm thick base materials, the squareness of diffused magnets is as high as 0.92 after HRE diffused along hard magnetization direction. Longer diffusion time and more usage amount of HRE are needed for original magnet. Furthermore, compared with original magnets, the grain boundary modified magnet show better magnetic properties at 150 ℃.

Keywords: Grain boundary; Nd-Fe-B alloy; Magnetic property; Heavy rare earth element

Thermodynamic Calculation and Experimental Investigation of NiCr-Cr$_3$C$_2$ Coating Prepared by Detonation Spray

Song QIN, Donghua LIU*, Sen HAN, Jianqiao ZHOU

Hunan Research Institute of Metallurgy and Materials Co., Ltd., Changsha, 410129, China

Abstract: In this study, the NiCr-Cr$_3$C$_2$ hardcoating was prepared on GH864 by means of the Detonation Spray (DS) technique. SEM and XRD were employed to analyze the microscopic morphology and phase composition of the coating. Microhardness tester, surface roughness tester, universal testing machine, and contact angle measuring instrument were used to characterize the hardness, porosity, bonding strength, roughness, thermal shock performance, and surface free energy of the coating, respectively. The thermodynamic calculation (Thermo-Calc 2023b, TCNi6 database) was studied in detail, and a developed database of CALPHAD approach was established to guide the design of coating and predict the phase composition and performances. The results showed that the NiCr-Cr$_3$C$_2$ hardcoating was complete, compact, and well-bonded. The thermodynamic calculation was successful to a certain extent, agreeing with the experimental characterization well in terms of phasecomposition and the major enriched element of phases. As the oxygen-fuel ratio increased, the hardness of the coating increased monotonously, and the porosity of the coating decreased first and then increased. When the oxygen-fuel ratio was 1.5, the quality of the as-prepared coating was the best, with a microhardness of 810 HV, a porosity of 0.58%, a bonding strength of 72.1 MPa, a surface roughness of 4.78 μm, and a surface free energy of 25.43 mJ/m^2. After being kept at 800 ℃ for 10 min and quenched with water at room temperature for 20 cycles, the coating had no warping, cracking, or peeling. Besides, the oxygen-fuel ratio changed the surface free energy and its components of the coating, thus having an impact on the surface characteristics of the coating.

Keywords: Detonation spray; Coating; CALPHAD method; Thermodynamic; Performance

Influence of Feed Rate on the Abradability of AlSi/PHB Sealing Coatings

Xinwo ZHAO[1,2,3,*], Jiangang SUN[1,2,3], Deming ZHANG[1,2,3]

1 Beijing General Research Institute of Mining and Metallurgy Group Co., Ltd., Beijing, 100160, China
2 Beijing Engineering Technology Research Centre of Surface Strengthening and Repairing of Industry Parts, Beijing, 102206, China
3 Beijing Key Laboratory of Special Coating Material and Technology, Beijing, 102206, China

Abstract: Abradable seal coating is a material used in aero engines to improve its efficiency. In this paper, using the method of simulated working conditions, the self-developed BGRIMM-ATR high-temperature and ultra-high-speed abradability test machine is used to conduct abradability tests. In this paper, using the method of simulated working conditions, the self-developed BGRIMM-ATR high-temperature and ultra-high-speed abradability test machine is used to conduct abradability experiments on the AlSi/PHB seal coating at different feed rates. Experiments on the AlSi/PHB seal coating at different feed rates to study the effect of feed rate changes on AlSi/the influence of PHB seal coating on abradability, the macroscopic and microscopic morphology of the coating were analysed by stereo microscope and scanning electron microscope (SEM), and a comprehensive evaluation of the abradability was carried out by a variety of characterization methods. The results show that at room temperature, with a linear velocity of 350 m/s and a feed depth of 500 μm, the abradability of the coating was reduced. When the feed rate increases from low to high, the abradability of the AlSi/PHB coating shows a trend that first increases and then decreases. The change of the feed rate has a significant effect on the relative abradability of AlSi/PHB. The abradability of the coating is excellent at the medium feed rate, which is mainly manifested by the cutting mechanism; at the low feed rate and the high feed rate, the abradability of the coating is reduced, which is mainly manifested as deep furrows, cracks, and chipping of the coating. The main mechanisms are ploughing, melting, and collision tearing.

Keywords: Sealing coatings; Abrasion behaviour; Abradability; Abradability tester

A New Method of Preparing Hydrophobic Photocatalytic Composite Coatings Based on Nano-TiO$_2$ and Soluble Polytetrafluoroethylene Prepared by Suspension Plasma Spraying

Chunyan HE[1], Xiujuan FAN[2,*], Shuanjian LI[1], Sainan CUI[2], Jie MAO[2], Chunming DENG[2], Min LIU[2], Kesong ZHOU[2]

1 Guangdong University of Technology, Guangzhou, 510006, China
2 Guangdong Academy of Sciences, Guangdong Institute of New Materials, National Engineering Laboratory for Modern Materials Surface Engineering Technology, The Key Lab of Guangdong for Modern Surface Engineering Technology, Guangzhou, 510651, China

Abstract: In this study, an environmentally friendly method was utilized to prepare a hydrophobic surface with photocatalytic features which can be applied on large surfaces without the complexity of the chemical methods. In this regard, the suspension of anatase titanium dioxide nanoparticles powder and soluble polytetrafluoroethylene (PFA) dispersion suspension were applied on 304 stainless steel using suspension plasma spraying. In addition, the PFA mass accounted for 0%, 5%, 15% and 25% of the total mass of PFA/TiO$_2$ suspension, and the corresponding prepared coatings were named as: 0 PT, 5 PT, 15 PT and 25 PT, respectively. The visible light absorption performance and photocatalytic activities of the prepared coatings were also explored by examining the degradation of methylene blue (MB) under UV-visible irradiation. The results showed that the PFA/TiO$_2$ coatings present many circular and ellipsoidal nanoparticles, flocculent porous micron-nano structure due to the PFA. It is important that the PFA is not carbonization and defluorination in suspension plasma spraying. Also, PFA evenly distributed in the coating. It led to formation of a hydrophobic photocatalytic surface with static contact angles above 90° in PFA/TiO$_2$ coatings, in particular, 85 wt.% anatase phases in 25 PT coating with static contact angles 134.1°. The PFA/TiO$_2$ coatings present better hydrophobicity and photocatalytic performance compared to the TiO$_2$ coating. The photocatalytic performance under visible light is gradually increasing in the 0 PT, 5 PT, 15 PT and 25 PT coatings, and the degradation rate of MB is up to 95% in 25 PT.

Keywords: Hydrophobic; Photocatalytic; Suspension plasma spraying; PFA/TiO$_2$

Challenges and Development Progress of Mn-based Cathode Materials

Jun WANG, Xuequan ZHANG, Yafei LIU*, Yanbin CHEN*

BGRMM Technology Group, Beijing Easpring Materials Technology Co. Ltd., Beijing, 100029, China

Abstract: Current bottlenecks in cobalt supply have negatively impacted commercial lithium-ion battery application and inspired the development of Co-free cathodes with high energy density. Li-Mn-rich layered oxides (LMRs) are promising to be next generation cathodes for lithium-ion batteries due to their high specific capacity (>250 mA·h/g) and low cost, but the voltage decay and capacity fading during cycling are the main challenge for their commercialization. Herein, Co-containing and Co-free LMRs were synthesized for investigating the function of Co on the electrochemical performance. It is found that Co can promote the oxidation of O^{2-} and contribute more capacity above 4.5 V, which leads to oxygen release and irreversible structural transformations. Moreover, this report will review the research progress of LMRs in the understanding of precursor process route selection, modification effect and mechanism of bulk doping, surface coating, liquid phase post-treatment, and the design of new special structures.

Keywords: Li-Mn-rich layered oxides; Cathode materials; Lithium-ion battery

富锰正极材料的挑战和开发进展

王俊，张学全，刘亚飞*，陈彦彬*

北京矿冶科技集团，北京当升材料科技股份有限公司，北京，100029，中国

摘要：目前钴供应的瓶颈限制了锂离子电池的进一步商业化应用，促使高能量密度无钴正极材料成为发展方向。层状富锂锰基材料凭借其高比容量(>250 mA·h/g)和低成本等优点，有望成为新一代锂离子电池用正极材料，但循环过程中的电压和容量衰减制约了其商业化应用进程。本报告研究了钴对富锂锰基材料电化学性能的影响，发现钴可以促进晶格 O^{2-} 的氧化，在4.5 V以上充电电压下提供更多的容量，但导致氧气的释放和不可逆的结构转变。此外，本报告将阐述近年来富锂锰基材料在前驱体工艺路线选择、体相掺杂、表面包覆、液相后处理的作用效果和改性机理，以及新型特殊结构设计等方面的研究进展。

关键词：层状富锂锰基氧化物；正极材料；锂离子电池

Activation of Peroxymonosulfate by Single Atom Co-N-C Catalysts for High-efficient Removal of Phenol: Performance, Mechanism and Stability

Jianqun WU[1,2,*], Shangqian ZHAO[1]

1 GRINM (Guangdong) Institute for Advanced Materials and Technology, Foshan, 528000, China
2 School of Civil Engineering and Architecture, East China Jiaotong University, Nanchang, 330013, China

Abstract: Biomass carbon-based catalysts have the advantages of biological cleaning, eco-friendly and cost-effective in water treatment. Meanwhile, nitrogen doped biochar promotes the development of non-radical peroxymonosulfate (PMS) activation in environmental remediation. Herein, we have designed biomass carbon-based catalysts with anchoring sites for single cobalt atoms in a defined $Co-N_3$ coordination structure (SA Co-N-C). A PMS activation system employing the SA Co-N-C as a high-efficiency catalyst was demonstrated, which can efficiently degrade phenol (Ph). The synergistic effect of radicals and non-radicals leads to the removal of Ph in SA Co-N-C/PMS system, while the adsorption effect played a negligible role. The non-radical pathways are prominent in the Ph decomposition process according to the quenching experiments, electron paramagnetic resonance, electrochemical analysis and electron-coupled oxidation system analysis, in which the contribution of electron-transfer was dominant. The density functional theory (DFT) calculations and experimental results showed that $Co-N_3$ site served as the main active site for PMS activation in SA Co-N-C/PMS system. Compared to the N-C and C-C sites, PMS is more inclined to adsorb on the $Co-N_3$ sites. In addition, SA Co-N-C/PMS system had excellent efficiencies in oxidative degradation of various organic pollutants, and this system tends to efficiently degrade organic pollutants containing strong electron-withdrawing group. This work dedicates to providing new insights into the non-radical pathway-catalyzed AOPs and efficient pathway for the effective degradation of organic pollutants.

Keywords: Single-atom; $Co-N_3$ sites; Peroxymonosulfate; Non-radical reaction; Electron-transfer

Figure 1 TEM image of SA Co-N-C

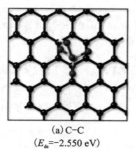

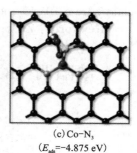

(a) C-C (E_{ds}=-2.550 eV)   (b) $C-N_3$ (E_{ads}=-2.580 eV)   (c) $Co-N_3$ (E_{ads}=-4.875 eV)

Figure 2 DFT calculation of PMS/C-C (a), PMS/$C-N_3$ (b), and PMS/$Co-N_3$ (c) stable structures

Defect-regulated Synthesis of ZrB$_2$ Powder by Carbon Thermal Reduction and Its Hot-pressing Densification Mechanism

Yuyang LIU[1,2,3,4,*], Xiaoning LI[1,2,3,4], Xingqi WANG[1,2,3,4]

1. National Engineering Research Center of Environment-friendly Metallurgy in Producing Premium Non-ferrous Metals, China GRINM Group Co., Ltd., Beijing, 101407, China
2. GRINM Resources and Environment Tech. Co., Ltd., Beijing, 101407, China
3. General Research Institute for Nonferrous Metals, Beijing, 100088, China
4. Beijing Engineering Research Center of Strategic Nonferrous Metals Green Manufacturing Technology, Beijing, 101407, China

Abstract: Based on defect modulation, ZrB$_2$ powder was controllably synthesized by carbon thermal reduction (CTR) method with B powder as a novel additive. Thermodynamic data of the ZrO$_2$-B$_2$O$_3$-C system in the presence of boron were calculated by HSC thermodynamic software and TG-DSC thermal analysis combined with XRD was used to investigate the reaction process. Transmission electron microscopy (TEM) and scanning electron microscopy (SEM) were used to analyze the defect structure and grain surface characteristics of the reaction products. The results show that the reaction of B with ZrO$_2$ and C exothermally changes the nucleation environment of ZrB$_2$ grains, and the generated helical dislocations act as a source of crystal growth steps to promote the growth of smooth interfaces, which leads to a transformation of the ZrB$_2$ grain growth mechanism. Quantitative calculations of the dislocation density of ZrB$_2$ powders based on the convolutional multiple whole profile fitting (CMWP-fit) suggest that the use of boron as an additive can modulate the defect concentration of the products. ZrB$_2$ powders with different dislocation densities were obtained by controlled synthesis, followed by hot-pressing densification. The results showed that the relative density of ZrB$_2$ ceramics increased from 83.9% to 96.7% as the dislocation density of the powder increased from 2.7×10^{13} m^{-2} to 7.6×10^{14} m^{-2} with the increase of boron content from 0 wt.% to 3 wt.%. In addition, the defects provide additional sintering-driving forces that prompt a change in the densification mechanism from grain boundary diffusion to dislocation slip during hot-pressing sintering, increasing the apparent activation energy from 105 kJ/mol to 210 kJ/mol.

Keywords: ZrB$_2$; Defect regulation; Caculation of dislocation density; Densification mechanism

缺陷调控的碳热还原法制备 ZrB$_2$ 粉体及其热压致密化性能研究

刘宇阳[1,2,3,4,*], 李小宁[1,2,3,4], 王星奇[1,2,3,4]

1. 中国有研科技集团有限公司 高品质有色金属绿色特种冶金国家工程研究中心, 北京, 101407, 中国
2. 有研资源环境技术研究院(北京)有限公司, 北京, 101407, 中国
3. 北京有色金属研究总院, 北京, 100088, 中国
4. 战略性有色金属绿色制造技术北京市工程研究中心, 北京, 101407, 中国

摘要: 基于缺陷调控, 以 B 粉为添加剂, 采用碳热还原(CTR)法可控合成 ZrB$_2$ 粉体。通过 HSC 热力学软件计算 B 粉作用下 ZrO$_2$-B$_2$O$_3$-C 体系热力学数据, 并采用 TG-DSC 热分析与 XRD 结合研究反应过程。采用透射电镜(TEM)和扫描电镜(SEM)分析产物的缺陷结构和晶粒表面特征, 结果表明, B 与 ZrO$_2$ 和 C 反应大量放热, 改变了 ZrB$_2$ 晶粒的形核环境, 产生的螺旋位错作为晶体生长台阶源促进光滑界面生长, 从而使得 ZrB$_2$ 晶粒生长机制发生转变。基于多重卷积整体轮廓拟合(CMWP-fit)方法定量计算 ZrB$_2$ 粉体位错密度, 结果表明, ZrB$_2$ 粉体合成过程中, 以 B 粉为添加剂可调控合成产物的缺陷浓度。通过可控合成得到不同位错密度的 ZrB$_2$ 粉体, 然后进行热压致密化, 结果表明, 随着 B 粉含量由 0 wt.%增加到 3 wt.%, 产物位错密度由 2.7×10^{13} m^{-2} 增加到 7.6×10^{14} m^{-2}, ZrB$_2$ 陶瓷的相对密度由 83.9%增加为 96.7%。此外, 缺陷提供额外的烧结驱动力, 在热压烧结过程中促使致密化机制由晶界扩散转变为位错滑移, 表观激活能由 105 kJ/mol 增加为 210 kJ/mol。

关键词: ZrB$_2$; 缺陷调控; 位错密度计算; 致密化机制

Determination of Elemental Impurities in Nickel-based Superalloys by Glow Discharge Mass Spectrometry

Fangfei HU[1,*], Jidong LI[2], Yingxin ZHANG[1], Pengyu LIU[2]

1 Guobiao (Beijing) Testing & Certification Co., Ltd., Beijing, 101407, China
2 China United Test & Certification Co., Ltd., Beijing, 101407, China

Abstract: A method for the determination of impurities in nickel-based superalloys by high-resolution glow discharge mass spectrometry (GDMS) was described. The optimum discharge conditions were investigated to obtain stable glow discharge and good sensitivity. Most of the interferences were separated by selecting appropriate isotopes or in high resolution mode of the instrument. All isotopes of only a few elements are severely affected by mass spectrometry interference, such as Zr. Select Zr 91 with relatively light interference and eliminate the interference through proportional deduction. The calibration relative sensitivity factors (RSF) for 17 elements were obtained with 4 matrix-matched certified reference materials. And the linear relationship of the standard curve is better than 0.99. A matrix-matched nickel-based superalloy sample was analyzed by using the calibration RSFs (quantitative analysis) and the standard RSFs in the instrument (semi-quantitative analysis), including C, N, O and S. The relative standard deviation of the 17 element determination results was less than 15%. The results of the verification methods, such as inductively coupled plasma optical emission spectrometry (ICP-OES), inductively coupled plasma mass spectrometry (ICP-MS), inductively coupled plasma tandem mass spectrometry (ICP-MS/MS), high frequency combustion infrared absorption and inert gas pulse infrared thermal conductivity, were closer to quantitative analysis results. The t-test results indicate that there was no significant difference between the results of GDMS and the verification methods. The validation revealed that the proposed analytical approach achieves reliable results for the rapid determination of several impurities, such as metals as well as non-metals, even the gas elements C, N, O and S.

Keywords: Determination of elemental impurities; Nickel-based superalloys; Glow discharge mass spectrometry; Relative sensitivity factors

Effect of A-EMS Melt Treatment on Microstructure and Mechanical Properties of Al-Zn-Mg-Cu Alloy Castings

Y. T. XU[1,2], Z. F. ZHANG[1,2], Z. H. GAO[1,2], Y. B. WANG[1,2], C. S. CHEN[1,2], J. Z. FAN[1,2]

1　GRINM Metal Composites Technology Co., Ltd., Beijing, 101407, China
2　General Research Institute for Non-Ferrous Metals, Beijing, 100088, China

Abstract: The high-end manufacturing fields such as aerospace, rail transit, national defense, and military industry have increasingly urgent demands for large, complex, high-strength, and high-tough aluminum alloy structural parts. Al-Zn-Mg-Cu alloys with low density, high strength and high stiffness have been sucessfully applied in the aerospace industry, but serious segregation and hot cracking susceptibility for such alloys in the casting process are not avoided especially for large-sized castings due to its high alloying contents and large density difference. Past research work showed that homogeneous composition distribution, uniformly fine microstructure during solidification, and the defects-related properties of these alloys could be effectively achieved by melt treatment including both chemical and physical methods, leading to a superior combination of higher strength levels and acceptable ductility even in the casting condition.

In this study, based on rheo-casting developed by authors, a new Al-Zn-Mg-Cu-Zr-Re cast alloy and typical castings with complicated profile were chosen and effects of melt treatment processes by combining micro-alloying and A-EMS on microstructure and mechanical properties of casting were studied. The results revealed that, an appropriate addition of micro-alloys not only benifited for the grain refinement, but also could effectivley restrain the formation of unfavorable phases; the macro-segregation and hot-cracking in the casting process were significantly alleviated by A-EMS melt treatment; and the strength and ductility of the alloy casting were found to be comparable to those of conventionally forged 7000 series alloys, indicating a good industrial application prospects.

Keywords: A-EMS; Melt treatment; Al-Zn-Mg-Cu alloy; Microstructure; Mechanical property; Rheo-casting

Interface Modulation for Inorganic Perovskite Solar Cell with Efficiency over 16%

Zhenyun ZHANG, Haoyue MA, Hongling ZHANG, Lin CAO, Bo WANG,
Chenyang ZHU, Peng SHANG*, Hongchun SHI, Yuanfei MA

GRINM Guojing Advanced Material Co., Ltd. General Research Institute for Nonferrous Metals, Langfang, 065000, China

Abstract: The hybrid organic-inorganic perovskite materials have attracted considerable interest due to their salient optical and electronic properties, such as strong light absorption ability, small exciton binding energy, long carrier diffusion length, high carrier mobility. The photovoltaic cell with perovskite material has become a very promising new type of solar cell due to its high efficiency, low cost and simple preparation. However, the serious interface recombination extremely restrict the development of perovskite solar cells (PSCs). This study focused on the generation and separation of photogenerated charges in PSCs.

Modification of the ETL/inorganic perovskite absorption layer interface with europium improved the morphology of the perovskite, promoted the efficient carriers separation efficiently at the interface, which effectively improved the photoelectric performance of the PSCs. In order to optimize charge accumulation at the ETL/inorganic perovskite interface, europium nitrate was choosed to treat tin dioxide, and investigate its regulation mechanism to improve the interface. On the one hand, the morphology of the perovskite is improved, the grain size grows bigger, the grain boundaries are reduced, the defects are reduced, and the carrier recombination inside the perovskite is reduced. On the other hand, the interface defects are passivated, the recombination of photo-generated charges at the interface is inhibited, and the efficient transfer of photo-generated charges is promoted. By finely optimizing the doping concentration, the highest efficiency of the PSCs can reach 16.83%.

Keywords: Perovskite; Interface modulation; Defect passivation

NiFe Layered Double Hydroxide Supported on Ni Fiber Felt for Oxygen Evolution at High Current Density

Qinglin LIU*, Yiwen TANG, Man LUO, Lijun JIANG

GRINM (Guangdong) Institute for Advanced Materials and Technology, Foshan, 528000, China

Abstract: To develop self-supporting oxygen evolution electrode with excellent activity and stability is a great strategy to solve the bottleneck in the application of alkaline electrolysis of water. To develop large size of electrode for industrial application, ambient synthesize method is necessary. NiFe layered double hydroxide (NiFe LDH) is an ideal catalyst for oxygen evolution reaction, with large specific surface area and impressive activity. However, the catalyst without substrate is hard to exhibit large current density during the electrochemical reaction. Hence, it is urgent to self-supporting NiFe LDH via a mild preparation recipe. Herein, needle-like NiFe layered double hydroxide supporting on Ni fiber felt (NiFe LDH/Ni fiber) was synthesized by one-step impregnation at room temperature, which is also possible to synthesize meter-sized oxygen evolution electrode. In a three-electrode system, the NiFe LDHs/Ni fiber electrode showed excellent OER performance with an overpotential of 300 mV at a current density of 600 mA/cm^2 in 30 wt.% KOH at 80 ℃. Besides, NiFe LDHs/Ni fiber expresses excellent stability. After chronoamperometric test at 600 mA/cm^2 for 14 days under the industrial catalytic conditions, the working potential of NiFe LDHs/Ni fiber//Raney Ni electrodes simply increases less than 1%. These experimental results imply that the NiFe LDHs/Ni fiber electrode synthesized by the one-step impregnation method is promising for industrial application.

Keywords: NiFe layered double hydroxide; Ni fiber felt; Oxygen evolution reaction; One-step impregnation; High current density

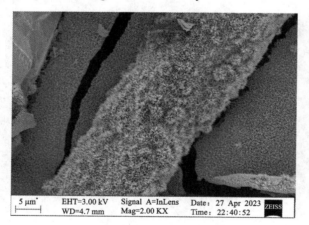

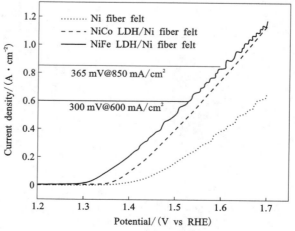

Study on Heat Transfer Performance of Mg Based Composite Hydrogen Storage Materials under Working Condition

Liyu ZHANG[1,2,4,*], Jianhua YE[1,2,4], Zhinian LI[1,3,4], Lijun JIANG[1,2,4]

1. GRINM Group Co., Ltd., National Engineering Research Center of Nonferrous Metals Materials and Products for New Energy, Beijing, 100088, China
2. GRINM (Guangdong) Institute for Advanced Materials and Technology, Foshan, 528000, China
3. GRIMAT Engineering Institute Co., Ltd., Beijing, 101407, China
4. General Research Institute for Nonferrous Metals, Beijing, 100088, China

Abstract: The problem of lacking safe and efficient hydrogen storage and transportation technology was one of the key factors restricting the large-scale commercial application of hydrogen energy. High-capacity hydrogen storage materials were the effective means to solve this problem. Mg-based hydrogen storage materials have become one of the most attractive hydrogen storage materials due to its high hydrogen storage capacity, abundant resources, and low price. However, due to the high enthalpy of the hydrogen absorption and desorption reaction of the Mg-based hydrogen storage material and the poor thermal conductivity of the powder, it is urgent to improve the heat and mass transfer performance of the bed to meet the application needs. At present, the calculation models for the heat and mass transfer performance of Mg-based hydrogen storage materials simplified the effective thermal conductivity of the Mg-based hydrogen storage material bed to a constant, resulting in deviation between the calculated value and the actual results under application conditions, which affects the accuracy of the model calculation. Therefore, there was an urgent need to systematically study the effective thermal conductivity of the Mg-based hydrogen storage material beds with working condition parameters, such as hydrogen pressure, temperature and hydrogen content to provide key data support for the optimization of heat and mass transfer in the system. Based on the above problems, the in-situ hydrogenation reaction ball milling method was used to achieve the 200 g per batch preparation of Mg-3 wt.%$Ti_{0.16}Cr_{0.24}V_{0.6}$ hydrogen storage composites. The hydrogenated Mg-3 wt.%$Ti_{0.16}Cr_{0.24}V_{0.6}$ composite sample (MH) was mixed with expanded graphite (EG) in a certain proportion, such as 90 wt.% MH and 10 wt.% EG. The MH/10 wt.% EG compact with 40 mm in diameter was prepared by uniaxial die pressing method under a pressure of 351 MPa. The transient plane heat source method was used to measure the thermal conductivity of the hydrogen storage compact samples under different atmospheres, hydrogen pressure, temperature and hydrogen content. The results showed that in the MH/EG compact, the EG was arranged in the layers perpendicular to the pressing direction, which forms the heat conduction pathways. The thermal conductivity value could reach 6.66 W/(m·K) at room temperature and 0.1 MPa hydrogen pressure. The MH/EG compact had higher thermal conductivity in hydrogen atmosphere, and the thermal conductivity showed logarithmic increasing trend with increasing hydrogen pressure, and linear decreasing trend with increasing temperature. And it showed S type trend with hydrogen content. As the hydrogen content decreased, the thermal conductivity increased slightly firstly, and when the dehydrogenation amount was about 3 wt.%, it significantly increased to a stable value. Then, we proposed an empirical formula for the relationship between the thermal conductivity of the MH/EG compact with the working conditions such as temperature, hydrogen pressure and hydrogen content and discussed the corresponding mechanism.

Keywords: Mg-based hydrogen storage materials; Thermal conductivity; Transient plane heat source method; Working conditions

The Effect of Garnet Type Oxide Solid State Electrolyte Coating on the Properties of Separators and the Electrical and Safety Performances of Pouch Cells

Bo WANG, Tianhang ZHANG[#], Zenghua CHANG,
Rennian WANG, Xiaopeng QI, Jiantao WANG, Rong YANG[*]

China Automotive Battery Research Institute Co., Ltd., Beijing, China

Abstract: Thermal runaway (TR) is a critical factor hindering the development of automotive power batteries. Particularly, mechanical abuse, such as puncturing/crushing, leads to internal short circuit (electrical abuse) in power battery, which causes massive heat realease, melting separator and a directly contact between cathode and anode materials, resulting in thermal abuse and eventually explosion. In order to enhance the safety and electrical performance of lithium-ion batteries, this study investigates the impact of coating the separator with a garnet type solid state oxide electrolyte, namely $Li_{6.5}La_3Zr_{1.5}Ta_{0.5}O_{12}$ (LLZTO), on the separator's properties, as well as the electrical and safety performance of lithium-ion pouch cells. Extensive tests were conducted using $LiNi_{0.9}Co_xMn_xO_2/C + SiO_x$ pouch cells, featuring a commendable capacity of 2 A·h and a high energy density of approximately 300 W·h/kg. The results demonstrate that the LLZTO solid-state electrolyte coating significantly improves the stability of the separator. After rigorous hot-box testing, the separator coated with LLZTO successfully suppresses severe internal short-circuits, maintaining a reliable voltage and minimizing the occurrence of internal short-circuits. Moreover, the LLZTO electrolyte coating helps to prevent separator shrinking and avoid direct contact between cathode and anode materials under mechanical abuse conditions, thereby preventing thermal runaway. The lithium-ion pouch cells with the LLZTO-coated separator exhibit satisfactory electrical performance. After 500 cycles at $1C$ rate, the capacity retention remains as high as 86.16%. The findings of this study underscore the potential benefits of the LLZTO solid-state electrolyte coating in enhancing the safety and electrical performance of lithium-ion batteries. This research provides a vital theoretical and practical foundation for the development of safer and more reliable energy storage solutions for electric transportation using lithium-ion batteries.

Keywords: Lithium-ion battery; Solid state oxide electrolyte; Separator

---

#: co-first author

The Microstructural Evolution and Mechanical Properties after the Solution Treatment of GH4099 Produced by Laser Powder Bed Fusion

Jiahao LIU[1,3], Wenqian GUO[2], Yonghui WANG[2], Shaoming ZHANG[3], Qiang HU[1,3,*]

1 Industrial Research Institute for Metal Powder Material, GRINM Group, Beijing, 101407, China
2 GRINM Additive Manufacturing Technology Co., Ltd., Beijing, 101407, China
3 General Research Institute for Nonferrous Metals, Beijing, 100088, China

Abstract: Owing to its ultrafine microstructures, GH4099 superalloy prepared by Laser Powder Bed Fusion exhibits an excellent combination of strength and plasticity but the brittleness at 900 ℃, which limits its performance improvement compared to the wrought counterpart. In this study, the optimized processing parameters are used to obtain GH4099 superalloy (As-built sample) with minimal defects. Furthermore, solid solution heat treatment (SST) is conducted to tune up its microstructure and mechanical properties at room temperature and 900 ℃. The characteristic results show that LPBF GH4099 mainly consists of cellular dendrite structure and columnar grains. Upon SST, the substructures of as-built sample are eliminated and the columnar grains transform the equiaxed grains due to static recrystallization. The $M_{23}C_6$-carbide in the alloy first precipitates and subsequently dissolves with the SST temperature increasing from 1140 ℃ to 1205 ℃. In addition, the residual stress is released and the density of dislocation drops. As a result, the SST-1140 sample has the highest yield strength (YS) of about 847 MPa and ultimate tensile strength (UTS) of about 1147 MPa with a desirable elongation of about 38.5% due to the precipitation strengthening of about 1 μm $M_{23}C_6$-carbide particles and dislocation strengthening. The fracture surface of the as-built and SST sample displays fine and dense dimples and tearing ridges, which feature ductile fracture. However, at 900 ℃, it was observed that the plastic deformation of either the as-built sample or SST sample is influenced by an intergranular brittle fracture, leading to the elevated temperature ductility loss, where cleavage facets and rock-candy intergranular fracture surfaces are observed. This means that the grain boundaries at the elevated temperature are weak regions and plastic deformation never occurs.

Keywords: GH4099 Superalloy; Laser powder bed fusion; Microstructural evolution; Mechanical properties

固溶处理对激光粉末床熔融制备 GH4099 合金的微观组织与力学性能演变

刘嘉豪[1,3]，郭文倩[2]，王永慧[2]，张少明[3]，胡强[1,2,3,*]

1 金属粉体材料产业技术研究院，有研科技集团有限公司，北京，101407，中国
2 有研增材技术有限公司，有研科技集团有限公司，北京，101407，中国
3 北京有色金属研究总院，北京，100088，中国

摘要：激光粉末床熔融制备的 GH4099 高温合金由于其独特的组织，具有良好的强度和塑性结合，但在 900 ℃时脆性较大，限制了其性能的提高。在本研究中，利用优化后的工艺参数获得了较少缺陷的 GH4099 高温合金（沉积态试样）。通过固溶热处理(SST)对其进行室温和 900 ℃的微观组织和力学性能调整。表征结果表明，沉积态 GH4099 主要由胞状枝晶结构和柱状晶粒组成。经固溶热处理后，试样的亚结构被消除，由静态再结晶柱状晶转变为等轴晶。随着 SST 温度从 1140 ℃升高到 1205 ℃，合金中的 $M_{23}C_6$ 碳化物先析出，然后溶解。残余应力得到释放，位错密度下降。因此，SST-1140 试样获得最大的屈服强度(YS)约为 847 MPa、极限抗拉强度(UTS)约为 1147 MPa 以及伸长率约为 38.5%，这主要来自于尺寸约 1 μm 的 $M_{23}C_6$ 碳化物颗粒的析出强化和位错强化。沉积态试样和 SST 试样的断口表面呈现出细小而致密的韧窝和撕裂脊，具有韧性断裂特征。然而，在 900 ℃时，观察到无论是沉积态试样还是 SST 试样的塑性变形都受到晶间脆性断裂的影响，导致高温塑性损失，出现解理面和冰糖状的晶间断口。这意味着在高温下晶界是弱区，未能发生塑性变形。

关键词：GH4099 高温合金；激光粉末床熔融；微观组织演变；力学性能

Core-shell FeCo@SiO$_2$ Nanocomposites with Controllable Shell Thickness and Tunable Electromagnetic Properties

Longxia YANG[1,2,3], Haicheng WANG[1,2,3,*]

1. State Key Laboratory of Advanced Materials for Smart Sensing, China GRINM Group Co. Ltd., Beijing, 100088, China
2. GRIMAT Engineering Institute Co. Ltd., Beijing, 101407, China
3. General Research Institute for Nonferrous Metals, Beijing, 100088, China

Abstract: A facile sol-gel method was utilized to coat uniform SiO$_2$ shell layer onto the surface of FeCo nanocubes, and core-shell structured FeCo@SiO$_2$ nanocomposites were obtained. It is found that, by adjusting the concentration of tetraethyl orthosilicate (TEOS), the thickness of the silica shell can be easily controlled in the range of 20-36 nm. When the amount of TEOS is 75 μL, the average shell thickness of SiO$_2$ is 22 nm. With the amount of TEOS gradually increasing to 100 and 150 μL, nearly 25 nm and 30 nm uniform shells could be achieved, respectively. Significantly, with the change of silica shell thickness, the electromagnetic parameters of FeCo@SiO$_2$ nanocomposites could be tuned elaborately. When the average silica shell thickness is 22 nm, FeCo@SiO$_2$ nanocomposites exhibit the largest value of tanδ_μ, which means the best magnetic loss and may contribute to electromagnetic wave absorption. Furthermore, the introduction of SiO$_2$ is capable of improving the antioxidant properties of FeCo nanoparticles and also beneficial to the improvement of impedance matching. Hopefully, it is expected that FeCo@SiO$_2$, or FeCo@SiO$_2$ based nanocomposites could be promising candidates for advanced electromagnetic materials with tunable properties, such as electromagnetic wave absorbing, or electromagnetic wave shielding materials.

Keywords: FeCo@SiO$_2$ nanocomposites; Core-shell; Controllable thickness; Electromagnetic parameters; Electromagnetic properties.

核壳 FeCo@SiO$_2$ 纳米复合材料及其壳层厚度和电磁性能调控

杨龙霞[1,2,3]，王海成[1,2,3,*]

1. 中国有研科技集团有限公司智能传感功能材料国家重点实验室，北京，100088，中国
2. 有研工程技术研究院有限公司，北京，101407，中国
3. 北京有色金属研究总院，北京，100088，中国

摘要：通过温和的溶胶-凝胶法在 FeCo 纳米立方体表面包覆了均匀的 SiO$_2$ 壳层，获得了一种核壳结构的 FeCo@SiO$_2$ 纳米复合材料。研究发现，只需调节 TEOS 的浓度，就能轻松地将 SiO$_2$ 壳层的厚度控制为 20~36 nm。当 TEOS 的用量分别为 75 μL、100 μL 和 150 μL 时，SiO$_2$ 的平均厚度为 22 nm、25 nm 和 30 nm。通过改变 SiO$_2$ 的壳层厚度，电磁参数可以得到精细调控。当 SiO$_2$ 的平均厚度为 22 nm 时，FeCo@SiO$_2$ 纳米复合材料的 tanδ_μ 值最大，即磁损耗最大，有助于电磁波的吸收。此外，SiO$_2$ 的引入能够改善 FeCo 纳米颗粒的抗氧化性能，也有利于改善阻抗匹配。因此，FeCo@SiO$_2$ 及其纳米复合材料有望成为性能可调的先进电磁材料，如电磁波吸收或电磁波屏蔽材料。

关键词：FeCo@SiO$_2$ 纳米复合材料；核壳结构；可控厚度；电磁参数；电磁性能

In-situ EBSD Investigation on Deformation Mechanism of Room Temperature Superplasticity of Bulk Recrystallized Molybdenum

Wenshuai CHEN[1,2,3], Yan LI[2], Xueliang HE[2], Zenglin ZHOU[1,2,3,*]

1 State Key Laboratory of Advanced Materials for Smart Sensing, China GRINM Group Co., Ltd., Beijing, 100088, China
2 GRIMAT Engineering Institute Co., Ltd., Beijing, 101407, China
3 General Research Institute for Nonferrous Metals, Beijing, 100088, China

Abstract: Body centered cubic refractory metals materials have excellent high temperature strength, but intercrystalline brittle fracture often occurs after recrystallization at high temperature. We use powder metallurgy Y-type hot rolling to prepare a pure molybdenum material (99.9%) that exhibits superplasticity after recrystallization. After annealing at 1200 ℃, it reaches a maximum room temperature total elongation of 113.5% and a tensile strength of 513 MPa. The deformation mechanism of room temperature superplasticity of molybdenum was studied through in-situ EBSD. The results showed that during the deformation process, the texture dominated by $\langle 110 \rangle$//RD did not change significantly. Entering the plastic deformation stage, high strain caused dislocation rearrangement, and the proportion of small angle grain boundaries sharply increased. Analysis showed that the room temperature superplasticity of molybdenum was mainly attributed to the excellent plastic deformation ability of $\langle 110 \rangle$//RD oriented grains. Other oriented grains rely on the formation of high lattice gradients internally undergo coordinated crystal rotation deformation, with different directions of grain rotation in different regions.

Keywords: Pure molybdenum; Superplastic deformation; Orientation; Grain boundary

再结晶钼材料室温超塑性变形机制的原位 EBSD 研究

陈文帅[1,2,3],李艳[2],何学良[2],周增林[1,2,3,*]

1 中国有研科技集团有限公司智能传感功能材料国家重点实验室,北京,100088,中国
2 有研工程技术研究院有限公司,北京,101407,中国
3 北京有色金属研究总院,北京,100088,中国

摘要:体心立方难熔金属材料具有优异的高温强度,但经高温服役再结晶后多发生晶间脆性断裂。我们利用粉末冶金-Y型热轧制备一种再结晶后呈现超塑性的纯钼材料(99.9%),经1200 ℃退火后达最高113.5%的室温总伸长率,抗拉强度达513 MPa。通过原位 EBSD 研究其室温超塑性的变形机制,结果表明:在变形过程中,以$\langle 110 \rangle$//RD 为主导的织构并没有明显变化,进入塑性变形阶段,高应变引起位错重新排列,小角度晶界比例急剧增加,分析发现钼的室温超塑性主要得益于$\langle 110 \rangle$//RD 取向晶粒优异的塑性变形能力,其他取向晶粒依靠内部形成高的晶格梯度发生晶体旋转协调变形,不同区域的晶粒旋转方向不同。

关键词:纯钼;超塑性变形;取向;晶界分布